Gold Chemistry

*Edited by
Fabian Mohr*

Further Reading

Hashmi, A. S. K., Toste, D .F. (Eds.)

Modern Gold Catalyzed Synthesis

2009
ISBN: 978-3-527-31952-7

Laguna, A. (Ed.)

Modern Supramolecular Gold Chemistry

Gold-Metal Interactions and Applications

2009
ISBN: 978-3-527-32029-5

Oro, L. A., Claver, C. (Eds.)

Iridium Complexes in Organic Synthesis

2009
Hardcover
ISBN: 978-3-527-31996-1

Gold Chemistry

Applications and Future Directions in the Life Sciences

Edited by
Fabian Mohr

WILEY-VCH

WILEY-VCH Verlag GmbH & Co. KGaA

The Editor

Prof. Dr. Fabian Mohr
Bergische Universität Wuppertal
Fachbereich C, Anorganische Chemie
Gaußstraße 20
42119 Wuppertal
Germany

All books published by Wiley-VCH are carefully produced. Nevertheless, authors, editors, and publisher do not warrant the information contained in these books, including this book, to be free of errors. Readers are advised to keep in mind that statements, data, illustrations, procedural details or other items may inadvertently be inaccurate.

Library of Congress Card No.: applied for

British Library Cataloguing-in-Publication Data
A catalogue record for this book is available from the British Library.

Bibliographic information published by the Deutsche Nationalbibliothek
The Deutsche Nationalbibliothek lists this publication in the Deutsche Nationalbibliografie; detailed bibliographic data are available on the Internet at http://dnb.d-nb.de.

© 2009 WILEY-VCH Verlag GmbH & Co. KGaA, Weinheim

All rights reserved (including those of translation into other languages). No part of this book may be reproduced in any form – by photoprinting, microfilm, or any other means – nor transmitted or translated into a machine language without written permission from the publishers. Registered names, trademarks, etc. used in this book, even when not specifically marked as such, are not to be considered unprotected by law.

Cover Design Adam-Design, Weinheim
Typesetting Thomson Digital, Noida, India
Printing betz-druck GmbH, Darmstadt
Binding Litges & Dopf GmbH, Heppenheim

Printed in the Federal Republic of Germany
Printed on acid-free paper

ISBN: 978-3-527-32086-8

Contents

Preface *XI*
Foreword *XIII*
List of Contributors *XV*

1	**Gold(I) Nitrogen Chemistry** *1*	
	Hanan E. Abdou, Ahmed A. Mohamed, and John P. Fackler Jr	
1.1	Introduction *1*	
1.2	Tetra-, Tri-, and Dinuclear Gold(I) Amidinate Complexes *3*	
1.3	Oxidative-Addition Reactions to the Dinuclear Gold(I) Amidinate Complex *9*	
1.4	Mercury(II) Cyanide Coordination Polymer *13*	
1.5	Formation of Mixed-Ligand Tetranuclear Gold(I) Nitrogen Clusters *15*	
1.6	Solvent Influences on Oxidation and Nuclearity of Gold Guanidinate Derivatives *21*	
1.7	Cyclic Trinuclear Gold(I) Nitrogen Compounds *24*	
1.8	Oxidative-Addition Reactions to the Cyclic Trinuclear Gold(I)-Nitrogen Compounds *28*	
1.9	Supramolecular Entities of Trinuclear Gold(I) Complexes Sandwiching Small Organic Acids *30*	
1.10	Gold(I) and Silver(I) Mixed-Metal Trinuclear Complexes *33*	
1.11	CO Oxidation Over Au/TiO$_2$ Prepared from Gold Nitrogen Complexes *36*	
1.12	Miscellaneous Observations *37*	
	References *37*	
2	**Chemistry of Gold(III) Complexes with Nitrogen and Oxygen Ligands** *47*	
	Maria Agostina Cinellu	
2.1	Introduction *47*	
2.2	Nitrogen Donor Ligands *47*	
2.2.1	Complexes with Neutral Monodentate Ligands *47*	
2.2.2	Complexes with Anionic Monodentate Ligands *51*

Gold Chemistry: Applications and Future Directions in the Life Sciences. Edited by Fabian Mohr
Copyright © 2009 WILEY-VCH Verlag GmbH & Co. KGaA, Weinheim
ISBN: 978-3-527-32086-8

2.2.3	Complexes with Multidentate Ligands	53
2.2.4	Complexes with Multidentate Ligands Containing Anionic N-Donors	59
2.2.5	Complexes with Polyazamacrocyclic Ligands	63
2.3	Oxygen Donor Ligands	65
2.3.1	Hydroxo Complexes	65
2.3.2	Oxo Complexes	69
2.3.3	Alkoxo Complexes	72
2.3.4	Complexes With Other O-donor Ligands	76
2.3.5	Complexes With Mixed N/O Ligands	78
	References	80
3	**Pentafluorophenyl Gold Complexes**	**93**
	A. Luquin, E. Cerrada, and M. Laguna	
3.1	Introduction	93
3.2	Pentafluorophenylgold(I) Derivatives	94
3.2.1	Neutral Pentafluorophenylgold(I) Derivatives	95
3.2.1.1	Neutral Pentafluorophenylgold(I) Derivatives with C-Donor Ligands	97
3.2.1.2	Neutral Pentafluorophenylgold(I) Derivatives with N-Donor Ligands	100
3.2.1.3	Neutral Pentafluorophenylgold(I) Derivatives with P-Donor Ligands	101
3.2.1.4	Neutral Pentafluorophenylgold(I) Derivatives with S-Donor Ligands	102
3.2.2	Anionic Pentafluorophenylgold(I) Derivatives	103
3.2.2.1	Reactivity of the Anionic Pentafluorophenylgold(I) Complexes Q[Au(C_6F_5)X]	105
3.2.3	Cationic Pentafluorophenylgold(I) Derivatives	107
3.2.4	Di and Polynuclear Pentafluorophenylgold(I) Derivatives	108
3.2.5	Heteronuclear Pentafluorophenylgold(I) Complexes	117
3.2.5.1	Heteronuclear Pentafluorophenylgold(I) Bismuth Complexes	117
3.2.5.2	Heteronuclear Pentafluorophenylgold(I) Tin Complexes	117
3.2.5.3	Heteronuclear Pentafluorophenylgold(I) Thallium Complexes	117
3.2.5.4	Heteronuclear Pentafluorophenylgold(I) Silver Complexes	119
3.2.5.5	Heteronuclear Pentafluorophenylgold(I) Copper Complexes	122
3.2.5.6	Heteronuclear Pentafluorophenylgold(I) Palladium Complexes	122
3.2.5.7	Heteronuclear Pentafluorophenylgold(I) Rhodium Complexes	123
3.2.5.8	Heteronuclear Pentafluorophenylgold(I) Iron Complexes	123
3.2.5.9	Heteronuclear pentafluorophenylgold(I) manganese complexes	126
3.2.5.10	Heteronuclear Pentafluorophenylgold(I) Chromium, Gold Molybdenum and Gold Tunsten Complexes	126
3.3	Pentafluorophenylgold(III) Derivatives	127

3.3.1	Mononuclear Pentafluorophenylgold (III) Derivatives	130
3.3.2	Di and Polynuclear Pentafluorophenylgold(III) Derivatives	138
3.3.2.1	C-Donor Ligands	138
3.3.2.2	Diamine and Carbene Bridges	143
3.3.2.3	N-, P- and As-Donor Ligands	143
3.3.2.4	S- and Se-Donor Ligands	147
3.3.3	Polynuclear Pentafluorophenylgold Derivatives with Au(I)-Au(III) Bond	148
3.3.4	Heteropolynuclear Pentafluorophenylgold(III) Derivatives	149
3.3.5	Pentafluorophenylgold(III) Derivatives as Catalyst	150
3.4	Gold Clusters	155
3.5	Pentafluorophenylgold(II) Derivatives	156
3.5.1	Dinuclear Pentafluorophenylgold(II) Complexes	156
3.5.2	Polynuclear Pentafluorophenylgold(II) Complexes	159
3.6	Outlook and Future Trends	162
	References	163

4	**Theoretical Chemistry of Gold – From Atoms to Molecules, Clusters, Surfaces and the Solid State**	**183**
	Peter Schwerdtfeger and Matthias Lein	
4.1	Introduction	183
4.2	The Origin of the Relativistic Maximum at Gold Along the 6th Period of Elements in the Periodic Table	186
4.3	Calculations on Atomic Gold	189
4.4	Relativistic Methods for Molecular Calculations and Diatomic Gold Compounds	194
4.5	Calculations on Inorganic and Organometallic Gold Compounds	203
4.6	Calculations on Gold Clusters	212
4.7	Calculations on Infinite Systems: from Surfaces to the Solid State of Gold	216
4.8	Summary	220
	References	221

5	**Luminescence and Photophysics of Gold Complexes**	**249**
	Chi-Ming Che and Siu-Wai Lai	
5.1	Introduction	249
5.2	Spectroscopic Properties of Gold(I) Complexes	250
5.2.1	Mononuclear Gold(I) Complexes	250
5.2.2	Di- and Polynuclear Gold(I) Complexes with Metal–Metal Interactions	252
5.2.3	High-Energy $^3[5d\sigma^*, 6p\sigma]$ Excited State Versus Visible Metal-Anion/Solvent Exciplex Emission	262
5.2.4	Chemosensory Applications	267
5.3	Spectroscopic Properties of Gold(III) Complexes	270
5.4	Photoinduced Electron Transfer Reactions of Gold Complexes	273

| 5.5 | Concluding Remarks 274 |
| | References 276 |

6	**Gold Compounds and Their Applications in Medicine** 283
	Elizabeth A. Pacheco, Edward R.T. Tiekink, and Michael W. Whitehouse
6.1	Introduction 283
6.2	The Aqueous Chemistry of Gold Compounds 283
6.2.1	Structures of Gold(I) and Gold(III) Complexes 284
6.2.2	Oxidation–Reduction Reactions 284
6.2.3	Ligand Exchange Mechanisms 285
6.3	Medicinally Important Gold Complexes, Their Analogs and Reactions 287
6.3.1	Oligomeric Gold(I) Thiolates and Analogs 287
6.3.2	Bis(thiolato)Gold(I) Species 290
6.3.3	Dithiocarbamates 291
6.3.4	Auranofin and Other Phosphine(Thiolato)Gold(I) Species 291
6.3.5	Au–S Bond Length Comparisons 292
6.3.6	Complexes of Diphosphine Ligands 293
6.3.7	Gold(I) Cyanide Complexes 294
6.3.8	Heterocyclic Carbenes 295
6.4	Gold–Protein Reactions and Complexes 295
6.4.1	Serum Albumin 295
6.4.2	Metallothioneins 297
6.4.3	Selenium-Dependent Glutathione Peroxidase 300
6.4.4	Gold(III) Oxidation of Insulin and Ribonuclease 301
6.4.5	Enzyme Inhibition 301
6.4.6	Zinc finger Proteins 302
6.4.7	Hemoglobin and Interprotein Gold Transfer ("Transauration") 303
6.4.8	Mitochondrial Thioredoxin Reductase 303
6.5	Physiological and Cellular Biochemistry 304
6.5.1	Biological Ligand Exchange 304
6.5.2	The Sulfhydryl-Shuttle Model 305
6.5.3	Equilibration of Intra- and Extracellular Gold 305
6.5.4	Cytotoxicity and Antitumor Activity of Gold Complexes 306
6.5.5	Oxidation States *in Vivo* 307
6.5.6	Immunochemical Consequences of Gold(III) 308
6.5.7	Anti-HIV Activity 308
6.5.8	Gold Nanoparticles 309
6.6	Conclusions 309
	References 310

7	**Nanoscience of Gold and Gold Surfaces** 321
	M.B Cortie and A. McDonagh
7.1	Introduction 321
7.2	Forms of Gold at the Nanoscale 323

7.2.1	Clusters and Nanoparticles of Less than 5 nm Diameter *323*
7.2.2	Nanospheres *325*
7.2.3	Nanoshells *325*
7.2.4	Nanorods *327*
7.2.5	Other Nanoparticle Shapes *327*
7.2.6	Mesoporous Sponges *327*
7.2.7	Thin Films *329*
7.2.8	Assemblages of Nanoparticles *330*
7.2.8.1	Disordered Aggregates *330*
7.2.8.2	Colloidal Crystals *330*
7.3	Onset of New Phenomena *331*
7.3.1	Optical *331*
7.3.1.1	The Plasmon Resonance *331*
7.3.1.2	Manipulation of the Plasmon Resonance *332*
7.3.1.3	Fluorescence and Luminescence *333*
7.3.2	Physical *334*
7.3.2.1	Depression of the Melting Point *334*
7.3.3	Chemical *334*
7.3.3.1	Heterogeneous Catalysis *334*
7.4	Surface Chemistry of Gold *335*
7.4.1	Hydrogen *335*
7.4.2	Halogens *336*
7.4.3	Oxygen *337*
7.4.4	Sulfur *338*
7.4.4.1	Isothiocyanates *339*
7.4.5	Selenium and Tellurium *339*
7.4.6	Nitrogen *340*
7.4.7	Phosphorus, Arsenic, Antimony *341*
7.4.8	Carbon *342*
7.5	Conclusions *342*
	References *343*

8	**Liquid Crystals Based on Gold Compounds** *357*
	Silverio Coco and Pablo Espinet
8.1	Introduction *357*
8.1.1	A Few General Concepts in Liquid Crystals *357*
8.1.2	Gold, an Ideal Metal for LC Studies *360*
8.2	Pyridine Complexes *361*
8.3	Dithiobenzoate Complexes *361*
8.4	Isocyanide Complexes *362*
8.4.1	Isocyanide-Halide Complexes *362*
8.4.2	Isocyanide-Alkynyl Complexes *370*
8.4.3	Isocyanide-Fluorophenyl Derivatives *372*
8.4.4	Ionic bis(isocyanide) Derivatives *379*
8.4.5	Mixtures: Liquid Crystalline "Molecular Alloys" *383*

8.5	Carbene Complexes *384*	
8.6	Complexes Containing Pyrazole-Type Ligands *387*	
8.6.1	Trinuclear Gold Pyrazolate Rings: Metallacrowns *387*	
8.6.2	"Mononuclear" Complexes *388*	
8.7	Ionic Imidazolium Derivatives *388*	
8.8	Liquid Crystalline Gold Nanoparticles *388*	
8.9	Conclusions *390*	
	References *392*	

Index *397*

Preface

Aurum, Gold, Oro, Złoto, Or, 金 are names in various languages for a metal which has fascinated, inspired and accompanied humankind for thousands of years. Its golden shine and glitter, its resistance to corrosion and its rarity made gold a very precious and highly sought after material. Gold jewellery, ornaments and decorations adorned royalty, clergy and the wealthy in civilizations all around the world. The desire for more and more gold was the driving force for the exploration of hitherto unknown corners of the globe as exemplified by the Spanish exploration of the New World in search of the "*el dorado*" and also by gold rushes in California, Alaska, South Africa and Australia. In medicine too, gold containing formulations have been prescribed to patients with various ailments for thousands of years. The chemistry of gold, however, is the newest chapter in the thousand year old history of this element. Only in the last few centuries, chemists have discovered the unique properties and reactivity of this metal and have found applications in a variety of fields including medicine and the materials sciences. Several books covering the chemistry of gold have previously been published: The first monograph by Richard Puddephatt appeared in 1978, which was followed in 1999 by a comprehensive treatise on gold chemistry, biochemistry and technology by Hubert Schmidbaur. Since then, the field of gold chemistry has continued to advance and head into completely new directions. The purpose of this book is to present these new directions and to update and highlight important applications of gold complexes. The book is divided into two parts: the first part outlining new chemistry of gold which includes chapters by Maria Cinellu and John Fackler Jr. focusing on the chemistry of gold(I) and gold(III) with nitrogen ligands, respectively. The first part also contains a chapter by Mariano Laguna dedicated to the rich and sometimes unusual chemistry of gold complexes containing the C_6F_5 group. Last, but by no means least, the chapter by Peter Schwerdtfeger showcases the advances that theoretical chemistry of gold has made in the last few years and how such computational work has helped to solve many puzzling questions in gold chemistry. The second part of this book deals with current and future applications of gold complexes and includes chapters on the photophysics of gold complexes by Chi-Ming Che, nanotechnology by Michael Cortie, the use of gold compounds in medicine by

Edward Tiekink and gold complexes in liquid crystals by Pablo Espinet. Homogenous catalysis by gold complexes is not included here since a book dedicated especially to this topic edited by Stephen Hashmi will also appear in 2009 from this publishing house.

Firstly, I wish to thank the team at Wiley-VCH, in particular Manfred Köhl, Rainer Münz and Martin Graf for their enthusiasm and support they have given me from the very beginning of this project. I also sincerely wish to thank all the authors for their excellent contributions, for sticking to the stylistic guidelines and for keeping (more or less) within the deadlines. I am indebted to members of my research group at the University of Wuppertal for their support with proof reading and with helping me keep my sanity during the final stages of this project.

Wuppertal, December 2008 *Soli Deo Gloria!*
Fabian Mohr

Foreword

For few elements the chemistry has recently undergone such an explosive growth as for gold. This is the more surprising since the chemistry of this element had been dormant during most of the 19th and 20th century when the chemistry of most other elements developed rapidly or at least with a steady pace. In the last three decades, however, the number of reports on significant advances in gold chemistry is increasing at such a rate that scientists interested in this field are already in a need of periodical reviews which critically summarize and highlight the most important contributions. Several publishers have reacted to this situation and a series of overviews dedicated to special subjects, and even special issues of periodicals and books with a more general scope have appeared at shorter and shorter intervals.

Meanwhile a growing share of these reviews is dedicated to specific applications of the many new findings in gold chemistry, which naturally have been the most powerful incentives for the worldwide research activities. At least in number, these application-oriented investigations have overruled already the curiosity-driven fundamental scientific studies. The symbiotic and synergistic nature of "pure and applied gold chemistry" turned out to be extremely fruitful and successful. The discovery of the unexpected activity of particulate gold in heterogeneous catalysis – observed e. g. in important reactions like the low-temperature oxidation of carbon monoxide or in the activation of olefins – was soon followed by work which provided clear evidence for a similar efficacy of gold salts in homogeneous catalysis. There is now a plethora of new protocols in organic synthesis with key steps based on a wide range of highly active and selective gold catalysts, which still keeps growing rapidly.

Another research area where gold and its compounds have widened the scope very considerably is the chemistry of complexes with specific photophysical properties. From early observations it had been known that various mono- and polynuclear gold complexes, aggregates and clusters are often strongly luminescent. Physicochemical scrutiny has since provided a much better understanding of the underlying effects and thus this chemistry could be developed into a promising source of components for LEDs, OLEDs and other devices. Even the transition from photonic towards electrogenerated chemiluminescence has also recently been accomplished. In

Gold Chemistry: Applications and Future Directions in the Life Sciences. Edited by Fabian Mohr
Copyright © 2009 WILEY-VCH Verlag GmbH & Co. KGaA, Weinheim
ISBN: 978-3-527-32086-8

addition, this work has been closely related to research activities dedicated to the development on gold chemistry-based meso-phases (liquid crystals) and NLO materials, where this metal can offer many advantages over established systems.

Simple gold flakes and more sophisticated gold preparations have played a role in pharmacy and medicine for millennia, and some well-defined complexes have much later indeed been established as potent drugs for the treatment of rheumatoid and arthritic deceases. Following the observations in platinum chemistry, in more recent years there have been positive results also regarding the usage of gold complexes for the treatment of cancer and other deseases, and pertinent investigations are consequently still or again an area of active research.

The most rapid growth of gold chemistry is currently observed in nanoparticle technology. Gold nanoparticles in all shapes and sizes are employed in many areas, as diverse as electrochemistry and electronics, photophysics, biochemistry, biology, and medicine. Moreover, self-assembly of monolayers of gold compounds on surfaces has provided surface scientists with an ideal playground for all sorts of investigations at interfaces.

The present volume reflects most of these active areas of research in articles written by some of the leading scientist who themselves have been pioneers in, or contributed to their particular fields of interest. It is very fortunate that articles have also been included which give status reports on the underlying fields of current fundamental research. From these it can be easily extracted that theoretical chemistry has meanwhile contributed enormously to a better understanding of the special effects that are characteristic of gold chemistry. It is also becoming obvious that in coordination chemistry the interest has slowly shifted from standard phosphorus and sulfur ligands to components with carbon and nitrogen donor centers, *viz.* carbenes, arenes and N-heterocycles, respectively. After a long period during which gold(I) complexes were in the focus of research, renewed attention has also recently been paid to compounds of gold in its higher oxidation states, mainly owing to advantages in catalytic activity and to novel photophysical effects.

This collection will therefore be a very valuable source of information and inspiration for all those who proudly call themselves gold chemists already, but also will attract others to this still adolescent division of chemistry.

Hubert Schmidbaur

List of Contributors

Hanan E. Abdou
Laboratory for Molecular Structure
& Bonding
Department of Chemistry
Texas A&M University
College Station, TX 77843
USA

Elena Cerrada
Instituto de Ciencia de Materiales
de Aragón
Consejo Superior de Investigaciones
Científicas – Universidad de Zaragoza
Plaza S. Francisco s/n
50009 Zaragoza
Spain

Chi-Ming Che
Department of Chemistry and
HKU-CAS Joint Laboratory on
New Materials
The University of Hong Kong
Pokfulam Road
Hong Kong SAR
P.R. China

Maria Agostina Cinellu
Dipartimento di Chimica
Università di Sassari
via Vienna 2
07100 Sassari
Italy

Silverio Coco
IU CINQUIMA/Química Inorgánica
Facultad de Ciencias
Universidad de Valladolid
47071 Valladolid
Spain

Michael B. Cortie
Institute for Nanoscale Technology
University of Technology Sydney
PO Box 123
Broadway NSW 2007
Australia

Pablo Espinet
IU CINQUIMA/Química Inorgánica
Facultad de Ciencias
Universidad de Valladolid
47071 Valladolid
Spain

John P. Fackler Jr
Laboratory for Molecular Structure
& Bonding
Department of Chemistry
Texas A&M University
College Station, TX 77843
USA

Gold Chemistry: Applications and Future Directions in the Life Sciences. Edited by Fabian Mohr
Copyright © 2009 WILEY-VCH Verlag GmbH & Co. KGaA, Weinheim
ISBN: 978-3-527-32086-8

Mariano Laguna
Instituto de Ciencia de Materiales
de Aragón
Consejo Superior de Investigaciones
Científicas – Universidad de Zaragoza
Plaza S. Francisco s/n
50009 Zaragoza
Spain

Siu-Wai Lai
Department of Chemistry and
HKU-CAS Joint Laboratory on
New Materials
The University of Hong Kong
Pokfulam Road
Hong Kong SAR
P.R. China

Matthias Lein
Center of Theoretical Chemistry
and Physics (CTCP)
The New Zealand Institute
for Advanced Study
Massey University Auckland
Private Bag 102904
North Shore City
0745 Auckland
New Zealand

Asunción Luquin
Instituto de Ciencia de Materiales
de Aragón
Consejo Superior de Investigaciones
Científices – Universidad de Zaragoza
Plaza S. Francisco s/n
50009 Zaragoza
Spain

Andrew McDonagh
Institute for Nanoscale Technology
University of Technology Sydney
PO Box 123
Broadway NSW 2007
Australia

Ahmed A. Mohamed
Laboratory for Molecular Structure
& Bonding
Department of Chemistry
Texas A&M University
College Station, TX 77843
USA

Elizabeth A. Pacheco
Department of Chemistry
The University of Texas at San Antonio
San Antonio, TX 78249-0698
USA

Peter Schwerdtfeger
Center of Theoretical Chemistry and
Physics (CTCP)
The New Zealand Institute
for Advanced Study
Massey University Auckland
Private Bag 102904
North Shore City
0745 Auckland
New Zealand

Edward R.T. Tiekink
Department of Chemistry
The University of Texas at San Antonio
San Antonio, TX 78249-0698
USA

Michael W. Whitehouse
Eskitis Institute for Cell and Molecular
Therapies
School of Biomolecular & Physical
Science
Griffith University
Queensland 41111
Australia

1
Gold(I) Nitrogen Chemistry

Hanan E. Abdou, Ahmed A. Mohamed, and John P. Fackler Jr

1.1
Introduction

Nitrogen ligands have rarely been used with gold(I) and hardly any chemistry has been described using anionic, bridging nitrogen ligands. Dinuclear gold(I) complexes containing either one or two bridging ligands, such as ylides and thiolates, and their oxidative-addition products have been attracting considerable attention for many years [1, 2]. Gold(I) complexes with N-donor ligands are much less common than those with P-donor ligands. However, the affinity of gold for nitrogen can be increased if a phosphine ligand is attached to gold, because of the efficient π-acceptor nature of the phosphine [2]. Therefore the majority of gold(I) complexes with anionic N-donor ligands (L$^-$) are complexes of the type R$_3$PAuL, such as Ph$_3$PAu(bis(trimethylsilyl)amide) and Ph$_3$PAu(4-nitro-anilide). Other complexes with L$^-$ corresponding to substituted pyrazoles, imidazoles and benzylimidazoles are also known [2]. The gold(I) amidinate complexes reviewed here are symmetrical with the Au atom bonded to two N atoms. Some work with gold(I) carbeniates (N,C) and benzylimidazolates (N,C) and pyrazolates (N,N) is also included in this review. Gold(III) complexes with nitrogen ligands are covered in Chapter 2 of this book.

Gold(I) with its [Xe]4f^{14}5d^{10} electronic configuration is often described as a soft metal ion [3] and therefore might be expected to have a preference for soft donor ligands such as sulfur and carbon over hard donor ligands such as those bonding through nitrogen or oxygen [4]. For example, when bifunctional ligands with two different donor atoms are used, the gold atom will bind to these ligands through the atom with the higher donor strength according to the sequence [2]: Si $\sim$ P > C > S > Cl > N > O > F.

The bifunctional ligands, for example 2-pyridylphophines, thioamides and the 1,1-dicyanoethylene-2,2-dithiolate, are coordinated to gold through the P and S atoms but not N, since P and S atoms are better "soft" donor atoms than N [5]. Therefore, it was generally assumed that gold(I) will not effectively coordinate to a donor nitrogen atom [2]. However, the interesting chemistry of the anionic bridging ligands, amidines, ArNHC(H)NAr, to be described here, does not bear this out.

Gold Chemistry: Applications and Future Directions in the Life Sciences. Edited by Fabian Mohr
Copyright © 2009 WILEY-VCH Verlag GmbH & Co. KGaA, Weinheim
ISBN: 978-3-527-32086-8

Cotton's group was able to exploit the amidine ligands for the synthesis of a variety of complexes spanning the transition elements [6]. Previous trials by his group to use the anionic bridging ligand amidines with gold(I) indicated that in the case of [M$_2$(ArNC(H)NAr)$_2$] compounds, Ar=-C$_6$H$_4$-4-Me and M=Ag, Cu, Au, the stability series [7] must be Cu ∼ Ag ≫ Au, since they were unable to isolate a gold compound. The gold amidinate complexes reported here are synthesized in open air at room temperature and are stable at room temperature for several months.

Theoretical studies by Pyykkö in 1998 for [M$_2$(NHCHNH)$_2$] systems, M=Cu, Ag, and Au, predicted the M–M distances at the MP2 level [8]. Experimentally, systems containing amidinate ligands were known with Cu and Ag but unknown with Au. The results for the models containing silver and copper are close to the X-ray structures of [M$_2$(ArNC(H)NAr)$_2$], Ar=C$_6$H$_4$-4-Me and M=Ag, Cu. The Ag–Ag distance is 2.705 and 2.712 Å and the Cu–Cu distance is 2.497 and 2.528 Å at the experimental and theoretical level, Table 1.1. The hypothetical dinuclear gold(I) amidinate compound was calculated to have an Au–Au distance at the MP2 level of 2.728 Å [8]. The dinuclear gold(I) amidinate complex now known proves the predicted Au-Au distance to be rather good.

Only very few examples of gold(II) nitrogen compounds are known. There is only one gold(II) nitrite complex, Au$_2$(ylide)$_2$(NO$_2$)$_2$, reported in the book, Gold Progress in Chemistry, Biochemistry, and Technology, which was synthesized by Fackler and co-workers [2]. The great majority of compounds with gold nitrogen bonds occur with gold in the oxidation states +I and +III with the electronic configuration [Xe] 4f^{14}5d^{10}6s^{0}6p^0 and [Xe]4f^{14}5d^{8}6s^{0}6p^0 respectively [3]. There is a strong tendency for disproportionation from Au(II) to Au(I) and Au(III) in mononuclear complexes [9] because the odd electron in d^9 metal complexes is in the antibonding d$_{x2-y2}$ σ orbital, strongly overlapping with the ligand orbitals (octahedral, tetragonally distorted or square planar). Surprisingly, several gold(II) amidinate complexes have been produced. However, they are dinuclear species. The Au(II) amidinate complexes also have halides or pseudo-halides coordinated to the Au(II) and are stable at room temperature.

Table 1.1 Optimized geometries at MP2 level and selected experimental structural parameters for [M$_2$(NHCHNH)$_2$] [8].

System	M-M (Å)	M-N (Å)
[Au$_2$(NHCHNH)$_2$]	2.728	2.005
[Ag$_2$(NHCHNH)$_2$]	2.712	2.043
[Cu$_2$(NHCHNH)$_2$]	2.528	1.834
Experimental structural parameters for [M$_2$(NHCHNH)$_2$], M = Ag and Cu		
[Cu$_2$(ArNC(H)NAr)$_2$], Ar = C$_6$H$_4$-4-Me	2.497	1.886
[Ag$_2$(ArNC(H)NAr)$_2$], Ar = C$_6$H$_4$-4-Me	2.705	2.116

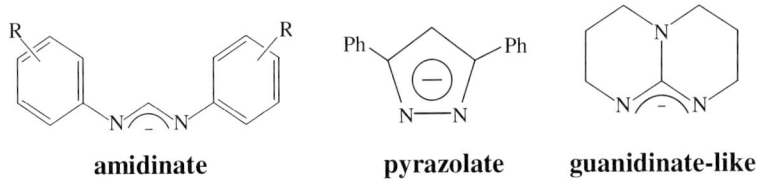

Figure 1.1 Some of the nitrogen ligands discussed in this review.

amidinate   pyrazolate   guanidinate-like

The anionic amidinate ligands, Figure 1.1, are known for their remarkable ability to bridge between the metal ions, facilitating the formation of short metal–metal distances and for their flexible coordination modes, leading to various molecular arrangements [10]. The use of these amidinate ligands in the coordination chemistry of the transition metals has produced complexes with extraordinarily short M–M distances [6]. These short distances are due, at least in part, to the ability of the amidinate anion to delocalize the negative charge while strongly donating sigma electron density to the metal atoms, supporting bond formation [11]. Previous studies have shown that the amidinate ligands form dinuclear Ag(I) and Cu(I) complexes [7, 12, 13]. Placing alkyl and aryl substituents on the amidinate NCN carbon atom influences the formation of tetranuclear and trinuclear structural motifs with Ag(I) [14]. Clearly the substituents play a role in determining the nuclearity and molecular arrangement of the complexes.

The nitrogen ligand chemistry with pyrazolate ligands, Figure 1.1, has produced mainly trinuclear complexes of group 11 with the structure, [M(μ-3,5-Ph$_2$Pz)]$_3$, M=Cu(I), Ag(I), Au(I). The hexanuclear gold complex [Au(μ-3,5-Ph$_2$Pz)]$_6$ has also been obtained, although in low yield [15]. Tetranuclear gold(I) pyrazolates were isolated only with a 3,5-di-isobutyl substituted pyrazolate [16]. The guanidinate-type anion ligand hpp (1,3,4,6,7,8-hexahydro-H-pyrimido[1,2-a]pyrimidine), Figure 1.1, forms complexes with extra short M–M bonding distances and stabilizes metals in high oxidation states [6]. Gold(I) and gold(II) complexes were isolated with the hpp ligand [17, 18].

1.2 Tetra-, Tri-, and Dinuclear Gold(I) Amidinate Complexes

The structural arrangement of group 11 amidinate complexes is determined by the substituents on the amidinate aryl groups as well as on the NCN carbon [14, 19]. The electronic vs. steric effect of the substituents on the molecular arrangement of gold(I) amidinate complexes have been studied in detail in the Fackler laboratory.

A series of symmetrical diaryl substituted amidinate ligands, ArNH(CH)NAr, has been synthesized, Figure 1.2. The substituents on the NCN aryl group vary from electron withdrawing groups such as -C$_6$F$_5$, 3-CF$_3$-C$_6$H$_4$, 3,5-Cl-C$_6$H$_3$ to donating groups such as 4-OMe-C$_6$H$_4$, 4-Me-C$_6$H$_4$, -C$_{10}$H$_7$. Ligands with sterically bulky groups in the ortho positions such as 2,6-Me$_2$-C$_6$H$_3$ as well as on the NCN carbon, NC(Me)N and NC(Ph)N, have also been prepared. The amidine ligands are readily

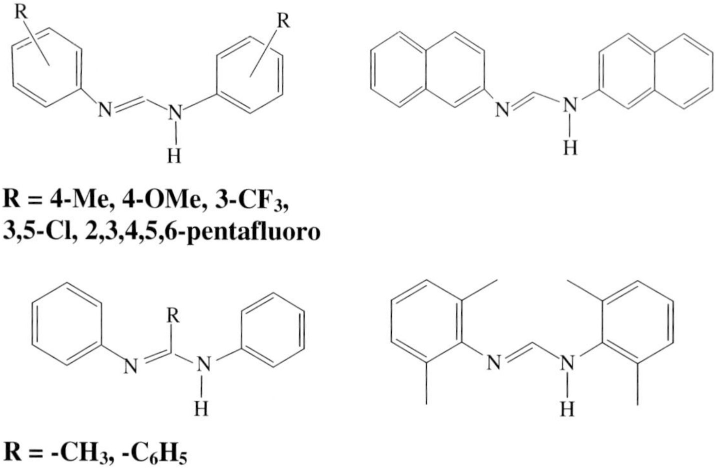

R = 4-Me, 4-OMe, 3-CF$_3$,
3,5-Cl, 2,3,4,5,6-pentafluoro

R = -CH$_3$, -C$_6$H$_5$

Figure 1.2 Amidine ligands used in the synthesis of terta-, tri-, and dinuclear gold complexes.

synthesized using modified literature procedures [14, 20]. The aniline derivative and triethylorthoformate (orthoester) are mixed and the reaction mixture heated to 140–160 °C to form the imido ester which later forms the amidine ligand, Figure 1.3.

Tetranuclear gold(I) amidinate complexes are synthesized by the reaction of Au(THT)Cl with the potassium or sodium salt of the amidinate ligand in THF, Figure 1.4. Syntheses involving various substituted amidinates resulted in tetranuclear gold(I) clusters, [Au$_4$(ArNC(H)NAr)$_4$]. The C-functionalized substituted amidine ligands, ArNC(Ph)NHAr and ArNC(Me)NHAr, Ar=-C$_6$H$_5$, were synthesized and reacted with Au(THT)Cl after deprotonation. Only tetranuclear clusters were isolated.

In each tetranuclear Au(I) aryl amidinate complex studied, [Au$_4$(ArNC(H)NAr)$_4$], the NC bond length in NCN is ~1.3 Å, indicating delocalization across the amidinate bridge. The four gold atoms are located at the corners of a rhomboid with the amidinate ligands bridged above and below the near plane of the four gold(I) atoms, Figures 1.5 and 1.6. The average Au···Au distance is ~3.0 Å, typical of Au(I)···Au(I)

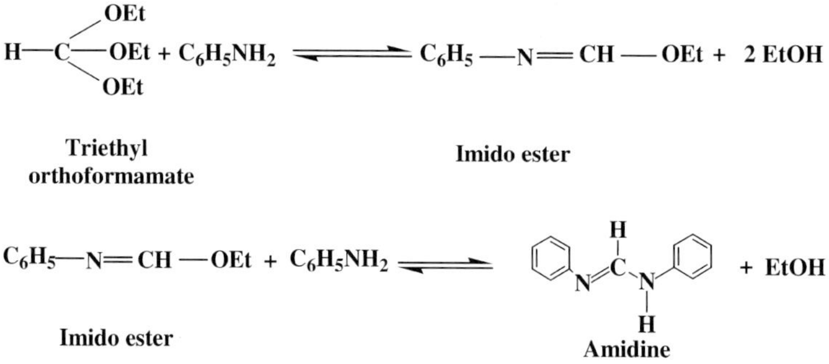

H—C(OEt)$_3$ + C$_6$H$_5$NH$_2$ ⇌ C$_6$H$_5$—N=CH—OEt + 2 EtOH

Triethyl orthoformamate   Imido ester

C$_6$H$_5$—N=CH—OEt + C$_6$H$_5$NH$_2$ ⇌ Amidine + EtOH

Imido ester

Figure 1.3 Synthesis of the amidine ligands.

1.2 Tetra-, Tri-, and Dinuclear Gold(I) Amidinate Complexes

R = H, Ar = C_6H_4-4-Me, C_6H_4-4-OMe, C_6H_3-3,5-Cl, C_6H_4-3-CF_3, -$C_{10}H_7$, -C_6F_5
R = -C_6H_5, Ar = -C_6H_5
R = -CH_3, Ar = C_6H_5

Figure 1.4 Schematic representation of the reaction between amidinate ligands and Au(THT)Cl.

aurophilic interactions. The four gold atoms are arranged in a near square (Au···Au···Au = 87–92°) in the tetranuclear structure [Au_4(ArNC(H)NAr)$_4$], Ar=-C_6F_5. The packing diagram shows weak F···F (~2.44 Å), Au···F (~3.14 Å intramolecular), and Au···F (~3.51 Å, intermolecular) interactions. Figures 1.7 and 1.8 are thermal ellipsoid plots of [Au_4(ArNC(Ph)NAr)$_4$] and [Au_4(ArNC(CH_3)NAr)$_4$], Ar= -C_6H_5. The average Au···Au distance is 2.94 Å, typical of Au(I)···Au(I) aurophilic interactions. The gold atoms are arranged in a near square (Au···Au···Au = 88–91°)

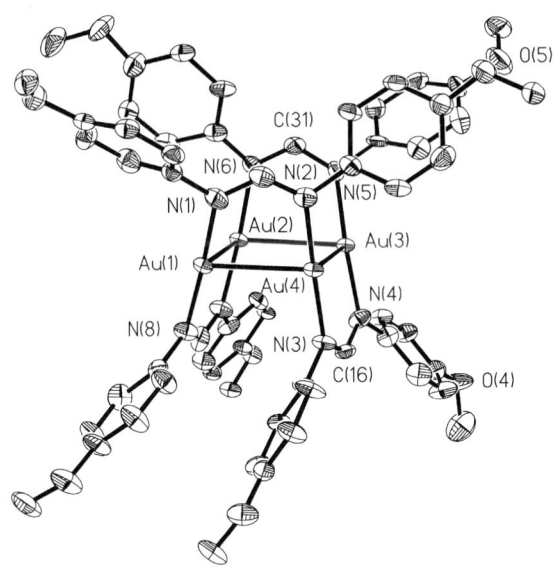

Figure 1.5 Thermal ellipsoid plot of [Au_4(ArNC(H)NAr)$_4$], Ar = 4-OMe-C_6H_4.

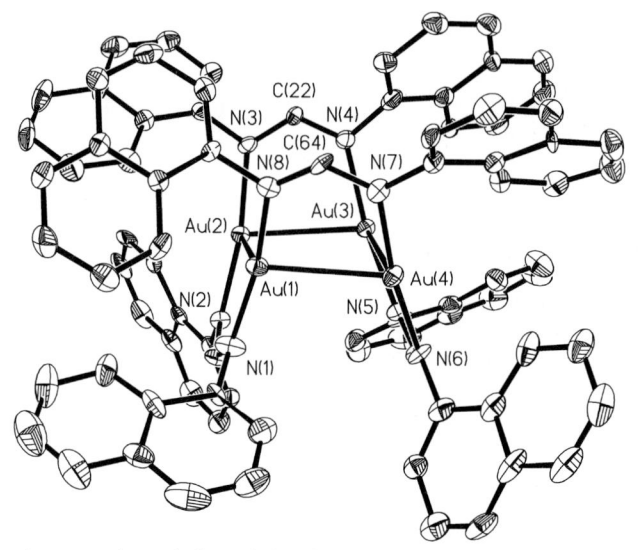

Figure 1.6 Thermal ellipsoid plot of [Au$_4$(ArNC(H)NAr)$_4$], Ar = C$_{10}$H$_7$.

in [Au$_4$(C$_6$H$_5$NC(CH$_3$)NC$_6$H$_5$)$_4$] and a distorted square (Au···Au···Au = 82–97°) in [Au$_4$(C$_6$H$_5$NC(Ph)NC$_6$H$_5$)$_4$].

Table 1.2 gives the Au···Au distances, and Au···Au···Au and N–Au–N angles for several homobridged tetranuclear Au(I) complexes and tetranuclear gold amidinate complexes. Similar structural arrangements have been found in the tetrameric 1,3-diphenyltriazenidogold(I) complex, [Au(PhNNNPh)]$_4$ (Au···Au = 2.85 Å) [21],

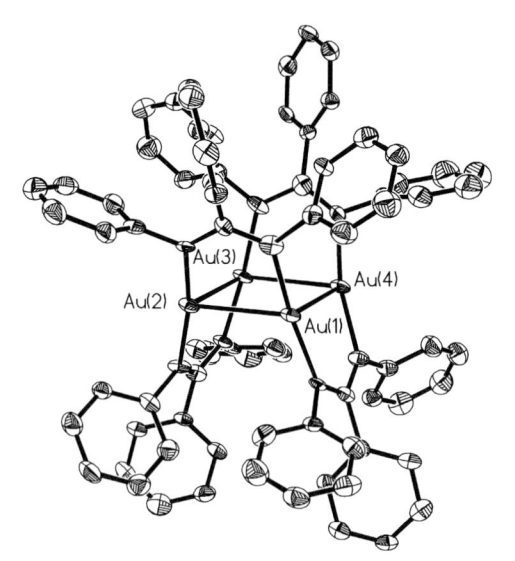

Figure 1.7 Thermal ellipsoid plot of [Au$_4$(C$_6$H$_5$NC(Ph)NC$_6$H$_5$)$_4$].

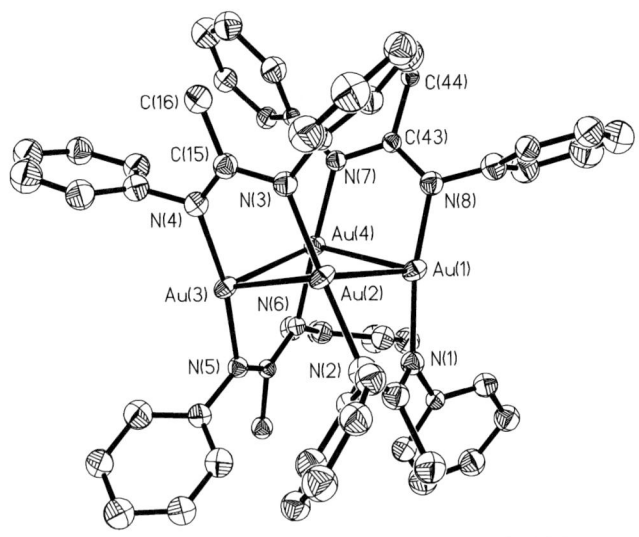

Figure 1.8 Thermal ellipsoid plot of [Au$_4$(C$_6$H$_5$NC(CH$_3$)NC$_6$H$_5$)$_4$].

[Au$_4$(CH$_3$CS$_2$)$_4$] (Au$\cdots$Au = 3.01 Å) [22] and the tetranuclear gold pyrazolate complex [Au(3,5-t-Bu-pz)]$_4$ (Au$\cdots$Au = 3.11 Å) [16]. The Au(I) atoms bridged by the more flexible amidinate ligands show shorter Au$\cdots$Au distances than those bridged by the rigid pyrazolate ligands (i.e., 2.9 Å versus 3.1 Å) [13].

Using sterically bulky groups in the ortho positions of the phenyl rings in ArNC(H)NHAr, such as Ar = 2,6-Me$_2$-C$_6$H$_3$, led to formation of dinuclear and trinuclear complexes. This suggests that steric factors can prevent the formation

Table 1.2 Average Au$\cdots$Au distances (Å) and Au$\cdots$Au$\cdots$Au angles (°) of tetranuclear gold(I) amidinate and related clusters.

Complex	Au$\cdots$Au	Au(1)$\cdots$Au(2)$\cdots$Au(3)	N-Au-N	Ref
[Au(PhNNNPh)]$_4$	2.85	89.92	176	[21]
[Au(CH$_3$CS$_2$)]$_4$	3.01	89.95	167	[22]
[Au(3,5-t-Bu-pz)]$_4$	3.11		175	[16]
Amidinate clusters [Au$_4$(ArNC(H)NAr)$_4$], Ar =				
C$_6$H$_4$-4-OMe	2.94	70.87, 109.12	174	[19]
C$_6$H$_3$-3,5-Cl	2.91	88.30, 91.53	177	[19]
C$_6$H$_4$-4-Me	3.03	63.59, 116.4	172	[19]
C$_{10}$H$_7$	2.98	68.52, 110.88	170	[19]
C$_6$F$_5$	2.96	92.3, 87.5	169	[19]
C$_6$H$_4$-3-CF$_3$	2.92	84.6, 95.3	176	[19]
Amidinate clusters [Au$_4$(PhNC(R)NPh)$_4$], R=				
C$_6$H$_5$	2.94	82.86, 97.66	173	[19]
CH$_3$	2.93	88.47, 91.14	168	[19]

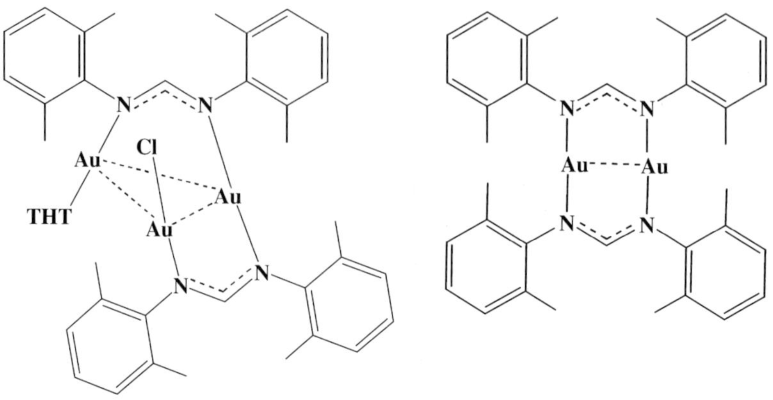

Figure 1.9 Structure of the trinuclear and dinuclear gold amidinate complexes.

of tetranuclear gold(I) amidinates, Figure 1.9 [23]. Models show that formation of the tetranuclear species is blocked by ligand-ligand interactions. Previous work indicated the formation of a dinuclear product, Au···Au = 2.646 Å, when Me_3Si was bonded to the N atoms, $\{Au_2[(Me_3SiN)_2C(Ph)]_2\}$ [4]. The formation of the dinuclear Au(II) guanidinate complex $[Au_2(hpp)_2Cl_2]$ also implicates the possible presence of a dinuclear Au(I) species with this sterically uncrowded ligand [17, 18].

The trinuclear species $[Au_3(2,6-Me_2-form)_2(THT)Cl]$ and the dinuclear $[Au_2(2,6-Me_2Ph-form)_2]$ were isolated by the reaction of the potassium salt of the corresponding amidinate ligand with (THT)AuCl, Figure 1.10. The structure of the trinuclear gold complex shows a short Au···Au distance of ~3.01 Å and a longer Au···Au distance of 3.66 Å, Figure 1.11. The Au···Au distance in the dinuclear complex $[Au_2(2,6-Me_2Ph-form)_2]$ is 2.711(3) Å, and the N-Au-N angle is 170.2(3)°. To our knowledge, the only other example of a symmetrically bridged dinuclear gold(I) nitrogen complex is $\{Au_2[(Me_3SiN)_2C(Ph)]_2\}$, which has an Au···Au distance of 2.646 Å [24]. The Au···Au distance in $[Au_2(2,6-Me_2Ph-form)_2]$ is 2.711 Å, Figure 1.12, close to the distance suggested by Pyykkö [8] for the $[Au_2(NHCHNH)_2]$. This is shorter than in the xanthate $[Au_2(^nBu\text{-xanthate})_2$ (2.849 Å) [25], the dithiophosphinate $[AuS_2PPh_2]_2$ (3.085 Å) [26], ylide $[Au(CH_2)_2PPh_2]_2$ (2.977 Å) [27], and dithiophosphonate $[AuS_2PPh(OEt)]_2$ (3.042 Å) [28], but somewhat closer to the observed Au···Au distances in the dithiolates $[PPN]_2[Au_2(\mu^2\text{-}\eta^2\text{-}CS_3)_2]$ (2.799 Å) [29], $[n\text{-}Bu_4N][Au(S_2C=C(CN)_2]_2$ (2.796 Å) [30], and $[Au(S_2C-N(C_5H_{11})_2)]_2$ (2.769 Å) [31].

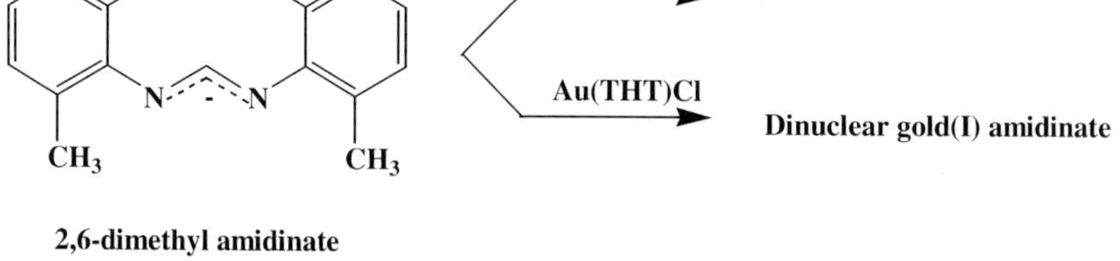

2,6-dimethyl amidinate

Figure 1.10 Synthesis of dinuclear and trinuclear gold(I) amidinates.

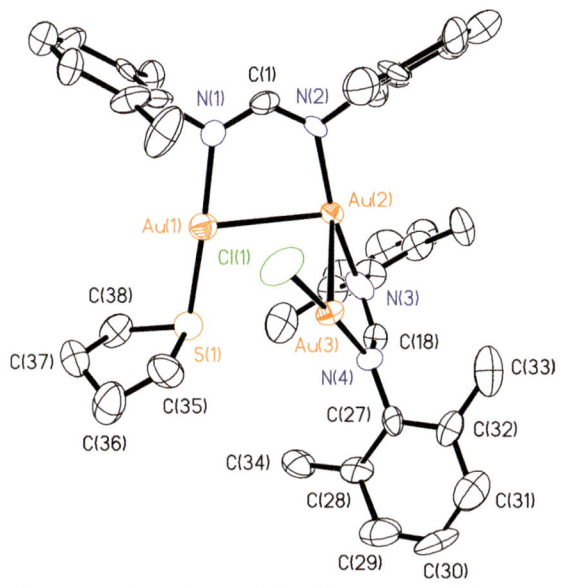

Figure 1.11 Thermal ellipsoid plot of [Au$_3$(2,6-Me$_2$Ph-form)$_2$(THT)Cl].

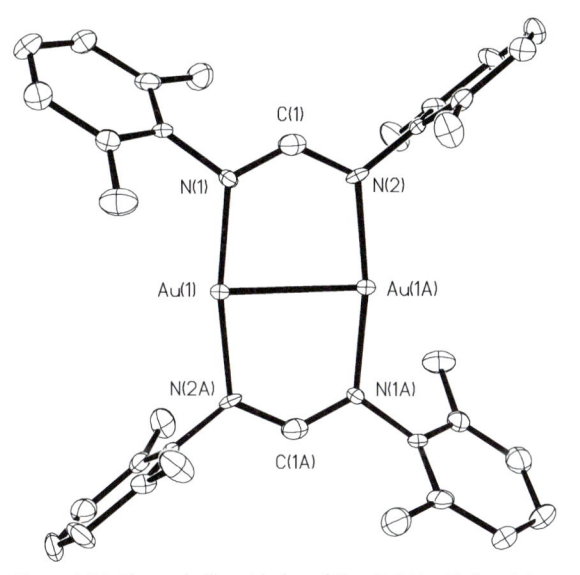

Figure 1.12 Thermal ellipsoid plot of [Au$_2$(2,6-Me$_2$Ph-form)$_2$].

1.3
Oxidative-Addition Reactions to the Dinuclear Gold(I) Amidinate Complex

Oxidative-addition reactions have been widely studied with bridged dinuclear metal complexes [1, 2, 5, 32]. Earlier work with the ylides and sulfur bonded ligands

Figure 1.13 Synthesis of gold(II) amidinate complexes by oxidative-addition to the dinuclear gold(I) amidinate.

(*vide infra*) has led to the formation of many Au(II)-Au(II) bonded complexes. No stable organometallic alkyl halide addition products of dinuclear Au(I) complexes have been characterized with ligands other than the ylides [32]. Oxidative-addition reactions to the dinuclear gold(I) amidinate complex, [Au$_2$(2,6-Me$_2$Ph-form)$_2$], result in the formation of Au(II) complexes. The Au(II) amidinate complexes are the first gold(II) species isolated with nitrogen ligands [23]. The complexes are stable at room temperature. Various reagents such as Cl$_2$, Br$_2$, I$_2$, benzoyl peroxide and CH$_3$I add to the dinuclear gold(I) amidinate complex to form oxidative-addition gold(II) products, [Au$_2$XY(2,6-Me$_2$Ph-form)$_2$], Figure 1.13 [23, 33, 34]. The methyl iodide addition product is the only organometallic Au(II) species formed to date with amidinate ligands [33].

The reaction of the dinuclear complex, [Au$_2$(2,6-Me$_2$Ph-form)$_2$], with the halogenated solvents, CH$_2$X$_2$, XCH$_2$CH$_2$X, CX$_4$ (X=Cl, Br, I) also forms Au(II) products. With the iodide derivatives the reaction occurs at the time of mixing. The analogous reactions with chloride and bromide derivatives take approximately 2–3 days, and 7 days with CH$_2$Cl$_2$ in order to oxidize all the Au(I) material. A crystalline product in which there are equal amounts of oxidized and unoxidized complexes in the same unit cell, [Au$_2$(2,6-Me$_2$Ph-form)$_2$X$_2$][Au$_2$(2,6-Me$_2$Ph-form)$_2$], X=Cl and X=Br, Figure 1.14, is isolated when the reaction is stopped after 3–4 h of stirring. In the reaction of the haloalkyls CH$_n$X$_m$, the qualitative order of reactivity with the dinuclear

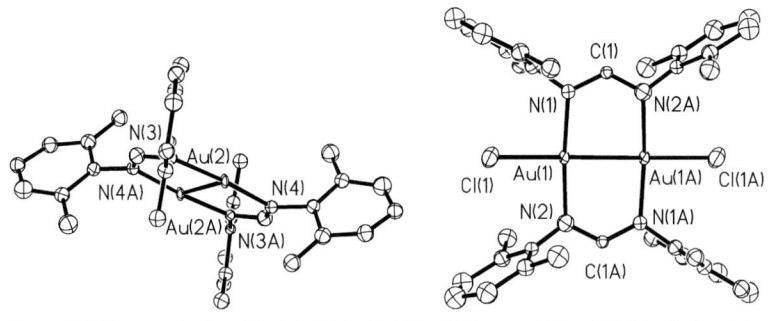

Figure 1.14 Thermal ellipsoid plot of [Au$_2$(2,6-Me$_2$Ph-form)$_2$Cl$_2$][Au$_2$(2,6-Me$_2$Ph-form)$_2$].

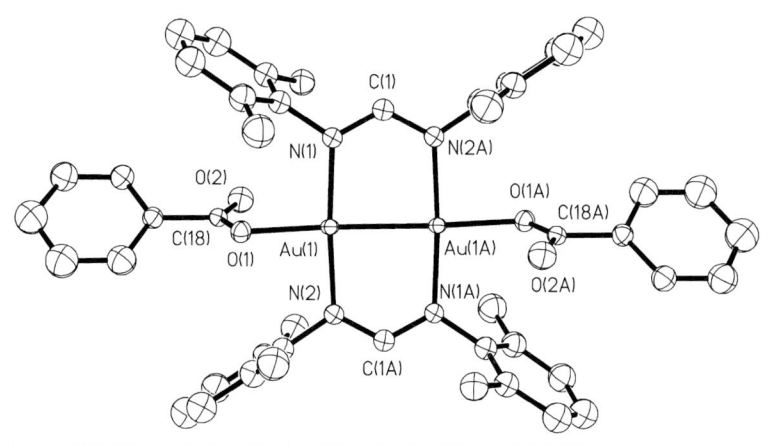

Figure 1.15 Thermal ellipsoid plot of [Au$_2$(2,6-Me$_2$Ph-form)$_2$(PhCO)$_2$].

gold complex (I > Br > Cl) follows inversely the order of carbon-halogen bond dissociation energy, C–Cl > C–Br > C–I.

The oxidative-addition of benzoyl peroxide, (PhCOO)$_2$, leads to the isolation of the first stable dinuclear gold(II) nitrogen complex also possessing Au–O bonds, [Au$_2$(2,6-Me$_2$Ph-form)$_2$(PhCO$_2$)$_2$] [34]. An analogous ylide complex, [Au$_2$((CH$_2$)$_2$PPh$_2$)$_2$(PhCO$_2$)$_2$] is known [33]. The benzoate amidinate product [Au$_2$(2,6-Me$_2$Ph-form)$_2$(PhCO)$_2$] was obtained, Figure 1.15, by adding an equivalent amount of benzoyl peroxide to a toluene solution of the dinuclear Au(I) amidinate. Infra-red spectroscopic studies of the gold(II) benzoate complex show two intense bands at 1628 and 1578 cm^{-1} due to v(C=O) and at 1320–1295 cm^{-1} due to v(C–O) frequencies. The separation between the two bands is $\sim$300 cm^{-1}, typical of unidentate benzoate bonding, a "pseudo ester" character. The bonding of the benzoates to the dinuclear gold(II) amidinate is similar to the unidentate bonding observed in the ylide complexes, which adopt an *anti* geometry [2, 35].

The oxidative-addition of benzoyl peroxide to the dinuclear gold(I) ylide complex formed a gold(II) complex with the shortest Au$\cdots$Au distance observed, 2.56–2.58 Å [2,35], for the dinuclear Au(II) ylide complexes, Table 1.3. Similarly, the Au$\cdots$Au distance in the oxidized product [Au$_2$(2,6-Me$_2$Ph-form)$_2$(PhCO)$_2$], 2.48 Å [34], is the shortest Au$\cdots$Au distance in the Au(II) amidinate complexes. The short Au$\cdots$Au distance in these complexes is due to the weak *trans*-directing and sigma covalent bonding ability of the carboxylate ligands to gold(II) compared with the halides [34].

A facile replacement of the benzoate groups in [Au$_2$(2,6-Me$_2$Ph-form)$_2$(PhCOO)$_2$] by chloride or bromide is achieved by adding equivalent amounts of PhICl$_2$ or [Bu$_4$N]Br, Figure 1.16. The replacement of the bromide in [Au$_2$(2,6-Me$_2$Ph-form)$_2$Br$_2$] by chloride is achieved by adding 1 mol of PhICl$_2$ to 1 mol in polar solvent such as CH$_3$CN.

The X-ray crystallography of the gold(II) amidinate complexes shows a decrease in the Au$\cdots$Au distance from 2.71 Å in the starting dinuclear complex to 2.51–2.57 Å in the oxidized species. The Au–X distances are Au–Cl = 2.36 Å, Figure 1.17, Au–Br = 2.47 Å, and Au–I = 2.68 Å, Figure 1.18. The Au–N distances decreases from

Table 1.3 Dinuclear Au(II) ylide and amidinate complexes characterized by X-ray studies.

Complex	d(Au^{II}···Au^{II})	d(Au-X)	d(Au-R)	Ref
[ClAu(CH$_2$PPh$_2$CH$_2$)$_2$AuCl]	2.600(1)	2.388(8)		[9]
[BrAu(CH$_2$PPh$_2$CH$_2$)$_2$AuBr]	2.614(1)	2.516(1)		[9]
[IAu(CH$_2$PPh$_2$CH$_2$)$_2$AuI]	2.650	2.693(8)		[9]
[(CH$_3$)Au(CH$_2$PPh$_2$CH$_2$)$_2$AuI]	2.695(4)			[9]
[(CH$_3$)Au(CH$_2$PMe$_2$CH$_2$)$_2$AuI]	2.695(4)	2.894(5)	2.13(5)	[9]
[PhCO$_2$Au(CH$_2$PPh$_2$CH$_2$)$_2$AuO$_2$CPh]	2.561(2)	2.117(13)		[9]
[ClAu(2,6-Me$_2$Ph-form)$_2$AuCl]	2.517(7)	2.356(2)		[34]
[BrAu(2,6-Me$_2$Ph-form)$_2$AuBr]	2.525(15)	2.470(2)		[34]
[IAu(2,6-Me$_2$Ph-form)$_2$AuI]	2.579(4)	2.682(4)		[346]
[(CH$_3$)Au$_2$(2,6-Me$_2$Ph-form)$_2$AuI]	2.529(11)	2.50	2.12	[33, 34]
[PhCO$_2$Au(2,6-Me$_2$Ph-form)$_2$AuO$_2$CPh]	2.489(10)	2.045(8)		[34]

2.035(7) Å in the dinuclear complex to 2.00–2.004 Å in the oxidative-addition products. The Au atoms have a nearly square-planer coordination geometry.

The reaction of methyl iodide, CH$_3$I, with [Au$_2$(2,6-Me$_2$Ph-form)$_2$] in ether generates [CH$_3$Au(2,6-Me$_2$Ph-form)$_2$AuI] in quantitative yield under nitrogen at 0 °C, in the absence of light [33, 34]. While the Au(II) atoms and the amidinate ligand atoms refine well, unfortunately, the structure has a disorder in the CH$_3$ and iodide positions since the spatial volume occupied by CH$_3$ and I is approximately identical. As a result their positions remain uncertain with regard to their exact distances from the Au(II) atoms. The Au–CH$_3$ and Au–I distances appear to be 2.12 Å and 2.50 Å, while in the dinuclear gold(I) ylide, [(CH$_3$)Au((CH$_2$)2PMe$_2$)$_2$AuI], the Au–CH$_3$ and Au–I distances are 2.13(5) and 2.894(5) Å, respectively, and the Au(II) –Au(II) distance is 2.695(4) Å [9]. Surprisingly, the Au–I distance in the Au(II) amidinate complex appears to be shorter than found in the ylide, [IAu((CH$_2$)$_2$PMe$_2$)$_2$AuI].

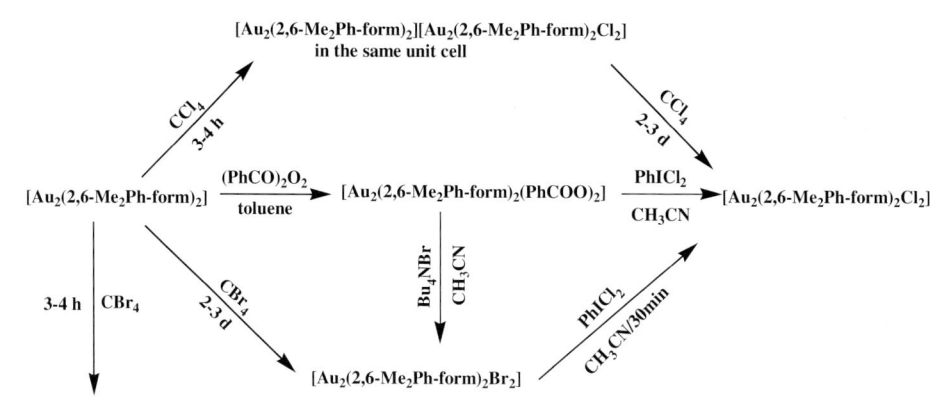

Figure 1.16 Schematic representation of the replacement reactions of [Au$_2$(2,6-Me$_2$Ph-form)$_2$].

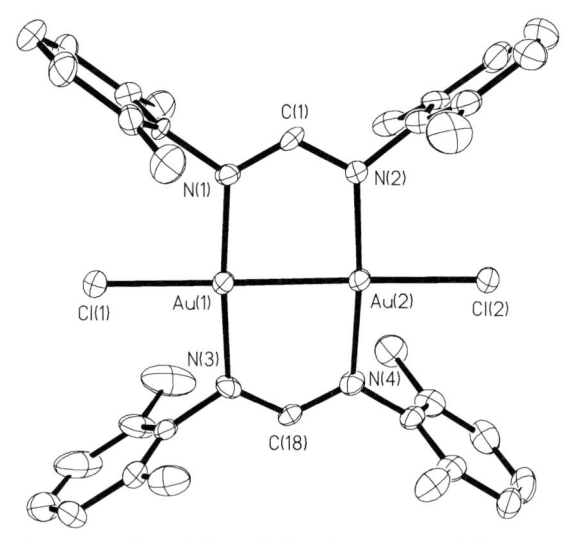

Figure 1.17 Thermal ellipsoid plot of [Au$_2$(2,6-Me$_2$Ph-form)$_2$Cl$_2$].

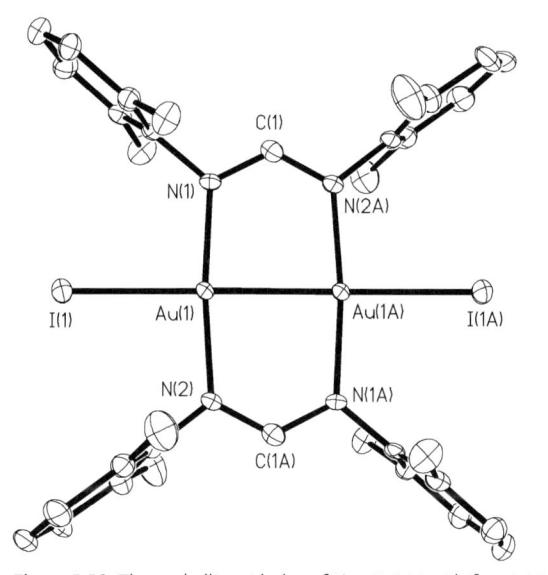

Figure 1.18 Thermal ellipsoid plot of [Au$_2$(2,6-Me$_2$Ph-form)$_2$I$_2$].

1.4
Mercury(II) Cyanide Coordination Polymer

The reaction of the dinuclear gold(I) amidinate complex, [Au$_2$(2,6-Me$_2$Ph-form)$_2$], with Hg(CN)$_2$ (1:2 stoichiometry) in THF forms a 2D coordination polymer, [Au$_2$(2,6-Me$_2$Ph-form)$_2$]·2Hg(CN)$_2$·2THF, not the expected oxidative-addition product of the type formed with the ylides. White crystals and a yellow powder are formed.

1 Gold(I) Nitrogen Chemistry

The white crystals change to yellow powder upon grinding, presumably with loss of THF and possibly some AuCN formation. Thermal gravimetric analysis of [Au$_2$(2,6-Me$_2$Ph-form)$_2$]·2Hg(CN)$_2$·2THF showed the release of THF gradually at >120 °C followed by decomposition at >200 °C. The powder diffraction pattern of the yellow residue after heating above 265 °C showed a pattern typical of AuCN (IR 2236 cm^{-1}) as confirmed by comparison with the powder diffraction pattern of a sample of AuCN obtained from the Aldrich Chemical Co.

The behavior of Hg(CN)$_2$ toward the dinuclear gold(I) amidinate complexes requires comment. In the case of the dinuclear gold(I) ylide, oxidation of the Au(I) to Au(II) resulted in the formation of a reduced mercury(0) product, Figure 1.19(a) [36]. In the mercury(II) cyanide reaction with the dinuclear gold(I) dithiophosphinate, Figure 1.19(b), the stability of the gold(I)-carbon bond compared

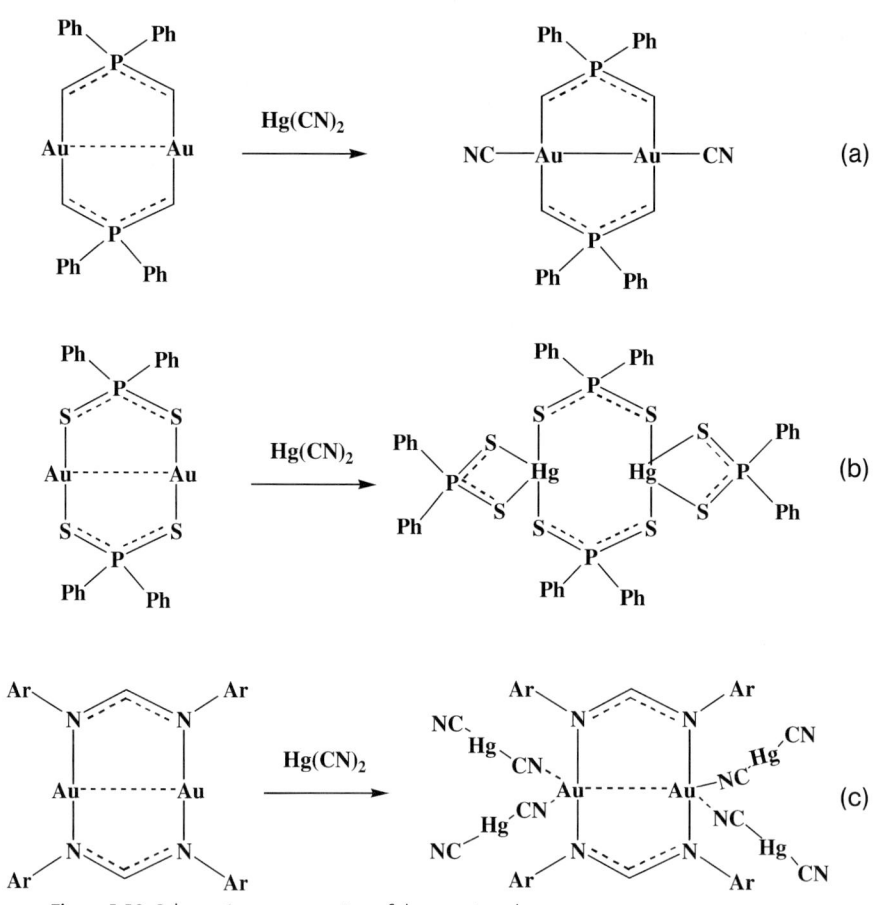

Figure 1.19 Schematic representation of the reactions between Hg(CN)$_2$ and the (a) dinuclear gold(I) ylide with loss of Hg(0), (b) dinuclear gold(I) dithiophosphinate with los of AuCN, and with (c) dinuclear Au(I) 2,6-Me$_2$formamidinate complexes.

with that of mercury(II)-carbon bond and the strength of the Hg(II)–S bonds compared with the labile Au(I)–S bonds appear to lead to the metathesis products observed. With the adduct to the amidinate ligand complex, Figure 1.19(c), the cyanide IR stretching frequency shifts from 2192 cm^{-1} in Hg(CN)$_2$ to ~2147 cm^{-1}, a value very near to the CN stretching frequency found (2145 cm^{-1}) in the dinuclear Au(II) ylide dicyanide [37]. However, the oxidation of the dinuclear Au(I) amidinate by the Hg(CN)$_2$ is much more difficult than the oxidation of the dinuclear Au(I) ylide. Cyclic voltammetric studies bear this out (*vide infra*) [38].

The differences observed in the chemistry of these dinuclear gold(I) amidinate complexes compared with dinuclear gold(I) complexes with sulfur and carbon ligands may be understood by examining their respective highest occupied molecular orbital (HOMO)s and lowest unoccupied molecular orbital (LUMO)s of the species. In the ylide complexes the HOMO is a metal-metal σ^* antibonding orbital, and the LUMO is a bonding σ orbital directed along the metal-metal axis. In the dinuclear gold(I) amidinates the HOMO is δ^* with regard to the π orbitals of the N ligands [39] and the LUMO is also largely ligand based.

Gold(I) ylides are oxidized in 0.1 M [Bu$_4$N]BF$_4$/THF at low potentials of +0.11 and +0.23 V vs. Ag/AgCl (quasi-reversible). The dinuclear amidinate oxidizes under the same conditions at +1.24 V vs. Ag/AgCl (reversible). These large differences in chemical character of the dinuclear gold(I) complexes appear to explain the widely different behavior of these compounds and especially toward the reaction with mercury cyanide.

The adduct formation of Hg(CN)$_2$ to the [Au$_2$(2,6-Me$_2$Ph-form)$_2$] increases the Au···Au distance from 2.7 Å in the dinuclear complex to 2.9 Å in the adduct, Figure 1.20. The gold centers are coordinated by four nitrogen atoms with Au–N distances in the range 2.09–2.51 Å. The N-Au-N angles associated with the amidinate ligands decreased from ~170° in the dinuclear starting material to ~161°. The N-Au-N angles are in the range 95–100° (angles from the cyanide groups). The 2D lattice contains two THF solvent molecules in the large cavities (~10.2 × 13.7 Å), Figure 1.20(b). The Hg–O distance is ~4.33 Å indicating that the interaction between the Hg centers and THF is not significant.

1.5
Formation of Mixed-Ligand Tetranuclear Gold(I) Nitrogen Clusters

Density functional theory (DFT) modeling calculations show that a dinuclear gold(I) amidinate complex is less stable than the tetranuclear gold(I) amidinate cluster, [Au$_4$(HNC(H)NH)$_4$]. However, replacing C by Si in the backbone reduces ring strain and makes the energies similar, Figures 1.21 and 1.22 [39].

Attempts to introduce less bulky anionic ligands to the dinuclear complex [Au$_2$(2,6-Me$_2$Ph-form)$_2$] cause the dinuclear gold(I) amidinate complex to rearrange and form tetranuclear gold(I) amidinate complexes [40]. The ligand exchange of the sterically bulky ligand, 2,6-Me$_2$Ph-form, in the dinuclear gold(I) amidinate complex with less bulky anionic ligands such as [ArNC(H)NAr]$^-$, Ar=C$_6$H$_4$-4-Me, Ar=C$_6$H$_4$-4-OMe,

16 | *1 Gold(I) Nitrogen Chemistry*

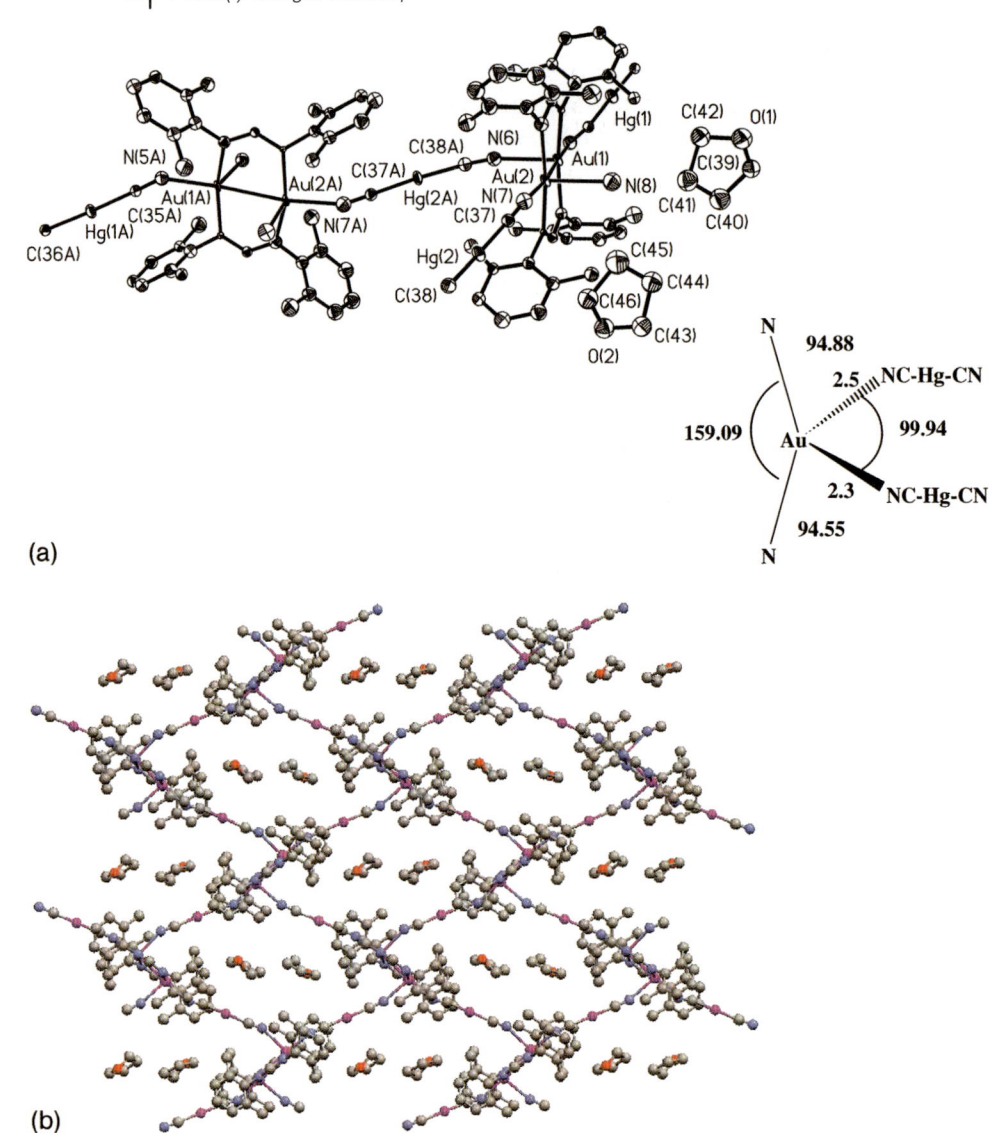

(a)

(b)

Figure 1.20 (a) Thermal ellipsoid plot and bond distances and angles of [Au$_2$(2,6-Me$_2$Ph-form)$_2$]·2Hg(CN)$_2$·2THF. (b) 2D of [Au$_2$(2,6-Me$_2$Ph-form)$_2$]·2Hg(CN)$_2$·2THF showing the THF solvent in the voids.

and 3,5-diphenylpyrazolate, to form mixed-ligand species provides a facile procedure for the synthesis of mixed-ligand complexes along with the increased nuclearity, Figure 1.23. Tetranuclear gold(I) complexes with amidinate and pyrazolate ligands can be formed, such as [Au$_4$(3,5-Ph$_2$pz)$_2$(2,6-Me$_2$Ph-form)$_2$] and [Au$_4$(3,5-Ph$_2$pz)$_3$(2,6-Me$_2$Ph-form)]. This result was extended to the synthesis of tetranuclear

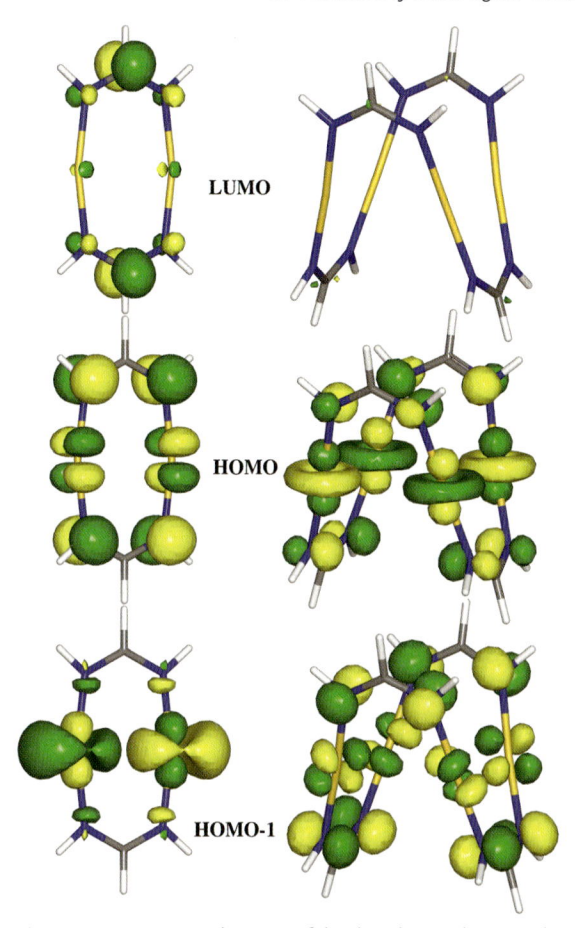

Figure 1.21 HOMO and LUMO of the dinuclear and tetranuclear gold amidinate species.

mixed-metal Au–Ag complexes with pyrazolate and amidinate ligands [Au$_2$(3,5-Ph$_2$pz)$_2$ Ag$_2$(2,6-Me$_2$Ph-form)$_2$] [40]. This complex is the only tetranuclear amidinate complex observed with the two bulky amidinate ligands facing each other. Apparently, the long Au$\cdots$Ag distances, $\sim$3.3 Å, allow the bulky amidinate ligands to be in this *syn* arrangement.

Reacting the amidinate salt, K[4-MePh-form], with the dinuclear gold(I) complex, [Au$_2$(2,6-Me$_2$Ph-form)$_2$], in a 1 : 1 stoichiometry in THF forms the dinuclear-tetranuclear complex [Au$_2$(2,6-Me$_2$Ph-form)$_2$][Au$_4$(4-MePh-form)$_4$]·2THF, Figure 1.24, with one tetranuclear and one dinuclear molecule in the same unit cell. Adjusting the reaction ratio to 2 : 1 formed the tetranuclear complex [Au$_4$(4-MePh-form)$_4$].

The reaction of the diphenylpyrazolate salt, Na[3,5-Ph$_2$pz], with the dinuclear gold(I) complex [Au$_2$(2,6-Me$_2$Ph-form)$_2$] in a 1 : 1 stoichiometric ratio resulted in the formation of two tetranuclear products, observed as blocks, [Au$_4$(3,5-Ph$_2$pz)$_2$ (2,6-Me$_2$Ph-form)$_2$]·2THF, Figure 1.25, and as needles, [Au$_4$(3,5-Ph$_2$pz)$_3$(2,6-Me$_2$Ph-form)]·THF, Figure 1.26. Adjusting the reaction ratio to 1.5 : 1 resulted in the

Gaussian 98 B3LYP Au (Stuttgart)
C,N (cc-pVDZ) H (D95)

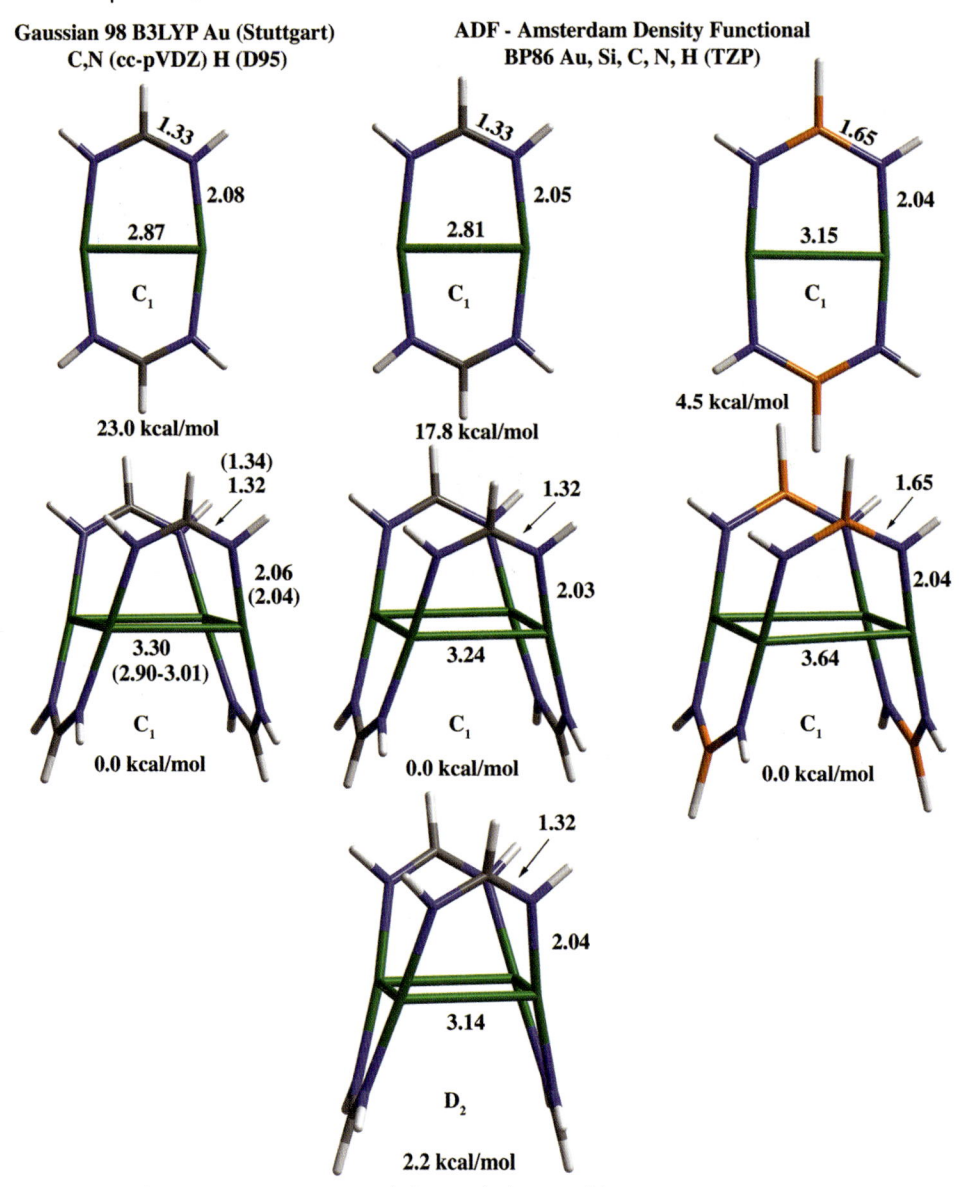

ADF - Amsterdam Density Functional
BP86 Au, Si, C, N, H (TZP)

1.33, 2.08, 2.87, C_1, 23.0 kcal/mol

1.33, 2.05, 2.81, C_1, 17.8 kcal/mol

1.65, 2.04, 3.15, C_1, 4.5 kcal/mol

(1.34) 1.32, 2.06 (2.04), 3.30 (2.90-3.01), C_1, 0.0 kcal/mol

1.32, 2.03, 3.24, C_1, 0.0 kcal/mol

1.65, 2.04, 3.64, C_1, 0.0 kcal/mol

1.32, 2.04, 3.14, D_2, 2.2 kcal/mol

Figure 1.22 Density Functional Theory calculations of the tetranuclear and dinuclear amidinate complexes at both the Gaussian 98 and ADF levels.

isolation of the tetranuclear mixed-ligand complex [Au$_4$(3,5-Ph$_2$pz)$_3$(2,6-Me$_2$Ph-form)]·THF. [40] These exchange reactions of the bulky amidinate ligand, [(2,6-Me$_2$-form)]$^-$, by the less steric ligands [(4-MePh-form)]$^-$ and [(3,5-Ph$_2$pz)]$^-$, are irreversible, Figure 1.23. These results validate the calculations which indicate that

1.5 Formation of Mixed-Ligand Tetranuclear Gold(I) Nitrogen Clusters | 19

```
                         K[(4-MePh-form)]
[Au₂(2,6-Me₂Ph-form)₂]  ─────────────────→  [Au₂(2,6-Me₂Ph-form)₂][Au₄(4-MePh-form)₄]
```

$[Au_2(2,6\text{-}Me_2Ph\text{-}form)_2]$ — K[(4-MePh-form)] → $[Au_2(2,6\text{-}Me_2Ph\text{-}form)_2][Au_4(4\text{-}MePh\text{-}form)_4]$

K[(4-MePh-form)], excess ⟶ (crossed out arrow back)

+ K[(2,6-Me₂Ph-form)]

K[3,5-Ph₂pz] ↓   ↓ K[(4-MePh-form)]

[Au₄(3,5-Ph₂pz)₂(2,6-Me₂Ph-form)₂]   [Au₄(4-MePh-form)₄]
[Au₄(3,5-Ph₂pz)₃(2,6-Me₂Ph-form)]

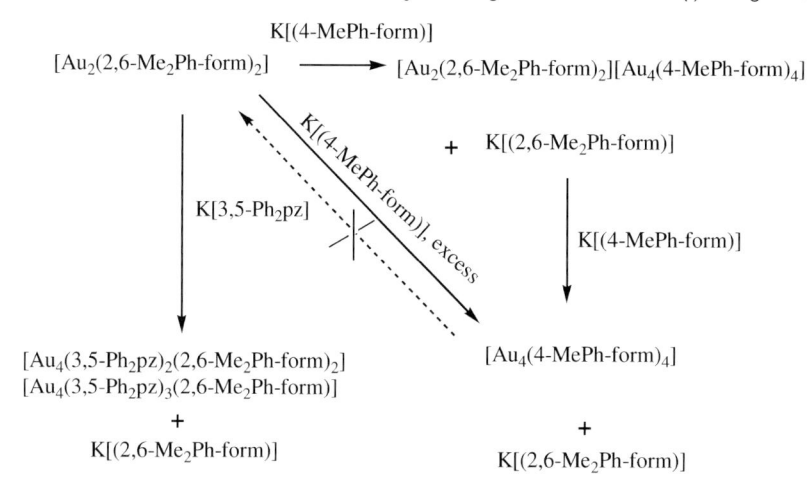

+   +
K[(2,6-Me₂Ph-form)]   K[(2,6-Me₂Ph-form)]

Figure 1.23 Schematic representation of the exchange reactions.

the tetranuclear structure is favored over the dinuclear arrangement with these nitrogen ligands.

The structures of the mixed ligand, dinuclear-tetranuclear Au₂Au₄ complexes show similar bond distances and angles, Au···Au = ~2.7 Å, (dinuclear) and ~3.0 Å (tetranuclear), to their parent complexes [19, 23, 34]. In the complexes [Au₄(3,5-Ph₂pz)₂ (2,6-Me₂-form)₂] and [Au₄(3,5-Ph₂pz)₃(2,6-Me₂-form)], the Au···Au dis-

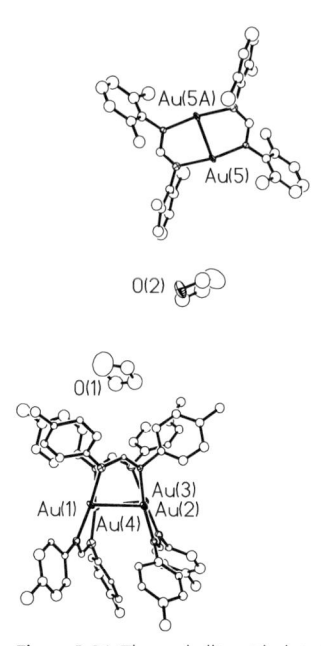

Figure 1.24 Thermal ellipsoid plot of [Au₂(2,6-Me₂Ph-form)₂][Au₄(4-Me-form)₄]·2THF.

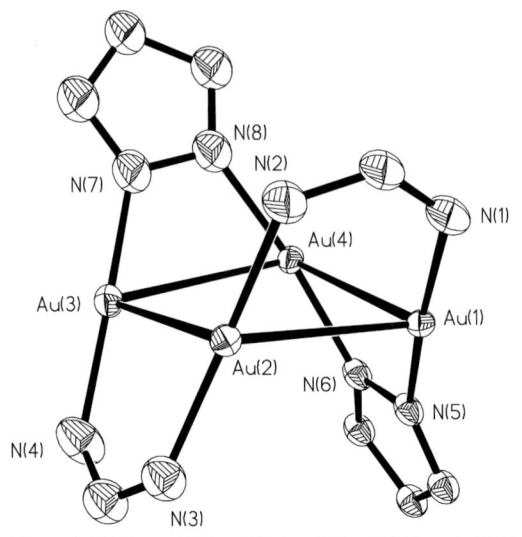

Figure 1.25 Thermal ellipsoid plot of [Au$_4$(3,5-Ph$_2$pz)$_2$(2,6-Me$_2$-form)$_2$].

tances range from 3.02–3.20 Å. Each pyrazolate ring in [Au$_4$(3,5-Ph$_2$pz)$_2$(2,6-Me$_2$Ph-form)$_2$]·2THF is facing an amidinate ligand, that is, *anti*. This avoids the steric bulk of the amidinate ligand, Figure 1.25. In the tetranuclear gold(I) pyrazolate complex, [Au$_4$(*t*-Bu-pz)$_4$], the Au···Au distance ranges from 3.11 to 3.18 Å [16]. The Au···Au distances linked by the pyrazolate ligands are slightly longer than those linked by the amidinate ligands, as expected from the required orientation of the Au–N sigma bonds of the different ligands.

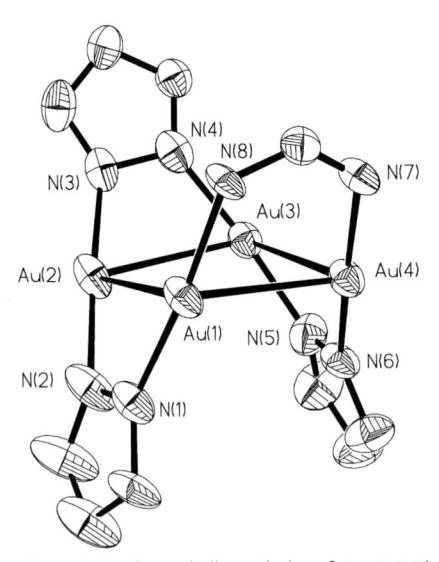

Figure 1.26 Thermal ellipsoid plot of [Au$_4$(3,5-Ph$_2$pz)$_3$(2,6-Me$_2$-form)].

1.6
Solvent Influences on Oxidation and Nuclearity of Gold Guanidinate Derivatives

Work with the Hhpp ligand was pioneered by Cotton and co-workers who showed that the di-metal complexes with Cr(II), Mo(II), or W(II), ionize readily, the latter more readily than cesium [41]. Recent work with the anionic hpp ligand has produced the compound [Au$_2$(hpp)$_2$Cl$_2$] with a short Au–Au (2.47 Å) distance [17, 18].

Although it is known that the direction of the lone pairs of electrons on nitrogen ligands can influence the nuclearity of the complexes formed, as for example, with the 3,5-diphenylpyrazolate ligand which normally forms trinuclear complexes with Au(I) or Ag(I), Figure 1.27, Raptis and coworkers [16] were able to obtain a tetranuclear Au(I) pyrazolate complex by using bulky groups at the 3 and 5 positions of the ring.

While we have been unsuccessful with many attempts by direct synthesis or reduction to isolate the gold(I) [Au$_2$(hpp)$_2$], we have discovered that solvent conditions determine whether oxidation to the dinuclear Au(II) species, [Au$_2$(hpp)$_2$Cl$_2$], occurs or a tetranuclear Au(I) species, [Au$_4$(hpp)$_4$], forms. It appears that the nuclearity of the gold(I) hpp compound depends on factors such as the disproportionation rate of the Au(I) in a given solvent and the presence of oxidants. The short ligand N$\cdots$N distance should promote tetranuclear product formation over dinuclear species but in the presence of oxidizing solvents and solvents supporting rapid disproportionation and in the presence of coordinating ligands like chloride, a gold(II) product is isolated.

This solvent role regarding the formation of a dinuclear Au(II) or a tetranuclear Au(I) product is noted when Na[hpp] is reacted with (THT)AuCl. In THF the product is the dinuclear Au(II) species, [Au$_2$(hpp)$_2$Cl$_2$], along with gold metal. In oxidizing solvents such as the chlorocarbon dichloromethane, [Au$_2$(hpp)$_2$Cl$_2$] is produced in high yield without Au(0) formation, Figure 1.28. If ethanol is used as the solvent, the product is the tetranuclear Au(I) species, [Au$_4$(hpp)$_4$]. A plausible rationalization of the different behavior in the two solvents is that ethanol solvates both the [(THT)Au]$^+$ and the Cl$^-$, reducing the potential of the cation for oxidation and allowing solvation of the sodium chloride. Several reducing agents were used in attempts to reduce the Au(II) complex to form the Au(I) product, including reagents such as KC$_8$ and K, but each produced gold metal. Using silver benzoate in a CH$_3$CN/THF solution to

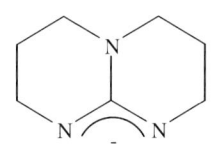

1,3,4,6,7,8-hexahydro-2H-pyramido[1,2-a]-pyramidinate

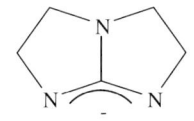

[hpp]$^-$

1,3,6-triazabicyclo[3.3.0]oct-4-ene

[tbo]$^-$

Figure 1.27 Structure of the anionic, bidentate nitrogen ligands [hpp]$^-$ and [tbo]$^-$.

22 | 1 Gold(I) Nitrogen Chemistry

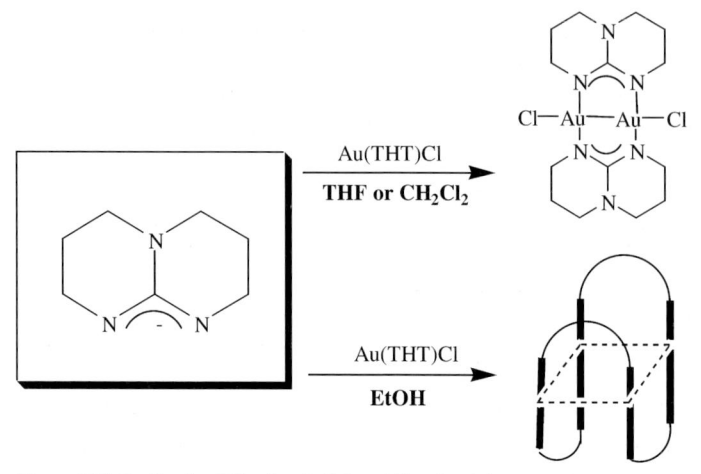

Figure 1.28 Synthesis of [Au$_2$(hpp)$_2$Cl$_2$] and [Au$_4$(hpp)$_4$].

remove the chlorides formed, a green hexanuclear product with mixed metals and ligands was achieved. The gold(II)-silver(I) complex [(PhCOO)$_6$Au$_4$(hpp)$_2$Ag$_2$] crystallized and was structurally characterized. It has a very short Au–Au distance, 2.4473(19) Å, and normal Au–Ag, 3.344(3) Å, and Ag–Ag, 2.771(6) Å distances [18].

The X-ray crystal structure of [Au$_2$(hpp)$_2$Cl$_2$] revealed a Au(II)–Au(II) distance of 2.4752(9) Å, Figure 1.29, which is shorter than the Au–Au distance observed in the amidinate, [Au$_2$(2,6-Me$_2$Ph-form)$_2$Cl$_2$] (2.617 Å) [23, 24]. This decrease in distance is

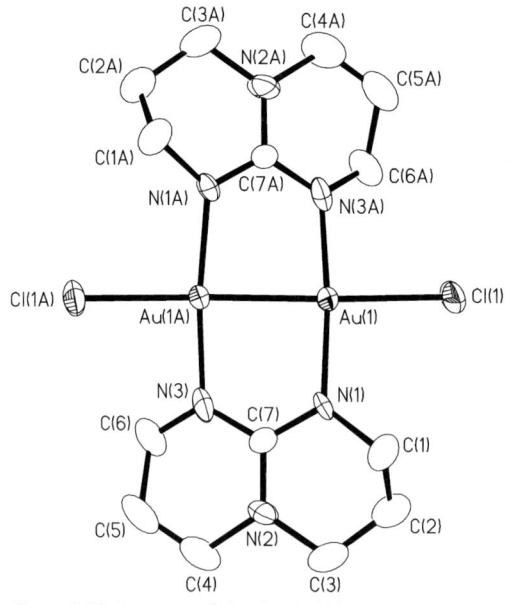

Figure 1.29 Structure of [Au$_2$(hpp)$_2$Cl$_2$].

1.6 Solvent Influences on Oxidation and Nuclearity of Gold Guanidinate Derivatives

dramatic, and results in a stable d^9-d^9 system with the Au(II) atoms in a square planar arrangement. The coordination angles range from 86.07–94.25° and sum to 359.42°. The dihedral angle between the N(1)-Au(1)-N(3A) and N(1A)-Au(1A)-N(3) plane in [Au$_2$(hpp)$_2$Cl$_2$] is 13.3°. The structure is puckered with deviation from the mean plane of 0.35 Å.

The molecular structures of Mo$_2$(hpp)$_4$ and W$_2$(hpp)$_4$ studied by Cotton show M–M distances of 2.067(1) and 2.162(1) Å, respectively [41]. These two complexes contain the shortest Mo$_2^{4+}$ or W$_2^{4+}$ quadruple bonds known. The ready oxidation of these complexes with the electron-rich bicyclic guanidinate ligand, hpp$^-$, and of the dinuclear gold species, clearly shows that ligands which favor short metal-metal distances promote reduction of the electron density between the metal atoms by electron loss.

The gold-gold distances in the [Au$_4$(hpp)$_4$] complex, range from 2.8975(5)–2.9392(6) Å, and are similar to those found in the tetranuclear gold amidinate complexes, Figure 1.30. The hpp ligand apparently shows a different behavior with different group 11 elements, forming a tetranuclear complex with gold and silver but to date only a dinuclear complex of copper(I) has been reported [42]. With the related smaller ring guanidinate [Au$_4$(tbo)$_4$] the average Au$\cdots$Au distance is 3.16 Å, Figure 1.31. The angles at Au$\cdots$Au$\cdots$Au in this complex are acute 66.03(3)–66.12(3)° and obtuse 111.92.64(3)–115.82(3)°.

Density functional theory and MP2 calculations on [Au$_2$(hpp)$_2$Cl$_2$] show that the HOMO is predominately hpp and chlorine-based with some Au-Au δ^* character and that the LUMO has metal-to-ligand (M-L) and metal-to-metal (M-M) σ^* character (approximately 50% hpp/chlorine, and 50% gold). DFT calculations on [Au$_4$(hpp)$_4$]

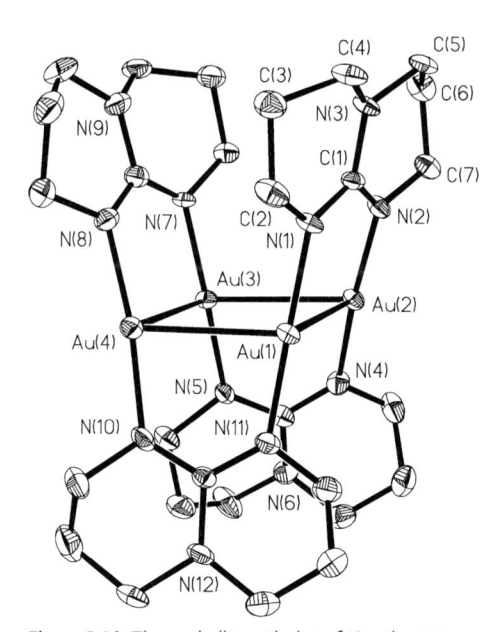

Figure 1.30 Thermal ellipsoid plot of [Au$_4$(hpp)$_4$].

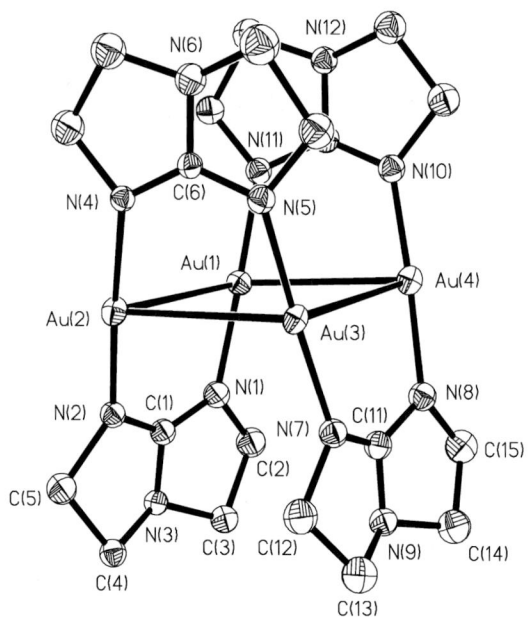

Figure 1.31 Thermal ellipsoid plot of [Au$_2$(tbo)$_4$].

show that the HOMO and HOMO-1 are a mixture of metal-metal antibonding character and metal-ligand antibonding character and that the LUMO is predominately metal based s character (85% Au and 15% hpp).

The calculated thermodynamics for the reduction of [Au$_2$(hpp)$_2$Cl$_2$] to [Au$_2$(hpp)$_2$] and Cl$_2$(g) suggest the reaction is endothermic with a $\Delta H°$ of 50.0, 48.6, 45.8, and 69.5 kcal mol^{-1} and a $\Delta G°$ of 38.3, 37.5, 35.3, and 57.8 kcal mol^{-1} depending upon the level of theory used [18]. The fact that the reaction is thermodynamically unfavorable is consistent with the difficulty in obtaining the Au(I) compound [Au$_2$(hpp)$_2$].

1.7
Cyclic Trinuclear Gold(I) Nitrogen Compounds

Trinuclear 9-membered rings can be formed by the reaction of gold(I) ions with exobidentate C,N or N,N monoanionic ligands. They are generally slightly irregular and puckered unless the metallocycle is imposed by intramolecular crystallographic symmetry. Gold-gold intramolecular interactions are always present and the complexes exhibit a roughly C$_{3h}$ or with symmetrical ligands a D$_{3h}$ symmetry. Crystal structures of these trinuclear complexes demonstrate formation of individual complexes, dimers, supramolecular columnar species or more complex supramolecular aggregates, Table 1.4 [43]. Dimers and supramolecular structures are held together by aurophilic intermolecular gold-gold interactions. Bulky substituents on the ligands can prevent intermolecular metal-metal interactions and the formation of supramolecular architectures.

1.7 Cyclic Trinuclear Gold(I) Nitrogen Compounds

Table 1.4 Trimeric cyclic gold(I) compounds[a].

Complex	Ref
[μ-N^1,C^2-pyAu]$_3$	[44, 45]
[μ-C(OMe)=N(C$_6$H$_{11}$)Au]$_3$	[46]
[μ-C(OMe)=N(Me)Au]$_3$	[47, 48]
[μ-C(OEt)=N(C$_6$H$_4$p-Me)Au]$_3$	[49, 50]
[μ-N^1,C^2-bzimAu]$_3$	[51]
[μ-N,N-3,5-Ph$_2$pzAu]$_3$	[52]
[μ-N,N-3,5(CF$_3$)pzAu]$_3$	[53, 54]
[μ-N,N-3,5(4-MeOPh)$_2$pzAu]$_3$	[55, 56]
[μ-N,N-pzAu]$_3$	[57]
[μ-N,N-4-MepzAu]$_3$	[57]

[a]py = pyridinate; bzim = 1-benzylimidazolate; pz = pyrazolate.

The first cyclic trinuclear compound of gold(I) was reported by Vaughan in 1970 [44]. The complex [μ-N^1,C^2-pyAu]$_3$ was obtained in a very good yield by adding triphenylarsine gold(I) chloride to a THF solution of 2-pyridyllithium at −40 °C, Figure 1.32.

Some other cyclic trinuclear gold(I) pyridine complexes, CTCs, were obtained by the same procedure using various substituted pyridines [44]. All the complexes except [μ-N^1,C^2-pyAu]$_3$ have a very low solubility in common organic solvents. In 1972, the synthesis of another CTC gold(I) complex [μ-C(OCH$_3$)=N(C$_6$H$_{11}$)Au]$_3$ was reported. The complex was obtained by the reaction of chloro(triphenylphosphine)gold(I) with cyclohexyl isocyanide in a methanolic potassium hydroxide solution, Figure 1.33 [46]. Using the same or similar synthetic approaches, many other analogous carbeniate cyclic gold(I) complexes have been described [47, 49]. They have the general formula [μ-C,N-carbAu]$_3$ (carb is C(OR)=NR') where R' is an aliphatic, alicyclic, alkyl aromatic or aromatic group, Figure 1.33.

Another family of gold(I) CTCs, having a C-Au-N environment, was described in which the bridging ligand between gold atoms is an alkyl-2-imidazolate anion (alkyl group = CH$_3$ or CH$_2$Ph) [51]. A typical reaction is carried out at −40 °C in THF using Vaughan's method, Figure 1.34, but in this case the crude brown solid is extracted

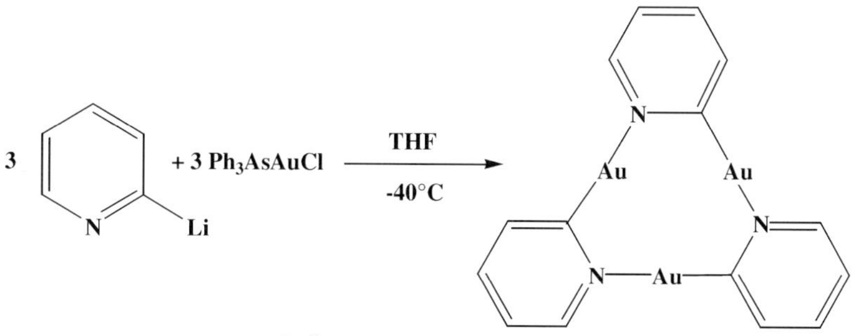

Figure 1.32 Synthesis of [μ-N^1,C^2-pyAu]$_3$.

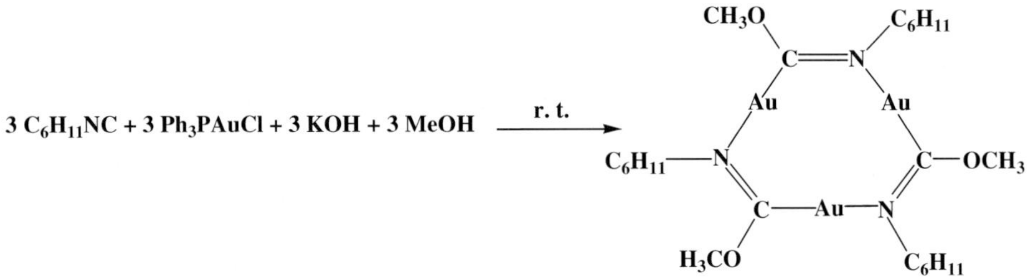

$3\ C_6H_{11}NC + 3\ Ph_3PAuCl + 3\ KOH + 3\ MeOH \xrightarrow{r.t.}$

Figure 1.33 Synthesis of $[\mu\text{-}C(OCH_3)=N(C_6H_{11})Au]_3$.

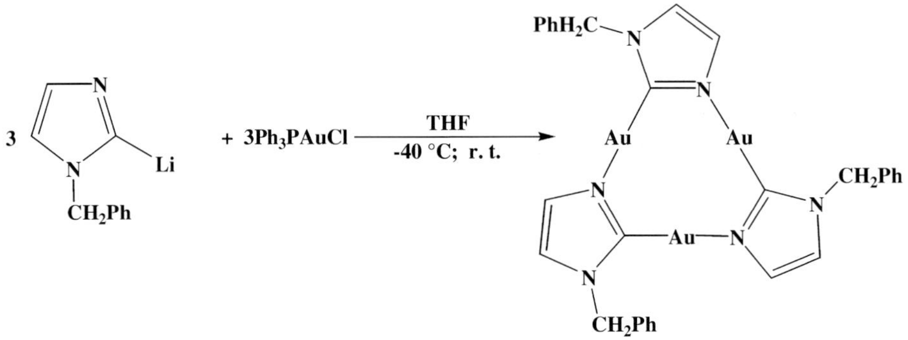

Figure 1.34 Synthesis of $[\mu\text{-}N^1,C^2\text{-}bzimAu]_3$.

overnight at room temperature with hexane. The reaction can be carried out using $(CH_3)_2SAuCl$ as a starting material instead of Ph_3PAuCl. In this way the CTCs $[\mu\text{-}C(OEt)=N(C_6H_4\text{-}p\text{-}Me)Au]_3$ and $[\mu\text{-}N^1,C^2\text{-}bzimAu]_3$ are immediately formed in a good yield, but the reaction is delicate and slow and often colloidal gold(0) is formed. The methyl analog of $[\mu\text{-}N^1,C^2\text{-}bzimAu]_3$ is quite soluble in the common organic solvents.

Gold(I) CTCs having a N-Au-N environment are also known and have the general formula $[\mu\text{-}N,N\text{-}pzAu]_3$ (pz$^-$ = pyrazolate or various ring substituted pyrazolates) [15b,52,53,55]. In these compounds the bidentate anion ligands bridging the gold atoms are obtained by deprotonation of a pyrazole ring with a base such as KOH or NaH, Figure 1.35 [15b]. It is noteworthy that when Na[3,5-Ph$_2$pz] and AgO$_2$CPh are added to a THF solution of Ph$_3$PAuCl a hexanuclear gold cycle having a 18-atom ring is formed [15b].

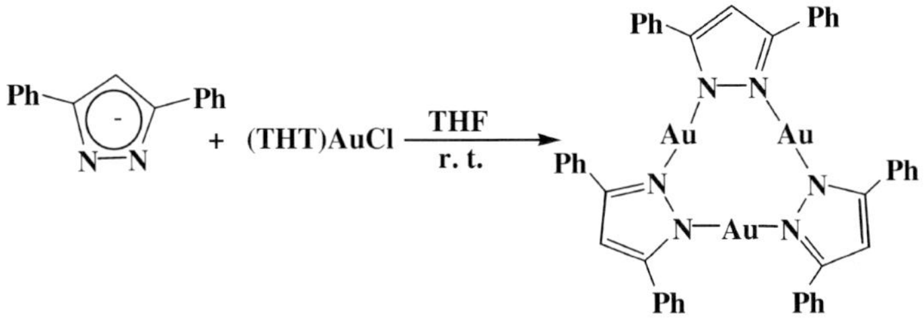

Figure 1.35 Synthesis of $[\mu\text{-}N,N\text{-}3,5\text{-}Ph_2pzAu]_3$.

1.7 Cyclic Trinuclear Gold(I) Nitrogen Compounds

The cyclic trimeric nature of the carbeniate Au(I) complexes was realized when the structure of [μ-C(OEt)=N(C$_6$H$_4$-p-Me)Au]$_3$ was reported [50]. This structure is a 9-membered ring formed by three carbeniate ligands bridging the gold(I) atoms through the N and C atoms. The deviation of the C-Au-N angles from the linearity as well as the puckering of the rings are due to the presence of intramolecular (average Au···Au = 3.272(1) Å) and intermolecular gold-gold interactions. Only two short Au···Au contacts of 3.244(1) Å are found between two CTCs, and the dimer is arranged in the crystal structure to give a characteristic Au$_6$ chair. The packing diagram of [μ-C(OEt)=N(C$_6$H$_4$-p-Me)Au]$_3$ shows additional weak Au(I)···Au(I) interactions at 3.824 Å.

The Au–Au distance in [μ-C(OMe)=N(Me)Au]$_3$ of 3.308(2) Å suggests an intramolecular interaction between the metal atoms [48]. The most important and unique feature of the complex [μ-C(OEt)=N(C$_6$H$_4$-p-Me)Au]$_3$ is the ability of the CTCs to aggregate in the solid state along the c axis to form ordered and disordered columnar stacks. In each unit cell, the two types of stacks occur in a 2:1 ratio. In the ordered stacks the intermolecular Au···Au distance is 3.346(1) Å and the gold centers are arranged to form an infinite trigonal prismatic array. In the disordered stacks there are two sets of positions for each gold triangle. The ability of the complex [μ-C(OEt)=N(C$_6$H$_4$-p-Me)Au]$_3$ to aggregate through gold-gold intermolecular interactions forming these supramolecular arrays confers to its extraordinary luminescent properties which have been described by Balch as solvoluminescence [48].

The complex [μ-N^1,C^2-pyAu]$_3$ shows an interesting and unusual crystal structure. It is located on a crystallographic mirror plane that lies perpendicular to the molecular plane. Each CTC is planar with gold-gold intramolecular interactions of 3.309(2) and 3.346(3) Å. The intermolecular Au···Au contacts are shorter than the intramolecular Au···Au contacts and fall in the range 3.105(2)-3.143(3) Å. Both CTC complexes [μ-N^1,C^2-pyAu]$_3$ and [μ-C(OMe)=N(C$_6$H$_{11}$)Au]$_3$ form dimers with a chair conformation of the gold atoms, but complex [μ-N^1,C^2-pyAu]$_3$ is further assembled by the apical gold atoms of the chairs to form extended stepwise chains.

Recently, gold pyrazolate CTCs have been described which produce room-temperature columnar mesophases [55]. These complexes have long chain substituents in the 3 and 5 positions of the pyrazolate ring. X-ray powder diffraction measurements have demonstrated that the supramolecular columnar arrangement is present in the crystalline solids as well as in the mesomorphic phase. The X-ray crystal structure of complex [μ- N,N-3,5(4'-MeOPh)$_2$pzAu]$_3$ which has an anisole unit on the pyrazolates, yields a unit cell which contains two independent CTCs. They are slightly different in the twist about the central metallocycle core and more markedly in the relative conformations of the phenyl substituents [56]. In the complex [μ-N,N-3,5(4'-MeOPh)$_2$pzAu]$_3$ the intramolecular Au···Au average distance is 3.3380(7) Å. The intermolecular Au···Au distance is greater than 4.252 Å with a mean stacking separation between two consecutive trimers of 4.54 Å. The packing mode observed appears to be controlled by van der Waals forces (i.e., no Au···Au interactions).

A more complex supramolecular architecture has been discovered for the complexes [μ-N,N-pzAu]$_3$, Figure 1.36, and [μ-N,N-4-MepzAu]$_3$ [57]. Intramolecular

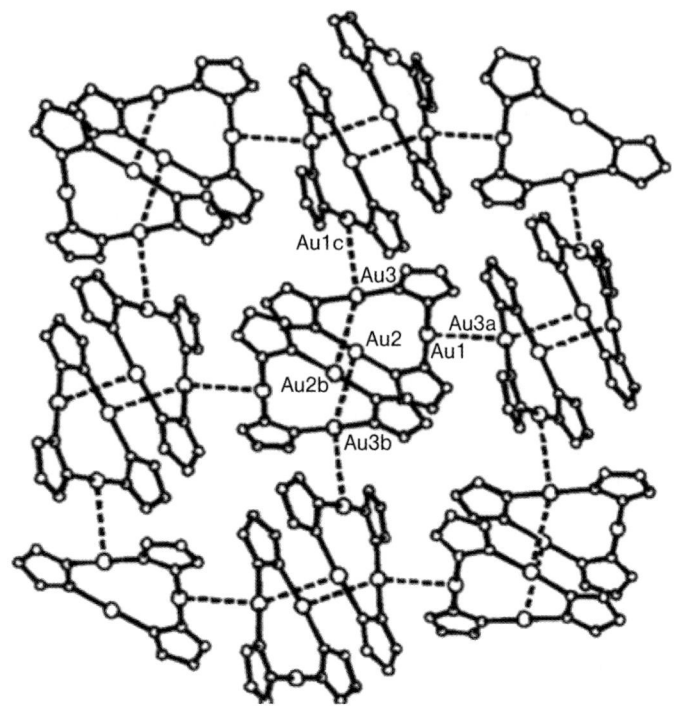

Figure 1.36 Two dimensional structure of [μ-N,N-pzAu]$_3$.

aurophilic gold-gold interactions are present with Au···Au distances of 3.372(1)-3.401(1) Å. Complex [μ-N,N-pzAu]$_3$ forms a two-dimensional network by self-assembly of the CTCs through intermolecular aurophilic interactions. Each [μ-N,N-pzAu]$_3$ forms a dimer such as those found in other CTCs, with two gold-gold interactions of 3.313(1) Å. Moreover, each dimer interacts with four other dimers through a single Au···Au contact 3.160(1) Å to form a 2-dimensional net.

1.8
Oxidative-Addition Reactions to the Cyclic Trinuclear Gold(I)-Nitrogen Compounds

Gold CTCs undergo oxidative-addition reactions of halogens at the metal centers [58]. There is evidence that electronic factors influence the reactivity of the gold atoms in these compounds. In fact, except for [μ-C(OMe)=N(Me)Au]$_3$, only one metal center appears to be oxidized to give mixed-valence Au$_2^I$/Au^{III} metallocycles. Surprisingly, aqua regia also fails to give complexes beyond the Au$_2^I$/Au^{III} oxidation state with the unoxidized pyrazolates. Thus an unusual stability of the d^{10}d^{10}d^8 configuration for these pyrazolate gold CTCs is observed. The electronic communication between the gold atoms may be the origin of this effect. The oxidation of the first gold atom may improve the π-acceptor ability of the two ligands coordinated to it so that they remove sufficient electron density from the remaining two Au^I atoms and prevent

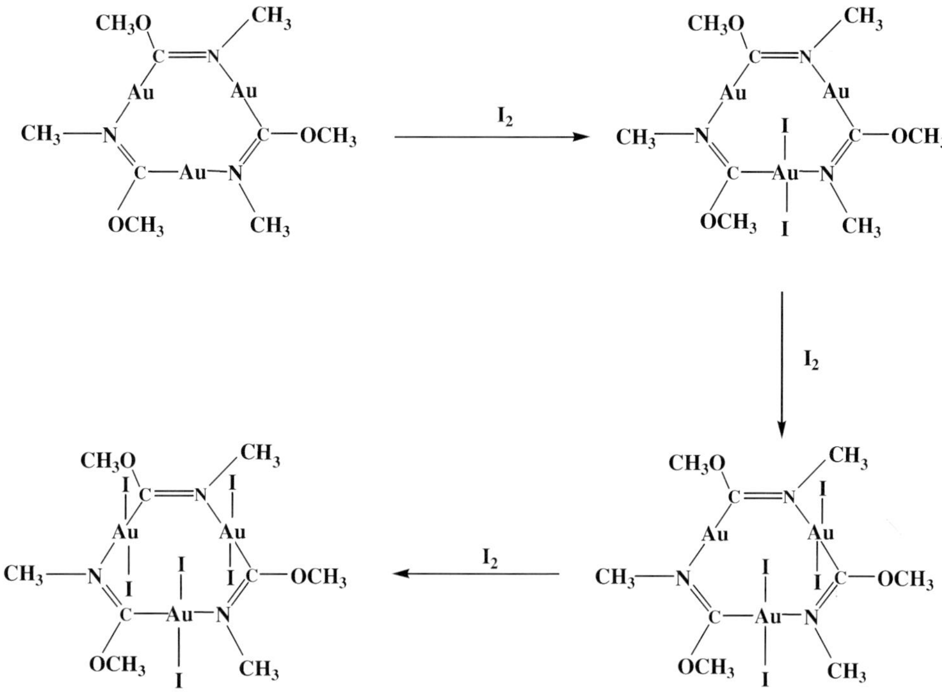

1.8 Oxidative-Addition Reactions to the Cyclic Trinuclear Gold(I)-Nitrogen Compounds

Figure 1.37 Synthesis of the mixed-valence or completely oxidized complexes [μ-C(OMe)=N(Me)Au]$_3$I$_n$ ($n = 2$–6).

their oxidation. However this hypothesis is not corroborated by the crystallographic data since change is not observed in the gold-ligand bond lengths. However the effect may involve Au-N pi interactions with subsequently little atom movement. The complex [μ-C(OMe)=N(Me)Au]$_3$ seems to be unique in the family of the gold CTCs and to date it is the only CTC of gold(I) that gives the stepwise addition of halogens, resulting in the formation of either mixed-valent or completely oxidized trinuclear gold complexes, Figure 1.37. The X-ray structures of these derivatives were recently reported many years later after their synthesis [58].

Crystallographic studies of the iodine oxidized carbeniate [μ-C(OMe)=N(Me)Au]$_3$ confirms the structures originally proposed [58]. All the structures retain the frame of the starting complex [μ-C(OMe)=N(Me)Au]$_3$, Figure 1.38. The variation in the intramolecular Au···Au separation is small. However, there is a trend toward increased Au···Au distance as more iodine is added to the complex. The Au–I distances fall in the range of 2.614(6)-2.633(7) Å. As a consequence of the repulsive intramolecular I···I contacts, the I-Au-I angles deviate significantly from linearity. They become smaller and smaller with increased number of iodide atoms bonded to the gold centers. The structure of [μ-C(OMe)=N(Me)Au]$_3$I$_6$.CH$_2$Cl$_2$ shows the formation of columns with short intermolecular I···I interactions ranging from 3.636(2) to 3.716(2) Å. The interaction between terminal iodide ligands appears to have a directional component.

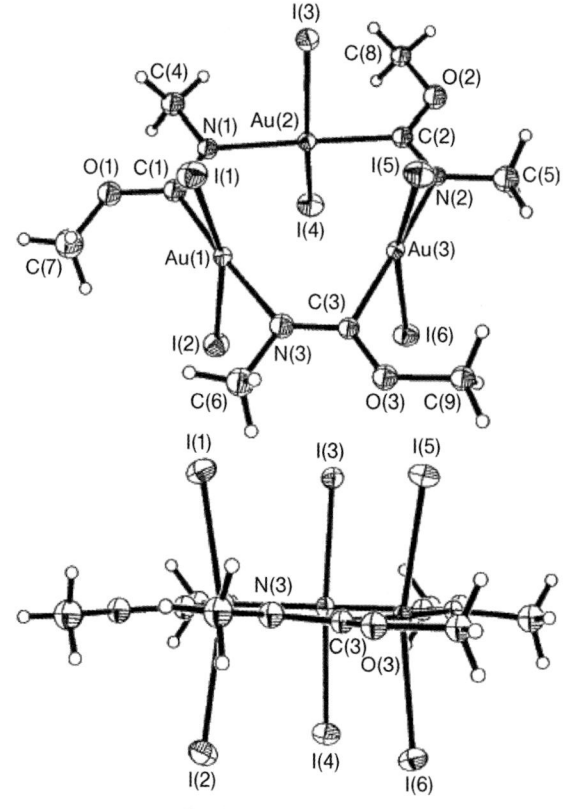

Figure 1.38 Top and side views of the completely oxidized complex [μ-C(OMe)=N(Me)Au]$_3$I$_6$.

Oxidative-addition of iodine also was investigated for the complex [μ-N^1, C^2-bzimAu]$_3$. This complex behaves like most of the CTCs since it adds iodine at only one gold center to yield [μ-N^1,C^2-bzimAu]$_3$I$_2$ [59]. The X-ray structure shows that it consists of discrete trinuclear units with the three gold atoms bridged by 1-benzylimidazolates.

1.9
Supramolecular Entities of Trinuclear Gold(I) Complexes Sandwiching Small Organic Acids

Extended linear chain inorganic compounds have special chemical and physical properties [60, 61]. This has led to new developments in fields such as supramolecular chemistry, acid-base chemistry, luminescent materials, and various optoelectronic applications. Among recent examples are the developments of a vapochromic light emitting diode from linear chain Pt(II)/Pd(II) complexes [62], a luminescent switch consisting of an Au(I) dithiocarbamate complex that possesses a luminescent linear

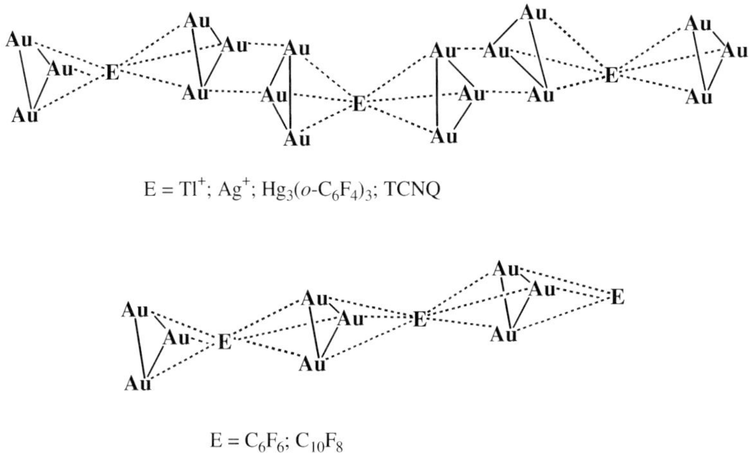

1.9 Supramolecular Entities of Trinuclear Gold(I) Complexes Sandwiching Small Organic Acids

$E = Tl^+; Ag^+; Hg_3(o-C_6F_4)_3; TCNQ$

$E = C_6F_6; C_{10}F_8$

Figure 1.39 Structural arrangements of the cyclic trinuclear Au^I compounds, CTCs, with various electrophilic adducts in ABBA and ABA chains.

chain in the presence of vapors of organic solvents [31], mixed-metal (Ag/Au or Tl/Au) [63] compounds that exhibit different colors and emissions when different organic solvents are introduced or removed, and the discovery of solvoluminescence [48] in a Au(I) CTC with an extended chain structure.

Recent results have demonstrated that the electron-rich trinuclear Au(I) complexes can interact with neutral electron-acceptor entities such as C_6F_6, $C_{10}F_{14}$, TCNQ, or $Hg_3(\mu-C_6F_4)_3$, and cation species such as Ag(I) and Tl(I) to produce infinite linear chain complexes, Figure 1.39 [64–66]. Balch and co-workers also demonstrated that trinuclear Au(I) compounds with alkyl-substituted carbeniate bridging ligands can interact with the large organic acceptors nitro-9-fluorenones [67]. DFT calculations clearly show that the donor regions in the trinuclear Au(I) compounds are located at the center of the 9-membered ring and that they extend to regions in space above and below the ring plane [63].

The TCNQ molecule in [TR(bzim)]$_2$·TCNQ is sandwiched between two units of [μ-N^1,C^2-bzimAu]$_3$ in a face-to-face manner so that it is best represented by the formula (π-[μ-N^1,C^2-bzimAu]$_3$)(μ-TCNQ)(π-[μ-N^1,C^2-bzimAu]$_3$). The cyanide groups clearly are not coordinated to the gold atoms. The distance between the centroid of TCNQ to the centroid of the Au$_3$ unit is 3.964 Å. The packing of [TR(bzim)]$_2$·TCNQ shows a stacked linear-chain structure with a repeat pattern of -(Au$_3$)(Au$_3$)(μ-TCNQ) (Au$_3$)(Au$_3$)(μ-TCNQ)- an ABBABB repeat. The complex [TR(bzim)]$_2$·TCNQ contains two very short intermolecular Au$\cdots$Au distances of 3.152 Å (identical for the two aurophilic bonds). The intermolecular Au$\cdots$Au distance is even shorter than the intramolecular distances in the starting compound, which are 3.475, 3.471, and 3.534 Å. The adjacent Au$_3$ units in [TR(bzim)]$_2$·TCNQ form a chair-type structure rather than the face-to-face (nearly eclipsed) pattern reported in Balch's studies of the nitro-9-fluorenones adducts with the trinuclear Au(I) alkyl-substituted carbeniate complexes.

The shortened intermolecular Au–Au distances in [TR(bzim)]$_2$·TCNQ may be associated with charge-transfer from the electron-rich Au center to the known electron acceptor TCNQ. A partial oxidation of the Au(I) atoms leads to the observed shortening of Au–Au distances. In the limit of complete oxidation to Au(II), a gold–gold single bond forms with [TCNQ]$^-$. The presence of [TCNQ]$^-$ impurity in the crystals may be the origin of the dark color of this complex since [TR(bzim)]$_2$ itself is colorless while TCNQ is light orange. Thin crystals and films are dark green as are solutions. It remains possible that charge transfer is the cause of the dark color, with adduct formation remaining intact in solution.

The crystal structure of [TR(carb)]·C$_6$F$_6$ shows a columnar stack consisting of alternating C$_6$F$_6$ and [TR(carb)] molecules [65]. The C$_6$F$_6$ molecule is sandwiched between two units of [TR(carb)], in a face-to-face manner so that a molecule of [TR(carb)]·C$_6$F$_6$ is best represented by the formula (π-[TR(carb)])$_{0.5}$(μ-C$_6$F$_6$)(π-[TR(carb)])$_{0.5}$. The distance between the centroid of C$_6$F$_6$ to the centroid of the Au$_3$ unit is 3.565 Å. The packing of [TR(carb)]·C$_6$F$_6$ shows a stacked linear-chain structure with a repeat pattern of ·(Au$_3$)(μ-C$_6$F$_6$)(Au$_3$)(μ-C$_6$F$_6$)· an ABAB pattern. The crystal structure of [TR(carb)] by itself shows a dimeric structure with intermolecular Au·Au bonds. Therefore, the C$_6$F$_6$ Lewis acid disrupts the intermolecular aurophilic bonding in [TR(carb)] chain with loss of visible luminescence [65]. These pi-acid, pi-base results with gold CTCs are similar to but opposite from the interactions reported by Gabbaï and co-workers, between the Lewis acid Hg$_3$(μ-C$_6$F$_4$)$_3$ and benzene, in which benzene acts as a Lewis base coordinating in a μ-6 manner to six Hg centers, three from each side [68, 69]. The two Au$_3$ units interacting with C$_6$F$_6$ in [TR(carb)]·C$_6$F$_6$ are eclipsed with respect to each other (nearly D_{3h}) whereas the two Hg$_3$ units in Hg$_3$(μ-C$_6$F$_4$)$_3$·benzene are nearly staggered (D_{3d}).

The Lewis acid Hg$_3$(μ-C$_6$F$_4$)$_3$ also forms a pi-acid/pi-base interaction with TR(carb). In addition to the crystal structure demonstrating the ABBABB pattern observed in other stacked materials which retain the aurophilic Au–Au interactions between four of the six basic Au(I) atoms of the BB moieties, studies have shown that the oligomeric acid/base interection is retained in solution. Pulsed gradient diffusion NMR studies [70] suggesting the oligomeric sizes and ^{13}C-^{19}F coupling between units demonstrate that the interactions are stronger than solvation of the CDCs.

The nucleophilic trinuclear Au(I) ring complex Au$_3$(p-tolN=COEt)$_3$, Figure 1.40, forms sandwich adducts with the organic Lewis acid octafluoronaphthalene, C$_{10}$F$_8$ [66]. The Au$_3$(p-tolN=COEt)$_3$·C$_{10}$F$_8$ adduct has a supramolecular structure consisting of columnar interleaved 1 : 1 stacks in which the Au$_3$(p-tolN=COEt)$_3$ π-base molecules alternate with the octafluoronaphthalene π-acid molecules with a distance between the centroid of octafluoronaphthalene to the centroid of Au$_3$(p-tolN=COEt)$_3$ of 3.458 and 3.509 Å. The stacking with octafluoronaphthalene completely quenches the blue photoluminescence of Au$_3$(p-tolN=COEt)$_3$, which is related to inter-ring Au–Au bonding, and leads to the appearance of a bright yellow emission band observed at room temperature. The structured profile, the energy, and the lifetime indicate that the yellow emission of the Au$_3$(p-tolN=COEt)$_3$·C$_{10}$F$_8$ adduct is due to gold influenced phosphorescence of the octafluoronaphthalene.

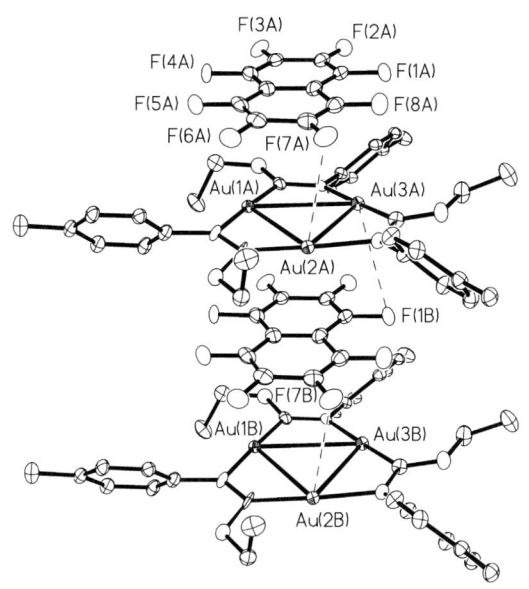

Figure 1.40 Thermal ellipsoid drawing of the stacked octafluoronaphthalene with $Au_3(p\text{-tolN}=COEt)_3$.

The 3.5 ms lifetime of the yellow emission of $Au_3(p\text{-tolN}=COEt)_3 \cdot C_{10}F_8$ is two orders of magnitude shorter than the lifetime of the octafluoronaphthalene phosphorescence which is observed at low temperature, thus indicating a gold heavy-atom effect.

1.10
Gold(I) and Silver(I) Mixed-Metal Trinuclear Complexes

Attention has been given to the synthesis of bimetallic silver-gold clusters [71] due to their effective catalytic properties, resistance to poisoning, and selectivity [72]. Recently molecular materials with gold and silver nanoclusters and nanowires have been synthesized. These materials are considered to be good candidates for electronic nanodevices and biosensors [73].

Based on the fact that pi-acids interact with the trinuclear gold(I) pi-bases, TR(carb) and TR(bzim), the trinuclear 3,5-diphenylpyrazolate silver(I) complex was reacted with each. Mixing $[Au_3(carb)_3]$ or $[Au_3(bzim)_3]$ with $[Ag_3(\mu\text{-}3,5\text{-}Ph_2pz)_3]$ in CH_2Cl_2 in stoichiometric ratios of 1:2 and 2:1 produced the mixed metal/mixed ligand complexes in the same gold-silver ratios. The crystalline products were not the expected acid-base adducts. It is suspected that the lability of the M–N bond (M=Au, Ag) in these complexes results in the subsequent cleavage of the cyclic complexes to produce the products statistically expected from the stoichiometry of materials used [74]. As a result of the lability of Au–N and Ag–N bonds, and the stability of

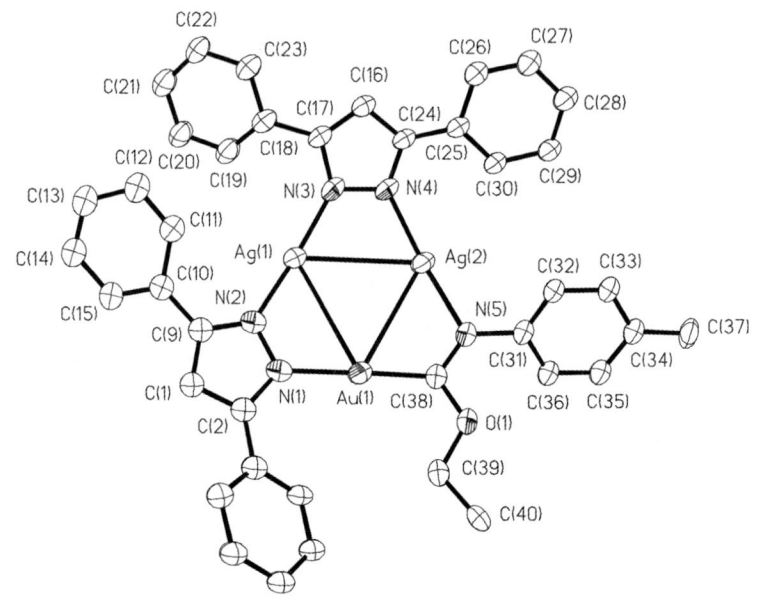

Figure 1.41 ORTEP diagram of [Au(carb)Ag$_2$(μ-3,5-Ph$_2$pz)$_2$].

Au–C bonds, mixed metal gold-silver dimers of planar, trinuclear complexes are readily formed by mixing gold(I) carbeniates and gold(I) benzylimidazolates with silver(I) pyrazolates in stoichiometric ratios. The complexes retain the ligands associated with the metal atoms of the starting materials.

The two trinuclear moieties of the dimer of [Ag$_3$(μ-3,5-Ph$_2$pz)$_3$] are rotated *anti* to each other [75], but this arrangement is less apparent in [Au(carb)Ag$_2$(μ-3,5-Ph$_2$pz)$_2$], Figure 1.41. The shortest Ag$\cdots$Ag interactions within the metallocycle rings of the dimer of [Ag$_3$(μ-3,5-Ph$_2$pz)$_3$] are about 3.4 Å, while between the trinuclear units the Ag$\cdots$Ag distance is 2.9712(14) Å. The Au$\cdots$Ag distances between trinuclear units in [Au(carb)Ag$_2$(μ-3,5-Ph$_2$pz)$_2$] are 3.311(2) Å and 3.082(2) Å. The metallocycles in [Au(carb)Ag$_2$(μ-3,5-Ph$_2$pz)$_2$] are irregular and puckered similar to those in the dimer of [Ag$_3$(μ-3,5-Ph$_2$pz)$_3$]. The structure of [Au$_2$(carb)$_2$Ag(μ-3,5-Ph$_2$pz)], Figures 1.42 and 1.43, shows one intermolecular interaction between the trinuclear gold units, with a Au$\cdots$Au distance of 3.33 Å. This is slightly longer than Au$\cdots$Au distances, 3.224–3.299 Å, in the irregular and puckered nine-membered ring of the dimer of [Au$_3$(carb)$_3$]. The Au$\cdots$Ag distances in [Au$_2$(carb)$_2$Ag(μ-3,5-Ph$_2$pz)] are 3.22–3.28 Å. The average distance of the two closest Au atoms between the trinuclear units of each dimer is 3.2 Å. A packing diagram shows a Au$\cdots$Au interaction, 3.857 Å, between the dimer units, similar to the distance observed in [Au$_3$(carb)$_3$], 3.824 Å. The inter-trinuclear Au$\cdots$Ag interactions in [Au(bzim)Ag$_2$(μ-3,5-Ph$_2$pz)$_2$] is 3.1423(8). The intermolecular distances, Au$\cdots$Ag, 3.53 and 3.38 Å and Ag$\cdots$Ag 3.35 Å are longer than those in the dimer of [Au(carb)Ag$_2$(μ-3,5-Ph$_2$pz)$_2$].

A few additional structural comparisons between the homonuclear gold and silver complexes and the mixed gold and silver complexes are of interest. In the dimer of the

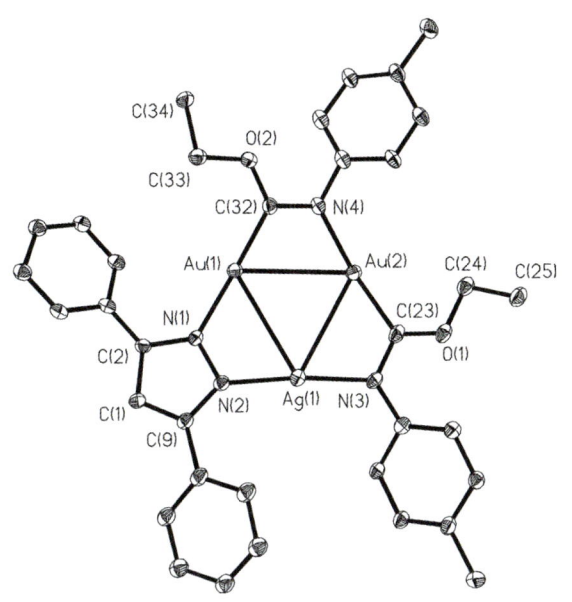

Figure 1.42 ORTEP diagram of [Au$_2$(carb)$_2$Ag(μ-3,5-Ph$_2$pz)].

trinuclear silver(I) 3,5-diphenylpyrazolate, [Ag$_3$(μ-3,5-Ph$_2$pz)$_3$], the six silver atoms are arranged as two triangles connected by only one short interaction. This drastically changes when a gold atom is introduced into the trinuclear unit as in [Au(carb)Ag$_2$(μ-3,5-Ph$_2$pz)$_2$]. An irregular square is formed by two Ag and two Au atoms with M-M distances in the range 3.08–3.40 Å. The other two silver atoms are above and below the plane of the square. A metallophilicity is observed in [Au(carb)Ag$_2$(μ-3,5-Ph$_2$pz)$_2$] in which each of the two gold atoms interact with three silver atoms. Three

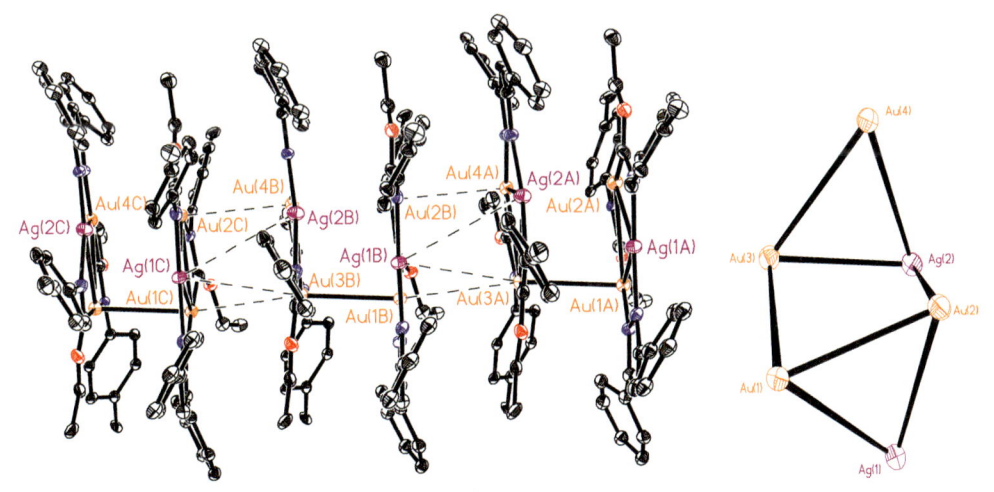

Figure 1.43 Packing diagram of [Au$_2$(carb)$_2$Ag(μ-3,5-Ph$_2$pz)]$_2$.

Au atoms and one Ag atom form a nearly regular square with distances range 3.21–3.42 Å.

Ligand bridged metal-metal distances display longer M–M distances than in the unbridged complexes. In the compounds studied, the non-bridged intermolecular M-M distances follow the order: Ag–Ag < Au–Ag < Au–Au while the intra-metallo-cycle M–M distances with bridging ligand bonding follow the order: Au–Au < Au–Ag < Ag–Ag. The synthesis of these mixed gold-silver compounds represents a new approach to cluster mixed metal synthesis with potential use in mixed-metal catalysis.

1.11
CO Oxidation Over Au/TiO$_2$ Prepared from Gold Nitrogen Complexes

Metal-organic and organometallic complexes have been widely used in the synthesis of catalysts, however, the use of metal-organic or organometallic gold complexes as catalyst precursors has been limited [76]. Gates and coworkers have reported that a supported mononuclear gold complex is active for ethylene hydrogenation at 353 K [77].

A series of Au/TiO$_2$ catalysts have been prepared from precursors of various metal-organic gold complexes (Au$_n$, $n = 2$–4) and their catalytic activity for CO oxidation studied. The Au/TiO$_2$ catalyst synthesized from a tetranuclear gold complex shows the best performance for CO oxidation with the TEM image of this catalyst indicating an average gold particle size of 3.1 nm.

Several factors may contribute to the high activity of our Au/TiO$_2$ catalysts. First, the use of metal-organic complexes as precursors can avoid the use of chloride. HAuCl$_4$ is widely used as a gold precursor in catalytic studies, invariably leaving a chloride residue in the catalyst after preparation. Recently, both experimental and theoretical studies have shown that chloride can poison the catalytic performance of gold catalysts for CO oxidation. Oh et al. have shown that chloride residue on a catalyst can promote agglomeration of Au particles during heat treatment, and can inhibit the catalytic activity by poisoning the active site [78]. Density functional calculations show that chloride can act as a poison by weakening the adsorption of O$_2$ and lowering the stability of the CO·O$_2$ intermediate complex [79]. Clearly metal-organic precursors provide an attractive route for the preparation of chloride-free gold catalysts. Another explanation for the high activity of our Au/TiO$_2$ catalysts also relates to the use of metal-organic precursor complexes. Upon deposition onto the oxide support, these complexes interact with the surface OH groups and become less mobile compared with gold atoms deposited using HAuCl$_4$. The catalyst particles appear to form at defect sites on the oxide as established by studies with MgO as the oxide surface. The defect sites may serve as calcination sites for the metal-organic catalyst precursors and perhaps inhibit agglomeration of gold particles during calcination. Factors preventing the sintering of gold lead to a narrow particle size distribution compared to the deposition-precipitation method of catalyst formation.

1.12
Miscellaneous Observations

The nearly simultaneous observation by the Schmidbaur [80] and Fackler [81] groups of the easy transmetallation of gold(I) with tetraphenylborate, which can be done in water, has caused Gray to develop this chemistry [82] in a general way with boronic acids. The considerable interest in the use of gold compounds as homogeneous catalysts [83] has prompted these studies. Recently Gray has described boronic acid transammination procedures to synthesize 3-coordinate azadipyrromethene complexes of gold(I) in order to examine their low energy absorption and emission properties. While the emission in the reported Au–N compound is comparable to what had been observed with the free azadipyrromethene ligand, the quantum yield is much lower. However, as Gray states [84], "The controlled auration of aromatic molecules affords access to broad classes of triplet-state luminophores and provides opportunities in materials design and organometallic photochemistry."

Acknowledgement

The Robert A. Welch Foundation of Houston, Texas is acknowledged for financial support of this work.

References

1 Fackler, J.P. (2002) Forty-five years of chemical discovery including a golden quarter-century. *Inorganic Chemistry*, **41**, 6959–6972.
2 (a) Schmidbaur, H. (ed.) (1999) *Gold Progress in Chemistry, Biochemistry, and Technology*, John Wiley & Sons, Ltd, Chichester, UK; (b) Grohman, A. and Schmidbaur, H. (1995) *Comprehensive Organometallic Chemistry II*, Vol. 3, (eds E.W. Abel, F.G. Stone and G. Wilkinson), Pergamon, Oxford, UK, pp. 1–56.
3 Puddephatt, R. (1978) *The Chemistry of Gold*, Elsevier, Oxford, UK.
4 Cotton, F.A. (1995) *Basic Inorganic Chemistry*, Vol. 3, John Wiley & Sons, New York.
5 Gimeno, M.C. and Laguna, A. (2004) *Comprehensive Coordination Chemistry II*, Vol. 6, Silver and Gold (eds J.A. McCleverty, T.J. Meyer and D.E. Fenton), Elsevier Pergamon, Oxford, UK.
6 (a) Clerac, R., Cotton, F.A., Dunbar, R.A., Murillo, C.A. and Wang, X. (2001) Dinuclear and Heteropolynuclear Complexes Containing Mo_2^{4+} Units. *Inorganic Chemistry*, **40**, 420–426; (b) Cotton, F.A., Lin, C. and Murillo, C.A. (2000) A Reliable Method of Preparation for Diiridium Paddlewheel Complexes: Structures of the First Compounds with Ir_2^{5+} Cores. *Inorganic Chemistry*, **39**, 4574–4578; (c) Cotton, F.A., Daniels, L.M., Murillo, C.A. and Schooler, P. (2000) Chromium(II) complexes bearing 2-substituted N,N′-di(aryl)formamidinate ligands. *Journal of the Chemical Society, Dalton Transactions*, (13), 2007–2012; (d) Cotton, F.A., Daniels, L.M., Murillo, C.A. and Schooler, P. (2000) Chromium(II)

complexes bearing 2,6-substituted N,N'-di(aryl)formamidinate ligands. *Journal of the Chemical Society, Dalton Transactions*, (13), 2001–2005; (e) Cotton, F.A., Daniels, L.M., Matonic, J.H. and Murillo, C.A. (1997) Highly distorted diiron(II, II) complexes containing four amidinate ligands. A long and a short metal–metal distance. *Inorganica Chimica Acta*, **256**, 277–282.

7 Cotton, F.A., Feng, X., Matusz, M. and Poli, R. (1988) Experimental and theoretical studies of the copper(I) and silver(I) dinuclear N,N'-Di-p-tolylformamidinato complexes. *Journal of the American Chemical Society*, **110**, 7077–7083.

8 Pyykkö, P. and Mendizabal, F. (1998) Theory of d^{10}-d^{10} closed-shell attraction. III. Rings. *Inorganic Chemistry*, **37**, 3018–3025.

9 Laguna, A. and Laguna, M. (1999) Coordination chemistry of gold(II) complexes. *Coordination Chemistry Reviews*, **193–195**, 837–856.

10 Barker, J. and Kilner, M. (1994) The coordination chemistry of the amidine ligand. *Coordination Chemistry Reviews*, **133**, 219–300.

11 Murray, H.H., Briggs, D.A., Guillermo, Garzon, Raptis, R., Porter, L.C. and Fackler, J.P. Jr (1987) Structural characterization of a linear [Au···Pt···Au] complex, $Au_2Pt(CH_2P(S)Ph_2)_4$, and its oxidized linear metal-metal bonded [Au-Pt-Au] product, $Au_2Pt(CH_2P(S)Ph_2)_4Cl_2$. *Organometallics*, **6**, 1992.

12 Ren, T., Lin, C., Amalberti, P., Macikenas, D., Protasiewicz, J.D., Baum, J.C. and Gibson, T.L. (1998) Bis(μ-N,N'-η^2-N,O-η^2-N',O'-di(o-methoxyphenyl)formamidinato)disilver(I): an interesting coordination geometry for silver(I) and room temperature fluorescence. *Inorganic Chemistry Communications*, **1**, 23–26.

13 Archibald, S.J., Alcock, N.W., Busch, D.H. and Whitcomb, D.R. (1999) Synthesis and characterization of functionalized N,N'-diphenylformamidinate silver(I) dimers: solid-state structures and solution properties. *Inorganic Chemistry*, **38**, 5571–5578.

14 Archibald, S.J., Alcock, N.W., Busch, D.H. and Whitcomb, D.R. (2000) Synthesis and characterization of silver(I) complexes with C-Alkyl functionalized N,N'-Diphenylamidinates: Tetrameric and trimeric structural motifs. *Journal of Cluster Science*, **11**, 261–283.

15 (a) Raptis, R.G., Murray, H.H. and Fackler, J.P. Jr (1987) The synthesis and crystal structure of a novel gold(I)-Pyrazolate hexamer containing an 18-Membered inorganic ring. *Journal of the Chemical Society, Chemical Communications*, (10), 737–739; (b) Murray, H.H., Raptis, R.G. and Fackler, J.P. Jr (1988) Syntheses and X-ray structures of group 11 pyrazole and pyrazolate complexes. X-ray crystal structures of Bis(3,5-diphenylpyrazole)copper(II) Dibromide, Tris(3,5-diphenylpyrazolato-N,N')trisilver(I)-2-Tetrahydrofuran, Tris(3,5-diphenylpyrazolato-N,N')trigold(I), and Hexakis(:-3,5-diphenylpyrazolato- N,N') hexagold(I). *Inorganic Chemistry*, **27**, 26–33; (c) Raptis, R.G. and Fackler, J.P. Jr (1988) Structure of Tris(3,5-diphenylpyrazolato-N,N')tricopper(I). Structural comparisons with the Silver(I) and Gold(I) pyrazolate trimers. *Inorganic Chemistry*, **27**, 4179–4182.

16 Yang, G. and Raptis, R.G. (2003) Synthesis, structure and properties of tetrameric gold(I) 3,5-di-tert-butyl–pyrazolate. *Inorganica Chimica Acta*, **352**, 98–104.

17 Irwin, M.D., Abdou, H.E., Mohamed, A.A. and Fackler, J.P. Jr (2003) Synthesis and X-ray structures of silver and gold guanidinate-like complexes. A Au(II) complex with a 2.47 Å Au-Au distance. *Chemical Communications*, (23), 2882–2883.

18 Mohamed, A.A., Mayer, A., Abdou, H.E., Irwin, M.D., Perez, L. and Fackler, J.P. Jr (2007) Dinuclear and tetranuclear gold-nitrogen complexes. Solvent influences on oxidation and nuclearity of gold

guanidinate derivatives. *Inorganic Chemistry*, **46**, 11165–11172.

19 (a) Mohamed, A.A., Abdou, H.E., Irwin, M.D., Lopez-de-Luzuriaga, J.M., and Fackler, J.P. Jr (2003) Gold(I) formamidinate clusters: The structure, luminescence, and electrochemistry of the tetranuclear. Base-free [Au$_4$(ArNC(H)NAr)$_4$]. *Journal of Cluster Science*, **14**, 253–266; (b) Abdou, H.E., Mohamed, A.A., López-de-Luzuriaga, J.M. and Fackler, J.P. Jr (2004) Tetranuclear gold(I) clusters with nitrogen donor ligands: luminescence and X-ray structure of gold(I) naphthyl amidinate complexes. *Journal of Cluster Science*, **15**, 397–411; (c) Abdou, H.E., Mohamed, A.A. and Fackler, J.P. Jr (2007) Synthesis, characterization, luminescence, and electrochemistry of the tetranuclear gold(I) amidinate clusters, precursors to CO oxidation catalysts: Au$_4$[(ArNC(H)NAr)]$_4$, Au$_4$[(PhNC(Ph)NPh)]$_4$ and Au$_4$[PhNC(CH$_3$)NPh)$_4$]. *Journal of Cluster Science*, **18**, 630–641; (d) Abdou, H.E., Mohamed, A.A. and Fackler, J.P. Jr (2007) Synthesis, characterization, luminescence, and electrochemistry of the tetranuclear gold(I) amidinate clusters, precursors to CO oxidation catalysts: Au$_4$[(ArNC(H)NAr)]$_4$, Au$_4$[(PhNC(Ph)NPh)]$_4$ and Au$_4$[PhNC(CH$_3$)NPh)$_4$]. *Journal of the Chinese Chemical Society*, **54**, 1107–1113.

20 Patai, S. (1975) *The Chemistry of Amidines and Imidates*, Vol. 1, John Wiley & Sons, New York.

21 Beck, J. and Strahle, J. (1986) Synthesis and structure of 1,3-Diphenyltriazenidogold (I), a tetrameric molecule with short gold-gold distances. *Angewandte Chemie (International Edition in English)*, **25**, 95–96.

22 Chiari, B., Piovesana, O., Tarantelli, T. and Zanazzi, P.F. (1985) Gold dithiocarboxylates. *Inorganic Chemistry*, **24**, 366–371.

23 Abdou, H.E., Mohamed, A.A. and Fackler, J.P. Jr (2005) Synthesis and X-ray structures of dinulear and trinuclear gold(I) and dinuclear Gold(II) amidinate complexes. *Inorganic Chemistry*, **44**, 166–168.

24 Fenske, D., Baum, G., Zinn, A. and Dehnicke, K. (1990) Ag$_2$[Ph-C(NsiMe$_3$)$_2$]$_2$ and Au$_2$[Ph-C(NsiMe$_3$)$_2$]$_2$ – amidinato complexes with short metal-metal distances. *Zeitschrift für Naturforschung B*, **45**, 1273–1278.

25 Mohamed, A.A., Kani, I., Ramirez, A.O. and Fackler, J.P. Jr (2004) Synthesis, characterization, and luminescent properties of dinuclear gold(I) xanthate complexes: X-ray structure of [Au$_2$(nBu-xanthate)$_2$]. *Inorganic Chemistry*, **43**, 3833–3839.

26 Van Zyl, W.E., López-de-Luzuriga, J.M., Fackler, J.P. Jr and Staples, R.J. (2001) Dithiophosphinates of Gold(I). Oxidative addition of C1$_2$ to a neutral, dinuclear gold(I) dithiophosphinate complex. X-ray crystal structures of [AuS$_2$P(C$_2$H$_5$)$_2$]$_2$, [AuS$_2$PPh$_2$]$_2$, Au$_2$(CH$_2$)$_2$PMe$_2$(S$_2$PPh$_2$), and Au$_2$C1$_2$[(CH$_2$)$_2$PMe$_2$][S$_2$PPh$_2$]. *Canadian Journal of Chemistry*, **79**, 896–903.

27 King, C., Wang, J.-C., Khan, M.N.I. and Fackler, J.P. Jr (1989) Luminescence and metal-metal interactions in binuclear Gold(I) compounds. *Inorganic Chemistry*, **28**, 2145–2149.

28 (a) Van Zyl, W.E., Staples, R.J. and Fackler, J.P. Jr (1998) Dinuclear gold(I) dithiophosphonate complexes: formation, structure and reactivity. *Inorganic Chemistry Communications*, **1**, 51–54; (b) Van Zyl, W.E., Lopez-de-Luzuriaga, J.M. and Fackler, J.P. Jr (2000) Luminescence Studies of dinuclear gold(I) phosphor-1,1-dithiolate complexes. *Journal of Molecular Structure*, **516**, 99–106.

29 Herrero, G.P. and Jones, P.G. (1995) Synthesis of the first trithiocarbonatogold complex: [N(PPh$_3$)$_2$]$_2$[Au$_2$(μ^2-η^2-CS$_3$)$_2$]. First crystal structure of a μ^2-η^2-bridging trithiocarbonato complex. *Journal of the Chemical Society, Chemical Communications*, (7), 745–746.

30 Nazrul, Md., Khan, I., Wang, S. and Fackler, J.P. Jr (1989) Synthesis and structural characterization of [n-Bu$_4$N]$_2$[Au$_2$(i-MNT)$_2$](i-MNT=1,1-Dicyanoethylene-2,2-dithiolate) and its oxidative-addition products [Ph$_4$As]$_2$[Au$_2$(i-MNT)$_2$Cl$_2$], [n-Bu$_4$N]$_2$[Au$_2$(i-MNT)$_2$Br$_2$], and [n-Bu$_4$N][Au(i-MNT)$_2$]. Spectral studies of the disproportionation of [n-Bu$_4$N]$_2$[Au$_2$(i-MNT)$_2$X$_2$] (X=Cl-, Br-, I-) into [n-Bu$_4$N][AuX$_2$] and [n-Bu$_4$N][Au(i-MNT)$_2$]. *Inorganic Chemistry*, **28**, 3579–3588.

31 Mansour, M.A., Connick, W.B., Lachicotte, R.J., Gysling, H.J. and Eisenberg, R. (1998) Linear chain Au(I) dimer compounds as environmental sensors: A luminescent switch for the detection of volatile organic compounds. *Journal of the American Chemical Society*, **120**, 1329–1330.

32 Fackler, J.P. (1997) Polyhedron report No. 63, metal-metal bond formation in the oxidative addition to dinuclear Gold(I) species. Implications from dinuclear and trinuclear gold chemistry for the oxidative addition process generally. *Polyhedron*, **16**, 1–17.

33 Abdou, H.E., Mohamed, A.A. and Fackler, J.P. Jr (2004) Oxidative addition of methyl iodide to dinuclear Gold(I) amidinate complex: schmidbaur's breakthrough reaction revisited with amidinates. *Zeitschrift für Naturforschung B. A Journal of Chemical Sciences*, **59**, 1480–1482.

34 Abdou, H.E., Mohamed, A.A. and Fackler, J.P. Jr (2007) Oxidative-addition to the dinuclear Au(I) amidinate complex, [Au$_2$(2,6-(CH$_3$)$_2$Ph-form)$_2$]. Syntheses and characterization of the Au(II) amidinate complexes. The first dinuclear Gold(II) nitrogen complex possessing bonds to oxygen. *Inorganic Chemistry*, **46**, 9692–9699.

35 Porter, L.C. and Fackler, J.P. Jr (1986) Structure of the first example of an organometallic dinuclear gold(II) complex possessing bonds to oxygen. *Acta Crystallographica*, **42**, 1128–1131.

36 Mohamed, A.A., Abdou, H.E. and Fackler, J.P. Jr (2006) Mercury(II) cyanide coordination polymer with dinuclear gold (I) amidinate. Structure of the 2-D [Au$_2$(2,6-Me$_2$-formamidinate)$_2$]·2Hg (CN)$_2$·2THF complex. *Inorganic Chemistry*, **45**, 11–12.

37 Murray, H.H., Mazany, A.M. and Fackler, J.P. Jr (1985) Molecular structures of [Au(CH$_2$)$_2$PPh$_2$]$_2$(CN)$_2$ and [(CH$_2$)Au(CH$_2$)$_2$PPh$_2$]$_2$(CN)$_2$. The first ylide dimer possessing Gold(III) centers bonded only to carbon. *Organometallics*, **4**, 154–157.

38 Mohamed, A.A., Bruce, A.E. and Bruce, M.R.M. (1999) The electrochemistry of gold and silver complexes, in *Chemistry of Organic Derivatives of Gold and Silver* (ed. S. Patai), John Wiley and Sons, New York.

39 Abdou, H. (2006) PhD. Thesis New Chemistry with Gold-Nitrogen Complexes: Synthesis and Characterization of Tetra-, Tri-, and Dinuclear Gold(I) Amidinate Complexes. Oxidative-Addition to the Dinuclear Gold (I) Amidinate, A&M University, Texas.

40 Abdou, H.E., Mohamed, A.A. and Fackler, J.P. Jr (2007) Syntheses of mixed-ligand tetranuclear Gold(I)-Nitrogen clusters by ligand exchange reactions with the dinuclear Gold(I) formamidinate complex Au$_2$(2,6-Me$_2$Ph-form)$_2$. *Inorganic Chemistry*, **46**, 141–146.

41 (a) Cotton, F.A., Gruhn, N.E., Gu, J., Huang, P., Lichtenberger, D.L., Murillo, C.A., Van Dorn, L.O. and Wilkinson, C.C. (2002) Most easily-ionized, closed-shell molecules known; easier than the cesium atom. *Science*, **298**, 1971–1974; (b) Wilkinson, Chad (2005) PhD. Thesis, A&M University, Texas; (c) Soria, D.B., Grundy, J., Coles, M.P. and Hitchcock, P.B. (2005) Stabilisation of high oxidation-state niobium using "electron-rich" bicyclic-guanidinates. *Journal of Organometallic Chemistry*, **690**, 2278–2284; (d) Cotton, F.A., Durivage, J.C., Gruhn Lichtenberger, D.L., Murillo, C.A., Van, L.O. and Wilkinson, C.W. (2006) Photoelectron

spectroscopy and DFT calculations of easily ionized quadruply bonded $Mo_2 4+$ compounds and their bicyclic guanidinate precursors. *The Journal of Physical Chemistry B*, **110**, 19793–19798.

42 Cotton, F.A., Feng, X. and Timmons, D.J. (1998) Further study of very close non-bonded Cu^I–Cu^I contacts. molecular structure of a new compound and density functional theory calculations. *Inorganic Chemistry*, **37**, 4066–4069.

43 Burini, A., Mohamed, A. and Fackler, J.P. (2003) Cyclic trinuclear Gold(I) compounds: synthesis. structures and supramolecular acid-base π-stacks. *Comments on Inorganic Chemistry*, **24**, 253–280.

44 Vaughan, L.G. (1970) Organogold chemistry. III. 2-Pyridylgold(I). *Journal of the American Chemical Society*, **11**, 730–731.

45 Hayashi, A., Olmstead, M.M., Attar, S. and Balch, A.L. (2002) Crystal chemistry of the Gold (I) trimer, $Au_3(NC_5H_4)_3$: formation of hourglass figures and self-association through aurophilic attraction. *Journal of the American Chemical Society*, **124**, 5791–5795.

46 Bonati, F. and Minghetti, G. (1972) Trimeric 1-(Cyclohexylimino) methoxymethylgold(I), A new type of organometallic compound. *Angewandte Chemie International Edition*, **11**, 429.

47 Parks, J.E. and Balch, A.L. (1974) Gold carbene complexes: preparation, oxidation, and ligand displacement. *Journal of Organometallic Chemistry*, **71**, 453–463.

48 (a) Vickery, J.C., Olmstead, M.M., Fung, E.Y. and Balch, A.L. (1997) Solvent-stimulated luminescence from the supramolecular aggregation of a trinuclear Gold(I) complex that displays extensive intermolecular Au···Au interactions. *Angewandte Chemie (International Edition in English)*, **36**, 1179–1181; (b) Vickery, J.C., Olmstead, M.M., Fung, E.Y. and Balch, A.L. (1998) Glowing gold rings: solvoluminescence from planar trigold(I) complexes. *Coordination Chemistry Reviews*, **171**, 151–159; (c) Gade, L.H. (1997) "Hyt was of Gold, and Shon so Bryghte…": Luminescent Gold(I) Compounds. *Angewandte Chemie (International Edition in English)*, **36**, 1171–1173.

49 Minghetti, G. and Bonati, F. (1974) Trimeric (alkoxy)(alkylimino)methylgold(I) compounds, $[(RO)(R'N=)CAu]_3$. *Inorganic Chemistry*, **13**, 1600–1602.

50 Tiripicchio, A., Tiripicchio Camellini, M. and Minghetti, G. (1979) The crystal structure of tris-μ-[(ethoxy)(N-p-tolylimino)methyl-N, C]trigold(I), $[(EtO)(MeC_6H_4N=)CAu]_3$. *Journal of Organometallic Chemistry*, **171**, 399–406.

51 Bonati, F., Burini, A., Pietroni, B.R. and Bovio, B. (1989) Reactions of C-imidazolyllithium derivatives with Group Ib compounds: Tris[μ-(1-alkylimidazolato-N^3,C^2)]tri-gold(I) and -silver(I). Crystal structure of bis (1-benzylimidazolin-2-yliden)gold(I) chloride. *Journal of Organometallic Chemistry*, **375**, 147–160.

52 Bonati, F., Burini, A., Pietroni, B.R. and Galassi, R. (1993) Gold(I) Derivatives of Furan, Thiophene, 2-mercaptopyridine, and of some Pyrazoles – Mass-spectroscopic evidence of tetranuclear Gold(I) compounds. *Gazzetta Chimica Italiana*, **123**, 691.

53 Banditelli, G., Bandini, A.L., Bonati, F. and Goel, R.G. (1982) Some New Gold(I) complexes with bulky ligands. *Gazzetta Chimica Italiana*, **112**, 539.

54 Bovio, B., Bonati, F. and Banditelli, G. (1984) X-ray crystal structure of tris[μ-3,5-bis(trifluoromethyl)pyrazolato-N,N'] trigold(I), a compound containing an inorganic nine-membered ring. *Inorganica Chimica Acta*, **87**, 25–33.

55 Barberà, J., Elduque, A., Gimenez, R., Oro, L.A. and Serrano, J.L. (1996) Pyrazolate Golden Rings: trinuclear complexes that form columnar mesophases at room temperature. *Angewandte Chemie (International Edition in English)*, **35**, 2832–2836.

56 Barberà, J., Elduque, A., Gimenez, R., Lahoz, F.J., Oro, L.A. and Serrano, J.L. (1998) (Pyrazolato)gold complexes showing room-temperature columnar mesophases. Synthesis, properties, and structural characterization. *Inorganic Chemistry*, **37**, 2960–2967.

57 Yang, G. and Raptis, R.G. (2003) Supramolecular assembly of trimeric Gold(I) pyrazolates through aurophilic attractions. *Inorganic Chemistry*, **42**, 261–263.

58 (a) Balch, A.L. and Doonan, D.J. (1977) Mixed valence gold chemistry: Stepwise oxidation of a cyclic trigold(I) complex. *Journal of Organometallic Chemistry*, **131**, 137; (b) Minghetti, G., Banditelli, G. and Bonati, F. (1979) Metal derivatives of azoles. 3. The pyrazolato anion (and homologs) as a mono- or bidentate ligand: preparation and reactivity of tri-, bi-, and mononuclear gold(I) derivatives. *Inorganic Chemistry*, **18**, 658–663; (c) Raptis, R.G. and Fackler, J.P. Jr (1990) Synthesis and crystal structure of a mixed-valence, AuI-AuIII, pyrazolato complex stable in aqua regia. X-ray photoelectron study of homo- and heterovalent gold-pyrazolato trimers. *Inorganic Chemistry*, **29**, 5003; (d) Bonati, F., Burini, A., Pietroni, B.R. and Bovio, B. (1991) Reactions of symmetric C-imidazolylgold(I) leading to Au^I carbene complexes or mixed valence or Au^{III} imidazolyl derivatives. Crystal structure of [1-benzyl-3-(carboethoxy)imidazolin-2-yliden]chlorogold(I). *Journal of Organometallic Chemistry*, **408**, 271–280; (e) Raptis, R.G., Murray, H.H. and Fackler, J.P. Jr (1988) The structure of [Au{3,5-(C$_6$H$_5$)$_2$C$_3$HN$_2$}]$_3$Cl$_2$: a trinuclear mixed-valence gold pyrazolate complex. *Acta Crystallographica Section C: Crystal Structure Communications*, **44**, 970; (f) Bovio, B., Burini, A. and Pietroni, B.R. (1993) Reactions of trimeric 1-benzyl-2-gold(I)imidazole leading to Au^I carbene complexes. Crystal structure of [1-benzyl-3-benzoyl-imidazolin-2-yliden]chlorogold (I). *Journal of Organometallic Chemistry*,
452, 287–291; (g) Vickery, J.C. and Balch, A.L. (1997) X-ray crystallographic studies of the products of oxidative additions of iodine to cyclic trinuclear Gold(I) complexes: directional effects for Au-I···I-Au interactions. *Inorganic Chemistry*, **36**, 5978–5983.

59 Burini, A., Pietroni, B.R., Bovio, B., Calogero, S. and Wagner, F.E. (1994) A ^{197}Au mössbauer study of reaction products of trimeric 1-benzyl-2-gold(I)-imidazole leading to Au^I carbene or Au^I imidazoline complexes and trinuclear Au^{III} imidazolyl derivatives. X-Ray crystal structure of [{(μ-1-benzylimidazolato-N^3, C^2)Au}$_3$I$_2$]. *Journal of Organometallic Chemistry*, **470**, 275–283.

60 Miller, J.S. (ed.) (1982) *Extended Linear Chain Compounds*, Vol. 1–3, Plenum Press, New York.

61 Hoffmann, R. (1987) How chemistry and physics meet in the solid state. *Angewandte Chemie (International Edition in English)*, **26**, 846–878.

62 Kunugi, Y., Mann, K.R., Miller, L.L. and Exstrom, C.L. (1998) A vapochromic LED. *Journal of the American Chemical Society*, **120**, 589–590.

63 (a) Fernández, E.J., López-de-Luzuriaga, J.M., Monge, M., Olmos, M.E., Pérez, J., Laguna, A., Mohamed, A.A. and Fackler, John P. Jr (2003) {Tl[Au(C$_6$Cl$_5$)$_2$]}$_n$: A vapochromic complex. *Journal of the American Chemical Society*, **125**, 2022–2023; (b) Fernández, E.J., López-de-Luzuriaga, J.M., Monge, M., Olmos, M.E., Pérez, J., Laguna, A., Mohamed, A.A. and Fackler, John P. Jr (2004) A Detailed study of the vapochromic behavior of {Tl[Au(C$_6$Cl$_5$)$_2$]}$_n$. *Inorganic Chemistry*, **43**, 3573–3581; (c) Fernández, E.J., Laguna, A., López-de-Luzuriaga, J.M., Monge, M., Montiel, M., Olmos, M.E. and Rodríguez-Castillo, M. (2006) Mesitylgold(I) and silver(I) perfluorocarboxylates as precursors of supramolecular Au/Ag systems. *Organometallics*, **25**, 4307–4315.

64 Burini, A., Fackler, J.P. Jr, Galassi, R., Grant, T.A., Omary, M.A.,

Rawashdeh-Omary, M.A., Pietroni, B.R. and Staples, R.J. (2000) Supramolecular chain assemblies formed by interaction of a B molecular acid complex of mercury with B-Base trinuclear gold complexes. *Journal of the American Chemical Society*, **122**, 11264.

65 Rawashdeh-Omary, M.A., Omary, M.A., Fackler, J.P. Jr, Galassi, R., Pietroni, B.R. and Burini, A. (2001) Chemistry and optoelectronic properties of stacked supramolecular entities of trinuclear Gold(I) complexes sandwiching small organic acids. *Journal of the American Chemical Society*, **123**, 9689.

66 (a) Mohamed, A.A., Manal, A., Rawashdeh-Omary, M.A., Omary, M.A. and Fackler, J.P. Jr (2005) External heavy-atom effect of gold in a supramolecular acid-base pi stack. *Dalton Transactions*, (15), 2597; (b) Omary, M.A., Mohamed, A.A., Rawashdeh-Omary, M.A. and Fackler, J.P. Jr (2005) Photophysics of supramolecular binary stacks consisting of electron-rich trinuclear Au(I) and organic electrophiles Coord. *Chemical Reviews*, **249**, 1372.

67 Olmstead, M.M., Jiang, F., Attar, S. and Balch, A.L. (2001) Alteration of the aurophilic interactions in trimeric Gold(I) compounds through charge transfer. behavior of solvoluminescent $Au_3(MeN=COMe)_3$ in the presence of electron acceptors. *Journal of the American Chemical Society*, **123**, 3260.

68 Gabbaï, F.P., Schier, A., Riede, J. and Tschinkl, M.T. (1999) Micropore decoration with bidentate lewis acids: spontaneous assembly of 1,2-Bis(chloromercurio) tetrafluorobenzene. *Angewandte Chemie International Edition*, **38**, 3547.

69 Tsunoda, M. and Gabbaï, F.P. (2000) $\mu_6\text{-}\eta^2{:}\eta^2{:}\eta^2{:}\eta^2{:}\eta^2{:}\eta^2$ As a new bonding mode for benzene. *Journal of the American Chemical Society*, **122**, 8335–8336.

70 Burini, A., Fackler, J.P. Jr, Galassi, R., Macchioni, A., Omary, M.A., Rawashdeh, M.A.-O., Pietroni, B.R., Sabatini, S. and Zuccaccia, C. (2002) ^{19}F, ^1H-HOESY and PGSE NMR studies of neutral trinuclear complexes of Au1 and HgII: evidence for acid-base stacking in solution. *Journal of the American Chemical Society*, **124**, 4570.

71 (a) Teo, B.K. and Keating, K. (1984) Novel triicosahedral structure of the largest metal alloy cluster: hexachlorododecakis (triphenylphosphine)-gold-silver cluster $[(Ph_3P)_{12}Au_{13}Ag_{12}Cl_6]m+$. *Journal of the American Chemical Society*, **106**, 2224; (b) Teo, B.K., Zhang, H. and Shi, X. (1990) Molecular architecture of a novel vertex-sharing biicosahedral cluster $[(p\text{-}Tol_3P)_{10}Au_{13}Ag_{12}Br_8](PF_6)$ containing a staggered-staggered-staggered configuration for the 25-atom metal framework. *Inorganic Chemistry*, **29**, 2083–2091; (c) Teo, B.K., Shi, X. and Zhang, H. (1991) Cluster of clusters. Structure of a novel gold-silver cluster $[(Ph_3P)_{10}Au_{13}Ag_{12}Br_8](SbF_6)$ containing an exact staggered-eclipsed-staggered metal configuration. Evidence of icosahedral units as building blocks. *Journal of the American Chemical Society*, **113**, 4329–4331; (d) Teo, B.K. and Zhang, H. (1991) Cluster of clusters. Structure of a new cluster $[(p\text{-}Tol_3P)_{10}Au_{13}Ag_{12}Cl_7](SbF_6)_2$ containing a nearly staggered-eclipsed-staggered metal configuration and five doubly-bridging ligands. *Inorganic Chemistry*, **30**, 3115–3116; (e) Teo, B.K. and Zhang, H. (1992) Molecular machines: molecular structure of $[(p\text{-}Tol_3P)_{10}Au_{13}Ag_{12}Cl_8](PF_7)$ – a Cluster with a biicosahedral rotorlike metal core and an unusual arrangement of bridging ligands. *Angewandte Chemie (International Edition in English)*, **31**, 445–447; (f) Teo, B.K., Shi, X. and Zhang, H. (1992) Cluster rotamerism of a 25-metal-atom cluster $[(Ph_3P)_{10}Au_{13}Ag_{12}Br_8]^+$ monocation: a molecular rotary unit. *Journal of the Chemical Society, Chemical Communications*, (17), 1195–1196; (g) Teo, B.K., Dang, H., Campana, C.F. and Zhang, H. (1998) Synthesis, structure, and

charactreization of (MePh$_2$P)$_{10}$Au$_{12}$Ag$_{13}$Br$_9$: The first example of a neutral bi-icosahedral Au⋯Ag cluster with a novel seven-membered satellite ring of bridging ligands. *Polyhedron*, **17**, 617–621;
(h) Usón, R., Laguna, A., Laguna, M., Manzano, B.R., Jones, P.G. and Sheldrick, G.M. (1984) Synthesis and reactivity of bimetallic Au–Ag polyfluorophenyl complexes; crystal and molecular structures of [{AuAg(C$_6$F$_5$)$_2$(SC$_4$H$_8$)}$_n$] and [{AuAg(C$_6$F$_5$)$_2$(C$_6$H$_6$)}$_n$]. *Journal of the Chemical Society, Dalton Transactions*, (2), 285; (i) Usón, R., Laguna, A., Laguna, M., Usón, R., Laguna, A., Laguna, M., Jones, P.G. and Erdbrügger, C.F. (1987) Bimetallic phosphorus ylide gold-silver complexes. *Organometallics*, **6**, 1778–1780;
(j) Tran, N.T., Powell, D.R. and Dahl, L.F. (2004) Generation of AuPd$_{22}$/Au$_2$Pd$_{21}$ analogues of the high-nuclearity Pd$_{23}$(CO)$_{20}$(PEt$_3$)$_{10}$ cluster containing 19-atom centered hexacapped-cuboctahedral (v_2-octahedral) metal fragment: structural-to-synthesis approach concerning formation of Au$_2$Pd$_{21}$(CO)$_{20}$(PEt$_3$)$_{10}$. *Dalton Transactions*, (2), 209–216;
(k) Tran, N.T., Powell, D.R. and Dahl, L.F. (2004) Nanosized Au$_2$Pd$_{41}$(CO)$_{27}$(PEt$_3$)$_{15}$ containing two geometrically unprecedented 13-coordinated Au-centered (μ_{13}-Au)Pd$_{13}$ polyhedra connected by triangular face-sharing and three interpenetrating 12-coordinated Pd-centered (μ_{12}-Pd)Au$_2$Pd$_{10}$ icosahedra: geometrical change in centered polyhedra induced by Au/Pd electronegativity-mismatch. *Dalton Transactions*, (2), 217–223; (l) Catalano, V.J. and Horner, S.J. (2003) Luminescent Gold(I) and Silver(I) Complexes of 2-(Diphenylphosphino)-1-methylimidazole (dpim): Characterization of a Three-Coordinate Au(I)-Ag(I) Dimer with a Short Metal-Metal Separation. *Inorganic Chemistry*, **42**, 8430–8438;
(m) Rawashdeh-Omary, M.A., Omary, M.A. and Fackler, J.P. Jr (2002) Argento-aurophilic Bonding in Organosulfur Complexes. The Molecular and Electronic Structures of the Heterobimetallic Complex AgAu(MTP)$_2$. *Inorganica Chimica Acta*, **334**, 376.

72 (a) Rousset, J.L., Aires, J.C.S., Sekhar, R., Mélinon, P., Prevel, B. and Pellarin, M. (2000) Comparative X-ray photoemission spectroscopy study of Au, Ni, and AuNi clusters produced by laser vaporization of bulk metals. *The Journal of Physical Chemistry B*, **104**, 5430–5435; (b) Rainer, D.R., Xu, C., Holmblad, P.M. and Goodman, D.W. (1997) Pd, Cu, and Au particles on Al$_2$O$_3$ thin films: An infrared reflection absorption spectroscopy study of monometallic and bimetallic planar model supported catalysts. *Journal of Vacuum Science & Technology A, Vacuum Surfaces and Films*, **15**, 1653–1662; (c) Baddeley, C.J., Tikhov, M., Hardacre, C., Lomas, J.R. and Lambert, R.M. (1996) Ensemble effects in the coupling of acetylene to benzene on a bimetallic surface: A study with Pd{111}/Au. *The Journal of Physical Chemistry*, **100**, 2189–2194; (d) Reifsnyder, S.N. and Lamb, H.H. (1999) Characterization of silica-supported Pd-Au clusters by X-ray absorption spectroscopy. *The Journal of Physical Chemistry B*, **103**, 321–329.

73 (a) Andres, R.P., Bein, T., Dorogi, M., Feng, S., Henderson, J.I., Kubiak, C.P., Mahoney, W., Osifchin, R.G. and Reifenberger, R. (1996) "Coulomb Staircase" at Room Temperature in a Self-Assembled Molecular Nanostructure. *Science*, **272**, 1323–1325; (b) Mirkin, C.A., Letsinger, R.L., Mucic, R.C. and Storhoff, J.J. (1996) A DNA-based method for rationally assembling nanoparticles into macroscopic materials. *Nature*, **382**, 607–609; (c) Alivisatos, A.P., Johnson, K.P., Peng, X., Wilson, T.E., Loweth, C.J., Bruchez, M.P. Jr and Schultz, P.G. (1996) Organization of 'nanocrystal molecules' using DNA. *Nature*, **382**, 609–611; (d) Hong, B.H., Bae, S.C., Lee, C.W., Jeong, S. and Kim, K.S. (2001) Ultrathin single-crystalline silver nanowire arrays formed in an ambient solution phase. *Science*, **294**,

348–351; (e) Hong, B.H., Lee, J.Y., Lee, C.W., Kim, J.C., Bae, S.C. and Kim, K.S. (2001) Self-assembled arrays of organic nanotubes with infinitely long one-dimensional H-Bond chains. *Journal of the American Chemical Society*, **123**, 10748–10749.

74 (a) Mohamed, A.A., Burini, A. and Fackler, J.P. Jr (2005) Mixed-metal triangular trinuclear complexes: dimers of gold-silver mixed-metal complexes from Gold(I) carbenieates and Silver(I) 3,5-Diphenylpyrazolates. *Journal of the American Chemical Society*, **127**, 5012; (b) Mohamed, A.A., Galassi, R., Fabrizio, P., Burini, A. and Fackler, J.P. Jr (2006) Gold(I) and Silver(I) mixed-metal trinuclaer complexes: dimeric products from the reaction of Gold(I) carbeniates or benzylimidazolates with Silver(I) 3,5-Diphenylpyrazolate. *Inorganic Chemistry*, **45**, 7770–7776.

75 Mohamed, A.A., Perez, L.M. and Fackler, J.P. Jr (2005) Unsupported intermolecular argentophilic interaction in the dimer of trinuclear silver(I) 3,5-diphenylpyrazolates. *Inorganica Chimica Acta*, **358**, 1657–1662.

76 (a) Yan, Z., Chinta, S., Mohamed, A.A., Fackler, J.P. Jr and Goodman, D.W. (2005) The role of F-centers in catalysis by Au supported on MgO. *Journal of the American Chemical Society*, **127**, 1604–1605; (b) Yan, Z., Chinta, S., Mohamed, A.A., Fackler, J.P. Jr and Goodman, D.W. (2006) CO Oxidation over Au/TiO2 prepared from metal-organic gold complexes. *Catalysis Letters*, **111**, 15–18.

77 Guzman, J. and Gates, B.C. (2003) Structure and reactivity of a mononuclear gold-complex catalyst supported on magnesium oxide. *Angewandte Chemie International Edition*, **42**, 690–693.

78 Oh, H.-S., Yang, J.H., Costello, C.K., Wang, Y.M., Bare, S.R., Kung, H.H. and Kung, M.C. (2002) Selective catalytic oxidation of CO: Effect of chloride on supported au catalysts. *Journal of Catalysis*, **210**, 375–386.

79 Broqvist, P., Molina, L.M., Grönbecka, H. and Hammer, B. (2004) Promoting and poisoning effects of Na and Cl coadsorption on CO oxidation over MgO-supported Au nanoparticles. *Journal of Catalysis*, **227**, 217–226.

80 Sladek, A., Hofreiter, S., Paul, M. and Schmidbaur, H. (1995) Sodium tetraphenylborate as a phenylating agent for gold(I) complexes. *Journal of Organometallic Chemistry*, **501**, 47–51.

81 Forward, J.M., Fackler, J.P. and Staples, R.J. (1995) Synthesis and structural characterization of the luminescent Gold(I) Complex [(MeTPA)$_3$AuI]I$_3$. Use of NaBPh$_4$ as a phenyl transfer reagent to form [(MeTPA)AuPh](BPh$_4$) and (TPA) AuPh. *Organometallics*, **14**, 4194–4198.

82 Partyka, D.V., Zeller, M., Hunter, A.D. and Gray, T.G. (2006) Relativistic functional groups: Aryl carbon-gold bond formation by selective transmetalation of boronic acids. *Angewandte Chemie International Edition*, **45**, 8188–8191.

83 There are several recent reviews: (a) Hashmi, A.S.K. (2007) Gold-catalyzed organic reactions. *Chemical Reviews*, **107**, 3180–3211; (b) Gorin, D.J. and Toste, D.F. (2007) Relativistic effects in homogeneous gold catalysis. *Nature*, **446**, 395–403; (c) Hashmi, A.S.K. and Hutchings, G.J. (2006) Gold Catalysis. *Angewandte Chemie International Edition*, **45**, 7896–7936.

84 Gray, T.G. (2007) Gilded organometallics. *Comments on Inorganic Chemistry*, **28**, 181–212.

2
Chemistry of Gold(III) Complexes with Nitrogen and Oxygen Ligands
Maria Agostina Cinellu

2.1
Introduction

The chemistry of gold has advanced considerably during the past decade. A large variety of gold compounds have found applications in various fields including catalysis [1], medicinal chemistry [2], photophysics and photochemistry [3] just to mention some of the most recently emerging areas. A renewed interest in these fields has prompted many researchers to synthesize new gold complexes with specific properties. In particular, gold(III) complexes with nitrogen and oxygen donor ligands occupy a predominant place. The versatility of nitrogen ligands [4], particularly heterocycles, has been exploited to give rise, *inter alia*, to a flourishing class of cycloaurated derivatives [5], most of which display promising antitumor activity. Polydentate nitrogen ligands and their derivatives have also been prepared for pharmacological and photophysical applications. Furthermore, several known complexes have been re-evaluated in the light of their possible catalytic and/or pharmacological activity and, in some cases, modifications (e.g., substitution of the ancillary ligands) have been made to vary their properties.

In spite of the low affinity for binding to oxygen, gold(III) alkoxo, hydroxo and even oxo complexes have been obtained [6, 7]. These are valuable models for Au-O(H) species which are likely to be involved in oxidation reactions catalyzed by metal-oxide-supported gold [8]. All these complexes have displayed interesting chemical reactivity and, in some cases, remarkable catalytic activity.

2.2
Nitrogen Donor Ligands

2.2.1
Complexes with Neutral Monodentate Ligands

Au(N)Cl$_3$ adducts are readily obtained by reaction of Au$_2$Cl$_6$ or [AuCl$_4$]$^-$ with a variety of substituted heterocyclic ligands including 2-phenylpyridine [9], 2-benzylpyridines [10],

Gold Chemistry: Applications and Future Directions in the Life Sciences. Edited by Fabian Mohr
Copyright © 2009 WILEY-VCH Verlag GmbH & Co. KGaA, Weinheim
ISBN: 978-3-527-32086-8

Scheme 2.1 Production of cyclometallated derivatives [Au(N,C)Cl$_2$] from 2-benzylpyridines by refluxing in MeCN/H$_2$O.

R = R' = H
R = H, R' = Me

2-anilinopyridine [11], 2-phenoxypyridines [11b, 12], 2-phenylsulfanylpyridine [11b], 2-benzoylpyridine [13], 2-thienylpyridyne [14], 2-phenylthiazole [15] and 1-ethyl-2-phenylimidazole [16]. All these adducts undergo facile activation of an *ortho* C—H bond of the aryl substituent to give cyclometallated derivatives [Au(N,C)Cl$_2$], by refluxing suspensions of the adducts in MeCN/H$_2$O. As an example, this process with 2-benzylpyridines is illustrated in Scheme 2.1.

Under comparable conditions, no *ortho*-metallated species are obtained from the 1-phenyl-pyrazole (1-Phpz) adduct, Au(N)Cl$_3$, formed by reaction of 1-Phpz with Na[AuCl$_4$] [17].

Coordination of only one pyridine, to give Au(N)Cl$_3$ adducts, is also observed in the case of 6-substituted 2,2'-bipyridines such as 6-phenyl- (6-Phbipy) [18], 6-(2''-thienyl)- (6-thbipy) [19], 6-alkyl- (6-Rbipy) (R = Me, CH$_2$Me, CHMe$_2$, CMe$_3$, CH$_2$CMe$_3$) [20] and 6-benzyl-2,2'-bipyridines (6-Bnbipy) (Bn = CH$_2$Ph, CHMePh, CMe$_2$Ph) [20]. This behavior contrasts with that of the unsubstituted ligand which forms the cationic adduct [Au(N,N)Cl$_2$]$^+$ [21]. An analogous situation is found in the case of 2,9-dimethyl-1,10-phenanthroline (2,9-Me$_2$phen) and 1,10-phenanthroline (phen) which give neutral Au(N,N)X$_3$ (X = Cl, Br) [22] and cationic [Au(N,N)Cl$_2$]$^+$ adducts, respectively [23]. The X-ray structure of the 6-methylbenzyl-2,2'-bipyridine adduct (Figure 2.1) [20] shows that the gold atom is bound to the nitrogen atom of the unsubstituted pyridine ring with an Au—N distance of 205 pm. The Au—N distance to the other nitrogen atom is 276 pm, which, although longer than a typical Au—N single bond, is well below the sum of the van der Waals radii of Au and N (220 + 150 = 370 pm). The overall coordination of the gold atom is square-planar with only a very

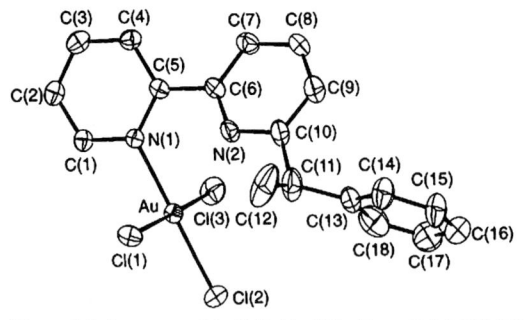

Figure 2.1 Structure of Au(6-Bnbipy)Cl$_3$ (Bn = CHMePh) [20].

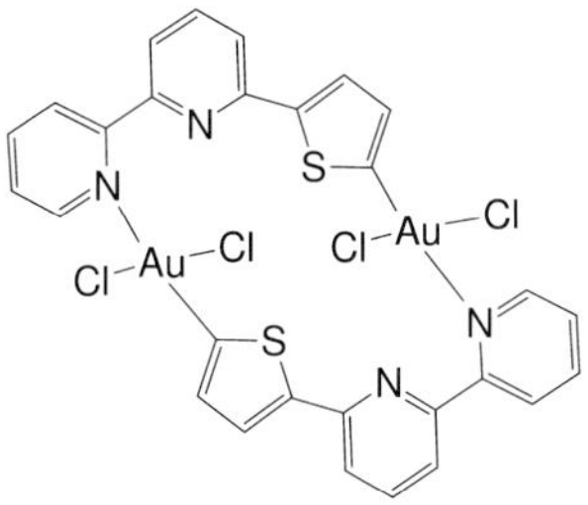

Figure 2.2 Products arising from C-H bond activation of Au(N)Cl$_3$ adducts (N = mono-ligated 6-substituted 2,2′-bipyridine).

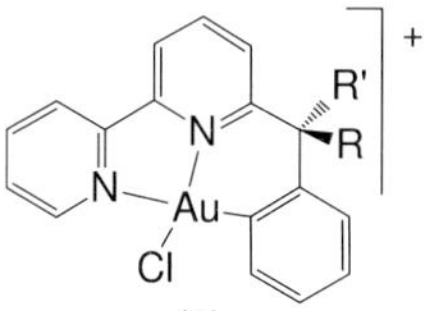

slight pyramidal distortion, while the 2,9-Me$_2$phen derivative, with Au–N distances of 209 and 258 pm, is best described as intermediate between trigonal and square pyramidal [22].

As in the case of 2-substituted pyridine adducts, activation of an aromatic C–H bond is observed when the adducts of 6-thbipy and of 6-Bnbipy are heated under reflux in aqueous media. A cyclic dimer (1) with bridging N,C ligands is obtained in the first case [24], while cyclometallated derivatives [Au(N,N,C)Cl]$^+$ (2) are formed in the latter [20] (Figure 2.2).

Under comparable reaction conditions, no C–H bond activation is observed for adducts of 6-Phbipy and 6-Rbipy. Nevertheless, [Au(N,N,C)Cl]$^+$ derivatives can be obtained with 6-Phbipy [18] and with 6-tBubipy (tBu = CMe$_3$) [20]. The former is obtained by a transmetallation reaction of the arylmercury(II) derivative with [AuCl$_4$]$^-$, while activation of a C(sp^3)–H bond of the *tert*-butyl substituent is accomplished by reaction of the Au(N)Cl$_3$ adduct **3** (N = 6-tBubipy) with AgBF$_4$ in the presence of excess ligand (Scheme 2.2).

In the latter case, a dinuclear oxo bridged complex [Au$_2$(N,N)$_2$(μ-O)$_2$][BF$_4$]$_2$ is also formed in small amounts. The same oxo complexes are obtained in high yields, as the

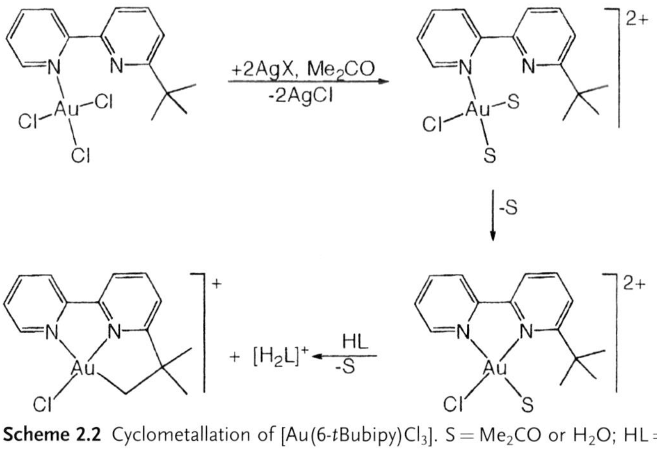

Scheme 2.2 Cyclometallation of [Au(6-tBubipy)Cl$_3$]. S = Me$_2$CO or H$_2$O; HL = 6-tBubipy.

PF_6^- salts, with all the 6-Rbipy ligands by reaction of the Au(6-Rbipy)Cl$_3$ adducts **3** (R = Me, Et, iPr or neoPn) with AcONa and excess KPF_6 in aqueous media [25] (Equation 2.1).

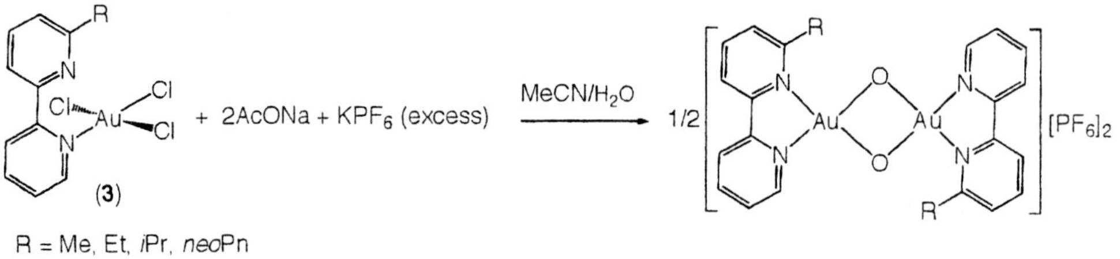

$$[\text{Au(6-Rbipy)Cl}_3] + 2\text{AcONa} + \text{KPF}_6 \text{ (excess)} \xrightarrow{\text{MeCN/H}_2\text{O}} 1/2\,[\text{Au}_2(\text{6-Rbipy})_2(\mu\text{-O})_2][\text{PF}_6]_2$$

(3)

R = Me, Et, iPr, neoPn

(2.1)

Nucleobases, including 9-methyl-, 9-ethyl-, 1,9-dimethyl-guanine and 2-amino-6-methoxy-9-methylpurine, form complexes of the type Au(N)Cl$_3$ when reacted with [AuCl$_4$]$^-$ in water at pH 3–4. Binding of a AuCl$_3$ unit to the N(7) position of the purine ring was confirmed by X-ray crystallography [26].

A few gold(III) complexes of monodentate N-ligands are sufficiently stable within a physiological environment to display interesting pharmacological activity. Among these, the trichorogold(III) derivatives of alkylimidazoles and dimethylbenzoxazole [27] and streptonigrin [28], exhibit cytotoxic activity against selected tumor cell lines. 1-[(2-chlorophenyl)diphenylmethyl]-1H-imidazole (CTZ) adduct [29] is active against cultures of *Trypanosoma cruzi*, the causative agent of Chagas disease, and the Au(N)$_2$Cl$_2$ adduct of chloroquine displays antimalarial activity [30].

The reaction of anhydrous gold(III) chloride [AuCl$_3$]$_2$ with a variety of aromatic hydrocarbons (ArH) and subsequent treatment with 2,6-lutidine (lut) affords the stable arylgold(III) complexes *trans*-[(lut)AuArCl$_2$] (**4**) [31]. These complexes undergo stoichiometric coupling with terminal alkynes to give Sonogashira type products in near quantitative yield (Scheme 2.3).

Gold(III) complexes with monodentate nitrogen ligands have also been prepared starting from (pentafluorophenyl)gold(III) compounds. Reaction of 2-amino-4,5-dihydrothiazole (2-amt) with [Au(C$_6$F$_5$)$_3$(tht)] (tht = tetrahydrothiophene) affords the complex [Au(C$_6$F$_5$)$_3$(2-amt)] (**5**) [32]. Crystal structure determination shows that 2-amt is coordinated through the ring nitrogen atom with an Au–N distance of 2.07 Å, while the short amino C–N bond length indicates some electron delocalization.

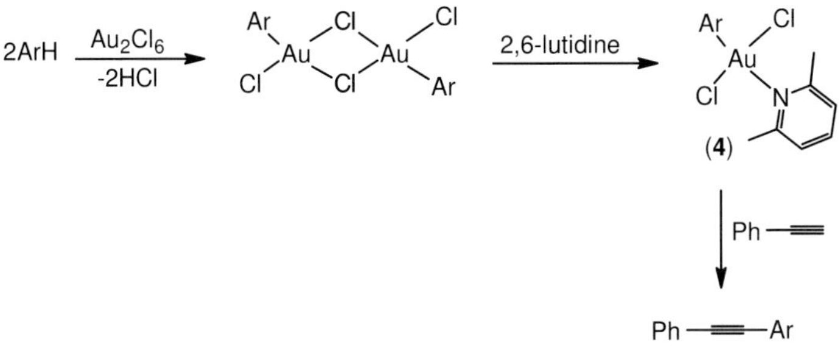

$$2\text{ArH} \xrightarrow[-2\text{HCl}]{\text{Au}_2\text{Cl}_6} [\text{Ar(Cl)Au}(\mu\text{-Cl})_2\text{Au(Cl)Ar}] \xrightarrow{\text{2,6-lutidine}} trans\text{-}[(\text{lut})\text{AuArCl}_2]\ (\mathbf{4}) \xrightarrow{\text{Ph}-\!\!\equiv} \text{Ph}-\!\!\equiv\!\!-\text{Ar}$$

Scheme 2.3 Production and reaction of *trans*-[(lut)AuArCl$_2$] (**4**).

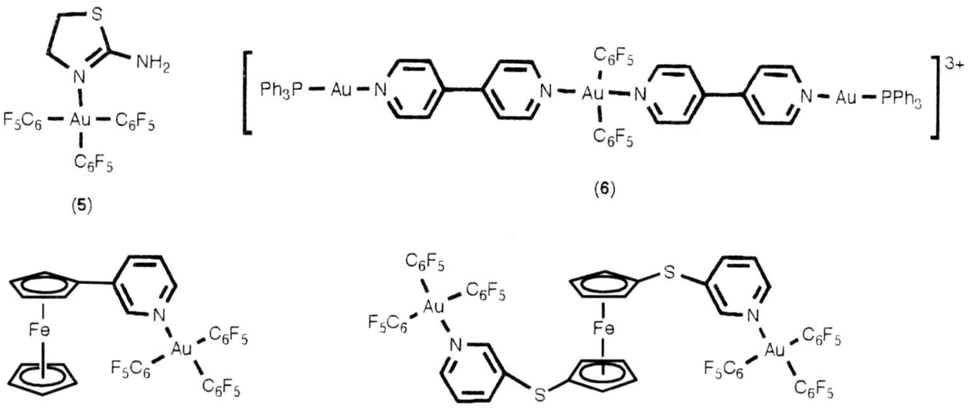

Figure 2.3 Selected examples of (pentafluorophenyl)gold(III) complexes with monodentate N-ligands.

The bifunctional ligand 4,4′-bipyridine (4,4′-bipy) links gold(I) and gold(III) ions in [Au^{III}(C$_6$F$_5$)$_2${(4,4′-bipy)Au^{I}PPh$_3$)}$_2$](NO$_3$)$_3$ (6) [33]. Hetero polynuclear complexes [Au(C$_6$F$_5$)$_3$(FcPy)] (7), [Au(C$_6$F$_5$)$_2$(FcPy)$_2$]ClO$_4$ (FcPy = 3-ferrocenylpyridine) [34] and [{Au(C$_6$F$_5$)$_3$}$_2${Fc(Spy)$_2$}] (8) [Fc(Spy)$_2$ = 1,1′-bis(2-pyridylthio)ferrocene] [35] are obtained from [Au(C$_6$F$_5$)$_3$(OEt$_2$)] or [Au(C$_6$F$_5$)$_2$(OEt$_2$)$_2$]ClO$_4$. Selected examples are shown in Figure 2.3.

Oxidative addition of PhICl$_2$ to gold(I) precursors has been used to prepare the acetimine derivatives [Au(NH=CMe$_2$)Cl$_3$] and [Au(NH=CMe$_2$)$_2$Cl$_2$]ClO$_4$ [36].

2.2.2
Complexes with Anionic Monodentate Ligands

Highly sensitive ammonium and methylammonium tetraazidoaurates(III) have been prepared according to different procedures depending on the ammonium counter ion [37] (Equations 2.2 and 2.3).

$$H[AuCl_4] \xrightarrow[H_2O/EtOH]{i) \text{ excess } NaN_3 / ii) [Me_4N]Br} [Me_4N][Au(N_3)_4] \quad (2.2)$$

$$[R_2NH_2][AuCl_4] \xrightarrow[MeOH]{\text{excess } AgN_3} [R_2NH_2][Au(N_3)_4] \quad (2.3)$$
$$R=Me, H$$

The red-orange crystalline solids exhibit increasing sensitivity on decreasing carbon content. [NH$_4$][Au(N$_3$)$_4$] is a promising candidate for the deposition of gold on surfaces, which can serve as a hydrogenation catalyst. The crystal structure of [Me$_4$N][Au(N$_3$)$_4$] shows a "whirl-wind" configuration of the tetraazidogold(III) anion, similar to that found in [Ph$_4$As][Au(N$_3$)$_4$] [38], except that in the former a polymeric stacking of the anions with weak Au···Au interactions of 351 and 358 pm is observed. This is one rare example of aurophilic interaction found in gold(III)

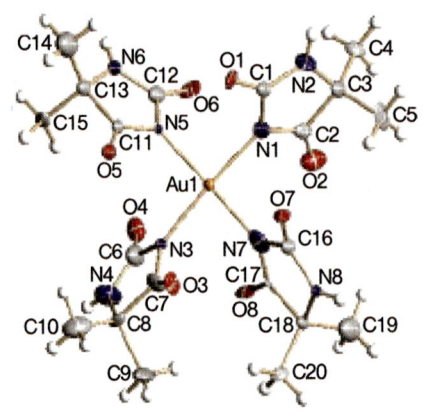

Figure 2.4 Structure of [Au(dmh)$_4$]$^-$. Reprinted with permission from [42]. Copyright (2005) American Chemical Society.

chemistry. The azido complexes M[Au(N$_3$)$_4$] (M = Li, Na), generated *in situ* from the tetrachloride and excess MN$_3$ in water, react with CNR (R = *t*Bu, Cy) to give gold-carbon bonded tetrazolato complexes [39].

A few stable gold(III) imidate complexes have been authenticated by X-ray structure determinations. Most of the claimed gold(III) imidate complexes including those containing succinimide, phthalimide, saccharin or N-methylidantoin derivatives [40] were subsequently reformulated as gold(I) species [41].

The square planar tetrahydantoinate complex Na[Au(dmh)$_4$] (dmhH = 5,5-dimethyl hydantoin) (Figure 2.4) is inert and very stable in alkaline solutions because of the favorable soft acid–base interactions [42]. Gold deposition from [Au(dmh)$_4$]$^-$ is observed at potentials more negative than −0.2 V *vs* Ag/AgCl by cyclic voltammetry. The saccharinate complexes [Au(sacc)$_2$(H$_2$O)$_2$]Cl and [Au(sacc)$_3$(H$_2$O)] [saccH = 1,2-benzisothiazol-3(2H)-one] are readily obtained by reaction of gold(III) chloride with Na[sacc] in water [43]. Replacement of the chloride ions of the cyclometallated complexes [Au(N,C)Cl$_2$] [N,CH = 2-benzylpyridine (2-Bnpy) or Me$_2$NCH$_2$C$_6$H$_5$] with deprotonated saccharine (sacc), phthalimide (phth) or isatin affords a series of gold (III) bis(imidate) complexes [Au(N,C)(imidate)$_2$] [44]. The X-ray crystal structures of [Au(2-Bnpy-H)(sacc)$_2$] and [Au(2-Bnpy-H)(phth)$_2$] show that the planar imidate ligands are perpendicular to the gold coordination plane; Au−N bond distances are 201 and 212 pm for the N atom *trans* to N and to C, respectively. The complexes show low to medium antitumor activity towards the P388 murine leukaemia cell line and modest anti-microbial activity.

A series of amido complexes [45] of the type [Au(N,N,C)(NRR′)](PF$_6$) (N,N,CH 6-benzyl-2,2′-bipyridines), [Au(N,C)(AcO)(NHAr)] (N,CH = 2-benzyl-pyridine) and [Au(N,N)(NHAr)$_2$](PF$_6$) (N,N = 2,2′-bipyridine) join the still small number of gold (III) derivatives of deprotonated primary and secondary amines [46]. These are obtained by σ-ligand metathesis of the corresponding hydroxo, methoxo, or acetato complexes [Au(N,N,C)(OR″)](PF$_6$) (R″ = H or Me), [Au(N,C)(AcO)$_2$] and [Au(N,N) (OH)$_2$](PF$_6$), with both aryl- (R = H, R′ = Ar = C$_6$H$_4$NO$_2$-4, C$_6$H$_4$Me-4 or C$_6$H$_3$Me$_2$-

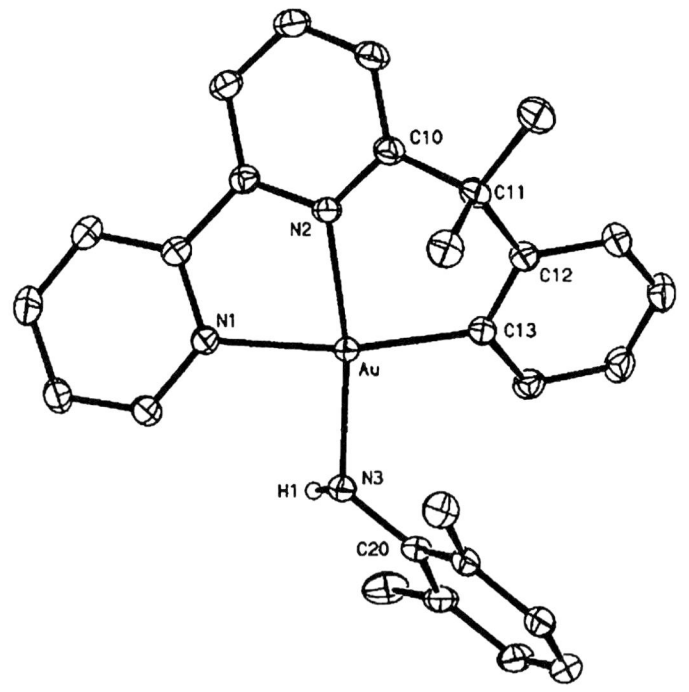

Figure 2.5 Structure of [Au(N,N,C)(NHC$_6$H$_3$Me$_2$-2,6)]$^+$ [N,N,CH = 6(1,1-dimethylbenzyl)-2,2'-bipyridine] [45b].

2,6) and alkyl- (R = H, R' = CH$_2$CHMe$_2$ or CHMeEt; R = R' = Et) amines. The Au–N bond distance of the coordinated amide in [Au(N,N,C)(NHC$_6$H$_3$Me$_2$-2,6)](PF$_6$) [N,N, CH = 6-(1,1-dimethylbenzyl)-2,2'-bipyridine] (Figure 2.5) is 202 pm [45b]. The slight elongation of the Au–N distance (206 pm) *trans* to the amide ligand, with respect to that found in the corresponding chloro complex (201 pm) [20] indicates a stronger *trans* influence of the amide ligand (Table 2.1).

2.2.3
Complexes with Multidentate Ligands

Gold(III) complexes with polydentate N-ligands are currently under investigation for pharmaceutical applications because they show good stability under physiological conditions. The enhanced stability toward reduction of various polyamine gold(III) complexes is shown by their electrochemical behavior in aqueous solution [47, 48]. Cyclic voltammetry shows that the formal potential of Au^{III}/Au^0 becomes more negative in the series [AuCl$_4$]$^-$, [Au(en)Cl$_2$]$^+$, [Au(dien)Cl]$^{2+}$, [Au(en)$_2$]$^{3+}$, [Au(cyclam)]$^{3+}$ (en = 1,2-ethylendiamine; dien=diethylentriamine; cyclam = 1,4,8, 11-tetraazacyclotetradecane) [47, 48]. Significant cytotoxic activities are found for [Au(en)$_2$]Cl$_3$ and [Au(dien)Cl]Cl$_2$, when tested against the human tumor cell line (A2780), either sensitive (A2780/S) or resistant (A2780/R) to cisplatin. In contrast, the encapsulated gold(III) derivative [Au(cyclam)](ClO$_4$)$_2$Cl is virtually

Table 2.1 Bond distances (pm) in selected complexes with monodentate ligands.

Complex	r(Au—N)	X[a]	r(Au-X)	References
Neutral ligands				
Au(6-Bnbipy)Cl$_3$[b]	204.7(4)	Cl	226.4(1)	[20]
	275.8(4)			
Au(1,9-Me$_2$guanine)Cl$_3$	202(3)	Cl	227(8)	[26]
Au(CTZ)Cl$_3$[c]	203.5(6)	Cl	223.1(2)	[29]
trans-[Au(lut)(2,5-Me$_2$C$_6$H$_3$)Cl$_2$][d]	216.0(4)	C	201.3(4)	[31]
Au(2-amt)(C$_6$F$_5$)$_3$[e]	206.6(2)	C	201.6(2)	[32]
Au(Fcpy)(C$_6$F$_5$)$_3$[f]	210.1(3)	C	201.7(3)	[34]
Anionic ligands				
[NH$_4$][Au(N$_3$)$_4$]	201.5(6)	N	207.1(7)	[37]
	199.6(9)	N	204(1)	
Na[Au(dmh)$_4$][g]	198(1)	N	196(2)	[42]
	198(1)	N	200(1)	
[Au(2-Bnpy-H)(sacc)$_2$][h]	211.8(4)	C	202.1(4)	[44]
	201.5(4)	N-py	204.1(4)	
[Au(6-Bnbipy-H)(NHC$_6$H$_3$Me$_2$-2,6)][i]	201.5(2)	N-py	205.6(2)	[45b]
[AuMe$_2$(NHMe)]$_2$	218.7(5)	C	203.9(8)	[46a]
	214.0(5)	C	203.9(9)	
[AuMe$_2$(NMe$_2$)]$_2$	214.0(4)	C	205.8(6)	[46b]
	214.0(5)	C	205.4(6)	

[a] X = atom in *trans*.
[b] 6-Bnbipy = 6-(1-methylbenzyl)-2,2'-bipyridine.
[c] CTZ = 1-[(2-chlorophenyl)diphenylmethyl]-1H-imidazole.
[d] lut = 2,6-dimethylpyridine.
[e] 2-amt = 2-amino-4,5-dihydrothiazole.
[f] Fcpy = 3-ferrocenylpyridine.
[g] Hdmh = 5,5-dimethylhydantoine.
[h] (2-Bnpy-H) = cyclometalated 2-benzylpyridine, saccH = saccharine.
[i] (6-Bnbipy-H) = cyclometalated 6-(1,1-dimethylbenzyl)-2,2'-bipyridine.

inactive [47, 49]. The relevant cell-killing properties of these compounds are unambiguously ascribed to the presence of a gold(III) center, while the higher cytotoxic activities displayed by [Au(terpy)Cl]Cl$_2$ and [Au(phen)Cl]Cl$_2$ [23], also tested against the same tumor cell line, seems to be due to the even more cytotoxic ligands (Table 2.2) [47].

The weak and reversible binding of these complexes to calf-thymus DNA (*ct* DNA) suggests a dominant electrostatic mode of interaction; nevertheless, relevant conformational distortions of the double helix are caused [50]. A multinuclear NMR study of the reactivity of [Au(en)Cl$_2$]Cl and [Au(en)$_2$]Cl$_3$ with guanosine 5'-monophosphate (5'-GMP) reveals that in an aqueous solution only [Au(en)Cl$_2$]Cl binds very weakly to 5'-GMP via N(7) to give a 1:1 adduct [48].

The 4'-substituted terpy complex [Au(4-MeOPhterpy)Cl](ClO$_4$)$_2$ (4-MeOPhterpy 4'-(4-methoxyphenyl)-2,2',6,6''-terpyridine) forms a strong adduct with *ct* DNA with a binding constant of 2.1×10^4 dm^3 mol^{-1} [51], manifesting a high degree of base

2.2 Nitrogen Donor Ligands | 55

Table 2.2 IC$_{50}$ values of selected gold(III) complexes against the human ovarian carcinoma A2780 cell lines sensitive (S) or resistant (R) to cisplatin. Adapted from Ref. [47].

Complex	IC$_{50}$ (µM) (±SE)^a		R/S
	A2780/S	A2780/R	
[Au(en)$_2$]Cl$_3$	8.36 ± 0.77	17.0 ± 4.24	2.03
[Au(dien)Cl]Cl$_2$	8.2 ± 0.93	18.7 ± 2.16	2.28
[Au(cyclam)]Cl$_3$	99.0	>120.0	
en, dien, cyclam	>120.0	>120.0	
[Au(terpy)Cl]Cl$_2$	0.2	0.37 ± 0.032	1.23
[Au(phen)Cl$_2$]Cl	3.8 ± 1.1	3.49 ± 0.91	0.92
terpy	0.125	0.36	2.88
phen	3.66 ± 1.52	2.73 ± 0.16	0.75
cisplatin	1.22 ± 0.43	14.16 ± 2.72	11.6

^{a}Expressed as mean ± SE of at least three determinations or mean of two determinations.

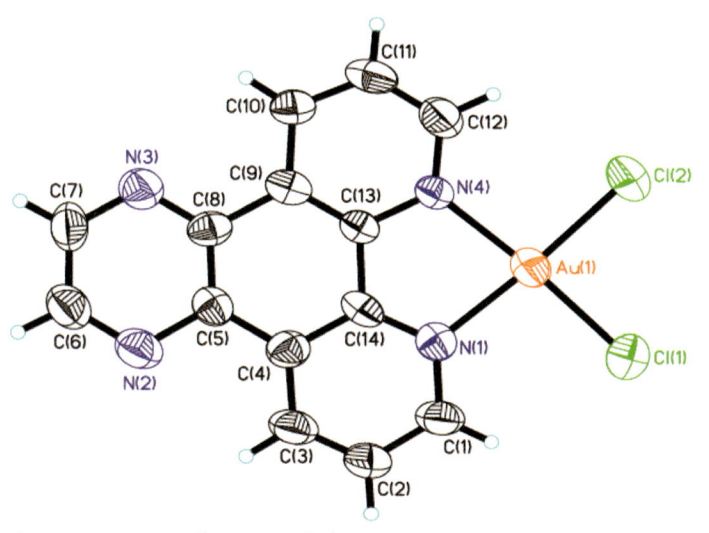

Figure 2.6 Structure of [Au(DPQ)Cl$_2$]$^+$ [52].

specificity at adenine-thymine sites. This complex also displays an intense emission at 480 nm (lifetime = 0.25 µs, quantum yield = 2.4 × 10^{-3}) when measured in 6–7% MeOH-tris buffer solution at room temperature. Important cytotoxic effects are also displayed by the square-planar gold(III) complexes containing the polypyridyl ligands dipyrido[3,2-f : 2′,3′-h]quinoxaline (DPQ), dipyrido[3,2-a : 2′,3′-c]phenazine (DPPZ) and dipyrido[3,2-a : 2′,3′-c](6,7,8,9-tetrahydro)phenazine (DPQC), with IC$_{50}$ values against A2780 being comparable to that of cisplatin [52]. Bond lengths and angles of the structurally characterized complex [Au(DPQ)Cl$_2$](PF$_6$) (Figure 2.6) are very similar to those of [Au(phen)Cl$_2$]Cl [23] (see Table 2.3).

Table 2.3 Bond distances (pm) in selected complexes with multidentate ligands.

Complex	r(Au—N)	X[a]	r(Au-X)	References
Neutral Ligands				
[Au(bipy)Cl$_2$]$^+$	203(1)	Cl	225.2(4)	[21]
	204.6(8)	Cl	225.2(4)	
[Au$_2$(6-*neo*Pnbipy)$_2$(μ-O)$_2$]$^{2+\,b}$	201.1(4)	O	197.6(3)	[25]
	210.0(3)	O	196.1(3)	
[Au(bipy)(mes)$_2$]$^{+\,c}$	212.0(4)	C	202.0(5)	[55]
	213.1(5)	C	202.9(6)	
[Au(4,4'-*t*Bu$_2$bipy)(tdt)]$^{+\,d}$	208(1)	S	226.4(3)	[56]
	206.8(9)	S	226.9(3)	
[Au(phen)Cl$_2$]$^+$	203.3(8)	Cl	226.3(3)	[23]
	205.6(8)	Cl	226.6(3)	
[Au(phen)(CH$_2$SiMe$_3$)$_2$]$^+$	212.8(6)	C	203.1(9)	[55]
	216.9(6)	C	203.9(9)	
[Au(DPQ)Cl$_2$]$^{+\,e}$	202(1)	Cl	223.9(4)	[52]
	206(1)	Cl	225.8(4)	
[Au(terpy)Cl]$^{2+}$	202.9(6)	N	201.8(6)	[54]
	193.1(7)	Cl	226.9(2)	
[Au(4-MeOPh-terpy)Cl]$^{2+\,f}$	202.0(7)	N	204.7(7)	[51]
	192.4(6)	Cl	225.6(2)	
[Au(4-MeS-terpy)Cl]$^{2+\,g}$	202.5(8)	N	201.8(8)	[53]
	194.5(7)	Cl	225.9(3)	
[Au(bimm)Cl$_2$]$^{+\,h}$	201.4(5)	Cl	226.0(2)	[59]
	202.0(5)	Cl	225.9(2)	
[Au(en)Cl$_2$]$^+$	202.9(4)	Cl	228.4(1)	[48]
	203.0(3)	Cl	228.1(1)	
[Au(dien)Cl]$^{2+}$	197(2)	N	198(2)	[68]
	205(2)	Cl	227.3(8)	
[Au(BPMA)Cl]$^{2+\,i}$	200.6(8)	N	200.6(8)	[65]
	200.8(7)	Cl	226.5(3)	
Anionic Ligands				
[Au(HDMG)$_2$]$^{+\,j}$	196(1)	N	202(1)	[58]
	201(1)	N	200(1)	
[Au(HBpz$_3$)Cl$_2$]	200.8(8)	Cl	225.9(3)	[61]
	201.9(7)	Cl	226.5(3)	

[a] X = atom in *trans*.
[b] 6-*neo*Pnbipy = 6-*neo*pentyl-2,2'-bipyridine.
[c] mes = mesityl.
[d] tdt = 3,4-toluenedithiolate.
[e] DPQ = dipyrido[3,2-f:2',3'-h]quinoxaline.
[f] 4-MeOPh-terpy = 4'-(4-methoxyphenyl)-2,2',6'.2"-terpyridine.
[g] 4-MeS-terpy = 4'-(4-methylsulfanyl-2,2',6'.2"-terpyridine.
[h] bimm = bis(1-methyl-2-imidazolyl)methoxymethane.
[i] BPMA = HN(CH$_2$-2-C$_5$H$_4$N)$_2$.
[j] H$_2$DMG = dimethylglyoxime.

2.2 Nitrogen Donor Ligands

The synthesis of a stable gold(III) complex with the bifunctional ligand 4′-methylsulfanyl terpyridine (4′-MeSterpy) [53] paves the way for new radiotherapeutic agents when the radioisotope ^{199}Au is used in place of the inactive ^{197}Au nucleus. The 4′ pendant group is able to covalently bind to a biological target. The bond distances and angles observed about the gold(III) atom in [Au(4′-MeSterpy)Cl](OTf)$_2$ are nearly identical to those found in [Au(terpy)Cl](Cl)(ClO$_4$) [54] (see Table 2.3).

The photophysical properties of the gold(III) diimine complexes [Au(N,N)Cl$_2$]$^+$ (N,N = 2,2′-bipyridines or 1,10-phenanthrolines) can be modified by replacing the chloride ligands with better σ-donors, as in the case of [Au(N,N)R$_2$](ClO$_4$) (N,N = bipy, phen or 4,7-Ph$_2$phen; R = mesityl or CH$_2$SiMe$_3$) where the introduction of organic functionalities significantly improves the relative luminescence quantum yields [55]. In the case of [Au(4,4′-tBu$_2$bipy)(tdt)](PF$_6$) the chelating 3,4-toluenedithiolate (tdt) ligand imparts mild solvatochromism to the diimine complex, while the corresponding dichloro precursor luminesces in low-temperature glass matrix [56]. In the solid state, the [Au(4,4′-tBu$_2$bipy)(tdt)]$^+$ cations are arranged in stacks with alternating intermolecular Au⋯Au separations of 360 and 375 pm.

A general metathesis route to double-salt compounds has been successfully applied to afford a series of bipy derivatives [Au(bipy)X$_2$][AuX′$_n$] (X = X′ = Cl or Br, n = 4; X = Cl, X′ = Br, n = 4 or X′ = CN, n = 2) featuring aurophilic interactions between cations and anions [57]. The geometry of the two square-planar metal centers in [Au(bipy)X$_2$][AuBr$_4$] (X = Cl or Br) is staggered with an Au^{II}⋯Au^{III} distance of 352 pm (Figure 2.7).

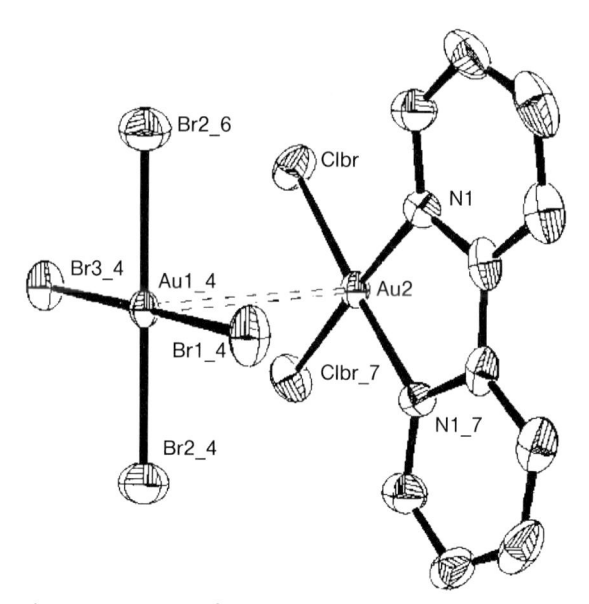

Figure 2.7 Structure of [Au(bipy)Cl$_2$][AuBr$_4$]. Reprinted with permission from [57]. Copyright (2006) American Chemical Society.

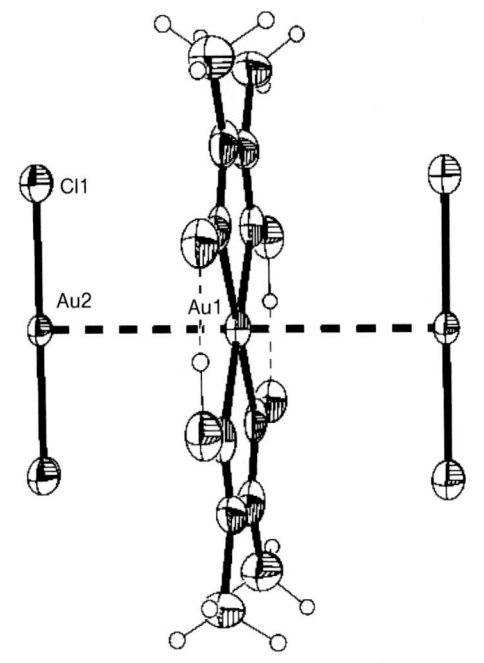

Figure 2.8 Structure of [Au(HDMG)$_2$][AuCl$_2$] [58].

In contrast, no Au^{III}···Au^{III} interactions are found in the dimethylglyoxime (H$_2$DMG) derivative [Au(HDMG)$_2$][AuCl$_4$] [58]. Here, one of the two square planar cations found is nearly orthogonal to the [AuCl$_4$] anion, and a short Au···Cl distance of 332 pm is found between the gold atom of the cation and one of the four chlorine atoms of the anion. In the second pair, the AuN$_4$ and AuCl$_4$ fragments form a dihedral angle of 8.1° with a Au···Au distance of 389 pm. Under slightly different reaction conditions, the same cationic complex is obtained as the [AuCl$_2$]$^-$ salt; here cation and anion are associated by short aurophilic interactions to give a polymeric chain with Au^{III}···Au^I contacts of 325 pm [58] (Figure 2.8).

The imidazole-containing chelate ligands bis(1-methyl-2-imidazolyl)ketone (bik), bis(1-methyl-2-imidazolyl)methoxymethane (bimm) and bis(1-methyl-2-imidazolyl) hydroxymethane (bihm) react with [AuCl$_4$]$^-$ in methanol or THF to give the cationic complexes [Au(N,N)Cl$_2$]X (X = Cl, N,N = bik, bihm; X = AuCl$_2$, N,N = bimm) [59]. In all three square planar complexes the six-membered chelate rings adopt a distorted boat conformation; DFT calculations support this coordination mode of the ligands. The LUMO of [Au(bik)Cl$_2$]$^+$ is calculated as Au–Cl centered in agreement with electrochemical and electron paramagnetic resonance (EPR) results. The known tris(pyrazol-1-yl)borate complex [Au{k^2-N,N-BH(pz)$_3$}Cl$_2$] [60] is best obtained by reaction of Na[BH(pz)$_3$] with Na[AuCl$_4$] in aqueous solution [61]; reaction of this complex with Na[BH(pz)$_3$], in the presence of NaClO$_4$, affords the bis-chelate species [Au{k^2-N,N-BH(pz)$_3$}$_2$](ClO$_4$). In these, as well as in the heteroleptic complexes obtained by reaction of Na[B(pz)$_4$] with the cyclometallated derivatives [Au(N,C)Cl$_2$]

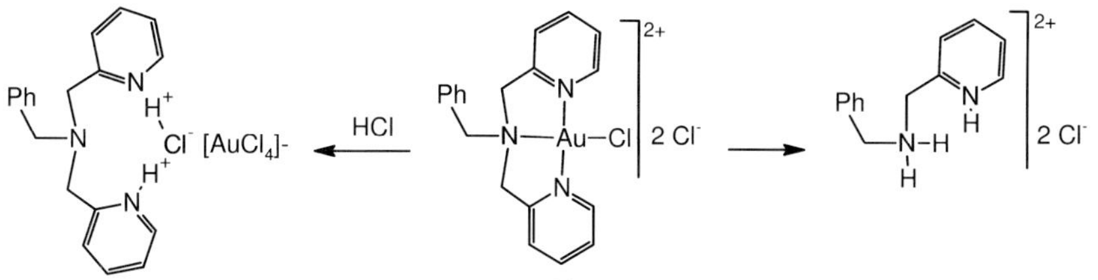

Scheme 2.4 Decomposition of the dicationic complex [Au(N,N′,N)Cl]Cl$_2$ [N,N′,N = PhCH$_2$N(CH$_2$-2-C$_5$H$_4$N)$_2$] (9).

{N,C = C$_6$H$_4$CH$_2$NMe$_2$-2 [62] or C$_6$H$_3$(N = NC$_6$H$_4$Me-4′)-2-Me-5 [63]}, the poly(pyrazolyl)borate ligands coordinate to gold in a bidentate mode forming six-membered chelate rings [61]. Reaction of aminopyridines such as N-isopropyl-N-2-methyl- and N-isopropyl-N-2-ethylpyridine with aqueous solutions of Na[AuCl$_4$] affords cationic complexes of the type [Au(N,N′)Cl$_2$][AuCl$_4$] (N = pyridine, N′ = amine nitrogen) [64]. The dicationic complex [Au(N,N′,N)Cl]Cl$_2$ [N,N′,N = PhCH$_2$N(CH$_2$-2-C$_5$H$_4$N)$_2$] (9) is easily decomposed in aqueous solution by cleavage of a C−N bond or, in dilute HCl solution, by protonation of the ligand to give [HN,N′,NH]Cl[AuCl$_4$] (10) (Scheme 2.4) [65]; coordination to gold(III) is clearly involved in the ligand-cleavage reaction.

Formation of a 2 : 1 L-histidine:Au(III) complex is suggested on the basis of multi-instrumental techniques; the N1 of the imidazole ring and the nitrogen of the α-amino group are likely involved in the coordination [66].

2.2.4
Complexes with Multidentate Ligands Containing Anionic N-Donors

The pyridylamidogold(III) complexes [Au(N,N′)Cl$_2$] [N,N′H = 2-(3,5-diphenyl-1H-pyrrol-2-yl)pyridine] (11) [64] and [Au(N,N′,N)Cl][X] [N,N′H,N = HN(CH$_2$-2-C$_5$H$_4$N)$_2$ (BPMA); X = BF$_4$, PF$_6$, or AuCl$_2$] [65] are readily obtained by spontaneous deprotonation of the corresponding pyridylamines in neutral aqueous Na[AuCl$_4$] solutions. Reaction of 11 with SnMe$_4$ affords the air-stable and thermally robust dimethylgold(III) derivative [Au(N,N′)Me$_2$] (12) (Figure 2.9) [64].

Gold(III)···gold(III) (354 and 373 pm) and gold(III)···gold(I) (334 and 349 pm) aurophilic interactions are observed in the salts [Au(BPMA-H)Cl](PF$_6$) (13-PF$_6$) and [Au(BPMA-H)Cl](AuCl$_2$) (13-AuCl$_2$), respectively [65]. Protonation of the amido nitrogen atom with HOTf leads to the isolation of the corresponding amino complex [Au(BPMA)Cl](OTf)$_2$ [14-(OTf)$_2$] whose pK_a is estimated at 3.5 (Equation 2.4). Comparison of the structures of the two complexes indicates that there is little p$_\pi$-d$_\pi$ bonding in the amido−gold bond and that the amide exerts a stronger *trans* influence than the amine group, as shown by the Au−Cl bond distances 234 and 227 pm, respectively (see Table 2.3 and Table 2.4). Analogous results were found for the couple [Au(dien)Cl]$^{2+}$/[Au(dien-H)Cl]$^+$ [67]; here the Au−Cl bond lengths are 227 and 233 pm, respectively [68].

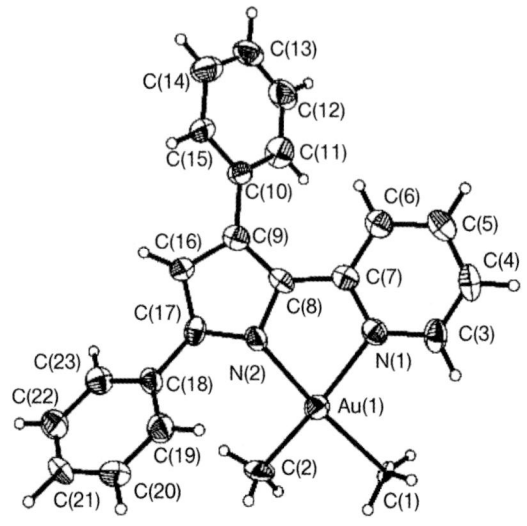

Figure 2.9 Structure of [Au(N,N')Me$_2$] (**12**) [64].

$$pK_a = 3.5 \text{ in } 0.5 \text{ mol dm}^{-3} \text{ [ClO}_4\text{]}^- \text{ [65]} \quad (2.4)$$

$$pK_a = 4.0 \text{ in } 0.5 \text{ mol dm}^{-3} \text{ [ClO}_4\text{]}^- \text{ [66]} \quad (2.5)$$

Reaction of trietylenetetraamine (trien) with H[AuCl$_4$] at low pH leads to the water soluble dinuclear complex [Au$_2$(trien-H)Cl$_3$]Cl$_2$, which features a deprotonated amino nitrogen bridging the two gold centers (Figure 2.10) [69].

A variety of gold(III) complexes of carboxamido substituted heterocyclic ligands are obtained by reaction of [AuCl$_4$]$^-$ with the appropriate ligands, these include [Au(N,N')Cl$_2$] (HN,N' = picolinamide) (**15**) [70], [Au(N,N',N'')Cl]Cl [N,N'H,N'' = N-(8-quinolyl)pyridine-2-carboxamide (**16**), N-(8-quinolyl)glycine-2-carboxamide (**17**) or N-(8-quinolyl)-L-alanine-2-carboxamide (**18**)] [71] (Figure 2.11).

These complexes show considerable *in vitro* cytotoxic effects against various tumor cell lines [70, 71]. Moreover, the cationic complexes [Au(N,N',N'')Cl]Cl are able to intercalate into ct DNA [71]. Gold(III) amidate complexes of histidine containing

Table 2.4 Bond distances (pm) in selected complexes with multidentate ligands containing anionic N-donors.

Complex	r(Au—N)	X[a]	r(Au-X)	References
Bidentate Ligands				
[Au{2-(Ph$_2$pyr)py}Cl$_2$][b]	208(1) amido	Cl	228.0(8)	[64]
	197(2) py	Cl	226.1(8)	
[Au{2-(Ph$_2$pyr)py}Me$_2$]	206.4(7) amido	C	213.5(6)	[64]
	214.9(7) py	C	205.1(9)	
[Au(pla)Cl$_2$][c]	196.9(5) amidate	Cl	229.7(1)	[70a]
	204.7(5) py	Cl	226.1(2)	
[Au(2-Bnpy-H)(1,2-amd$_2$Ar)][d]	208.9(3) amidate	C	226.3(3)	[75]
	202.2(3) amidate	Npy	205.8(3)	
Terdentate Ligands				
[Au(BPMA-H)Cl][BF$_4$][e]	199.4(5) amido	Cl	234.1(2)	[65]
	201.4(6) py	Npy	202.3(6)	
[Au(Quingly)Cl]$^+$ [f]	195.3(6) amidate	Cl	227.7(2)	[71]
	202.7(6) quin	Namine	202.4(6)	
[Au(gly-L-his-H)Cl]$^+$	194(1) amidate	Cl	227.3(3)	[72a]
	199.1(8) imidazole	Namine	200.2(9)	
[Au(amd$_4$EDTA-H)Cl]$^+$ [g]	200(2) amidate	Namine	210.5(9)	[74]
	206(1) amine	Cl	227.4(3)	
[Au(dien-H)Cl]$^+$	205(4) amine	Namine	201(4)	[68]
	206(4) amido	Cl	233(1)	
Tetradentate and Macrocyclic Ligands				
[Au(gly-gly-L-his-2H)]$^+$	203.8(9) imidazole	Namidate	194.1(9)	[73]
	201(1) amidate	Namine	205(1)	
[Au$_2$(trien-H)Cl$_3$]$^{2+}$	202.4(9) amine	Cl	227.5(3)	[69]
	207.4(8) amido	Cl	228.2(3)	
	201.7(9) amine	Cl	228.7(3)	
	204.5(9) amine			
[Au(cyclam)]$^{3+}$	204[h]			[77]
[Au(Me$_4$C$_{10}$N$_4$)]$^+$ [i]	197.6(4), 198.0(5)			[78c]
[Au(TPP)]$^+$ [j]	203.2(5), 203.3(5)			[91a]

[a]X = atom in *trans*.
[b]2-(Ph$_2$pyrH)py = 2-(3,5-diphenyl-1H-pyrrol-2-yl)pyridine.
[c]plaH = picolinamide.
[d](2-Bnpy-H) = cyclometalated 2-benzylpyridine, 1,2-amd$_2$Ar = [1,2-{NC(O)Me}$_2$C$_6$H$_4$]$^{2-}$.
[e]BPMA-H = [N(CH$_2$-2-C$_5$H$_4$N)$_2$]$^-$.
[f]Quingly = N-(8-quinolyl)pyridine-2-carboxamidate.
[g]amd$_4$EDTA = 1,2-diaminoethane-N,N,N',N'-tetra-(N-methylacetamide).
[h]Average distance.
[i]Me$_4$C$_{10}$N$_4$ = 5,7,12,14-tetramethyl-1,4,8,11-tetraazamacrocyclotetradeca-4,6,11,13-tetraenato.
[j]TPP = *meso*-tetraphenylporphyrin.

peptides, such as glycyl-L-histidine dipeptide (Hgly-L-his) [Au(gly-L-his)Cl]$^+$ (**19**) [72] and glycylglycyl-L-histidine tripeptide (H$_2$glygly-L-his) [Au(glygly-L-his)]$^+$ (**20**) [73], or of amides of EDTA, compound **21**, and PDTA [74], are valuable model systems for studying gold(III)-peptide/protein interactions (Figure 2.12). Complex **19** is highly

Figure 2.10 $[Au_2(trien-H)Cl_3]^{2+}$.

Figure 2.11 Complexes of carboxamido substituted heterocyclic ligands.

Figure 2.12 Amidate complexes of histidine containing peptides, (19) and (20), and of amides of EDTA, (21).

cytotoxic towards A2780S and A2780R ovarian carcinoma human cell lines and is also able to bind DNA [72b].

Several gold(III) complexes containing chelating bis(amidate) ligands are obtained by reaction of the corresponding dichloride complexes [Au(N,C)Cl$_2$] [N,CH = 2-benzylpyridine, 2-anilinopyridine, 2-(p-tolyl)pyridine or dimethylbenzylamine] or [Au(N,N′)Cl$_2$] (N,N′H = picolinamide) with 1,2-C$_6$H$_4$(NHCOMe)$_2$ and silver(I) oxide, or with C$_2$H$_4$(NHSO$_2$Tol)$_2$ (Tol = p-tolyl) or 1,2-C$_6$H$_4$(NHSO$_2$Tol)$_2$ and trimethylamine [75] (examples are shown in Figure 2.13).

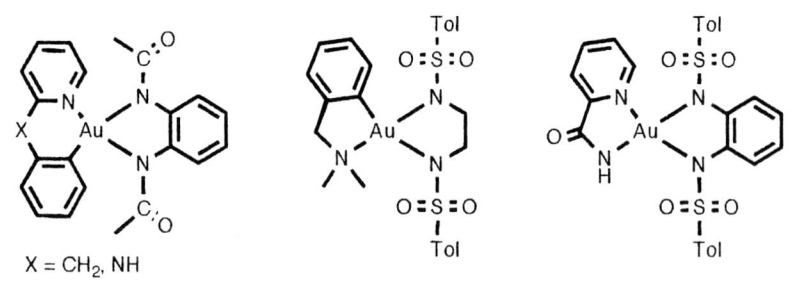

X = CH$_2$, NH

Figure 2.13 Selected examples of gold(III) complexes containing chelating bis(amidate) ligands.

Most of these compounds display medium to high antitumor activity against P388 murine leukaemia cells; the N,N'-bis(p-toluensulfonyl)ethylenediamide derivatives being the most active with IC$_{50}$ values in the range of 0.33–4.35 µM. Some of the bis-acetamido derivatives also show a broad spectrum activity against a range of bacteria and fungi.

2.2.5
Complexes with Polyazamacrocyclic Ligands

Cyclam (1,4,8,11-tetraazacyclotetradecane), monooxocyclam and their phenol and pyridyl substituted derivatives form various types of macrocyclic gold(III) polyamine complexes [76] (Figure 2.14). The complexes are stable in acidic aqueous solutions, but unstable in neutral to alkaline solutions. Dissociation of a proton from one of the cyclam units occurs even at neutral pH with pK_a values of 5.0–5.4 at 25 °C and $I = 0.1$ (NaClO$_4$). The X-ray crystal structure of [Au(cyclam)]$^{3+}$ shows a four-coordinate square-planar N$_4$ geometry with the $RRSS$ configuration of cyclam and an average Au−N bond distance of 204 pm [77]. The reduction potential for Au(III) to Au(I) in [Au(cyclam)]$^{3+}$ varies from −0.16 V (vs SCE), at pH 1.0, to −0.55 V at pH 9.0, according to the degree of cyclam NH deprotonation [77].

Bis(ethylenediamine)gold(III) chloride reacts with a variety of β-diketonates in aqueous base, via Schiff base condensation, to form 14-membered tetraaza 12π macrocyclic species [78]. The parent member of the series [AuL^1]$^+$ being **22** (where H$_2$L^1 = 5,7,12,14-tetramethyl-1,4,8,11-tetraazacyclotetradeca-4,6,11,13-tetraene) (Figure 2.15). The X-ray structure shows the cation to be nearly planar. The observed pattern of C−C and C−N distances indicates delocalization of π-electrons within the six-membered β-diiminate rings. Open-chain tetraaza ligand complexes [AuL^2]$^{2+}$, in which condensation of only one β-diketonate has occurred,

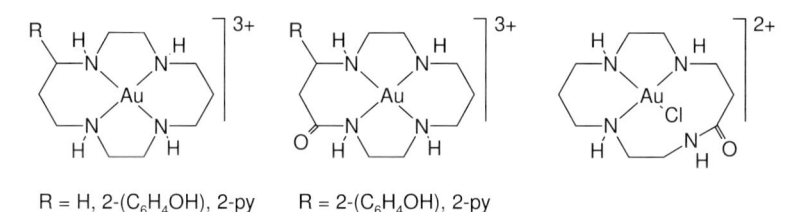

R = H, 2-(C$_6$H$_4$OH), 2-py   R = 2-(C$_6$H$_4$OH), 2-py

Figure 2.14 Complexes with cyclam derivatives.

(22)   (23)

Figure 2.15 Complexes derived by from the reaction of [Au(en)$_2$]Cl with β-diketonates.

can be isolated as intermediates in this reaction. These may be used for further condensation with a different β-diketonate. Oxidation of this complex with trityl tetrafluoroborate introduces a double bond in position C2-C3 [78b, 79]. The tetraazamacrocyclic complex $[AuL^1]^+$ in aqueous solution undergoes two-stage protonation to form singly and doubly protonated forms of the complex, $[AuHL^1]^{2+}$ and $[AuH_2L^1]^{3+}$, respectively [80]; protonation occurs at the central carbon atoms of the six-membered rings. The open-chain tetraaza ligand derivative $[AuL^2]^{2+}$ is likewise protonated to give $[AuHL^2]^{3+}$ [81]. Other complexes with tetraaza macrocycles have been prepared by reaction of $[Au(en)_2]Cl_3$, ethanolamine, or nitroethane, and formaldehyde, although with nitroethane an acyclic species is also obtained (**23** in Figure 2.15) [82]. A gold(III) complex with the hexaaza macrocycle 1,8-dimethyl-1,3,6,8,10,13-hexaazacyclotetradecane (L^3) has been obtained by transmetallation reaction from the nickel compound $[NiL^3]^{2+}$ by reaction with $[AuCl_4]^-$ [83].

A variety of gold(III) porphyrins have been synthesized and their unique chemical, physical and biological properties studied (examples are shown in Figure 2.16).

A new method for the incorporation of gold(III) into porphyrins, based on the disproportionation of $[Au(THT)_2]BF_4$ under mild conditions, was developed [84]. Most gold(III) porphyrin units are used as acceptors in porphyrin dyads [85] and triads [85c, e, 86] because they can easily be reduced, either chemically or photochemically. Recent electrochemical experiments on $[Au(P)]PF_6$ [H_2P = 5,10,15,20-tetrakis(3,5-di-*tert*-butylphenyl)porphyrin] (**24**-PF_6) in a variety of non-aqueous solvents have shown that gold(III) porphyrins are not electrochemically inert but can be rapidly and reversibly converted to an Au(II) form of the compound [87]. Nevertheless, substituents at one of the eight β-pyrrolic positions of the macrocycles can drive the first electron transfer either to the gold center or to the porphyrin ring to yield, in this case, an Au(III) porphyrin π-anion radical [88]. The complex tetrakis(1-pyrenyl)porphyrinatogold(III) acetate (**25**-OAc) is prepared by metallation of the atropisomeric mixture of the free-base porphyrin ligand with $[AuCl_4]^-$ in glacial acetic acid [89]. Optical studies have been carried out and have shown that this system has the potential to serve as a novel type of catalytic photonucleobase for long-wavelength sensitized cleavage of DNA and other guanine-containing nucleic acids [90]. A series of tetraarylporphyrinatogold(III) complexes (**26**) is found to exert much higher potency than cisplatin in killing a variety of human cancer cells,

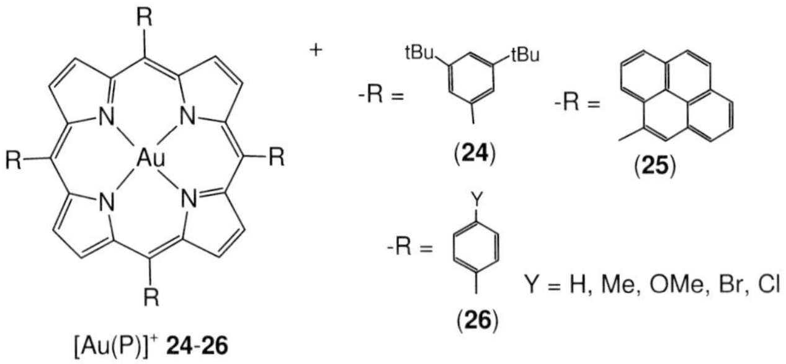

Figure 2.16 Selected examples of gold(III) porphyrins.

including the drug-resistant variants; the gold-induced cytotoxicity occurs through an apoptotic pathway [91]. The same complexes **26** inhibit the human immunodeficiency virus type-1 (HIV-1) reverse transcriptase [92]. One-electron reduction of these complexes occurs at negative potential (-0.96 to -1.02 V vs $Cp_2Fe^{+/0}$) indicating the high stabilization of Au(III) by the dianionic porphyrin ligand [91a]. These features make gold(III) porphyrins robust Lewis acid catalysts, as shown by the cycloisomerization of allenones to furans [93]: yields up to 98% and substrate conversion >99% are obtained with [Au(TPP)]X [$H_2TPP = $ meso-tetraphenylporphyrin (**26**, Y = H) or meso-tetrakis(pentafluorophenyl)porphyrin; X = Cl or OTf] as catalyst; the recyclable catalytic system affords a total product turnover number of 8300. The same gold(III)-porphyrin complex catalyzes efficiently the hydroamination and hydration of phenylacetylene [93].

2.3
Oxygen Donor Ligands

2.3.1
Hydroxo Complexes

Gold(III) hydroxo complexes are generally obtained by reaction of the corresponding precursors with a base in aqueous solution. Most of the reported species are polynuclear complexes with bridging hydroxides: one of the best known examples being dimethylgold(III) hydroxide [94] which has a dimeric structure in aqueous solution but in the solid state or in benzene solution exists a tetramer [95]. It is prepared by treatment of dimethylgold(III) iodide with silver salts and sodium hydroxide and then precipitated by addition of nitric acid. The dimethylgold(III) aquo ion **27** (stable in aqueous solution) is a weak acid and transfers one proton in dilute solution over the pH range 5–7 to give the dimeric hydroxo complex **28** (Equation 2.6) [96].

$$2 cis\text{-}[AuMe_2(OH_2)_2]^+ \rightleftharpoons [Au_2Me_4(\mu\text{-}OH)_2] + 2H_3O^+ \quad (2.6)$$
$$\quad\quad (27) \quad\quad\quad\quad\quad\quad\quad (28)$$

The dimer **28** is the predominant soluble hydrolysis product: polymerization to give the water-insoluble tetramer **29** (Figure 2.17) occurs either in neutral aqueous solution of the hydroxide or in organic solvents. The hydroxide is also soluble in alkaline solution where the hydroxoaurate species **30** is formed (Equation 2.7).

$$[Au_2Me_4(\mu\text{-}OH)_2] + 2OH^- \rightleftharpoons 2 cis\text{-}[AuMe_2(OH)_2]^- \quad (2.7)$$
$$\quad (28) \quad\quad\quad\quad\quad\quad\quad (30)$$

Adventitious water is responsible for the formation of the dimeric hydroxo complex **31** obtained by reaction of $AuCl_3$ with 1,4-dilithiotetraphenylbutadiene in ether solution [97]. The hydroxo-bridged complex $[Au(C_6H_4NO_2\text{-}2)_2(\mu\text{-}OH)]_2$ (**32**) was obtained either by reaction of $Na[Au(C_6H_4NO_2\text{-}2)_2(OPh)_2]$ with traces of water in CH_2Cl_2/n-hexane solution or by treatment of the dichloroaurated complex with NaOH [98]. The crystal structure of **32**·2Et$_2$O shows that it is a centrosymmetric

Figure 2.17 Hydroxo bridged gold(III) complexes.

dimer, as for complex **31** [97b], and that the central Au$_2$O$_2$ ring is thus exactly planar with Au–O distances 207 and 208 pm and an Au–O–Au angle of 98.8(2)°; the transannular Au···Au distance is 315 pm.

Monomeric gold(III) hydroxo complexes have been described quite recently; all of them feature a polypyridine as co-ligand (Figure 2.18). The terpy derivative [Au(terpy)(OH)][ClO$_4$]$_2$ (**33**) (terpy = 2,2′:6′,2″-terpyridine) was isolated in the course of equilibrium and kinetic studies of gold(III) complexes with terpy carried out in aqueous solution [99]. The highly charged aquo complex **34**, formed according to Equation 2.8, behaves as a strong acid ($K_a \geq 0.8$ mol dm^{-3}) and dissociates completely into the corresponding hydroxo species **33**, which can be isolated in the solid state as its perchlorate. The crystal structure of **33** consists of square planar [Au(terpy)(OH)]$^{2+}$ cations having Au–N distances of 201, 201 and 195 pm and an Au–OH distance of 200 pm. The square planar geometry is expanded to distorted tetragonal bipyramidal by linking the two perchlorate anions with Au–O distances of 302 and 307 pm, which are intermediate between bonding and van der Waals interactions (Table 2.5).

$$[Au(terpy)Cl]^{2+} + H_2O \rightleftharpoons \underset{(34)}{[Au(terpy)(OH_2)]^{3+}} + Cl^- \qquad (2.8)$$

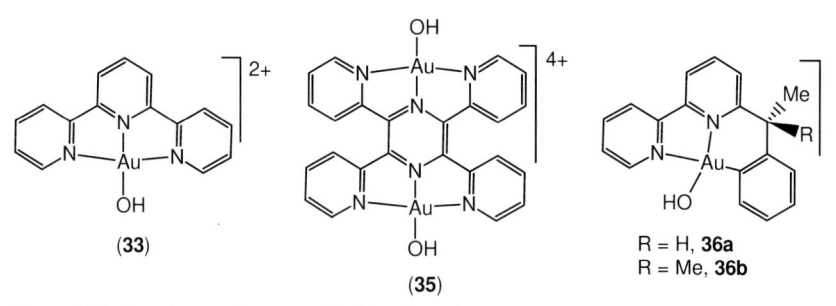

Figure 2.18 Complexes with terminal hydroxo ligands.

Table 2.5 Bond distances (pm) in selected complexes with O-donor ligands.

Complex	r(Au—O)	X[a]	r(Au-X)	References
Hydroxo Complexes				
[Au$_4$Me$_8$(μ-OH)$_4$]	215[b]	C	205[b]	[95]
[Au$_2$(Ph$_4$C$_4$)$_2$(μ-OH)$_2$][c]	223(2)	C	209(2)	[97b]
	226(2)	C	213(3)	
	215(2)	C	189(3)	
	222(2)	C	211(3)	
[Au$_2$(C$_6$H$_4$NO$_2$-2)$_2$(μ-OH)$_2$]	207.3(4)	C	199.2(5)	[98]
	207.5(4)	C	199.5(5)	
[Au(terpy)(OH)]$^{2+}$	200.0(4)	N	194.9(4)	[99]
Oxo Complexes				
[Au$_2$(bipy)$_2$(μ-O)$_2$]$^{2+}$	197.1(5)	N	200.0(4)	[109]
	195.7(6)	N	201.5(4)	
[Au$_2$(6,6'-Me$_2$bipy)$_2$(μ-O)$_2$]$^{2+}$[d]	195.5(5)	N	206.5(6)	[109]
[Au$_2$(6-Bnbipy-H)$_2$(μ-O)]$^{2+}$[e]	197.1(5)	N	202.9(5)	[102]
	195.6(5)	N	204.0(5)	
[Au(O)(OH$_2$)P$_2$W$_{18}$O$_{68}$]$^{17-}$	176(2)	O$_w$	229(4)	[7]
[Au(O)(OH$_2$)P$_2$W$_{20}$O$_{70}$(OH$_2$)$_2$]$^{9-}$	177(4)	O$_w$	232(6)	[7]
Alkoxo Complexes				
cis-[AuMe$_2$(OPh)(PPh$_3$)]	209(1)	C	203(3)	[118]
[Au(bipy)(OMe)$_2$]$^+$	197.1(4)	N	203.9(5)	[104]
	196.0(6)	N	203.2(5)	
Complexes with other O-anions				
Zn[Au(AcO)$_4$]$_2$	198[b]			[129a]
Sr[Au(AcO)$_4$]$_2$	198[b]			
ClO$_2$[Au(ClO$_4$)$_4$]	187, 206[b]			[130]
[Au$_2$Me$_4$(μ-OAc)$_2$]	212.2(4)	C	203.2(6)	[132a]
	212.3(4)	C	201.6(6)	
	213.9(4)	C	202.2(7)	
	211.7(5)	C	202.3(7)	

[a]X = atom in *trans*.
[b]Average values.
[c]Ph$_4$C$_4$ = tetraphenyl-ciclopentadienyl.
[d]6,6'-Me$_2$bipy = 6,6'-dimethyl-2,2'-bipyridine.
[e]6-Bnbipy-H = cyclometalated 6-(1,1-dimethylbenzyl)-2,2'-bipyridine.

$$[\text{Au(terpy)(OH}_2)]^{3+} \rightleftharpoons [\text{Au(terpy)(OH)}]^{2+} + \text{H}^+ \quad (2.9)$$
$$(34)(33)$$

The closely related dinuclear complex [Au$_2$(tppz)(OH)$_2$]Cl$_4$ (**35**) [tppz = 2,3,5,6-tetrakis(2-pyridinyl)pyrazine] is easily obtained by reaction of the ligand with K[AuCl$_4$]·2H$_2$O in refluxing methanol [100]. In cyclic voltammetry complex **35** undergoes two main irreversible reduction processes: the first one, a four-electron process, leads to the formation of a gold(I) species in solution, the second is a two electron process giving metallic gold.

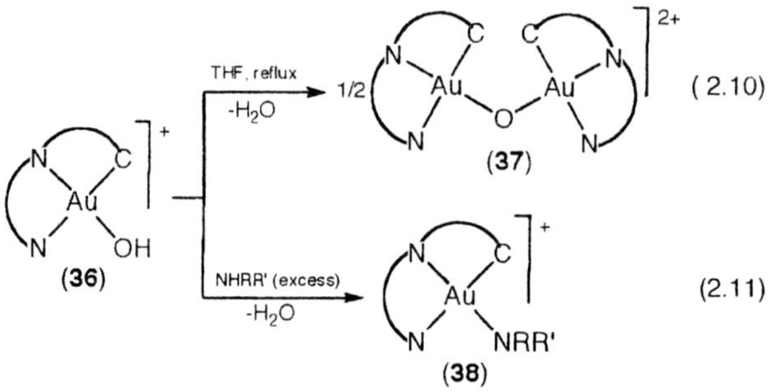

Scheme 2.5 Reactions of complexes **36**. In Equation 2.11: R = H, R′ = C$_6$H$_4$NO$_2$-4, p-tol, o-Xyl, i-Bu, s-Bu or R = R′ = Et.

Treatment of the cyclometallated complexes [Au(N,N,C)Cl][PF$_6$] [N,N,CH = 6-methylbenzyl- (**a**) or 6-(1,1-dimethylbenzyl)-2,2′-bipyridine (**b**)] [20] with KOH or Ag$_2$O in aqueous media affords the hydroxo complexes [Au(N,N,C)(OH)][PF$_6$] (**36**) in fairly good yields [45b,101]: these are air-stable white solids, quite soluble in water and in many organic solvents. When refluxed in anhydrous THF they condense to give the oxo-bridged complexes [Au$_2$(N,N,C)$_2$(μ-O)]$^{2+}$ (**37**) (Equation 2.10 in Scheme 2.5) which, in turn, can be obtained by a different route [102] (see Section 3.2); the reaction can be reversed by refluxing the oxo complex in water.

Complexes **36** undergo a proton exchange reaction with protic reagents; for example, a series of monomeric amido complexes are easily prepared using this route [45] (Equation 2.11 in Scheme 2.5).

The water-soluble compound **36b**, very stable within a physiological buffer at 37 °C, exhibits cytotoxic effects on ovarian A2780 human cancer cells and promotes apoptosis to a greater extent than platinum drugs [103].

Other isolated monomeric hydroxo complexes are the two bipyridyl derivatives [Au(N,N)(OH)Cl][PF$_6$] (**39**) and [Au(N,N)(OH)$_2$][PF$_6$] (**40**) (N,N = 2,2′-bipyridine (bipy) [104] or 4,4′-di-*tert*-butyl-2,2′-bipyridine (4,4′-*t*Bu$_2$bipy) [105]) which have been obtained by hydrolysis of the corresponding dichlorides in aqueous solution. Partial hydrolysis to give complexes **39** takes place in AcONa solution while complete hydrolysis to **40** requires the presence of Ag$_2$O. Complexes **39** and **40** (N,N=bipy) undergo proton exchange in ROH (R = Me, Et, or *i*-Pr) solution to give the corresponding alkoxides **41** and **42** [104] (Equations 2.10 and 2.11).

$$[\text{Au(bipy)(OH)Cl}]^+ + \text{ROH} \rightleftharpoons [\text{Au(bipy)(OR)Cl}]^+ + \text{H}_2\text{O} \quad (2.12)$$
$$\quad\quad (39) \quad\quad\quad\quad\quad\quad\quad\quad\quad (41)$$

$$[\text{Au(bipy)(OH)}_2]^+ + 2\text{ROH} \rightleftharpoons [\text{Au(bipy)(OR)}_2]^+ + 2\text{H}_2\text{O} \quad (2.13)$$
$$\quad\quad (40) \quad\quad\quad\quad\quad\quad\quad\quad\quad (42)$$

R = Me, Et, *i*-Pr

The bis amido complex [Au(bipy)(NHC$_6$H$_4$NO$_2$-4)$_2$][PF$_6$] (**43**) has been similarly obtained by reaction of **40** with *p*-nitroaniline in acetone solution (Equation 2.14 in Scheme 2.6) [45b]. Complex **40** promotes the stoichiometric oxidation of various amines different from *p*-nitroaniline. Azotoluene is the main organic product of the

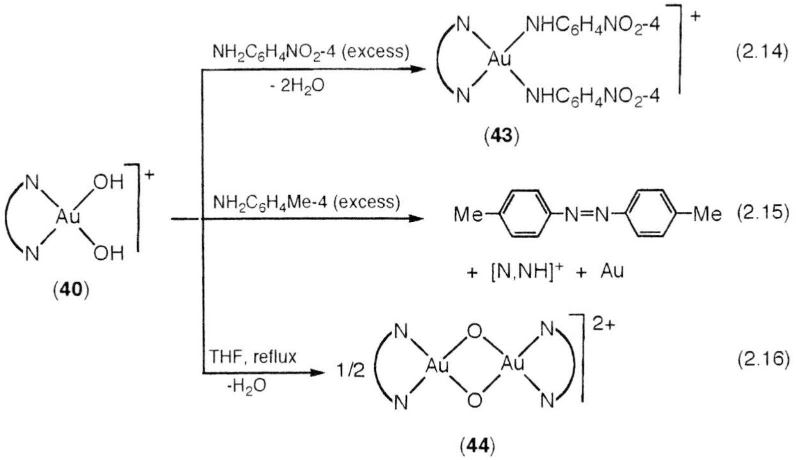

Scheme 2.6 Reactions of complex 40. N,N = bipy.

reaction with p-toluidine, other products being metallic gold and the protonated ligand [bipyH][PF$_6$] (Equation 2.15 in Scheme 2.6) [45b].

As observed for complex **36**, condensation of complexes **40** also occurs in refluxing THF to give the oxo-bridged complexes [Au$_2$(N,N)$_2$(μ-O)$_2$][PF$_6$]$_2$ (**44**) (N,N = bipy [104] or 4,4'-tBu$_2$bipy [105] (Equation 2.16 in Scheme 2.6). Complexes **44**, which can be most conveniently obtained by a different route (see Section 3.2), are rehydrated in boiling water to give the hydroxo complexes **40**. The electrochemical behavior of the hydroxo complexes **39** and **40** (N,N = bipy) in cyclic voltammetry and controlled-potential electrolysis in MeCN is compared with that of the dichloro precursor [Au(bipy)Cl$_2$][PF$_6$] [106]. Both hydroxo complexes undergo an easy, irreversible one-electron reduction process. Notably, substitution of Cl$^-$ with OH$^-$ ligands renders the electroreduction more difficult ($E_{\text{p,c,peakA}} = -0.01$ V, -0.54 V, -0.88 V vs Cp$_2$Fe$^{+/0}$ for dichloro, monohydroxo, and dihydroxo complexes, respectively). Complex **40**, although less stable in biological media and less active than the cyclometallated complex **36**, also shows important cell-killing effects with IC$_{50}$ values in the micromolar range [103a].

2.3.2
Oxo Complexes

Gold oxo species are likely to be involved in oxidation processes catalyzed by metal-oxide-supported gold [8]. The first gold(III) oxo-bridged complexes [25, 102] have been reported about 20 years after the first gold(I) oxo complex [(AuPPh$_3$)$_3$(μ-O)]$^+$ [107]. All of them are stabilized by chelating nitrogen ligands. The cyclometallated derivative [Au$_2$(N,N,C)$_2$(μ-O)][X]$_2$ (**37**) [N,N,CH = 6-methylbenzyl- (**a**) or 6-(1,1-dimethylbenzyl)-2,2'-bipyridine (**b**); X = PF$_6$ or BF$_4$) [102] is a rare example of metal oxo complex with an unsupported M-O-M bridge (M = late transition metal) [108]. Complex **37** can be prepared as the BF$_4^-$ salt by reaction of [Au(C,N,N)Cl][BF$_4$] with AgBF$_4$ in acetone, while the PF$_6^-$ salt is prepared by condensation of the hydroxo complex **36**-PF$_6$ in

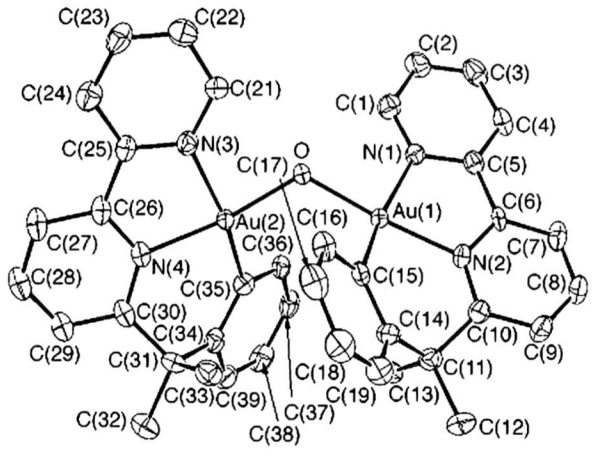

Figure 2.19 Structure of **37b** [102].

boiling THF. The crystal structure of **37b**-(BF$_4$)$_2$ has been determined by X-ray diffraction (Figure 2.19) [102]. The cation displays an idealized C2 symmetry, with the twofold axis passing through the oxygen atom and the midpoint of the Au(1)···Au (2) vector; the Au—O—Au angle is 121.3(2)° and the average Au—O distance is 196 pm. The distance Au(1)···Au(2) is 342 pm, too long to be considered bonding although slightly shorter than the sum of the estimated van der Waals radii.

The oxo complexes [Au$_2$(N,N)$_2$(μ-O)$_2$][PF$_6$]$_2$ [**44**-(PF$_6$)$_2$] [N,N = bipy (**a**) or 4,4′-*t*Bu$_2$bipy (**b**)] can also be prepared by reaction of [Au(N,N)Cl$_2$][PF$_6$] with an aqueous solution of KOH in acetonitrile [104, 105]. Analogous complexes with a variety of 6-substituted 2,2′-bipyridines (6-Rbipy) [R = Me (**c**) Et (**d**), *i*-Pr (**e**), *t*-Bu (**f**), CH$_2$CMe$_3$ (*neo*-Pn) (**g**) or C$_6$H$_3$Me$_2$-2,6 (*o*-Xyl) (**h**)] and 6,6′-dimethyl-2,2′-bipyridine (6,6′-Me$_2$bipy) (**i**) are obtained either as BF$_4^-$ salts, by reaction of the adducts [Au(N) Cl$_3$] **3** (N = monodentate 6-Rbipy) with AgBF$_4$ in acetone, or as PF$_6^-$ salts, by reaction of the adducts with AcONa and excess KPF$_6$ in MeCN/H$_2$O (see Equation 2.1), the latter method being by far the most convenient [25]. The structure in the solid state of compounds **44** (for N,N = bipy, 6-Mebipy, 6-*neo*-Pnbipy, 6-*o*-Xylbipy and 6,6′-Me$_2$-bipy) has been determined by X-ray diffraction (Figure 2.20) [25, 109]. All compounds contain a common structural motif consisting of an Au$_2$O$_2$ "*diamond core*" linked to two bipyridyl ligands in a roughly planar arrangement. The average Au—O bond length (197 pm) is similar to that found for the monooxobridged complex **37b** and can be compared with the 193 pm Au—O distance observed in polymeric gold(III) oxide involving a bridging oxygen atom [110]. The central Au$_2$O$_2$ ring is exactly planar with a rather short transannular Au···Au distance, the average value being 301 pm.

The hydroxo complexes [Au(6-Rbipy)(OH)$_2$]$^+$, postulated as intermediates in the formation of **44c–i**, are not isolated; nevertheless, in all the mass spectra (FAB conditions) a weak peak is found corresponding to these species. Unchanged **44c–i** are quantitatively recovered from the reaction with aqueous solutions of HX (X = BF$_4$ or PF$_6$): neither dihydroxo complexes similar to those observed in the vapor phase, nor hydroxo bridged dimers are obtained [25].

2.3 Oxygen Donor Ligands | 71

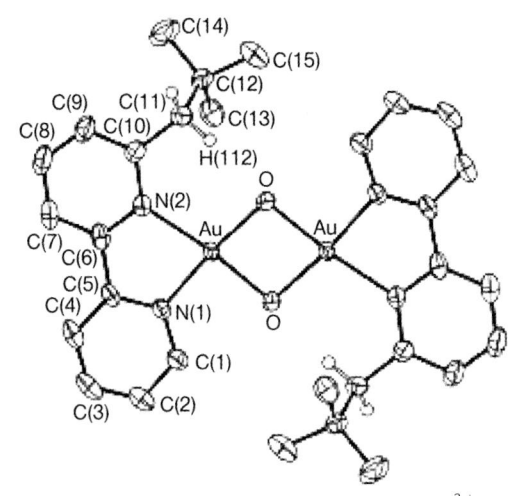

Figure 2.20 Structure of [Au$_2$(6-neoPnbipy)$_2$(μ-O)$_2$]$^{2+}$ **44g** [25].

Introduction of different kinds of alkyl or aryl substituents at position(s) 6 (and 6′) of the bipyridine ligands leads to small structural changes in the complexes which are responsible of their different chemical behavior in solution [109] as well as of their biological activity [111]. Notably, complex **44i** (N,N = 6,6′-Me$_2$bipy) which presents the largest structural deviation, with respect to **44a** (N,N = bipy), shows the highest oxidizing power, the least thermal stability and the greatest cytotoxic activity.

Complexes **44c–i** are able to transfer oxygen to PPh$_3$ and olefins [112]. From the reaction with PPh$_3$, OPPh$_3$ is obtained together with the gold(I) complex [Au(PPh$_3$)$_2$]$^+$ and free 6-Rbipy [112a]. Reaction with a variety of linear [112a, c] and cyclic [112b] olefins results in the formation of oxygenated olefin derivatives and gold(I) olefin complexes [Au(N,N)(η^2-olefin)]$^+$ or [Au$_2$(N,N)$_2$(μ-η^2:η^2-diolefin)]$^{2+}$. From the reaction with norbornene unprecedented oxaauracyclobutane complexes [Au(N,N)(κ^2-O,C-oxynorbornyl)]$^+$ (**45**) are also formed (Scheme 2.7) [112b]. Reaction of **45a** (N,N = 6-Mebipy) with excess norbornene leads to the

Scheme 2.7 Formation of oxaauracyclobutane complexes [Au(N,N)(κ^2-O,C-oxynorbornyl)]$^+$ (**45**).

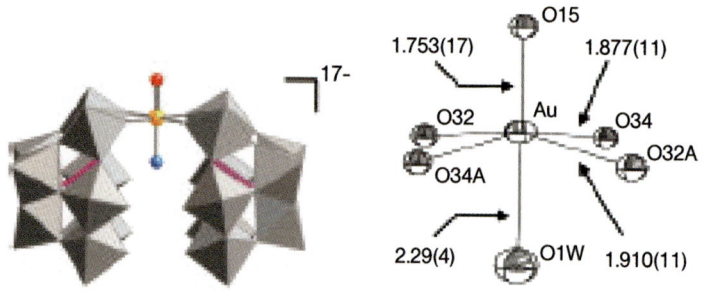

Figure 2.21 Structure of polyanion [Au(O)(OH$_2$)P$_2$W$_{18}$O$_{68}$]$^{17-}$ (46). Reprinted in part with permission from [7]. Copyright (2007) American Chemical Society.

$$Au^I + H_2O_2 \longrightarrow Au^I\text{-OOH} + H^+ \longrightarrow Au^{III}\text{=O} + H_2O$$

$$Au^{III}\text{=O} + RH \longrightarrow Au^{II}\text{-OH} + R^{\cdot}$$

$$R^{\cdot} + O_2 \longrightarrow ROO^{\cdot}$$

$$ROO^{\cdot} + Au^{II}\text{-OH} \longrightarrow ROOH + Au^{III}\text{=O}$$

Scheme 2.8 Oxidation of alkanes with hydrogen peroxide catalyzed by gold(III) and gold(I) complexes.

formation of *exo*-2,3-epoxynorbornane and [Au(N,N)(η^2-norbornene)]$^+$; an analogous reaction pathway is thus assumed for the formation of the other olefin complexes and of the oxygenated olefin derivatives. These results provide evidence for the intermediacy of metallaoxetanes in the oxygen-transfer reaction to olefins catalyzed by late transition metals [113]. Oxaauracycle species have also been proposed as intermediates for styrene oxidation on Au(III) [114].

Finally, two unique terminal gold-oxo molecular complexes have been obtained and structurally characterized [7]. Complexes K$_{15}$H$_2$[Au(O)(OH$_2$)P$_2$W$_{18}$O$_{68}$]·25H$_2$O (46) (Figure 2.21) and K$_7$H$_2$[Au(O)(OH$_2$)P$_2$W$_{20}$O$_{70}$(OH$_2$)$_2$]·27H$_2$O (47) have been synthesized by reaction of AuCl$_3$ with tungsten-oxide cluster ligands, [PW$_9$O$_{34}$]$^{9-}$, that model redox-active metal oxide surfaces. The very short (*circa* 176 pm) Au-O$_{oxo}$ distances account for multiply bonded gold-oxo species.

Au^{III}=O species are postulated, *inter alia*, as active intermediates in the oxidation of alkanes with hydrogen peroxide catalyzed by gold(III) and gold(I) complexes [115]. The reaction sequence is proposed in Scheme 2.8.

2.3.3
Alkoxo Complexes

Gold(III) alkoxo complexes, stabilized by a variety of ancillary ligands, are known with the same alkoxo ligands that also form stable gold(I) complexes [6]. An early example is the siloxane compound [Me$_2$AuOSiMe$_3$]$_2$ (48) which features a four-membered ring structure with silyloxide bridges between the square planar gold

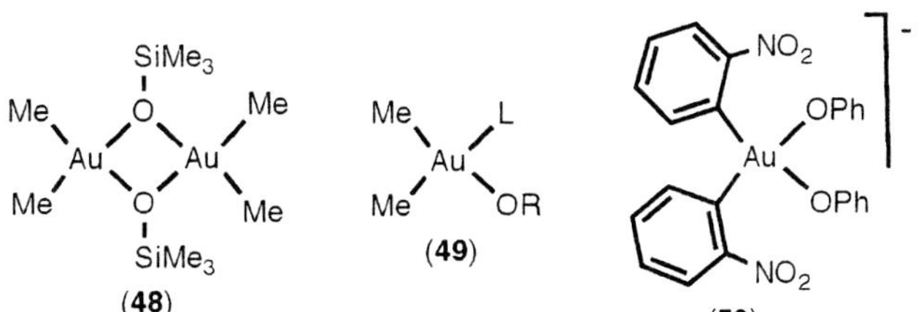

49: R = Ph, p-Tol, CH$_2$CF$_3$ or CH(CF$_3$)$_2$;
L = Ph, PCy$_3$, PMe$_2$Ph, PMe$_3$ or PEt$_3$

Figure 2.22 Stable gold(III) alkoxides stabilized by C-donor ligands.

centers (Figure 2.22) [116]. This complex has found application as precursor for chemical vapor deposition (CVD) [117].

Thermally stable aryloxo and fluoroalkoxo complexes having the general formula cis-[AuMe$_2$(OR)(PR'$_3$)] (**49**) (R = Ph **a**, p-tol **b**, CH$_2$CF$_3$ **c** CH(CF$_3$)$_2$ **d**; R' = Ph) have been synthesized by metathesis of cis-[AuMe$_2$I(PR'$_3$)] with KOR in THF [118]. The crystal structure of the phenoxo derivative shows a typical square-planar configuration with normal bond distances (Au–O 209 pm) and angles. Reaction of cis-[AuMe$_2$(OR)(PR'$_3$)] (R=CH$_2$CF$_3$, CH(CF$_3$)$_2$, Ph; R' = Ph, Cy) with metal hydrides MH of the type [CoH(CO)$_4$], [MnH(CO)$_5$], [MoHCp(CO)$_3$], [WHCp(CO)$_3$] or [WHCp(CO)$_2$(PMe$_3$)] proceeds with liberation of the free alcohol, ethane, and the gold(I) derivatives (R$_3$P)Au-M [119]. The reaction with active methylene compounds, such as H$_2$C(CN)$_2$, H$_2$C(CN)(COOR) (R=Me or Et) or HC$_2$Ph in non-polar solvents results in hydrogen abstraction to give the free alcohol and the corresponding organogold derivatives (Equation 2.17) [118]. Although less active than the corresponding gold(I) alkoxides, the fluoroalkoxides cis-[AuMe$_2$(OR)(PPh$_3$)] [R=CH$_2$CF$_3$, CH(CF$_3$)$_2$] can act as catalysts for the Knoevenagel condensation reaction (Equation 2.18) [120]. In contrast to analogous alkoxogold(I) complexes, only the aryloxides cis-[AuMe$_2$(OPh)L] (**49a**) (L=PMePh$_2$, PMe$_2$Ph, PMe$_3$, PEt$_3$) are capable of cleaving the C–Si bond of Me$_3$SiCF$_3$ (Equation 2.19) [121]. A large steric effect of the ancillary phosphine is observed in the case of gold(III), thus suggesting that the gold center is also taking part in the rate determining step of the C–Si bond cleavage reaction. A four-center concerted intermediate has been proposed.

$$\text{cis-[AuMe}_2\text{(OR)(PPh}_3\text{)]} + \text{R'H} \longrightarrow \text{cis-[AuMe}_2\text{R'(PPh}_3\text{)]} + \text{ROH}$$
(**49**)

(2.17)

$$\text{H}_2\text{C}\genfrac{}{}{0pt}{}{X}{Y} + \text{O=CH(Ph)} \xrightarrow{49c,d} \genfrac{}{}{0pt}{}{X}{Y}\text{C=CH(Ph)} + \text{H}_2\text{O}$$

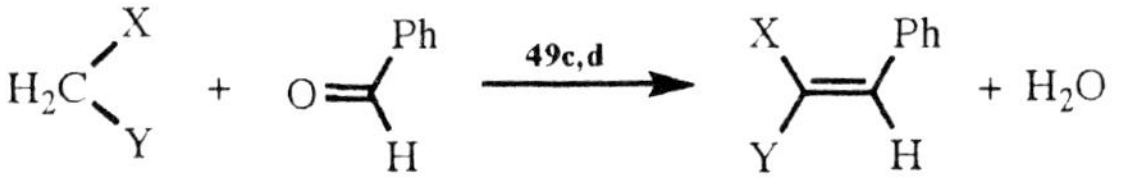

(2.18)

$$\text{cis-[AuMe}_2\text{(OPh)L]} + \text{Me}_3\text{SiCF}_3 \longrightarrow \text{cis-[AuMe}_2\text{(CF}_3\text{)L]} + \text{Me}_3\text{SiOPh}$$
(**49a**)

Eq. 2.17: R' = H$_2$C(X)(Y), C$_2$Ph
Eq. 2.18: X = Y = CN; X = CN, Y = C(O)OMe or C(O)OEt

(2.19)

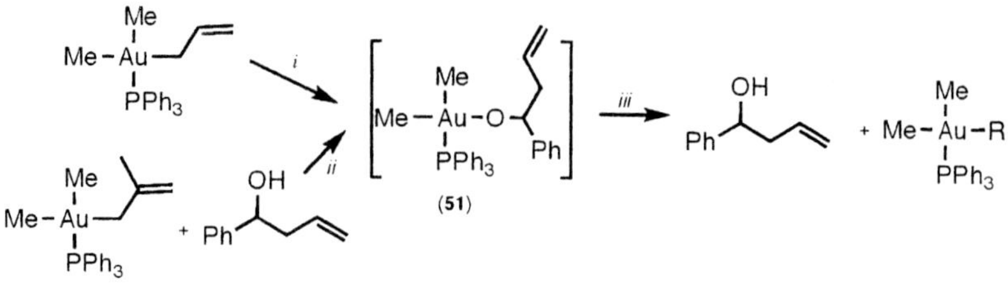

Scheme 2.9 Formation and reaction of alkoxogold(III) intermediate cis-[AuMe$_2$\{OCH(Ph)CH$_2$CH=CH$_2$\}(PPh$_3$)] (**51**); (i) + PhCHO; (ii) – Me$_2$C=CH$_2$; (iii) + RH [R=CH$_2$(CN)$_2$ or PhCHO].

Alkoxo complexes cis-[AuMe$_2$(OR)(PPh$_3$)] (**50**)(R=Me **a**, Et **b**, i-Pr **c**), although not isolated, have been prepared *in situ* from cis-[AuMe$_2$I(PPh$_3$)] and NaOR in alcohol or by alcoholysis of cis-[AuMe$_2$\{CH$_2$CH=CHMe\}(PPh$_3$)] in methanol. These highly reactive alkoxides readily insert carbon monoxide to give alkoxocarbonyl complexes cis-[AuMe$_2$(COOR)(PPh$_3$)] [122]. Similarly, the alkoxogold(III) intermediate cis-[AuMe$_2$\{OCH(Ph)CH$_2$CH=CH$_2$\}(PPh$_3$)] (**51**), formed from the insertion of benzaldehyde into the Au-allyl bond in cis-[AuMe$_2$\{CH$_2$CH=CH$_2$\}(PPh$_3$)], has been detected by ^{1}H NMR spectroscopy. Complex **51** can be obtained by alcoholysis of the methallylgold(III) complex cis-[AuMe$_2$\{CH$_2$C(Me)=CH$_2$\}(PPh$_3$)] with the homoallylalcohol, PhCH(OH)CH$_2$CH=CH$_2$ (Scheme 2.9) [123]. The alkoxo intermediate **51** reacts with active methylene compounds to give free homoallyl alcohol and C-bonded gold(III) enolates.

The anionic diphenoxide Na[Au(C$_6$H$_4$NO$_2$-2)$_2$(OPh)$_2$] (**52**), obtained by metathesis of the chloride Me$_4$N[Au(C$_6$H$_4$NO$_2$-2)$_2$Cl$_2$] with NaOPh, is quite stable in the solid state but in dichloromethane solution it reacts with traces of water to give the neutral dimeric hydroxo complex **6** [98]. Very stable catecholate complexes of the type [Au(N,C)(O$_2$Ar)] [N,CH=dimethylbenzylamine, 2-benzylpyridine or 2-anilinopyridine; Ar(OH)$_2$=catechol, tetrachlorocatechol or 3,5-tBu$_2$catechol] have been shown to possess both high antitumor and antibacterial activity [124].

Although primary and secondary alcohols including MeOH, EtOH, i-PrOH or ArCH$_2$OH are easily oxidized by [AuCl$_4$]$^-$ [125], stable alkoxogold(III) complexes have been isolated with 6-benzyl-2,2'-bipyridines and 2,2'-bipyridine as co-ligands. The cyclometallated derivatives [Au(N,N,C)(OR)][PF$_6$] (**53**) (C,N,NH=6-methylbenzyl- or 6-(1,1-dimethylbenzyl)-2,2'-bipyridine; R=Me **a**, Et **b**, i-Pr **c**, CH$_2$C$_6$H$_4$NO$_2$-3 **d**, Ph **e**, p-tol **f**) have been prepared according to different routes which depend on both the C,N,N and the alkoxo ligands [45a, 101]. For example, with the dimethylbenzyl substituted ligand both methoxo and ethoxo derivatives can be prepared by metathesis of the chloride [Au(N,N,C)Cl][PF$_6$] with NaOR (R=Me, Et) in ROH, whereas with the methylbenzyl substituted bipy the same alkoxides are only accessible by proton exchange of the acetato complex [Au(N,N,C)(OAc)][PF$_6$] with

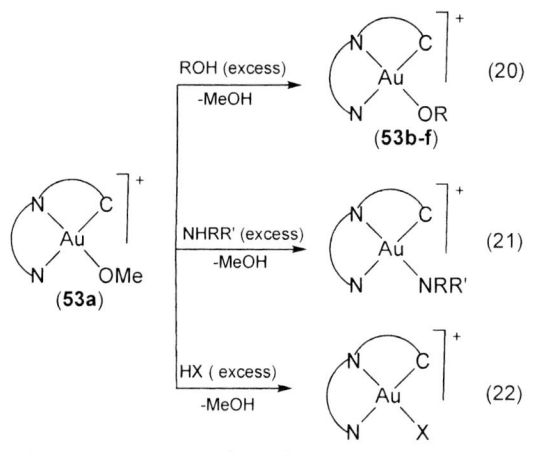

Scheme 2.10 Reactions of complexes **53a**. Equation 2.20: R = Et, i-Pr, $CH_2C_6H_4NO_2$-3, Ph or p-Tol. Equation 2.21: R = H, R' = $C_6H_4NO_2$-4, p-Tol, o-Xyl, i-Bu, s-Bu or R = R' = Et. Equation 2.22: X = C_2Ph, SC_6H_4Me-4.

ROH, since the benzylic C-H bond of the substituent in the chloride complex is activated by the strong base NaOR [45a].

All the other alkoxides **53c–f** are readily prepared using proton exchange between the methoxo complex and ROH under mild conditions (Equation 2.20 in Scheme 2.10) [101]. The methoxo derivatives are also the best starting materials for the synthesis of a number of complexes of general formula [Au(N,N,C)X][PF$_6$], where X is an anionic C-, N- or S-donor ligand, by proton exchange with HX (Equations 2.21 and 2.22 in Scheme 2.10) [45]. The bipy derivatives [Au(bipy)(OR)Cl][PF$_6$] (**41**) (R = Me, Et) and [Au(bipy)(OR)$_2$][PF$_6$] (**42**) (R = Me **a**, Et **b**, i-Pr **c**) are the only examples of gold alkoxides stabilized by nitrogen ligands only [104]. While complexes **41** are exclusively prepared by proton exchange reaction of the hydroxide **39** with ROH (Equation 2.10), complexes **42** can be obtained either by this route, from complex **40** (Equation 2.11), from the diacetato derivative [Au(bipy)(OAc)$_2$][PF$_6$], or by metathesis of the dichloro [Au(bipy)Cl$_2$][PF$_6$] with NaOR (R = Me, Et) in ROH [104]. The crystal structure of the dimethoxo complex **42a** (Figure 2.23) displays a square-planar coordination at the gold atom with typical bond distances and angles; the Au−O distances of 197 and 196 pm, are the same as those of the oxo complexes.

The methoxo complex [Au(sp)(OMe)(Cl)I] (**54**) (sp = diphenyl-2-styrylphosphine), possessing fairly good stability, has been fully characterized in solution [126]. It is the final product of a rearrangement induced by the substitution of a chloride with an iodide in the cyclometallated complex [Au(sp-OMe)Cl$_2$] (sp-OMe = 2-CH-(CH$_2$OMe)C$_6$H$_4$PPh$_2$) (Scheme 2.11). In contrast, the analogous diiodo derivative [Au(sp)(OMe)I$_2$] (**55**), although detected in solution, undergoes spontaneous reduction to [Au(sp)I] and, presumably, IOMe which further decomposes to HI and HCH=O.

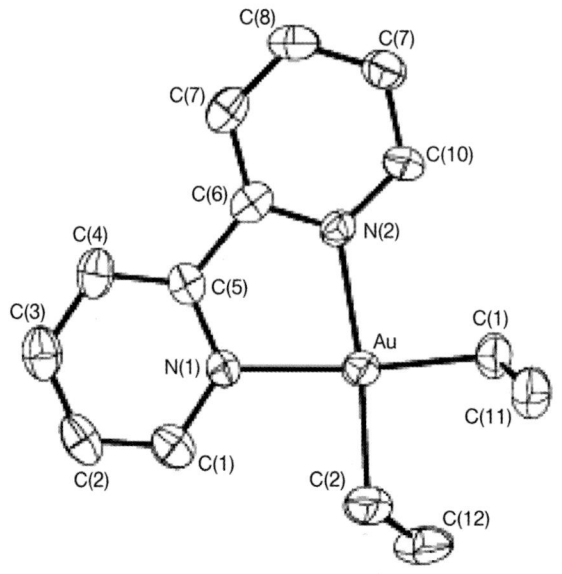

Figure 2.23 Structure of [Au(bipy)(OMe)$_2$]$^+$ (**42a**) [104].

Scheme 2.11 Formation and reactivity of methoxo intermediates **54** and **55**.

2.3.4
Complexes With Other O-donor Ligands

Mixed-metal salts with carboxylate, sulfate, or acetate have been synthesized. The gold(III) sulfates M[Au(SO$_4$)$_2$] (M = Na, K, Rb) are obtained by evaporation of Au(OH)$_3$ and Na$_2$SO$_4$ in concentrated sulfuric acid [127]. The gold atoms are coordinated by four oxygen atoms of different [SO$_4$]$^{2-}$ groups, forming an infinite chain [127]; in the caesium salt there are 2D layers [128]. The salts M[AuX$_4$]$_2$ (X = O$_2$CMe, O$_2$CEt; M = Pb, Zn, Ca, Sr) are prepared by reaction of M(OH)$_2$ with HAuCl$_4$, followed by treatment with the carboxylic acid, or by reaction of aurates M[Au(OH)$_4$]$_2$ with RCO$_2$H [129]. The structure of Zn[Au(OAc)$_4$]$_2$·2H$_2$O is a layered polymer while those of Pb[Au(OAc)$_4$]$_2$ and Sr[Au(OAc)$_4$]$_2$ are molecular units. Reaction of Cl$_2$O$_6$ with Au metal, AuCl$_3$ or HAuCl$_4$·nH$_2$O yields the well-defined chloryl salt ClO$_2$Au(ClO$_4$)$_4$ whose crystal structure displays discrete ClO$_2^+$ ions lying in channels formed by Au(ClO$_4$)$_4^-$ stacks [130]. Au is located in a square planar environment with Au–O bond lengths of 187 and 206 pm.

The water soluble acetato complex [Au(N,C)(OAc)$_2$] (N,C = 2-C$_6$H$_4$CH$_2$NMe$_2$) exhibits selective *in vitro* cytotoxicity and shows *in vivo* antitumor activity against human carcinoma xenografts [131]. Also active are the analogous bis-acetato and bis-carboxylato complexes supported by cyclometallated 2-phenylpyridine [9b] and 2-(1,1-dimethylbenzyl)-pyridine [103b]. The stable dimethylgold(III) carboxylates [AuMe$_2$(O$_2$CR)]$_2$ (R = Me, tBu, CF$_3$ or Ph) have been structurally characterized: they are dimers with bridging carboxylate ligands with Au···Au distances in the range of 298–308 pm and Au–O bond lengths of 212–217 pm (Figure 2.24) [132]. The dinuclear molecules form infinite chains with Au···Au contacts in the range of 346 (R = CF$_3$) –364 pm (R = Me). These complexes are viable precursors for gold CVD due to their sufficient volatility and thermal stability [133].

The well known complex [AuMe$_2$(acac)] (acacH = acetylacetone) [134] and several other substituted analogs [135] have found application as precursors for the CVD of gold [136], in laser-directed metal deposition of high-purity gold [137] and, quite recently, also in the preparation of mononuclear gold complexes supported on metal oxide powder [138]. The supported complexes AuMe$_2${OMg}$_2$ (where the {OMg} ligands are part of the support), obtained by adsorption of [AuMe$_2$(acac)] on calcined MgO, are identified as the catalytically active species in the hydrogenation of ethylene.

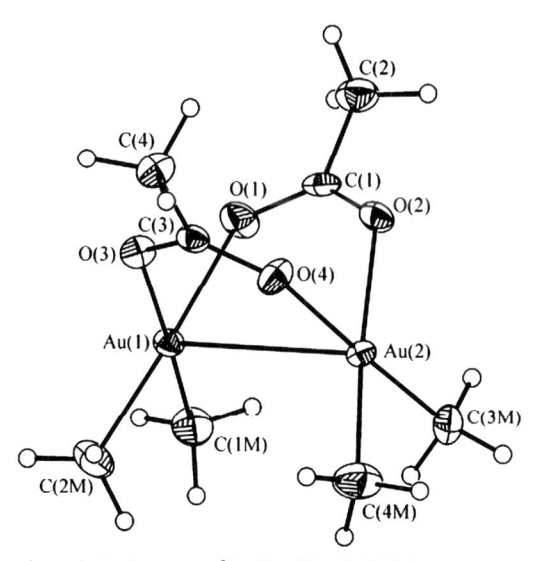

Figure 2.24 Structure of [AuMe$_2$(OAc)]$_2$ [132a].

2.3.5
Complexes With Mixed N/O Ligands

A number of gold(III) complexes with chelating N,O⁻ ligands have been reported. In these ligands activation of the O−H bond to give the cyclic alkoxogold derivative is made easier by the coordination of a heterocyclic nitrogen atom, as in the case of the 8-hydroxyquinoline (N,OH) derivatives [Au(N,C)(N,O)][BF$_4$] (**56**) (N,CH = dimethylbenzyl amine) [139] and [AuMe$_2$(N,O)] (**57**) [140], and of the 2-pyridylmethanol and 2-(2-pyridyl)ethanol derivatives [Au(N,O)Cl$_2$] (**58a, 58b**) [141], or of an iminic nitrogen, as in the case of the Schiff bases derivatives [Au(N,O)Cl$_2$] (**59**) and [Au(N,O)$_2$][AuX$_4$] (N,OH = 2-HOC$_6$H$_4$CH=NR; R = Me **a**, Et **b**, *i*Pr **c**, *n*Bu **d**, CH$_2$C$_6$H$_5$ **e**, C$_6$H$_{11}$ **f**; X = Cl, Br) [141b]. Other [Au(N,O)X$_2$] (X = Cl or Br) (**60**) or [Au(N,O)$_2$]$^+$ derivatives are obtained by reaction of [AuX$_4$]$^-$ with pyridine-2-carboxylic acids (Figure 2.25) [141b,142]. The N-coordinated [Au(N)Cl$_3$] adduct initially formed by reaction of pyridine-2-carboxaldehyde with KAuCl$_4$ in water adds a molecule of H$_2$O to the aldehydic group to form the corresponding diolic derivative which, by elimination of HCl, affords the [Au(N,O)Cl$_2$] derivative **61** (Scheme 2.12) [143]. Successive treatment with ROH (R = Me, Et or CH$_2$CF$_3$) gives the corresponding N,O-chelated hemiacetal complexes **62**. A kinetic study shows that displacement of the Au-alkoxide oxygen by chloride ions takes place only in the presence of perchloric acid [143b].

Some of the [Au(N,O)Cl$_2$] derivatives, namely **58a**, **59a** and **59b**, which have been tested for cytotoxic activity against various human tumor cell lines, have shown significant effects [144]. Compound **57** is a potential alternative to dimethylgold (III) β-diketonates and can be used as a starting material for gold coatings by the CVD method [140]. Recent studies have shown that the long-term stability of these

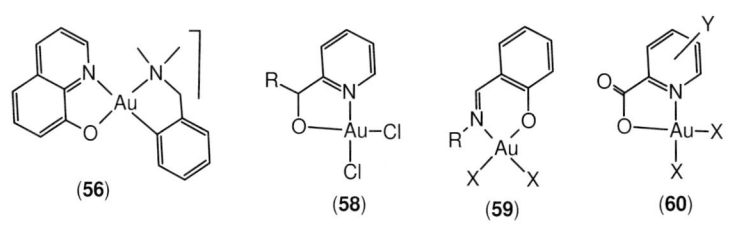

Figure 2.25 Complexes with chelating N,O⁻ ligands.

Scheme 2.12 Formation and reaction of the [Au(N,O)Cl$_2$] derivative **61**.

Figure 2.26 Complexes with phenolic Schiff bases.

complexes makes them efficient precatalysts. In particular, the 2-pyridylmethoxo complex **58a** has been used in the synthesis of highly substituted phenols from furans [145]. Highly stable and efficient catalysts are also the 2-pyridyl carboxylato derivatives [Au(N,O)Cl$_2$] (**60**) (N,OH = pyridine-2-carboxylic acid, pyridine-2,4-dicarboxylic acid or 3-hydroxypyridine-2-carboxylic acid) [146]. Gold(III) complexes of the type [Au(O,N,N,O)]$^+$, containing tetradentate Schiff base ligands (**63** in Figure 2.26), are conveniently obtained by reaction of the ligands with [Bu$_4$N][AuCl$_4$]: radiochemical studies with Au-198 show that these complexes are also formed at the tracer level [147]. A range of gold(III) phenolic Schiff base complexes (examples are illustrated in Figure 2.26) have been shown to be efficient and selective catalysts in a variety of transformations including olefin hydrogenation [148], self-coupling of aryl boronic acids [149] and hydrosilylation of a variety of functionalities [150].

Reaction of HAuCl$_4$ with the dipeptide glycylalanine (H-Gly-Ala-OH) and the tripeptide glycylalanylalanine (H-Gly-Ala-Ala-OH) gives the corresponding complexes [Au(Gly-Ala-O)Cl] (**66**) and [Au(Gly-Ala-Ala-O)] (**67**) where the peptides are coordinated to gold through $-$NH$_2$, N$^-$ amide(s) and CO$_2^-$ groups in a near square planar environment (Figure 2.27) [151].

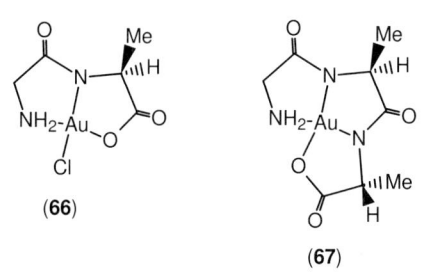

Figure 2.27 Polypeptide derivatives.

References

1 Hashmi, A.S.K. and Hutching, G.J. (2006) Gold catalysis. *Angewandte Chemie International Edition*, **45**, 7896; Bond, G.C., Louis, C. and Thompson, D.T. (2006) *Catalytic Science Series*, Vol. 6, Catalysis by Gold (ed. G.J. Hutchings), Imperial College Press, London, UK; Hashmi, A.S.K. (2007) Gold-catalyzed organic reactions. *Chemical Reviews*, **107**, 3180.

2 Guo, Z. and Sadler, P.J. (1999) Metals in medicine. *Angewandte Chemie International Edition*, **38**, 1512; Shaw, C.F. III (1999) Gold-based therapeutic agents. *Chemical Reviews*, **99**, 2589; Zhang, C.X. and Lippard, S.J. (2003) New metal complexes as potential therapeutics. *Current Opinion in Chemical Biology*, **7**, 481; Tiekink, E.R.T. (2002) Gold derivatives for the treatment of cancer. *Critical Reviews in Oncology/Hematology*, **42**, 225; Tiekink, E.R.T. (2003) Gold compounds in medicine: potential anti-tumour agents. *Gold Bulletin*, **36**, 117; Messori, L. and Marcon, G. (2004) Gold complexes as antitumour agents, in *Metal Ions in Biological Systems*, Vol. 42, (eds A. Sigel and H. Sigel), Marcel Dekker, Inc., New York, p. 385; Messori, L. and Gabbiani, C. (2005) Recent trends in antitumor gold(III) complexes: Innovative cytotoxic metallodrugs for cancer treatment, in *Metal Compounds in Cancer Chemotherapy* (eds J.M. Pérez, M.A. Fuertes and C. Alonso), Research Signpost, Kerala, India, p. 355; Kostova, I. (2006) Gold coordination complexes as anticancer agents. *Anti-Cancer Agents in Medicinal Chemistry*, **6**, 19.

3 Sibley, S., Thompson, M.E., Burrows, P.E. and Forrest, S.R. (1999) Electroluminescence in molecular materials, in *Optoelectronic Properties of Inorganic Compounds* (eds D.M. Roundhill and J.P. Jr.Fackler), Plenum Press, New York, pp. 29–54; Fun, E.Y., Olmstead, M.M., Vickery, J.C. and Balch, A.L. (1998) Glowing gold rings: solvoluminescence from planar trigold(I) complexes. *Coordination Chemistry Reviews*, **171**, 151; Chen, C.H. and Shi, J. (1998) Metal chelates as emitting materials for organic electroluminescence. *Coordination Chemistry Reviews*, **171**, 161; Vogler, A. and Kunkely, H. (2001) Photoreactivity of gold complexes. *Coordination Chemistry Reviews*, **219–221**, 489; Fernández, E.J., Laguna, A., López-de-Luzuriaga, J.M., Monge, M., Montiel, M., Olmos, M.E., Pérez, J. and Rodríguez-Castillo, M. (2007) Pyridine gold complexes. An emerging class of luminescent materials. *Gold Bulletin*, **40**, 172.

4 Togni, A. and Venanzi, L.M. (1994) Nitrogen donors in organometallic chemistry and homogeneous catalysis. *Angewandte Chemie (International Edition in English)*, **33**, 497; Reedijk, J. (1987) Heterocyclic nitrogen-donor ligands, in *Comprehensive Coordination Chemistry*, Vol 2, Ligands (eds G. Wilkinson, F.G.A. Stone and E.W. Abel), Pergamon, Oxford, UK, pp. 73–98.

5 Henderson, W. (2006) The Chemistry of cyclometallated gold(III) complexes with C,N-donor ligands. *Advances in Organometallic Chemistry*, **54**, 207.

6 Cinellu, M.A. and Minghetti, G. (2002) Gold(I) and gold(III) complexes with anionic oxygen donor ligands: hydroxo, oxo and alkoxo complexes. *Gold Bulletin*, **35**, 11.

7 Cao, R., Anderson, T.M., Piccoli, P.M.B., Schultz, A.J., Koetzle, T.F., Geletii, Y.V., Slonkina, E., Hedman, B., Hodgson, K.O., Hardcastle, K.I., Fang, X., Kirk, M.L., Knotteenbelt, S., Kögerler, P., Musaev, D.G., Morokuma, K., Takahashi, M. and Hill, C.L. (2007) Terminal gold-oxo complexes. *Journal of the American Chemical Society*, **129**, 11118.

8 Haruta, M. (1997) Size and support-dependency in the catalysis of gold. *Catalysis Today*, **36**, 153; Bond, G.C. and Thompson, D.T. (2000) Gold-catalysed oxidation of carbon monoxide. *Gold Bulletin*, **33**, 41; Monnier, J.R. (2001) The direct epoxidation of higher olefins using molecular oxygen. *Applied Catalysis A: General*, **221**, 73; Fu, L., Wu, N.Q., Yang, J.H., Qu, F., Johnson, D.L., Kung, M.C., Kung, H.H. and Dravid, V.P. (2005) Direct evidence of oxidized gold on supported gold catalysts. *The Journal of Physical Chemistry B*, **109**, 3704; Min, B.K., Alemozafar, A.R., Pinnaduwage, D.S. and Friend, C.M. (2006) Efficient CO oxidation at low temperature on Au(111). *The Journal of Physical Chemistry B*, **110**, 19833.

9 (a) Constable, E.C. and Leese, T.A. (1989) Cycloaurated derivatives of 2-phenylpyridine. *Journal of Organometallic Chemistry*, **363**, 419; (b) Fan, D., Yang, C.-T., Ranford, J.D., Lee, P.F. and Vittal, J.J. (2003) Chemical and biological studies of the dichloro(2-phenylpyridine)gold(III) complex and its derivatives. *Dalton Transactions*, (13), 2680; (c) Fan, D., Yang, C.-T., Ranford, J.D., Vittal, J.J. and Lee, P.F. (2003) Synthesis, characterization, and biological activities of 2-phenylpyridine gold(III) complexes with thiolate ligands. *Dalton Transactions*, (17), 3376.

10 Cinellu, M.A., Zucca, A., Stoccoro, S., Minghetti, G., Manassero, M. and Sansoni, M. (1995) Synthesis and characterization of gold(III) adducts and cyclometallated derivatives with 2-substituted pyridines. Crystal structure of [Au{$NC_5H_4(CMe_2C_6H_4)$-2}Cl_2]. *Journal of the Chemical Society, Dalton Transactions*, (17), 2865.

11 (a) Nonoyama, M., Nakajima, K. and Nonoyama, K. (1997) Direct cycloauration of 2-anilinopyridine (Hanp) with tetrachloroaurate(III) and the X-ray crystal structure of [$AuCl_2$(anp)]. *Polyhedron*, **16**, 4039; (b) Fuchita, Y., Ieda, H., Kayama, A., Kinoshita-Nagaoka, J., Kawano, H., Kameda, S. and Mukuriya, M. (1998) Cycloauration of 2-substituted pyridine derivatives. Synthesis, structure and reactivity of six-membered cycloaurated complexes of 2-anilino-, 2-phenoxy- and 2-(phenylsulfanyl)-pyridine. *Journal of the Chemical Society, Dalton Transactions*, (24), 4095.

12 Zhu, Z., Cameron, B.R. and Skerlj, R.T. (2003) Cycloauration of substituted 2-phenoxypyridine derivatives and X-ray crystal structure of gold, dichloro[2-[[5-[(cyclopentylamino)carbonyl]-2-pyridinyl-|N]oxy]phenyl-|C]-, (SP-4-3)-. *Journal of Organometallic Chemistry*, **677**, 57.

13 Fuchita, Y., Ieda, H., Tsunemune, Y., Kinoshita-Nagaoka, J. and Kawano, H. (1998) Synthesis, structure and reactivity of a new six-membered cycloaurated complex of 2-benzoylpyridine [$AuCl_2$(pcp-C^1,N)] [pcp = 2-(2-pyridylcarbonyl)phenyl]. Comparison with the cycloaurated complex derived from 2-benzylpyridine. *Journal of the Chemical Society, Dalton Transactions*, (5), 791.

14 (a) Constable, E.C. and Sousa, L.R. (1992) Metal-ion dependent reactivity of 2-(2′-thienyl)pyridine (Hthpy). *Journal of Organometallic Chemistry*, **427**, 125; (b) Fuchita, Y., Ieda, H., Wada, S., Kameda, S. and Mukuriya, M. (1999) Organogold(III) complexes derived from auration reactions of thienyl-substituted pyridine derivatives. *Journal of the Chemical Society, Dalton Transactions*, (24), 4431.

15 Ieda, H., Fujiwara, H. and Fuchita, Y. (2001) Synthesis and reactivity of the five-membered cycloaurated complex of phenylthiazole. *Inorganica Chimica Acta*, **319**, 203.

16 Fuchita, Y., Ieda, H. and Yasutake, M. (2000) First intramolecular aromatic substitution by gold(III) of a ligand other than pyridine derivatives. Synthesis and crystal structure of the novel five-membered cycloaurated complex of 1-ethyl-2-phenylimidazole.

Journal of the Chemical Society, Dalton Transactions, (3), 271.

17 Minghetti, G., Cinellu, M.A., Pinna, M.V., Stoccoro, S., Zucca, A. and Manassero, M. (1998) Gold(III) derivatives with C(4)-aurated 1-phenylpyrazole. *Journal of Organometallic Chemistry*, **568**, 225.

18 Constable, E.C., Henney, R.P.G., Leese, T.A. and Tocher, D.A. (1990) Cyclometallation reactions of 6-phenyl-2,2′-bipyridine; a potential C,N,N-donor analogue of 2,2′:6′,2″-terpyridine. Crystal and molecular structure of dichlorobis(6-phenyl-2,2′-bipyridine)ruthenium(II). *Journal of the Chemical Society, Dalton Transactions*, (2), 443.

19 Constable, E.C., Henney, R.P.G. and Leese, T.A. (1989) The direct cycloauration of 6-(2″-thienyl)-2,2′-bipyridine. *Journal of Organometallic Chemistry*, **361**, 277.

20 Cinellu, M.A., Zucca, A., Stoccoro, S., Minghetti, G., Manassero, M. and Sansoni, M. (1996) Synthesis and characterization of gold(III) adducts and cyclometallated derivatives with 6-benzyl- and 6-alkyl-2,2′-bipyridines. *Journal of the Chemical Society, Dalton Transactions*, (22), 4217.

21 McInnes, E.J.L., Welch, A.J. and Yellowleese, L.J. (1995) 2,2(-Bipyridinedichlorogold(III) tetrafluoroborate. *Acta Crystallographica Section C: Crystal Structure Communications*, **51**, 2023.

22 Robinson, W.T. and Sinn, E. (1975) Synthesis, structure, and properties of trichloro- and tribromo-(2,9-dimethyl-1,10-phenanthroline)gold(III). *Journal of the Chemical Society, Dalton Transactions*, (8), 726.

23 Abbate, F., Orioli, P., Bruni, B., Marcon, G. and Messori, L. (2000) Crystal structure and solution chemistry of the cytotoxic complex 1,2-dichloro(o-phenanthroline)gold(III) chloride. *Inorganica Chimica Acta*, **311**, 1.

24 Constable, E.C., Henney, R.P.G., Raithby, P.R. and Sousa, L.R. (1992) Cyclometallation reactions of 6-(thienyl)-2,2′-bipyridine with d^8 transition metal ions. *Journal of the Chemical Society, Dalton Transactions*, (14), 2251.

25 Cinellu, M.A., Minghetti, G., Pinna, M.V., Stoccoro, S., Zucca, A., Manassero, M. and Sansoni, M. (1998) μ-Oxo and alkoxo complexes of gold(III) with 6-alkyl-2,2′-bipyridines. Synthesis, characterization and X-ray structures. *Journal of the Chemical Society, Dalton Transactions*, (11), 1735.

26 Schimanski, A., Freisinger, E., Erxleben, A. and Lippert, B. (1998) Interactions between $[AuX_4]^-$ (X = Cl, CN) and cytosine and guanine model nucleobases: salt formation with (hemi-) protonated bases, coordination, and oxidative degradation of guanine. *Inorganica Chimica Acta*, **283**, 223.

27 Cossu, F., Mativic, Z., Radanovic, D. and Ponticelli, G. (1994) Cytotoxic activity of some gold(III) complexes. *Il Farmaco*, **49**, 301.

28 Moustatih, A., and Garnier-Suillerot, A. (1989) Bifunctional antitumor compounds: synthesis and characterization of a gold(III)-streptonigrin complex with thiol-modulating properties. *Journal of Medicinal Chemistry*, **32**, 1426.

29 Sánchez-Delgado, R.A., Navarro, M., Lazardi, K., Atencio, R., Capparelli, M., Vargas, F., Urbina, J.A., Bouillez, A., Noels, A.F. and Masi, D. (1998) Toward a novel metal based chemotherapy against tropical diseases. Synthesis and characterization of new metal-clotrimazole complexes and evaluation of their activity against *Trypanosoma cruzi*. *Inorganica Chimica Acta*, **275–276**, 528.

30 Navarro, M., Vásquez, F., Sánchez-Delgado, R.A., Pérez, H., Sinou, V. and Schrével, J. (2004) Toward a Novel Metal-Based Chemotherapy against Tropical Diseases. 7. Synthesis and *in Vitro* Antimalarial Activity of New Gold-Chloroquine Complexes. *Journal of Medicinal Chemistry*, **47**, 5204.

31 Fuchita, Y., Utsunomiya, Y. and Yasutake, M. (2001) Synthesis and reactivity of arylgold(III) complexes from aromatic hydrocarbons *via* C—H bond activation. *Journal of the Chemical Society, Dalton Transactions*, (16), 2330.

32 Bardají, M., Laguna, A., Pérez, M.R. and Jones, P.G. (2002) Unexpected ring-opening reaction to a new cyanamide-thiolate ligand stabilized as a dinuclear gold complex. *Organometallics*, **21**, 1877.

33 Byabartta, P. (2007) Gold(I)-gold(III)-4,4'-bpy-phosphine complexes: synthesis and multinuclear NMR study. *Transition Metal Chemistry*, **32**, 314.

34 Barranco, E.M., Crespo, O., Gimeno, M.C., Jones, P.G., Laguna, A. and Villacampa, M.D. (1999) Synthesis, structure and redox behaviour of gold and silver complexes with 3-ferrocenylpyridine. *Journal of Organometallic Chemistry*, **592**, 258.

35 Barranco, E.M., Crespo, O., Gimeno, M.C., Jones, P.G., Laguna, A. and Sarrocca, C. (2001) 1,1'-Bis(2-pyridylthio) ferrocene: a new ligand in gold and silver chemistry. *Journal of the Chemical Society, Dalton Transactions*, (17), 2523.

36 Vicente, J., Chicote, M.T., Guerrero, R., Saura-Llamas, I.M., Jones, P.G. and Ramírez de Arellano, M.C. (2001) The first acetimine gold(I) and gold(III) complexes and the first acetonine complexes. *Chemistry - A European Journal*, **7**, 638.

37 Klapötke, T.M., Krumm, B., Galvez-Ruiz, J.-C. and Nöth, H. (2005) Highly sensitive ammonium tetraazidoaurates(III). *Inorganic Chemistry*, **44**, 9625.

38 Beck, W. and Nöth, H. (1984) Röntgenstruktur von tetraphenylarsonium-tetraazidoaurat(III). *Chemische Berichte*, **117**, 419; Beck, W., Klapötke, T.M., Klüfers, P., Kramer, G. and Rienäcker, C.M. (2001) X-Ray crystal structures and quantum chemical calculations of tetraphenyl-arsonium tetraazidoaurate(III) and azido (triphenylphosphine)gold(I). *Zeitschrift fur Anorganische und Allgemeine Chemie*, **627**, 1669.

39 Wehlan, M., Thiel, R., Fuchs, J., Beck, W. and Fehlhammer, W.P. (2000) New C-tetrazolato complexes of rhodium(III), palladium(II) and gold(III). *Journal of Organometallic Chemistry*, **613**, 159.

40 (a) Pope, W.J. (1929) UK Patent 338506; (b) Karasch, M.S. and Isbell, H.S. (1931) The chemistry of organic gold compounds. IV. Gold imide compounds. *Journal of the American Chemical Society*, **53**, 3059.

41 (a) Tyabji, A.M., and Gibson, C.S. (1952) Gold imides. *Journal of the Chemical Society*, 450; (b) Malik, N.A., Sadler, P.J., Neidle, S. and Taylor, G.L. (1978) X-ray crystal structure of sodium bis(N-methylhydantoinato)gold(I) tetrahydrate; a linear, planar complex of pharmacological interest stabilised by two nitrogen ligands. *Journal of the Chemical Society, Chemical Communications*, (16), 711.

42 Oyaizu, K., Ohtani, Y., Shiozawa, A., Sugawara, K., Saito, T. and Yuasa, M. (2005) Highly stable gold(III) complex with a hydantoin ligand in alkaline media. *Inorganic Chemistry*, **44**, 6915.

43 Teleb, S.M. (2004) Spectral and thermal studies of saccharinato complexes. *The Journal of the Argentine Chemical Society*, **92**, 31.

44 Kilpin, K.J., Henderson, W. and Nicholson, B.K. (2007) Synthesis, characterisation and biological activity of cycloaurated organogold(III) complexes with imidate ligands. *Polyhedron*, **26**, 204.

45 (a) Cinellu, M.A., Minghetti, G., Pinna, M.V., Stoccoro, S., Zucca, A. and Manassero, M. (1999) Replacement of the chloride ligand in [Au(C,N,N)Cl][PF$_6$] cyclometallated complexes by C, N, O and S donor anionic ligands. *Journal of the Chemical Society, Dalton Transactions*, (16), 2823; (b) Cinellu, M.A., Minghetti, G., Pinna, M.V., Stoccoro, S., Zucca, A. and

Manassero, M. (2003) Synthesis and characterization of mononuclear amidogold(III) complexes - Crystal structure of [Au(N$_2$C$_{10}$H$_7$(CMe$_2$C$_6$H$_4$)-6](NHC$_6$H$_3$Me$_2$-2,6)][PF$_6$] - Oxidation of 4-methylaniline to azotoluene. *European Journal of Inorganic Chemistry*, 2304.

46 (a) Adams, H.-N., Grässle, U., Hiller, W. and Strähle, J. (1983) Gold-stickstoff-heterocyclen. Synthese, eigenschaften und struktur von [(CH$_3$)$_2$AuNH$_2$]$_4$ und [(CH$_3$)$_2$AuN(CH$_3$)$_2$]$_2$. *Zeitschrift für Anorganische und Allgemeine Chemie*, **504**, 7; (b) Grässle, U., Hiller, W. and Strähle, J. (1985) Gold-stickstoff-heterocyclen. 2. Synthese, eigenschaften und struktur von N-methylamido-dimethylgold(III), [(CH$_3$)$_2$AuNHCH$_3$]$_2$. *Zeitschrift für Anorganische und Allgemeine Chemie*, **529**, 29.

47 Messori, L., Abbate, F., Marcon, G., Orioli, P., Fontani, M., Mini, E., Mazzei, T., Carotti, S., O'Connell, T. and Zanello, P. (2000) Gold(III) complexes as potential antitumor agents: solution chemistry and cytotoxic properties of some selected gold(III) compounds. *Journal of Medicinal Chemistry*, **43**, 3541.

48 Zhu, S., Gorski, W., Powell, D.R. and Walmsley, J.A. (2006) Synthesis, structures, and electrochemistry of gold(III) ethylenediamine complexes and interactions with guanosine 5′-monophosphate. *Inorganic Chemistry*, **45**, 2688.

49 Carotti, S., Guerri, A., Mazzei, T., Messori, L., Mini, E. and Orioli, P. (1998) Gold(III) compounds as potential antitumor agents: Cytotoxicity and DNA binding properties of some selected polyamine-gold(III) complexes. *Inorganica Chimica Acta*, **281**, 90.

50 Messori, L., Orioli, P., Tempi, C. and Marcon, G. (2001) Interactions of selected gold(III) complexes with calf thymus DNA. *Biochemical and Biophysical Research Communications*, **281**, 352.

51 Liu, H.-Q., Cheung, T.-C., Peng, S.-M. and Che, C.-M. (1995) Novel luminescent cyclometaiated and terpyridine gold(III) complexes and DNA binding studies. *Journal of the Chemical Society, Chemical Communications*, (17), 1787.

52 Palanichamy, K., and Ontko, A.C. (2006) Synthesis, characterization, and aqueous chemistry of cytotoxic Au(III) polypyridyl complexes. *Inorganica Chimica Acta*, **359**, 44.

53 Sampath, U.-S., Putnam, W.C., Osiek, T.A., Touami, S., Xie, J., Cohen, D., Cagnolini, A., Droege, P., Klug, D., Barnes, C.L., Modak, A., Bashkin, J.K. and Jurisson, S.S. (1999) Terpyridyl derivatives as bifunctional chelates: synthesis and crystal structures of 4′-[2-(1,3-dioxolan-2-yl)ethylsulfanyl]-2,2′:6′,2″-terpyridine and chloro(4′-methylsulfanyl-2,2′:6′,2″-terpyridine) gold(III) bis(trifluoromethanesulfonate). *Journal of the Chemical Society, Dalton Transactions*, (12), 2049.

54 Hollis, L.S., and Lippard, S.J. (1983) Aqueous chemistry of (2,2′,2″-terpyridine)gold(III). Preparation and structures of chloro(2,2′,2″-terpyridine) gold dichloride trihydrate ([Au(terpy)Cl]Cl$_2$·3H$_2$O) and the mixed valence gold(I)-gold(III) salt bis[chloro(2,2′,2″-terpyridine)gold] tris(dichloroaurate) tetrachloroaurate ([Au(terpy)Cl]$_2$[AuCl$_2$]$_3$[AuCl$_4$]). *Journal of the American Chemical Society*, **105**, 4293.

55 Yam, V.W.-W., Choi, S.W.-K., Lai, T.-F. and Lee, W.K. (1993) Syntheses, crystal structures and photophysics of organogold(III) diimine complexes. *Journal of the Chemical Society, Dalton Transactions*, (6), 1001.

56 Mansour, M.A., Lachicotte, R.J., Gysling, H.J. and Eisenberg, R. (1998) Syntheses, molecular structures, and spectroscopy of gold(III) dithiolate complexes. *Inorganic Chemistry*, **37**, 4625.

57 Hayoun, R., Zhong, D.K., Rheingold, A.L. and Doerrer, L.H. (2006) Gold(III) and platinum(II) polypyridyl double salts and

a general metathesis route to metallophilic interactions. *Inorganic Chemistry*, **45**, 6120.

58 Simonov, Y., Bologa, O., Bourosh, P., Gerbelu, N., Lipkowski, J. and Gdaniec, M. (2006) Synthesis and structural characterization of gold(III) dioximates with anions [AuCl$_4$]$^-$ and [AuCl$_2$]$^-$. *Inorganica Chimica Acta*, **359**, 721.

59 Bulak, E., Sarper, O., Dogan, A., Lissner, F., Schleid, T. and Kaim, W. (2006) Dichlorogold(III) complexes of bis(1-methyl-2-imidazolyl)ketone and related ligands: Geometrical and electronic structures. *Polyhedron*, **25**, 2577.

60 Borkett, N.F., Bruce, M.I. and Walsh, J.D. (1980) Chemistry of the group 1B Metals. XIV Some poly(pyrazolyl)borate derivatives containing gold. *Australian Journal of Chemistry*, **33**, 949.

61 Vicente, J., Chicote, M.T., Guerrero, R. and Herber, U. (2002) Synthesis of tris- and tetrakis(pyrazol-1-yl)borate gold(III) complexes. Crystal structures of [Au{k^2-N,N'-BH(Pz)$_3$}Cl$_2$] (pz = pyrazol-1-yl) and [Au{k^2-N,N'-B(Pz)$_4$}{k^2-C,N-C$_6$H$_4$CH$_2$NMe$_2$-2)]ClO$_4$·CHCl$_3$, *Inorganic Chemistry*, **41**, 1870.

62 Vicente, J., Chicote, M.T. and Bermudez, M.D. (1984) 2-[(Dimethylamino)methyl] phenylgold(III) complexes. *Journal of Organometallic Chemistry*, **268**, 191.

63 Vicente, J., and Chicote, M.T. (1981) Synthesis and reactivity of dichloro 2-(phenylazo)phenyl gold(III). *Inorganica Chimica Acta*, **54**, L259.

64 Schouteeten, S., Allen, O.R., Haley, A.D., Ong, G.L., Jones, G.D. and Vicic, D.A. (2006) A robust dimethylgold(III) complex stabilized by a 2-pyridyl-2-pyrrolide ligand. *Journal of Organometallic Chemistry*, **691**, 4975.

65 Cao, L., Jennings, M.C. and Puddephatt, R.J. (2007) Amine-amide equilibrium in gold(III) complexes and a gold(III)-gold(I) aurophilic bond. *Inorganic Chemistry*, **46**, 1361.

66 Cuadrado, J.A., Zhang, W., Hang, W. and Majidi, V. (2000) Speciation of gold (III)–L-histidine complex: a multi-instrumental approach. *Journal of Environmental Monitoring*, **2**, 355.

67 Baddley, W.H., Basolo, F., Gray, H.B., Nölting, C. and Poë, A.J. (1963) Acidodiethylenetriaminegold(III) complexes: preparation, solution chemistry, and electronic structure. *Inorganic Chemistry*, **2**, 921.

68 Nardin, G., Randaccio, L., Annibale, G., Natile, G. and Pitteri, B. (1980) Comparison of structure and reactivity of bis-(2-aminoethyl)amine- and bis(2-aminoethyl)amido-chlorogold(III) complexes. *Journal of the Chemical Society, Dalton Transactions*, (2), 220.

69 Messori, L., Abbate, F., Orioli, P., Tempi, C. and Marcon, G. (2002) Au$_2$trien: a dinuclear gold(III) complex with unprecedented structural features. *Chemical Communications*, (6), 612.

70 (a) Fan, D., Yang, C.-T., Ranford, J.D. and Vittal, J.J. (2003) Chemical and biological studies of gold(III) complexes with uninegative bidentate N-N ligands. *Dalton Transactions*, (35), 4749; (b) Hill, D.T., Burns, K., Titus, D.D., Girard, G.R., Reiff, W.M. and Mascavage, L.M. (2003) Dichloro(pyridine-2-carboxamido-N1, N2)gold(III), a bis-nitrogen aurocycle: syntheses, gold-197 Mossbauer spectroscopy, and X-ray crystal structure. *Inorganica Chimica Acta*, **346**, 1.

71 Yang, T., Tu, C., Zhang, J., Lin, L., Zhang, X., Liu, Q., Ding, J., Xu, Q. and Guo, Z. (2003) Novel Au(III) complexes of aminoquinoline derivatives: crystal structure, DNA binding and cytotoxicity against melanoma and lung tumour cells. *Dalton Transactions*, (24), 3419.

72 (a) Wienken, M., Lippert, B., Zangrando, E. and Randaccio, L. (1992) Gold(III) glycyl-L-histidine dipeptide complexes. Preparation and x-ray structures of monomeric and cyclic tetrameric species. *Inorganic Chemistry*, **31**, 1983; (b) Carotti, S., Marcon, G., Marussich, M., Mazzei, T., Messori, L., Mini, E. and Orioli, P. (2000) Cytotoxicity and DNA binding properties

of a chloro glycylhistidinate gold(III) complex (GHAu). *Chemico-Biological Interactions*, **125**, 29.

73 Best, S.L., Chattopadhyay, T.K., Djuran, M.I., Palmer, R.A., Sadler, P.J., Sóvágó, I. and Varnagy, K. (1997) Gold(III) and palladium(II) complexes of glycylglycyl-L-histidine: crystal structures of [Au^{III}(Gly-Gly-L-His-H$_{-2}$)]Cl·H$_2$O and [Pd^{II}(Gly-Gly-L-His-H$_{-2}$)]·1.5H$_2$O and HisεNH deprotonation. *Journal of the Chemical Society, Dalton Transactions*, (15), 2587.

74 Cornejo, A.A., Castineiras, A., Yanovsky, A.I. and Nolan, K.B. (2003) Gold(III) complexes of amides of EDTA (L, L^1) and PDTA (L^2, L^3) as model systems for gold (III)–peptide/protein interactions.: Synthesis of [AuLCl][AuCl$_4$], [AuL$_2$Cl]Cl and related complexes and the crystal and molecular structures of [Au(LH$_{-1}$)Cl] [AuCl$_4$] containing a deprotonated amide ligand and of (LH$_2$)$_2$[AuCl$_4$]$_{3.5}$ [AuCl$_2$]$_{0.5}$·2H$_2$O. *Inorganica Chimica Acta*, **349**, 91.

75 Kilpin, K.J., Henderson, W. and Nicholson, B.K. (2007) Organogold(III) complexes containing chelating bis (amidate) ligands: Synthesis, characterisation and biological activity. *Polyhedron*, **26**, 434.

76 Kimura, E., Kurogi, Y. and Takahashi, T. (1991) The first gold(III) macrocyclic polyamine complexes and application to selective gold(III) uptake. *Inorganic Chemistry*, **30**, 4117.

77 Kimura, E., Kurogi, Y., Koike, T., Shionoya, M. and Iitaka, Y. (1993) Gold (III)-cyclam complexes. X-ray crystal structure and a useful macrocyclic effect on the reduction of gold(III). *Journal of Coordination Chemistry*, **28**, 33.

78 (a) Brawner, S.A., Lin, I.J.B., Kim, J.-H. and Everett, G.W. (1978) Synthesis of β-diiminate chelates by condensation of 2,4-pentanedione with Pt(NH$_3$)$_6$Cl$_4$, Pt(en)$_3$Cl$_4$, and Au(en)$_2$Cl$_3$. Crystal and molecular structure of *trans*-[Pt(NH$_3$)$_2$ (2,4-pentanediiminate)$_2$](ClO$_4$)$_2$. *Inorganic Chemistry*, **17**, 1304;

(b) Kim, J.-H. and Everett, G.W. (1979) Synthesis of macrocyclic complexes of gold(III) by condensation of bis (ethylenediamine)gold(III) chloride with β-diketones. *Inorganic Chemistry*, **18**, 3145; (c) Kim, J.-H. and Everett, G.W. (1981) Crystal and molecular structure of a macrocyclic complex of gold(III). *Inorganic Chemistry*, **20**, 853.

79 Park, C.H., Lee, B. and Everett, G.W. (1982) Crystal and molecular structure of a 14-π-electron macrocyclic gold(III) complex. *Inorganic Chemistry*, **21**, 1681.

80 Afanasyeva, V.A. and Mironov, I.V. (2001) Gold(III) macrocyclic complexes in aqueous solution. *Russian Journal of Coordination Chemistry*, **27**, 878; Afanasyeva, V.A., Glinskaya, L.A., Klevtsova, R.F. and Sheludyakova, L.A. (2005) Crystal and molecular structure of [Au(C$_{14}$H$_{24}$N$_4$)][H$_3$O](ClO$_4$)$_4$. *Journal of Structural Chemistry*, **46**, 131.

81 Afanasyeva, V.A., Glinskaya, L.A., Klevtsova, R.F. and Mironov, I.V. (2005) Protonation of the acyclic tetraaza metallocomplex of gold(III) [Au (C$_9$H$_{19}$N$_4$)]$^{2+}$ in aqueous solution. Synthesis and crystal and molecular structure of [Au(C$_9$H$_{20}$N$_4$)](H$_5$O$_2$) (ClO$_4$)$_4$. *Journal of Structural Chemistry*, **46**, 876.

82 (a) Suh, M.P., Kim, I.S., Shim, B.Y., Hong, D. and Yoon, T.-S. (1996) Extremely Facile Template Synthesis of Gold(III) Complexes of a saturated azamacrocycle and crystal structure of a six-coordinate gold(III) complex. *Inorganic Chemistry*, **35**, 3595; (b) Rossignoli, M., Bernhardt, P.V., Lawrance, G.A. and Maeder, M. (1997) Gold(III) template synthesis of a pendant-arm macrocycle. *Journal of the Chemical Society, Dalton Transactions*, (3), 323.

83 Bang, H., Lee, E.J., Lee, E.Y., Suh, J. and Suh, M.P. (2000) Transmetallation of nickel(II) ion in a nickel(II) azamacrocyclic complex with gold (III) ion. *Inorganica Chimica Acta*, **308**, 150.

84 Chambron, J.C., Heitz, V. and Sauvage, J.-P. (1997) Incorporation of gold(III) into porphyrins by disproportionation of [Au(tht)$_2$]BF$_4$. *New Journal of Chemistry*, **21**, 237.

85 (a) Brun, A.M., Harriman, A., Heitz, V. and Sauvage, J.-P. (1991) Charge-transfer across oblique bisporphyrins - 2-center photoactive molecules. *Journal of the American Chemical Society*, **113**, 8657; (b) Kilså, K., Kajanus, J., Macpherson, A.N., Mårtensson, J. and Albisson, B. (2001) Bridge-dependent electron transfer in porphyrin-based donor-bridge-acceptor systems. *Journal of the American Chemical Society*, **123**, 3069; (c) Fukuzumi, S., Ohkubo, K., Shao, J., Kadish, K.M., Hutchison, J.A., Ghiggino, K.P., Sintic, P.J. and Crossley, M.J. (2003) Metal-centered photoinduced electron transfer reduction of a gold(III) porphyrin cation linked with a zinc porphyrin to produce a long-lived charge-separated state in nonpolar solvents. *Journal of the American Chemical Society*, **125**, 14984; (d) Flamigni, L., Barigelletti, F., Armaroli, N., Collin, J.-P., Sauvage, J.-P. and Williams, J.A.G. (1998) Photoinduced processes in highly coupled multicomponent arrays based on a ruthenium(II) bis(terpyridine) complex and porphyrins. *Chemistry - A European Journal*, **4**, 1744; (e) Flamigni, L., Barigelletti, F., Armaroli, N., Ventura, B., Collins, J.-P., Sauvage, J.-P. and Williams, J.A.G. (1999) Triplet-triplet energy transfer between porphyrins linked via a ruthenium(II) bisterpyridine complex. *Inorganic Chemistry*, **38**, 661.

86 Flamigni, L., Dixon, I.M., Collins, J.-P. and Sauvage, J.-P. (2000) A Zn(II) porphyrin–Ir(III) bis-terpyridine–Au(III) porphyrin triad with a charge-separated state in the microsecond range. *Chemical Communications*, (24), 2479; Dixon, I.M., Collin, J.-P., Sauvage, J.-P., Barigelletti, F. and Flamigni, L. (2000) Charge separation in a molecular triad consisting of an iridium(III) - bis-terpy central core and porphyrins as terminal electron donor and acceptor groups. *Angewandte Chemie International Edition*, **39**, 1292.

87 Kadish, K.M., Wenbo, E., Ou, Z., Shao, J., Sintic, P.J., Ohkubo, K., Fukuzumi, S. and Crossley, M.J. (2002) Evidence that gold (III) porphyrins are not electrochemically inert: facile generation of gold(II) 5,10,15,20-tetrakis(3,5-di-*tert*-butylphenyl)porphyrin. *Chemical Communications*, (4), 356.

88 Ou, Z., Kadish, K.M., Wenbo E., Shao, J., Sintic, P.J., Ohkubo, K., Fukuzumi, S. and Crossley, M.J. (2004) Substituent effects on the site of electron transfer during the first reduction for gold(III) porphyrins. *Inorganic Chemistry*, **43**, 2078.

89 Knör, G. (2000) Molecular design of cationic sensitizers carrying redox active intercalator substituents: Towards recognition and photocatalytic cleavage of DNA. *Journal of Information Recording*, **25**, 111.

90 Knör, G. (2001) Intramolecular charge transfer excitation of *meso*-tetrakis (1-pyrenyl) porphyrinato gold(III) acetate. Photosensitized oxidation of guanine. *Inorganic Chemistry Communications*, **4**, 160.

91 (a) Che, C.-M., Sun, R.W.-Y., Yu, W.-Y., Ko, C.-B., Zhu, N. and Sun, H. (2003) Gold(III) porphyrins as a new class of anticancer drugs: cytotoxicity, DNA binding and induction of apoptosis in human cervix epitheloid cancer cells. *Chemical Communications*, (14), 1718; (b) Wang, Y., He, O.-Y., Sun, R.W.-Y., Che, C.-M. and Chiu, J.-F. (2007) Cellular pharmacological properties of gold(III) porphyrin 1a, a potential anticancer drug lead. *European Journal of Pharmacology*, **554**, 113.

92 Sun, R.W.-Y., Yu, W.-Y., Sun, H. and Che, C.-M. (2004) *In vitro* inhibition of human immunodeficiency virus type-1 (HIV-1) reverse transcriptase by gold(III) porphyrins. *ChemBioChem*, **5**, 1293.

93 Zhou, C.-Y., Chan, P.W.H. and Che, C.-M. (2006) Gold(III) porphyrin-catalyzed

cycloisomerization of allenones. *Organic Letters*, **8**, 325.

94 Miles, M.G., Glass, G.E. and Tobias, R.S. (1966) Structure of dimethylgold(III) compounds. Spectroscopic studies on the aquo ion and several coordination compounds. *Journal of the American Chemical Society*, **88**, 5738.

95 Glass, G.E., Konnert, J.H., Miles, M.G., Britton, D. and Tobias, R.S. (1968) Crystal and molecular structure and the solution conformation of dimethylgold(III) hydroxide, inorganic intermediate ring compound. *Journal of the American Chemical Society*, **90**, 1131.

96 Harris, S.J. and Tobias, R.S. (1969) Inorganic condensation reactions. Hydrolysis of dimethylgold(III) and the growth of tetrakis(dimethylgold hydroxide). *Inorganic Chemistry*, **8**, 2259.

97 (a) Braye, E.H., Hübel, W. and Caplier, I. (1961) New unsaturated heterocyclic systems. I. *Journal of the American Chemical Society*, **83**, 4406; (b) Peteau-Boisdenghien, M., Meunier-Piret, J. and van Meerssche, M. (1975) 1-Hydroxy-2,3,4,5-tetraphenylaura-cyclopentadiene dimer; $C_{56}H_{42}Au_2O_2$. *Acta Crystallographica Section C: Crystal Structure Communications*, **4**, 375.

98 Vicente, J., Bermúdez, M.D., Carrión, F.J. and Jones, P.G. (1996) Synthesis of some μ-hydroxo-, phenoxo- and O,O-acetylacetonato-arylgold(III) complexes. Crystal structure of $[Au(C_6H_4NO_2-2)_2(\mu-OH)]_2 \cdot 2Et_2O$. *Journal of Organometallic Chemistry*, **508**, 53.

99 Pitteri, B., Marangoni, G., Visentin, F., Bobbo, T., Bertolasi, V. and Gilli, P. (1999) Equilibrium and kinetic studies of (2,2':6',2''-terpyridine)gold(III) complexes. Preparation and crystal structure of $[Au(terpy)(OH)][ClO_4]_2$. *Journal of the Chemical Society, Dalton Transactions*, (5), 677.

100 Bortoluzzi, M., De Faveri, E., Daniele, S. and Pitteri, B. (2006) Synthesis of a new tetrakis(2-pyridinyl)pyrazine complex of gold(III) and its computational, spectroscopic and electrochemical characterization. *European Journal of Inorganic Chemistry*, 3393.

101 Pinna, M.V. (2000) PhD. Thesis, University of Sassari, Italy.

102 Cinellu, M.A., Minghetti, G., Pinna, M.V., Stoccoro, S., Zucca, A. and Manassero, M. (1998) The first gold(III) dinuclear cyclometallated derivatives with a single oxo bridge. *Chemical Communications*, (21), 2397.

103 (a) Marcon, G., Carotti, S., Coronnello, M., Messori, L., Mini, E., Orioli, P., Mazzei, T., Cinellu, M.A. and Minghetti, G. (2002) Gold(III) complexes with bipyridyl ligands: solution chemistry, cytotoxicity, and DNA binding properties. *Journal of Medicinal Chemistry*, **45**, 1672; (b) Coronnello, M., Mini, E., Caciagli, B., Cinellu, M.A., Bindoli, A., Gabbiani, C. and Messori, L. (2005) Mechanisms of cytotoxicity of selected organogold(III) compounds. *Journal of Medicinal Chemistry*, **48**, 6761.

104 Cinellu, M.A., Minghetti, G., Pinna, M.V., Stoccoro, S., Zucca, A. and Manassero, M. (2000) Gold(III) derivatives with anionic oxygen ligands: mononuclear hydroxo, alkoxo and acetato complexes. Crystal structure of $[Au(bpy)(OMe)_2][PF_6]$. *Journal of the Chemical Society, Dalton Transactions*, (8), 1261.

105 Cocco, F.S. (2007) PhD Thesis, University of Sassari.

106 Sanna, G., Pilo, M.I., Minghetti, G., Cinellu, M.A., Spano, N. and Seeber, R. (2000) Electrochemical properties of gold(III) complexes with 2,2'-bipyridine and oxygen ligands. *Inorganica Chimica Acta*, **310**, 34.

107 Nesmeyanov, A.N., Perevalova, E.G., Struchkov, Yu.T., Antipin, M.Yu., Grandberg, K.I. and Dyadchenko, V.P. (1980) Tris(triphenylphosphinegold) oxonium salts. *Journal of Organometallic Chemistry*, **201**, 343.

108 Sharp, P.R. (1999) Late transition metal oxo and imido chemistry. *Comments on Inorganic Chemistry*, **21**, 85.

109 Gabbiani, C., Casini, A., Messori, L., Guerri, A., Cinellu, M.A., Minghetti, G., Corsini, M., Rosani, C., Zanello, P. and Arca, M. (2008) Structural characterization, solution studies, and DFT calculations on a series of binuclear gold(III) oxo complexes: relationships to biological properties. *Inorganic Chemistry*, **47**, 2368.

110 Jones, P.G., Rumpel, H., Schwarzmann, E., Sheldrick, G.M. and Paulus, H. (1979) Gold(III) oxide. *Acta Crystallographica Section B: Structural Science*, **35**, 1435.

111 Casini, A., Cinellu, M.A., Minghetti, G., Gabbiani, C., Coronnello, M., Mini, E. and Messori, L. (2006) Structural and solution chemistry, antiproliferative effects, and DNA and protein binding properties of a series of dinuclear gold(III) compounds with bipyridyl ligands. *Journal of Medicinal Chemistry*, **49**, 5524.

112 (a) Cinellu, M.A., Minghetti, G., Stoccoro, S., Zucca, A. and Manassero, M. (2004) Reaction of gold(III) oxo complexes with alkenes. Synthesis of unprecedented gold alkene complexes, [Au(N,N)(alkene)][PF$_6$]. Crystal structure of [Au(bipy^{ip})(2-CH$_2$CHPh)][PF$_6$] (bipy^{ip} = 6-isopropyl-2,2-bipyridine). *Chemical Communications*, (14), 1618; (b) Cinellu, M.A., Minghetti, G., Cocco, F., Stoccoro, S., Zucca, A. and Manassero, M. (2005) Reactions of gold(III) oxo complexes with cyclic alkenes. *Angewandte Chemie International Edition*, **44**, 6892; (c) Cinellu, M.A., Minghetti, G., Cocco, F., Stoccoro, S., Zucca, A., Manassero, M. and Arca, M. (2006) Synthesis and properties of gold alkene complexes. Crystal structure of [Au(bipy^{oXyl})(η^2-CH$_2$=CHPh)](PF$_6$) and DFT calculations on the model cation [Au(bipy)(η^2-CH$_2$=CH$_2$)]$^+$. *Dalton Transactions*, (48), 5703.

113 Jørgensen, K.A. and Schiøtt, B. (1990) Metallaoxetanes as intermediate in oxygen-transfer reactions - reality or fiction? *Chemical Reviews*, **90**, 1483.

114 Deng, X., and Friend, C.M. (2005) Selective oxidation of styrene on an oxygen-covered Au(111). *Journal of the American Chemical Society*, **127**, 17178.

115 Shul'pin, G.B., Shilov, A. and Süss-Fink, G. (2001) Alkane oxygenation catalysed by gold complexes. *Tetrahedron Letters*, **42**, 7253.

116 Schmidbaur, H., and Bergfeld, M. (1966) Trimethylsiloxydimethylgold. *Inorganic Chemistry*, **5**, 2069.

117 Uchida, H., Saito, N., Sato, M., Take, M. and Ogi, K. (1994) (Mitsubishi Material Corp) Manufacture of gold thin film by chemical vapor deposition, Jpn. Kokai Tokkyo Koho JP 06,101,046.

118 Sone, T., Iwata, M., Kasuga, N. and Komiya, S. (1991) Dimethylgod(III) aryloxides and alkoxides having a triphenylphosphine ligand. *Chemistry Letters*, **20**, 1949.

119 Usui, Y., Hirano, M., Fukuoka, A. and Komiya, S. (1997) Hydrogen abstraction from transition metal hydrides by gold alkoxides giving gold-containing heterodinuclear complexes. *Chemistry Letters*, **26**, 981.

120 Komiya, S., Sone, T., Usui, Y., Hirano, M. and Fukuoka, A. (1996) Condensation reactions of benzaldehyde catalysed by gold alkoxides. *Gold Bulletin*, **29**, 131.

121 Usui, Y., Noma, J., Hirano, M. and Komiya, S. (2000) C–Si bond cleavage of trihalomethyltrimethylsilane by alkoxo- and aryloxogold or -copper complexes. *Inorganica Chimica Acta*, **309**, 151.

122 (a) Komiya, S., Ishikava, M. and Ozaki, S. (1988) Isolation of *cis*- dimethyl (methoxycarbonyl)(triphenylphosphine) gold(III). *Organometallics*, **7**, 2238; (b) Komiya, S., Sone, T., Ozaki, S., Ishikava, M. and Kasuga, N. (1992) Synthesis, structure and properties of dimethyl(alkoxycarbonyl)gold(III) complexes having a triphenylphosphine ligand. *Journal of Organometallic Chemistry*, **428**, 303.

123 Sone, T., Ozaki, S., Kasuga, N.C., Fukuoka, A. and Komiya, S. (1995) Synthesis, structure, and reactivities of *cis*-dimethyl(η^1-allyl)gold(III) complex.

Diastereoselective allylation of aromatic aldehyde. *Bulletin of the Chemical Society of Japan*, **68**, 1523.

124 Goss, C.H.A., Henderson, W., Wilkins, A.L. and Evans, C. (2003) Synthesis, characterisation and biological activity of gold(III) catecholate and related complexes. *Journal of Organometallic Chemistry*, **679**, 194.

125 Pal, B., Sen, P.K. and Sen Gupta, K.K. (2001) Reactivity of alkanols and aryl alcohols towards tetrachloroaurate(III) in sodium acetate-acetic acid buffer medium. *Journal of Physical Organic Chemistry*, **14**, 284.

126 Parish, R.V., Boyer, P., Fowler, A., Kahn, T.A., Cross, W.I. and Pritchard, R.G. (2000) Gold complexes derived from diphenyl-2-styrylphosphine. *Journal of the Chemical Society, Dalton Transactions*, (14), 2287.

127 Wickleder, M.S., and Büchner, O. (2001) The gold sulfates $MAu(SO_4)_2$ (M = Na, K, Rb). *Zeitschrift für Naturforschung B. A Journal of Chemical Sciences*, **56b**, 1340.

128 Wickleder, M.S. and Esser, K. (2002) Synthese und kristallstruktur von $CsAu(SO_4)_2$. *Zeitschrift für Anorganische und Allgemeine Chemie*, **628**, 911.

129 (a) Jones, P.G., Schelbach, R., Schwarzmann, E., Thörne, C. and Vielmäder, A. (1988) Darstellung und kristallstruktur einiger tetracarboxylatoaurate(III). *Zeitschrift für Naturforschung B. A Journal of Chemical Sciences*, **43b**, 807; (b) Jones, P.G. (1984) Digold(III) strontium octaacetate dihydrate, $Au_2Sr(C_2H_3O_2)_8 \cdot 2H_2O$. *Acta Crystallographica Section C: Crystal Structure Communications*, **40**, 804.

130 Cunin, F., Deudon, C., Favier, F., Mula, B. and Pascal, J.L. (2002) First anhydrous gold perchlorato complex: $ClO_2Au(ClO_4)_4$. Synthesis and molecular and crystal structure analysis. *Inorganic Chemistry*, **41**, 4173.

131 Buckley, R.G., Elsome, A.M., Fricker, S.P., Henderson, G.R., Theobald, B.R.C., Parish, R.V., Howe, B.P. and Kelland, L.R. (1996) Antitumor properties of some 2-[(dimethylamino)methyl]phenyl-gold(III) complexes. *Journal of Medicinal Chemistry*, **39**, 5208.

132 (a) Bessonov, A.A., Baidina, I.A., Morozova, N.B., Semyannikov, P.P., Trubin, S.V., Gel'fond, N.V. and Igumenov, I.K. (2007) Synthesis, crystal structure, and thermal behavior of dimethylgold(III) acetate. *Journal of Structural Chemistry*, **48**, 282; (b) Bessonov, A.A., Morozova, N.B., Kurat'eva, N.V., Baidina, I.A., Gel'fond, N.V. and Igumenov, I.K. (2008) Synthesis and crystal structure of dimethylgold(III) carboxylates. *Russian Journal of Coordination Chemistry*, **34**, 70.

133 Bessonov, A.A., Morozova, N.B., Gel'fond, N.V., Semyannikov, P.P., Trubin, S.V., Shevtsov, Yu.V., Shubin, Yu.V. and Igumenov, I.K. (2007) Dimethylgold(III) carboxylates as new precursors for gold CVD. *Surface & Coatings Technology*, **201**, 9099.

134 Brain, F.H. and Gibson, G.S. (1939) The organic compounds of gold. Part VII. Methyl and ethyl compounds. *Journal of the Chemical Society*, 762.

135 Komiya, S., and Kochi, J.K. (1977) Reversible linkage isomerisms of β-diketonato ligands. Oxygen-bonded and carbon-bonded structures in gold(III) acetylacetonate complexes induced by phosphines. *Journal of the American Chemical Society*, **99**, 3695.

136 Puddephatt, R.J. (1999) Gold metal and gold alloys in electronics and thin film technology, in *Gold: Progress in Chemistry, Biochemistry and Technology* (ed. H. Schmidbaur), John Wiley & Sons, Ltd, Chichester, UK, pp. 237–256.

137 Baum, T.H. (1987) Laser chemical vapor deposition of gold. *Journal of the Electrochemical Society*, **134**, 2616.

138 Guzman, J., and Gates, B.C. (2004) A mononuclear gold complex catalyst supported on MgO: spectroscopic characterization during ethylene

hydrogenation catalysis. *Journal of Catalysis*, **226**, 111.

139 Vicente, J., Chicote, M.T., Bermudez, M.D., Jones, P.G., Fittschen, C. and Sheldrick, G.M. (1986) Some attempts to prepare five-co-ordinated gold(III) complexes. Crystal and molecular structures of [Au(κ-*C,N*- $C_6H_4CH_2NMe_2$-2)(phen)(PPh$_3$)][BF$_4$]$_2 \cdot$CH$_2$Cl$_2$, [Au(κ-*C,N*-$C_6H_4CH_2NMe_2$-2)(κ-*N,O*-NC$_9$H$_6$O)]BF$_4$, and [Au(κ-*C,N*-$C_6H_4CH_2NMe_2$-2)(κ-*N,S*-H$_2$NC$_6$H$_4$S)]ClO$_4$. *Journal of the Chemical Society, Dalton Transactions*, (11), 2361.

140 Bessonov, A.A., Morozova, N.B., Semyannikov, P.P., Trubin, S.V., Gel'fond, N.V. and Igumenov, I.K. (2008) Thermal properties of dimethylgold(III) 8-hydroxyquinolinate and 8-mercaptoquinolinate. *Russian Journal of Coordination Chemistry*, **34**, 186.

141 (a) Canovese, L., Cattalini, L., Marangoni, G. and Tobe, M.L. (1985) Displacement of pyridine-2-methanol from dichloro (pyridine-2-methanolato)gold(III) in acidic solution. Ring opening at Oxygen. *Journal of the Chemical Society, Dalton Transactions*, (4), 731; (b) Dar, A., Moss, K., Cottrill, S.M., Parish, R.V., McAuliffe, C.A., Pritchard, R.G., Beagley, B. and Sandbank, J. (1992) Complexes of gold(III) with mononegative bidentate N,O-ligands. *Journal of the Chemical Society, Dalton Transactions*, (12), 1907.

142 Annibale, G., Canovese, L., Cattalini, L., Marangoni, G., Michelon, G. and Tobe, M.L. (1984) Displacement by chloride of pyridine-2-carboxylate from dichloro(pyridine-2-carboxylato)gold(III) in acidic solution; the position of ring opening. *Journal of the Chemical Society, Dalton Transactions*, (8), 1641.

143 (a) Bertolasi, V., Annibale, G., Marangoni, G., Paolucci, G. and Pitteri, B. (2003) Syntheses, characterization and X-ray structures of gold(III) complexes obtained by addition of bases to gold(III)(2-pyridinecarboxaldehyde). *Journal of Coordination Chemistry*, **56**, 397; (b) Marangoni, G., Pitteri, B., Annibale, G. and Bortoluzzi, M. (2006) Ring opening and displacement by chloride of the bidentate chelate ligand from dichloro [pyridine-2- (methoxymethanolato)]gold (III) - A kinetic and mechanistic study. *European Journal of Inorganic Chemistry*, **29**, 765.

144 Calamai, P., Carotti, S., Guerri, A., Messori, L., Mini, E., Orioli, P. and Speroni, G.P. (1997) Biological properties of two gold(III) complexes: AuCl$_3$ (Hpm) and AuCl$_2$ (pm). *Journal of Inorganic Biochemistry*, **66**, 103; Calamai, P., Carotti, S., Guerri, A., Mazzei, T., Messori, L., Mini, E., Orioli, P. and Speroni, G.P. (1998) Cytotoxic effects of gold(III) complexes on established human tumor cell lines sensitive and resistant to cisplatin. *Anti-Cancer Drug Design*, **13**, 67; Calamai, P., Guerri, A., Messori, L., Orioli, P. and Speroni, G.P. (1999) Structure and DNA binding properties of the gold(III) complex [AuCl$_2$(esal)]. *Inorganica Chimica Acta*, **285**, 309.

145 Hashmi, A.S.K., Rudolph, M., Weyrauch, J.P., Wölfe, M., Frey, W. and Bats, J.W. (2005) Gold catalysis: proof of arene oxides as intermediates in the phenol synthesis. *Angewandte Chemie International Edition*, **44**, 2798.

146 Hashmi, A.S.K., Weyrauch, J.P., Rudolph, M. and Kurpejovic, E. (2004) Gold catalysis: the benefits of N and N,O ligands. *Angewandte Chemie International Edition*, **43**, 6545.

147 Barnholtz, S.L., Lyndon, J.D., Huang, G., Venkatesh, M., Barnes, C.L., Ketring, A.R. and Jurisson, S.S. (2001) Syntheses and characterization of gold(III) tetradentate Schiff base complexes. X-ray crystal structures of [Au(sal$_2$pn)]Cl$\cdot$2.5H$_2$O and [Au (sal$_2$en)]PF$_6$. *Inorganic Chemistry*, **40**, 972.

148 Comas-Vives, A., González-Arellano, C., Corma, A., Iglesias, M., Sánchez, F. and Ujaque, G. (2006) Single-site homogeneous and heterogenized gold(III) hydrogenation catalysts: mechanistic

implications. *Journal of the American Chemical Society*, **128**, 4756.
149 González-Arellano, C., Corma, A., Iglesias, M. and Sánchez, F. (2005) Homogeneous and heterogenized Au(III) Schiff base-complexes as selective and general catalysts for self-coupling of aryl boronic acids. *Chemical Communications*, (15), 1990.
150 Corma, A., González-Arellano, C., Iglesias, M. and Sánchez, F. (2007) Gold nanoparticles and gold(III) complexes as general and selective hydrosilylation catalysts. *Angewandte Chemie International Edition*, **46**, 7820.
151 Kolev, Ts., Koleva, B.B., Zareva, S.Y. and Spiteller, M. (2006) Au(III)-complexes of the alanyl-containing peptides glycylalanine and glycylalanylalanine – Synthesis, spectroscopic and structural characterization. *Inorganica Chimica Acta*, **359**, 4367.

3
Pentafluorophenyl Gold Complexes
A. Luquin, E. Cerrada, and M. Laguna

3.1
Introduction

The pentafluorophenyl group bonds to a metal center mainly and, in the case of gold, exclusively through the ipso carbon atom affording a carbon-gold bond. Thus, if we are refering to pentafluorophenyl gold derivatives, we need to discuss organometallic gold chemistry which began a century ago [1, 2]. As with many other parts of the chemistry of gold, the development of this area began very slowly but it was from the 1970–80s to the end of the century when gold and organogold chemistry increased steadily, with the seminal works and reviews [3–6] of Schmidbaur, Puddephatt and Uson. From that time onwards work on organogold chemistry has been increasing exponentially.

The development of arylgold chemistry is very similar, starting in the 1930s [7] with the synthesis of the first arylgold derivative that was only fully characterized in 1959 [8] and showing a progressive increase in activity from the 1970–80s up to the present. The amount of interest in the arylgold derivatives is important, as is evident from a number of different reviews either general surveys [9–12] or specific ones in gold(I) [13], arylgold chemistry [14] or more recently in arylgold(I) chemistry [15]. In some of the reviews mentioned it is of note that the use of polyhalophenyl groups in general, and the polyfluorophenyl group in particular, is an important reason for the success of this area. That is why we present here a chapter dedicated to the polyfluorophenyl group in gold chemistry taking into account all the oxidation states.

The first thing to point out is that the use of a pentafluorophenyl C_6F_5 group with late transition metal confers on the complexes great stability, both thermodynamic and kinetic. This general fact, that is also true in gold chemistry, can be explained by different factors:

1. The strong electron-withdrawing effect of the C_6F_5 group that is estimated to have a Hammett constant σ_p of 0.4 [16, 17] and a Brown constant σ^- of 0.99 [18, 19]. The Lewis acid strength of $B(C_6F_5)_3$ is between BF_3 and BCl_3 which means that the electron-withdrawing effect of the C_6F_5 group should be comparable with that of these two halogens [20].

Gold Chemistry: Applications and Future Directions in the Life Sciences. Edited by Fabian Mohr
Copyright © 2009 WILEY-VCH Verlag GmbH & Co. KGaA, Weinheim
ISBN: 978-3-527-32086-8

2. Unlike alkyl-gold bonds the Au—C bonds in pentafluorophenyl compounds are more resistant to cleavage by protic acids, giving more chemical integrity to the complexes.

3. The pentafluorophenyl group imparts greater crystallinity to the complexes and as a result many complexes have been studied by X-ray crystallography. Although with other metal centers C_6F_5–C_6F_5 or C_6F_5–C_6H_5 π–π stacking interactions are observed [21, 22], there are not many examples in gold chemistry and they have been shown very recently [23].

The explanation for the rich gold chemistry shown around the pentafluorophenyl ligand is not only the stability that the C_6F_5 group confers on the complexes but also the synthesis of very suitable starting materials. They are weakly coordinated ligands that permit, by replacement reactions, the preparation of a great number of new complexes. Neutral, such as AsPh$_3$, or anionic, such as OClO$_3$, were the first neutral or anionic groups used for these purposes as we will see below. The extension to safer chemistry leading to the use of ligands such as tetrahydrothiophene (tht), triflate (CF$_3$SO$_3$) or μ-chloride, particularly the former, was a big step in the preparation of new gold complexes. In fact the synthesis of complexes with the tht ligand is the most important factor in the progress of gold chemistry. Thus the preparation of [Au(C_6F_5)(tht)] [24], [Au(C_6F_5)$_3$(tht)] [25], cis-[Au(C_6F_5)$_2$(μ-Cl)]$_2$ [26], trans-[Au(C_6F_5)$_2$(tht)$_2$](CF$_3$SO$_3$) [27, 28] [Au(C_6F_5)$_3$OEt$_2$] [29, 172], and [Au(C_6F_5)$_2$(OEt$_2$)$_2$]ClO$_4$ [29, 186] are the bases of the development of pentafluorophenyl gold chemistry.

This review covers the literature up to the end of 2007 and it is arranged according to the oxidation state of gold. Although such ordering should begin with cluster chemistry with the oxidation state of gold between 0 and I, followed by I, II and III, we will start with gold(I) chemistry, which chronologically is the starting point of pentafluorophenylgold chemistry. We will continue with gold(III) derivatives, in which we will cover all the polynuclear gold (III)/(I) complexes and we will finish with the known examples in cluster gold chemistry and gold(II) derivatives.

3.2
Pentafluorophenylgold(I) Derivatives

The first synthesized pentafluorophenylgold(I) complex was [Au(C_6F_5)PPh$_3$] obtained by the reaction of [AuClPPh$_3$] and C_6F_5MgBr [30–32]. This complex was at the same time the first characterized [33] by X-ray studies showing the common nearly linear distribution of the phosphorous and the *ipso* carbon atom (178(1)°). Nowadays, there are many methods of obtaining this complex, in many cases as a by-product of some reaction and, in other cases, its formation is the driving force of the reaction. We will see examples of both in this chapter.

Treatment of [Au(C_6F_5)$_2$ClPPh$_3$] with hydrazine [34] leads to formation of [Au(C_6F_5)PPh$_3$]. Similarly, the reaction between [Au(PPh$_3$)$_2$]ClO$_4$ and NBu$_4$[Au(C_6F_5)$_2$] or NEt$_4$[Au(C_6F_5)Cl] affords [Au(C_6F_5)PPh$_3$], the former being a good method for its preparation. An example of an undesired by-product is in the reaction of [(μ-Cl)(AuPPh$_3$)$_2$]ClO$_4$ with Ag(C$_6$F$_5$) [35] where instead of a pentafluorophenyl bridge the neutral Au(I) complex is formed. Attempts to oxidize [AuX(PPh$_3$)] with [Tl(C$_6$F$_5$)$_2$Br] (X = C$_6$F$_5$, NO$_3$, CH$_3$COO, SCN, C$_6$H$_5$ [36]) yields both [Tl(C$_6$F$_5$)BrX] and [Au(C$_6$F$_5$)PPh$_3$]. The exchange of the pentafluorophenyl group of the thallium derivative with the X group of the gold complex occurs but there is no oxidation.

Although it is not the intention here to give many details of infra-red and NMR data for each complex, it is important to note that the IR [30–32] of the pentafluorophenyl group in [Au(C$_6$F$_5$)PPh$_3$] shows signals at 1500, 1050, 950 and 790 cm^{-1}. The latter presents the behavior of the carbon-metal component and is sensitive to the oxidation state of the gold center and to the numbers and disposition of the pentafluorophenyl groups. The ^{19}F NMR spectra [30–32] show three resonances for each different pentafluorophenyl group and, in the case of [Au(C$_6$F$_5$)PPh$_3$], they resonate at 116.0 ppm corresponding to the *ortho* fluorine, 162.4 ppm due to the *meta* fluorine atoms and at 158.3 ppm assignable to the *para* fluorine atoms. The former two are complex multiplets, whereas the latter usually appears as a triplet.

A similar complex is the derivative containing triphenylarsine [Au(C$_6$F$_5$)AsPh$_3$] which can be obtained by reaction of [AuCl(AsPh$_3$)] with (C$_6$F$_5$)MgBr [37] in analogous manner to the preparation of the PPh$_3$ complex. In this case too, there are some reactions that give [Au(C$_6$F$_5$)AsPh$_3$] as undesired by-product for example the reaction of [Au(C$_6$F$_5$)$_2$Cl(AsPh$_3$)] with AgNO$_3$ [38].

The first complex described with an Au–S [39] bond was [Au(C$_6$F$_5$)(tht)] (tht = tetrahydrothiophene) that was synthesized by reacting [AuCl(tht)] with LiC$_6$F$_5$. The complex can also be prepared by the reaction of [Au(tht)$_2$]ClO$_4$ with NBu$_4$[Au(C$_6$F$_5$)$_2$] [35].

This complex is very interesting from a synthetic point of view because the tht ligand is easily replaced [40] which allows the preparation of many [Au(C$_6$F$_5$)L] complexes. The crystal structure of [Au(C$_6$F$_5$)(tht)] was studied by X-ray diffraction in 2007 [41] and it shows the asymmetric unit containing two independent monomers with similar paddle-like geometries that form infinite chains along the *a* axis of the crystal through alternating long and two different short aurophilic interactions (3.306, 3.191 and 3.128 Å). Identical pairs of molecules are altenatingly rotated by 180° (Figure 3.1).

This Au(I) complex was tested as co-catalyst in palladium-catalyzed alkynylation reactions but this derivative is much less active than the analogous chloride complex [AuCl(tht)] needing over twice as long to fully convert the starting material [42].

3.2.1
Neutral Pentafluorophenylgold(I) Derivatives

The reaction between [Au(C$_6$F$_5$)AsPh$_3$] and [Au(C$_6$F$_5$)(tht)], mainly the latter, with C-, N-, P- As- or S-donor ligands leads to the corresponding [Au(C$_6$F$_5$)L] complexes as shown in Table 3.1. The reaction carried out with BiPh$_3$ causes decomposition to metallic gold.

3 Pentafluorophenyl Gold Complexes

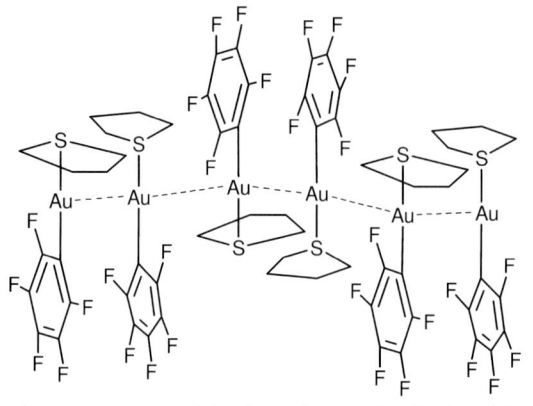

Figure 3.1 Drawing of the identical pairs of molecules of the complex [Au(C$_6$F$_5$)(tht)] alternatingly rotated.

Table 3.1 [Au(C$_6$F$_5$)L] complexes.

Reagent	Complex	Ref
CH$_2$(S$_2$CNMe$_2$)$_2$	[{Au(C$_6$F$_5$)}{CH$_2$(S$_2$CNMe$_2$)$_2$}]	[43]
CH$_2$(S$_2$CNEt$_2$)$_2$	[{Au(C$_6$F$_5$)}{CH$_2$(S$_2$CNEt$_2$)$_2$}]	[43]
CH$_2$(S$_2$CN^{n}Bu$_2$)$_2$	[{Au(C$_6$F$_5$)}{CH$_2$(S$_2$CN^{n}Bu$_2$)$_2$}]	[43]
CH$_2$(S$_2$CNBz$_2$)$_2$	[{Au(C$_6$F$_5$)}{CH$_2$(S$_2$CNBz$_2$)$_2$}]	[43]
CNC$_6$H$_4$O(O)CC$_6$H$_4$OC$_{10}$H$_{21}$	[Au(C$_6$F$_5$)CNC$_6$H$_4$O(O)CC$_6$H$_4$OC$_{10}$H$_{21}$]	[44]
NEt$_3$	[Au(C$_6$F$_5$)NEt$_3$]	[45]
H$_2$NCH$_2$NH$_2$	[Au(C$_6$F$_5$)H$_2$NCH$_2$NH$_2$]	[46]
H$_2$NCH$_2$CH$_2$NH$_2$	[Au(C$_6$F$_5$)H$_2$NCH$_2$CH$_2$NH$_2$]	[46]
2-amino-4,5-dihydrothiazole	[Au(C$_6$F$_5$)(2-amt)]	[47]
N(H)=CPh$_2$	[Au(C$_6$F$_5$)(N(H)=CPh$_2$)]	[48]
C$_3$H$_5$NS$_2$	[Au(C$_6$F$_5$)(C$_3$H$_5$NS$_2$)]	[49]
C$_4$H$_4$N$_2$S	[Au(C$_6$F$_5$)(C$_4$H$_4$N$_2$S)]	[49]
C$_5$H$_5$NS	[Au(C$_6$F$_5$)(C$_5$H$_5$NS)]	[49]
C$_7$H$_5$NS$_2$	[Au(C$_6$F$_5$)(C$_7$H$_5$NS$_2$)]	[49]
C$_7$H$_6$N$_2$S	[Au(C$_6$F$_5$)(C$_7$H$_6$N$_2$S)]	[49]
Ph$_2$CN$_2$	[Au(C$_6$F$_5$)(Ph$_2$C=N−N=PPh$_2$)]	[50]
PEt$_3$	[Au(C$_6$F$_5$)PEt$_3$]	[45]
PPh$_3$	[Au(C$_6$F$_5$)PPh$_3$]	[45]
PCy$_3$	[Au(C$_6$F$_5$)PCy$_3$]	[45]
phen	[Au(C$_6$F$_5$)phen]	[45]
PPhH$_2$	[Au(C$_6$F$_5$)PPhH$_2$]	[51]
PPy$_3$	[Au(C$_6$F$_5$)PPy$_3$]	[52]
PPhPy$_2$	[Au(C$_6$F$_5$)PPhPy$_2$]	[52]
PPh$_2$CCH	[Au(C$_6$F$_5$)(PPh$_2$CCH)]	[53]
(Ph$_2$PC(S)N(H)Me)	[Au(C$_6$F$_5$){Ph$_2$PC(S)N(H)Me}]	[54]
PhP(O)H	[Au(C$_6$F$_5$)PhP(O)H]	[55]
C$_6$F$_4$NCSPPh$_2$	[Au(C$_6$F$_5$)(C$_6$F$_4$NCSPPh$_2$)]	[56]
P(SCH$_3$)$_3$	[Au(C$_6$F$_5$)P(SCH$_3$)$_3$]	[57]
P(SPh$_3$)$_3$	[Au(C$_6$F$_5$)P(SPh$_3$)$_3$]	[57]
1,3,5-triaza-7-phosphaadamantane	[Au(C$_6$F$_5$)PTA]	[28]

Table 3.1 (Continued)

Reagent	Complex	Ref
TPPMS	[Au(C$_6$F$_5$)(TPPMS)]	[58]
TPPDS	[Au(C$_6$F$_5$)(TPPDS)]	[58]
TPPTS	[Au(C$_6$F$_5$)(TPPTS)]	[58]
CH(PPh$_2$)$_3$	[CH{PPh$_2$Au(C$_6$F$_5$)}$_3$]	[59]
AsPh$_3$	[Au(C$_6$F$_5$)AsPh$_3$]	[45]
S=CSCH$_2$CH$_2$S	[Au(C$_6$F$_5$)(S=CSCH$_2$CH$_2$S)]	[60]
SC(SMe)NH(p-MeC$_6$H$_4$)	[Au(C$_6$F$_5$)SC(SMe)NH(p-MeC$_6$H$_4$)]	[61]
SC(SMe)NH(o-MeC$_6$H$_4$)	[Au(C$_6$F$_5$)SC(SMe)NH(o-MeC$_6$H$_4$)]	[61]
SC(SMe)NH(p-MeOC$_6$H$_4$)	[Au(C$_6$F$_5$)SC(SMe)NH(p-MeOC$_6$H$_4$)]	[61]
SC(SMe)NH(3,5-Me$_2$C$_6$H$_3$)	[Au(C$_6$F$_5$)SC(SMe)NH(3,5-Me$_2$C$_6$H$_3$)]	[61]

3.2.1.1 Neutral Pentafluorophenylgold(I) Derivatives with C-Donor Ligands

The addition of [Au(C$_6$F$_5$)(tht)] to a dichloromethane solution of CNC$_6$H$_4$O(O)CC$_6$H$_4$OC$_{10}$H$_{21}$ gave the complex [44] shown in the Figure 3.2. This complex has very interesting behavior because it is a liquid crystal that displays a Sm$_A$ mesophase that has been identified in optical microscopy by its typical oily streaks and homeotropic textures, giving focal-conic textures at temperatures close to the clearing point. It was shown that the perhalophenyl group produced lower transition temperatures, shorter mesogenic ranges and an enhancement of the Sm$_A$ phase according to the expected lower lateral intermolecular interaction as a consequence of the greater molecular width.

The reaction of methylene bis(dialkyldithiocarbamates) with [Au(C$_6$F$_5$)(tht)] leads to the complexes [{Au(C$_6$F$_5$)}$_2${CH$_2$(S$_2$CNR$_2$)$_2$}] [43] (R = Me, Et, n-Bu, Bz). The ^{1}H NMR resonances of the bridging methylene groups in the complexes are observed at lower field than that in the free ligand and some modification of the C=S stretching is observed.

Gold(I) isocyanide complexes [62, 63] are prepared as shown in Equation 3.1.

$$[Au(C_6F_5)(tht)] + CNR \rightarrow [Au(C_6F_5)(CNR)] + tht \qquad (3.1)$$

Where R = Tol, Ph, ^{t}Bu, Cy, Me [63].

Depending on the R group, this reaction could lead to the formation of gold(I) isocyanide complexes that behave as liquid crystals. Thus, complexes [Au(C$_6$F$_5$)(C≡N(C$_6$H$_4$)OC$_{10}$H$_{21}$-p)] and [Au(C$_6$F$_5$)(C≡N(C$_6$H$_4$)OC$_n$H$_{2n+1}$-p)] [64] where n = 4, 6, 8, 10 and 12 show this behavior. All of these complexes are mesomorphic and behave as liquid crystals showing a nematic (N) phase when the isocyanide has a

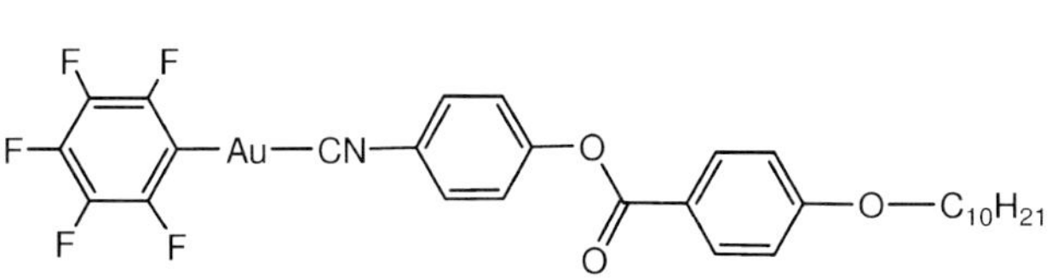

Figure 3.2 Complex that behaves as a liquid crystal.

short tail ($n = 4$), N and smectic A (Sm_A) phase when they have and intermediate tail ($n = 6, 8$) and only Sm_A phase for longer chains. As the length of the chain increases, the range of Sm_A phase increases and the range of the N phase decreases. The transition temperatures decrease when increasing lengths.

The reaction of the gold(I) pentafluorophenyl isocyanide complexes with primary and secondary amines as well as alcohols leads to the corresponding gold(I) [62, 65] carbenes (Table 3.2). The addition of amines leads to the corresponding carbenes

Table 3.2 [Au(C_6F_5)(Carbene)] complexes.

CNR^1	$NH_{3-x}R_x^2$ or R^2 OH	Ref
CNTo	$MeNH_2$	[62]
CNTo	$^tBuNH_2$	[62]
CNTo	$CyNH_2$	[62]
CNTo	$C_6H_5CH_2NH_2$	[62]
CNTo	$C_6H_5CH_2CH_2NH_2$	[62]
CNTo	NH_3	[62]
CNTo	$C_6H_5NH_2$	[62]
CNTo	p-$FC_6H_4NH_2$	[62]
CNTo	p-$CH_3OC_6H_4NH_2$	[62]
CNTo	p-$NO_2C_6H_4NH_2$ (no reaction)	[62]
CNTo	Et_2NH	[62]
CNTo	Pr_2NH	[62]
CNTo	$(C_6H_5CH_2)_2NH$ (no reaction)	[62]
CNTo	MeOH	[62]
CNTo	EtOH	[62]
CNPh	CH_3NH_2	[65]
CNPh	$C_6H_5NH_2$	[65]
CNPh	$C_6H_{11}NH_2$	[65]
CNPh	p-$FC_6H_4NH_2$	[65]
CNPh	$C_6H_5CH_2NH_2$	[65]
CNPh	p-$CH_3C_6H_4NH_2$	[65]
CNPh	p-$CH_3OC_6H_4NH_2$	[65]
CNPh	$C_6H_5CH_2CH_2NH_2$	[65]
CN^tBu	CH_3NH_2	[65]
CN^tBu	$C_6H_5CH_2NH_2$	[65]
CNCy	CH_3NH_2	[65]
CNCy	t-$C_4H_9NH_2$	[65]
CNCy	$C_6H_{11}NH_2$	[65]
CNCy	$C_6H_5CH_2NH_2$	[65]
CNPh	CH_3OH	[65]
CNPh	CH_3CH_2OH	[65]
CNCy	CH_3OH	[65]
CNPh	$NH_2CH_2CH_2NH_2$	[46]
CNPh	$NH_2CH_2CH_2CH_2NH_2$	[46]
CNTo	$NH_2CH_2CH_2NH_2$	[46]
CNTo	$NH_2CH_2CH_2CH_2NH_2$	[46]
CNMe	$MeNH_2$	[63]

under milder conditions than those required for alcohols; this is due to the better nucleophilicity of amines compared with alcohols. When the isocyanide group is attached to an electron-withdrawing substituent the process occurs readily. For example, the reaction of cyclohexylisocyanide pentafluorophenyl Au(I) proceeds only very slowly and no reaction is observed with ethanol. These results are likely due to kinetic reasons. The accepted mechanism of this type of reaction goes through a nucleophilic attack (Equation 3.2 and 3.3).

$$[Au(C_6F_5)(CNR^1)] + NH_{3-x}R^2_x \rightarrow [Au(C_6F_5)[C(NHR^1)(NH_{2-x}R^2_x)] \quad (3.2)$$

$$[Au(C_6F_5)(CNR^1)] + R^2OH \rightarrow [Au(C_6F_5)[C(NHR^1)(OR^2)] \quad (3.3)$$

The cyclic carbene complex shown in equation 3.4 was studied by X-ray diffraction [66], it shows a linear complex (angle C−Au−C 178.6(4)°) and the gold aryl bond distance is 1.993(10) Å which is in accordance with such bonds in other known pentafluorophenyl complexes. The gold carbene carbon distance is 1.961(9) Å, the dihedral angle between the planes formed by the two organic ligands is 5.35° and the shortest intermolecular Au−Au distance is 3.95 Å.

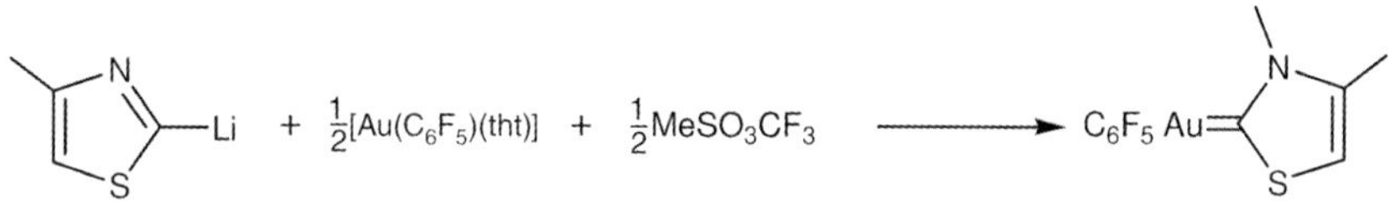

$$\text{thiazole-Li} + \tfrac{1}{2}[Au(C_6F_5)(tht)] + \tfrac{1}{2}MeSO_3CF_3 \longrightarrow C_6F_5\,Au\text{-thiazole} \quad (3.4)$$

When the Au(I) complex [Au(C_6F_5)(tht)] reacts with lithium 1-methylimidazol-2-yls the organic compound is protonated or alkylated and the carbene complexes are obtained [67] (see Figure 3.3).

In the complex with two methylimidazol units one of them is N-coordinated and the other a C-coordinating carbene ligand. It seems that in the synthesis of this complex a homoleptic rearrangement of the C_6F_5 precursor complex is accompanied by a rapid migration of the proton nitrogen to the initially coordinated carbon.

When 4',4'-dimethyl-2'-(2-thienyl)-2'-oxazoline lithiated at C^3 is reacted with [Au(C_6F_5)(tht)] the organogold complex analogous to the previous example was obtained.

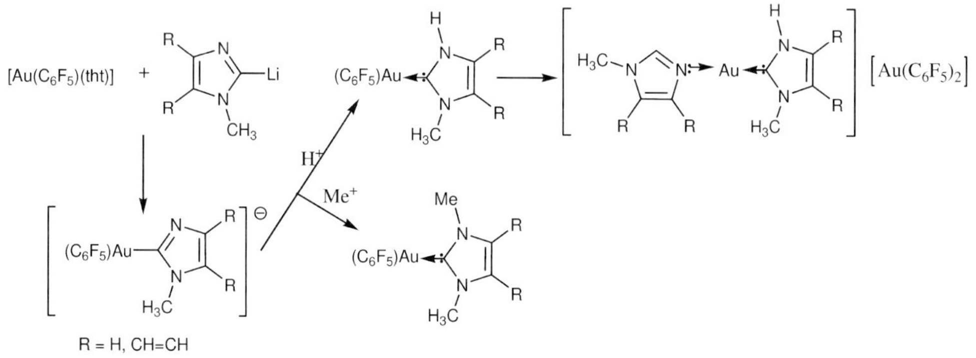

R = H, CH=CH

Figure 3.3 Synthesis of different carbene derivatives.

In contrast, this complex does not undergo homoleptic rearrangement in solution. It was the first diorganocarbene gold complex synthesized [68].

The Au(I) derivative $N^nBu_4[Au(C_6F_5)_2]$ reacts with $[Au(CH_2PPh_3)_2]ClO_4$ to give the neutral complex $[Au(C_6F_5)(CH_2PPh_3)]$ [69]. This reaction is a general reaction for several Au(I) derivatives of the type $NBu_4[AuX_2]$.

Complexes containing the ylide ligand can be also obtained by reaction of the yellow solution of ylides (prepared by addition of butyllithium to tetrahydrofurane or diethylether solutions of the phosphonium salt) with $[Au(C_6F_5)(tht)]$. The weak ligand tht is replaced and the ylide complex is obtained (Equation 3.5) [70].

$$Au(C_6F_5)(tht) + RHC\text{-}ER'_3 \rightarrow [Au(C_6F_5)(CHRER'_3)]$$
$$R = H, \ ER' = PPh_3, \ PPh_2Me, \ PPhMe_2, \ AsPh_3 \quad (3.5)$$
$$R = Me, \ Et \ or \ Ph, \ ER' = PPh_3 \ [71].$$

3.2.1.2 Neutral Pentafluorophenylgold(I) Derivatives with N-Donor Ligands

These complexes are obtained by reaction of the nitrogen ligand with the Au(I) starting material $[Au(C_6F_5)(tht)]$. The ketazine complex $[Au(C_6F_5)(Ph_2C=N-N=PPh_2)]$ [50] was also obtained this way (by reaction with Ph_2CN_2 and shows resonances in the aromatic region of the ^1H NMR spectrum. An almost linear arangement (175.8(2)°) of the ligands has been observed. The stability of the complex is attributable to the presence of the (C_6F_5) ligand because attempts to obtain other ketazine complexes produce the free azine even at low temperature [72].

The reaction of the complex $[Au(C_6F_5)(Ph_2C=N-N=CPh_2)]$ with 2,6-dimethyl-phenylisocyanide (CNXy) leads to the formation of $[Au(C_6F_5)(CNXy)]$ [50] by displacement of the ligand. The IR shows an absorption at 2214 cm^{-1} that is due to the $v_{(CN)}$ of the coordinated isocyanide.

The reaction of the $[Au(C_6F_5)(tht)]$ with 2-amino-4,5-dihydrothiazole (2-amt) yields the complex $[Au(C_6F_5)(2\text{-}amt)]$ [47]. When $[Au(C_6F_5)(tht)]$ reacts with $N(H)=CPh_2$ the complex $[Au(C_6F_5)\{N(H)=CPh_2\}]$ [48] is obtained. The structure of this complex shows that two molecules of the complex are associated in an antiparallel manner with Au(I)-Au(I) interactions of 3.5884(7) Å.

When 4′,4′-dimethyl-2′-(2-thienyl)-2′-oxazoline lithiated at C^5 is reacted with $[Au(C_6F_5)(tht)]$ the neutral nitrogen-coordinated pentafluorophenylgold complex and not the expected monocarbene complex [68] is obtained. 4-Methylthiazole, 4-methyl-2-mehylsulfanylthiazole and piperidine react with $[Au(C_6F_5)(tht)]$ forming the complexes shown in Figure 3.4 [73].

The ^1H and ^{13}C NMR spectra of the thiazole and thioether complexes does not provide any clue to identify the coordinated heteroatom although the signal of the

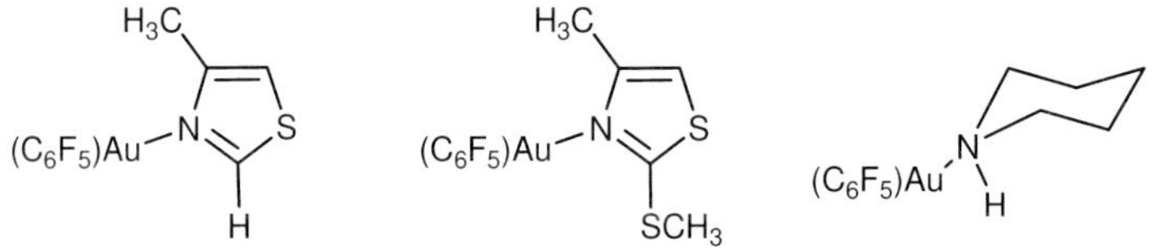

Figure 3.4 Complexes obtained by reaction of [Au(C$_6$F$_5$)(tht)] with different N-ligands.

NCS-carbon is shifted dowfield in the two first complexes. The amine complex shows a shift of the N—H signal which appears at 2.08 ppm in the free ligand and at 5.17 ppm when it is coordinated. The structure of the thiazole complex was studied by X-ray diffraction, confirming the almost linear coordination about the gold atom and showing that the ligands are almost co-planar.

^{15}N NMR experiments where undertaken to determine the site of coordination of the two other complexes. The thioether complex showed an upfield shift for the N-atom in the thioether gold complex and a similar result was obtained in the case of the thiazole compound, confirming imine coordination.

3.2.1.3 Neutral Pentafluorophenylgold(I) Derivatives with P-Donor Ligands

When the Au(I) pentafluorophenyl complex [Au(C_6F_5)(tht)] reacts with a phosphine the tht ligand is easily displaced to give the corresponding phosphine complexes. This was the method employed to obtain all the following complexes.

Using the phosphine PPhH$_2$ the complex [Au(C_6F_5)(PPhH$_2$)] [51] is obtained which in the ^{31}P {^{1}H} NMR spectrum shows a broad singlet at −46.0 ppm. When the phosphines employed are PPy$_3$ or PPhPy$_2$ the corresponding complexes [Au(C_6F_5)P] [52] where P=PPy$_3$ or PPhPy$_2$ are obtained. The ^{31}P{^{1}H} NMR spectrum shows a singlet at 42.5 ppm for the first complex and a singlet at 41.6 ppm for the latter.

The reaction with PPh$_2$CCH leads to the formation of [Au(C_6F_5)(PPh$_2$CCH)] [53] whose ^{31}P{^{1}H} NMR spectrum shows a singlet at 17.2 ppm, in the ^{1}H NMR spectrum the resonance of the C≡CH proton is observed at 3.46 ppm. The IR spectrum shows, besides the pentafluorophenyl absorptions, a band at 3271 cm^{-1} due to the $v_{(C≡CH)}$ and another absorption at 2056 cm^{-1} for the asymmetric C≡C stretch. The structure of this complex was studied by X-ray diffraction, the Au(I) atom is an almost linearly coordinated and the Au—C and Au—P distances are in the range of the values found for similar complexes. The excitation and the emission data in the solid state at 77 K are 331 and 445 nm.

The ligand diphenylphosphinothioformamide Ph$_2$PC(S)N(H)Me reacts in the same way to give the complex [Au(C_6F_5){Ph$_2$PC(S)N(H)Me}] [54]. The ^{1}H NMR spectrum shows the presence of the amine protons and the ^{31}P{^{1}H}NMR has a singlet at 56.0 ppm more than 41 ppm downfield compared to that of the free ligand. A curious characteristic of this complex is that when dissolved in chloroform it suffers a progressive discoloration that is complete in approximately one hour. The ^{31}P{^{1}H}NMR spectrum of the colorless solution shows a singlet at 50.7 ppm, the signals due to the amine proton in the proton NMR decrease till disappearance; the same happens with the signals in the ^{19}F{^{1}H}NMR spectrum. This fact seems to be in accordance with a process of simultaneous deprotonation of the amine protons of two complexes by the pentafluorophenyl rings of the adjacent unit, as was shown before in the complex [Au(C_6F_5)PhP(O)H] [55].

When the reaction of [Au(C_6F_5)(tht)] is carried out with C_6F_4NCSPPh$_2$ the complex [56] shown in Figure 3.5 was obtained and its structure was determined by X-ray diffraction. The benzothiazole ligands are known to display photophysical properties with an improvement of the photostability and emission intensity in conjugated systems as well as in the complexes that contain them, and the presence of

Figure 3.5 Complex obtained by reaction of [Au(C$_6$F$_5$)(tht)] with C$_6$F$_4$NCSPPh$_2$.

Au(I) in such systems should enhance the phosphorescence. The complex [Au(C$_6$F$_5$)(C$_6$F$_4$NCSPPh$_2$)] is luminescent in the solid state at room temperature and at 77 K also in glass media at 77 K but neither in deoxygenated dichloromethane nor in acetonitrile solutions.

The reaction of the trimethyl tritiophosphite and triphenyl tritiophosphite with the gold derivative [Au(C$_6$F$_5$)(tht)] leads to the gold(I) complexes [57] shown in Figure 3.6. The crystal structure of the trimethyl tritiophosphite gold (I) complex was studied by X-ray diffraction and two different polymorphs were discovered.

The reaction of the gold(I) complex [Au(C$_6$F$_5$)(tht)] with the water soluble phosphine 1,3,5-triaza-7-phosphaadamantane (PTA) gives the complex [Au(C$_6$F$_5$)PTA] [28].

Some [Au(C$_6$F$_5$)(L)] complexes have been tested as catalysts or co-catalysts in some organic processes. For example, [Au(C$_6$F$_5$)(TPPXS)] where TPPXS is triphenylphosphine mono-, di- or trisulfonated, was tested as a catalyst in the hydration of phenylacetylene. The [Au(C$_6$F$_5$)(TPPXS)] complexes were synthesized by reacting [Au(C$_6$F$_5$)(tht)] with the appropriate phosphine in equimolecular amounts. It was shown that the pentafluorophenyl complexes displayed similar conversions and turnovers to the chloro analogs, but at lower catalyst concentration [Au(C$_6$F$_5$)(TPPTS)] displayed poorer results than [AuCl(TPPTS)] [58].

3.2.1.4 Neutral Pentafluorophenylgold(I) Derivatives with S-Donor Ligands

The complex [Au(C$_6$F$_5$)(S=CSCH$_2$CH$_2$S)] [60] was obtained in a very low yield (20%) by a simple substitution reaction of the tht group of the [Au(C$_6$F$_5$)(tht)] by S=CSCH$_2$CH$_2$S. In the ^{1}H NMR spectrum a singlet assigned to the CH$_2$CH$_2$ protons is seen at 4.22 ppm and the ^{13}C NMR spectrum shows a singlet at 49.6 ppm from the CH$_2$ carbons of the thione ligand, and multiplets in the region of 137–154 ppm for the pentafluorophenyl group but no signal was observed for the C=S group.

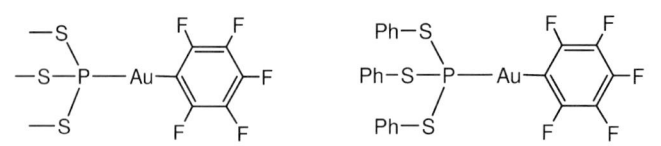

Figure 3.6 Complexes obtained by reaction of [Au(C$_6$F$_5$)(tht)] with trimethyl tritiophosphite and triphenyl tritiophosphite.

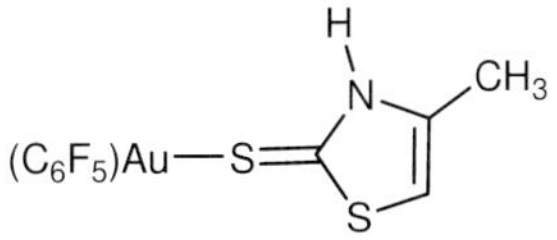

Figure 3.7 Complex obtained by reaction of [Au(C$_6$F$_5$)(tht)] with 4-methyl-3H-thiazole-2-thione.

Neutral Au(I) pentafluorophenyl complexes with heterocyclic thiones can be prepared by displacement of the tht ligand in [Au(C$_6$F$_5$)(tht)] with a thione [49] as shown in Equation 3.6.

$$[Au(C_6F_5)(tht)] + HL \rightarrow [Au(C_6F_5)(HL)] + tht$$
$$HL = C_3H_5NS_2, C_4H_4N_2S, C_5H_5NS, C_7H_5NS_2, C_7H_6N_2S. \quad (3.6)$$

The IR spectra of these complexes shows no band due to the SH absorption and two bands, the first at approximately 3300 ($\nu_{(NH)}$) and the second one at 1620 cm^{-1} ($\delta_{(NH)}$).

The reaction of [Au(C$_6$F$_5$)(tht)] with 4-methyl-3H-thiazole-2-thione leads to the thione complex shown in Figure 3.7 [73].

The ^{1}H and ^{13}C NMR spectrum of the thione complex does not provide any clue to identify the coordinated heteroatom although the signal of the NCS-carbon is shifted upfield in the complex. The structure of the complex was studied by X-ray diffraction, confirming the almost linear coordination around the gold atom and showing that the ligands are almost co-planar.

The reaction of methyl dithiocarbamates with [Au(C$_6$F$_5$)(tht)] leads to the formation of the neutral complexes [Au(C$_6$F$_5$)SC(SMe)NHR] [61], R = p-MeC$_6$H$_4$, o-MeC$_6$H$_4$, p-MeOC$_6$H$_4$ or 3,5-Me$_2$C$_6$H$_3$. The ^{1}H NMR shows that the dithiocarbamate is S-coordinated and the IR spectra shows the absorption of the $\nu_{(NH)}$.

The reactivity of these complexes with amines, that are typical nucleophilic agents, was studied and the formation of the neutral complexes [Au(C$_6$F$_5$)SC(NHR')NHR], R = p-MeC$_6$H$_4$ R' = n-Bu, Cy, CH$_2$CH$_2$C$_6$H$_5$; R = p-MeOC$_6$H$_4$, R' = n-Bu was achieved although in a low yield. The ^{1}H NMR spectra show the resonances due to the R'NH group; two different NH resonances are noticed in every product: one from each amine group. In the IR spectra two different NH absorptions are shown. With primary amines there was no reaction [61].

3.2.2
Anionic Pentafluorophenylgold(I) Derivatives

The gold(I) complex [Au(C$_6$F$_5$)(tht)] has been used as starting material to synthesize numerous anionic Au(I) pentafluorophenyl derivatives as shown in Table 3.3.

The complex PPN[Au(C$_6$F$_5$)Cl] was used as a catalyst in cycloaddition reactions [76] and, although the ratio endo/exo was practically unchanged, the reaction time was reduced from 96 h to 72 h (using 10 mol% of catalyst).

The anionic complex K[Au(C$_6$F$_5$)(CN)] reacts with BzPh$_3$PClO$_4$ to give the complex BzPh$_3$P[Au(C$_6$F$_5$)(CN)] that is easily crystallizable thanks to the bulky cation [63]. This complex reacts with different substances such as [Au(C$_6$F$_5$)(tht)] that lead to the

Table 3.3 Anionic pentafluorophenylgold(I) complexes.

Reagent	Complex	Ref
[PPN]Cl	PPN[Au(C_6F_5)Cl]	[74, 75]
[PPN]I	PPN[Au(C_6F_5)I]	[74]
[PPN]SCN	PPN[Au(C_6F_5)SCN]	[74]
[PPN]N_3	PPN[Au(C_6F_5)N_3]	[74]
[NBu_4]I	NBu_4[Au(C_6F_5)I]	[74]
[NBu_4]CN	NBu_4[Au(C_6F_5)CN]	[74]
KCN	K[Au(C_6F_5)(CN)]	[63]
HCl + NEt_4Cl	[NEt_4][Au(C_6F_5)Cl]	[39, 40]
HBr + NBu_4Br	[NBu_4][Au(C_6F_5)Br]	[39, 40]
Me_4NCl	NMe_4[Au(C_6F_5)Cl]	[75]
LiC_6F_5 + NBu_4Br	[NBu_4][Au(C_6F_5)$_2$]	[40]
o-phenylenebis(dimethylarsine) (pdma)	[Au(pdma)$_2$][Au(C_6F_5)$_2$]	[45]
$SbPh_3$	[Au($SbPh_3$)$_4$][Au(C_6F_5)$_2$]	[45]
(CH_3PPh_3)Br	[CH_3PPh_3][AuBr(C_6F_5)]	[71]
($MeCH_2PPh_3$)Br	[$MeCH_2PPh_3$][AuBr(C_6F_5)]	[71]
($EtCH_2PPh_3$)Br	[$EtCH_2PPh_3$][AuBr(C_6F_5)]	[71]
($PhCH_2PPh_3$)Cl	[$PhCH_2PPh_3$][Au(C_6F_5)Cl]	[71]
(CH_3AsPh_3)I	[CH_3AsPh_3][Au(C_6F_5)I]	[71]
$LiC_6F_3H_2$ + [NBu_4]	NBu_4[Au(C_6F_5)($C_6F_3H_2$)]	
[Au(C_6F_5)$_3$(μ-PPhH)Au(PPh$_3$)] + PPN(acac)	PPN[Au(C_6F_5)$_3$(μ_3-PPh){Au(PPh$_3$)}{Au(C_6F_5)}]	[51]
PPN[{Au(C_6F_5)$_3$}$_2$(μ-PPhH)] + PPN(acac)	(PPN)$_2$[{Au(C_6F_5)$_3$}$_2$(μ_3-PPh){Au(C_6F_5)}]	[51]

production of BzPh$_3$P[Au(C_6F_5)CNAu(C_6F_5)] [63]. The last complex can also be obtained by reaction of BzPh$_3$P[Au(C_6F_5)(CN)] with a silver salt such as AgClO$_4$. The complex BzPh$_3$P[Au(C_6F_5)CN] reacts with triethylamine and an acid, for example HCl or HBF$_4$, and leads to formation of the complex [NHEt$_3$][Au(C_6F_5)CN] [63].

The electrochemistry of the complex [NBu$_4$][Au(C_6F_5)$_2$] shows that the reduction peak is <−2.6 V [77].

The structure of the complex [Au(pdma)$_2$][Au(C_6F_5)$_2$] was studied by X-ray crystallography and was the first full X-ray analysis of a four-coordinate Au(I) complex and also the first of an [AuR$_2$] anion where R is an organic group.

The complex [Au(SbPh$_3$)$_4$][Au(C_6F_5)$_2$] [45] can be obtained in two different ways, the first is the reaction between [Au(C_6F_5)(tht)] and SbPh$_3$ and the second synthesis can be via metathesis of NBu$_4$[Au(C_6F_5)$_2$] with [Au(SbPh$_3$)$_4$][ClO$_4$].

The crystal structure of the complex [PhCH$_2$PPh$_3$][Au(C_6F_5)Cl] has been determined by X-ray diffraction; it consists of an anion-cation pair where the anion shows a two coordinate Au(I) atom bonded to a Cl atom and a C_6F_5 ring and the coordination is almost linear (molecules with two different angles were found but the differences between the two molecules in the asymmetric unit are not significant: 177.8(5) and 179.8(4) Å) [78].

Complexes of general formulae $[RCH_2EPh_3][Au(C_6F_5)X]$ E = P, As; R = H, Me, Et or Ph; X = Cl, Br or I [71] can be used as precursors for the synthesis of gold ylide derivatives via deprotonation of the RCH_2 group of the cation and elimination of the halide attached to the metal center (Equation 3.7).

$$[RCH_2EPh_3][Au(C_6F_5)X] + NaH \rightarrow [Au(C_6F_5)(CHREPh_3)] + NaX + H_2$$
$$E = P; R = H, Me, Et \text{ or } Ph; E = As; R = H [71].$$

(3.7)

The ylide derivatives can also be obtained by using TlC_5H_5 instead of NaH, but this reaction does not work when E = As. When using the first method the yellow color characteristic of the free ylide ($RCH=PPh_3$) can only be observed when R = Ph, E = P; in the other cases, no change in color has been observed even when the reaction takes place. When the starting material is TlC_5H_5 no free ylide can be perceived; in these cases, except when R = Ph, a little decomposition to metallic gold and a small amount of $[Au(C_6F_5)_2]^-$ can be detected by IR spectroscopy of the crude mixture. It is possible to evoke two different mechanisms: the first one the attack upon the cation under formation of ylide and subsequent substitution of the halide linked to the gold atom, and the second one the substitution of the halogen of the organoaurate and the interaction with the cation under hydrogen abstraction leads to the formation of the gold ylide derivative. The second mechanism seems to be preferred in all cases, except for R = Ph where the first mechanism is significant.

The reaction between $[Au(C_6F_5)(tht)]$, $LiC_6F_3H_2$ and $[NBu_4]Br$ leads to the formation of $NBu_4[Au(C_6F_5)(C_6F_3H_2)]$ but this is not a general process because the reaction with LiC_6H_5 and NaC_5H_5 leads to a partial decomposition to metallic gold and to $NBu_4[Au(C_6F_5)_2]$ and $NBu_4[Au(C_6F_5)Br]$ as the only isolable products.

The alkynyl gold(I) complex $PPN[Au(C_6F_5)(C\equiv CH)]$ [79] can be obtained by reaction of Tl(acac) with $PPN[Au(C_6F_5)Cl]$ followed by bubbling of C_2H_2 through the mixture. The IR spectrum shows the typical signals for the pentafluorophenyl group bound to Au(I) as well as the signals corresponding to the $v_{(\equiv CH)}$ at 3268 and the $v_{(C\equiv C)}$ at 1970 cm^{-1}. The $^{13}C\{^1H\}$ NMR spectra shows the signals of the $\equiv CH$ proton at 65.77 and $\equiv CAu$ at 88.22 ppm. Treatment of this complex with bases to obtain the dinuclear complex leads to formation of mixtures of products. This complex was one of the first ethynylgold complexes characterized by X-ray diffraction; the gold atom is in a quasi linear environment (176.9(2)°) and the Au−C≡C angle is also slightly bent (176.9(2)°).

As we will show later in this chapter, the most important anionic derivative, judged by its stability and reactivity is $NBu_4[Au(C_6F_5)_2]$, prepared by a one pot synthesis from $[AuCl(tht)]$, LiC_6F_5 and $[NBu_4]Br$. It is a very nice reaction which affords beautiful colorless crystals in very good yield. The electrochemistry of the complex $NBu_4 [Au(C_6F_5)_2]$ shows that the reduction peak is <-2.6 V [77]

3.2.2.1 Reactivity of the Anionic Pentafluorophenylgold(I) Complexes Q[Au(C$_6$F$_5$)X]

When the anionic complexes $Q[Au(C_6F_5)X]$ react with $AgClO_4$ complexes of the type $Q[(C_6F_5)AuXAu(C_6F_5)]$ where Q = Ph_3BzP, PPN and X = SCN, N_3 are obtained [80].

The reaction of the anionic gold(I) complex PPN[Au(C$_6$F$_5$)Cl] with [MeSi{Me$_2$SiN(p-tol)}$_3$SnLi(OEt$_2$)] leads to formation of an anionic complex with a Sn—Au bond, shown in Equation 3.8, whose ^{31}P{^{1}H}NMR spectrum shows a singlet at 47.5 ppm [81].

$$\text{(3.8)}$$

The reaction of the anionic gold(I) complex NBu$_4$[Au(C$_6$F$_5$)Br] with Na(SR) (SR 4-oxo-2-thioxopyrimidines) gives NBu$_4$[Au(C$_6$F$_5$)(SR)] [82]. Complexes Q[Au(C$_6$F$_5$)Cl] react with H$_2$S to form complexes of the type Q[Au(C$_6$F$_5$)SH] or Q$_2$[{Au(C$_6$F$_5$)}$_3$(μ$_3$-S)] in which the chlorine atom is replaced by the SH group, depending on the reaction conditions and also on the cation used [75]. Thus while complexes containing PPN or NBu$_4$ give the hydrosulfido complexes, the complex where Q = NMe$_4$ always forms the sulfido complex, but when Q = NEt$_4$ the complex obtained depends on the reaction conditions. Although Q[Au(C$_6$F$_5$)SH] and Q$_2$[{Au(C$_6$F$_5$)}$_3$(μ$_3$-S)] smell of H$_2$S, the PPN salts are stable and can be stored at room temperature for several weeks; however, the NBu$_4$[Au(C$_6$F$_5$)SH] complex must be kept at −20 °C otherwise it decomposes and NEt$_4$[Au(C$_6$F$_5$)SH] decomposes in a few days even when stored at −20 °C. The ^{19}F{^{1}H} NMR spectrum of the decomposed product indicates the presence of NEt$_4$[{Au(C$_6$F$_5$)}$_3$(μ$_3$-S)]. This may be due to the stability of the Q[Au(C$_6$F$_5$)SH] complexes which decreases with the size of the cation. The crystal structure of NEt$_4$[{Au(C$_6$F$_5$)}$_3$(μ$_3$-S)]) was studied by X-ray diffraction. The environment around the sulfur atom is a distorted trigonal pyramid with short Au⋯Au contacts, which means narrow Au—S—Au angles. The aurophilic interactions are similar to those of the analogous complex containing PPh$_3$ instead of pentafluorophenyl groups, and the gold atoms are in linear environments (176.6(4), 177.0(3) and 177.7(3)°).

The reaction of 4,4'-diphenyltetrathiafulvalene with the gold(I) derivatives NBu$_4$[Au(C$_6$F$_5$)$_2$] and BzPPh$_3$[Au(C$_6$F$_5$)Cl] leads to formation of the anionic complexes (TTFPh$_2$)$_2$[Au(C$_6$F$_5$)$_2$] and (TTF)[Au(C$_6$F$_5$)Cl] [83], the crystal structure of the former has been studied by X-ray diffraction. The results show that the anion is linear at the gold atom and that each gold is involved in four short Au⋯S contacts to four different TTFPh$_2$ units of two different stacks. TTFPh$_2$ stacks are formed parallel to the x-axis, the two independent cations alternate and are not parallel to each other. The anions are arranged in the interstices with the C—Au—C axis approximately perpendicular to the stack. The conductivity of the complex at room-temperature is σ = 150 Sm^{-1} and the dependence of the conductivity with the temperature is typical of semiconductor behavior.

3.2.3
Cationic Pentafluorophenylgold(I) Derivatives

The gold(I) complex [Au(C_6F_5)(tht)] reacts with [$Ph_2PCH_2PPh_2Me$]ClO_4 to give the compound [Au(C_6F_5)($Ph_2PCH_2PPh_2Me$)]ClO_4 [84] whose IR spectrum displays the bands due to the anion at 1100 and 625 cm^{-1} and also two bands at 950 and 795 cm^{-1} due to the pentafluorophenyl group linked to the gold(I) center. When the neutral gold(I) complex is treated with [$MePh_2PCHPPh_2Me$]CF_3SO_3 the ylide complex [Au{CH(PPh_2Me)$_2$}(C_6F_5)]CF_3SO_3 [85] is obtained. In the IR and in the $^{19}F\{^1H\}$ NMR spectra the pentafluorophenyl group bonded to the Au(I) is seen and in the $^{31}P\{^1H\}$ NMR spectrum a singlet at 22.7 ppm is observed. In the ^1H NMR spectrum a doublet at 2.22 ppm for the methyl group and a triplet at 4.28 ppm due to the methylide bridge are observed in addition to the signals of the phenyl groups. The structure of this complex was studied by X-ray diffraction. The gold atom is in an almost linear environment and it is bonded to two kinds of carbon atoms: one of a pentafluorophenyl group (2.029(6) Å) similar to the distances found in complexes where the Au-C_6F_5 is *trans* to the ylidic carbon) and the other from an ylidic ligand (2.112(6) Å): the two Au—C distances are very different. There are no Au—Au interactions because of the bulky ligand.

The complex [Au(C_6F_5)($Ph_2PCH_2PPh_2Me$)] can be treated with acids such as HCl or HBF_4 to give the cationic complexes [Au(C_6F_5)($Ph_2PCH_2PPh_2Me$)]X (X = Cl, BF_4) which behave similarly to the compound [$Ph_2PCH_2PPh_2Me$]ClO_4. When the reaction is carried out with the same starting material and [AuCl(tht)], apart from displacement of the tht ligand, a rearrangement takes place and a cationic ligand is obtained. This ligand can also be obtained by reaction with [Au(tht)$_2$]ClO_4 instead of [AuCl(tht)] (Equation 3.9)

$$[Au(C_6F_5)(Ph_2PCH_2PPh_2Me)] + [AuCl(tht)] \rightarrow \begin{array}{c} MePh_2\text{-}CH\text{-}PPh_2\text{-}Au\text{-}C_6F_5 \\ | \\ Au \\ | \\ MePh_2\text{-}CH\text{-}PPh_2\text{-}Au\text{-}C_6F_5 \end{array} \quad [AuCl_2] \quad (3.9)$$

By treating [Au(C_6F_5)($Ph_2PCH_2PPh_2Me$)]ClO_4 with [Au(acac)(PPh_3)] in a molar ratio 1:2, the methylene protons are substituted and the trinuclear complex [Au(C_6F_5){(PPh_2C(Au(PPh_3))$_2PPh_2Me$}]ClO_4 [86] is obtained. This complex shows three resonances in the $^{31}P\{^1H\}$ NMR spectrum with relative intensities 2:1:1 (37.6 ppm PPh_3; 44.9 ppm $AuPPh_2$ and 23.7 ppm PPh_2Me). The $^{19}F\{^1H\}$ NMR spectrum displays the typical pattern for the pentafluorophenyl groups bonded to Au(I).

The neutral complex [Au(C_6F_5)($Ph_2PCHPPh_2Me$)] [84] is obtained by reacting the cationic Au(I) complex [Au(C_6F_5)($Ph_2PCH_2PPh_2Me$)]ClO_4 with NaH. The ^{31}P NMR spectrum shows a multiplet at 25.0 ppm that corresponds to the P atom coordinated to the gold atom, the other P atom gives a signal at 17.5 ppm, $J_{(PP)}$ is 70.9 Hz. In the crystal structure of this complex the gold(I) center has an approximately linear coordination, with an angle P—Au—C = 174.8(2)°, the distances and angles between the carbon atom and the phosphorus in the phosphine ligand are similar to the angles and distances found in complexes containing the bidentate bis(diphenylphosphino)methanide.

3.2.4
Di and Polynuclear Pentafluorophenylgold(I) Derivatives

Two different strategies can be used for the synthesis of di- and trinuclear pentafluorophenylgold(I) derivatives using [Au(C$_6$F$_5$)(tht)] as one of the starting materials.

(a) Reaction of [Au(C$_6$F$_5$)(tht)] with potential polydentate ligands in the appropriate ratio to form the corresponding polinuclear derivatives.

(b) Reaction of [Au(C$_6$F$_5$)(tht)] with a gold complex containing either a coordination site in the molecule (a vacant donor atom) or a center that can be transformed into a donor site through a simple reaction.

We present examples of both methods in Tables 3.4 and 3.5, respectively. characterization of the new complexes by X-ray diffraction studies shows many examples of short intramolecular Au−Au contacts, in addition to intermolecular Au−Au contacts as observed in mononuclear pentafluorophenylgold(I) derivatives. ^{19}F{^{1}H}NMR spectroscopy is a very good tool to obtain data about the structures. As there are not many F atoms, the NMR spectra are clean and usually three groups of resonances are observed for one C$_6$F$_5$ group. Thus, the presence of six resonances, or better two *p*-F signals, means that two different C$_6$F$_5$ groups are present in the molecule. Occasionally, there are five signals for one C$_6$F$_5$ group, which indicates that free rotation about the *ipso* carbon is not possible and the *m*-F and *o*-F are not equivalent. In addition, a phosphorous atom *trans* to a C$_6$F$_5$ group should lead to broadening of the signal due to coupling with *o*-F.

The synthesis of di- or tri-nuclear pentafluorophenyl derivatives by the use of polydentate ligands with [Au(C$_6$F$_5$)(tht)] is quite a general procedure as shown in Table 3.4. Many of these complexes have been characterized by X-ray diffraction and show short Au···Au contacts as in [(C$_6$F$_5$)Au(SdppmS)Au(C$_6$F$_5$)] [88] of 3.163(1) Å or in the dinuclear complex [57] shown in Figure 3.8 of 3.4671(9) Å.

One example of the aforementioned ^{19}F{^{1}H}NMR spectra with five fluorine resonances for one C$_6$F$_5$ group is the complex [(C$_6$F$_5$)Au(μ-S$_2$CPR$_3$)Au(C$_6$F$_5$)] [89] (R$_3$ = (C$_6$H$_{11}$)$_3$, Et$_3$, Bu$_3$, PhMe$_2$ or PhEt$_2$). The rotation about the *ipso* carbon is hindered and so the five F atoms are inequivalent.

When (PPh$_2$CH$_2$)$_3$CCH$_3$ reacts with [Au(C$_6$F$_5$)(tht)] two possible products can be obtained depending on the molar ratio of the starting materials, if a ratio 1 : 3 is used the trinuclear complex [{Au(C$_6$F$_5$)}$_3${(PPh$_2$CH$_2$)$_3$CCH$_3$}] is obtained, but when the ratio is 1 : 2 only two phosphorus atoms are coordinated to the gold atoms and the third one is oxidized [{Au(C$_6$F$_5$)}$_2${(PPh$_2$CH$_2$)$_2$C(CH$_3$)CH$_2$PPh$_2$O}], even when working under inert atmosphere [91].

Another property that should be present in this type of complex is photoluminescence; in recent decades this has attracted the attention of many gold researchers. It may well be possible, that many older complexes might also be photoluminescent but have simply not been studied. [{Au(C$_6$F$_5$)}$_2${(P^{i}Pr$_2$)$_2$CH$_2$}] [93] in the solid state at room temperature shows an excitation maximum at 335 nm and an emission

Table 3.4 Di and trinuclear pentafluorophenylgold(I) derivatives using as starting materials polydentate ligands.

Ligand	Polynuclear complex	Ref
4,4′-bipy	[(C_6F_5)Au(4,4′-bipy)Au(C_6F_5)]	[46]
CH(PPh$_2$)$_3$	[CH{PPh$_2$Au(C_6F_5)}$_3$]	[59]
bis(diphenylarsino)methane (dpam)	[(C_6F_5)Au(μ-dpam)Au(C_6F_5)]	[29]
bis(diphenyphosphino)amine (dppa)	[(C_6F_5)Au(μ-dppa)Au(C_6F_5)]	[87]
bis(diphenyphosphino)methane disulfide (dppmS$_2$)	[(C_6F_5)Au(μ-SdppmS)Au(C_6F_5)]	[88]
S$_2$CP(C$_6$H$_{11}$)$_3$	[(C_6F_5)Au(μ-S$_2$CP(C$_6$H$_{11}$)$_3$)Au(C_6F_5)]	[89]
S$_2$CPEt$_3$	[(C_6F_5)Au(μ-S$_2$CPEt$_3$)Au(C_6F_5)]	[89]
S$_2$CPBu$_3$	[(C_6F_5)Au(μ-S$_2$CPBu$_3$)Au(C_6F_5)]	[89]
S$_2$CPPhMe$_2$	[(C_6F_5)Au(μ-S$_2$CPPhMe$_2$)Au(C_6F_5)]	[89]
S$_2$CPPhEt$_2$	[(C_6F_5)Au(μ-S$_2$CPPhEt$_2$)Au(C_6F_5)]	[89]
bis(diphenylphosphino)ferrocene (dppf)	[Au$_2$(C_6F_5)$_2$(dppt)]	[90]
(PPh$_2$CH$_2$)$_3$CCH$_3$ ratio 1:3	[{Au(C_6F_5)}$_3${(PPh$_2$CH$_2$)$_3$CCH$_3$}]	[91]
(PPh$_2$CH$_2$)$_3$CCH$_3$ ratio 1:2	[{Au(C_6F_5)}$_2${(PPh$_2$CH$_2$)$_2$C(CH$_3$)CH$_2$PPh$_2$O}]	[91]
(PPh$_2$)$_2$C=CH$_2$	[{Au(C_6F_5)}$_2${(PPh$_2$)$_2$C=CH$_2$}]	[92]
(P^{i}Pr$_2$)$_2$CH$_2$	[{Au(C_6F_5)}$_2${(P^{i}Pr$_2$)$_2$CH$_2$}]	[93]
Ph$_2$PC≡CPPh$_2$ (dppa)	[{Au(C_6F_5)}$_2$(dppa)$_2$]	[94]
Ph$_2$PCH$_2$PPhCH$_2$PPh$_2$ (dpmp)	[{Au(C_6F_5)}$_3$(μ-dpmp)]	[95]
3,4-(NHPPh$_2$)$_2$MeC$_6$H$_3$	[MeC$_6$H$_3$(NHPPh$_2$AuC$_6$F$_5$)$_2$]	[96]
1,2-(NHPPh$_2$)$_2$C$_6$H$_4$	[C$_6$H$_4$(NHPPh$_2$AuC$_6$F$_5$)$_2$]	[96]
Ph$_2$PCH$_2$CH$_2$PPhCH$_2$CH$_2$PPh$_2$ (triphos)	[Au(C_6F_5)}$_3$(μ-triphos)]	[97]
Py$_2$NH	[Au$_2$(C_6F_5)$_2$(μ-Py$_2$NH)]	[98]
1,2-bis(1,3,2-dithiaphospholan-2-ylthio)ethane		[57]
		[99]
bis(3,5-dimethylpyrazol-1-yl)methane		

maximum at 445 nm and when the temperature drops to 77 K the excitation maximum is at 327 nm and the emission maximum at 440 nm.

The crystal structure of the dinuclear complex [(C_6F_5)Au(Ph$_2$PC≡CPPh$_2$)Au(C_6F_5)] [94] was also studied and it shows that there are discrete molecules without

Table 3.5 Di and trinuclear pentafluorophenylgold(I) derivatives using gold complexes as starting materials.

Starting complex	Polynuclear complex	Ref
[AuC$_3$H$_4$NS$_2$(PPh$_3$)]	[(C$_6$F$_5$)AuC$_3$H$_4$NS$_2$Au(PPh$_3$)]	[49]
[AuC$_5$H$_4$NS(PPh$_3$)]	[(C$_6$F$_5$)AuC$_5$H$_4$NSAu(PPh$_3$)]	[49]
[SCNAuPPh$_3$]	[(C$_6$F$_5$)AuSCNAuPPh$_3$]	[80]
[N$_3$AuPPh$_3$]	[(C$_6$F$_5$)AuN$_3$AuPPh$_3$]	[80]
[AuCl$_x$(C$_6$F$_5$)$_{3-x}$(Ph$_2$PCH$_2$PPh$_2$Me)] ($x=0, 1$)	[Cl$_x$(C$_6$F$_5$)$_{3-x}$Au{Ph$_2$PCH$_2$PPh$_2$Me}Au(C$_6$F$_5$)] ($x=0, 1$)	[84]
Ph$_3$BzP[(C$_6$F$_5$)AuSCN]	[Ph$_3$BzP][(C$_6$F$_5$)AuSCNAu(C$_6$F$_5$)]	[80]
Ph$_3$BzP[(C$_6$F$_5$)AuN$_3$]	[Ph$_3$BzP][(C$_6$F$_5$)AuN$_3$Au(C$_6$F$_5$)]	[80]
PPN[(C$_6$F$_5$)AuSCN]	PPN[(C$_6$F$_5$)AuSCNAu(C$_6$F$_5$)]	[8]
[S(Au$_2$dppf)]	[S(AuC$_6$F$_5$)(Au$_2$dppf)]	[100]
[Au{(PPh$_2$)$_2$C$_2$B$_9$H$_{10}$}(PPh$_3$)]	[Au$_2${(PPh$_2$)$_2$C$_2$B$_9$H$_{10}$}(C$_6$F$_5$)(PPh$_3$)]	[101]
[Au$_3$(S$_2$CNMe$_2$)(μ-S$_2$CNMe$_2$){μ$_3$-(PPh$_2$)$_3$CH}]ClO$_4$	[Au$_3$(μ$_3$-{(PPh$_2$)$_3$CH})(μ-S$_2$CNMe$_2$)(C$_6$F$_5$)]ClO$_4$ (structural formula shown)	[102], [103]
[Au$_3$(μ-dpmp)(S$_2$CNMe$_2$)$_2$Cl]	Trinuclear structure with Ph$_2$P–PPh–PPh$_2$ bridges, Au–Au–Au, S,S-CNMe$_2$ bridge and C$_6$F$_5$	[103]
Trinuclear complex with Ph$_2$P–PPh–PPh$_2$, Au–Cl–Au–Au, S,S-CNMe$_2$, C$_6$F$_5$	Cationic trinuclear complex [Ph$_2$P–PPh–PPh$_2$/Au–Au–Au/S,S-CNMe$_2$/C$_6$F$_5$][CF$_3$SO$_3$]	[104]
Dinuclear Au complex: Cl, Cl–Au–Au–N(Pr^i)H with ortho-phenylene-PPh$_2$	Analogous complex with C$_6$F$_5$ replacing one Cl	
[(C$_6$F$_5$)AuNH$_2$(CH$_2$)$_2$NH$_2$]	[(C$_6$F$_5$)AuNH$_2$(CH$_2$)$_2$NH$_2$AuC$_6$F$_5$]	[46]
[(C$_6$F$_5$)AuNH$_2$(CH$_2$)$_3$NH$_2$]	[(C$_6$F$_5$)AuNH$_2$(CH$_2$)$_3$NH$_2$AuC$_6$F$_5$]	[46]

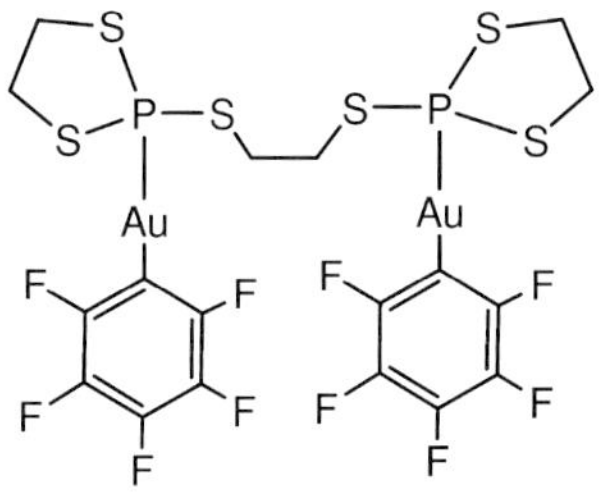

Figure 3.8 Complex whose structure has been studied by X-ray diffraction.

short Au···Au intramolecular contacts. The complex emits weakly at room temperature but has an intense emission at 77 K (room temperature: λ_{exc} 370 nm and λ_{emis} 516 nm; 77 K: λ_{exc} 355 nm and λ_{emis} 529 nm).

The trinuclear complex [Au$_3$(μ-triphos)(C$_6$F$_5$)$_3$] [97] in dichloromethane shows an absorption around 270 nm and in the solid state at room temperature the complex does not emit, even using an excitation frequency below 300 nm. At lower temperature (77 K) the complex emits with a maximum at 450 nm. Thus luminescence properties can be dramatically influenced by the pentafluorophenyl group which indicates its important contribution to the energy levels involved in the electronic transitions.

The complex shown in Figure 3.9 [104] is luminescent in the solid state at 77 K with three emission maxima at 431, 448 and 460 nm. The excitation maxima are at 305 and 370 nm. The origin of the luminescence has been attributed to intraligand transitions with contributions of charge transfer character.

In addition to the simple reaction of a polydentate ligand with [Au(C$_6$F$_5$)(tht)] the other versatile synthesis of di- and trinuclear derivatives is the use of gold complexes as ligands.

The anionic complex Ph$_3$BzP[Au(C$_6$F$_5$)X] reacts with the Au(I) derivative [Au(C$_6$F$_5$)(tht)] to give the dinuclear complex [Ph$_3$BzP][(C$_6$F$_5$)AuXAu(C$_6$F$_5$)] X=SCN or N$_3$ [80]. This reaction gives very low yields of the binuclear complexes because the reaction does not go to completion and the resulting product is always contaminated with the starting materials. They can however be separated because they have different solubilities. The similar anionic complex PPN[Au(C$_6$F$_5$)SCN] also reacts with [Au(C$_6$F$_5$)(tht)] forming the complex PPN[(C$_6$F$_5$)AuSCNAu(C$_6$F$_5$)] [80].

The excess of electron density on the methanide carbon of complex [Au(C$_6$F$_5$)(Ph$_2$PCHPPh$_2$Me)] [84] causes it to act as a C-donor nucleophile, producing a

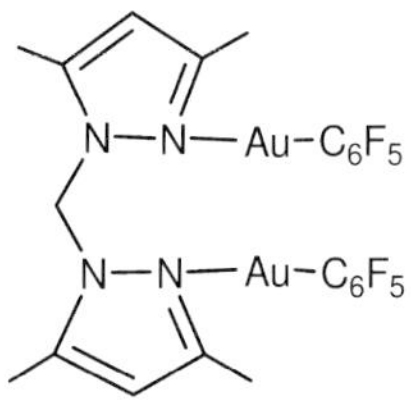

Figure 3.9 Luminescent dinuclear gold(I) complex.

Figure 3.10 Trinuclear complex [(Au$_2$S$_2$CNR$_2$)]{(PPh$_2$)$_3$CH}Au(C$_6$F$_5$)].

binuclear complex [(C$_6$F$_5$)Au{Ph$_2$PCHPPh$_2$Me}Au(C$_6$F$_5$)]. The crystal structure of this complex was determined by X-ray diffraction and it shows the expected linear geometry for the gold centers, but this time the geometry of the bidentate methanide ligand is significantly different to the starting material; this time P—C bonds are much longer (1.790(10), 1.801(11) Å) and P—C—P angle much smaller (114.9(6)°) which is consistent with the electron donation to the second gold atom. This donation reduces the electron delocalization over the methanide ligand.

In the preparation of binuclear complexes, the deprotonated N atom is a potential donor ligand. By adding Na$_2$CO$_3$ to a mixture of NBu$_4$[AuBr(C$_6$F$_5$)] or [AuCl(PPh$_3$)] and [Au(C$_6$F$_5$)(C$_5$H$_5$NS)] the binuclear complexes NBu$_4$[(C$_6$F$_5$)Au(C$_5$H$_5$NS)Au(C$_6$F$_5$)] [49] and [(C$_6$F$_5$)Au(C$_5$H$_5$NS)Au(PPh$_3$)] [49] are obtained.

The reaction between [S(Au$_2$dppf)] (dppf = 1,1'-bis(diphenylphosphino)ferrocene) and [Au(C$_6$F$_5$)(tht)] gives the neutral derivative [S(AuC$_6$F$_5$)(Au$_2$dppf)] [100], the ^{31}P{^{1}H} NMR spectrum of which shows a singlet at 25.3 ppm.

The crystal structure of the neutral complex [Au$_2${(PPh$_2$)$_2$C$_2$B$_9$H$_{10}$}(C$_6$F$_5$)(PPh$_3$)] [101] was studied by X-ray diffraction and it was the first example of a structure where the anionic diphosphine acts as a bridging ligand. The Au(I) atom shows a distorted linear coordination geometry (P—Au—P 164.69(7)° and P—Au—C 166.6(2)°) and it has short Au···Au contacts (2.9885(8) Å).

The trinuclear complex [(Au$_2$S$_2$CNR$_2$)]{(PPh$_2$)$_3$CH}Au(C$_6$F$_5$)] [102] (Figure 3.10) has an IR spectra that confirms the presence of the pentafluorophenyl group bonded to the Au(I) atom. The ^{31}P{^{1}H} NMR spectrum shows a doublet at 43.6 ppm and a triplet at 41.2 ppm while the ^{19}F{^{1}H} NMR spectrum shows the typical pattern for the pentafluorophenyl group bonded to the Au(I) atom.

The two trinuclear complexes [(Au$_2$S$_2$CNR$_2$)(PPh$_2$CH$_2$PPh$_2$CH$_2$PPh$_2$)Au(C$_6$F$_5$)] [103] (Figure 3.11) are synthesized by treatment of [Au$_3$(μ-dpmp)(S$_2$CNMe$_2$)$_2$Cl]

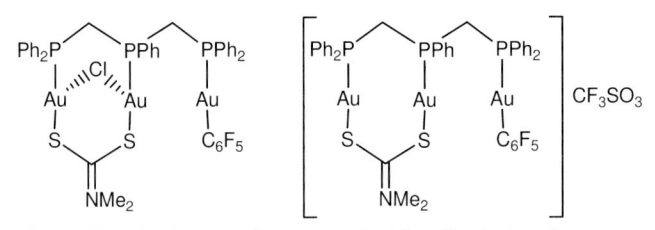

Figure 3.11 Trinuclear complexes synthesized from [Au$_3$(μ-dpmp)(S$_2$CNMe$_2$)$_2$Cl] by treatment with [Au(C$_6$F$_5$)(tht)] and AgCF$_3$SO$_3$.

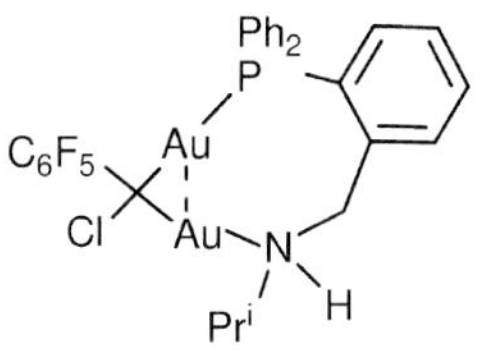

Figure 3.12 Complex [(C_6F_5)Au(PPh$_2C_6H_5$CH$_2$N^{i}PrH)AuCl].

with [Au(C_6F_5)(tht)] and subsequently the neutral complex thus obtained affords the cationic complex upon treatment with AgCF$_3$SO$_3$. Both complexes are luminescent at room temperature and the solid state emission and excitation spectra at 298 and 77 K have been determined.

The reaction to synthesize the complex [(C_6F_5)Au(PPh$_2C_6H_5$CH$_2$N^{i}PrH)AuCl] shown in Figure 3.12 [104] is carried out due to the selective displacement of the N-bound Au—Cl fragment by Au(C_6F_5), leaving the already P-bound AuCl fragment intact. It is the first example of an AuCl fragment being substituted by an organogold(I) fragment and, taking into account that only one of the two AuCl fragments is substituted, the relative P-Au(I) and N-Au(I) bond strengths can be compared. It should be possible to replace the remaining chlorine atom by treatment with Li(C_6F_5). The structure of this complex was studied by X-ray diffraction, and shows intramolecular Au$\cdots$Au interactions of 3.1013(10) Å.

The weakly attached ligand of the starting complexes [Au(C_6F_5)(tht)] or [AuCl(tht)] can be displaced by amine groups in such a way that the nitrogen atom is coordinated to different gold centers [(C_6F_5)AuNH$_2$(CH$_2$)$_n$NH$_2$AuX] ($n = 2, 3$) [46]. The same reaction can be carried out using another Au(I) complex [Au(OClO$_3$)PPh$_3$] as starting material to obtain the dinuclear cationic derivative [(C_6F_5)AuNH$_2$(CH$_2$)$_n$NH$_2$-AuPPh$_3$]ClO$_4$.

There are other methods of obtaining di- and tri-nuclear derivatives that do not use [Au(C_6F_5)(tht)] and which fit into neither of the previously discussed categories. One of them uses [Au(C_6F_5)(SMe$_2$)] [79] but that does not offer better results compared to the tht derivative. The arylation of [ClAu(dppe)AuCl] with (C_6F_5)MgBr affording [(C_6F_5)Au(dppe)Au(C_6F_5)] [105] is one example of how to make the complex the complicated way! The synthesis of [Au(C_6F_5)(tht)] is very easy and displacement of the tht with dppe cleanly affords [(C_6F_5)Au(dppe)Au(C_6F_5)] without the need of a Grignard reagent and hydrolysis procedures.

Of the two methods generally used, one is the use of Q[Au(C_6F_5)X] (Q = large anion such as PBzPh$_3$, PPh$_3$Me, NBu$_4$, PPN; X = Cl, Br, I). Thus, the reaction of the anionic gold(I) derivative Ph$_3$BzP[(C_6F_5)AuX] with AgClO$_4$ leads to formation of the anionic dinuclear complexes [Ph$_3$BzP][(C_6F_5)AuXAu(C_6F_5)] [80] X = SCN or N$_3$. PPh$_3$Me[Au(C_6F_5)Br] reacts with 1,2-(SH)$_2C_2B_{10}H_{10}$ to give the anionic derivative PPh$_3$Me[Au(C_6F_5)$_2$(μ-S$_2C_2B_{10}H_{10}$)] [106]. When the complex [Au(C_6F_5)(PPh$_2$CCH)] reacts with NaOMe/MeOH in the presence of PPN[Au(C_6F_5)Cl] the alkynyl derivative PPN[Au(C_6F_5)(PPh$_2$CC)Au(C_6F_5)] [53] is formed. The reaction of sodium metal with 4,5-bis(benzoylthio)-1,3-dithiole-2-selenone followed by the addition of PPN[Au(C_6F_5)Cl] leads to the dinuclear complex (PPN)$_2$[Au$_2$(dmise)(C_6F_5)$_2$] [107]. The reaction of Na$_2$dmit (dmit = 2-thioxo-1,3-dithiole-4,5-dithiolate) with Q[Au(C$_6$F$_5$)X] (Q = NBu$_4$, PPN; X = Cl, Br) gives the anionic complexes Q$_2$[Au$_2$(μ-dmit)(C$_6$F$_5$)$_2$] [108] which show characteristic bands of dmit and two bands due to the pentafluorophenyl groups at 950 and 970 cm^{-1} in the IR spectrum. When these complexes react with (TTF)$_3$(BF$_4$)$_2$ (TTF = tetrathiafulvalene) the highly insoluble complex (TTF)[Au$_2$(dmit)(C$_6$F$_5$)$_2$] is obtained. Elemental analysis of this complex is the basis for the proposed stoichiometry. The IR spectrum shows a band at 1350 cm^{-1} due to the C$_3$S$_5^-$ group.

Other methods to synthesize this type of complexes involve the use of acetylacetonate gold complexes as deprotonating agent. This is used in the preparation of [Au(C$_6$F$_5$){PPh$_2$CH(AuPPh$_3$)PPh$_2$}Au(C$_6$F$_5$)] [86] from the reaction of [Au(C$_6$F$_5$)(PPh$_2$CH$_2$PPh$_2$)Au(C$_6$F$_5$)] with [Au(acac)(PPh$_3$)] (acac = acetylacetonate). The reaction of [Au(C$_6$F$_5$)(PPh$_2$CCH)] with [Au(acac)PPh$_3$] gives a mixture of [Au(C$_6$F$_5$)PPh$_3$] and [Au(C$_6$F$_5$)(PPh$_2$CC)AuPPh$_3$] [53].

When complexes such as [RC$_6$H$_3$(NHPPh$_2$AuC$_6$F$_5$)$_2$] are reacted with NBu$_4$(acac) the anionic derivatives [NBu$_4$][RC$_6$H$_3${NHPPh$_2$AuC$_6$F$_5$)}{NPPh$_2$AuC$_6$F$_5$)}] [96] (R = Me, H) are obtained.

Complexes [Au$_2$(C$_6$F$_5$)$_2$(μ-2-C$_6$H$_4$PR$_2$)$_2$] (R = Ph and Et) are not thermodynamically stable and when a solution of these complexes (in d^8-toluene) is heated to 90 °C for 48 h the complexes are transformed into [Au$_2$(C$_6$F$_5$)$_2$(μ-2,2′-R$_2$PC$_6$H$_4$C$_6$H$_4$PR$_2$)] [109, 110] (R = Ph and Et) as the main product and [(C$_6$F$_5$)$_2$Au^{III}(μ-2-C$_6$H$_4$PR$_2$)$_2$Au^I] (R = Ph and Et), respectively. The structure of the two main products has been studied by X-ray diffraction, the gold(I) atoms are in a slightly distorted linear arrangement and the Au—C and Au—P distances are in the range of distances for similar complexes. The ^{31}P{^{1}H} NMR spectra shows in both cases a triplet (36.3 ppm for R = Ph and 32.8 ppm for R = Et with J = 6.3 Hz in both cases) and the ^{19}F{^{1}H} NMR spectra show the typical pattern for the pentafluorophenyl group bonded to the Au(I) atom. The Au···Au distances suggest that there are aurophilic interactions in the molecule when R = Ph while for R = Et these interactions are not observed. Orthometallated complexes with AsPh$_2$ show similar behavior and the dinuclear complex shown in Figure 3.13 [111] was obtained and characterized.

The dinuclear gold(I) pentafluorophenyl carbene derivatives [(C$_6$F$_5$)Au{C(NHR)(NH(CH$_2$)$_n$NH$_2$)}AuC$_6$F$_5$] [46] can be obtained by reacting the Au(I) complex [(C$_6$F$_5$)AuNH$_2$(CH$_2$)$_n$NH$_2$] with [(C$_6$F$_5$)AuCNR] (n = 2,3 and R = Ph, Tolyl).

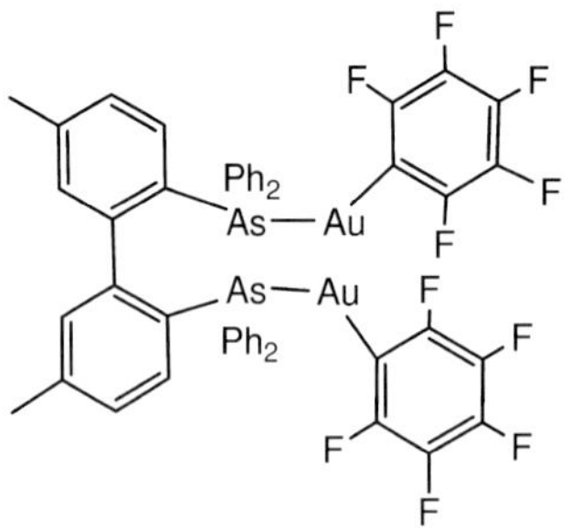

Figure 3.13 Dinuclear complex with AsPh$_2$.

Polynuclear Au(I) pentafluorophenyl complexes can also be obtained by reaction of [Au(C$_6$F$_5$)(tht)] with polydentate ligands or with gold complexes that still have different coordination sites or can be created before the subsequent reaction. The ligands or complexes that react with [Au(C$_6$F$_5$)(tht)] and the complexes obtained are listed in Table 3.6.

During the reaction to obtain the unexpected pentanuclear Au(I) complex [Au$_5$(C$_6$F$_5$){(SPPh$_2$)$_2$C}$_2$PPh$_3$] [112] the only side product identified was [Au(C$_6$F$_5$)(PPh$_3$)] and the yield, based on gold, was high (see Figure 3.14). The structure was confirmed by X-ray diffraction although the precision was poor.

The molecule can be seen as a cluster of three gold atoms bridged by two (Ph$_2$PS)$_2$C^{2-} ligands although distances are longer than found in gold clusters (Au(1)-Au(2) 2.989(6); Au(1)-Au(3) 3.226(7) and Au(2)-Au(3) 3.182(7) Å). The oxidation state of all the gold atoms in this molecule should be +1.

The tetranuclear Au derivative [{Au(C$_6$F$_5$)}$_4$(PPh$_2$)$_4$TTF] [113] shown in Figure 3.15, has a ^{31}P{^{1}H} NMR spectrum showing a singlet at 21.3 ppm. The UV/Vis spectrum

Table 3.6 Polynuclear complexes.

Ligand or starting complex	Complex	Ref
[(SPPh$_2$)$_2$C(AuPPh$_3$)$_2$]	[Au$_5$(C$_6$F$_5$){(SPPh$_2$)$_2$C}$_2$PPh$_3$]	[112]
(PPh$_2$)$_4$TTF (TTF = tetrathiafulvalene)	[{Au(C$_6$F$_5$)}$_4$(PPh$_2$)$_4$TTF]	[113]
Ph$_2$PC(S)N(H)Me	(structure: H-N and H-N groups with Ph$_2$P-S and S-PPh$_2$, bridging Au----Au---Au----Au with C$_6$F$_5$ groups)	[54]
[Au{Ph$_2$PC(S)NMe}]$_2$	(structure: Au–C$_6$F$_5$ with N, Ph$_2$P, S, Au----Au, S, PHPh$_2$, N, C$_6$F$_5$–Au)	[54]
[Ph$_2$PCH(PPh$_2$AuPPh$_2$)$_2$CHPPh$_2$][ClO$_4$]$_2$	[{(C$_6$F$_5$)Au}Ph$_2$PCH(PPh$_2$AuPPh$_2$)$_2$CHPPh$_2${Au(C$_6$F$_5$)}][ClO$_4$]$_2$	[59]
[Ph$_2$PC(PPh$_2$AuPPh$_2$)$_2$CPPh$_2$]	[{(C$_6$F$_5$)Au}Ph$_2$PC(PPh$_2$AuPPh$_2$)$_2$CPPh$_2${Au(C$_6$F$_5$)}]	[59]
[X(Ph$_2$PAuPPh$_2$)$_2$X] X = CH o N	[(C$_6$F$_5$)AuX(Ph$_2$PAuPPh$_2$)$_2$XAu(C$_6$F$_5$)]	[114]

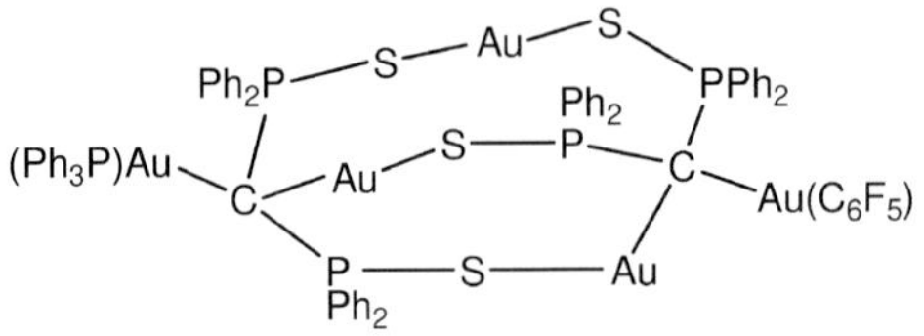

Figure 3.14 Structure of the [Au$_5$(C$_6$F$_5$){(SPPh$_2$)$_2$C}$_2$PPh$_3$] complex.

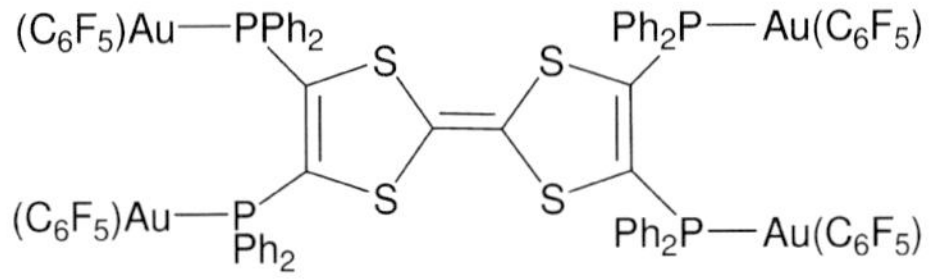

Figure 3.15 Tetranuclear derivative.

shows λ_{max} (nm) (ε, M^{-1}cm^{-1}) at 212(10699), 272 (27577), 325 (14712) and 485(668) and the cyclic voltamogram shows the $E_{1/2}^1 = 0.76$ V and $E_{1/2}^2 = 1.16$ V.

The ^{19}F{^{1}H} NMR spectrum of the complex [54] in Figure 3.16, shows two inequivalent pentafluorophenyl groups meaning that there are two types of magnetically different Au(C$_6$F$_5$) units in the molecule. The ^{31}P{^{1}H} NMR spectrum shows a singlet at 60.0 ppm.

The crystal structure of this complex was determined by X-ray diffraction and the sulfur coordination to the Au atom was confirmed. In this structure all the distances are in the ranges found for similar molecules but inter- and intramolecular Au$\cdots$Au contacts are longer that those in other similar molecules.

A tetranuclear derivative with the same type of ligand is depicted in Figure 3.17 [54]. In the IR spectrum the vibration due to the C=N bond is observed at 1592 cm^{-1}. The ^{31}P{^{1}H} NMR spectrum shows a singlet at 48.0 ppm.

The PPh$_2$ groups of complexes [Ph$_2$PCH(PPh$_2$AuPPh$_2$)$_2$CHPPh$_2$][ClO$_4$]$_2$ and [Ph$_2$PC(PPh$_2$AuPPh$_2$)$_2$CPPh$_2$] can react with gold(I) derivatives such as [Au(C$_6$F$_5$)(tht)] to give the cationic [{(C$_6$F$_5$)Au}Ph$_2$PCH(PPh$_2$AuPPh$_2$)$_2$CHPPh$_2${Au(C$_6$F$_5$)}][ClO$_4$]$_2$ or the neutral [{(C$_6$F$_5$)Au}Ph$_2$PC(PPh$_2$AuPPh$_2$)$_2$CPPh$_2${Au(C$_6$F$_5$)}] tetranuclear complexes, respectively [59].

The reaction of [Au(C$_6$F$_5$)(tht)] with the dinuclear derivative [Au{PPh$_2$CH(CSS)PPh$_2$Me)$_2$}][ClO$_4$] [115] affords the tetranuclear complex [Au{PPh$_2$CH(CSS{Au(C$_6$F$_5$)})PPh$_2$Me}$_2$][ClO$_4$] [116]. This black solid changes color to orange when exposed to vapors of acetone, ethanol, methanol, tetrahydrofuran and others. This

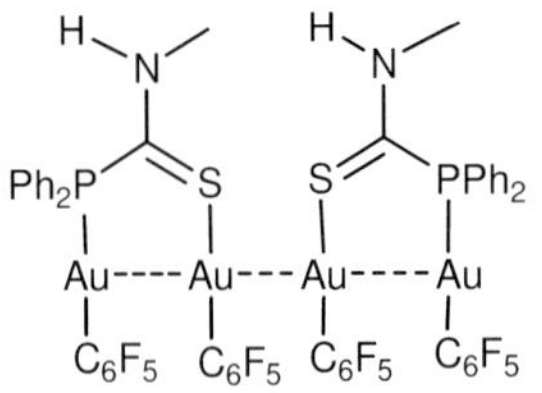

Figure 3.16 Tetranuclear complex with inequivalent pentafluorophenyl groups.

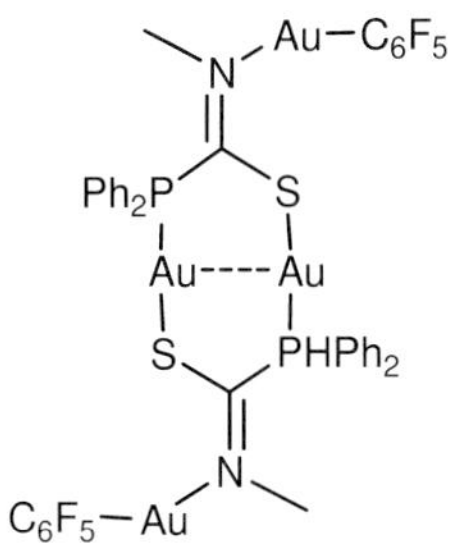

Figure 3.17 Tetranuclear derivative.

vapochromic behavior is reversible within seconds when exposure to solvent vapor ceases. This property has been utilized for the development of a fiber optical sensor for the detection of vapors of organic compounds (VOCs) [117, 118]. The structure of both complexes has been established by X-ray diffraction studies [115, 116].

3.2.5
Heteronuclear Pentafluorophenylgold(I) Complexes

3.2.5.1 Heteronuclear Pentafluorophenylgold(I) Bismuth Complexes
A transmetallation reaction was used to synthesize the complex [Au(C_6F_5)$_2$][Bi($C_6H_4CH_2NMe_2$-2)$_2$] [119] and the crystal structure of the complex was studied by X-ray diffraction. This is the first example of metallophilic interactions of the Bi(III) atom and this is the most interesting characteristic of this complex, the presence of Au$\cdots$Bi distances of 3.7284(5) Å.

3.2.5.2 Heteronuclear Pentafluorophenylgold(I) Tin Complexes
The tin compounds [SnR$_2$(SC$_6$H$_4$PPh$_2$)$_2$] (R = Me, Ph, ^{t}Bu) react with [Au(C$_6$F$_5$)(tht)] to give the trinuclear complexes [SnR$_2${(SC$_6$H$_4$PPh$_2$)Au(C$_6$F$_5$)}$_2$] [120] which show a broad singlet in their ^{31}P{^{1}H} NMR spectra (R = Me, 37.0 ppm, R = ^{t}Bu, 35.6 ppm, R = Ph, 36.2 ppm). The molecular structure of the complex where R = ^{t}Bu was studied by X-ray diffraction showing a linear coordination of the Au(I) center (177.5(3) and 176.0(3)°).

3.2.5.3 Heteronuclear Pentafluorophenylgold(I) Thallium Complexes
The reaction of Li[Au(C$_6$F$_5$)$_2$] with TlNO$_3$ in the presence of OPPh$_3$ leads to the formation of the mixed-metal complex [Tl(OPPh$_3$)$_2$][Au(C$_6$F$_5$)]$_2$ [121, 122] which was the first example of a extended unsupported Au-Tl linear chain. The same reaction with the tetrabutylammonium salt instead of the lithium salt does not give any product, only the starting materials are recovered. However, the product can be obtained by treatment of [Au(C$_6$F$_5$)$_2$Cl(PPh$_3$)] with thallium acetylacetonate. The structure of the complex was studied by X-ray diffraction; it shows that the two independent gold atoms lie on an inversion center and display an exact linear coordination to the pentafluorophenyl groups. The Tl center is bonded to two OPPh$_3$ ligands with Tl—O distances of 2.483(3) and 2.550(4) Å. The geometry

around the gold atom is almost square planar, including the metal-metal interactions, and that at the thallium is distorted trigonal bipyramidal. Tl—Au distances are 3.0358(8) and 3.0862(8) Å, nearly equal and similar to the sum of the Tl and Au metallic radii (3.034 Å). The complex is luminescent both at room temperature and at 77 K in the solid state [123].

The complex [Tl(bipy)]$_2$[Au(C$_6$F$_5$)$_2$]$_2$ [122, 124] was obtained by treating NBu$_4$[Au(C$_6$F$_5$)$_2$] with TlPF$_6$ and 4,4'-bipyridine. The structure of the complex was studied by X-ray diffraction and the asymmetric unit shows a thallium atom bonded to two bipyridine ligands and a [Au(C$_6$F$_5$)$_2$]$^-$ unit. The Au—Tl interaction (3.0161(2)Å) is shorter than the sum of thallium and gold metallic radii and there are also Au$\cdots$Au contacts of 3.4092(3) Å that forms linear tetranuclear moieties with an unusual Tl—Au—Au—Tl arrangement. The structure of the complex can be described as an infinite three-dimensional polymer. In the solid state it shows very intense luminescence both at room temperature and at 77 K.

The reaction of NBu$_4$[Au(C$_6$F$_5$)$_2$] with TlPF$_6$ leads to the complex [AuTl(C$_6$F$_5$)$_2$] [125] which can react with different ligands in a similar manner than the gold-silver complexes discussed later. Thus, it can react with Tl(acac) to give the complex [AuTl$_3$(acac)$_2$(C$_6$F$_5$)$_2$] [122, 125]. The crystal structure of this complex was studied by X-ray diffraction. It contains Tl$_2$(acac)$_2$ and [AuTl(C$_6$F$_5$)$_2$] units linked via two Tl—O bonds (2.577(3) Å); there are also unsupported Tl$\cdots$Tl contacts (3.7200(4) and 3.7607(4) Å) and the Tl$_2$(acac)$_2$ unit links only one [AuTl(C$_6$F$_5$)$_2$] fragment forming a double chain unidimensional polymer with a Tl$\cdots$Au distance 3.0653(4) Å. The complex is luminescent at room temperature in the solid state showing a single emission at 429 nm (exc 364 nm) but at 77 K it displays two independent emissions with two different excitation profiles at 427 and 507 nm.

When [AuTl(C$_6$F$_5$)$_2$]$_n$ reacts with DMSO the complex [Tl$_2${Au(C$_6$F$_5$)$_2$}$_2${μ-DMSO}$_3$]$_n$ [126] is obtained. The crystal structure of this complex shows unsupported Au$\cdots$Tl interactions that range from 3.2225(6) to 3.5182(8) Å but there are no Tl$\cdots$Tl interactions. There are Au$\cdots$Au interactions of 3.733 Å and the gold centers are almost linearly coordinated to two pentafluorophenyl groups. The complex is strongly luminescent both at room temperature (emits at 440 nm (exc.390 nm)) and at 77 K (emits at 460 nm (exc. 360 nm)).

The complex [AuTl(C$_6$F$_5$)$_2$]$_n$ reacts with ethylenediamine to give the complex shown on the left in Figure 3.18. This complex reacts in tetrahydrofuran with 1 equivalent of acetone to give the condensation product (shown on the right in Figure 3.18). With 2 equivalents of acetone the bis-condensation product is formed. Similarly, two equivalents of acetophenone also gives the bis-condensation product as shown in Figure 3.18 [127]. It seems that the acidity of the thallium center has some influence on the mechanism of each condensation. The crystal structure of the bis-acetone condensation product was studied by X-ray diffraction and shows polymeric chains where Au(I) centers are in a square-planar environment. The ethylenediamine complex reacts in the solid state with ketone vapors at room temperature displaying an irreversible color change from green to yellow. The same behavior is observed in the other two complexes. All these complexes are brightly luminescent in the solid state at room temperature.

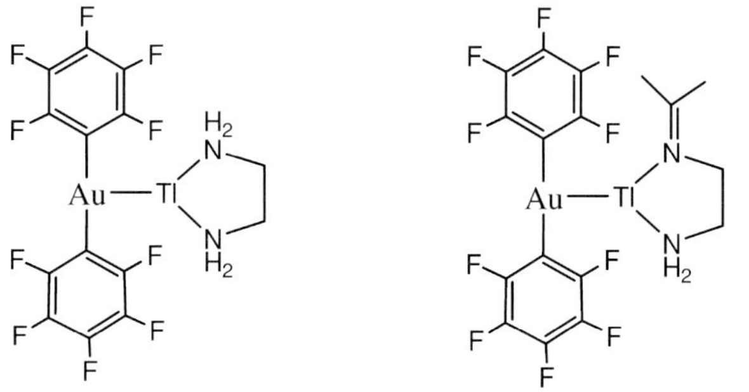

Single emission at 505 (exc 430)   Single emission at 625 (exc 515)

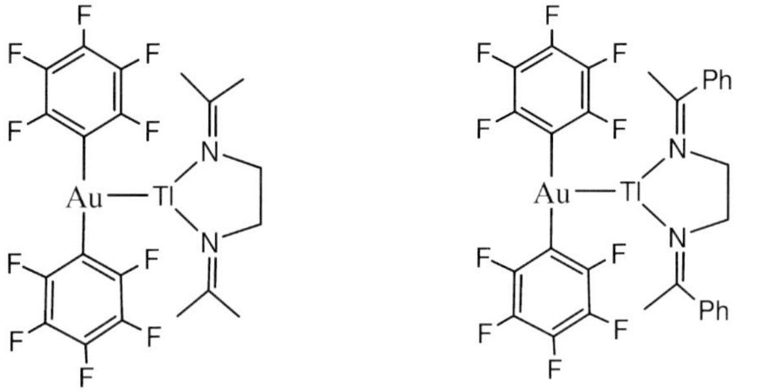

Single emission at 675 (exc 455)   Single emission at 575 (exc 465 nm)

Figure 3.18 Au—Tl complexes obtained by condensation reactions and their respective luminescent behavior.

3.2.5.4 Heteronuclear Pentafluorophenylgold(I) Silver Complexes

Many heteronuclear complexes have been obtained, as was shown for polynuclear gold derivatives, by reaction of the appropriate gold complex with silver salts and the silver center bonds to one free donor center in the starting complexes. That is the case in the reaction of [Au(C$_6$F$_5$)(Ph$_2$PCH$_2$SPh)] with AgCF$_3$SO$_3$ (in the ratio 2 : 1) or with AgCF$_3$CO$_2$ (in the ratio 1 : 1 and 1 : 2) that leads to the formation of the complexes [Au$_2$Ag(C$_6$F$_5$)$_2$(Ph$_2$PCH$_2$SPh)$_2$](CF$_3$SO$_3$), [Au$_2$Ag$_2$(C$_6$F$_5$)$_2$(CF$_3$CO$_2$)$_2$(Ph$_2$PCH$_2$SPh)$_2$] and [AuAg$_2$(C$_6$F$_5$)(CF$_3$CO$_2$)$_2$(Ph$_2$PCH$_2$SPh)] [128], respectively. The ^{19}F{^{1}H}NMR spectrum shows the pattern of the pentafluorophenyl group bonded to the Au(I) atom and a singlet around −77 ppm for the CF$_3$SO$_3$ group and −73 ppm for the CF$_3$CO$_2$ group. The crystal structure of the complex [Au$_2$Ag$_2$(C$_6$F$_5$)$_2$(CF$_3$CO$_2$)$_2$ (Ph$_2$PCH$_2$SPh)$_2$] was studied by X-ray diffraction, the gold atom is in an approximately linear environment (170.16(9)°) and the Au—C and Au—P distances are similar to those in other complexes.

The reaction of silver perchlorate with the complex [Au(C$_6$F$_5$)(Ph$_2$PCHPPh$_2$Me)] gives the trinuclear heterometallic cationic complex [{(C$_6$F$_5$)AuPh$_2$PCH(PPh$_2$Me)}$_2$Ag] ClO$_4$ [84].

The polynuclear derivatives [{AuAg(C$_6$F$_5$)$_2$L}$_n$] [129, 130] were prepared by reacting the gold derivatives NBu$_4$[Au(C$_6$F$_5$)$_2$] with Ag[ClO$_4$], the polymeric complex obtained this way [{AuAg(C_6F_5)$_2$}$_n$] reacts in dichloromethane with a neutral ligand (L) to give the final product. Examples with many different ligands including tetrahydrothiophene, pyridine, 2,2′-bipyridyl, 1,10-phenantroline, ethylenediamine, 1,3-propylenediamine, pyridine N-oxide, OPPh$_3$, SPPh$_3$, PPh$_2$Me, C_8H_{14}, C_8H_8, C_8H_{12}, C_7H_8, C_6H_{10}, C_2Ph_2, OCMe$_2$, C_6H_6, $C_6H_5CH_3$, OCH$_3$ have been prepared. Complexes with the ligands OPPh$_3$, SPPh$_3$, PPh$_2$Me do not precipitate but evaporation of the reaction mixture and subsequent extraction of the residue with Et$_2$O followed by removal of QClO$_4$ and concentration of the filtrate, leads to the solid complexes. On the other hand, the lability of the C_2Ph_2 ligand (Equation 3.10) can be used for syntheses of arene complexes:

$$[\{AuAg(C_6F_5)_2 0.5 C_2Ph_2\}_n] + \text{arene} \rightarrow [\{AuAg(C_6F_5)_2(\text{arene})\}_n] + 0.5\ C_2Ph_2$$
(3.10)

The crystal structure of some of these complexes (L=tht, benzene, acetone and acetonitrile) [129–131, 140] has been studied by X-ray diffraction. All of them are polymeric chains that repeat the structural unit shown in Figure 3.19 through short Au···Au contacts, but the magnitude of these contacts depends on the nature of the ligands (see Table 3.7); such contacts are well known for Au(I) complexes although this distance for the complex [{AuAg(C_6F_5)$_2$(tht)}$_n$] is the shortest observed in the absence of any bridging ligands.

The short Au···Ag contacts may be due to some degree of metal-metal bonding which means that these were the first reported Au–Ag bonds. The complexes with acetone and acetonitrile shown are also the first reported examples in gold chemistry in which the pentafluorophenyl ligands act as a bridge between Au and Ag centers. This type of behavior is generally more common in Pt chemistry.

Another process that leads to the formation of some of the complexes described above is the addition of [Au(C_6F_5)L] to etheral solutions of [Ag(C_6F_5)]. Because of the easy preparation and stability of NBu$_4$[Au(C_6F_5)$_2$] which can be stored without

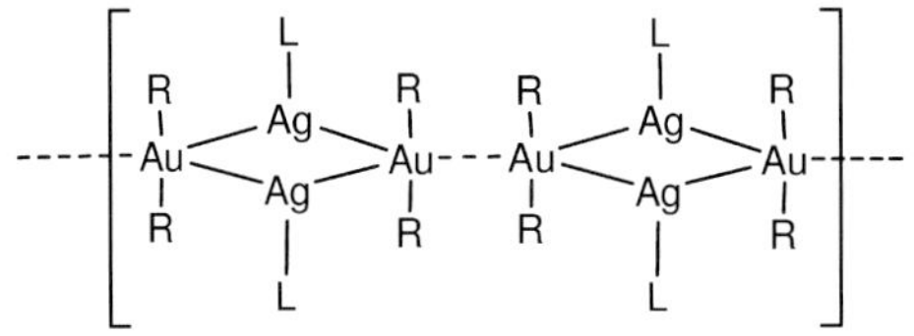

Figure 3.19 Short gold contacts in the polymeric complexes.

Table 3.7 Au···Au and Au···Ag contacts of the polymeric derivatives.

Complex	Au···Au contacts (Å)	Au···Ag contacts (Å)
[{AuAg(C_6F_5)$_2$(tht)}$_n$]	2.889	2.726 and 2.718
[{AuAg(C_6F_5)$_2$(C_6H_6)}$_n$]	3.013	2.702 and 2.792
[Au$_2$Ag$_2$(C_6F_5)$_4$(OCMe$_2$)$_2$]$_n$	3.2674(11)	2.7903(9) and 2.7829(9)
[Au$_2$Ag$_2$(C_6F_5)$_4$(N≡CCH$_3$)$_2$]$_n$	2.8807(4)	2.7577(5) and 2.7267(5)

decomposition at room temperature for months, the former method is however more convenient for the preparation of these polynuclear complexes.

The complex [Au$_2$Ag$_2$(C$_6$F$_5$)$_4$(OCMe$_2$)$_2$]$_n$ [131] luminesces both at room temperature and at 77 K. Even though the excitation profile is complicated, there is a maximum at 468 nm that leads to a maximum emission band at 546 nm at room temperature. When the temperature is 77 K the maximum emission is shifted to 554 nm. The lifetimes are on the nanosecond time scale ($\tau_1 = 68(5)$ ns, $\tau_2 = 298(19)$ ns). Taking into account the lifetime values and the small Stokes shift, it can be said that the emission is fluorescence and it is believed to originate from a gold-centered spin allowed transition. The blue shift observed with increasing temperature is consistent with an increase in the Au–Au separation as a result of thermal expansion. The optical behavior changes when the solid is dissolved in acetone, giving a colorless solution. In this case, the excitation peak appears at 332 nm and the emission at 405 nm; the emission probably arises from the pentafluorophenyl localized $\pi\pi^*$ excited states, although π-MMCT transitions cannot be excluded. This behavior suggests that in dilute solutions the Au–Au interactions are lost. It was also observed that changes in the concentration of the solution induces changes in the excitation and emission wavelengths and as a consequence a deviation from the Lambert–Beer law.

The addition of a further equivalent of neutral ligand to a suspension of [{AuAg(C$_6$F$_5$)$_2$L}$_n$] in dichloromethane leads to a colorless solution which upon addition of n-hexane precipitates complexes of the type [AgLL′][Au(C$_6$F$_5$)$_2$] with L = Py, L′ = py or L = phen, bipy, L′ = PPh$_3$. Taking into account all available data, the probable formation of these complexes is shown in the Equation 3.11:

$$[\{AuAg(C_6F_5)_2L\}_n] + nL' \rightarrow n[AgLL']^+[Au(C_6F_5)_2]^- \qquad (3.11)$$

Similarly, in the presence of acetone or another solvent, the equilibrium shown in Equation 3.12 should be present.

$$[\{AuAg(C_6F_5)_2L\}_n] + nS \rightarrow n[L\text{-}Ag\text{-}S]^+[Au(C_6F_5)_2]^- \qquad (3.12)$$

In accordance with that, the silver derivative reacts with [NBu$_4$][Au(C$_6$F$_5$)$_2$] in the presence of pyridine to give the complex [Ag(py)$_3$][Au(C$_6$F$_5$)$_2$]·py [132].

All these complexes change their color when dissolved and some of them have been used to develop optical fiber sensors [116] for the detection of VOCs. Thus, [{AuAg(C$_6$F$_5$)$_2$(phen)}$_n$] was used to develop an optical fiber sensor [133] able to detect different concentrations of acetone vapor using a laser source of 1550 nm. This sensor was also able to detect acetone in water solution. A study of the change in the mass of the solid complex when exposed to acetone vapor was carried out. The results show a mass increase in the presence of acetone vapor and a gradual loss of weight when the vapor source is removed. In order to prepare these complexes more safely Ag(CF$_3$SO$_3$) may be used instead of the potentially explosive AgClO$_4$, without reduction of yields.

The complex where L = C$_6$H$_5$C≡CC$_6$H$_5$ [134] was also used to develop an optical fiber sensor as the complex changes color from dark yellow to pale yellow. The sensor was tested at different laser wavelengths and also with an 850 nm LED and for three different solvents (methanol, ethanol and acetic acid) and five concentrations of those solvents using the LED source.

With the complex where L = pyridine an optical nanosensor was developed [135–137], the method used to fix the vapochromic material to the optical fiber was the electrostatic self assembling method (ESA) and the light source used was an 850 nm LED. The sensor was tested for two different alcohols (ethanol and methanol) and it was possible to distinguish between different concentrations. It was also possible to discriminate between the two different alcohols.

The complex obtained when L = 2,2'-bipyridine was used in the development of an optical fiber sensor that works in the reflection mode [138]. The sensor response for three different VOCs with four different concentrations each was studied and it was shown that the sensor responded to the three VOCs and that it was possible to distinguish between the different concentrations.

3.2.5.5 Heteronuclear Pentafluorophenylgold(I) Copper Complexes

The reaction of the complex $[Au(C_6F_5)(Ph_2PCH_2SPh)]$ with $[Cu(CH_3CN)_4]CF_3SO_3$ leads to formation of the complex [128] $[Au_2Cu(C_6F_5)_2(Ph_2PCH_2SPh)_2](CF_3SO_3)$.

The unsupported Au(I)···Cu(I) interactions were first described in the complex $[Cu\{Au(C_6F_5)_2\}(N\equiv CCH_3)(\mu_2\text{-}C_4H_4N_2)]_n$ [139] which was obtained by reaction of $[Au_2Ag_2(C_6F_5)_4(N\equiv CCH_3)_2]_n$ with CuCl and pyrimidine. The crystal structure of the complex shows that there are no Au···Au interactions (the distance is longer than 6 Å) and the Au···Cu distance is 2.8216(6) Å. The complex is strongly luminescent both at room temperature and at 77 K in the solid state with a broad unstructured emission at 525 nm (max excitation 390 nm) and 529 nm (max excitation 371 nm), respectively and it also shows some degree of rigidochromic effect. In solution the complex is also luminescent but the behavior depends on the solvent.

The complex $[Au_2Ag_2(C_6F_5)_4(N\equiv CCH_3)_2]_n$ [140] was treated with CuCl and the analogous complex $[Au_2Cu_2(C_6F_5)_4(N\equiv CCH_3)_2]_n$ [140] was obtained. The crystal structure of both complexes was studied, and as in the other complexes of this kind, they have Au(I)···M(I) interactions as well as Au···Au interactions. Both complexes are brightly luminescent in the solid state at room temperature and at 77 K. In acetonitrile solution a gradual red shift is observed. The luminescence in solution depends on the concentration, the temperature and the solvent used; the lack of emission in donor solvents of the complex with Cu can be interpreted in terms of a quenching effect by coordination of donor solvent molecules to the copper center.

3.2.5.6 Heteronuclear Pentafluorophenylgold(I) Palladium Complexes

Gold palladium binuclear complexes [141] can be synthesized by reaction of $[Au(C_6F_5)(tht)]$ with the palladium derivative trans-$[pdRR'(dppm)_2]$ where R = R' = C_6F_5; R = R' = C_6Cl_5; R = C_6F_5, R' = Cl. It was however not possible to distinguish the two structures for these complexes shown in Figure 3.20:

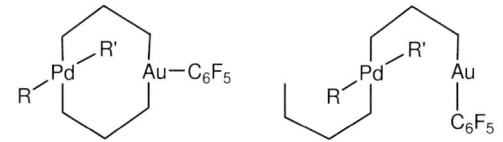

Figure 3.20 Possible structures of the PdAu complex.

Figure 3.21 Trinuclear PdAu$_2$ complex.

When the same reaction is carried out in a 1:2 ratio the trinuclear complexes where R = C$_6$F$_5$ (Figure 3.21) and C$_6$Cl$_5$ are obtained. The crystal structure of the former was solved by X-ray diffraction. The complex where R = C$_6$F$_5$ can also be obtained by reaction of [Au(C$_6$F$_5$)(dppm)] with *trans*-[Pd(C$_6$F$_5$)(tht)$_2$]

The reaction between [Au(ClO$_4$)PPh$_3$] and *trans*-[Pd(C$_6$F$_5$)$_2$(dppm)$_2$AuC$_6$F$_5$] does not give a trinuclear complex, rather migration of a C$_6$F$_5$ group occurs giving [Au(C$_6$F$_5$)PPh$_3$] and *trans*-[Pd(C$_6$F$_5$)$_2$(dppm)$_2$Au]ClO$_4$.

3.2.5.7 Heteronuclear Pentafluorophenylgold(I) Rhodium Complexes

The neutral complexes [Au(C$_6$F$_5$)(PPhPy$_2$)] and [Au(C$_6$F$_5$)(PPy$_3$)] were used to prepare the monocationic complexes [142] shown in Figure 3.22 by reaction with [Rh$_2$(μ-Cl$_2$)L$_2$] [L = 1,5-cyclooctadiene (COD), 5,6,7,8-tetrafluoro-1,4-ethenonaphthalene (TFB)]. The crystal structure of the complex with R = PPhPy$_2$ and TFB as the diolefin was studied by X-ray diffraction.

3.2.5.8 Heteronuclear Pentafluorophenylgold(I) Iron Complexes

All known pentafluorophenylgold(I) iron complexes are derivatives of ferrocene. In fact, they are obtained mainly by reaction of [Au(C$_6$F$_5$)(tht)] with modified ferrocenes

Figure 3.22 Monocationic gold rhodium complex.

containing donor atoms, resulting in complexes with the Au(C$_6$F$_5$) unit bonded to this donor center as represented by compounds **1–11** [143–154]. Some examples have been structurally characterized by X-ray diffraction and while some show Au···Au contacts, none show any form of gold-iron interaction.

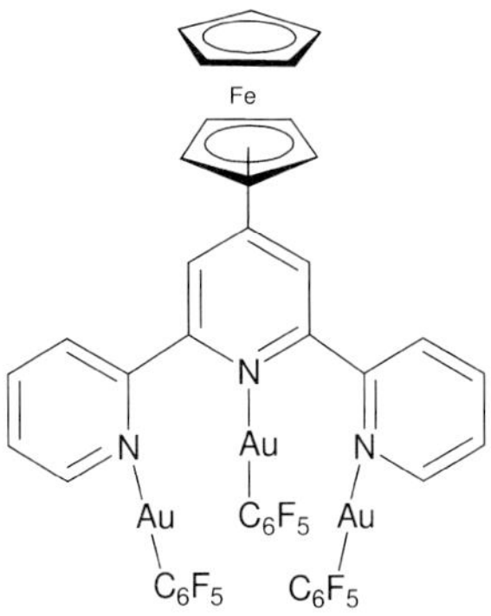

11

The complex obtained with ferrocenylmethylpyrazole (**9**) is noteworthy because the structure [151] shows the molecules associated into pairs via intermolecular Au···Au interactions of 3.1204(6) Å and these pairs form chains held together by H···Au and H···F interactions. The 3D supramolecular structure is formed through additional C−H···F hydrogen bonds. These types of H···F and H···Au interactions are not common in pentafluorophenylgold chemistry.

The reaction of [Au(C$_6$F$_5$)(tht)] with the ligands shown in Figure 3.23 also give the the corresponding Au−Fe complexes [152]. The ligands have several sites that can coordinate to a metal; the expected coordination is via the quinoline nitrogen because the 4-amino group donates electrons to the quinoline ring increasing the electron density at the quinoline nitrogen. However, the unambiguous characterization of these complexes is difficult because it is possible that the tertiary amine may also be involved in coordination.

These complexes were tested for *in vitro* antimalarial activity showing that the coordination to the Au increases the potency of chloroquine. The ferrocenyl ligands

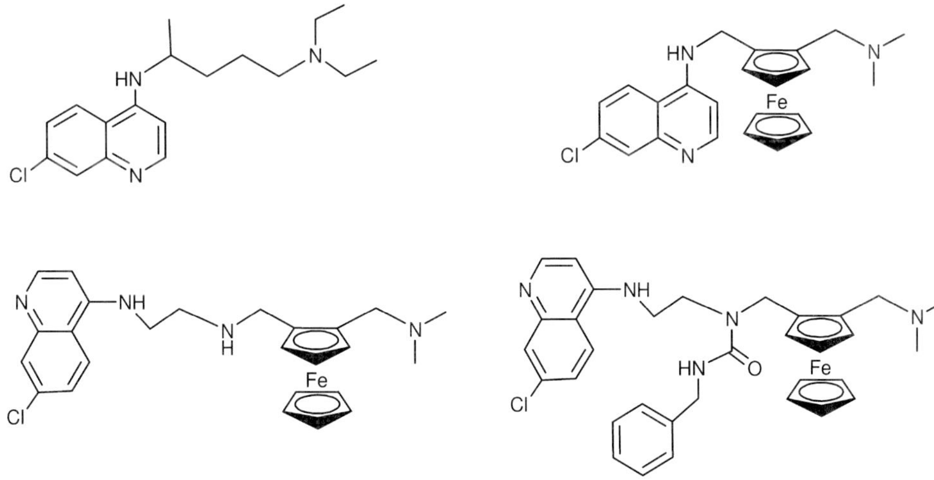

Figure 3.23 Ligands that reacts with [Au(C$_6$F$_5$)(tht)].

are more effective than the gold complexes of chloroquine, especially against the resistant strains of the parasite.

3.2.5.9 Heteronuclear pentafluorophenylgold(I) manganese complexes

Complexes [(S_2CPCy_3)(CO)$_3$Mn(μ-dppm)Au(C_6F_5)]PF$_6$ and [(η^5-C_5H_4Me)(CO)$_2$ MnSCPCy$_3$SAu(C_6F_5)] were synthesized by the reaction of [Au(C_6F_5)(tht)] with *fac*-[Mn(CO)$_3$(dppm){SC(S)-PCy$_3$}]PF$_6$ and [Mn(η^5-C_5H_4Me)(CO)$_2$(SC(S)PCy$_3$)], respectively. The coordination of the Au(C_6F_5) fragment is different in the two complexes. In the former coordination is through the diphosphine and in the latter through the S atom. This complex was the first example of an S_2CPR_3 group acting as a 4e bridge between two different metals.

3.2.5.10 Heteronuclear Pentafluorophenylgold(I) Chromium, Gold Molybdenum and Gold Tunsten Complexes

It is possible to obtain heterometallic chromium, molybdenum and tungsten gold complexes by reaction of [Au(C_6F_5)(tht)] with different metallic compounds including the deprotonated [M($\equiv$CR)(CO)$_2${R'C(pz)$_3$}][BF$_4$] (M = W, Mo; R = Me, C_6H_4Me-4, $C_6H_3Me_2$-2,6; R' = H, Me) derivatives, which gives the neutral alkylidyne metal complexes [M($\equiv$CR)(CO)$_2${(C_6F_5)AuC(pz)$_3$}] [155] in which the bridgehead carbon atom coordinates to a pentafluorophenylgold moiety. When the gold-molybdenum or gold-tungsten complexes obtained in this manner are treated with a further equivalent of the gold starting material or with [AuCl(tht)], complexes [WAuX(μ-CMe)(CO)$_2${(C_6F_5)AuC(pz)$_3$}] [155] [X = (C_6F_5), Cl] are obtained. The ^{13}C NMR spectrum of the complex with X = Cl could not be recorded because the compound is insoluble but the other complex shows a resonance at 288.3 ppm due to the μ-CMe carbon, which means that the ethylidyne ligand does not fully bridge the metal-metal bond (in that case the resonance should be appreciably more deshielded). The C$\equiv$W group, in both compounds, is relatively unperturbed by the bond of the AuX group, so the ethylidyne group is likely to remain much closer to the tungsten. In order to obtain species with metal-metal bonds, the complex where X = Cl was treated with metal carbonyl anions but only Na[Mn(CO)$_5$] afforded a stable complex although in very low yield.

The complex [WCo$_2$(μ_3-CR)(CO)$_8${(C_6F_5)AuC(pz)$_3$}] [155] (R = Me, C_6H_4Me-4) was obtained by reaction of [W($\equiv$CR)(CO)$_2${(C_6F_5)AuC(pz)$_3$}] (R = Me, C_6H_4Me-4) with [Co$_2$(CO)$_8$]. Reaction of [W($\equiv$CMe)(CO)$_2${(C_6F_5)AuC(pz)$_3$}] with [Pt(nb) (PMe$_2$Ph)$_2$] gives a mixture of the expected product ([WPt(μ-CMe)(CO)$_2$(PMe$_2$Ph)$_2$ {(C_6F_5)AuC(pz)$_3$}]) and the complex [WPtAu(C_6F_5)(μ_3-CMe)(CO)$_2$(PMe$_2$Ph)$_2$ {(C_6F_5) AuC(pz)$_3$}] which can be prepared by reaction with [Au(C_6F_5)(tht)].

[Au(C_6F_5)(tht)] reacts with [W{$\equiv$CC$_6$H$_4$Me-4}(CO)$_2$(cp)] and [W{$\equiv$CN(Et)Me} (CO)$_2$(cp)] to give [AuW(μ-CC$_6$H$_4$Me-4)(C_6F_5)(CO)$_2$(cp)] [156]. and [AuW{μ-CN(Et) Me}(C_6F_5)(CO)$_2$(cp)] [157], respectively. The signal for the μ-aminocarbyne carbon atom in the ^{13}C *NMR* spectrum appears at 259.6 ppm for the latter complex whilst the signal of the former appears low-field shifted (287.4 ppm).

The crystal structure shows that the Au(C_6F_5) fragment achieves the two coordinate 14-electron configuration by accepting an electron pair from the W–C π-orbital localized in the plane. The Au—W and Au—C distances are 2.727(1) Å and 2.13(2) Å, respectively. These distances can be compared with the corresponding values in the complex [AuWBr(bipy)(C_6F_5)(CO)$_2$(μ-CC$_6$H$_4$Me-4)] [156] Au—W = 2.752(1) Å and Au—C = 1.83(2) Å.

Neutral complexes [M(CO)$_4${(PPh$_2$)$_2$CHPPh$_2$Au(C_6F_5)}] [158] where M = Cr, Mo, W can be obtained by reaction of the complexes [M(CO)$_4${(PPh$_2$)$_2$CHPPh$_2$}] with [Au(C_6F_5)(tht)]. The crystal structure of the Mo complex was established by X-ray crystallography. The bite angle of the phosphine is 67.62(8)°, similar to that of the smallest value found for complexes containing chelated Ph$_2$PCH$_2$PPh$_2$. The molybdenum gold distance is 4.565 Å, too long to allow any significant bonding interaction.

The reaction of the previous complexes with [NBu$_4$][acac] leads to the deprotonation of the phosphine and formation of the methanide compounds [NBu$_4$][M(CO)$_4${(PPh$_2$)$_2$CPPh$_2$Au(C_6F_5)}] [158]. Again, the crystal structure of the molybdenum complex was determined by X-ray crystallography. The bite angle of the phosphine here is even smaller than that found in the previous complex: 65.14(5)°. The molybdenum-gold distance is far longer than in the neutral complex (6.172 Å).

When [Mo(CO)$_6$] or [Mo(CO)$_4$(NBD)] was treated with equimolecular amounts of [Au(C_6F_5)(PPy$_3$)], the dark purple complex fac-[Mo(CO)$_3$(Py$_3$PAuC_6F_5-N$_3$)] [52] was obtained. The ^{31}P{^{1}H} NMR shows a singlet at 43.3 ppm.

All the pentafluorophenylgold(I) complexes whose X-ray structure has been determined are collected in Table 3.8. Although there are many complexes and, probably for this reason, it is very difficult to extract any conclusion about the lengthening of the gold-C ipso distances depending on the atom being trans to the pentafluorophenyl group. In fact, the Au—C distances in many different [Au(C_6F_5)$_2$]$^-$ complexes vary from 2.040 to 2.060 Å, being even higher, 2.090 in the polynuclear complexes with silver, copper and thallium. In the later case some electronically deficient 3c-2e$^-$ bond has been proposed.

3.3
Pentafluorophenylgold(III) Derivatives

The first pentafluorophenyl gold(III) complexes formed by oxidation reactions were described in 1969. Thus, [Au(C_6F_5)Br$_2$(PPh$_3$)] [30, 35] was prepared by addition of bromine to the gold(I) derivative [Au(C_6F_5)(PPh$_3$)] and cis-[Au(C_6F_5)$_2$Cl(PPh$_3$)] [32, 34, 161] was obtained from the reaction of the dimeric bispentafluorophenylmonobromo thallium(III) [162] and the gold(I) complex [AuCl(PPh$_3$)]. The later reaction involves precipitation of TlBr and the oxidation of the gold(I) derivative as the consequence of the transfer of two C_6F_5 groups. The formation of a simple or double halogen bridge [36] (Figure 3.24) in a first step has been assumed for this oxidation process, followed by the transfer of the two C_6F_5 groups to the metal center via an inner sphere mechanism.

Subsequent halide metathesis reactions of MX (MX = KBr, LiBr, KI and NaI) with [Au(C_6F_5)$_2$Cl(PPh$_3$)] gave only decomposition products [163], although the corre-

Table 3.8 Au-C$_6$F$_5$ distances of pentafluorophenylgold (I) complexes.

Compound	Au-C$_{trans\ to\ (X)}$	Ref
[Au(C$_6$F$_5$)PPh$_3$]	2.07(2) (P)	[33]
[Au(C$_6$F$_5$)(Ph$_2$PCH$_2$PPh$_2$Me)]	2.057(6) (P)	[84]
[(C$_6$F$_5$)AuS(dppm)SAu(C$_6$F$_5$)]	2.063(12) 2.058(12) (S)	[159]
[Au(C$_6$F$_5$)(Ph$_2$C=N−N=PPh$_2$)]		[50]
[Au$_3$(C$_6$F$_5$)$_3$(η^3-Fcterpy)]	1.961(12) 1.982(11) 2.012(10) (C)	[154]
[Au(C$_6$F$_5$)CNMeCMe=CHS)]	1.993(10) (C)	[66]
[Au$_2${(PPh$_2$)$_2$C$_2$B$_9$H$_{10}$}(C$_6$F$_5$)(PPh$_3$)]	2.068(8) (P)	[101]
[Au$_5$(C$_6$F$_5$){(SPPh$_2$)$_2$C}$_2$PPh$_3$]	2.086(82) (C)	[112]
[Au(C$_6$F$_5$){Ph$_2$PCH(PPh$_2$Me)}Au(C$_6$F$_5$)]	2.053(15) 2.011(11) (P,C)	[84]
[PPh$_3$Bz][Au(C$_6$F$_5$)Cl]	2.054(17) 2.042(15) (Cl)	[78]
Et$_4$N[{Au(C$_6$F$_5$)}$_3$(μ$_3$-S)]	2.017(14) 2.042(13) 2.019(14) (S)	[75]
(TTFPh$_2$)$_2$[Au(C$_6$F$_5$)$_2$]	2.040(3) (C)	[83]
[{AuAg(C$_6$F$_5$)$_2$(C$_6$H$_6$)}$_n$]	2.063(11) 2.051(11) (C)	[130]
[{AuAg(C$_6$F$_5$)$_2$(tht)}$_n$]	2.057(12) 2.059(11) (C)	[129, 130]
trans-[Pd(C$_6$F$_5$)$_2${dppmAu(C$_6$F$_5$)}$_2$]	2.074(7) (P)	[141]
[AuW{μ-CN(Et)Me}(C$_6$F$_5$)(CO)$_2$(cp)]	2.07(2) (C,W)	[157]
[Mo(CO)$_4${(PPh$_2$)$_2$CHPPh$_2$Au(C$_6$F$_5$)}]	2.051(10) (P)	[158]
[NBu$_4$][Mo(CO)$_4${(PPh$_2$)$_2$CPPh$_2$Au(C$_6$F$_5$)}]	2.066(6) (P)	[158]
[Au(C$_6$F$_5$)(Fcpy)]	2.00(2) (N)	[144]
[Tl(OPPh$_3$)$_2$][Au(C$_6$F$_5$)$_2$]	2.058(5) 2.053(6) (C)	[121]
[SnR$_2${(SC$_6$H$_4$PPh$_2$)Au(C$_6$F$_5$)}$_2$]	2.052(10) 2.013(11) (S)	[120]
[Au{PPh$_2$C(PPh$_2$Me)C(X)S}$_2$][Au(C$_6$F$_5$)$_2$]	2.054(7) (C)	[115]
[Au$_2$(C$_6$F$_5$)$_2$(μ-dppm)]	2.039(12) (P)	[160]
[MeC$_6$H$_3$(NHPPh$_2$AuC$_6$F$_5$)$_2$]	2.056(9) 2.041(9) (P)	[96]
PPN[Au(C$_6$F$_5$)(C≡CH)]	2.049(5) (C)	[79]
[Au$_2$Ag$_2$(C$_6$F$_5$)$_4$(OCMe$_2$)$_2$]$_n$	2.086(8), 2.092(7) (C)	[131]
[Au$_2$Ag$_2$(C$_6$F$_5$)$_2$(CF$_3$CO$_2$)$_2$(Ph$_2$PCH$_2$SPh)$_2$]	2.056(3) (C)	[128]
[Au{CH(PPh$_2$Me)$_2$}(C$_6$F$_5$)]CF$_3$SO$_3$	2.029(6) (C)	[85]
[Au(C$_6$F$_5$)(PPh$_2$CH$_2$Fc)]	2.070(5) (P)	[147]
[Au$_2$(C$_6$F$_5$)$_2$(μ-2,2'-Ph$_2$PC$_6$H$_4$C$_6$H$_4$PPh$_2$)]	2.02(1), 2.04(1) (P)	[109]
[Au$_2$(C$_6$F$_5$)$_2$(μ-2,2'-Et$_2$PC$_6$H$_4$C$_6$H$_4$PEt$_2$)]	2.056(7) (P)	[109]
[Tl(bipy)]$_2$[Au(C$_6$F$_5$)$_2$]$_2$	2.048(3) 2.049(3) (C)	[124]
[Au(C$_6$F$_5$){N(H)=CPh$_2$}]	2.002(10) (N)	[48]
[Au(C$_6$F$_5$)(PPh$_2$CCH)]	2.059(8) (P)	[53]
[{Au(C$_6$F$_5$)}$_2${(P^{i}Pr2)$_2$CH$_2$}]	2.044(5), 2.046(6) (P)	[93]
(C$_6$F$_5$)Au—N(4-methylthiazole)	2.00(1) (N)	[73]
(C$_6$F$_5$)Au—S(4-methylthiazole-2-thione)	2.057(14) (S)	[73]

Table 3.8 (Continued)

Compound	Au-C*trans* to (X)	Ref
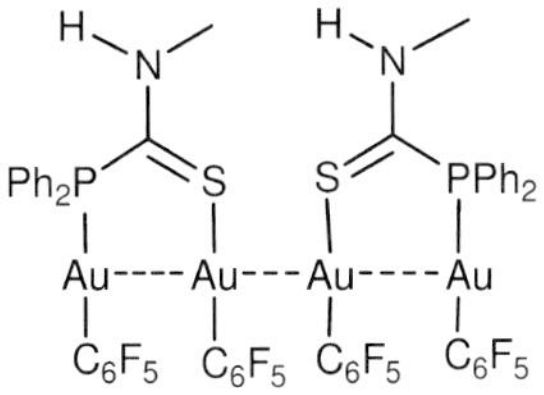 [structure: Ph₂P(=S)NHMe and MeNH(S=)PPh₂ bridging Au---Au---Au---Au chain, each Au bearing C₆F₅]	2.068(8) (P); 2.029(8) (S) 2.041(8) (S); 2.074(7) (P)	[54]
[AuTl₃(acac)₂(C₆F₅)₂] [structure: ferrocenyl-CH₂-NH-pyridyl-Au(C₆F₅)]	2.047(5) (C) 1.988(4) (N)	[125] [153]
[Cu{Au(C₆F₅)₂}(N≡CCH₃)(μ₂-C₄H₄N₂)]ₙ [{Au(C₆F₅)}₂(PPh₂CCPPh₂)] [structure: ferrocene with PPh₂-Au-C₆F₅ and SPh substituents]	2.056(5), 2.045(5) (C) 2.045(8) (P) 2.056(6) (P)	[139] [94] [150]
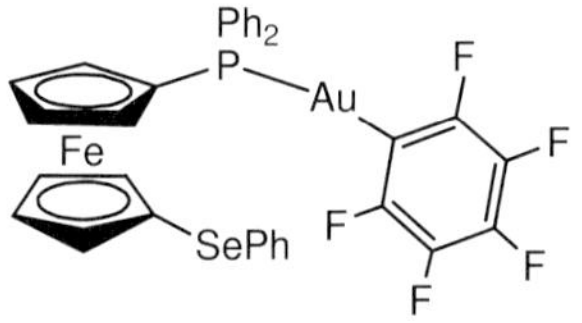 [structure: ferrocene with PPh₂-Au-C₆F₅ and SePh substituents]	2.049(5) 2.037(5)(P)	[150]
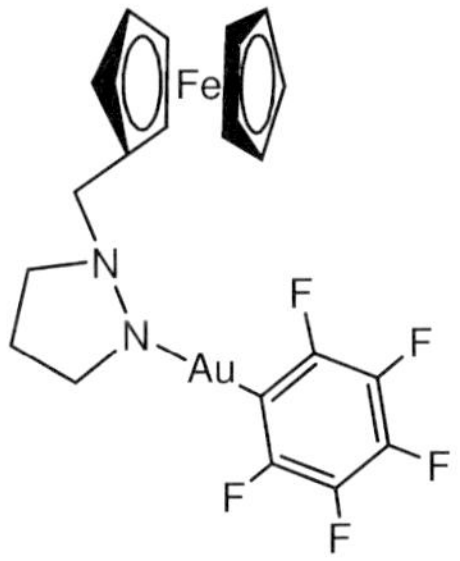 [structure: 1,1'-disubstituted ferrocene with CH₂ linked to pyrazolidine N, other N bonded to Au-C₆F₅]	2.023(7) (N)	[151]
[Au₂Tl₂(C₆F₅)₄(OSMe)₃] [structure: biphenyl with two AsPh₂-Au-C₆F₅ groups]	2.057(8), 2.060(8) (C) 2.038(5) (As)	[126] [111]

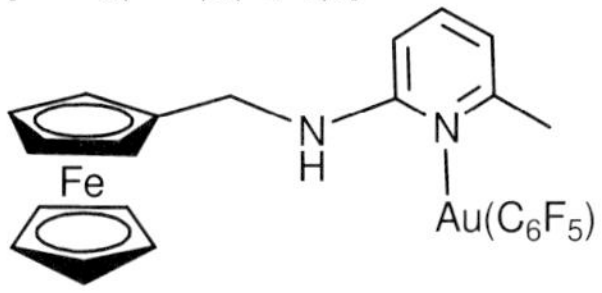

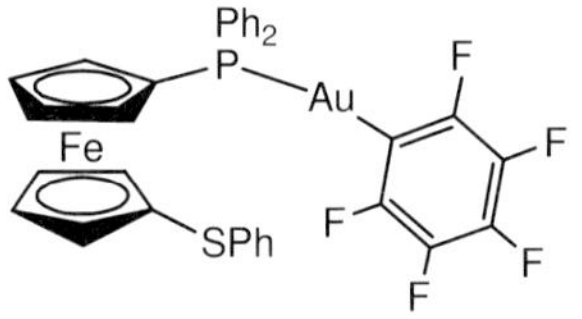

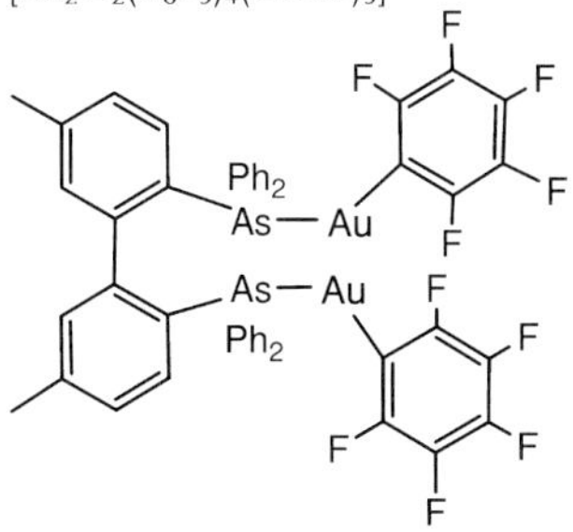

(Continued)

Table 3.8 (Continued)

Compound	Au-C$_{trans\ to\ (X)}$	Ref
(structure: benzothiazole-PPh$_2$-Au-C$_6$F$_5$)	2.054(3) (P)	[56]
[Au(C$_6$F$_5$)P(SMe)$_3$]	2.043(5), 2.049(6) (P)	[57]
[Au(C$_6$F$_5$)P(SPh)$_3$]	2.041(5) (P)	[57]
	2.063(10) (P)	[57]
(structure: bis-dithiaphospholane-Au-C$_6$F$_5$ dimer)		
[Au$_2$Ag$_2$(C$_6$F$_5$)$_4$(N≡CMe)$_2$]	2.055(6), 2.088(6) (C)	[140]
[Au$_2$Cu$_2$(C$_6$F$_5$)$_4$(N≡CMe)$_2$]	2.043(5), 2.090(4) (C)	[140]
[Au(C$_6$F$_5$)tht]	2.014(9), 2.03(1) (S)	[41]
		[104]
(structure: C$_6$F$_5$-Au-PPh$_2$-C$_6$H$_4$-CH$_2$-N(Pr^i)H-Au-Cl)		
[Au(C$_6$F$_5$)$_2$][Bi(C$_6$H$_4$CH$_2$NMe$_2$-2)$_2$]	2.056(9), 2.028(8) (C)	[119]

sponding bromo and iodo-derivatives can be prepared by oxidation reactions of bromo and iodo-triphenylphosphine gold(I) complexes with the C$_6$F$_5$ transfer agent [Tl(C$_6$F$_5$)$_2$Br]$_2$.

3.3.1
Mononuclear Pentafluorophenylgold(III) Derivatives

The most general method of preparing pentafluorophenylgold(III) derivatives is oxidative addition with the appropriate oxidant to the gold(I) complex. The choice of

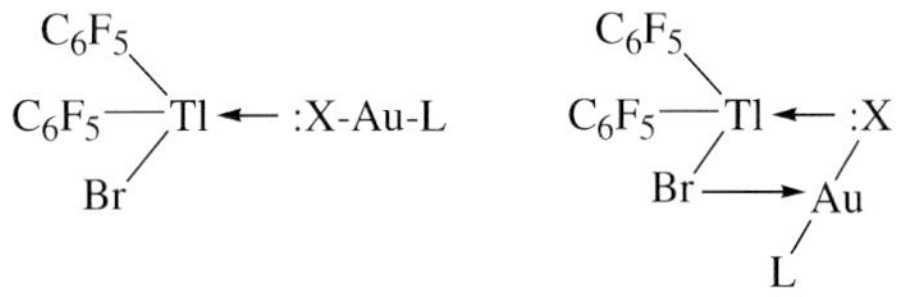

Figure 3.24 Formation of simple or double halogen bridges.

oxidant such as halogen, TlCl$_3$, [Tl(C$_6$F$_5$)$_2$Br]$_2$ or [Tl(C$_6$F$_5$)$_2$Cl]$_2$ [164] depends exclusively on the stoichiometry of the desired product.

Thus, complexes with the formula [Au(C$_6$F$_5$)X$_2$L] can be prepared by addition of the corresponding halogen X$_2$ or TlCl$_3$ to the starting gold(I) compound. It has been shown that the course of the halogenation is determined by the nature of the ligand L. In the case of the labile ligands (L = AsPh$_3$ [37] and tht [165]) or even with L = PEt$_3$ [166] the addition of X$_2$ (X = Cl or Br) gives [Au(C$_6$F$_5$)X$_2$L] with a *trans* disposition of the halogens; however oxidative addition of TlCl$_3$ affords the *cis* isomer [Au(C$_6$F$_5$)X$_2$(AsPh$_3$)]. However, when L is triphenylphosphine (PPh$_3$) [167] only chlorination with TlCl$_3$ gives the corresponding chloro-compound [Au(C$_6$F$_5$)Cl$_2$(PPh$_3$)] with the *cis* conformation. Direct halogenation with chlorine, even with a large excess, gives a mixture of the starting material and [AuCl(PPh$_3$)]. Nevertheless, [Au(C$_6$F$_5$)X$_2$(PPh$_3$)] (X = Br, I) can be prepared by direct halogenation of the gold(I) derivative with bromine or iodine [167].

Neutral ylide, isocyanide and carbene pentafluorophenyl gold(III) complexes with the general formulae *trans*-[Au(C$_6$F$_5$)X$_2$(CH$_2$PR$_3$)] (PR$_3$ = PPh$_3$, PPh$_2$Me, PPhMe$_2$), X = Cl or SCN) [168], *trans*-[Au(C$_6$F$_5$)X$_2$(CNR)] (CNR = CNPh, CN^{t}Bu, CNCy; X = Cl, Br, I)) [65] and *trans*-[Au(C$_6$F$_5$)Br$_2${C(NHR)(NHR′)}] (R = Ph, ^{t}Bu, C$_6$H$_{11}$; R′ = C$_6$H$_{11}$, CH$_2$C$_6$H$_5$, CH$_3$, X = Br, I) [65] are obtained by the simple oxidative addition of the corresponding halogen to the gold(I) compounds.

Only one anionic trihalopentafluorophenyl gold(III) complex with the formula [Au(C$_6$F$_5$)X$_3$]$^-$ has been described, NBu$_4$[Au(C$_6$F$_5$)Br$_3$] [39], obtained by the oxidation of NBu$_4$[Au(C$_6$F$_5$)Br] with bromine.

The incorporation of two or even more C$_6$F$_5$ units to the gold center can be achieved by using the thallium dimer [Tl(C$_6$F$_5$)$_2$X]$_2$ (X = Cl or Br) as oxidant. Thus, the complexes *trans*-[Au(C$_6$F$_5$)$_2$XL] with L = PPh$_3$ [34, 169], tht [169], CNPh [65] or ylide [170] (CH$_2$PR$_3$ = CH$_2$PPh$_3$, CH$_2$PPh$_2$M$_2$, CH$_2$PPhMe$_2$) and X = Cl are prepared by the reaction of [AuClL], [Au(isocyanide)Cl] or [Au(ylide)Cl] with the thallium arylating agent. When L = ylide [170] the reaction gives additional compounds such as [Tl(C$_6$F$_5$)$_3$] and [Au(CH$_2$PR$_3$)$_2$][Au(C$_6$F$_5$)$_4$]. If there are two pentafluorophenyl rings in the starting material —NBu$_4$[Au(C$_6$F$_5$)$_2$]— only the addition of the halogen or TlCl$_3$ is necessary to isolate complexes with the stoichiometry [Au(C$_6$F$_5$)$_2$X$_2$]$^-$. Consequently *trans*-NBu$_4$[Au(C$_6$F$_5$)$_2$X$_2$] (X = Cl, Br, I) [74, 171] are accessible by the reaction of NBu$_4$[Au(C$_6$F$_5$)$_2$] with the corresponding halogen and *cis*-NBu$_4$[Au(C$_6$F$_5$)$_2$Cl$_2$] [74] by addition of TlCl$_3$. The *trans* isomer can isomerize to the *cis* derivative with heating [74].

Oxidation of the gold(I) derivative [Au(C$_6$F$_5$)(tht)] with [Tl(C$_6$F$_5$)$_2$Cl]$_2$ gives the corresponding gold(III) complex [Au(C$_6$F$_5$)$_3$(tht)] [24, 25, 172] which is an excellent starting material for the preparation of a large number of trispentafluorophenyl gold(III) complexes with the general formula [Au(C$_6$F$_5$)$_3$L] (see below), by simple displacement of the labile tetrahydrothiophene ligand. Anionic [Au(C$_6$F$_5$)$_3$X]$^-$ and [Au(C$_6$F$_5$)$_4$]$^-$ derivatives are obtained from the oxidative addition of [Tl(C$_6$F$_5$)$_2$Br]$_2$ or [Tl(C$_6$F$_5$)$_2$Cl]$_2$ to NBu$_4$[Au(C$_6$F$_5$)Br] and NBu$_4$[Au(C$_6$F$_5$)$_2$], respectively giving the corresponding gold(III) complexes NBu$_4$[Au(C$_6$F$_5$)$_3$X] [39] and NBu$_4$[Au(C$_6$F$_5$)$_4$] [74]. Oxidative addition is not the only way to prepare the latter complex. The use of the arylating agent Ag(C$_6$F$_5$) [172] or the addition of a freshly prepared solution of Li(C$_6$F$_5$) [39, 40, 173] to gold(III) derivatives such as N(PPh$_3$)$_2$[Au(C$_6$F$_5$)$_3$Cl] or [AuCl$_3$(tht)] or even the addition of N(PPh$_3$)$_2$[Co(CO)$_4$] [174] to [Au(C$_6$F$_5$)$_3$(tht)] affords Q[Au(C$_6$F$_5$)$_4$] in good yields. The anion [Au(C$_6$F$_5$)$_3$Cl]$^-$ has been also obtained in the reaction of [{Au(C$_6$F$_5$)}$_2$(μ-dppm)] with [Tl(C$_6$F$_5$)$_2$Cl]$_2$ giving [Au{(dppm)AuCl}$_2$][Au(C$_6$F$_5$)$_3$X] as a consequence of the transfer of two C$_6$F$_5$ units to the Au(I) center [175].

Very recently, it was shown that a homoleptic rearrangement in the mononuclear organogold(III) derivative [Au(C$_6$F$_5$)$_3$(tht)] gives [Au(C$_6$F$_5$)$_4$][Au(C$_6$F$_5$)$_2$(tht)$_2$] which was structurally characterized [41].

Ag(C$_6$F$_5$) has also been used as arylating agent in the preparation of the diauracycle (μ-methylene)bisylide digold(III) [μ-(CH$_2$){Au(CH$_2$)$_2$PPh$_2$}$_2$(C$_6$F$_5$)$_2$] [176] using [μ-(CH$_2$){Au(CH$_2$)$_2$PPh$_2$}$_2$I$_2$] as starting material.

Complexes of the type [Au(C$_6$F$_5$)$_3$L] where L = PPh$_3$ or tht can be also obtained by using Ag(C$_6$F$_5$) as arylating agent [35] and cis-[AuCl(C$_6$F$_5$)$_2$PPh$_3$] [34] or trans-[AuCl$_2$(C$_6$F$_5$)(tht)] [165] as starting materials.

Mixed triaryl gold(III) derivatives [Au(C$_6$F$_5$)R'$_2$(tht)] [169] (R' = 2,3,4,6-C$_6$F$_4$H and 2,4,6-C$_6$F$_3$H$_2$) have also been prepared by oxidative addition of [TlR'$_2$Cl]$_2$ to the gold(I) complex [Au(C$_6$F$_5$)tht].

A third route for the preparation of pentafluorophenyl gold(III) derivatives involves the replacement of a labile ligand in a gold(III) complex with neutral or anionic molecules. In the case of substitution by neutral ligands, three different general formulae can be described: [Au(C$_6$F$_5$)X$_2$L], [Au(C$_6$F$_5$)$_2$L$_2$]$^+$ and [Au(C$_6$F$_5$)$_3$L]. Tetrahydrothiophene (tht) is one of the most commonly employed labile ligands in the chemistry of gold. In complexes such as [Au(C$_6$F$_5$)X$_2$(tht)] [165], cis-[Au(C$_6$F$_5$)$_2$(tht)$_2$]ClO$_4$ [177] and [Au(C$_6$F$_5$)$_3$(tht)] [24, 25, 172] the tht ligand is readily displaced by neutral C, N, P, As or Sb-donor ligands giving: [Au(C$_6$F$_5$)X$_2$L] —L = PPh$_3$, X = Cl; L = AsPh$_3$, X = I; L = pdma (o-phenylenebis(dimethyl)arsine), phen (1,10-phenantroline), X = Cl, Br; L = CNC$_6$H$_4$Me-p, X = Cl, Br, I— [165, 178], [Au(C$_6$F$_5$)$_2$L$_2$]ClO$_4$ —L$_2$ = dppmS$_2$ (bis(diphenylphosphine)methane disulfide) [88], dppf (1,1'-bis(diphenylphosphine)ferrocene) [90], PTA (1,3,5-triaza-7-phosphaadamantane)— [28] and [Au(C$_6$F$_5$)$_3$L] —L = PPh$_3$ [25, 172], PMePh$_2$ [172], POPh$_3$ [172], PTA [28], PPh$_2$H [179], AsPh$_3$ [25], SbPh$_3$ [172], PPh$_2$C$_6$H$_4$SH [180] (2-(diphenylphosphine)thiophenol), dppm [25, 172] (1,2-bis(diphenylphosphine)methane), dppe [172] (1,2-bis(diphenylphosphine)ethane), dppf [90], FcCH$_2$PPh$_2$ ((ferrocenylmethyl)phosphine) [147], (PPh$_2$)$_2$C=CH$_2$ (vinylidenebis(diphenylphosphine)) [92], pdma [172], diars (1,2-bis(dimethylarsine)benzene) [25] P^{i}Pr$_2$CH$_2$PPh$_2$ (diisopropylphosphine-(diphenylphosphine)) [93], NH$_3$ [172], py [172], 2-amt (2-amino-4,5-dihydrothiazole) [47], isocyanides and carbenes [172]: CNC$_6$H$_4$Me-p, C(NHC$_6$H$_4$Me-p)$_2$, C(NHC$_6$H$_4$Me-p)(NEt$_2$), C(NHC$_6$H$_4$Me-p)(NHCH$_2$CH$_2$NH$_2$), phosphonium dithiocarboxylates [89]: S$_2$C-PR$_3$

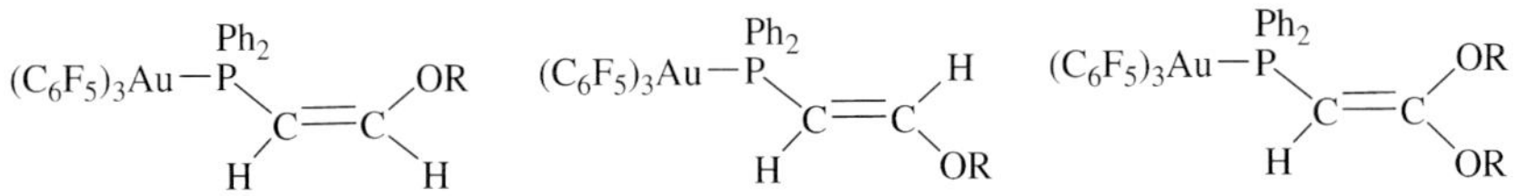

Figure 3.25 Single or double alcohol addition to the triple bond of PPh2(C≡CH).

($PR_3 = P(C_6H_{11})_3$, PEt_3, PBu_3, $PPhMe_2$, $PPhEt_2$) and ylides [70] (RCH-ER'_3) with $ER'_3 = PPh_3$, R = H, Me, Et; $ER'_3 = PPh_2Me$, PPh_2Me_2, $AsPh_3$; R = H−.

Phosphinoalkynes PR_2CCR' are potentially bifunctional ligands, able to coordinate as a phosphine, as an acetylene, or a combination of both. In the particular case of PPh_2CCH (where R′ = H) the ligand can also act as an acetylide by deprotonation of the C≡CH group. This ligand displaces tht in [Au(C_6F_5)$_3$(tht)] giving [Au(C_6F_5)$_3$(PPh_2CCH)]. Unexpectedly, the deprotonation of the alkyne with sodium alkoxide gives a single and a double alcohol addition to the triple bond leading to new phosphines coordinated to the Au(C_6F_5)$_3$ fragment: (2-methoxyethenyl)diphenylphosphine and (2,2-dimethoxyethenyl)diphenylphosphine, originating from a single or double anti-Markovnikov addition of methanol to the non-coordinated alkyne fragment [181] (Figure 3.25).

The displacement of tht with salts leads to the isolation of ionic complexes of the type trans-Q[Au(C_6F_5)X_2X'] (Q = BzPPh$_3$, X = Cl, Br, X′ = Cl; X = SCN, X′ = Br; X = X′ = Br) [165] and Q[Au(C_6F_5)$_3$X] with X = halide, pseudo-halide or dithiocarbamate —S_2CNR_2—: (Q = NEt$_4$, X = Cl, Q = NBu$_4$, BzPPh$_3$, X = SCN) [25], (Q = BzPPh$_3$, K, Et, X = CN) [63], (Q = NEt$_4$, NBu$_4$, BzPPh$_3$, N(PPh$_3$)$_2$, X = Cl, Br, I, SCN, N$_3$) [172], (Q = NBu$_4$, N(PPh$_3$)$_2$; X = S_2CNR_2 with R = Me, Et, CH$_2$Ph) [182]. With phosphonium salts QX where Q = PPh$_3$Me, MeCH$_2$PPh$_3$, EtCH$_2$PPh$_3$, PhCH$_2$PPh$_3$, MeAsPh$_3$ and X = Cl, Br, I [71] the corresponding complexes Q[Au(C_6F_5)$_3$X] give gold(III) ylide derivatives [Au(C_6F_5)$_3$(ylide)] (ylide = CH$_2$PPh$_3$, CH$_2$AsPh$_3$) [71] via deprotonation with NaH.

It has been shown that [Au(C_6F_5)$_2$(ClO$_4$)L] (L = PPh$_3$ and AsPh$_3$) [183] are excellent starting materials since the very poor coordination ability of the perchlorate ligand permits its replacement by other molecules. Hence, complexes with anionic ligands [Au(C_6F_5)$_2$X(PPh$_3$)] (X = N$_3$, HCO$_3$, CN, SCN) [38, 183, 184] and [Au(C_6F_5)$_2$X(AsPh$_3$)] (X = CN, SCN, N$_3$, OOCC$_6$H$_5$), neutral monodentate ligands [Au(C_6F_5)$_2$LL′]ClO$_4$ (L = OPPh$_3$, OAsPh$_3$, ONC$_5$H$_5$, ONC$_9$H$_7$, NC$_9$H$_7$, PEt$_3$, PBu$_3$, PPh$_2$Me, PPh$_3$, AsPh$_3$, py, H$_2$O and L′ = PPh$_3$, AsPh$_3$) [183, 184] or neutral bidentate ones ([Au(C_6F_5)$_2$(L-L)(PPh$_3$)] [184], L-L = dppe, diars have been described.

A different way to prepare pentafluorophenyl gold(III) complexes is depicted by replacement of the halogen atom from the starting material as MCl or AgCl. Some examples are illustrated by the following equations (Equations 3.13–3.19):

$$[Au(C_6F_5)X_2(L\text{-}L)] + AgClO_4 \rightarrow AgX + [Au(C_6F_5)X(L\text{-}L)]ClO_4 [178] \quad (3.13)$$

(X = Cl, Br; L-L = pdma, phen)

$$[Au(C_6F_5)_2Cl(AsPh_3)] + AgX \rightarrow AgCl + [Au(C_6F_5)_2X(AsPh_3)][38] \quad (3.14)$$

(X = NO$_3$, NO$_2$, CH$_3$COO, CF$_3$COO, ClO$_4$)

$$Q[Au(C_6F_5)_2Cl_2] + 2\, MX \rightarrow 2MCl + \textit{trans-}Q[Au(C_6F_5)_2X_2]\, [63, 26] \quad (3.15)$$

(Q = NBu$_4$, X = CN, SCN, Q = Et$_3$NH, X = CN)

$$\textit{trans-}NBu_4[Au(C_6F_5)_2Cl_2] + MX \rightarrow 2MCl + NBu_4[[Au(C_6F_5)_2$$
$$(S_2C_2B_{10}H_{10})]\,[185] \quad (3.16)$$

MX = Na$_2$S$_2$C$_2$B$_{10}$H$_{10}$

$$\textit{cis-}NBu_4[Au(C_6F_5)2Cl_2]^{75} + 2AgClO_4 \xrightarrow[-2AgCl]{OEt_2} \textit{cis-}[Au(C_6F_5)_2(OEt_2)_2]ClO_4\,[186] \quad (3.17)$$

$$NBu_4[Au(C_6F_5)_3Br] + AgClO_4 \xrightarrow{OEt_2} QClO_4 + AgBr + [Au(C_6F_5)_3(OEt_2)]\,[172] \quad (3.18)$$

$$NBu_4[Au(C_6F_5)_3Br] + MX \rightarrow MCl + NBu_4[Au(C_6F_5)_3(SCB_{10}H_{10}CMe)]\,[187] \quad (3.19)$$

MX = Na SCB$_{10}$H$_{10}$CMe

When reaction (3.13) is carried out in equimolecular amounts [Au(C$_6$F$_5$)$_2$Cl(OEt$_2$)] [29, 87] is obtained. And the same reaction without diethylether leads to a dinuclear compound with bridging halogen atoms [Au(μ-X)(C$_6$F$_5$)$_2$]$_2$ (X = Cl, Br) [26] which is an excellent starting material in pentafluorophenyl gold(III) chemistry. Such halogen bridges are readily cleaved by the addition of neutral ligands giving cis-[Au(C$_6$F$_5$)$_2$ClL] (L = PPh$_3$, py, dppm) as a result of a symmetric cleavage or the ionic complexes cis,cis-[Au(C$_6$F$_5$)$_2$(L-L)][Au(C$_6$F$_5$)$_2$Cl$_2$] (L-L = bipy, phen, pdma) [26]. The cleavage of the halogen bridge by anionic ligands provides neutral complexes such as [Au(C$_6$F$_5$)$_2$(acac)] [26], [Au(C$_6$F$_5$)$_2${(SePPh$_2$)$_2$N}] [188] and [Au(C$_6$F$_5$)$_2$(S$_2$CNR$_2$)] (R = Me, Et, CH$_2$Ph) [182] also available via oxidative addition of [Tl(C$_6$F$_5$)$_2$Cl]$_2$ to the bis(dithiocarbamate) digold(I) [Au$_2$(S$_2$CNR$_2$)$_2$]. Pseudo-halogen bridges with the formula [Au(C$_6$F$_5$)$_2$X]$_2$ (X = SCN, N$_3$CF$_3$COO) [26] are obtained by reaction of MX with [Au(μ-X)(C$_6$F$_5$)$_2$]$_2$.

Addition of vinylidenbis(diphenylphosphine) or vinylidenbis(diphenylphosphine) disulfide to this dinuclear [Au(μ-Cl)(C$_6$F$_5$)$_2$]$_2$ gives the isolation of mononuclear gold (III) derivatives with mono-coordination through P or S atom: [Au(C$_6$F$_5$)$_2$Cl{PPh$_2$C(=CH$_2$)PPh$_2$}] [189], [Au(C$_6$F$_5$)$_2$Cl{SPPh$_2$C(=CH$_2$)PPh$_2$ = S}] [190] and [Au(C$_6$F$_5$)$_2${(SPPh$_2$)$_2$C=CH$_2$}]ClO$_4$ [190]. Such coordination to the metal center activates the double bond in the vinylidenbis(diphenylphosphine) in such a way that various nucleophiles can be added by Michael-type additions giving formulae: [Au(C$_6$F$_5$)$_2${(PPh$_2$)$_2$CCH$_2$Nu}] (Nu = CCPh, C$_5$H$_5$, CH(COMe)$_2$, OEt, SPh, S$_2$CNR$_2$, with R = Me, Et, Bz) and [Au(C$_6$F$_5$)$_2${(SPPh$_2$)$_2$CCH$_2$Nu}] (Nu = CH(COMe)$_2$, CN, OEt, C$_5$H$_5$). Treatment of [Au(C$_6$F$_5$)$_2$Cl{PPh$_2$C(=CH$_2$)PPh$_2$}] [189] with oxygen-based nucleophiles such as Ag$_2$O or NaOEt afford the complexes [{Au(C$_6$F$_5$)$_2${(PPh$_2$)$_2$CCH$_2$}$_2$O] and [{Au(C$_6$F$_5$)$_2${(PPh$_2$)$_2$CCH$_2$OEt}] [189].

Since tetrahydrothiophene is more strongly bonded to Au(III) than to Au(I), complexes with attached OEt$_2$ molecules are better candidates for substitution reactions. Complexes of the type: [Au(C$_6$F$_5$)$_2$Cl(OEt$_2$)], [Au(C$_6$F$_5$)$_2$(OEt$_2$)$_2$]ClO$_4$ or [Au(C$_6$F$_5$)$_3$(OEt$_2$)] are the most often employed in this kind of reaction. As examples:

3.3 Pentafluorophenylgold(III) Derivatives

Scheme 3.1 Formation of cyclic carbenes.

[Au(C$_6$F$_5$)$_2$Cl(OEt$_2$)] or [Au(C$_6$F$_5$)$_3$(OEt$_2$)] can displace this poorly coordinated molecule by addition of S-, O-, N-, P-, As- donor ligands. As a result, [Au(C$_6$F$_5$)$_2$Cl(dpma)] [29], [Au(C$_6$F$_5$)$_3$(dpma)] [29], [Au(C$_6$F$_5$)$_3$(dppmS$_2$)] [88], [Au(C$_6$F$_5$)$_3$(L)] (L = THF, OCMe$_2$, dioxane, EtOH, SPPh$_3$, SPPh$_2$Me, OPPh$_3$, OAsPh$_3$, NCMe, NCPh, NCPr, NC(CH=CH$_2$), o-(CN)$_2$C$_6$H$_4$) [191] (L = dithiocarbamates, S(MeS)CNHR, R = p-MeC$_6$H$_4$, o-MeC$_6$H$_4$, p-MeOC$_6$H$_4$, 3,5-Me$_2$C$_6$H$_3$) [61], [Au(C$_6$F$_5$)$_3$(EPPh$_2$CH$_2$PPh$_2$R)]ClO$_4$ (E = S, O; R = Me, CH$_2$C$_6$F$_5$, CH$_2$COOMe, Bz; E = O, R = Bz) [191] or [Au(C$_6$F$_5$)$_3$(HL)] (HL = heterocyclic thiones: C$_3$H$_5$NS$_2$, C$_4$H$_4$N$_2$S, C$_5$H$_5$NS, C$_7$H$_5$NS$_2$, C$_7$H$_6$N$_2$S) [49] are described.

Addition of AgClO$_4$ to an acetone or diethylether solution of the dimer [Au(μ-Cl)(C$_6$F$_5$)$_2$]$_2$ gives cis-[Au(C$_6$F$_5$)$_2$(OCMe$_2$)$_2$]ClO$_4$ [177] and cis-[Au(C$_6$F$_5$)$_2$(OEt$_2$)$_2$]ClO$_4$, also described in Equation 3.17. Both complexes are excellent starting materials by displacement of the labile ligands OCMe$_2$ or OEt$_2$ by other more coordinating neutral monodentate, bidentate or anionic ligands, giving complexes of the type: cis-[Au(C$_6$F$_5$)$_2$L$_2$]ClO$_4$ (L = tht, PPh$_2$Me, L$_2$ = dppm, pdma, bipy) and cis-[Au(C$_6$F$_5$)$_2$L$_2$] (L$_2$ = acac [177], BiBzIm (2,2'-bidenzimidazole) [177], (CN)$_2$ [177] and (CNR)$_2$ (R = Ph, p-tolyl) [186]. The latter bis(isocyanide)bis(pentafluorophenyl) gold(III) complexes lead to the formation of cyclic carbenes as shown in Scheme 3.1.

Cyclic carbenes [46] are also described by the addition of a isocyanide to a diamine gold(III) complex (Scheme 3.2).

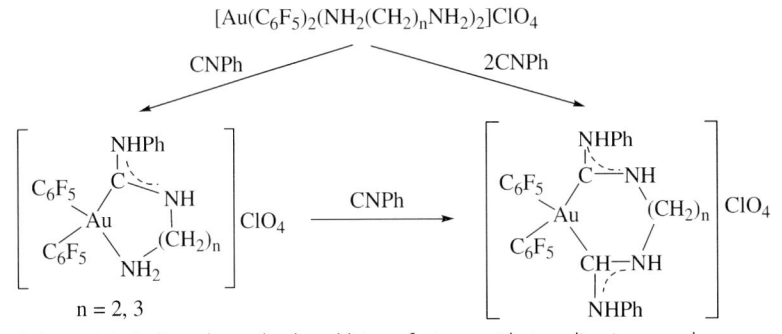

Scheme 3.2 Cyclic carbenes by the addition of a isocyanide to a diamine complex.

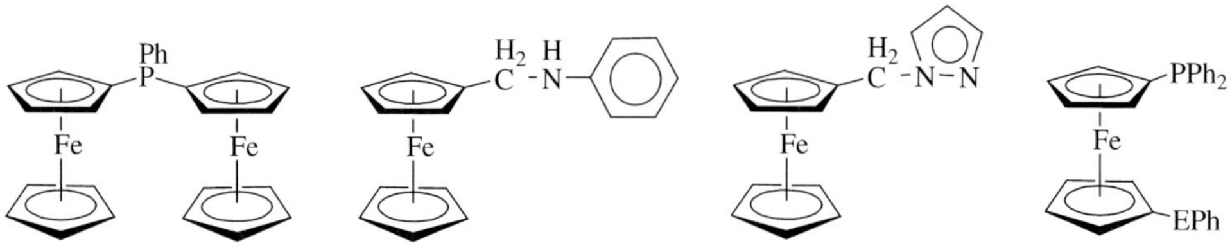

Figure 3.26 Ferrocenyl derivatives.

Many ferrocenyl derivatives have been used as N-, P-donor monodentate ligands (Figure 3.26) *via* displacement of the labile OEt$_2$ molecule or cleaving the chlorine bridge in [Au(μ-Cl)(C$_6$F$_5$)$_2$]$_2$. Thus, complexes of the type: [Au(C$_6$F$_5$)$_2$(Fcpy)$_2$]ClO$_4$ [144], [Au(C$_6$F$_5$)$_3$(L)] —L = Fcpy (Fc = Fe(η^5-C$_5$H$_5$)$_2$) (3-ferrocenylpyridine) [144], PFc$_2$Ph (diferrocenylphenylphosphine) [145], FcCH$_2$NHpyMe [153] and FcCH$_2$pz (ferrocenylmethylpyrazole)— [151] and [Au(C$_6$F$_5$)$_2$Cl(L)] —L = Fcpy [144], FcCH$_2$NHpyMe— [153]. Chalcogeno-phosphino-derivatives of ferrocene such as: 1-diphenylphosphino-1'-(phenylsulfanyl)ferrocene and 1-diphenylphosphino-1'-(phenylselenyl)ferrocene —Fc(EPh)PPh$_2$ (E = S, Se)— (Figure 3.26) can also act as mono- or bidentate ligands through the P atom or through the S, Se atoms, giving the mononuclear chelating complexes: [Au(C$_6$F$_5$)$_2${Fc(SePh)PPh$_2$}]ClO$_4$, [Au(C$_6$F$_5$)$_3${Fc(EPh)PPh$_2$}] (E = S, Se), where Fc(EPh)PPh$_2$ is coordinated through the P atom and the dinuclear chelated complex [{Au(C$_6$F$_5$)$_3$}$_2${Fc(SePh)PPh$_2$}] [150].

Elimination of the chlorine atom from [Au(C$_6$F$_5$)$_2$Cl(dpma)] [29] with AgClO$_4$ gives the corresponding chelate bis(diphenylarsine)methanegold derivative, which evolves to the methanide [Au(C$_6$F$_5$)$_2$(Ph$_2$AsCHAsPh$_2$)] under deprotonation reaction with NaH.

The ligands N-[bis(isopropoxy)thiophosphoryl]thiobenzamide and N-[bis(isopropoxy)thiophosphoryl]-N'-phenylthiourea can be deprotonated with the acetylacetonate gold(III) derivative [Au(C$_6$F$_5$)$_2$(acac)] [26] giving the corresponding pentafluorophenyl complexes with the ligand acting as a chelate one [192].

A transmetallation reaction with [Hg(C$_6$F$_5$)$_2$] can be used as another way to prepare pentafluorophenylgold(III) complexes as a result of the transference of the aryl groups from mercury to the gold center. Thus, Q[Au(C$_6$F$_5$)Cl$_3$] (Q = BzPPh$_3$, NMe$_4$), also available by a simple oxidation of the gold(I) derivative, can be obtained by reaction of Q[AuCl$_4$] and [Hg(C$_6$F$_5$)$_2$]. [Hg(C$_6$F$_5$)$_2$] does not transfer the C$_6$F$_5$ fragment to Q[Au(C$_6$F$_5$)Cl$_3$], however using as starting material a pentafluorophenylgold(III) derivative with a second aryl group such as 2-C$_6$H$_4$CH$_2$NMe$_2$ or C$_6$H$_4$N=NPh-2, the transmetallation is possible in the presence of an excess of NMe$_4$Cl, which favors the elimination of the mercury residue as (NMe$_4$)$_2$[Hg$_2$Cl$_6$], giving [Au(2-C$_6$H$_4$CH$_2$NMe$_2$)(C$_6$F$_5$)Cl] [193] (**12**) or [Au(C$_6$H$_4$N=NPh-2)(C$_6$F$_5$)Cl] [194] (**13**). The former complex **12** affords [Au(2-C$_6$H$_4$CH$_2$NMe$_2$)(C$_6$F$_5$)X] (X = O$_2$CMe, Br) *via* substitution reactions.

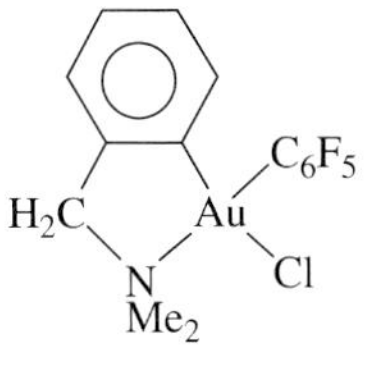

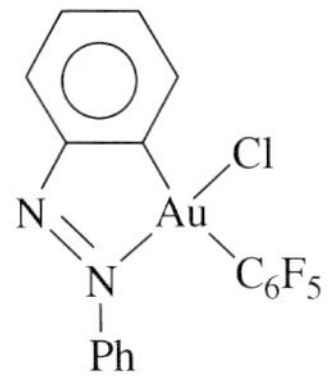

3.3 Pentafluorophenylgold(III) Derivatives

12 — structure showing H$_2$C, NMe$_2$, Au, C$_6$F$_5$, Cl

13 — structure showing N, N-Ph, Au, Cl, C$_6$F$_5$

Dithiolate tin or zinc complexes can also act as transmetallating agents towards gold centers, in this case transferring the dithiolate group. As an example [SnMe$_2$(S-S)] and Q$_2$[Zn(S-S)$_2$] (Q = NEt$_4$, N(PPh$_3$)$_2$; S-S = 1,2-S$_2$C$_6$H$_4$, 3,4-S$_2$C$_6$H$_3$Me, C$_3$S$_5$) transfer the S2-donor ligand to pentafluorophenyl gold(III) derivatives in very mild conditions leading to [Au(C$_6$F$_5$)(S-S)L] (L = PPh$_3$, PPh$_2$Me, AsPh$_3$) [195] and the polynuclear compounds [Au(C$_6$F$_5$)(S-S)]$_n$ [196] which in the case of S-S = 1,2-S$_2$C$_6$H$_4$, the X-ray analysis reveals a trinuclear nature (Figure 3.27). In the preparation of [Au(C$_6$F$_5$)(S$_2$C$_6$H$_4$)]$_3$ a small amount of [Au(C$_6$F$_5$)(S$_2$C$_6$H$_4$)(SC$_6$H$_4$SPPh$_3$)] is obtained which has been crystallographically characterized [196]. Substitution of the labile ligand AsPh$_3$ in [Au(C$_6$F$_5$)(S-S)(AsPh$_3$)] by diphosphines gives the mononuclear [Au(C$_6$F$_5$)(S-S)(L-L)] (L-L = dppm, dppe) with a free P atom, which can coordinate an additional metallic unit leading to the unsymmetrical digold (I,III) [Au(C$_6$F$_5$)(S-S)(μ-dppm)(AuX)] (X = Cl, C$_6$F$_5$) [160].

An additional route to prepare pentafluorophenyl gold(III) derivatives has been also described for gold(I) and consists of the atypical intermolecular ylide transfer reaction from an ylide-gold(I) to another gold(III) complex. Accordingly, [Au(CH$_2$PPh$_3$)$_2$]ClO$_4$ transfers the ylide fragment to [Au(C$_6$F$_5$)$_3$(OEt$_2$)] or [Au(μ-Cl)(C$_6$F$_5$)$_2$]$_2$ giving rise to [Au(C$_6$F$_5$)$_3$(CH$_2$PR$_3$)] (PR$_3$ = PPh$_3$, PPh$_2$Me) and [Au(C$_6$F$_5$)$_2$(Cl)(CH$_2$PPh$_3$)] respectively [69, 197].

Other oxidants such as TCNQ (7,7′,8,8′-tetracyanoquinodimethane) can be used to oxidize a pentafluorophenylgold(I) complex, as takes place in the case of Q$_2$[Au$_2$(μ-dmit)(C$_6$F$_5$)$_2$], which in the presence of TCNQ evolves in the oxidation of one gold atom giving a gold(III) [108] complex according to Equation 3.20

$$Q_2[Au_2(\mu\text{-dmit})(C_6F_5)_2] + TCNQ \rightarrow Q[Au(dmit)(C_6F_5)_2] + QTCNQ + Au°$$
(3.20)

Q = NBu$_4$ and N(PPh$_3$)$_2$

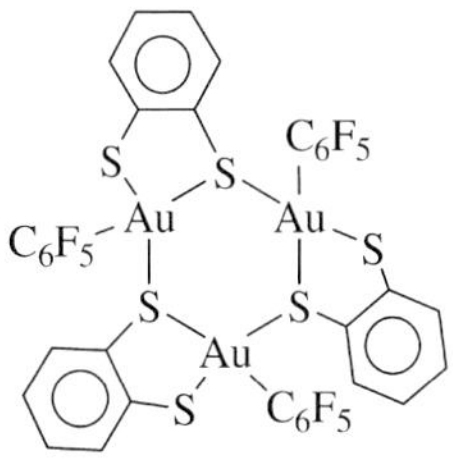

Figure 3.27 Trinuclear gold(III) complex.

Cation-radical salts with pentafluorophenyl gold(III) anions such as $(TTFPh_2)_{2.5}[Au(C_6F_5)_2Cl_2]$ and $(TTFPh_2)[Au(C_6F_5)_2I_2]$, where $TTFPh_2$ is the donor molecule 4,4′-diphenyltetrathiafulvalene, can be performed by electrocrystallization techniques [83].

3.3.2
Di and Polynuclear Pentafluorophenylgold(III) Derivatives

3.3.2.1 C-Donor Ligands

The presence of an excess of electron density on the carbon atom in methanide gold (III) complexes permits them to act as C-donor nucleophiles forming dinuclear or even polynuclear derivatives.

Thus, the abstraction of the chlorine atom in $[Au(C_6F_5)_2Cl(dppm)]$ [175] provides a mononuclear gold(III) derivative with the diphosphine acting as a chelate ligand, which is an excellent starting material in the synthesis of dinuclear or polynuclear methanide and methanidiide complexes after deprotonation of the dppm ligand (first deprotonation with HNa and the second with acetylacetonate gold(I) derivatives) [198–201] (Scheme 3.3)

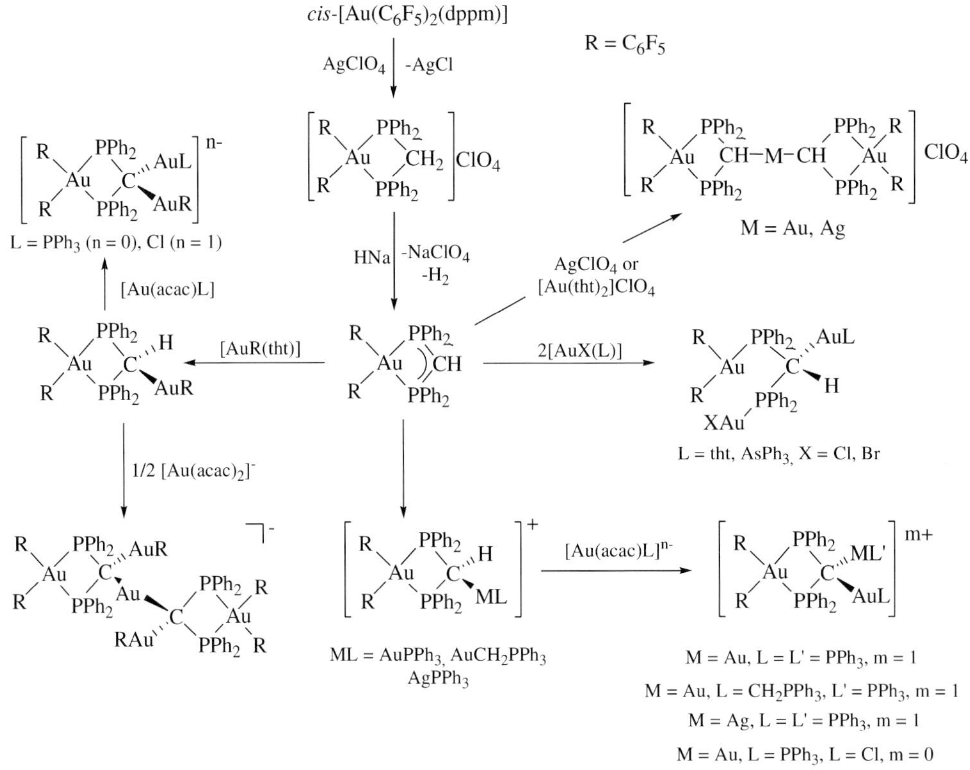

Scheme 3.3 Synthesis of cyclic methanide derivatives.

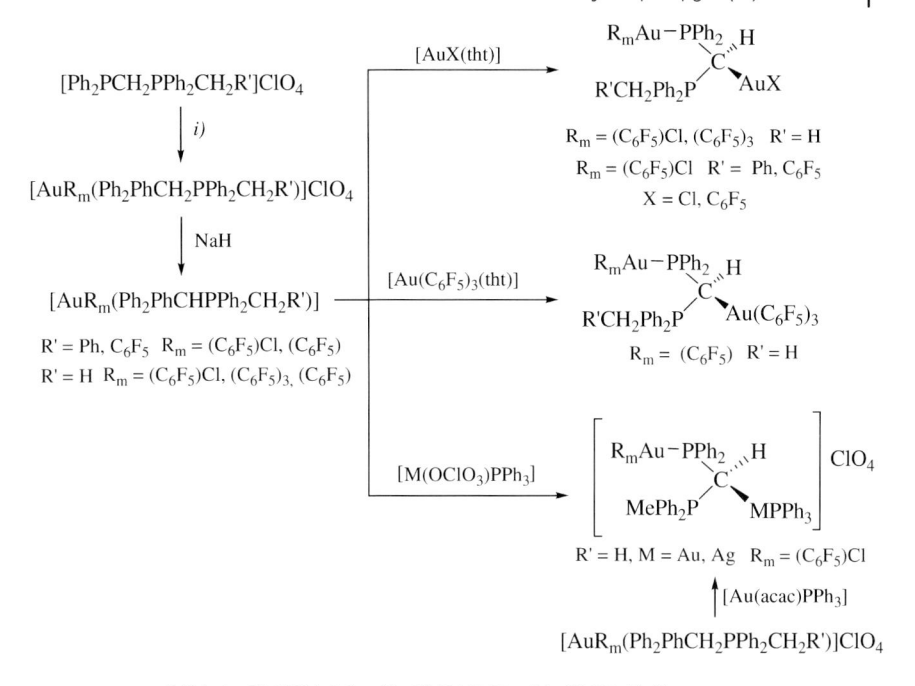

3.3 Pentafluorophenylgold(III) Derivatives

[Ph$_2$PCH$_2$PPh$_2$CH$_2$R']ClO$_4$

↓ i)

[AuR$_m$(Ph$_2$PhCH$_2$PPh$_2$CH$_2$R')]ClO$_4$

↓ NaH

[AuR$_m$(Ph$_2$PhCHPPh$_2$CH$_2$R')]

R' = Ph, C$_6$F$_5$ R$_m$ = (C$_6$F$_5$)Cl, (C$_6$F$_5$)
R' = H R$_m$ = (C$_6$F$_5$)Cl, (C$_6$F$_5$)$_3$, (C$_6$F$_5$)

[AuX(tht)] →

R$_m$Au−PPh$_2$ H
         \\C/
R'CH$_2$Ph$_2$P   AuX

R$_m$ = (C$_6$F$_5$)Cl, (C$_6$F$_5$)$_3$ R' = H
R$_m$ = (C$_6$F$_5$)Cl   R' = Ph, C$_6$F$_5$
X = Cl, C$_6$F$_5$

[Au(C$_6$F$_5$)$_3$(tht)] →

R$_m$Au−PPh$_2$ H
         \\C/
R'CH$_2$Ph$_2$P   Au(C$_6$F$_5$)$_3$

R$_m$ = (C$_6$F$_5$) R' = H

[M(OClO$_3$)PPh$_3$] →

[R$_m$Au−PPh$_2$ H
           \\C/
   MePh$_2$P   MPPh$_3$] ClO$_4$

R' = H, M = Au, Ag   R$_m$ = (C$_6$F$_5$)Cl

↑ [Au(acac)PPh$_3$]

[AuR$_m$(Ph$_2$PhCH$_2$PPh$_2$CH$_2$R')]ClO$_4$

i) [{Au(μ-Cl)(C$_6$F$_5$)$_2$}$_2$] or [Au(C$_6$F$_5$)(tht)] or [Au(C$_6$F$_5$)$_3$(tht)]

Scheme 3.4 Synthesis of non-cyclic methanide derivatives.

Cleavage of the chlorine bridges in the dinuclear complex [Au(μ-Cl)(C$_6$F$_5$)$_2$]$_2$ by the addition of the phosphino phosphonium ligand [Ph$_2$PCH$_2$PPh$_2$Me]ClO$_4$ [84] and [Ph$_2$PCH$_2$PPh$_2$CH$_2$R']ClO$_4$ (R' = Ph, C$_6$F$_5$) [202] or the displacement of the labile tht in [Au(C$_6$F$_5$)$_3$(tht)] by the same salts leads to mononuclear [AuR$_m$(Ph$_2$PCH$_2$PPh$_2$CH$_2$R')]ClO$_4$ (R$_m$ = (C$_6$F$_5$)Cl, (C$_6$F$_5$)$_3$; R' = H, Ph or C$_6$F$_5$). The presence of the positive charge on the quaternary phosphorus atom increases the acidity of the methylenic protons. Therefore subsequent deprotonation with NaH [84, 202] provides the preparation of methanide gold(III) derivatives able to attach a new metallic fragment to the CH unit, giving mixed-valence dinuclear compounds (Scheme 3.4). The use of acetylacetonate gold complexes as deprotonating agents such as [Au(acac)PPh$_3$] affords the deprotonation and the coordination of the AuPPh$_3$ unit in one step [86]. This procedure leads to the synthesis of non-cyclic phosphino-methanide derivatives, since the presence of the quaternary phosphorus atom avoids the cyclization.

These non-cyclic phosphino methanide derivatives can also be available by using the dinuclear complex [Au(C$_6$F$_5$)$_3$(μ-PPh$_2$CH$_2$PPh$_2$)Au(C$_6$F$_5$)] [172] in the presence of [Au(acac)PPh$_3$] giving directly the trinuclear methanide derivatives [Au(C$_6$F$_5$)$_3${PPh$_2$CH(AuPPh$_3$)PPh$_2$}Au(C$_6$F$_5$)] and N(PPh$_3$)$_2$[Au(C$_6$F$_5$)$_3$.{PPh$_2$CH(AuX)PPh$_2$}Au(C$_6$F$_5$)] (X = Cl, C$_6$F$_5$) as a result of a deprotonation with N(PPh$_3$)$_2$(acac) in the first place followed by displacement of tht in [AuX(tht)]. A pentanuclear mixed-valence gold is additionally described by reaction between [Au(C$_6$F$_5$)$_3$(μ-PPh$_2$CH$_2$PPh$_2$)Au(C$_6$F$_5$)] and N(PPh$_3$)$_2$[Au(acac)$_2$] [86].

Another way to describe this non-cyclic phosphino methanide derivatives consists of the use of monodentate complexes with the mono-oxide [203] or monosulfide [204] dppm giving the methanide or methanediide derivatives (**14**) after addition of the appropriate amounts of [Au(acac)PPh$_3$]. A mixed-valence Au(I)-Au(III) can be isolated by coordination of the fragment AuPPh$_3$ to the free sulfur atom in the mono-oxide starting material [204].

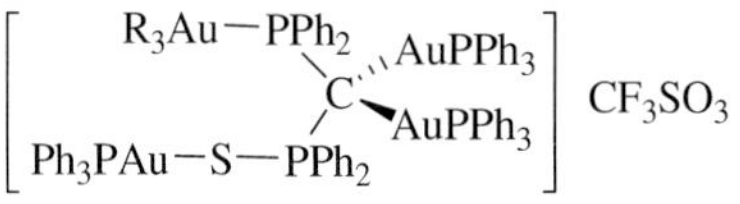

$$\left[\begin{array}{c} R_3Au-PPh_2\diagdown_{\diagup\!\!\!\!AuPPh_3} \\ C \\ Ph_3PAu-S-PPh_2{\diagup}^{\diagdown AuPPh_3} \end{array}\right] CF_3SO_3$$

14

In the case of the phosphonium salt [PPh$_2$CH$_2$CH$_2$COOMe]ClO$_4$ and the corresponding sulfide [SPPh$_2$CH$_2$CH$_2$COOMe]ClO$_4$, deprotonation with NaH leads to mononuclear cyclic methanides (**15** and **17**) that coordinate additional fragments to the CH group giving dinuclear mixed valence gold(I)-gold(III) derivatives (**16** and **18**) [205, 206]. A second deprotonation in **16** leads to methanediide auracycles which show basic nature, coordinating a second metallic center (**19**) [207].

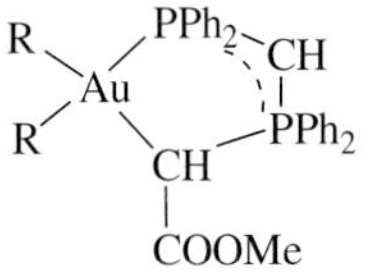

15 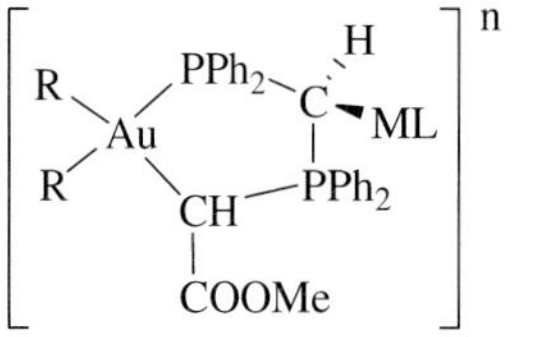 **16** 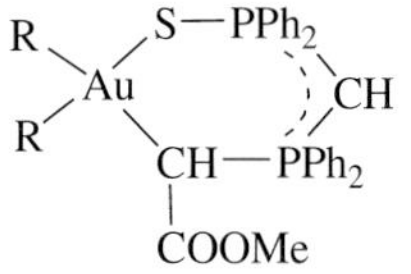 **17** 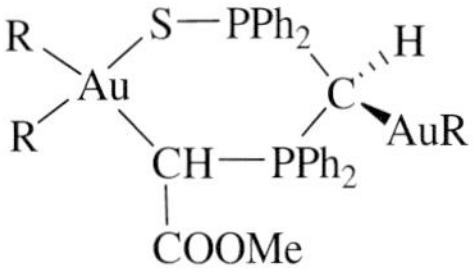 **18**

ML = AuR, AuCl, n = 0; AgPPh$_3$, n = -1   R = C$_6$F$_5$

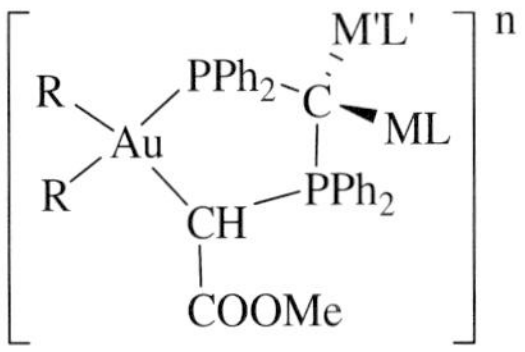

19

ML = M'L' = AuPPh$_3$ or AgPPh$_3$, n = 1

ML = AuPPh$_3$ or AgPPh$_3$, M'L' = AuC$_6$F$_5$, n = 0

Deprotonation of the six-membered bis(diphenylphosphine)methane disulfide auracycle [Au(C$_6$F$_5$)$_2$(dppmS$_2$)]ClO$_4$ [88] with NaH gives a four-membered methanide cycle (**20**) with a structure similar to that of [(Et$_3$P)ClPtSPPh$_2$CHPPh$_2$S] published elsewhere [208]. Using as deprotonationg agent an acetylacetonate gold(I) complex such as [Au(acac)L] (L = PPh$_3$, AsPh$_3$) the six-membered cycle structure is maintained at the same time that a mono- or di-deprotonation takes place, affording di- or tri-nuclear methanide or methanidide auracycles (**21**) [204]. Addition of [Au(acac)PPh$_3$] to **20** gives the corresponding methanide auracycle [Au(C$_6$F$_5$)$_2${SPPh$_2$C(AuPPh$_3$)PPh$_2$S}] with the same four-membered cyclic structure [204].

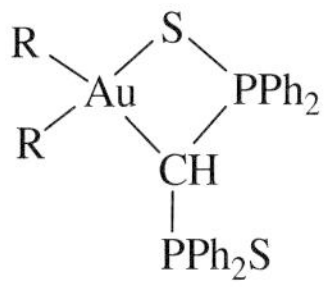

3.3 Pentafluorophenylgold(III) Derivatives

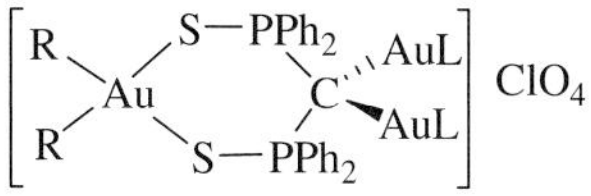

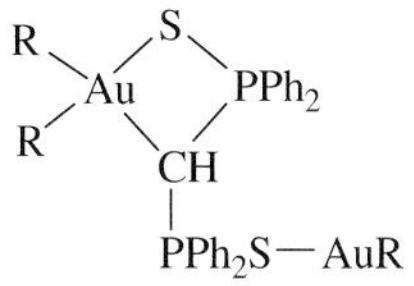

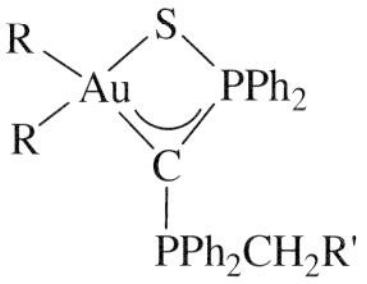

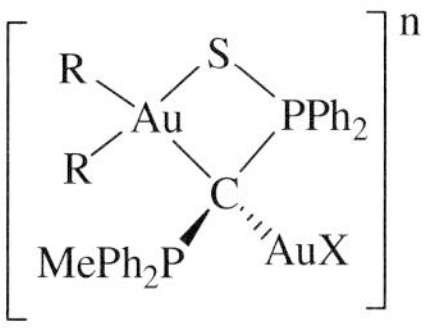

20, 21, 22, 23, 24 (structural diagrams)

Coordination of an additional metal center to the free sulfur atom in **20** affords dinuclear (**22**) or trinuclear [(C$_6$F$_5$)$_2$Au(S-PPh$_2$CHPPh$_2$-S)M(S-PPh$_2$CHPPh$_2$-S)Au(C$_6$F$_5$)$_2$]ClO$_4$ (M = Au, Ag) derivatives. A double deprotonation of [Au(C$_6$F$_5$)$_2$Cl(SPPh$_2$CH$_2$PPh$_2$CH$_2$R)]ClO$_4$ (R = H, Ph) [209] gives the methanide auracycles (**23**) similar in structure to **20**, which incorporate new gold(I) fragments affording dinuclear complexes (**24**).

A five-membered methanide auracycle [Au(C$_6$F$_5$)$_2$(SPPh$_2$CH$_2$PPh$_2$)]ClO$_4$ (**25**) is described with the monosulfonated dppm, obtained after chlorine elimination of [Au(C$_6$F$_5$)$_2$Cl(SPPh$_2$CH$_2$PPh$_2$)] with a silver salt. After deprotonation of the methylene group (**26**) and later coordination of additional metal centers affords dinuclear and trinuclear methanide derivatives (**27, 28**) [210].

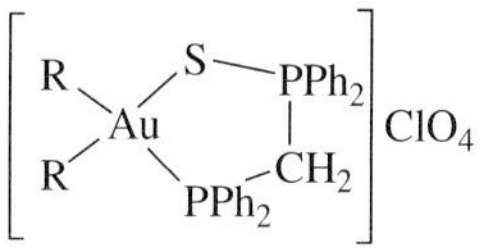

25, 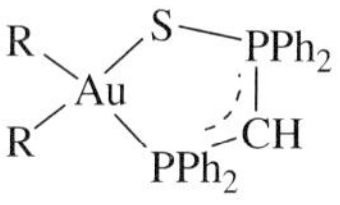 26, 27, 28 (structural diagrams)

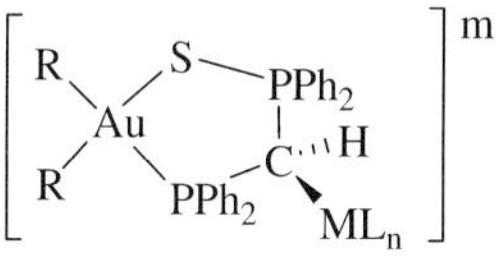

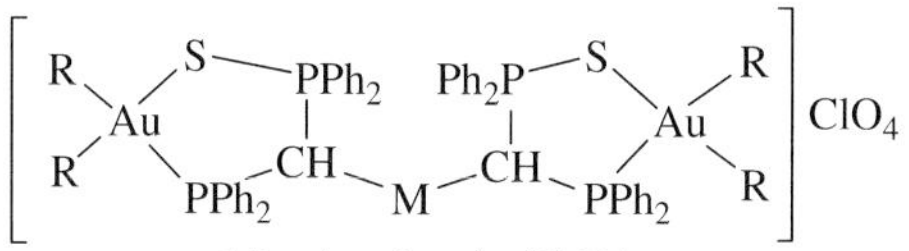

As shown above, the use of acetylacetonate complexes such as Q[Au(acac)$_2$] or [Au(acac)(PPh$_3$)] as deprotonating agents allows the formation of Au–C bonds very easily [86, 204, 211]. Thus, the mono-deprotonation of the trinuclear [{Au(C$_6$F$_5$)$_3$(PPh$_2$CH$_2$PPh$_2$)}$_2$Au]ClO$_4$ [211] using NBu$_4$(acac)$_2$ leads to the bis(methanide)

trinuclear gold(III) NBu$_4$[{Au(C$_6$F$_5$)$_3$(PPh$_2$CHPPh$_2$)}$_2$Au] that affords the pentanuclear gold(I)-gold(III) derivative after coordination of an Au-X fragment NBu$_4$[{Au(C$_6$F$_5$)$_3$(PPh$_2$CH(AuX)PPh$_2$)}$_2$Au] (X = Cl, C$_6$F$_5$) (**29**). Using tetranuclear gold(I) complexes as starting materials with the acac ligand [(acac)AuCH(PPh$_2$AuPPh$_2$)$_2$CHAu(acac)] in the presence of the phosphane gold(III) derivatives [Au(C$_6$F$_5$)$_2$(dppm)] [199] and [Au(C$_6$F$_5$)$_3$(dppm)] [172] leads to the deprotonation of the CH$_2$ group of the dppm ligand and formation of the hexanuclear methanide complexes (**30** and **31**) [212].

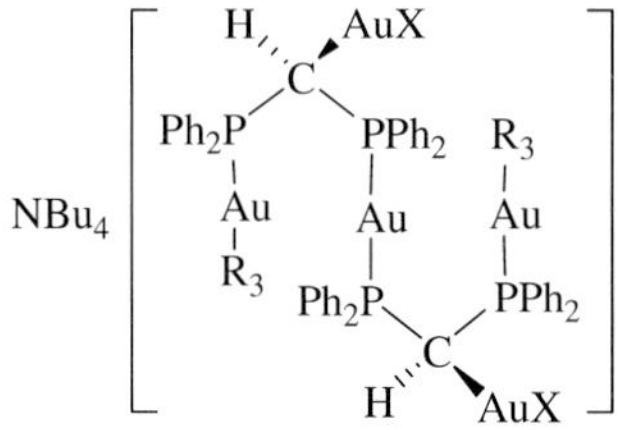

29

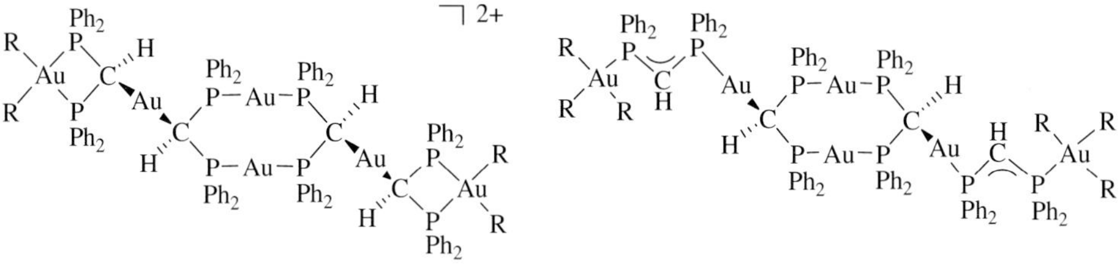

30 **31**

Since the deprotonation of the alkyne in PPh$_2$C≡CH with sodium alkoxide gives a single and a double alcohol addition to the triple bond (see above), the addition of acetylacetonate complexes gives the deprotonation at the same time as the coordination of additional metallic fragments to the C atom of the phosphinoalkyne: [(C$_6$F$_5$)$_3$Au(PPh$_2$CC)(MPPh$_3$)] (M = Au, Ag), [(C$_6$F$_5$)$_3$Au(PPh$_2$CC)Au(μ-dppe)Au(PPh$_2$CC)Au(C$_6$F$_5$)$_3$] and [(C$_6$F$_5$)$_3$Au(PPh$_2$CC)Au(PPh$_2$CC)Au(C$_6$F$_5$)$_3$] [181].

A zwitterionic complex [(C$_6$F$_5$)$_2$Au^{III}(μ-2-C$_6$H$_4$PPh$_2$)$_2$Au^I] [109] structurally characterized, has been described as a result of the evolution of the thermodynamically unstable gold(II) derivative [Au^{II}$_2$(C$_6$F$_5$)$_2$(μ-2-C$_6$H$_4$PPh$_2$)$_2$] via an intermediate [(C$_6$F$_5$)Au(μ-2-Ph$_2$PC$_6$H$_3$-6-Me)Au(C$_6$F$_5$){η^2-(6-MeC$_6$H$_3$-2-PPh$_2$)] also characterized by X-ray analysis [110] (Equation 3.21)

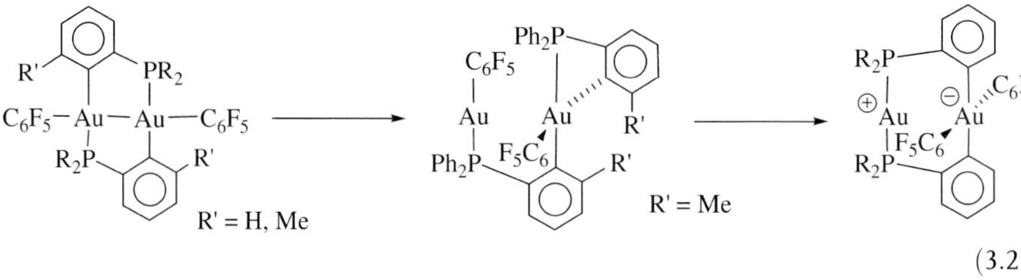

(3.21)

3.3.2.2 Diamine and Carbene Bridges

It is well known that amines react with gold isocyanide to give the corresponding carbenes [62, 213]. Similarly, diamines are able to use one of the $-NH_2$ groups giving the gold(III) carbene $[Au(C_6F_5)_2C\{NH(CH_2)_nNH_2)(NHR)\}]$ ($n=2,3$; $R=Ph$, C_6H_4Me-p).

Carbene gold(III) with diamine ligand can coordinate a second metallic center, since the N-ligand contains a free NH_2 group and can react with a new gold isocyanide under the formation of double dinuclear carbenes [46] (**32**).

$$(C_6F_5)_mAu-C\begin{matrix}ToHN\\NH\text{-}(CH_2)_n\text{-}NH\\NHTo\end{matrix}C-Au(C_6F_5)_m$$

$$m = 1 \text{ or } 3, n = 2, 3$$

32

3.3.2.3 N-, P- and As-Donor Ligands

The terminal N atom of the cyanide, isocyanide and azyde derivatives acts as a donor atom providing dinuclear complexes by coordination of an additional metallic fragment. As examples the reaction of pentafluorophenyl gold(III) $Q[Au(C_6F_5)_3CN]$, $Q[Au(C_6F_5)_2X_2]$ ($Q=NBu_4$, $X=CN$; $Q=BzPPh_3$, $X=SCN$)), $[Au(C_6F_5)_2X(PPh_3)]$ ($X=SCN$ [183], N_3 [184]) and gold(I) $Q[Au(C_6F_5)CN]$ complexes displace the labile ligand from $[Au(OClO_3)(PPh_3)]$, $[Au(C_6F_5)_2(OClO_3)(PPh_3)]$ or $[Au(C_6F_5)_3(tht)]$ giving $[(C_6F_5)_3Au(\mu\text{-}X)AuPPh_3]$ ($X=CN$ [63] and SCN [80]), $[(C_6F_5)Au(\mu\text{-}X)Au(C_6F_5)_2PPh_3]$ ($X=SCN$, N_3) [80], $[(C_6F_5)_3Au(\mu\text{-}SCN)Au(C_6F_5)_2PPh_3]$ [80], $[(C_6F_5)_2CNAu(\mu\text{-}CN)AuPPh_3]$ [63], $[(Ph_3P)(C_6F_5)_2Au(\mu\text{-}X)Au(C_6F_5)_2(PPh_3)]ClO_4$ [214] and $Q[(C_6F_5)Au(\mu\text{-}CN)Au(C_6F_5)_3]$ [63]. Reaction with perchlorate gold [63] or silver salts [63, 80] afford cyano-bridged dinuclear compounds or even polymeric species (Equations 3.22–3.24)

$$Q[Au(C_6F_5)_3X] + AgClO_4 \rightarrow AgCN + QClO_4 + Q[(C_6F_5)_3Au(\mu\text{-}X)Au(C_6F_5)_3] \quad (3.22)$$

$$Q = BzPPh_3, (PPh_3)_2N; \quad X = CN, SCN, N_3$$

$$Q[Au(C_6F_5)_2(CN)_2] + AgClO_4 \rightarrow QClO_4 + [(F_5C_6)_2(CN)Au(\mu\text{-}CN)Ag]_x \quad (3.23)$$

$$Q[Au(C_6F_5)_2(CN)_2] + [Au(OClO_3)(tht)] \rightarrow QClO_4 + [(C_6F_5)_2Au(CN)_2Au]_x \quad (3.24)$$

Coordination of pentafluorophenyl gold(III) to the two free N-donor atoms in 1,1'-bis(2-pyridylthio)ferrocene leads to the dinuclear complex $[\{Au(C_6F_5)_3\}_2\{Fc(Spy)_2\}]$ structurally characterized [149].

Coordination of an additional metallic fragment to a free donor atom of a bidentate diphosphine, dppm [172, 175], $(PPh_2)_2C=CH_2$ (vinylidenebis(diphenylphosphine) [92] and dppa (bis(diphenylphosphino)amine) [87] give mixed valence Au(I)-Au(III) complexes with the formulae: [M'(μ-dppm)Au(PPh$_3$)]ClO$_4$ (M' = AuCl(C$_6$F$_5$)$_2$, Au(C$_6$F$_5$)$_3$), [M'(μ-dppm)AuX] (M' = AuCl(C$_6$F$_5$)$_2$, Au(C$_6$F$_5$)$_3$, X = Cl, C$_6$F$_5$), [Au(C$_6$F$_5$)$_3$(μ-L-L)AuX] (LL = (PPh$_2$)$_2$C=CH$_2$ and dppa; X = Cl, C$_6$F$_5$) or [Au(C$_6$F$_5$)X$_2$(μ-dppa)Au(C$_6$F$_5$)] (X = Cl, Br) by reaction of [Au(C$_6$F$_5$)$_2$Cl(L-L)] or [Au(C$_6$F$_5$)$_3$(L-L)] (L-L = dppm, (PPh$_2$)$_2$C=CH$_2$, dppa) with [Au(OClO$_3$)PPh$_3$] or [AuX(tht)] (X = Cl, C$_6$F$_5$). Trinuclear derivatives with the formula: [{Au(C$_6$F$_5$)$_3$(μ-L-L)}$_2$M]ClO$_4$ [92, 211] (M = Au, Ag; LL = (PPh$_2$)$_2$C=CH$_2$ and dppm) are described from the reaction of [Au(C$_6$F$_5$)$_3$(L-L)] and [Au(tht)$_2$]ClO$_4$ or AgClO$_4$.

Acetylacetonato (acac) complexes can be employed as deprotonating agents (see above) and, in the particular case of [Au(C$_6$F$_5$)$_2$(acac)], deprotonation of tris(diphenylphosphino)methane (PPh$_2$)$_3$CH leads to the methanide gold(III) complex [Au(C$_6$F$_5$)$_2${(PPh$_2$)$_2$CPPh$_2$}] [215]. The displacement of labile ligands by the free phosphorus atom in the later complex affords dinuclear and trinuclear methanide derivatives, with formulae: [(C$_6$F$_5$)$_2$Au{(PPh$_2$)$_2$CPPh$_2$}AuL$_n$] (AuL$_n$ = AuCl, AuC$_6$F$_5$, Au(C$_6$F$_5$)$_2$Cl, Au(C$_6$F$_5$)$_3$), [{(C$_6$F$_5$)$_2$Au[(PPh$_2$)$_2$CPPh$_2$]}$_2$M]ClO$_4$ (M = Au, Ag) and [{(C$_6$F$_5$)$_2$Au[(PPh$_2$)$_2$CPPh$_2$]}$_2$Au(C$_6$F$_5$)$_2$]ClO$_4$.

The phosphine (Ph$_2$PCH$_2$)$_3$CMe (tdppme) can act as a mono-, bi- or even tridentate-chelating ligand, giving to the corresponding mono-, bi- or trinuclear derivatives. The monodentate complex can be the starting material for the preparation of higher nuclearities. The displacement of tht in [Au(C$_6$F$_5$)$_3$(tht)] or brigde-chlorine cleavage in [Au(C$_6$F$_5$)$_2$(μ-Cl)]$_2$ lead to the mononuclear products: [{Au(C$_6$F$_5$)$_3${Ph$_2$PCH$_2$C(Me)(CH$_2$PPh$_2$)$_2$}] and [{Au(C$_6$F$_5$)$_2${Ph$_2$PCH$_2$C(Me)(CH$_2$PPh$_2$)$_2$}]Cl, which after addition of a gold(I) or gold(III) fragments affords the di- and trinuclear complexes [91]: [{Au(C$_6$F$_5$)$_3${Ph$_2$PCH$_2$C(Me)(CH$_2$PPh$_2$)$_2$}AuL]ClO$_4$ (L = PPh$_3$, CH$_2$PPh$_3$), [{Au(C$_6$F$_5$)$_2${Ph$_2$PCH$_2$C(Me)(CH$_2$PPh$_2$)$_2$}Au(C$_6$F$_5$)$_3$]ClO$_4$, [{Au(C$_6$F$_5$)$_3${Ph$_2$PCH$_2$C(Me)(CH$_2$PPh$_2$)$_2$}$_2$Au]ClO$_4$ and [{Au(C$_6$F$_5$)$_3${Ph$_2$PCH$_2$C(Me)(CH$_2$PPh$_2$)$_2$}(AuX)$_2$] (X = Cl, C$_6$F$_5$).

The use of polyphosphine ligands favors the presence of metallic interactions and the presence of luminescence properties [216]. For instance, the tridentate ligand bis(diphenylphosphinemethyl)phenylphosphine (dpmp) −Ph$_2$PCH$_2$PPhCH$_2$PPh$_2$− in the presence of [Au(C$_6$F$_5$)(tht)] gives the trinuclear luminescent complex [{Au(C$_6$F$_5$)}$_3$](μ-dpmp)] [95] (see gold(I) section). However the mixed valence gold(III)-gold(I)-gold(III) [{Au(C$_6$F$_5$)$_3$}$_2$(μ-dpmp)(AuX)] (X = Cl, C$_6$F$_5$) obtained from the mononuclear [{Au(C$_6$F$_5$)$_3$}(dpmp)], which is a mixture of isomers with the gold(III) coordinated to the central or to the lateral phosphorus and [AuX(tht)], do not luminesce, probably because there is no Au(I)-Au(III) interaction. Curiously, longer separation in between the three phosphorus atoms, as takes place in the polyphosphine bis(2-diphenylphosphinoethyl)phenylphosphine (triphos) −Ph$_2$PCH$_2$CH$_2$PPhCH$_2$CH$_2$PPh$_2$−, gives luminescent dinuclear and trinuclear tris(pentafluorophenyl) gold(III) derivatives [{Au(C$_6$F$_5$)$_3$}$_n$(μ-triphos)] (n = 2 or 3) [97].

The secondary phosphane in [Au(C$_5$F$_6$)$_3$(PPh$_2$H)] [179] can be deprotonated with N(PPh$_3$)$_2$[Au(acac)$_2$] affording in the same reaction a trinuclear phosphido-bridged Au(III)-Au(I) derivative N(PPh$_3$)$_2$[{Au(C$_6$F$_5$)$_3$(μ-PPh$_2$)}$_2$Au] [179, 217]. Deprotonation with NBu$_4$(acac) and addition of [Au(C$_6$F$_5$)(tht)] or [MX(PPh$_3$)] (M = Au, Ag, X = Cl, TfO) gives the polynuclear complexes: NBu$_4$[Au(C$_6$F$_5$)$_3$(μ-PPh$_2$){Au(C$_6$F$_5$)}] and [Au(C$_6$F$_5$)$_3$(μ-PPh$_2$)(MPPh$_3$)] (M = Au and Ag) [217]. If [Au(C$_6$F$_5$)$_3$(tht)] is added to the deprotonated phosphine complex a dinuclear gold(III) phosphide with a single atom bridging the metallic centers NBu$_4$[{Au(C$_6$F$_5$)$_3$}$_2$(μ-PPh$_2$)] [217] is obtained. And when the diphenylphosphine is added to [Au(μ-Cl)(C$_6$F$_5$)$_2$]$_2$ the cleavage of the halogen bridges affords a doubly-bridged dimmeric gold(III) phosphide [Au$_2$(C$_6$F$_5$)$_4$(μ-PPh$_2$)$_2$] [218].

Trans-[Au(C$_6$F$_5$)$_2$(PPh$_2$H)$_2$]ClO$_4$ [217], obtained from [Au(C$_6$F$_5$)$_2$(OEt$_2$)$_2$]ClO$_4$ and the free phosphine PPh$_2$H contains two free P-donor atoms, which after deprotonation with an acac salt and coordination to a metallic center give phosphido-bridged mixed valence compounds of the type trans-NBu$_4$[Au(C$_6$F$_5$)$_2$(μ-PPh$_2$){Au(C$_6$F$_5$)}$_2$], cis-[Au(C$_6$F$_5$)$_2$(μ-PPh$_2$){Au(PPh$_3$)}$_2$]ClO$_4$ and [Au(C$_6$F$_5$)$_2$(μ-PPh$_2$)$_2$Au]$_2$ or the heteronuclear [Au(C$_6$F$_5$)$_2$(μ-PPh$_2$)$_2$M]$_2$ (M = Ag, Cu) [217].

The primary phosphine PPhH$_2$ with two hydrogen atoms can give phosphide or phosphodiide polynuclear derivatives by the substitution of one or the two H atoms. Using [Au(C$_6$F$_5$)$_3$(PPhH$_2$)] as starting material the mono or di-deprotonation in the presence of acetylacetonate gold derivatives or acetylacetonate salts and the subsequent addition of a metallic compound lead to the isolation of the dinuclear species: [Au(C$_6$F$_5$)$_3$(μ-PPhH)(AuPR$_3$)] (PR$_3$ = PPh$_3$, PMe$_3$), N(PPh$_3$)$_2$[{Au(C$_6$F$_5$)$_3$}$_2$(μ-PPhH)], the trinuclear compounds: (N(PPh$_3$)$_2$)$_m$[{Au(C$_6$F$_5$)$_3$}$_2$(μ-PPhH){AuL}] (m = 1, L = PPh$_3$; m = 2, L = C$_6$F$_5$), (N(PPh$_3$)$_2$)$_m$[{Au(C$_6$F$_5$)$_3$}$_2$(μ-PPhH){AuPPh$_3$}{AuL}] (m = 0, L = PPh$_3$ (33); m = 1, L = C$_6$F$_5$), N(PPh$_3$)$_2$[{Au(C$_6$F$_5$)$_3$}$_2$(μ-PPhH)M] (M = Au, Ag) and the pentanuclear (N(PPh$_3$)$_2$)$_3$[{Au(C$_6$F$_5$)$_3$}$_2$(μ3-PPhH)$_2$}$_2$Au] derivative(34) [51]. The reaction of this phosphine with the dimmer [Au(μ-Cl)(C$_6$F$_5$)$_2$]$_2$ gives the trinuclear [Au(C$_6$F$_5$)$_2$(μ-PPhH)]$_3$ derivative characterized by X-ray analysis, which undergoes a tetranuclear derivative by deprotonation and coordination of a AuPPh$_3$ unit with [Au(acac)(PPh$_3$)] [51].

R = C$_6$F$_5$

The reaction of the diphosphines [96] 3,4-(NHPPh$_2$)$_2$MeC$_6$H$_3$ and 1,2-(NHPPh$_2$)$_2$C$_6$H$_4$ with [Au(C$_6$F$_5$)$_3$(tht)] give the mononuclear gold(III) complex [Au(C$_6$F$_5$)$_3$(L-L)] (L-L = 3,4-(NHPPh$_2$)$_2$MeC$_6$H$_3$) and 1,2-(NHPPh$_2$)$_2$C$_6$H$_4$), which behave as excellent starting materials in the preparation of the dinuclear [{Au(C$_6$F$_5$)$_3$}$_2$(μ-L-L)] and the mono-deprotonated [{Au(C$_6$F$_5$)$_3$}$_2$(L-L′)] (L-L′ = 3,4-(NPPh$_2$)(NHPPh$_2$)MeC$_6$H$_3$) and 1,2-(NPPh$_2$)(NHPPh$_2$)C$_6$H$_4$). Addition of a AuPPh$_3$ fragment to the later mono-deprotonated derivatives afford the trinuclear complex [C$_6$H$_4${NHPPh$_2$Au(C$_6$F$_5$)$_3$}{N(AuPPh$_3$)PPh$_2$Au(C$_6$F$_5$)$_3$}] [96].

The phosphine 2-(diphenylphosphine)aniline (PNH$_2$) can be used as a P,N-donor as in [Au(C$_6$F$_5$)$_2$(PNH$_2$)]ClO$_4$ or simply P-donor ligand as in the compound [Au(C$_6$F$_5$)$_2$X(PNH$_2$)] (X = C$_6$F$_5$, Cl) [219]. Deprotonation of one or the two protons of the NH$_2$ group with acetylacetonate complexes afford mixed valence Au(I)-Au(III) dinuclear or trinuclear derivatives [Au(C$_6$F$_5$)$_2$(PNH(MPPh$_3$)}]ClO$_4$ (M = Au, Ag) and [Au(C$_6$F$_5$)$_2${PN(AuPPh$_3$)$_2$}]ClO$_4$, with weak gold(I)-gold(III) interactions in the latter complex as derived from the X-ray structure. The bis(diphenylphosphane)acetylene (Ph$_2$PC≡CPPh$_2$) can act as a bridging ligand affording dinuclear complexes as [{Au(C$_6$F$_5$)$_3$}$_2$(μ-Ph$_2$PC≡CPPh$_2$)] [94].

An important number of polynuclear pentafluorophenylgold(III) and mixed gold(I)-gold(III) derivatives have been described with the diphosphine 1,2-diphosphinobenzene, which contains four hydrogen atoms that can be removed and substituted by a metal atom (Scheme 3.5) [220].

The bis(arsine) dpma coordinated by only one donor atom has one free As donor atom capable of coordinating a new metallic fragment. As examples [Au(C$_6$F$_5$)$_3$(dpma)] reacts with [Au(C$_6$F$_5$)$_3$(OEt$_2$)], [Au(OClO$_3$)(PPh$_3$)] or AgClO$_4$ affording the di- and tri-nuclear derivatives with dpma as a bridge ligand: [(C$_6$F$_5$)$_3$Au(μ-dpma)

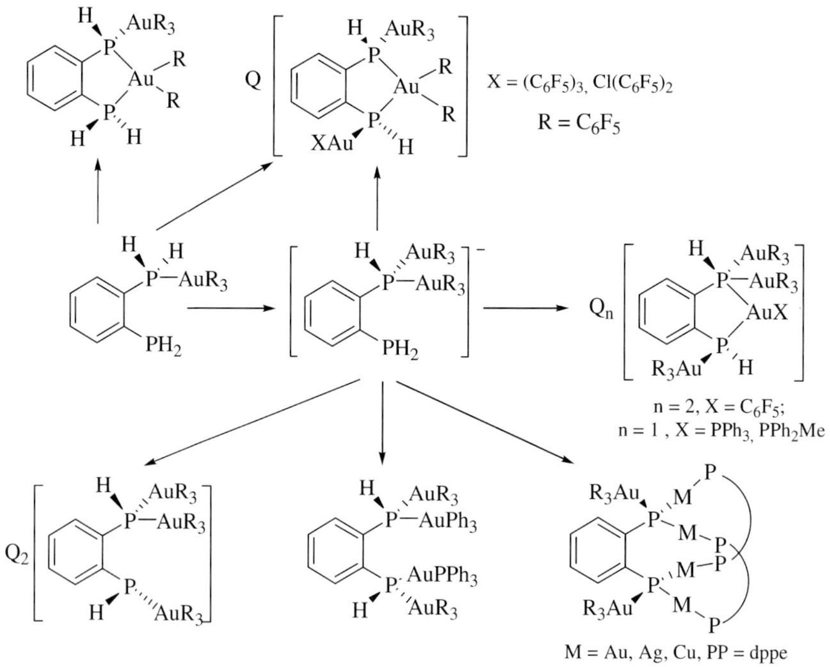

Scheme 3.5 Pentafluorophenylgold(III) or gold(I)/(III) complexes with 1,2- diphosphinobenzene.

Au(C₆F₅)₃], [(C₆F₅)₃Au(μ-dpma)(AuPPh₃)]ClO₄ and [(C₆F₅)₃Au(μ-dpma)-Ag(OClO₃)(μ-dpma)Au(C₆F₅)₃] respectively [29]. The reaction with [AuCl(tht)] in a molar ratio 1:2 or with [Au(tht)₂]ClO₄ gives not only the tht displacement but also a ligand rearrangement leading a trinuclear gold(III)-gold(I)-gold(III) derivative [(C₆F₅)₃Au(μ-dpma)Au(μ-dpma)Au(C₆F₅)₃][AuCl₂] (35). Dinuclear mixed valence gold derivatives can also be obtained through the selective oxidation of the dinuclear gold(I) complex [29] (Equation 3.25)

$$\begin{array}{c}Ph_2As\frown AsPh_2\\|\quad\quad|\\Au\quad Au\\|\quad\quad|\\C_6F_5\ C_6F_5\end{array} \xrightarrow{Cl_2} \begin{array}{c}Ph_2As\frown AsPh_2\\|\ \ \ R\ \ |\\Au\quad Au\\Cl{\diagup}|\quad|\\Cl\quad R\end{array} \quad\quad (3.25)$$

3.3.2.4 S- and Se-Donor Ligands

The mononuclear phosphonium dithiocarboxylate gold(III) [Au(C₆F₅)₃(S₂C-PR₃)] (PR₃ = P(C₆H₁₁)₃, PEt₃, PBu₃, PPhMe₂, PPhEt₂) with a free S-donor atom can add an additional gold(III) metallic unit giving rise to homo dinuclear derivatives with the formula [Au(C₆F₅)₃(S₂C-PR₃)Au(C₆F₅)₃] [89].

The methylene bis(dialkyldithiocarbamate) (CH₂(S₂CNR₂)₂ with R = Me, Et, n-Bu, Bz) [221], which in the case of n-butyl substituent is sold as a lubricant additive for use as an extreme pressure agent, displaces the diethylether ligand in [Au(C₆F₅)₃(OEt₂)] to give the dinuclear [{Au(C₆F₅)₃}₂{(CH₂(S₂CNR₂)₂}] [43], where both gold atoms are coordinated to one sulfur atom of the methylene bis(dialkyldithiocarbamate) acting as bridging ligand.

The reaction of NBu₄[AuBr(C₆F₅)₃] and NaSH gives the first hydrosulfido gold(III) complex NBu₄[Au(C₆F₅)₃(SH)] which adds a second Au(C₆F₅)₃ fragment via displacement of the labile diethylether in [Au(C₆F₅)₃(OEt₂)] giving the dinuclear compound NBu₄[{Au(C₆F₅)₃}₂(SH)]. It is well known that the AuPR₃ group is isolobal with the hydrogen atom and mostly uses the 6s orbital of gold for bonding. This isolobal analogy can also be extended to AgPPh₃; thus treatment of the later dinuclear complex with [M(OClO₃)PPh₃] (M = Au, Ag) in the presence of a base gives the trinuclear derivatives NBu₄[{Au(C₆F₅)₃}₂(SMPPh₃)] isolobal with the starting dinuclear compound [222].

1,1'-Bis(diethyldithiocarbamate)ferrocene ligand [Fc(S₂CNEt₂)₂] (Fc = η⁵-C₅H₄)₂Fe), and the dinuclear complex [Au₂(SC₆F₅)₂(μ-dppf)] (dppf = 1,1'-bis(diphenylphosphino)ferrocene) can act as a bidentate ligand through the sulfur atoms, giving dinuclear pentafluorophenyl gold(III) complexes [{AuR_m}₂Fc(S₂CNEt₂)₂] [223] (R_m (C₆F₅)₃, Cl(C₆F₅)₂) or the tetranuclear [Au₄(μ₃-SC₆F₅)₂(C₆F₅)₆(μ-dppf)] [224] by displacement of the labile OEt₂ from [Au(C₆F₅)₃(OEt₂)] or chlorine bridge cleavage of [Au(μ-Cl)(C₆F₅)₂]₂. Other molecules derived from ferrocene such as 1,1'-bis(diphenylthiophosphoryl)ferrocene (dptpf), 1,1'-bis(diphenylselenophosphoryl)-ferrocene (dpspf), 1,1'-bis(phenylthio)ferrocene −Fc(SPh)₂− and 1,1'-bis(phenylseleno)ferrocene −Fc(SePh)₂− (Figure 3.28) can act as bidentate ligands by coordination through the S or Se donor atoms to two metal centers, giving dinuclear pentafluorophenyl gold(III) derivatives with the formula [{AuR_m}₂(L-L)] (L-L = dptpf [148],

dptpf, dpspf

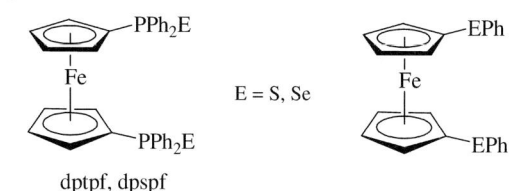

Figure 3.28 Molecules derived from ferrocene.

dpspf [225]; $R_m = (C_6F_5)_3$ and $(C_6F_5)_2Cl$) and [{Au$(C_6F_5)_3$}$_2$(L-L)] (L-L = Fc(SPh)$_2$, Fc(SePh)$_2$) [226].

[Au$(C_6F_5)_3$(OEt$_2$)] and [Au$(C_6F_5)_2$(OEt$_2$)$_2$]ClO$_4$ can displace the labile OEt$_2$ in the presence of S-, Se-centered phosphine gold complexes, such as [E(Au$_2$(dppf)$_2$)] (E = S, Se) and [E(AuPPh$_3$)$_2$] (E = S, Se) giving the corresponding trinuclear and tetranuclear mixed species [Se(Au$_2$dppf){Au$(C_6F_5)_3$}] [227] (**36**), [Se(AuPPh$_3$)$_2${Au$(C_6F_5)_3$}] [227], [E(Au$_2$dppf){Au$(C_6F_5)_3$}$_2$] (E = S [228], Se [227]), [E(AuPPh$_3$)$_2${Au$(C_6F_5)_3$}$_2$] (E = S [229], Se [227]), [S(Au$_2$dppf){Au$(C_6F_5)_2$}] [228, 229] and the pentanuclear [Se(AuPPh$_3$)$_2${Au$(C_6F_5)_2$}$_2$] [227]. The sulfur derivatives show weak Au(I)-Au(III) contact as established by crystal structure determination [228]. Trinuclear and tetranuclear selenolate gold complexes [230] with weak aurophilic Au(I)-Au(III) interactions have been described as the result of coordination of one or two Au$(C_6F_5)_3$ fragments to the gold(I) selenolate compound [Au$_2$(SePh)$_2$(μ-dppf)] (dppf 1,1′-bis(diphenylphosphino)ferrocene) with bridging selenolate ligands [Au$_3$$(C_6F_5)_3$(μ-SePh)(μ-dppf)] and [Au$_4$$(C_6F_5)_6$(μ-SePh)(μ-dppf).

36

The sulfide of the diphosphines 3,4-(NHPPh$_2$)$_2$MeC$_6$H$_3$ and 1,2-(NHPPh$_2$)$_2$C$_6$H$_4$ displace the tetrahydrothiophene ligand from [Au$(C_6F_5)_3$(tht)] giving the dinuclear gold(III) complexes [RC$_6$H$_3${NHPPh$_2$SAu$(C_6F_5)_3$}$_2$] (R = Me, H) [96].

Mononuclear and dinuclear gold(I) derivatives with 2-aminobenzenethiol [Au(2-SC$_6$H$_4$NH$_2$)(PPh$_3$)] and [{Au(2-SC$_6$H$_4$NH$_2$)}$_2$(μ-dppm)] can introduce an additional gold(III) fragment using the S-donor atom leading mixed valence Au(I)-Au(III) luminescent materials [(AuPPh$_3$){Au$(C_6F_5)_3$}(μ$_2$-2-SC$_6$H$_4$NH$_2$)] and [Au$(C_6F_5)_3$(μ$_2$-2-SC$_6$H$_4$NH$_2$)(AudppmAu)(μ$_2$-2-SC$_6$H$_4$NH$_2$)Au$(C_6F_5)_3$], where the luminescence properties can be attributed to the presence of gold-gold interactions [231].

3.3.3
Polynuclear Pentafluorophenylgold Derivatives with Au(I)-Au(III) Bond

The exclusive example of a direct formal gold(I)-gold(III) bond (2.769(1) Å) described [{Au(CH$_2$)$_2$PPh$_2$}$_2$Au$(C_6F_5)_3$] [12, 232, 233] can be prepared by the displacement of the diethylether molecule of [Au$(C_6F_5)_3$(OEt$_2$)] and a posterior coordination to the diauracycle [Au(CH$_2$)$_2$(PPh$_2$)]$_2$ [234, 235]. Addition of two moles of [Au(C$_6$F$_5$)$_3$(OEt$_2$)] affords the tetranuclear [{Au(CH$_2$)$_2$PPh$_2$}$_2${Au(C$_6$F$_5$)$_3$}$_2$].

3.3.4
Heteropolynuclear Pentafluorophenylgold(III) Derivatives

Many heteropolynuclear Au(III)/Ag derivatives have been described in the previous Section 3.2.

Heteronuclear pentafluorophenyl Pd-Au(I) derivatives are described in Chapter 3.2.5.6, dedicated to gold(I). Some examples of dinuclear and trinuclear derivatives containing Pd and Au(III) complexes have been obtained by reaction of palladium derivatives with gold(III) complexes of the type Q[Au(C$_6$F$_5$)$_3$X] [25] where X is a potentially bidentate ligand (CN or SCN) [236]: (NBu$_4$)[(C$_6$F$_5$)$_3$Au-X-Pd(C$_6$F$_5$)$_2$(tht)] and (NBu$_4$)$_2$[(C$_6$F$_5$)$_3$Au-X-Pd(C$_6$F$_5$)$_2$-X-Au(C$_6$F$_5$)$_3$].

A tetramer with four gold(III) and four Ag(I) (Figure 3.29) [177] atoms is isolated by reaction of cis-NBu$_4$[Au(C$_6$F$_5$)$_2$(CN)$_2$] with AgClO$_4$, as a consequence of the precipitation of NBu$_4$ClO$_4$ instead of AgCN.

Mixed gold(III) and molybdenum or tungsten derivatives can be obtained by ligand displacement or bridge-cleavage reactions. Thus, treatment of [M(CO)$_4$(tdppme)] (M = Mo, W) with [Au(C$_6$F$_5$)$_3$(tht)] or [Au(μ-Cl)(C$_6$F$_5$)$_2$]$_2$ leads to the neutral dinuclear products [(OC)$_4$M{Ph$_2$PCH$_2$)$_2$CMe(CH$_2$PPh$_2$)}Au(C$_6$F$_5$)$_2$X] (X = C$_6$F$_5$, Cl) [237], where tdppme is the tridentate-chelating ligand (Ph$_2$PCH$_2$)$_3$CMe. The reaction between [Au(C$_6$F$_5$)$_3$(dppm)] [172] and the corresponding acetylacetonate gold(I) derivative [(OC)$_4$Mo{Ph$_2$PCH$_2$)$_2$CMe(CH$_2$PPh$_2$)}Au(acac)] affords the trinuclear methanide complex [(OC)$_4$Mo{Ph$_2$PCH$_2$)$_2$CMe(CH$_2$PPh$_2$)}Au(Ph$_2$PCHPPh$_2$)Au(C$_6$F$_5$)$_3$] as a consequence of the deprotonation of the CH$_2$ group.

Oxidation of [Au(C$_6$F$_5$)$_3$(dppm)] with H$_2$O$_2$ gives the corresponding mono-oxide [Au(C$_6$F$_5$)$_3$(Ph$_2$PCH$_2$PPh$_2$O)], where the oxygen atom can act as ligand giving trinuclear derivatives Au(III) and Tl(III) [Tl(C$_6$F$_5$)$_2$X{Au(C$_6$F$_5$)$_3$(Ph$_2$PCH$_2$PPh$_2$O)}$_2$] (X = C$_6$F$_5$ or Cl) [203], in the presence of [Tl(C$_6$F$_5$)$_2$Cl]$_2$ [164] or [Tl(C$_6$F$_5$)$_3$(dioxane)] [238]. In this trinuclear complex the thallium center bridges two [Au(C$_6$F$_5$)$_3$(Ph$_2$PCH$_2$PPh$_2$O)] units via Tl−O bond.

The thermally stable tripodal tris(amido)methalates of the group 14 element can act as metal building blocks [239, 240] in the synthesis of heterometallic systems

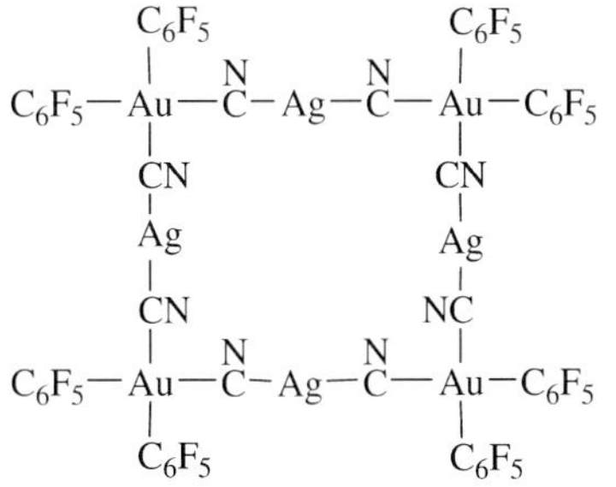

Figure 3.29 Tetramer with four Au(III) and four Ag(I).

containing unsupported metal-metal bonds. The first example of unsupported Sn-Au (III) bond N(PPh$_3$)$_2$[MeSi{Me$_2$SiN(p-tol)}$_3$SnAu(C$_6$F$_5$)$_3$] [81] (**37**) has been prepared by reaction of [MeSi{Me$_2$SiN(p-tol)}$_3$Sn(OEt$_2$)] with the organoaurate N(PPh$_3$)$_2$[Au (C$_6$F$_5$)$_3$Cl].

37   **38**

Trinuclear phosphido-bridged Au(III)-M derivatives with M = Ag and Cu (PPh$_3$)$_2$N [{Au(C$_6$F$_5$)$_3$(μ-PPh$_2$)}$_2$]M] have been described deprotonating [Au(C$_6$F$_5$)$_3$ (PPh$_2$H)] [179] with N(PPh$_3$)$_2$(acac) and the subsequent addition of the corresponding silver or copper salt. A pentanuclear mixed gold(III)-silver(I) phosphide with an unusual T-frame μ$_3$-Cl bridge is obtained (**38**) [241] in the reaction of [Au (C$_6$F$_5$)$_3$(PPh$_2$H)] with Ag(CF$_3$SO$_3$), [Ag(CF$_3$SO$_3$)(PPh$_3$)], [N(PPh$_3$)$_2$Cl] and [N(PPh$_3$)$_2$ (acac)] as deprotonating agent.

3.3.5
Pentafluorophenylgold(III) Derivatives as Catalyst

Some halogen-gold(III) derivatives containing at least one pentafluorophenyl ring such as *trans*-NBu$_4$[Au(C$_6$F$_5$)$_2$X$_2$] (X = Cl, Br), [Au(C$_6$F$_5$)$_2$(μ-Cl)]$_2$, BzPPh$_3$[Au (C$_6$F$_5$)Cl$_3$] and [Au(C$_6$F$_5$)Cl$_2$]$_2$ have shown catalytic activity in the addition of water and methanol to terminal alkynes (phenylacetylene and *n*-heptyne) [27]. The studies of a stoichiometric reaction with [Au(C$_6$F$_5$)$_2$(μ-Cl)]$_2$ and PhCCH provides a proposal for a mechanism, where a gold(III)-vinyl intermediate is suggested (Scheme 3.6).

Pentafluorophenyl gold(III) and gold(I) derivatives of the type [Au(μ-Cl)(C$_6$F$_5$)$_2$]$_2$, *trans*-[Au(C$_6$F$_5$)$_2$Cl$_2$], (PPh$_3$)$_2$N[Au(C$_6$F$_5$)Cl$_3$], [Au(C$_6$F$_5$)Cl$_2$(tht)] and (PPh$_3$)$_2$N[Au (C$_6$F$_5$)Cl] [76] are able to catalyze the 1,3-dipolar cycloaddition reaction of N-benzyl-C(2-pyridyl) nitrone and methyl acrylate decreasing the reaction time and favoring the formation of the *exo(cis)* isomer (Scheme 3.7). The corresponding gold(III) and gold(I) complexes with the formulae: [Au(C$_6$F$_5$)(2-PyBN)], [Au(C$_6$F$_5$)$_2$Cl(2-PyBN)] and [Au(C$_6$F$_5$)(2-PyBN)] (2-PyBN = N-benzyl-C(2-pyridyl) nitrone) do not react with methyl acrylate to give the cycloaddition products and display low catalytic activity, for this reason they can not be considered as plausible intermediates in the cycloaddition reaction.

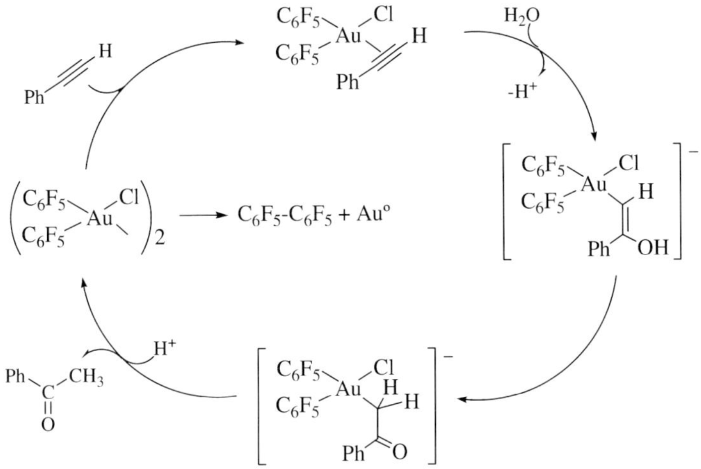

Scheme 3.6 Proposal of catalytic mechanism.

There are some differences in between the Au–C distances in the C_6F_5 units, which depends in most of the cases on the ligand in *trans* position to the pentafluorophenyl ring. The simplest cases are complexes with only one Au-C_6F_5 fragment, as illustrated in Table 3.9. The values of the distances reflect an increasing *trans* influence in the order $N < S < C(sp) < C(sp^2)$ for the donor atom of the ligand, except in the case of complex $[Au(C_6F_5)(S_2C_6H_4)]_3$ probably due to the rigid trinuclear structure. In the cases of mononuclear and polynuclear pentafluorophenyl complexes with $Au(C_6F_5)_2$ units (Tables 3.10 and 3.11 respectively) display slight differences in between both fragments when they have the same donor atom in *trans*. But an increasing *trans* influence is observed in most of the examples (with different donor atoms in *trans*) in the following order $N < Cl < S \sim Se < P < C$ donor ligands.

In mononuclear and polynuclear $Au(C_6F_5)_3$ compounds (Tables 3.12 and 3.13 respectively) the Au-C_6F_5 bond lengths in the fragments mutually in *trans* position are in the same range and the remaining $Au(C_6F_5)$ displays different distances depending

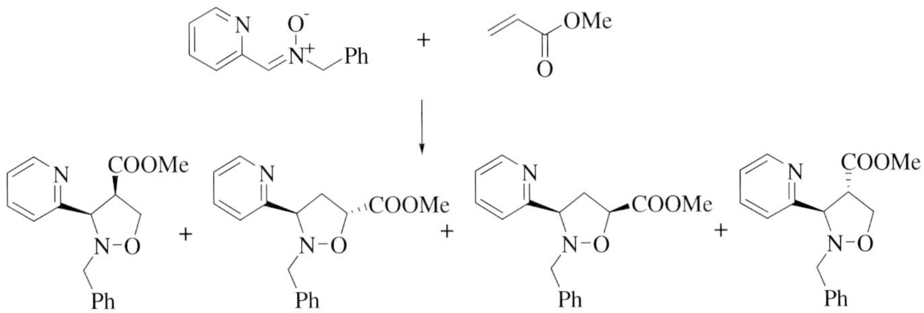

Scheme 3.7 1,3-dipolar cycloaddition reaction of N-benzyl-C(2-pyridyl) nitrone and methyl acrylate.

Table 3.9 Au-C_6F_5 distances in Au(C_6F_5) compounds.

Compound	Au-C$_{trans\ to\ (X)}$	Ref
[Au($C_6H_4CH_2NMe_2$-2)(C_6F_5)Cl]	2.022(10) (N)	[194]
[Au$_2$(μ-C_6F_5)(μ-C_6H_4-2-Me)(Ph$_2$PC$_6H_4$Me-2)$_2$] (see structure)	2.074(4) (C)	[110]
[Au(C_6F_5)($S_2C_6H_4$)(PPh$_3$)]	2.061(4) (S)	[195]
[Au(C_6F_5)($S_2C_6H_4$)]$_3$	2.035(7) (S)	[196]
[Au(C_6F_5)($S_2C_6H_4$)(SC$_6H_4$SPPh$_3$)]	2.07(2) (S)	[196]
[(μ-CH$_2$){Au(CH$_2$)$_2$PPh$_2$}$_2$(C_6F_5)$_2$]	2.104(7) (C)	[176]

Table 3.10 Au-C_6F_5 distances in mononuclear Au(C_6F_5)$_2$ compounds.

Compound	Au-C$_{trans\ to\ (X)}$	Ref
[Au(C_6F_5)$_2${CN(p-tolyl)}$_2$]	2.067(6), 2.097(5) (C)	[186]
[Au$_2$(μ-C_6F_5)(μ-C_6H_4-2-Me)(Ph$_2$PC$_6H_4$Me-2)$_2$]$^+$ (see structure)	2.097(10) (C)	[109]
[Au(C_6F_5)$_2$(Fcpy)$_2$]	2.011(9), 2.015(9) (N)	[144]
[Au(C_6F_5)$_2$(PPh$_2C_6H_4NH_2$)]ClO$_4$	2.019(6) (N), 2.073(7) (P)	[219]
[Au(C_6F_5)$_2$(PPh$_2C_6H_4$NH)]	2.041(4), 2.053(4) (N); 2.067(4), 2.050(5) (P)	[219]
[Au(C_6F_5)$_2$Cl(FcCH$_2$NHpyMe)]	2.012(3) (N), 2.028(2) (Cl)	[153]
[Au(C_6F_5)$_2$(Ph$_2$PCHPPh$_2$)]ClO$_4$	2.071(10) (P)	[175]
NBu$_4$[{Au(C_6F_5)$_2$}$_2$(μ-PPh$_2$)]	2.060(8), 2.055(7) (P)	[218]
[Au(C_6F_5)$_2${(PPh$_2$)$_2$CCH$_2$(SPh)}]	2.099(13), 2.087(10) (P)	[189]
[Au(C_6F_5)$_2${(PPh$_2$)$_2$CCH$_2$(S$_2$CNEt$_2$)}]	2.082(6), 2.086(7) (P)	[189]
[Au(C_6F_5)$_2${(Ph$_2$P)$_2$CPPh$_2$}]	2.10(2), 2.074(14) (P)	[215]
[Au(C_6F_5)$_2$(SPPh$_2$CHPPh$_2$)]	2.069(5) (P), 2.061(6) (S)	[210]
[Au(C_6F_5)$_2$Cl(PPh$_2C_6H_4NH_2$)]	2.038(4) (Cl), 2.068(4) (P)	[219]
[Au(C_6F_5)$_2$(S$_2$CN(CH$_2$Ph)$_2$]	2.049(6), 2.047(6) (S)	[182]
[Au(C_6F_5)$_2${(SPPh$_2$)$_2$CCH$_2$(C$_5H_4$)}]	2.032(3), 2.033(3) (S)	[189]
PPN[Au(C_6F_5)$_2$(μ-dmit)]	2.059(5), 2.061(6) (S)	[108]
[Au(C_6F_5)$_2$(SPPh$_2$O)(CH$_2$PPh$_2$Me)]	2.056(26) (S), 2.084(27) (C)	[209]
[Au(C_6F_5)$_2${(SePh$_2$)$_2$N]	2.052(2), 2.055(2) (Se)	[188]
[Au(C_6F_5)$_4$][Au(C_6F_5)$_2$Cl$_2$]	2.066(6), 2.051(6); 2.050(6), 2.056(6)	[41]

3.3 Pentafluorophenylgold(III) Derivatives

Table 3.11 Au-C$_6$F$_5$ distances in polynuclear Au(C$_6$F$_5$)$_2$ compounds.

Complex	Au-C$_{trans\ to\ (X)}$	Ref
[Au(C$_6$F$_5$)$_2$(μ-2-C$_6$H$_4$PPh$_2$)$_2$Au]	2.081(10), 2.097(10) (Csp)	[109]
[Au(C$_6$F$_5$)$_2${(PPh$_2$C$_6$H$_4$N(AuPPh$_3$)$_2$)}]ClO$_4$	2.046(3) (N), 2.088(4) (P)	[219]
[{Au(C$_6$F$_5$)$_2$}$_3$(μ-PPhH)$_3$]	2.052(8)–2.078(7) (P)	[51]
[Au(C$_6$F$_5$)$_2$(Ph$_2$PCHPPh$_2$)Au(C$_6$F$_5$)]	2.075(6) (P)	[198]
[Au(C$_6$F$_5$)$_2$(Ph$_2$PCHPPh$_2$)[Au(Ph$_2$PCHPPh$_2$){Au(C$_6$F$_5$)$_2$}]ClO$_4$	2.108 (10), 2.118(11) (P)	[198]
[{Au(C$_6$F$_5$)$_2$}$_2$(μ-PPh$_2$)$_2$]	2.060(8), 2.055(7) (P)	[218]
[Au(C$_6$F$_5$)$_2$(μ-PPh$_2$)$_2$Ag]$_2$	2.073(9), 2.078(9) (P) 2.069(9), 2.091(8)	[218]
[Au(C$_6$F$_5$)$_2${(Ph$_2$P)$_2$CPPh$_2$}(AuCl)]	2.097(8), 2.077(7) (P)	[215]
[Au(C$_6$F$_5$)$_2$(Ph$_2$PC(AuC$_6$F$_5$)(AuPPh$_3$)PPh$_2$CHCOOMe]	2.03(3), 2.08(3) (P)	[189]
[Au(C$_6$F$_5$)$_2${(PPh$_2$)$_2$C(AuPh$_3$)$_2$}]	2.054(3), 2.154(4) (P)	[200]
[Au(C$_6$F$_5$)$_2${(Ph$_2$PCHAu(C$_6$F$_5$)PPh$_2$CHCOOMe)}]	2.110(15) (P), 2.069(16) (C)	[207]
[Au(C$_6$F$_5$)$_2$Cl{Ph$_2$PCH(Aupy)PPh$_2$}AuCl]	2.037(18) (P), 2.052(20) (Cl)	[199]
[Au(C$_6$F$_5$)$_2${(Ph$_2$PCHAu(C$_6$F$_5$)PPh$_2$CHCOOMe)}]	2.098(14) (P), 2.111(13) (Csp^2)	[205]
trans-NBu$_4$[{Au(C$_6$F$_5$)$_3$}(1,2-PHC$_6$H$_4$PH){Au(C$_6$F$_5$)$_2$Cl}{μ-Au(C$_6$F$_5$)$_2$}]	2.074(7), 2.030(7) (P)	[220]
[Au(C$_6$F$_5$)$_2${(SPPh$_2$)$_2$C(AuAsPh$_3$)$_2$}]ClO$_4$	2.02(3), 2.08(4) (S)	[204]
[Se(AuPPh$_3$)$_2${Au(C$_6$F$_5$)$_2$}$_2$]	2.046(8), 2.070(7) (Se)	[227]
[{Au(C$_6$F$_5$)$_2$Cl}$_2$(μ-Fc(PPh$_2$S)]	2.008(8) (Cl), 2.040(8) (S)	[148]
[{S(Au$_2$dppf)}$_2${Au(C$_6$F$_5$)$_3$}]OTf	2.053(14), 2.062(14) (S)	[228]

Table 3.12 Au-C$_6$F$_5$ distances in mononuclear Au(C$_6$F$_5$)$_3$ compounds.

Compound	Au-C$_{mutually\ in\ trans}$	Au-C$_{trans\ to\ (X)}$	Ref
[Au(C$_6$F$_5$)$_3$(CH$_2$PPh$_2$Me)]	2.055(4), 2.069(3)	2.067(6) (C)	[197]
[Au(C$_6$F$_5$)$_3$(C$_3$H$_4$NSNH$_2$)]	2.063(2), 2.068(2)	2.016(2) (N)	[47]
[Au(C$_6$F$_5$)$_3$Cl][Au$_3$(μ-dppm)$_2$Cl$_2$]	2.063(2), 2.068(2)	2.016(2) (N)	[175]
[Au(C$_6$F$_5$)$_3$(FcCH$_2$NHpyMe)]	2.064(5), 2.071(5)	2.026(5) (N)	[153]
[Au(C$_6$F$_5$)$_3$(Fcpy)]	2.067(3), 2.068(3)	2.017(3) (N)	[144]
[Au(C$_6$F$_5$)$_3${PPh$_2$(CH$_2$CH(OMe)$_2$)}]	2.065(3), 2.068(3)	2.067(3) (P)	[181]
[Au(C$_6$F$_5$)$_3${PPh$_2$C≡C)(AuPPh$_3$)}]	2.063(8), 2.067(3)	2.082(7) (P)	[181]
[Au(C$_6$F$_5$)$_3$(PTA)]	2.069(2), 2.079(2)	2.066(2) (P)	[28]
[Au(C$_6$F$_5$)$_3$(Ph$_2$PC$_6$H$_4$SH)]	2.060(5), 2.074(4)	2.057(5) (P)	[180]
[Au(C$_6$F$_5$)$_3$(triphos)]	2.078(6), 2.082(7)	2.079(3) (P)	[95]
[Au(C$_6$F$_5$)$_3$(PPh$_2$CH$_2$Fc)]	2.080(4), 2.068(4)	2.067(3) (P)	[147]
[Au(C$_6$F$_5$)$_3$(pdma)]	2.065(11), 2.085(11)	2.104(12) (As)	[172]
[Au(C$_6$F$_5$)$_3$(Fc(SPh)$_2$)]	2.068(4), 2.073(4)	2.028(4) (S)	[226]
[Au(C$_6$F$_5$)$_3$(S$_2$CPEt$_3$)]	2.067(4), 2.076(4)	2.037(3) (S)	[89]
[Au(C$_6$F$_5$)$_3$(tht)]	2.068, 2.067; 2.056, 2.066	2.035(5), 2.031(3) (S)	[41]
[Au$_4$(PPh$_2$CH$_2$PPhCH$_2$PPh$_2$)$_2$Cl$_2$][Au(C$_6$F$_5$)$_3$Cl]	2.065(10), 2.071(10)	2.014(11) (Cl)	[243]

Table 3.13 Au-C$_6$F$_5$ distances in polynuclear Au(C$_6$F$_5$)$_3$ compounds.

Compound	Au-C$_{mutually\ in\ trans}$	Au-C$_{trans\ to\ (X)}$	Ref
[{Au(C$_6$F$_5$)$_3$}$_2${Fc(Spy)$_2$}]	2.058(6), 2.077(6)	2.005(5) (N)	[149]
PPN[{Au(C$_6$F$_5$)$_3$}$_2$(μ-PPhH)]	2.062(2), 2.064(2)	2.058(2) (P)	[51]
	2.062(2), 2.074(2)	2.063(2)	
[{Au(C$_6$F$_5$)$_3$}$_2$(μ-PPh){AuPPh$_3$}$_2$]	2.055(3), 2.074(3)	2.083(3) (P)	[51]
[{Au(C$_6$F$_5$)$_3$}$_2$(Me-C$_6$H$_3$){NHPPh$_2$}$_2$]	2.064(10), 2.071(10)	2.088(9) (P)	[96]
	2.062(10), 2.076(3)	2.070(10)	
[C$_6$H$_4${NHPPh$_2$Au(C$_6$F$_5$)$_3$}]	2.069(5), 2.070(5)	2.062(6) (P)	[96]
{N(AuPPh$_3$)PPh$_2$}Au(C$_6$F$_5$)$_3$]	2.074(5), 2.079(5)	2.061(6)	
trans-NBu$_4$[{Au(C$_6$F$_5$)$_3$} (1,2-PHC$_6$H$_4$PH) {Au(C$_6$F$_5$)$_2$Cl}{μ-Au(C$_6$F$_5$)$_2$}]	2.068(7), 2.078(7)	2.075(7) (P)	[220]
[{Au(C$_6$F$_5$)$_3$}(1,2-PC$_6$H$_4$P) {Au(C$_6$F$_5$)$_3$} {μ-Au(dppe)Au}$_2$]	2.064(9), 2.080(9)	2.104(8), 2.085(9) (P)	[220]
	2.058(10), 2.086(9)		
[{Au(C$_6$F$_5$)$_3$}(1,2-PC$_6$H$_4$P) {Au(C$_6$F$_5$)$_3$} {μ-Ag(dppe)Au}$_2$]	1.932(19), 2.051(19)	2.026(17), 2.096(16) (P)	[220]
	2.069(17), 2.084(16)		
NBu$_4$[{Au(C$_6$F$_5$)$_3$}$_2$(μ-PPh$_2$)]	2.070(9), 2.065(9)	2.063(8), 2.061(8) (P)	[218]
	2.061(8), 2.074(9)		
[Au(C$_6$F$_5$)$_3$(μ$_2$-2-SC$_6$H$_4$NH$_2$) (AudppmAu)(μ$_2$-2-SC$_6$H$_4$NH$_2$) Au(C$_6$F$_5$)$_3$]	2.001(14), 2.071(13)	2.050(13), 2.039(15) (P)	[231]
	2.017(14), 2.105(15)		
[Au(C$_6$F$_5$)$_3$(μ-PPh$_2$)(AuPPh$_3$)]	2.055(6), 2.074(6)	2.068(6) (P)	[217]
[Au(C$_6$F$_5$)$_3$(μ-PPh$_2$)(AgPPh$_3$)]	2.063(3), 2.064(3)	2.073(3) (P)	[217]
[Au(C$_6$F$_5$)$_3${μ-PPh$_2$(C=CH$_2$)PPh$_2$} (AuC$_6$F$_5$)]	2.050(3), 2.071(3)	2.059(3) (P)	[92]
[{Au(C$_6$F$_5$)$_3$}$_3$(μ-triphos)]	2.063(7), 2.079(7)	2.054(7), 2.057(7), 2.047(7) (P)	[97]
	2.073(7), 2.079(7)		
	2.075(7), 2.078(7)		
NBu$_4$[{Au(C$_6$F$_5$)$_3$PPh$_2$CHPPh$_2$}$_2$Au]	2.057(8), 2.063(8)	2.063(8), 2.080(8) (P)	[211]
	2.064(8), 2.069(9)		
[Tl(C$_6$F$_5$)$_2$Cl{Au(C$_6$F$_5$)$_3$ (PPh$_2$CH$_2$PPh$_2$(O))}$_2$]	2.075(10), 2.074(9)	2.059(10) (P)	[203]
PPN[{Au(C$_6$F$_5$)$_3$(μ-PPh$_2$)}$_2$Au]	2.052(5), 2.058(6)	2.073(6) (P)	[179]
PPN[{Au(C$_6$F$_5$)$_3$(μ-PPh$_2$)}$_2$Ag]	2.052(5), 2.068(5)	2.078(5) (P)	[179]
[{Au(C$_6$F$_5$)$_3$}(dpam)Ag(OClO$_3$) (dpam) {Au(C$_6$F$_5$)$_3$}]	2.075(7), 2.083(9)	2.067(8), 2.060(9) (As)	[29]
	2.067(8), 2.075(9)		
[Au$_4$(μ$_3$-SC$_6$F$_5$)$_2$(C$_6$F$_5$)$_6$(μ-dppf)]	2.049(7), 2.056(7)	2.048(7) (S)	[224]
[S(Au$_2$dppf){Au(C$_6$F$_5$)$_3$}]	2.061(14), 2.079(13)	2.02(2) (S)	[228]
[{Au(C$_6$F$_5$)$_3$}$_2$(Fc(SPh)$_2$)]	2.073(4), 2.068(4)	2.028(4) (S)	[226]
[{Au(C$_6$F$_5$)$_3$}$_2$(μ-S$_2$CPEt$_3$)]	2.086(15), 2.090(13)	2.048(16), 2.058(20) (S)	[89]
	2.062(14), 2.088(14)		
[{Au(C$_6$F$_5$)$_3$}$_2$(SH)]	2.080(14), 2.051(13)	2.042(13), 2.051(13) (S)	[222]
	2.102(14), 2.086(14)		

Table 3.13 (Continued)

Compound	Au-C$_{mutually\ in\ trans}$	Au-C$_{trans\ to\ (X)}$	Ref
[{Au(C$_6$F$_5$)$_3$}$_2$(SAgPPh$_3$)]	2.050(8), 2.058(7) 2.062(7), 2.058(7)	2.038(8), 2.046(7) (S)	[222]
[S(Au$_2$dppf){Au(C$_6$F$_5$)$_3$}$_2$]	2.067(10), 2.093(10) 2.057(11), 2.069(11)	1.979(13), 2.034(12) (S)	[229]
[S(AuPPh$_3$)$_2${Au(C$_6$F$_5$)$_3$}$_2$]	2.018(13), 2.064(13) 2.051(12), 2.09(2)	2.091(13), 2.01(2) (S)	[229]
[{Au(C$_6$F$_5$)$_3$}$_2$(SAuPPh$_3$)]	2.07(2), 2.05(2) 2.04(2), 2.05(2)	2.05(2), 2.00(2) (S)	[222]
[{Au(C$_6$F$_5$)$_3$}$_2$(μ-SePh)$_2$(μ-dppf)]	2.076(3), 2.066(3) 2.061(3), 2.054(3)	2.036(3), 2.051(4) (Se)	[230]
[Se{Au$_2$(μ-dppf)}{Au(C$_6$F$_5$)$_3$}]	2.079(11)	2.060(12) (Se)	[227]

on the donor atom of the ligand placed *trans* in the molecule. In these cases the observed *trans* influence exhibits the following order: N < Cl < S ~ Se < P < C < As.

To the best of our knowledge there are only four examples of [Au(C$_6$F$_5$)$_4$]$^-$ crystallographically characterized [41, 173, 174, 242] with Au-C$_6$F$_5$ distances ranging from 2.050 to 2.098(13) Å.

3.4
Gold Clusters

The oxidation states I and III are the most common in gold chemistry but there are also some polynuclear complexes of the general formulae [Au$_n$L$_m$]$^{a+}$ where the oxidation state of gold is between 0 and I, because a < n and, generally m < n. No organometallic cluster complexes of gold are known with the exception of [Au$_{10}$(C$_6$F$_5$)$_4$(PPh$_3$)$_5$] [244] obtained by reaction of [Au$_9$(PPh$_3$)$_8$](NO$_3$)$_3$ with NBu$_4$[Au(C$_6$F$_5$)$_2$] in 1:3 ratio. Complex shown in Figure 3.30 is separated by column chromatography in good yield and it was the first gold cluster analyzed by mass spectrometry (LSIMS)$^+$.

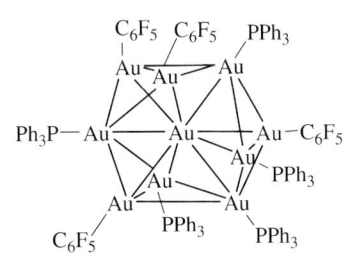

Figure 3.30 Structure of the cluster [Au$_{10}$(C$_6$F$_5$)$_4$(PPh$_3$)$_5$].

3.5
Pentafluorophenylgold(II) Derivatives

The number of gold(II) complexes is very scarce when we compare them with the more common gold(I) and gold(III) derivatives. The energy required to reach [245, 246] Au^{2+} from atomic gold is not very different to that required to form either Cu^{2+} or Ag^{2+} and to attain M^{3+} less energy is required for Au than for Cu and Ag. Therefore this argument is not enough to justify the lack of stability of the oxidation state II for gold. There is a strong tendency to disproportionation from Au^{2+} to give Au^+ and Au^{3+} because the odd electron in d^9 metal complexes is in the d_{x2-y2} orbital [245–247] (octahedral tetragonally distorted or square planar arrangement) which has much higher energy than the one with copper and can easily be ionized. The formation of a gold-gold bond gives more stable compounds and the Au_2^{4+} core derivatives are the more stable and abundant types of gold(II) complexes. In fact only the complexes containing C_6F_5 groups are of this kind and although mononuclear gold (II) complexes have been reported [248] only one has been X-ray characterized [249] [Au([9]aneS$_3$)$_2$] (BF$_4$)$_2$ and none contains the pentafluorophenyl group Figure 3.31.

3.5.1
Dinuclear Pentafluorophenylgold(II) Complexes

One of the reasons for the poor stability of gold(II) complexes must be the unfavorable energy of the odd electron. The formation of a metal-metal bond in binuclear gold(II) complexes gives an extra stability and avoids a decomposition pathway. Thus, the number of gold(II) complexes containing the Au_2^{4+} core has been increasing in the last two decades and different stoichiometries are known. The most abundant are those containing two equal bridges holding together the gold centers and keeping them in close proximity, giving rise to a diauracycle with a gold-gold bond across the cycle. But beside this class of derivatives, some dinuclear gold(II) complexes with two different bridges have been synthesized, as well as complexes with only one bridge holding the gold(II) centers, and the first Au_2^{4+} complex without any bridge was described in 1996 [250]. Nowadays we have examples of all these types in pentafluorophenylgold(II) complexes including this interesting new type of Au_2^{4+} without any bridge [41].

Table 3.14 shows all the known dinuclear gold(II) complexes containing the Au_2^{4+} core. As happened in gold(II) chemistry the most abundant are those containing a double bridge of the bis ylide $CH_2PPh_2CH_2$. The first derivative [(C_6F_5)Au

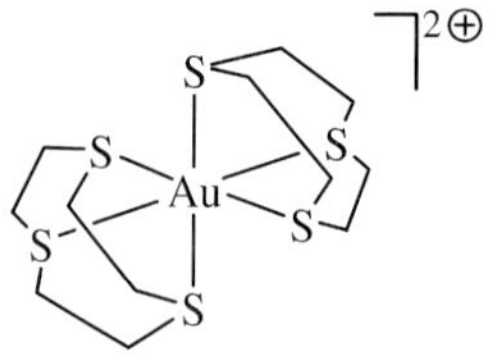

Figure 3.31 Structure of the gold (II) derivative [Au([9]aneS$_3$)$_2$](BF$_4$)$_2$.

3.5 Pentafluorophenylgold(II) Derivatives

Table 3.14 Dinuclear Gold(II) Au_2^{4+} complexes with at least one C_6F_5 group.

Complex	Ref.	d(Au-Au)	d(Au-X)	d(Au-R)
[(C_6F_5)Au(CH$_2$PPh$_2$CH$_2$)$_2$Au(C_6F_5)]	[174]	2.677(1)	—	2.151(1)
[(C_6F_5)Au(CH$_2$PPh$_2$CH$_2$)$_2$Au(NO$_3$)]	[251]			
[(C_6F_5)Au(CH$_2$PPh$_2$CH$_2$)$_2$AuCl]	[251]			
[(C_6F_5)Au(CH$_2$PPh$_2$CH$_2$)$_2$AuSCN]	[251]			
[(C_6F_5)Au(CH$_2$PPh$_2$CH$_2$)$_2$Au($C_6F_3H_2$)]	[251]			
[(C_6F_5)Au(CH$_2$PPh$_2$CH$_2$)$_2$Au(MeCOO)]	[251]			
[(C_6F_5)Au(CH$_2$PPh$_2$CH$_2$)$_2$Au(OClO$_3$)]	[253]			
[(C_6F_5)Au(CH$_2$PPh$_2$CH$_2$)$_2$Au(PPh$_3$)]ClO$_4$	[251]	2.661((8)	2.443(3)	2.078(12)
[(C_6F_5)Au(CH$_2$PPh$_2$CH$_2$)$_2$Au(tht)]ClO$_4$	[251]			
[(C_6F_5)Au(CH$_2$PPh$_2$CH$_2$)$_2$Au(AsPh$_3$)]ClO$_4$	[251]			
[(C_6F_5)Au(CH$_2$PPh$_2$CH$_2$)$_2$Au(PTol$_3$)]ClO$_4$	[251]			
[(C_6F_5)Au(CH$_2$PPh$_2$CH$_2$)$_2$Au(py)]ClO$_4$	[251]			
[(C_6F_5)Au(CH$_2$PPh$_2$CH$_2$)(S$_2$CNMe$_2$)Au(C_6F_5)]	[254]			
[(C_6F_5)Au(CH$_2$PPh$_2$CH$_2$)(S$_2$CPCy$_3$)Au(C_6F_5)](ClO$_4$)	[255]			
[(C_6F_5)Au(o-C_6H_4PPh$_2$)$_2$Au(C_6F_5)]	[109]	2.6139(4)	—	—
[Cl(C_6F_5)Au(PPh$_2$NHPPh$_2$)Au(C_6F_5)Cl]	[87]	2.576(2)	2.341(3)	—
[(C_6F_5)$_2$Au(PPh$_2$NHPPh$_2$)Au(C_6F_5)$_2$]	[87]			
[Br(C_6F_5)Au(PPh$_2$NHPPh$_2$)Au(C_6F_5)Br]	[87]			
[(C_6F_5)Au{(PPh$_2$NHPPh$_2$)$_2$Au(C_6F_5)](ClO$_4$)$_2$	[87]			
[Au$_2$(C_6F_5)$_4$(tht)$_2$]	[41]	2.5679(7)	2.418(3)	2.110(2) 2,078(2)

(CH$_2$PPh$_2$CH$_2$)$_2$Au(C_6F_5)] was synthesized by Facker et al. [174] by oxidation of the dinuclear derivative of gold(I) with [Tl(C_6F_5)$_2$Br]. Although in low yield, it was characterized by X-ray studies. Synthesis with higher yield was reported by Laguna and Laguna [251] either by oxidation of the dinuclear gold(I) derivative or by metathesis replacement with [Ag(C_6F_5)] of the corresponding dichloride gold(II), the latter being the best way (Equation 3.26).

$$Cl-Au(\mu\text{-}CH_2PPh_2CH_2)_2Au-Cl + 2Ag(C_6F_5) \longrightarrow C_6F_5-Au(\mu\text{-}CH_2PPh_2CH_2)_2Au-C_6F_5 + 2AgCl$$

(3.26)

This good yield from the preparation of [(C_6F_5)Au(CH$_2$PPh$_2$CH$_2$)$_2$Au(C_6F_5)] permitted the study of its reactivity and the synthesis of the complexes of Table 3.14 of general formula [(C_6F_5)Au(CH$_2$PPh$_2$CH$_2$)$_2$AuX] X = anionic ligands, and [(C_6F_5)Au (CH$_2$PPh$_2$CH$_2$) $_2$AuL]ClO$_4$ L = neutral ligands, both obtained by reaction of [(C_6F_5)

Au(CH$_2$PPh$_2$CH$_2$)$_2$Au(C$_6$F$_5$)] with the corresponding [XAu(CH$_2$PPh$_2$CH$_2$)$_2$AuX] or [LAu(CH$_2$PPh$_2$CH$_2$)$_2$AuL](ClO$_4$)$_2$. Other syntheses of these nonsymmetrical derivatives are the methathetical reactions starting from [(C$_6$F$_5$)Au(CH$_2$PPh$_2$CH$_2$)$_2$AuCl] with silver salts (Equation 3.27).

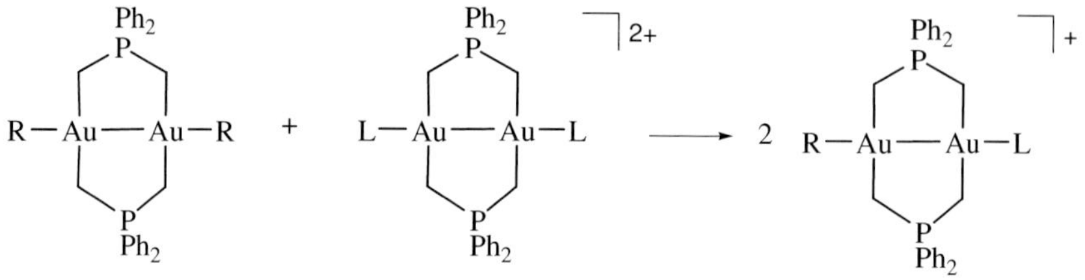

$$R-Au(\mu\text{-}CH_2PPh_2CH_2)_2Au-R + L-Au(\mu\text{-}CH_2PPh_2CH_2)_2Au-L^{2+} \longrightarrow 2\ R-Au(\mu\text{-}CH_2PPh_2CH_2)_2Au-L^{+}$$

(3.27)

Working this way the complexes [(C$_6$F$_5$)Au(CH$_2$PPh$_2$CH$_2$)$_2$Au(OClO$_3$)] and [(C$_6$F$_5$)Au(CH$_2$PPh$_2$CH$_2$)$_2$Au(tht)](ClO$_4$) become accessible. The tetrahydrothiophene (tht) and the perchlorate are weak ligands in gold(I) and gold(III) chemistry and also in gold(II) chemistry. They give enough stability to some gold(II) complexes and, in addition, they can be replaced easily by any other anionic or cationic ligand. Therefore gold(II) complexes containing tht or OClO$_3$ were the starting point for the preparation not only of a great variety of dinuclear gold(II) derivatives but also of polynuclear gold(II) complexes (see below).

The preparation of gold(II) complexes with two different bridges can be accomplished by metathetical reactions with the corresponding dichloride gold(II) complex [ClAu(CH$_2$PPh$_2$CH$_2$)(S$_2$CNMe$_2$)AuCl] and [ClAu(CH$_2$PPh$_2$CH$_2$)(S$_2$CPCy$_3$)AuCl] (ClO$_4$), although no studies by X-ray diffraction of this type of structures have been reported. Another interesting pentafluorophenylgold(II) complex is the ortho-metallated complex [(C$_6$F$_5$)Au(o-C$_6$H$_4$PPh$_2$)$_2$Au(C$_6$F$_5$)] reported by Bennett [109] synthesized by reaction with LiC$_6$F$_5$ over the corresponding chloride derivative.

All these gold(II) complexes are unstable in solution at room temperature and decompose to gold(III)/gold(I) derivatives as mentioned above. The two bridges holding the two gold centers are not enough to maintain this oxidation state, thus complexes with only one bridge should be less stable, as they are, and not many examples of this kind have been reported. Only three examples with bis(diphenylphosphino)amine and the pentafluorophenyl group have been reported including the X-ray structure of one of them.

If it is difficult to get the Au$_2^{4+}$ core maintained by two bridges and some examples with only one bridge have been reported, the existence of some examples without any bridge should be noticed. In fact the first example of this class of complexes was published in 1996 [250], although it was in 2007 that this type of complex containing C$_6$F$_5$ groups was reported [41]. This very important example was reported by Raubenheimer et al. by coproportionation reaction of gold(I) [Au(C$_6$F$_5$)(tht)] and gold(III) [Au(C$_6$F$_5$)$_3$(tht)] affording [Au$_2$(C$_6$F$_5$)$_4$(tht)$_2$] whose structure shows the Au$_2^{4+}$ core bonding four C$_6$F$_5$ groups.

The disproportionation reactions for dinuclear gold(II) complexes to give gold (III)/gold(I) derivatives are known and well represented. On the contrary only one previous copropotionation reaction to give gold(II) complexes starting from gold (III)/gold(I) has been reported [252].

3.5.2
Polynuclear Pentafluorophenylgold(II) Complexes

Since the most abundant gold(II) complexes belonging to the Au_2^{4+} formulation have been discussed above, now we will refer to nuclearity higher than two. This includes complexes with more than two gold atoms in the molecule or with other metals in their composition in addition to the gold centers (see Table 3.15).

The first polynuclear complexes containing the C_6F_5 group appear by substitution reactions of $[(C_6F_5)Au(CH_2PPh_2CH_2)_2Au(tht)](ClO_4)$ with NaS_2CNR_2 affording the tetranuclear complexes $[\{(C_6F_5)Au(CH_2PPh_2CH_2)_2Au\}_2(S_2CNR_2)]$ (R = Me, Et). On the base of the spectroscopic properties a structure with the dithiocarbamate ligand acting as a bridge between two eight-membered diauracycles of gold(II) has been proposed [256].

The synthesis of polynuclear gold(II) complexes, involving different kinds of gold-gold bonds, has been achieved by reaction of gold(II) derivatives with the Au_2^{4+} core with either gold(I) or gold(III) compounds. Some of us reported [242] the reaction of $[(C_6F_5)Au(CH_2PPh_2CH_2)_2Au(C_6F_5)]$ with $[Au(C_6F_5)_3(OEt_2)]$ which led serendipitously to the pentanuclear cationic complex $[\{Au_2(CH_2PPh_2CH_2)_2(C_6F_5)\}Au(C_6F_5)_2][Au(C_6F_5)_4]$. The structure of the cation is schematically shown in Equation 3.28. Its backbone is a linear chain of five gold atoms, all of which have square planar geometry. Two diauracycle units $[(C_6F_5)Au(CH_2PPh_2CH_2)_2Au]$ are bonded to a $Au(C_6F_5)_2$ moiety which lies nearly perpendicular to the two eight membered rings. The Au–Au distances of 2.755(1) and 2.640(1) Å are characteristic of metal-metal bonds, the former corresponding to the unsupported gold-gold bond.

Table 3.15 Polynuclear pentafluorophenylgold(II) complexes.

Complex	Ref.	d(Au-Au)
$[\{(C_6F_5)Au(CH_2PPh_2CH_2)_2Au\}_2(S_2CNMe_2)]$	[256]	—
$[\{(C_6F_5)Au(CH_2PPh_2CH_2)_2Au\}_2(S_2CNEt_2)]$	[256]	—
$[\{(C_6F_5)Au(CH_2PPh_2CH_2)_2Au\}_2\{S_2CN(CH_2Ph)_2\}]$	[256]	—
$[\{Au_2(CH_2PPh_2CH_2)_2(C_6F_5)\}Au(C_6F_5)_2][Au(C_6F_5)_4]$	[242]	2.640(1), 2.755(1)
$[\{Au_2(CH_2PPh_2CH_2)_2(C_6F_5)\}Au(C_6F_5)_2][(ClO_4)$	[253]	—
$[\{(C_6F_3H_2)Au(CH_2PPh_2CH_2)_2Au\}_2Au(C_6F_3H_2)_2]ClO_4$	[253]	—
$[\{(C_6F_5)Au(CH_2PPh_2CH_2)_2Au\}_2Au_2(CH_2PPh_2CH_2)_2]ClO_4$	[253]	—
$[\{(C_6F_3H_2)Au(CH_2PPh_2CH_2)_2Au\}_2Au_2(CH_2PPh_2CH_2)_2]ClO_4$	[253]	2.654(1), 2.737(1), 2.838(1)
$[Au_4(C_6F_5)_2\{(PPh_2)_2CH\}_2Cl_2]$	[260]	—
$[Au_4(C_6F_5)_2\{(PPh_2)_2CH\}_2Br_2]$	[260]	—
$[Au_4(C_6F_5)_2\{(PPh_2)_2CH\}_2(tht)_2]$	[260]	—
$[Au_4(C_6F_5)_2\{(PPh_2)_2CH\}_2(PPh_3)_2]$	[260]	2.730(1), 2.909(2)
$[Au_4(C_6F_5)_2\{(PPh_2)_2CH\}_2(PPh_2Me)_2]$	[260]	—
$[Au_4(C_6F_5)_2\{(PPh_2)_2CH\}_2\{P(tol)_3\}_2]$	[260]	—

$$R-Au\underset{\underset{Ph_2}{P}}{\overset{\overset{Ph_2}{P}}{\diamond}}Au-R \xrightarrow{[AuR_3OEt_2]} \left[R-Au\underset{\underset{Ph_2}{P}}{\overset{\overset{Ph_2}{P}}{\diamond}}Au\underset{R}{\overset{R}{-}}Au\underset{\underset{Ph_2}{P}}{\overset{\overset{Ph_2}{P}}{\diamond}}Au-R \right] [AuR_4]$$

R = C$_6$F$_5$

(3.28)

In accordance with extended Hückel calculations [233, 257] the pentanuclear chain can be better described as Au(III)-Au(I)-Au(I)-Au(I)-Au(III), so the central gold atom can be regarded as part of one $[Au(C_6F_5)_2]^-$ unit which acts as an electron donor to the dinuclear gold cation $[(C_6F_5)Au(CH_2PPh_2CH_2)_2Au]^+$. Complexes containing the organoaurates $[AuR_2]^-$ (R = C$_6$F$_5$ or 2,4,6-C$_6$F$_3$H$_2$) as nucleophilic centers have been described [129, 130] and cationic dinuclear gold complexes have been postulated as intermediates in oxidative additions [258].

The unit $[(C_6F_5)Au(CH_2PPh_2CH_2)_2Au]^+$ was nearly prepared from different gold (II) complexes above mentioned, such as $[(C_6F_5)Au(CH_2PPh_2CH_2)_2Au(tht)]ClO_4$ and $[(C_6F_5)Au(CH_2PPh_2CH_2)_2Au(OClO_3)]$. The tetrahydrothiophene or perchlorate ligands are easily displaced in gold chemistry, although they are enough to stabilize the gold(II) precursors. The reaction of these gold(II) complexes with either organoaurates NBu$_4$[AuR$_2$] (R = C$_6$F$_5$ or 2,4,6-C$_6$F$_3$H$_2$) or the diauracycle [Au(CH$_2$PPh$_2$CH$_2$)$_2$Au], which nucleophilic character was repeatedly manifested [258, 259], is the way to the preparation of these chains of gold centers.

Scheme 3.8 shows the results of these reactions. The pentanuclear chain of gold atoms is now accessible with ClO$_4^-$ as anion. A complex containing the trifluorophenyl group, $[\{(C_6F_3H_2)Au(CH_2PPh_2CH_2)_2Au\}_2Au(C_6F_3H_2)_2]ClO_4$, can be obtained using the starting trifluorophenyl derivatives instead. When the diauracycle [Au(CH$_2$PPh$_2$CH$_2$)$_2$Au] is used as electron donor, hexanuclear complexes [{RAu(CH$_2$PPh$_2$CH$_2$)$_2$Au}$_2$Au$_2$(CH$_2$PPh$_2$CH$_2$)$_2$](ClO$_4$)$_2$ (R = C$_6$F$_5$, 2,4,6-C$_6$F$_3$H$_2$ or CH$_3$) were obtained. The X-ray structure of the trifluorophenyl derivative has been determined. The whole dication consists of three Au(CH$_2$PPh$_2$CH$_2$)$_2$Au diauracycles in a nearly perpendicular disposition linked together through two unbridged gold-gold bonds, building up an almost linear six atom gold chain. The gold-gold distances in the pentanuclear and hexanuclear derivatives are very close despite the differences in the aryl group bonded to the ending gold atoms. In accordance with these data the bond in penta-, Au$_5^{9+}$, and hexa-nuclear, Au$_6^{10+}$, complexes can be explained as a donation of electron density from the central gold(I) center to two [RAu(CH$_2$PPh$_2$CH$_2$)$_2$Au]$^+$ fragments, which were gold(II) in the starting materials. Extended Hückel calculations [257] in the hexanuclear derivative show a better description as Au(III)-Au(I)-Au(I)-Au(I)-Au(I)-Au(III), very close to the description of pentanuclear gold complex and slightly different from the preparative procedures.

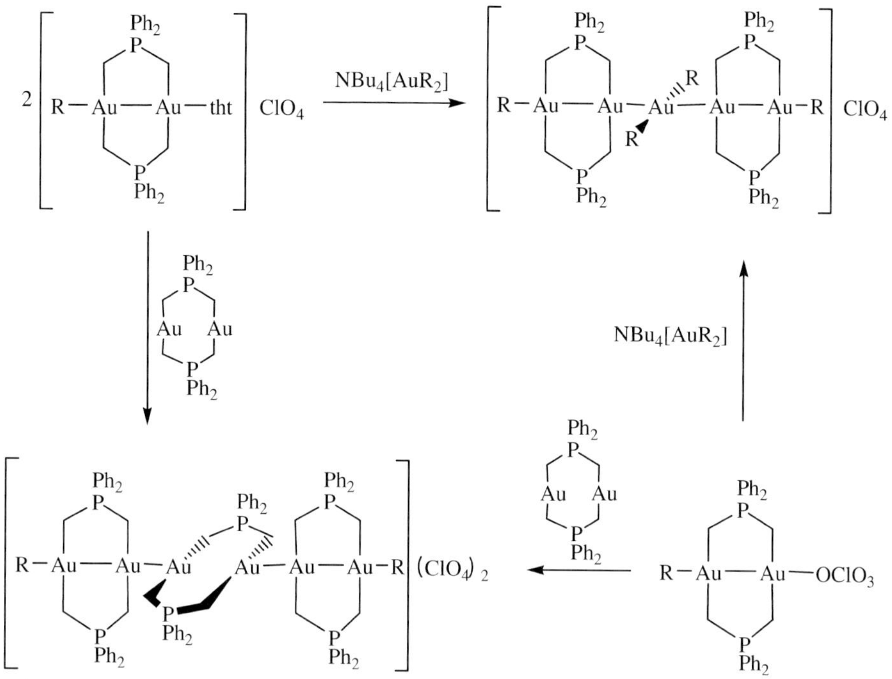

3.5 Pentafluorophenylgold(II) Derivatives

Scheme 3.8 R = C$_6$F$_5$, 2,4,6-C$_6$F$_3$H$_2$, Me.

The reaction of the tetranuclear complex [Au(PPh$_2$CH{Au(C$_6$F$_5$)}PPh$_2$)$_2$Au] [114] with chlorine or bromine does not oxidize the two gold centers of the diauracycle to give the usual gold(II) derivatives with an X–Au–Au–X backbone. Instead, a new type of reaction occurs, probably due to the presence of other gold(I) centers in close proximity [260]. Displacement reactions using silver salts, such as [Ag(OClO$_3$)(tht)] and [Ag(OClO$_3$)PR$_3$] afford cationic complexes which in the case of the blue PPh$_3$ derivative, the X-ray structure shows a linear chain of three gold atoms in a P–Au–Au–Au–P backbone (Scheme 3.9). The assignment of oxidation states in the linear chain is not straightforward. Because the sum must equal +5, the only two possibilities with integral oxidation states are Au(II)-Au(I)-Au(II) or Au(I)-Au(III)-Au(I) the former being the most appropriate and very close to those of heteronuclear gold(II)

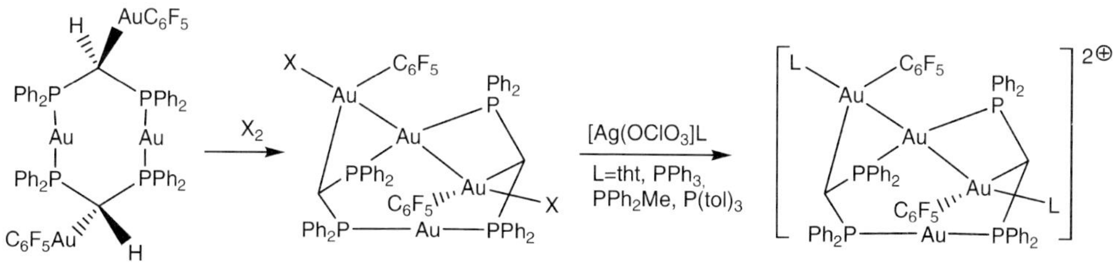

Scheme 3.9 Formation of the "Blue Gold" complexes.

derivative Au(II)-Pt(II)-Au(II) reported by Fackler [261]. Because of the intense blue color the oxidation state could be considered as fractional oxidation states Au_3^{5+} as happened in platinum [262] and rhodium and iridium [263] complexes which curiously present the same blue color, and they are called "Blue Platinum" [262], "Blue Rhodium and Iridium" [263] and by analogy we are in the presence of "Blue Gold" complexes.

3.6
Outlook and Future Trends

As we stated in the introduction the pentafluorophenylgroup, C_6F_5, has some properties that made it an important ligand for the stabilization of gold complexes. Its electron-withdrawing character, its resistance to cleavage against different agents, and the crystallinity that it confers on the complexes, made it an important ligand in the chemistry of transition metals, in general, and in gold chemistry in particular. The number of references of this chapter is in accordance with this.

The chemistry of the pentafluorophenylgold complexes covers all the oxidation states in one representation in accordance with the chemistry of gold. This shows that the pentafluorophenyl group has no preference for stabilization of any oxidation number. This general assertion is not totally true in cluster chemistry in which only one pentafluorophenylgold complex has been reported and in our opinion it corresponds more with a lack of reactions that with the properties of the pentafluorophenyl group. In fact the electron-withdrawing character of the C_6F_5 group should increase the stability of these complexes.

The oxidation number of pentafluorophenylgold derivatives, not highly represented is gold(II), although taking into account the scarce representation of this kind of gold derivative we must be optimistic. A close look at Section 5 shows us that the use of C_6F_5 group gives such different new bonding situations (new gold-gold bonds, chains of gold atoms in different oxidation states, fractional situations "Blue Gold," unsupported Au_2^{4+} derivatives and so on) that really is a invitation to develop new complexes in this area, because, in our opinion, not all the possibilities are yet manifested.

In the more widely represented pentafluorophenylgold(I) and gold(III) complexes the strength of the stability that the ligand confers on the complex should be at the same time its weakness regarding its chemistry. If we want to take advantage of this fact we need to use C_6F_5 groups to stabilize new bonding situations on the other coordination possibilities of the gold(I) or gold(III) centers. In order to minimize the lack of reactivity of C_6F_5 group we need, and this review is an invitation, to force the preparation of C_6F_5 complexes in which the group will act as electron-deficient donor between two metallic centers or those complexes in which the π-π stacking interaction between C_6F_5 group should be the reason for 3D macromolecular assembly and properties.

Applications of the pentafluorophenylgold complexes have been mentioned when appropriate throughout the chapter. Catalytic reactions, vapochromic behavior, components of VOCs detectors, liquid crystals, luminescence complexes and water solubility have been mentioned. We wish we could write a new chapter in the future in which the possible and real applications of this type should be the dominant argument, because pentafluorophenylgold complexes present the stability required for this purpose.

Acknowledgements

We wish to thank the Spanish Science and Innovation Ministry/FEDER grant number BQU2005-0899CO3-01.

References

1. Pope, W.J. and Gibson, C.S. (1907) The alkyl compounds of gold. *Journal of the Chemical Society*, **91**, 2061–2066.
2. Pope, W.J. and Gibson, C.S. (1908) The alkyl compounds of gold, Diethylauric bromide. *Proceedings of the Chemical Society*, **24**(23), 245.
3. Schmidbaur, H. (1976) Is Gold Chemistry A Topical Field Of Study. *Angewandte Chemie International Edition*, **15**(12), 728–740.
4. Uson, R., Laguna, A. and Vicente, J. (1977) Aryl-Gold Chemistry. *Synthesis and Reactivity in Inorganic and Metal-Organic Chemistry*, **7**(5), 463–496.
5. Puddephatt, R.H. (1978) *The Chemistry of Gold*, Elsevier Scientific, Amsterdam.
6. Schmidbaur, H. (1980) *Gmelin handbuch der Anorganishen Chemie Organogold Compounds*, Springer Verlag, Berlin.
7. Kharasch, M.S. and Isbell, H.S. (1931) The chemistry of organic gold compounds. III. Direct introduction of gold into the aromatic nucleus (Preliminary communication). *Journal of the American Chemical Society*, **53**(2), 3053–3059.
8. Calvin, G., Coates, G.E. and Dixon, P.S. (1959) Organic Derivatives Of Gold (I). *Chemistry & Industry*, (52), 1628–11628.
9. Schmidbaur, H. (1999) *Gold: Progress in Chemistry, Biochemistry and Technology*, John Wiley & Sons, Ltd, Chichester, UK, pp. 922.
10. Laguna, A. and Laguna, M. (1990) Bis (Diphenylphosphino)Methanide Or Bis (Diphenylphosphino)Methamide And Its Derivatives As Ligands In Gold Chemistry – A Review. *Journal of Organometallic Chemistry*, **394**(1–3), 743–756.
11. Laguna, A. and Laguna, M. (1999) Coordination chemistry of gold(II) complexes. *Coordination Chemistry Reviews*, **195**, 837–856.
12. Uson, R. (1989) Electron-rich metal centers (Au, Pt) as sources of organometallic complexes with unusual features. *Journal of Organometallic Chemistry*, **372**, 171–182.
13. Uson, R. and Laguna, A. (1986) Recent Developments In Arylgold Chemistry. *Coordination Chemistry Reviews*, **70**, 1–50.
14. Laguna, A. and Gimeno, M.C. (1994) New trends in aryl gold chemistry. *Trends in Organometallic Chemistry*, **1**, 231–261.
15. Fernandez, E.J., Laguna, A. and Olmos, M.E. (2005) Recent developments in arylgold(I) chemistry. *Advances in Organometallic Chemistry*, **52**, 77–141.
16. De Pascale, R.J. and Tamborski, C. (1967) The Reactions of Sodium

Pentafluorophenolate with Substituted Pentafluorobenzenes. *The Journal of Organic Chemistry*, **32**, 3163–3168.

17 Smith, M.B. and March, J. (2001) March's Advanced Organic Chemistry: Reactions, Mechanisms, and Structure, 5th edn, John Wiley & Sons, Inc., New York.

18 Ho, K.C. and Miller, J. (1966) Sn Mechanism In Aromatic Compounds. 32. Reactivity Of Some Perhalogenobenzenes. *Australian Journal of Chemistry*, **19**(3), 423–436.

19 Miller, J. and Yeung, H.W. (1967) Sn Mechanism In Aromatic Compounds.35. Comparative Reactivity Of Pentachlorofluorobenzene Hexafluorobenzene And Fluorobenzene. *Australian Journal of Chemistry*, **20**(2), 379–381.

20 Piers, W.E. (2005) The chemistry of Perfluoroaryl Boranes. *Advances in Organometallic Chemistry*, **52**, 1–77.

21 Casas, J.M., Diosdado, B.E., Falvello, L.R., Fornies, J., Martin, A. and Rueda, A.J. (2004) Hydrogen-bond mediation of supramolecular aggregation in neutral bis-(C_6F_5)Pt complexes with aromatic H-bond donating ligands. A synthetic and structural study. *Dalton Transactions*, (17), 2733–2740.

22 Uson, M.A. and Llanos, J.M. (2002) Synthesis of Group 10 polyfluorothiolate mono- and bi-nuclear complexes. Crystal structures of [Ni$(SC_6HF_4)_2$(dppe)], [(dppe)Ni$(\mu$-$SC_6HF_4)_2$Pd$(C_6F_5)_2$] and [(dppe)Ni$(\mu$-$SC_6F_5)_2$Pd$(C_6F_5)_2$]. *Journal of Organometallic Chemistry*, **663**(1–2), 98–107.

23 Cerrada, E., Laguna, M. and Lardíes, N. (2009) Bis(1,2,3-thiadiazoles) as precursors in the synthesis of bis (alkynethiolate)gold(I) derivatives. Unpublished results. *European Journal of Inorganic Chemistry*, 137–146.

24 Uson, R., Laguna, A. and Laguna, M. (1989) (Tetrahydrothiophene)gold(I) or gold(III) complexes. *Inorganic Syntheses*, **26**, 85–91.

25 Uson, R., Laguna, A., Laguna, M. and Fernandez, E. (1980) Tris (pentafluorophenyl)(tetrahydrothiophene) gold(III). *Inorganica Chimica Acta*, **45**(5), L177–L178.

26 Uson, R., Laguna, A., Laguna, M. and Abad, M. (1983) Synthesis and reactions of di-m-halo- or -pseudohalotetrakis (pentafluorophenyl)digold(III). *Journal of Organometallic Chemistry*, **249**(2), 437–443.

27 Casado, R., Contel, M., Laguna, M., Romero, P. and Sanz, S. (2003) Organometallic Gold(III) Compounds as Catalysts for the Addition of Water and Methanol to Terminal Alkynes. *Journal of the American Chemical Society*, **125**(39), 11925–11935.

28 Mohr, F., Cerrada, E. and Laguna, M. (2006) Organometallic Gold(I) and Gold (III) Complexes Containing 1,3,5-Triaza-7-phosphaadamantane (TPA): Examples of Water-Soluble Organometallic Gold Compounds. *Organometallics*, **25**(3), 644–648.

29 Laguna, A., Laguna, M., Fananas, J., Jones, P.G. and Fittschen, C. (1986) Mono-, bi- and trinuclear bis (diphenylarsino)methane gold complexes. Crystal and molecular structure of [{$(C_6F_5)_3$Au $(Ph_2AsCH_2AsPh_2)$}$_2$Ag$(OClO_3)$]. *Inorganica Chimica Acta*, **121**(1), 39–45.

30 Vaughan, L.G. and Sheppard, W.A. (1969) New organogold chemistry. I. Synthesis and fluorine-19 nuclear magnetic resonance studies of some fluoroorganogold compounds. *Journal of the American Chemical Society*, **91**(22), 6151–6156.

31 Vaughan, L.G. and Sheppard, W.A. (1970) Organogold chemistry. II. Tris (pentafluorophenyl)gold(III). *Journal of Organometallic Chemistry*, **22**(3), 739–742.

32 Royo, P. (1972) Synthesis, structure, and reaction of fluorophenylorganometallic complexes of transition metals. *Revista de la Academia de Ciencias Exactas, Fisicas, Quimicas y Naturales de Zaragoza*, **27**(2), 235–251.

33 Baker, R.W. and Pauling, P.J. (1972) Crystal structure of (pentafluorophenyl)(triphenylphosphine) gold(I). *Journal of the Chemical Society, Dalton Transactions*, (20), 2264–2266.

34 Nyholm, R.S. and Royo, P. (1969) Synthesis of perfluorophenyl metal complexes using $(C_6F_5)_2TlBr$ as oxidant. *Journal of the Chemical Society D: Chemical Communications*, (8), 421.

35 Uson, R., Laguna, A., Fernandez, E.J., Mendia, A. and Jones, P.G. (1988) (Polyhalophenyl)silver(I) complexes as arylating agents: crystal structure of $[(\mu\text{-}2,4,6\text{-}C_6F_3H_2)(AuPPh_3)_2]ClO_4$. *Journal of Organometallic Chemistry*, **350**(1), 129–138.

36 Uson, R., Royo, P. and Laguna, A. (1974) Reactions of complexes of gold(I) with bis(pentafluorophenyl)thallium(III) halides. *Journal of Organometallic Chemistry*, **69**(3), 361–365.

37 Uson, R., Laguna, A. and Vicente, J. (1975) Dihaloperhaloaryl(triphenylarsine)gold(III) complexes. *Journal of Organometallic Chemistry*, **86**(3), 415–421.

38 Uson, R., Laguna, A. and Buil, J. (1975) Neutral bis(pentafluorophenyl)triphenylarsinegold(III) complexes. *Journal of Organometallic Chemistry*, **85**(3), 403–408.

39 Uson, R., Laguna, A. and Vicente, J. (1976) Preparation and properties of stable salts containing mono- or bis-(pentafluorophenyl)aurate(I) and mono-, tris-, or tetrakis-(pentafluorophenyl)aurate(III) ions. *Journal of the Chemical Society, Chemical Communications*, (10), 353–354.

40 Uson, R., Laguna, A. and Vicente, J. (1977) Novel anionic gold(I) and gold(III) organocomplexes. *Journal of Organometallic Chemistry*, **131**(3), 471–475.

41 Coetzee, J., Gabrielli, W.F., Coetzee, K., Schuster, O., Nogai, S.D., Cronje, S. and Raubenheimer, H.G. (2007) Structural studies of gold(I, II, and III) compounds with pentafluorophenyl and tetrahydrothiophene ligands. *Angewandte Chemie International Edition*, **46**(14), 2497–2500.

42 Jones, L.A., Sanz, S. and Laguna, M. (2007) Gold compounds as efficient co-catalysts in palladium-catalyzed alkynylation. *Catalysis Today*, **122**(3–4), 403–406.

43 Gimeno, M.C., Jambrina, E., Laguna, A., Laguna, M., Murray, H.H. and Terroba, R. (1996) Gold and silver complexes of methylenebis(dialkyldithiocarbamates) and the X-ray structure of $CH_2[S_2CN(CH_2Ph)_2]_2$. *Inorganica Chimica Acta*, **249**(1), 69–77.

44 Ferrer, M., Mounir, M., Rodriguez, L., Rossell, O., Coco, S., Gomez-Sal, P. and Martin, A. (2005) Effect of the organic fragment on the mesogenic properties of a series of organogold(I) isocyanide complexes. X-ray crystal structure of $[Au(C \equiv CC_5H_4N)(CNC_6H_4O(O)CC_6H_4OC_{10}H_{21})]$. *Journal of Organometallic Chemistry*, **690**(9), 2200–2208.

45 Uson, R., Laguna, A., Vicente, J., Garcia, J., Jones, P.G. and Sheldrick, G.M. (1981) Preparation of three- and four-coordinate gold(I) complexes; crystal structure of bis[o-phenylenebis(dimethylarsine)]gold(I) bis(pentafluorophenyl)aurate. *Journal of the Chemical Society, Dalton Transactions*, (2), 655–657.

46 Uson, R., Laguna, A. and Villacampa, M.D. (1984) Mono and binuclear gold complexes with diamine and carbene ligands. *Inorganica Chimica Acta*, **81**(1), 25–31.

47 Bardaji, M., Laguna, A., Perez, M.R. and Jones, P.G. (2002) Unexpected Ring-Opening Reaction to a New Cyanamide-Thiolate Ligand Stabilized as a Dinuclear Gold Complex. *Organometallics*, **21**(9), 1877–1881.

48 Codina, A., Fernandez, E.J., Jones, P.G., Laguna, A., Lopez-de-Luzuriaga, J.M., Monge, M., Olmos, M.E., Perez, J. and Rodriguez, M.A. (2002) Do Aurophilic

Interactions Compete against Hydrogen Bonds? Experimental Evidence and Rationalization Based on ab Initio Calculations. *Journal of the American Chemical Society*, **124**(23), 6781–6786.

49 Uson, R., Laguna, A., Laguna, M., Jimenez, J., Gomez, M.P., Sainz, A. and Jones, P.G. (1990) Gold complexes with heterocyclic thiones as ligands. X-ray structure determination of [Au(C_5H_5NS)$_2$]ClO$_4$. *Journal of the Chemical Society, Dalton Transactions*, (11), 3457–3463.

50 Bordoni, S., Busetto, L., Cassani, M.C., Albano, V.G. and Sabatino, P. (1994) Reactions of [Au(C_6F_5)(SC_4H_8)] with diazoalkanes. Synthesis and molecular structures of [cyclic] [Au(C_6F_5)(Ph$_2$C:NN: CPh$_2$)]. *Inorganica Chimica Acta*, **222** (1–2), 267–273.

51 Blanco, M.C., Fernandez, E.J., Olmos, M.E., Perez, J., Crespo, O., Laguna, A. and Jones, P.G. (2004) Gold(III) Phenylphosphides and -phosphodiides. *Organometallics*, **23**(19), 4373–4381.

52 Casares, J.A., Espinet, P., Hernando, R., Iturbe, G., Villafane, F., Ellis, D.D. and Orpen, A.G. (1997) Poly(2-pyridyl) phosphines, PPynPh$_{3-n}$ (n = 2, 3), and Their P-Substituted Derivatives as Tripodal Ligands in Molybdenum(0) Carbonyl Complexes. *Inorganic Chemistry*, **36**(1), 44–49.

53 Bardaji, M., Jones, P.G. and Laguna, A. (2002) Acetylenephosphino gold(I) derivatives: structure, reactivity and luminescence properties. *Journal of the Chemical Society, Dalton Transactions*, (18), 3624–3629.

54 Crespo, O., Fernandez, E.J., Jones, P.G., Laguna, A., Lopez-de-Luzuriaga, J.M., Monge, M., Olmos, M.E. and Perez, J. (2003) Coordination modes of diphenylphosphinothioformamide in its neutral and deprotonated forms at gold(I). *Dalton Transactions*, (6), 1076–1082.

55 Hollatz, C., Schier, A. and Schmidbaur, H. (1997) Aggregation of a Neutral Gold (I) Complex through Cooperative Action of Hydrogen Bonding and Auriophilicity. *Journal of the American Chemical Society*, **119**(34), 8115–8116.

56 Fernandez, E.J., Laguna, A., Lopez-de-Luzuriaga, J.M., Monge, M., Montiel, M., Olmos, M.E. and Rodriguez-Castillo, M. (2006) Synthesis, coordination to Au(I) and photophysical properties of a novel polyfluorinated benzothiazolephosphine ligand. *Dalton Transactions*, (30), 3672–3677.

57 Strasser, C.E., Cronje, S., Schmidbaur, H. and Raubenheimer, H.G. (2006) The preparation, properties and X-ray structures of gold(I) trithiophosphite complexes. *Journal of Organometallic Chemistry*, **691**(22), 4788–4796.

58 Sanz, S., Jones, L.A., Mohr, F. and Laguna, M. (2007) Homogenous Catalysis with Gold: Efficient Hydration of Phenylacetylene in Aqueous Media. *Organometallics*, **26**(4), 952–957.

59 Fernandez, E.J., Gimeno, M.C., Jones, P.G., Laguna, A., Laguna, M. and Lopez-de-Luzuriaga, J.M. (1993) Di-, tri- and tetra-nuclear gold(I) complexes with tris (diphenylphosphino)-methane or -methanide as ligand. Crystal structures of two modifications of [(O)Ph$_2$PC (PPh$_2$AuPPh$_2$)$_2$CPPh$_2$(O)].4CH$_2$Cl$_2$. *Journal of the Chemical Society, Dalton Transactions*, (22), 3401–3406.

60 Raubenheimer, H.G., Otte, R., Linford, L., Van Zyl, W.E., Lombard, A. and Kruger, G.J. (1992) Synthesis and characterization of gold(I) and gold(III) thione complexes. *Polyhedron*, **11**(8), 893–900.

61 Bardaji, M., Laguna, A., Laguna, M. and Merchan, F. (1994) Methyl dithiocarbamate gold(I) and gold(III) complexes. Synthesis and reactivity with amines. *Inorganica Chimica Acta*, **215** (1–2), 215–218.

62 Uson, R., Laguna, A., Vicente, J., Garcia, J. and Bergareche, B. (1979) Preparation of pentahalophenyl p-tolylisocyanide complexes of gold(I) and their reactions with amines, ammonia and alcohols.

63 Uson, R., Laguna, A., Brun, P., Laguna, M. and Abad, M. (1981) Synthesis and reactions of cyanopentafluorophenyl aurate(I) and -(III). *Journal of Organometallic Chemistry*, **218**(2), 265–273.

64 Bayon, R., Coco, S., Espinet, P., Fernandez-Mayordomo, C. and Martin-Alvarez, J.M. (1997) Liquid-Crystalline Mono- and Dinuclear (Perhalophenyl) gold(I) Isocyanide Complexes. *Inorganic Chemistry*, **36**(11), 2329–2334.

65 Uson, R., Laguna, A., Vicente, J., Garcia, J., Bergareche, B. and Brun, P. (1978) Neutral isocyanide and carbene pentafluorophenyl complexes of gold(I) and gold(III). *Inorganica Chimica Acta*, **28**(2), 237–243.

66 Raubenheimer, H.G., Scott, F., Roos, M. and Otte, R. (1990) The synthesis of gold (I) carbene complexes using 4-methylthiazolyllithium. *Journal of the Chemical Society, Chemical Communications*, (23), 1722–1723.

67 Raubenheimer, H.G., Lindeque, L. and Cronje, S. (1996) Synthesis and characterization of neutral and cationic diamino carbene complexes of gold(I). *Journal of Organometallic Chemistry*, **511** (1–2), 177–184.

68 Desmet, M., Raubenheimer, H.G. and Kruger, G.J. (1997) Synthesis and Characterization of Thienyl Oligomeric, Carbene, and Nitrogen-Donor Complexes of Gold(I). *Organometallics*, **16**(15), 3324–3332.

69 Uson, R., Laguna, A., Laguna, M., Uson, A. and Gimeno, M.C. (1987) Novel phosphorus ylide transfer reactions from gold(I) to gold(I) or from gold(I) to gold (III) centers. *Organometallics*, **6**(3), 682–683.

70 Uson, R., Laguna, A., Laguna, M., Uson, A. and Gimeno, M.C. (1988) Synthesis of pentahalophenylgold complexes of ylides: a comparison of methods. *Synthesis and Reactivity in Inorganic and Metal-Organic Chemistry*, **18**(1), 69–82.

71 Uson, R., Laguna, A., Laguna, M. and Uson, A. (1983) A new route for the synthesis of gold(I) and gold(III) pentafluorophenyl(ylide) complexes. *Inorganica Chimica Acta*, **73**(1), 63–66.

72 Bertani, R., Michelin, R.A., Mozzon, M., Traldi, P., Seraglia, R., Busetto, L., Cassani, M.C., Tagliatesta, P. and D'Arcangelo, G. (1997) Mass Spectrometric Detection of Reactive Intermediates. Reaction Mechanism of Diazoalkanes with Platinum(0) and Gold (I) Complexes. *Organometallics*, **16**(14), 3229–3233.

73 Cronje, S., Raubenheimer, H.G., Spies, H.S.C., Esterhuysen, C., Schmidbaur, H., Schier, A. and Kruger, G.J. (2003) Synthesis and characterisation of N-coordinated pentafluorophenyl gold(I) thiazole-derived complexes and an unusual self-assembly to form a tetrameric gold(I) complex. *Dalton Transactions*, (14), 2859–2866.

74 Uson, R., Laguna, A., Garcia, J. and Laguna, M. (1979) Anionic perfluorophenyl complexes of gold(I) and gold(III). *Inorganica Chimica Acta*, **37**(2), 201–207.

75 Vicente, J., Chicote, M.-T., Gonzalez-Herrero, P., Gruenwald, C. and Jones, P.G. (1997) Complexes with S-Donor Ligands. 5. Syntheses of the First (Hydrosulfido)- and Anionic Sulfidoorganogold(I) Complexes. Crystal and Molecular Structure of $(Et_4N)_2[\{Au(C_6F_5)\}_3(\mu_3\text{-}S)] \cdot 0.5MeC(O)Et$. *Organometallics*, **16**(15), 3381–3387.

76 Ade, A., Cerrada, E., Contel, M., Laguna, M., Merino, P. and Tejero, T. (2004) Organometallic gold(III) and gold(I) complexes as catalysts for the 1,3-dipolar cycloaddition to nitrones: synthesis of novel gold-nitrone derivatives. *Journal of Organometallic Chemistry*, **689**(10), 1788–1795.

77 Koelle, U. and Laguna, A. (1999) Electrochemistry of Au-complexes. *Inorganica Chimica Acta*, **290**(1), 44–50.

78 Briggs, D.A., Raptis, R.G. and Fackler, J.P. Jr. (1988) The structure of benzyl (triphenyl)phosphonium chloro (pentafluorophenyl)aurate(I): an organometallic gold(I) pentafluorophenyl complex useful as an ylide precursor. *Acta Crystallographica, Section C: Crystal Structure Communications*, **C44**(7), 1313–1315.

79 Vicente, J., Chicote, M.-T., Abrisqueta, M.-D. and Jones, P.G. (2000) A Novel Type of Alkynylgold(I) Complex. Crystal and Molecular Structures of PPN[Au(Ar)(C≡CH)] [Ar = C_6F_5, $C_6H_2(NO_2)_3$-2,4,6]. *Organometallics*, **19**(13), 2629–2632.

80 Uson, R., Laguna, A., Laguna, M. and Perez, V. (1981) Neutral and anionic single-bridged binuclear complexes of gold(I) and gold(III). *Synthesis and Reactivity in Inorganic and Metal-Organic Chemistry*, **11**(4), 361–372.

81 Findeis, B., Contel, M., Gade, L.H., Laguna, M., Gimeno, M.C., Scowen, I.J. and McPartlin, M. (1997) Tris(amido) tingold Complexes in Different Oxidation States. First Structural Characterization of a Sn–Au–Au–Sn Linear Chain. *Inorganic Chemistry*, **36**(11), 2386–2390.

82 Gimeno, M.C., Jambrina, E., Fernandez, E.J., Laguna, A., Laguna, M., Jones, P.G., Merchan, F.L. and Terroba, R. (1997) Gold (I) complexes with biologically active thiolate ligands. *Inorganica Chimica Acta*, **258**(1), 71–75.

83 Cerrada, E., Laguna, M., Bartolome, J., Campo, J., Orera, V. and Jones, P.G. (1998) Cation-radical salts with organometallic gold anions. X-ray structure of [TTFPh]$_2$[Au(C_6F_5)$_2$]. *Synthetic Metals*, **92**(3), 245–251.

84 Uson, R., Laguna, A., Laguna, M., Lazaro, I., Morata, A., Jones, P.G. and Sheldrick, G.M. (1986) Synthesis of mono-, di-, and trinuclear gold complexes containing the (diphenylphosphino) methyldiphenylphosphoniomethanide ligand. Crystal structures of [(diphenylphosphino) methyldiphenylphosphoniomethanide-P]pentafluorophenylgold(I) and m-[(diphenylphosphino) methyldiphenylphosphoniomethanide-C (Au$_1$)P(Au$_2$)]-bis[pentafluorophenylgold (I)]. *Journal of the Chemical Society, Dalton Transactions*, (3), 669–675.

85 Romeo, I., Bardaji, M., Concepcion Gimeno, M. and Laguna, M. (2000) Gold(I) complexes containing the cationic ylide ligand bis(methyldiphenylphosphonio) methylide. *Polyhedron*, **19**(15), 1837–1841.

86 Fernandez, E.J., Gimeno, M.C., Laguna, A., Lopez-De-Luzuriaga, J.M. and Olmos, E. (1998) Two alternatives for the synthesis of non-cyclic phosphinomethanide derivatives of gold. *Polyhedron*, **17**(22), 3919–3925.

87 Uson, R., Laguna, A., Laguna, M., Fraile, M.N., Jones, P.G. and Sheldrick, G.M. (1986) Mono- and binuclear gold(I), gold (II), and gold(III) perhaloaryl complexes with the ligand bis(diphenylphosphino) amine. Crystal and molecular structure of m-[bis(diphenylphosphino)amine] dichlorobis(pentafluorophenyl)digold (II). *Journal of the Chemical Society, Dalton Transactions*, (2), 291–296.

88 Laguna, A., Laguna, M., Rojo, A. and Fraile, M.N. (1986) Gold(I) and gold(III) complexes containing bis (diphenylphosphine)methane disulfide or bis(diphenylphosphine)amine disulfide ligands. *Journal of Organometallic Chemistry*, **315**(2), 269–276.

89 Uson, R., Laguna, A., Laguna, M., Luz Castilla, M., Jones, P.G. and Fittschen, C. (1987) Mono- and binuclear gold(I) and gold(III) complexes with S_2C-PR3 ligands. X-ray crystal structures of tris (pentafluorophenyl) (triethylphosphoniodithioformate-S)gold (III) and m-(triethylphosphoniodithioformate-S,S′)bis[tris(pentafluorophenyl) gold(III)] dichloromethane (1/1). *Journal of the Chemical Society, Dalton Transactions*, (12), 3017–3022.

90 Gimeno, M.C., Laguna, A., Sarroca, C. and Jones, P.G. (1993) 1,1′-Bis (diphenylphosphino)ferrocene (dppf) complexes of gold(I) and gold(III). Crystal structures of [(dppf)AuPPh$_3$]ClO$_4$.CHCl$_3$ and [(dppf)Au(μ-dppf)Au(dppf)] (ClO$_4$)$_2$.2CH$_2$Cl$_2$. *Inorganic Chemistry*, **32** (26), 5926–5932.

91 Fernandez, E.J., Gimeno, M.C., Laguna, A., Laguna, M., Lopez-de-Luzuriaga, J.M. and Olmos, E. (1996) Different coordination modes of the 1,1,1-tris (diphenylphosphinomethyl)ethane ligand in gold(I) and gold(III) complexes. *Journal of Organometallic Chemistry*, **514** (1–2), 169–175.

92 Fernandez, E.J., Gimeno, M.C., Jones, P.G., Laguna, A., Lopez-de-Luzuriaga, J.M. and Olmos, E. (1997) Coordination compounds of coinage metals with vinylidenebis(diphenylphosphine) and its disulfide and their reactivity towards nucleophiles. *Chemische Berichte/Recueil*, **130**(10), 1513–1517.

93 Bardaji, M., Jones, P.G., Laguna, A., Villacampa, M.D. and Villaverde, N. (2003) Synthesis and structural characterization of luminescent gold(I) derivatives with an unsymmetric diphosphine. *Dalton Transactions*, (23), 4529–4536.

94 Bardaji, M., de la Cruz, M.T., Jones, P.G., Laguna, A., Martinez, J. and Villacampa, M.D. (2005) Luminescent dinuclear gold complexes of bis(diphenylphosphano) acetylene. *Inorganica Chimica Acta*, **358** (5), 1365–1372.

95 Bardaji, M., Laguna, A., Orera, V.M. and Villacampa, M.D. (1998) Synthesis, Structural Characterization, and Luminescence Studies of Gold(I) and Gold(III) Complexes with a Triphosphine Ligand. *Inorganic Chemistry*, **37**(20), 5125–5130.

96 Bella, P.A., Crespo, O., Fernandez, E.J., Fischer, A.K., Jones, P.G., Laguna, A., Lopez-de-Luzuriaga, J.M. and Monge, M. (1999) Gold complexes of 3,4-bis (diphenylphosphinoamino)toluene and 1,2-bis(diphenylphosphinoamino) benzene. A comparative study. *Journal of the Chemical Society, Dalton Transactions*, (22), 4009–4017.

97 Bardaji, M., Laguna, A., Vicente, J. and Jones, P.G. (2001) Synthesis of Luminescent Gold(I) and Gold(III) Complexes with a Triphosphine Ligand. *Inorganic Chemistry*, **40**(12), 2675–2681.

98 Burgos, M., Crespo, O., Gimeno, M.C., Jones, P.G. and Laguna, A. (2003) Gold, silver and palladium complexes with the 2,2′-dipyridylamine ligand. *European Journal of Inorganic Chemistry*, (11), 2170–2174.

99 Jose Calhorda, M., Costa, P.J., Crespo, O., Gimeno, M.C., Jones, P.G., Laguna, A., Naranjo, M., Quintal, S., Shi, Y.-J. and Villacampa, M.D. (2006) Group 11 complexes with the bis(3,5-dimethylpyrazol-1-yl)methane ligand. How secondary bonds can influence the coordination environment of Ag(I): The role of coordinated water in [Ag$_2$(μ-L)$_2$(OH$_2$)$_2$](OTf)$_2$. *Dalton Transactions*, (34), 4104–4113.

100 Canales, F., Gimeno, M.C., Laguna, A. and Jones, P.G. (1996) Aurophilicity at Sulfur Centers. Synthesis and Reactivity of the Complex [S(Au$_2$dppf)]; Formation of Polynuclear Sulfur-Centered Complexes. Crystal Structures of [S(Au$_2$dppf)].2CHCl$_3$, [(μ-Au$_2$dppf) {S(Au$_2$dppf)}2](OTf)$_2$.8CHCl$_3$, and [S(AuPPh$_2$Me)$_2$(Au$_2$dppf)] (ClO$_4$)$_2$.3CH$_2$Cl$_2$. *Journal of the American Chemical Society*, **118**(20), 4839–4845.

101 Crespo, O., Gimeno, M.C., Laguna, A. and Villacampa, M.D. (1998) μ-[7,8-Bis (diphenylphosphino)-7,8-dicarba-nido-undecaborato-1kP, 2kP′] (pentafluorophenyl-1kC) (triphenylphosphine-2kP)digold(I) dichloromethane solvate. *Acta Crystallographica, Section C: Crystal Structure Communications*, **C54**(2), 203–205.

102 Fernandez, E.J., Lopez-de-Luzuriaga, J.M., Monge, M., Olmos, E., Laguna, A., Villacampa, M.D. and Jones, P.G. (2000) Trinuclear gold(I) complexes with various coordination modes of N,N-dimethyldithiocarbamate. *Journal of Cluster Science*, **11**(1), 153–167.

103 Bardaji, M., Laguna, A., Jones, P.G. and Fischer, A.K. (2000) Synthesis, Structural Characterization of Luminescent Trinuclear Gold(I) Complexes with Dithiocarbamates. *Inorganic Chemistry*, **39**(16), 3560–3566.

104 Williams, D.B.G., Traut, T., Kriel, F.H. and van Zyl, W.E. (2007) Bidentate amino- and iminophosphine ligands in mono- and dinuclear gold(I) complexes: Synthesis, structures and AuCl displacement by AuC_6F_5. *Inorganic Chemistry Communications*, **10**(5), 538–542.

105 Uson, R., Laguna, A., Vicente, J. and Garcia, J. (1976) Preparation of organogold(III) complexes by oxidizing dichloro-, or bis(pentafluorophenyl)-μ-bis(diphenylphosphino)ethanedigold(I). *Journal of Organometallic Chemistry*, **104**(3), 401–406.

106 Crespo, O., Gimeno, M.C., Jones, P.G., Ahrens, B. and Laguna, A. (1997) Synthesis and Characterization of Gold(I) Complexes with the Ligand 1,2-Dithiolate-o-carborane. Crystal Structures of $[Au_2(\mu\text{-}S_2C_2B_{10}H_{10})(PPh_3)_2]$, $[Au_2(\mu\text{-}S_2C_2B_{10}H_{10})(\mu\text{-}\{(PPh_2)_2(CH:CH)\})]$, and $[Au_2(\mu\text{-}S_2C_2B_{10}H_{10})(\mu\text{-}\{(PPh_2)_2(C_6H_4)\})]$. *Inorganic Chemistry*, **36**(4), 495–500.

107 Cerrada, E., Laguna, M. and Sorolla, P.A. (1997) Gold complexes with 2-selenoxo-1,3-dithiole-4,5-dithiolato. *Polyhedron*, **17**(2–3), 295–298.

108 Cerrada, E., Jones, P.G., Laguna, A. and Laguna, M. (1996) Synthesis and Reactivity of Heteroleptic Complexes of Gold with 2-Thioxo-1,3-dithiole-4,5-dithiolate (dmit). X-ray Structure of $[Au_2(\mu\text{-dmit})(PPh_3)_2]$, $(NBu_4)[Au(dmit)(PPh_3)]$, and $(PPN)[Au(dmit)(C_6F_5)_2]$. *Inorganic Chemistry*, **35**(10), 2995–3000.

109 Bennett, M.A., Hockless, D.C.R., Rae, A.D., Welling, L.L. and Willis, A.C. (2001) Carbon-Carbon Coupling in Dinuclear Cycloaurated Complexes Containing Bridging 2-(Diphenylphosphino)phenyl or 2-(Diethylphosphino)phenyl. Role of the Axial Ligand and the Fine Balance between Gold(II)-Gold(II) and Gold(I)-Gold(III). *Organometallics*, **20**(1), 79–87.

110 Bennett, M.A., Bhargava, S.K., Mohr, F., Welling, L.L. and Willis, A.C. (2002) Synthesis and X-ray structure of a heterovalent, cycloaurated pentafluorophenylgold(I)/pentafluorophenylgold(III) complex. *Australian Journal of Chemistry*, **55**(4), 267–270.

111 Kitadai, K., Takahashi, M., Takeda, M., Bhargava, S.K., Priver, S.H. and Bennett, M.A. (2006) Synthesis, structures and reactions of cyclometalated gold complexes containing (2-diphenylarsino-n-methyl)phenyl (n = 5, 6). *Dalton Transactions*, (21), 2560–2571.

112 Gimeno, M.C., Laguna, A., Laguna, M., Sanmartin, F. and Jones, P.G. (1993) Reactions of $[Au(acac)PPh_3]$ with diphosphine derivatives: different coordination modes of gold to the ligand systems. X-ray structure of $[SPPh_2C(AuPPh_3)_2PPh_2CH(AuPPh_3)COOMe]ClO_4$ and $[Au_5(C_6F_5)\{(SPPh_2)_2C\}_2(PPh_3)]$. *Organometallics*, **12**(10), 3984–3991.

113 Cerrada, E., Diaz, C., Diaz, M.C., Hursthouse, M.B., Laguna, M. and Light, M.E. (2002) Tetrathiafulvalene-functionalized phosphine as a coordinating ligand. X-Ray structures of $(PPh_2)_4TTF$ and $[(AuCl)_4\{(PPh_2)_4TTF\}]$. *Journal of the Chemical Society, Dalton Transactions*, (6), 1104–1109.

114 Uson, R., Laguna, A., Laguna, M., Gimeno, M.C., Jones, P.G., Fittschen, C. and Sheldrick, G.M. (1986) Di- and tetranuclear complexes with bis(diphenylphosphino)amide and bis(diphenylphosphino)methanide as bi- and tridentate ligands. X-ray structures of $[(Ph_3P)(O_3ClO)AgN$

(Ph$_2$PAuPPh$_2$)$_2$NAg(OClO$_3$)(PPh$_3$)] and [(C$_6$F$_5$)AuCH(Ph$_2$PAuPPh$_2$)$_2$CHAu (C$_6$F$_5$)]. *Journal of the Chemical Society, Chemical Communications*, (7), 509–510.

115 Bardaji, M., Gimeno, M.C., Jones, P.G., Laguna, A., Laguna, M., Merchan, F. and Romeo, I. (1997) Carbon-Carbon Coupling via Nucleophilic Addition of a Gold(I) Methanide Complex to Heterocumulenes. *Organometallics*, **16**(5), 1083–1085.

116 Bariáin, C., Matías, I.R., Romeo, I., Garrido, J. and Laguna, M. (2000) Detection of volatile organic compound vapors by using a vapochromic material on a tapered optical fiber. *Applied Physics Letters*, **77**(15), 2274–2276.

117 Romeo, M.L.J.G.I. (2001) Compuestos vapocrómicos y su uso en sensores medio ambientales de vapores de compuestos orgánicos, Patent ES-2151424.

118 Laguna, M., Bariáin, C., Garrido, J., Matías, I.R. and Romeo, I. (2001) Optical fiber sensor for volatile organic vapors, Patent WO-047936.

119 Fernandez, E.J., Laguna, A., Lopez-de-Luzuriaga, J.M., Monge, M., Nema, M., Olmos, M.E., Perez, J. and Silvestru, C. (2007) Experimental and theoretical evidence of the first Au(I).Bi(III) interaction. *Chemical Communications*, (6), 571–573.

120 Fernandez, E.J., Hursthouse, M.B., Laguna, M. and Terroba, R. (1999) Phosphinethiolate tin(IV)-11 Group-metal derivatives. X-ray structure of [Au$_2$Sn(^{t}Bu)$_2$(C$_6$F$_5$)$_2$(SC$_6$H$_4$PPh$_2$)$_2$]. *Journal of Organometallic Chemistry*, **574**(2), 207–212.

121 Crespo, O., Laguna, A., Fernandez, E.J., Lopez-de-Luzuriaga, J.M., Mendia, A., Monge, M., Olmos, E. and Jones, P.G. (1998) [Tl(OPPh$_3$)$_2$][Au(C$_6$F$_5$)$_2$]: the first extended unsupported gold-thallium linear chain. *Chemical Communications*, (20), 2233–2234.

122 Fernandez, E.J., Laguna, A. and Lopez-de-Luzuriaga, J.M. (2005) Luminescence in complexes with Au(I)–Tl(I) interactions. *Coordination Chemistry Reviews*, **249**, 1423–1433.

123 Fernandez, E.J., Laguna, A. Lopez-de-Luzuriaga, J.M, Mendizabal, F., Monge, M. Olmos, M.E, and Perez, J. (2003) Theoretical and Photoluminescence Studies on the $d^{10}-S^2$ Au(I)–Tl(I) Interaction in Extended Unsupported Chains. *Chemistry - A European Journal*, **9**(2), 456–465.

124 Fernandez, E.J., Jones, P.G., Laguna, A., Lopez-de-Luzuriaga, J.M., Monge, M., Perez, J. and Olmos, M.E. (2002) Synthesis, Structure, and Photophysical Studies of Luminescent Two- and Three-Dimensional Gold-Thallium Supramolecular Arrays. *Inorganic Chemistry*, **41**(5), 1056–1063.

125 Fernandez, E.J., Laguna, A., Lopez-de-Luzuriaga, J.M., Monge, M., Montiel, M., Olmos, M.E. and Perez, J. (2004) Thallium(I) Acetylacetonate as Building Blocks of Luminescent Supramolecular Architectures. *Organometallics*, **23**(4), 774–782.

126 Fernandez, E.J., Laguna, A., Lopez-de-Luzuriaga, J.M., Montiel, M., Olmos, M.E. and Perez, J. (2005) Dimethylsulfoxide gold-thallium complexes. Effects of the metal-metal interactions on the luminescence. *Inorganica Chimica Acta*, **358**(14), 4293–4300.

127 Fernandez, E.J., Laguna, A., Lopez-de-Luzuriaga, J.M., Montiel, M., Olmos, M.E. and Perez, J. (2006) Easy Ketimine Formation Assisted by Heteropolynuclear Gold-Thallium Complexes. *Organometallics*, **25**(7), 1689–1695.

128 Fernandez, E.J., Lopez-De-Luzuriaga, J.M., Monge, M., Rodriguez, M.A., Crespo, O., Gimeno, M.C., Laguna, A. and Jones, P.G. (2000) Heteropolynuclear complexes with the ligand Ph$_2$PCH$_2$SPh: theoretical evidence for metallophilic Au-M attractions. *Chemistry - A European Journal*, **6**(4), 636–644.

129 Uson, R., Laguna, A., Laguna, M., Jones, P.G. and Sheldrick, G.M. (1981) Synthesis

and reactivity of bimetallic gold-silver complexes. X-ray structure of a chain polymer containing the moiety.$(F_5C_6)_2Au$ $(\mu\text{-}AgSC_4H_8)_2Au(C_6F_5)_2$. *Journal of the Chemical Society, Chemical Communications*, (21), 1097–1098.

130 Uson, R., Laguna, A., Laguna, M., Manzano, B.R., Jones, P.G. and Sheldrick, G.M. (1984) Synthesis and reactivity of bimetallic gold-silver polyfluorophenyl complexes: crystal and molecular structures of catena-di-μ-(tetrahydrothiophene)argentiobis[bis (perfluorophenyl)gold] and catena-di-μ-(η_2-benzene)argentiobis[bis (perfluorophenyl)gold]. *Journal of the Chemical Society, Dalton Transactions*, (2), 285–292.

131 Fernandez, E.J., Gimeno, M.C., Laguna, A., Lopez-de-Luzuriaga, J.M., Monge, M., Pyykkoe, P. and Sundholm, D. (2000) Luminescent Characterization of Solution Oligomerization Process Mediated Gold-Gold Interactions. DFT Calculations on $[Au_2Ag_2R_4L_2]_n$ Moieties. *Journal of the American Chemical Society*, **122**(30), 7287–7293.

132 Fernandez, E.J., Laguna, A., Lopez-De-Luzuriaga, J.M., Monge, M., Pyykko, P. and Runeberg, N. (2002) A study of the interactions in an extended unsupported gold-silver chain. *European Journal of Inorganic Chemistry*, (3), 750–753.

133 Luquin, A., Bariáin, C., Vergara, E., Cerrada, E., Garrido, J., Matías, I.R. and Laguna, M. (2005) New preparation of gold-silver complexes and optical fiber environmental sensors based on vapochromic $[Au_2Ag_2(C_6F_5)_4(phen)_2]_n$. *Applied Organometallic Chemistry*, **19**(12), 1232–1238.

134 Bariáin, C., Matías, I.R., Fdez-Valdivielso, C., Elosúa, C., Luquin, A., Garrido, J. and Laguna, M. (2005) Optical fiber sensors based on vapochromic gold complexes for environmental applications. *Sensors and Actuators B*, **108**, 535–541.

135 Elosúa, C., Bariáin, C., Matías, I.R., Arregui, F.J., Luquin, A. and Laguna, M. (2006) Volatile alcoholic compounds fiber optic nanosensor. *Sensors and Actuators B*, **115**, 444–449.

136 Luquin, A., Elosúa, C., Vergara, E., Estella, J., Cerrada, E., Bariáin, C., Matías, I.R., Garrido, J. and Laguna, M. (2007) Application of Gold Complexes in the Development of Sensors for Volatile Organic Compounds. *Gold Bulletin*, **40**(3), 225–233.

137 Elosúa, C., Bariáin, C., Matías, I.R., Arregui, F.J., Luquin, A., Vergara, E. and Laguna, M. (2008) Indicator immobilization on Fabry-Perot nanocavities towards development of fiber optic sensors. *Sensor and actuators B*, **130**, 158–163.

138 Casado, S., Elosúa, C., Bariáin, C., Segura, A., Matías, I.R., Fernández, A., Luquin, A., Garrido, J. and Laguna, M. (2006) Volatile-organic-compound optic fiber sensor using a gold-silver vapochromic complex. *Optical Engineering*, **45**(4), 044401.1–044401.7.

139 Fernandez, E.J., Laguna, A., Lopez-de-Luzuriaga, J.M., Monge, M., Montiel, M. and Olmos, M.E. (2005) Unsupported Gold(I)-Copper(I) Interactions through $\eta_1Au\text{-}[Au(C_6F_5)_2]$- Coordination to Cu^+ Lewis Acid Sites. *Inorganic Chemistry*, **44**(5), 1163–1165.

140 Fernandez, E.J., Laguna, A., Lopez-de-Luzuriaga, J.M., Monge, M., Montiel, M., Olmos, M.E. and Rodriguez-Castillo, M. (2006) Photophysical and Theoretical Studies on Luminescent Tetranuclear Coinage Metal Building Blocks. *Organometallics*, **25**(15), 3639–3646.

141 Uson, R., Laguna, A., Fornies, J., Valenzuela, I., Jones, P.G. and Sheldrick, G.M. (1984) Polynuclear palladium and gold perhalophenyl derivatives with dppm bridges. Crystal and molecular structure of trans- $C_6F_5Au(\mu\text{-}dppm)Pd(C_6F_5)_2(\mu\text{-}dppm)Au\, C_6F_5$ (dppm $= Ph_2PCH_2PPh_2$). *Journal of Organometallic Chemistry*, **273**(1), 129–139.

142 Casares, J.A., Espinet, P., Martin-Alvarez, J.M. and Santos, V. (2006) Neutral and

Cationic Complexes with P-Bonded 2-Pyridylphosphines as N-Donor Ligands toward Rhodium. Electrical Charge vs. Steric Hindrance on the Conformational Control. *Inorganic Chemistry*, **45**(17), 6628–6636.

143 Viotte, M., Gautheron, B., Kubicki, M.M., Mugnier, Y. and Parish, R.V. (1995) New Iron(II)- and Gold(I)-Containing Metallocenes. X-ray Structure of a Three-Coordinate Gold(I) Ferrocenophane-Type Representative. *Inorganic Chemistry*, **34** (13), 3465–3473.

144 Barranco, E.M., Crespo, O., Gimeno, M.C., Jones, P.G., Laguna, A. and Villacampa, M.D. (1999) Synthesis, structure and redox behavior of gold and silver complexes with 3-ferrocenylpyridine. *Journal of Organometallic Chemistry*, **592**(2), 258–264.

145 Gimeno, M.C., Jones, P.G., Laguna, A. and Sarroca, C. (1999) Synthesis of gold-(I) and -(III) complexes with diferrocenylphenylphosphine (PFc$_2$Ph). Crystal structure of [Au(PFc$_2$Ph)$_2$]ClO$_4$. *Journal of Organometallic Chemistry*, **579** (1–2), 206–210.

146 Barranco, E.M., Gimeno, M.C. and Laguna, A. (1999) Synthesis of gold ferrocenyl substituted ylides and methanides: crystal structure of [FcCH(AuPPh$_3$)PPh$_3$]ClO$_4$ [Fc = (η_5-C$_5$H$_5$)Fe(η_5-C$_5$H$_4$)]. *Inorganica Chimica Acta*, **291** (1–2), 60–65.

147 Barranco, E.M., Crespo, O., Gimeno, M.C., Laguna, A., Jones, P.G. and Ahrens, B. (2000) Gold and Silver Complexes with the Ferrocenyl Phosphine FcCH$_2$PPh$_2$ [Fc = (η_5-C$_5$H$_5$)Fe(h$_5$-C$_5$H$_4$)]. *Inorganic Chemistry*, **39**(4), 680–687.

148 Gimeno, M.C., Jones, P.G., Laguna, A. and Sarroca, C. (2000) 1,1′-Bis(diphenylthiophosphoryl)ferrocene complexes of gold(I) and gold(III). *Journal of Organometallic Chemistry*, **596**(1–2), 10–15.

149 Barranco, E.M., Crespo, O., Gimeno, M.C., Jones, P.G., Laguna, A. and Sarroca, C. (2001) 1,1′-Bis(2-pyridylthio)ferrocene: a new ligand in gold and silver chemistry. *Journal of the Chemical Society, Dalton Transactions*, (17), 2523–2529.

150 Aguado, J.E., Canales, S., Gimeno, M.C., Jones, P.G., Laguna, A. and Villacampa, M.D. (2005) Group 11 complexes with unsymmetrical P,S and P,Se disubstituted ferrocene ligands. *Dalton Transactions*, (18), 3005–3015.

151 Barranco, E.M., Gimeno, M.C., Laguna, A. and Villacampa, M.D. (2005) Gold and silver complexes with the ferrocenyl-pyrazolate ligand FcCH2pz. *Inorganica Chimica Acta*, **358** (14), 4177–4182.

152 Blackie, M.A.L., Beagley, P., Chibale, K., Clarkson, C., Moss, J.R. and Smith, P.J. (2003) Synthesis and antimalarial activity in vitro of new heterobimetallic complexes: Rh and Au derivatives of chloroquine and a series of ferrocenyl-4-amino-7-chloroquinolines. *Journal of Organometallic Chemistry*, **688**(1–2), 144–152.

153 Barranco, E.M., Crespo, O., Gimeno, M.C., Jones, P.G. and Laguna, A. (2004) The role of secondary interactions in group 11 metal complexes containing the ferrocene ligand FcCH$_2$NHpyMe in supramolecular structures. *European Journal of Inorganic Chemistry*, (24), 4820–4827.

154 Aguado, J.E., Calhorda, M.J., Gimeno, M.C. and Laguna, A. (2005) Unprecedented η3-M3 coordination mode in a terpyridine ligand. *Chemical Communications*, (26), 3355–3356.

155 Byers, P.K., Carr, N. and Stone, F.G.A. (1990) Chemistry of polynuclear metal complexes with bridging carbene or carbyne ligands. Part 106. Synthesis and reactions of the alkylidyne complexes [M($\equiv$CR)(CO)$_2${(C$_6$F$_5$)AuC(pz)$_3$}] (M = W or Mo, R = alkyl or aryl, pz = pyrazol-1-yl); crystal structure of [WPtAu(C$_6$F$_5$)(μ_3-CMe)(CO)$_2$(PMe$_2$Ph)$_2${(C$_6$F$_5$)AuC(pz)$_3$}]. *Journal of the Chemical Society, Dalton Transactions*, (12), 3701–3708.

156 Carriedo, G.A., Riera, V., Sanchez, G. and Solans, X. (1988) Synthesis of new heteronuclear complexes with bridging

carbyne ligands between tungsten and gold. X-ray crystal structure of (2,2'-bipyridine)bromo-dicarbonyl (pentafluorophenyl)-μ-(p-tolylmethylidyne)goldtungsten. *Journal of the Chemical Society, Dalton Transactions*, (7), 1957–1962.

157 Albano, V.G., Busetto, L., Cassani, M.C., Sabatino, P., Schmitz, A. and Zanotti, V. (1995) Synthesis of new heterodinuclear aminocarbyne complexes: crystal structures of [AuW{μ-CN(Et)Me}(C_6F_5)(CO)$_2$(η-C_5H_5)] and [{CuW{μ-CN(Et)Me}Cl(CO)$_2$(η-C_5H_5)}$_2$]. *Journal of the Chemical Society, Dalton Transactions*, (12), 2087–2093.

158 Fernandez, E.J., Gimeno, M.C., Jones, P.G., Laguna, A., Laguna, M. and Olmos, E. (1994) Synthesis of heterometallic species with (Ph$_2$P)$_3$CH or (Ph$_2$P)3C-. Crystal structures of [Mo(CO)$_4${(PPh$_2$)$_2$CHPPh$_2$AuCl}], [Mo(CO)$_4${(PPh$_2$)$_2$CHPPh$_2$Au(C_6F_5)}] and [NBu$_4$][Mo(CO)$_4${(PPh$_2$)$_2$CPPh$_2$Au(C_6F_5)}]. *Journal of the Chemical Society, Dalton Transactions*, (19), 2891–2898.

159 Jones, P.G. and Thoene, C. (1992) μ-[Bis(diphenylphosphino)methane]bis[pentafluorophenylgold(I)] at 178 K. *Acta Crystallographica, Section C: Crystal Structure Communications*, C48(7), 1312–1314.

160 Terroba, R., Fernandez, E.J., Hursthouse, M.B. and Laguna, M. (1998) Dithiolate diphosphine polynuclear gold complexes. X-ray structure of [Au$_2$(μ-dppm)(C_6F_5)$_2$($S_2C_6H_4$)]. *Polyhedron*, **17**(15), 2433–2439.

161 Baker, R.W. and Pauling, P. (1969) Molecular structure of chlorobis(pentafluorophenyl)triphenylphosphinegold(III). *Journal of the Chemical Society D: Chemical Communications*, (13), 745.

162 Deacon, G.B., Green, J.H.S. and Nyholm, R.S. (1965) Chemistry Of Organothallium Compounds.I. Preparations And Properties Of Bis(Pentalfuorophenyl)Thallium(3) Compounds. *Journal of the Chemical Society*, (MAY), 3411–3425.

163 Uson, R., Royo, P. and Laguna, A. (1971) Preparation of bromo and iodobis(pentafluorophenyl)triphenylphosphinegold(III). *Inorganic and Nuclear Chemistry Letters*, **7** (10), 1037–1040.

164 Uson, R. and Laguna, A. (1982) Chlorobis(Pentafluorophenyl)Thallium(III). *Inorganic Syntheses*, **21**, 71–74.

165 Uson, R., Laguna, A. and Bergareche, B. (1980) Dihalopentafluorophenyl-tetrahydrothiophenegold(III) complexes, their preparation and reactions. *Journal of Organometallic Chemistry*, **184**(3), 411–416.

166 Uson, R., Laguna, A. and Vicente, J. (1976) Dihaloperhalophenyl(triethylphosphine)gold(III) complexes. *Synthesis and Reactivity in Inorganic and Metal-Organic Chemistry*, **6**(4), 293–305.

167 Uson, R., Laguna, A. and Pardo, J. (1974) Preparation of dihaloperhaloaryl complexes of gold(III) by oxidizing perhaloaryl complexes of gold(I) with halogens of thallium(III) chloride. *Synthesis and Reactivity in Inorganic and Metal-Organic Chemistry*, **4**(6), 499–513.

168 Uson, R., Laguna, A., Laguna, M., Gimeno, M.C., De Pablo, A., Jones, P.G., Meyer-Baese, K. and Erdbruegger, C.F. (1987) Synthesis and reactivity of neutral complexes of the types [AuX$_3$(ylide)] and trans-[Au(C_6F_5)X$_2$(ylide)] (X = halide or pseudohalide). X-ray structure of [Au(SCN)$_3$(CH$_2$PPh$_3$)]. *Journal of Organometallic Chemistry*, **336**(3), 461–468.

169 Uson, R., Laguna, A. and Cuenca, T. (1980) Chlorobis(polyfluorophenyl)thallium(III) complexes and their reactions with gold(I) and tin(II) compounds. *Journal of Organometallic Chemistry*, **194**(3), 271–275.

170 Uson, R., Laguna, A., Laguna, M., Uson, A. and Gimeno, M.C. (1988) Reactions of pentafluorophenyl(ylide)silver(I) or -gold(I) complexes with chlorobis

(pentafluorophenyl)thallium(III). *Journal of the Chemical Society, Dalton Transactions*, (3), 701–703.

171 Jones, P.G. (1993) Crystal structure of tetra-n-butylammonium trans-dichlorobis(pentafluorophenyl)aurate (III), $(C_4H_9)_4N(Cl_2Au(C_6F_5)_2)$. *Zeitschrift für Kristallographie*, **208**(2), 362–365.

172 Uson, R., Laguna, A., Laguna, M., Fernandez, E., Jones, P.G. and Sheldrick, G.M. (1982) Tris(pentafluorophenyl)gold (III) complexes, *Journal of the Chemical Society, Dalton Transactions*, (10), 1971–1976.

173 Jones, P.G. and Bembenek, E. (1994) Crystal structure of bis (triphenylphosphine)iminium tetrakis (pentafluorophenyl)aurate(III), $[((C_6H_5)_3P)_2N] + [Au(C_6F_5)_4]$. *Zeitschrift für Kristallographie*, **209**(8), 690–692.

174 Murray, H.H., Fackler, J.P., Jr. Porter, L.C., Briggs, D.A., Guerra, M.A. and, Lagow, R.J. (1987) Synthesis and X-ray crystal structures of $[Au(CH_2)_2PPh_2]_2(CF_3)_2$, $[Au(CH_2)_2PPh_2]_2(C_6F_5)_2$, and $[PPN][Au(C_6F_5)_4]$: two dinuclear gold(II) ylide complexes containing alkyl and aryl ligands and a tetrakis (pentafluorophenylaurate(III) anion complex. *Inorganic Chemistry*, **26**(3), 357–363.

175 Uson, R., Laguna, A., Laguna, M., Fernandez, E., Villacampa, M.D., Jones, P.G. and Sheldrick, G.M. (1983) Mono-, bi-, and trinuclear bis (diphenylphosphino)methanegold complexes. Crystal and molecular structures of [bis(diphenylphosphino) methane]bis(pentafluorophenyl)gold(III) perchlorate and 1,2; 2,3-di-μ-[bis (diphenylphosphino)methane]-1,3-dichlorotrigold(I) chlorotris (pentafluorophenyl)aurate(III). *Journal of the Chemical Society, Dalton Transactions*, (8), 1679–1685.

176 Bardaji, M., Gimeno, M.C., Jimenez, J., Laguna, A., Laguna, M. and Jones, P.G. (1992) Neutral or cationic (μ-methylene) bisylide digold(III) complexes. *Journal of Organometallic Chemistry*, **441**(2), 339–348.

177 Uson, R., Laguna, A. and Arrese, M.L. (1984) A new approach to the synthesis of cationic, neutral or anionic diarylgold(III) complexes. *Synthesis and Reactivity in Inorganic and Metal-Organic Chemistry*, **14** (4), 557–567.

178 Uson, R., Laguna, A., De la Orden, M.U., Parish, R.V. and Moore, L.S. (1985) Synthesis and gold-197 Moessbauer spectroscopic studies of dihalo (pentafluorophenyl)(bidentate ligand) gold(III) complexes. *Journal of Organometallic Chemistry*, **282**(1), 145–148.

179 Blanco, M.C., Fernandez, E.J., Jones, P.G., Laguna, A., Lopez-de-Luzuriaga, J.M. and Olmos, M.E. (1998) [Au $(C_6F_5)_3(PPh_2H)$]: a precursor for the synthesis of gold(III) phosphide complexes. *Angewandte Chemie International Edition*, **37**(21), 3042–3043.

180 Jones, P.G., Terroba, R., Fernandez, E. and Laguna, M. (2002) Tris (pentafluorophenyl)[2-(diphenylphosphino)thiophenol-kP]gold (III) dichloromethane solvate. *Acta Crystallographica, Section E: Structure Reports Online*, **E58**(3), m90–m92

181 Bardaji, M., Laguna, A. and Jones, P.G. (2001) Reactivity of an Acetylenephosphine Gold(III) Precursor: Addition to the Triple Bond or Formation of Alkynyl Derivatives. *Organometallics*, **20**(18), 3906–3912.

182 Uson, R., Laguna, A., Laguna, M., Castilla, M.L., Jones, P.G. and Meyer-Baese, K. (1987) Pentafluorophenyl (dithiocarbamate)gold(III) complexes. Crystal and molecular structure of $(C_6F_5)_2Au\{S_2CN(CH_2Ph)_2\}$. *Journal of Organometallic Chemistry*, **336**(3), 453–460.

183 Uson, R., Royo, P. and Laguna, A. (1973) Two new ways for the synthesis of pentafluorophenyl-gold(III) complexes. *Synthesis in Inorganic and Metal-Organic Chemistry*, **3**(3), 237–244.

184 Uson, R., Laguna, A. and Sanjoaquin, J.L. (1974) Perchloratobis(pentafluorophenyl)

triphenylphosphinegold(III) as a precursor of new gold complexes. *Journal of Organometallic Chemistry*, 80(1), 147–154.

185 Crespo, O., Gimeno, M.C., Jones, P.G. and Laguna, A. (1997) Synthesis of gold (III) complexes with the 1,2-dithiolate-o-carborane ligand. Crystal structures of [N(PPh$_3$)$_2$][AuCl$_2$(S$_2$C$_2$B$_{10}$H$_{10}$)] and [AuCl(S$_2$C$_2$B$_{10}$H$_{10}$)(CH$_2$PPh$_3$)]. *Journal of Organometallic Chemistry*, 547(1), 89–95.

186 Uson, R., Laguna, A., Villacampa, M.D., Jones, P.G. and Sheldrick, G.M. (1984) Reactions of cis-diisocyanidebis (perfluorophenyl)gold(III) complexes with hydrazines. Crystal and molecular structure of bis(perfluorophenyl)[3-phenyl-1,4-bis(p-toluidino)-2,3-diazabut-1-en-1-yl-4-ylidene]gold. *Journal of the Chemical Society, Dalton Transactions*, (9), 2035–2038.

187 Crespo, O., Gimeno, M.C., Jones, P.G., Laguna, A. and Villacampa, M.D. (1997) Synthesis and characterization of gold-(I) and -(III) complexes with 1-methyl-2-sulfanyl-1,2-dicarba-closo-dodecaborate. *Journal of the Chemical Society, Dalton Transactions*, (17), 2963–2967.

188 Canales, S., Crespo, O., Gimeno, M.C., Jones, P.G., Laguna, A., Silvestru, A. and Silvestru, C. (2003) Gold and silver complexes with the diselenium ligand [Ph$_2$P(Se)NP(Se)Ph$_2$]. *Inorganica Chimica Acta*, 347, 16–22.

189 Fernandez, E.J., Gimeno, M.C., Jones, P.G., Laguna, A. and Olmos, E. (1997) Strong Activation of the Double Bond in (PPh$_2$)$_2$C:CH$_2$. Novel Synthesis of Gold (III) Methanide Complexes by Michael Addition Reactions. *Organometallics*, 16 (6), 1130–1136.

190 Fernandez, E.J., Gimeno, M.C., Jones, P.G., Laguna, A., Lopez-De-Luzuriaga, J.M. and Olmos, E. (1997) Synthesis of highly stable intermediates in Michael-type additions to the double bond in (SPPh$_2$)$_2$C:CH$_2$. *Journal of the Chemical Society, Dalton Transactions*, (19), 3515–3518.

191 Uson, R., Laguna, A., Laguna, M., Jimenez, J. and Durana, M.E. (1990) Tris (pentafluorophenyl)gold(III) complexes with oxygen-, nitrogen- and sulfur-donor ligands. *Inorganica Chimica Acta*, 168(1), 89–92.

192 Crespo, O., Brusko, V.V., Gimeno, M.C., Tornil, M.L., Laguna, A. and Zabirov, N.G. (2004) Gold and silver complexes with the ligands N-[bis(isopropoxy)-thiophosphoryl]thiobenzamide and N-[bis(isopropoxy)thiophosphoryl]-N'-phenylthiourea. *European Journal of Inorganic Chemistry*, 423–430.

193 Vicente, J., Bermudez, M.D., Chicote, M.T. and Sanchez-Santano, M.J. (1989) Attempts to prepare mixed diarylgold(III) complexes containing a pentafluorophenyl group. Synthesis of [2-{(dimethylamino)methyl}phenyl-C1N)-(pentafluorophenyl)gold(III) complexes. *Journal of Organometallic Chemistry*, 371 (1), 129–135.

194 Vicente, J., Bermudez, M.D., Escribano, J., Carrillo, M.P. and Jones, P.G. (1990) Synthesis of intermediates in the carbon-hydrogen bond activation of acetone with 2-phenylazophenylgold(III) complexes and in the carbon-carbon coupling of aryl groups from diarylgold(III) complexes. Crystal and molecular structures of [cyclic] [Au{C$_6$H$_3$(N:NC$_6$H$_4$Me-4')-2-Me-5}(acac-C)Cl] (acac = acetylacetonate), cis-[Au(C$_6$H$_4$N:NPh-2)Cl$_2$(PPh$_3$)], and [cyclic] [Au(C$_6$H$_4$CH$_2$NMe$_2$-2)(C$_6$F$_5$)Cl]. *Journal of the Chemical Society, Dalton Transactions*, (10), 3083–3089.

195 Cerrada, E., Fernandez, E.J., Gimeno, M.C., Laguna, A., Laguna, M., Terroba, R. and Villacampa, M.D. (1995) Synthesis of dithiolate gold(III) complexes by dithiolate transfer reactions. X-ray structure of [Au(C$_6$F$_5$)(S$_2$C$_6$H$_4$)(PPh$_3$)]. *Journal of Organometallic Chemistry*, 492 (1), 105–110.

196 Cerrada, E., Fernandez, E.J., Jones, P.G., Laguna, A., Laguna, M. and Terroba, R.

(1995) Synthesis and Reactivity of Trinuclear Gold(III) Dithiolate Complexes. X-ray Structure of [Au(C$_6$F$_5$)(S$_2$C$_6$H$_4$)]$_3$ and [Au(C$_6$F$_5$)(S$_2$C$_6$H$_4$)(SC$_6$H$_4$SPPh$_3$)]. *Organometallics*, **14**(12), 5537–5543.

197 Jones, P.G. (1992) [(Methyldiphenylphosphonio)methanide]tris(pentafluorophenyl)gold(III) acetone solvate. *Acta Crystallographica, Section C: Crystal Structure Communications*, **C48**(7), 1209–1211.

198 Uson, R., Laguna, A., Laguna, M., Manzano, B.R., Jones, P.G. and Sheldrick, G.M. (1984) Synthesis of polynuclear complexes containing the tridentate bis(diphenylphosphino)methanide ligand. Crystal structures of the compounds μ-[bis(diphenylphosphino)methanide]tris(pentafluorophenyl)digold(I)/(III) and μ-[bis(diphenylphosphino)methanide]-μ-[bis(diphenylphosphino)methanide]tetrakis(pentafluorophenyl)trigold(I)/(III)/(III) perchlorate. *Journal of the Chemical Society, Dalton Transactions*, (5), 839–843.

199 Uson, R., Laguna, A., Laguna, M., Manzano, B.R., Jones, P.G. and Sheldrick, G.M. (1985) Synthesis of trinuclear gold (I) and gold(III) complexes containing the tridentate bis(diphenylphosphino)methanide ligand. Crystal structure of μ$_3$-[bis(diphenylphosphino)methanido-C(Au$_3$)P(Au$_1$)P'(Au$_2$)]-1,2-dichloro-1,1-bis(pentafluorophenyl)-3-pyridinetrigold. *Journal of the Chemical Society, Dalton Transactions*, (11), 2417–2420.

200 Fernandez, E.J., Concepcion Gimeno, M., Jones, P.G., Laguna, A., Laguna, M. and Lopez-de-Luzuriaga, J.M. (1992) Synthesis and structural characterization of polynuclear complexes containing the eight-electron donor bis(diphenylphosphino)methanediide ligand. *Journal of the Chemical Society, Dalton Transactions*, (23), 3365–3370.

201 Laguna, A., Laguna, M., Gimeno, M.C., Fernandez, E.J., Jimenez, J., Bardaji, M., Lopez-de-Luzuriaga, J.M. and Jones, P.G. (1993) Polynuclear derivatives of gold; synthesis, structure, and reactivity. *Special Publication - Royal Society of Chemistry*, **131**, (Chemistry of the Copper and Zinc Triads), 168–171.

202 Uson, R., Laguna, A., Laguna, M., Lazaro, I. and Tarton, M.T. (1988) Synthesis of benzyl- or (pentafluorobenzyl)diphenylphosphonio(diphenylphosphino)methanide complexes of gold(I) and gold(III) and their use as carbon-donor ligands. *Journal of the Chemical Society, Dalton Transactions*, (1), 155–158.

203 Alvarez, B., Fernandez, E.J., Gimeno, M.C., Jones, P.G., Laguna, A., Laguna, M. and Lopez-de-Luzuriaga, J.M. (1996) Synthesis and structure of [Tl(C$_6$F$_5$)$_2$Cl{Au(C$_6$F$_5$)$_3$(PPh$_2$CH$_2$PPh$_2$(O))}$_2$]. *Journal of Organometallic Chemistry*, **525**(1–2), 109–113.

204 Alvarez, B., Fernandez, E.J., Gimeno, M.C., Jones, P.G., Laguna, A. and Lopez-De-Luzuriaga, J.M. (1998) Gold complexes with mono- or di-chalcogenides of bis(diphenylphosphino)methanide ligands. X-ray crystal structure of [Au(C$_6$F$_5$)$_2${(SPPh$_2$)$_2$C(AuAsPh$_3$)$_2$}]ClO$_4$. *Polyhedron*, **17**(11–12), 2029–2035.

205 Uson, R., Laguna, A., Laguna, M., Lazaro, I. and Jones, P.G. (1987) Four-, five-, or six-membered methanide auracycles: X-ray structure of [(C$_6$F$_5$)$_2$AuPPh$_2$CH(AuC$_6$F$_5$)PPh$_2$CHCOOMe]. *Organometallics*, **6**(11), 2326–2331.

206 Jones, P.G. (1988) The distortion of pentafluorophenyl groups in metal complexes: limitations of rigid-body refinement. *Journal of Organometallic Chemistry*, **345**(3), 405–411.

207 Gimeno, M.C., Jones, P.G., Laguna, A., Laguna, M. and Lazaro, I. (1993) Five-membered methanediide auracycles. Crystal structure of [Au(C$_6$F$_5$)$_2${Ph$_2$PC(AuPPh$_3$)(AuC$_6$F$_5$)PPh$_2$(CHCO$_2$Me)}]. *Journal of the Chemical Society, Dalton Transactions*, (14), 2223–2227.

208 Browning, J., Bushnell, G.W., Dixon, K.R. and Pidcock, A. (1983) [PtCl(PEt$_3$)(CH(PPh$_2$S)$_2$)], A Novel C,S-Bonded Chelate With Dynamic Stereochemistry Controlled By A Metal-Ligand Pivot. *Inorganic Chemistry*, **22**(16), 2226–2228.

209 Uson, R., Laguna, A., Laguna, M., Lazaro, I., Jones, P.G. and Fittschen, C. (1988) Mono- and bi-nuclear four-membered methanide auracycles; synthesis and reactivity. X-ray structure of cis-diphenylthiophosphinato [(methyldiphenylphosphonio)methanide] bis(pentafluorophenyl)gold(III). *Journal of the Chemical Society, Dalton Transactions*, (9), 2323–2327.

210 Uson, R., Laguna, A., Laguna, M., Fraile, M.N., Jones, P.G. and Erdbruegger, C.F. (1989) (Diphenylphosphinomethyl) diphenylphosphine sulfide (Ph$_2$PCH$_2$PPh$_2$S) and its methanide anion Ph$_2$PCHPPh$_2$S- as ligands in organogold chemistry. X-ray crystal structure of [(diphenylphosphino)(diphenylthiophosphinoyl)methanido-P,S]bis(pentafluorophenyl)gold(III). *Journal of the Chemical Society, Dalton Transactions*, (1), 73–77.

211 Fernandez, E.J., Gimeno, M.C., Jones, P.G., Laguna, A., Laguna, M. and Lopez-de-Luzuriaga, J.M. (1995) Synthesis, Structure, and Reactivity of the Anionic Trinuclear Methanide NBu$_4$[{Au(C$_6$F$_5$)$_3$(PPh$_2$CHPPh$_2$)}$_2$Au]. *Organometallics*, **14**(6), 2918–2922.

212 Fernandez, E.J., Gimeno, M.C., Jones, P.G., Laguna, A., Laguna, M., Lopez-de-Luzuriaga, J.M. and Rodriguez, M.A. (1995) Synthesis, structure and reactivity of [(acac)AuCH(PPh$_2$AuPPh$_2$)$_2$CHAu(acac)], a complex containing the tridentate ligand [HC(PPh$_2$)$_2$]. *Chemische Berichte*, **128**(2), 121–124.

213 Minghetti, G., Bonati, F. and Banditelli, G. (1976) Carbene Complexes Of Gold (III) And Reactions Of Coordinated Ligand. *Inorganic Chemistry*, **15**(7), 1718–1720.

214 Uson, R., Laguna, A. and Castrillo, M.V. (1979) Unusual cationic single-bridged binuclear complexes of gold(I) and gold (III). *Synthesis and Reactivity in Inorganic and Metal-Organic Chemistry*, **9**(4), 317–324.

215 Fernandez, E.J., Gimeno, M.C., Jones, P.G., Ahrens, B., Laguna, A., Laguna, M. and Lopez de Luzuriaga, J.M. (1994) Tris (diphenylphosphino)methanide gold(III) complexes. Crystal structures of [Au(C$_6$F$_5$)$_2${(Ph$_2$P)$_2$CPPh$_2$}] and [(F$_5$C$_6$)$_2$Au{(Ph$_2$P)$_2$CPPh$_2$}AuCl]. *Journal of the Chemical Society, Dalton Transactions*, (23), 3487–3492.

216 Balch, A.L. (1994) Construction Of Small Polynuclear Complexes With Trifunctional Phosphine-Based Ligands As Backbones, in *Progress In Inorganic Chemistry*, vol. **41** (ed. S.J. Lippard), John Wiley & Sons, Inc., New York, pp. 239–329.

217 Blanco, M.C., Fernandez, E.J., Lopez-De-Luzuriaga, J.M., Olmos, M.E., Crespo, O., Gimeno, M.C., Laguna, A. and Jones, P.G. (2000) Heteropolynuclear phosphide complexes: phosphorus as unique atom bridging coinage metal centers. *Chemistry - A European Journal*, **6**(22), 4116–4123.

218 Blanco, M.C., Fernandez, E.J., Fischer, A.K., Jones, P.G., Laguna, A., Olmos, M.E. and Villacampa, M.D. (2000) NBu$_4$[{Au(C$_6$F$_5$)$_3$}$_2$(μ-PPh$_2$)]: a gold(III) phosphide with a single atom bridging the metallic centers. *Inorganic Chemistry Communications*, **3**(4), 163–165.

219 Fernandez, E.J., Gil, M., Olmos, M.E., Crespo, O., Laguna, A. and Jones, P.G. (2001) Ability of a Au(III)-N Unit to Bond Two Aurophilically Interacting Gold(I) Centers. *Inorganic Chemistry*, **40**(13), 3018–3024.

220 Blanco, M.C., Fernandez, E.J., Olmos, M.E., Perez, J. and Laguna, A. (2006) From diphosphane to diphosphodiide gold(III) derivatives of 1,2-diphosphinobenzene. *Chemistry - A European Journal*, **12**(12), 3379–3388.

221 Cespedes, C. and Vega, J.C. (1994) Reactions Of Dichloromethane With Thioanions. 1. Preparation Of Bis(N,N-Dialkylthiocarbamoylthio)Methanes. *Phosphorus, Sulfur Silicon Related Elements*, **90**(1–4), 155–158.

222 Canales, F., Canales, S., Crespo, O., Gimeno, M.C., Jones, P.G. and Laguna, A. (1998) Synthesis and Reactivity of the First (Hydrosulfido)gold(III) Complex. Crystal Structure of the Derivatives NBu$_4$[{Au(C$_6$F$_5$)$_3$}$_2$SR] with the Isolobal Fragments R = H, AuPPh$_3$, AgPPh$_3$. *Organometallics*, **17**(8), 1617–1621.

223 Gimeno, M.C., Jones, P.G., Laguna, A., Sarroca, C., Calhorda, M.J. and Veiros, L.F. (1998) Gold(I) and gold(III) complexes with the 1,1′-bis(diethyldithiocarbamate)ferrocene ligand. *Chemistry - A European Journal*, **4**(11), 2308–2314.

224 Crespo, O., Canales, F., Gimeno, M.C., Jones, P.G. and Laguna, A. (1999) Synthesis of [Au$_2$(SC$_6$F$_5$)$_2$(μ-dppf)] and [Au$_2$(μ-SC$_6$F$_5$)(μ-dppf)] (dppf = 1,1′-Bis(diphenylphosphino)ferrocene). Reactivity toward Various Metallic Fragments. *Organometallics*, **18**(16), 3142–3148.

225 Canales, S., Crespo, O., Gimeno, M.C., Jones, P.G. and Laguna, A. (2000) Coordination chemistry of 1,1′-bis(diphenylselenophosphoryl)ferrocene (dpspf) towards Group 11 elements. Crystal structures of [Ag(dpspf){(SPPh$_2$)$_2$CH$_2$}]OTf and [Au(dpspf)][AuCl$_2$]. *Journal of Organometallic Chemistry*, **613**(1), 50–55.

226 Canales, S., Crespo, O., Fortea, A., Gimeno, M.C., Jones, P.G. and Laguna, A. (2002) Gold and silver complexes with the ligands Fc(SPh)$_2$ and Fc(SePh)$_2$ (Fc = (η$_5$-C$_5$H$_4$)$_2$Fe). *Journal of the Chemical Society, Dalton Transactions*, (10), 2250–2255.

227 Canales, S., Crespo, O., Gimeno, M.C., Jones, P.G., Laguna, A. and Mendizabal, F. (2001) Mixed Gold(I)-Gold(III) Complexes with Bridging Selenido Ligands. Theoretical Studies of the Gold (I)-Gold(III) Interactions. *Organometallics*, **20**(23), 4812–4818.

228 Calhorda, M.J., Canales, F., Gimeno, M.C., Jimenez, J., Jones, P.G., Laguna, A. and Veiros, L.F. (1997) Gold(I)-Gold(III) Interactions in Polynuclear Sulfur-Centered Complexes. Synthesis and Structural Characterization of [S(Au$_2$dppf){Au(C$_6$F$_5$)$_3$}] and [{S(Au$_2$dppf)}$_2${Au(C$_6$F$_5$)$_2$}]OTf (dppf = 1,1′-Bis(diphenylphosphino)ferrocene). *Organometallics*, **16**(17), 3837–3844.

229 Canales, F., Gimeno, M.C., Laguna, A. and Jones, P.G. (1996) Synthesis and Structural Characterization of Tetranuclear Sulfur-Centered Complexes with Mixed-Valent Gold Atoms: [S(Au$_2$dppf){Au(C$_6$F$_5$)$_3$}$_2$] (dppf = 1,1′-Bis(diphenylphosphino)ferrocene) and [S(AuPPh$_3$)$_2${Au(C$_6$F$_5$)$_3$}$_2$]. *Organometallics*, **15**(15), 3412–3415.

230 Canales, S., Crespo, O., Gimeno, M.C., Jones, P.G. and Laguna, A. (2004) Selenolate Gold Complexes with Aurophilic Au(I)-Au(I) and Au(I)-Au(III) Interactions. *Inorganic Chemistry*, **43**(22), 7234–7238.

231 Bardaji, M., Calhorda, M.J., Costa, P.J., Jones, P.G., Laguna, A., Perez, M.R. and Villacampa, M.D. (2006) Synthesis, Structural Characterization, and Theoretical Studies of Gold(I) and Gold(I)-Gold(III) Thiolate Complexes: Quenching of Gold(I) Thiolate Luminescence. *Inorganic Chemistry*, **45**(3), 1059–1068.

232 Uson, R., Laguna, A., Laguna, M., Tarton, M.T. and Jones, P.G. (1988) The first example of a direct formal gold(I)-gold(III) bond. Synthesis and structure of [[Au(CH$_2$)$_2$PPh$_2$]$_2$Au(C$_6$F$_5$)$_3$]. *Journal of the Chemical Society, Chemical Communications*, (11), 740–741.

233 Calhorda, M.J. and Veiros, L.F. (1994) Metal-metal bonds in a Au5 chain and other species. *Journal of Organometallic Chemistry*, **478**(1–2), 37–44.

234 Basil, J.D., Murray, H.H., Fackler, J.P., Tocher, J., Mazany, A.M.,

Trzcinskabancroft, B., Knachel, H., Dudis, D., Delord, T.J. and Marler, D.O. (1985) Experimental And Theoretical-Studies Of Dinuclear Gold(I) And Gold(II) Phosphorus Ylide Complexes – Oxidative Addition, Halide Exchange, And Structural-Properties Including The Crystal And Molecular-Structures Of [Au(CH$_2$)$_2$PPh$_2$]$_2$ And [Au(CH$_2$)$_2$PPh$_2$]$_2$(CH$_3$)Bri. *Journal of the American Chemical Society*, **107**(24), 6908–6915.

235 Schmidbaur, H., Hartmann, C., Riede, J., Huber, B. and Muller, G. (1986) Alkylation Of Methylene-Bridged And Ylide-Bridged Binuclear Gold(III) Complexes. *Organometallics*, **5**(8), 1652–1656.

236 Uson, R., Fornies, J., Laguna, A. and Valenzuela, J.I. (1982) Pseudohalo-bridged pentafluorophenyl gold-palladium complexes. *Synthesis and Reactivity in Inorganic and Metal-Organic Chemistry*, **12**(7), 935–946.

237 Fernandez, E.J., Gimeno, M.C., Jones, P.G., Laguna, A., Laguna, M. and Olmos, E. (1996) Synthesis of heteropolynuclear complexes with 1,1,1-tris(diphenylphosphinomethyl)ethane. Crystal structure of [(OC)$_4$Mo{(Ph$_2$PCH$_2$)$_2$CMe(CH$_2$PPh$_2$)}AuCl]. *Journal of the Chemical Society, Dalton Transactions*, (17), 3603–3608.

238 Uson, R., Laguna, A., Abad, J.A. and Dejesus, E. (1983) Preparation Of Monomeric Neutral Or Anionic Tris(Polyfluorophenyl)-Thallium(III) And Of Anionic Heteronuclear Tris(Polyfluorophenyl)-Thallium Metal-Carbonyl-Complexes. *Journal of the Chemical Society, Dalton Transactions*, (6), 1127–1130.

239 Hellmann, K.W., Friedrich, S., Gade, L.H., Li, W.S. and McPartlin, M. (1995) Tripodal Triamidostannates As Building-Blocks In The Generation Of Sn-M-Bonded Heterobimetallics (M = Fe, Ru). *Chemische Berichte*, **128**(1), 29–34.

240 Memmler, H., Kauper, U., Gade, L.H., Stalke, D. and Lauher, J.W. (1996) A chiral triamidostannate: Its structure and properties as a nucleophile. *Organometallics*, **15**(17), 3637–3639.

241 Blanco, M.C., Fernandez, E.J., Olmos, M.E., Crespo, O., Laguna, A. and Jones, P.G. (2002) A Pentanuclear Mixed Gold(III)-Silver(I) Phosphide with an Unusual T-Frame m3-Cl Bridge. *Organometallics*, **21**(12), 2426–2429.

242 Uson, R., Laguna, A., Laguna, M., Jimenez, J. and Jones, P.G. (1991) Mixed valence pentanuclear gold complex with a linear Au^{5+} chain. *Angewandte Chemie-International Edition*, **30**(2), 198.

243 Bardaji, M., Laguna, A. and Villacampa, M.D. (2000) Bis[bis(diphenylphosphinomethyl)phenylphosphine]dichlorotetragold(I) bis[chlorotris(pentafluorophenyl)aurate(III)]. *Acta Crystallographica, Section C: Crystal Structure Communications*, **C56**(11), E487–E488

244 Laguna, A., Laguna, M., Gimeno, M.C. and Jones, P.G. (1992) Synthesis and X-ray characterization of the neutral organometallic gold cluster [Au$_{10}$(C$_6$F$_5$)$_4$(PPh$_3$)$_5$]. *Organometallics*, **11**(8), 2759–2760.

245 Puddephatt, R.J. (1987) *Comprehensive Coordination Chemistry*, **5** (eds J. Wardell, E.W. Abel, F.G.A. Stone and G. Wilkinson), pp. 861.

246 Grohmann, H. and Schmidbaur, H. (1995) *Comprehensive Organometallic Chemistry II*, Vol. 3 (eds J. Wardell, E.W. Abel, F.G.A. Stone and G. Wilkinson), Pergamon Press, Oxford, pp. 1.

247 Schmidbaur, H. and Dash, K.C. (1982) Compounds Of Gold In Unusual Oxidation-States. *Advances in Inorganic Chemistry*, **25**, 239–266.

248 Beurskens, P.T., Blaauw, H.J.A., Cras, J.A. and Steggerda, J.J. (1968) Preparation Structure And Properties Of Bis(N,N-Di-N-Butyldithiocarbamato)Gold(III) Dihaloaurate(I). *Inorganic Chemistry*, **7**(4), 805–810.

249 Blake, A.J., Greig, J.A., Holder, A.J., Hyde, T.I., Taylor, A. and Schroder, M. (1990) Bis

(1,4,7-Trithiacyclononane)Gold Dication - A Paramagnetic, Mononuclear Au-Ii Complex. *Angewandte Chemie International Edition*, **29**(2), 197–198.

250 Yam, V.W.W., Choi, S.W.K. and Cheung, K.K. (1996) Synthesis, photophysics and thermal redox reactions of a [{Au(dppn)Cl}$_{(2)}$]$^{(2+)}$ dimer with an unsupported Au-II-Au-II bond. *Chemical Communications*, (10), 1173–1174.

251 Laguna, A., Laguna, M., Jimenez, J., Lahoz, F. and Olmos, E. (1992) Asymmetric monocationic or neutral gold(II) complexes: X-ray crystal structure of [(C$_6$F$_5$)Au(CH$_2$PPh$_2$CH$_2$)$_2$AuPPh$_3$] ClO$_4$·O.5CH$_2$Cl$_2$. *Journal of Organometallic Chemistry*, **435**(1–2), 235–247.

252 Mendez, L.A., Jimenez, J., Cerrada, E., Mohr, F. and Laguna, M. (2005) A family of alkynylgold(III) complexes [Au-I(μ-{CH$_2$}$_{(2)}$PPh$_2$)$_{(2)}$Au-III(C CR)$_{(2)}$)] (R = Ph, Bu-t, Me$_3$Si): Facile and reversible comproportionation of gold(I)/gold(III) to digold(II). *Journal of the American Chemical Society*, **127**(3), 852–853.

253 Laguna, A., Laguna, M., Jimenez, J., Lahoz, F.J. and Olmos, E. (1994) Mixed-valent linear chains of gold atoms. X-ray structure of [{(2,4,6-C$_6$F$_3$H$_2$)Au(CH$_2$PPh$_2$CH$_2$)$_2$Au}$_2$Au(CH$_2$PPh$_2$CH$_2$)$_2$Au](ClO$_4$)$_2$. *Organometallics*, **13**(1), 253–257.

254 Bardaji, M., Gimeno, M.C., Jones, P.G., Laguna, A. and Laguna, M. (1994) Dinuclear Gold(II) Complexes Containing Two Different Bridging Ligands. Crystal Structure of [Au$_2${μ-(CH$_2$)$_2$PPh$_2$}{μ-S$_2$CN(CH$_2$Ph)$_2$}Br$_2$]. *Organometallics*, **13**(9), 3415–3419.

255 Bardaji, M., Laguna, A. and Laguna, M. (1995) Phosphoniodithioformate gold derivatives. Synthesis of tricationic gold (II) complexes. *Journal of Organometallic Chemistry*, **496**(2), 245–248.

256 Bardaji, M., Blasco, A., Jiménez, J., Jones, P.G., Laguna, A., Laguna, M., Merchán, F., (1994) Di- and tetranuclear gold(II) complexes with dithiocarbamate and related ligands. X-ray structure of Au (CH$_2$PPh$_2$CH$_2$)$_2$ (S$_2$CNMe$_2$)$_2$). *Inorganica Chimica Acta*, **223**, 55–61.

257 Veiros, L.F. and Calhorda, M.J. (1996) How bridging ligands and neighboring groups tune the gold-gold bond strength. *Journal of Organometallic Chemistry*, **510** (1–2), 71–81.

258 Fackler, J.P. Jr. (1997) Metal-metal bond formation in the oxidative addition to dinuclear gold(I) species. Implications from dinuclear and trinuclear gold chemistry for the oxidative addition process generally. *Polyhedron*, **16**(1), 1–17.

259 Laguna, M. and Cerrada, E. (1999) *Metal Clusters in Chemistry* (eds P. Braunstein, L.A. Oro and P.R. Raithby), Wiley-VCH, Weinheim.

260 Gimeno, M.C., Jimenez, J., Jones, P.G., Laguna, A. and Laguna, M. (1994) Synthesis and Structure of [Au$_4$(C$_6$F$_5$)$_2${(PPh$_2$)$_2$CH}$_2$(PPh$_3$)$_2$] (ClO$_4$)$_2$·4CH$_2$Cl$_2$, a Complex with Two Direct Gold-Gold Bonds. *Organometallics*, **13**(6), 2508–2511.

261 Fackler, J.P., Galarza, E., Garzon, G., Mazany, A.M., Murray, H.H., Omary, M.A.R., Raptis, R., Staples, R.J., Van Zyl, W.E. and Wang, S.N. (2002) The diphenylmethylenethiophosphinate (MTP) ligand in gold(I), platinum(II), lead (II), thallium(I), and mercury(II) complexes, sym-Au-2(MTP)(2), (PPN)[Au (MTP)(2)], Au$_2$Pt(MTP)(4), Au$_2$Pb(MTP) (4), AuTl(MTP)(4), Hg(MTP)(2), Hg (MTP)(2)(AuCl)(2), and (HgAuI)-Au-II (MTP)(2)(AuCl$_4$)-Cl-III. *Inorganic Syntheses*, **33**, 171–180.

262 Hofmann, K.A. and Bugge, G. (1908) Platinum blue. *Berichte der Deutschen Chemischen Gesellschaft*, **41**, 312–314.

263 Ciriano, M.A., Tejel, C. and Oro, L.A. (1999) From Platinum blues to Rhodium and Iridium blues. *Chemistry - A European Journal*, **5**(4), 1131–1135.

4
Theoretical Chemistry of Gold – From Atoms to Molecules, Clusters, Surfaces and the Solid State

Peter Schwerdtfeger and Matthias Lein

4.1
Introduction

The theoretical and computational chemistry of heavy elements is currently undergoing enormous growth due to the ever-increasing computer power combined with the development of new algorithms and theoretical methodologies. In the case of gold and its compounds, the number of papers published in this field is currently growing at an exponential rate, Figure 4.1. The strong interest in gold chemistry and physics arose early on in the computational community, after Pekka Pyykkö and Jean-Paul Desclaux's surprising discovery in 1979 [1, 2] that neutral gold shows an unusually large relativistic 6s-orbital stabilization (contraction) compared with its neighboring atoms in the periodic table, which is now termed the *gold maximum of relativistic effects*, Figure 4.2. This distinct maximum is also found for the lighter Group 11 elements Cu and Ag as Figure 4.2 shows, but to a much smaller extent, and shifted to element 112 within the trans-actinide series, as the very large relativistic 7s shell contraction leads to a change in the ground state configuration for Röntgenium (Rg; $Z = 111$) from $d^{10}s^1$ to d^9s^2 [3, 4]. For gold, this rather large relativistic 6s stabilization results in a significant increase in the ionization potential (by 2.0 eV) and electron affinity (by 1.1 eV) [5, 6]. Using Mulliken's recipe [7, 8], this change results in a relativistic increase in the electronegativity of gold from 1.9 (as in Cu or Ag) to 2.4 [5]. With such a high electronegativity similar to that of iodine, gold is often seen as a pseudohalide. It is self-evident that such large relativistic effects for Au will significantly change any physical property, and thus may explain many of the anomalies observed for gold and its compounds when compared with copper and silver, which for some properties are summarized in Table 4.1.

Because of these large relativistic effects, gold exhibits some very unusual features compared with its lighter congeners [9], Table 4.1. This ranges from the stabilization of high oxidation states [10, 11], aurophilic interactions [12–14], unusual planar structures for bare gold clusters [15–17], the high tendency to form inter-metallic compounds [18], some unusual photochemistry [19] and chemiluminescent

Gold Chemistry: Applications and Future Directions in the Life Sciences. Edited by Fabian Mohr
Copyright © 2009 WILEY-VCH Verlag GmbH & Co. KGaA, Weinheim
ISBN: 978-3-527-32086-8

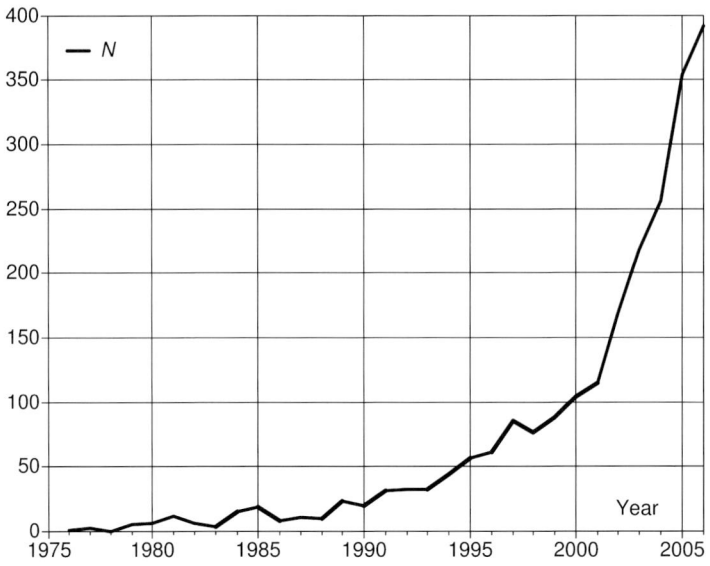

Figure 4.1 Number N of computational papers published per year involving gold compounds.

properties of Au(I) and Au(III) compounds [20], to the more recently explored novel catalytic activities in both homogeneous catalysis [21–24] involving Au(I) and Au(III) in solution, and heterogeneous catalysis involving gold nanoparticles on support surfaces [25]. In fact, since Haruta's discovery in 1987 that gold nanomaterials show unexpected catalytic activities [26–29], research in gold nanostructures has become

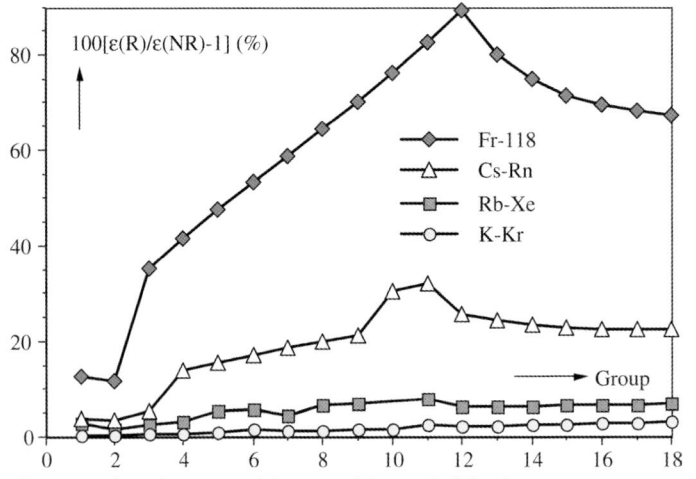

Figure 4.2 The relativistic stabilization of the ns shell for the elements K to Kr ($n=4$), Rb to Xe ($n=5$), Cs to Rn ($n=6$), and Fr to element 118 ($n=7$). Redrawn from the data of Desclaux [32] and Schwerdtfeger and Seth [33]. Pd is ground state $d^{10}s^0$ and is therefore missing in this graph.

Table 4.1 A comparison of properties of Group 11 elements [34].

Property	Cu	Ag	Au
Colour	bronze	silver	yellow
Specific resistivity (10^{-8} Wm)	1.72	1.62	2.4
Thermal conductivity (J cm^{-1} s^{-1} K^{-1})	3.85	4.18	3.1
Electronic heat capacity (10^{-4} J K^{-1} mol^{-1})	6.926	6.411	6.918
Melting point (°C)	1083	961	1064
Boiling point (°C)	2567	2212	3080
Atomic volume (cm^3 mol^{-1})	7.12	10.28	10.21
Electronegativity	1.9	1.9	2.4
Cohesive energy (kJ mol^{-1})	330	280	370
O_2-Chemisorption energy (eV)	5.4	6.0	3.6
Desorption temperature CO on metal surface (K)	190–210	40–80	170–180
Common oxidation states	I,II	I	I,III
Solid MF fluorides	unknown	AgF	unknown
Superconductors	many	rare	rare

the centre of immense activity in the past decade [30], not only on the experimental side with widespread applications, but also on the theoretical side ranging from cluster simulations to extended surface calculations including catalytic reactions [31].

A number of books have been edited on relativistic electronic structure theory [35–38], and three excellent textbooks were published only very recently by Dyall and Faegri [39], by Grant [40] and by Reiher and Wolf [41], which are all highly commended. We note the already extensive literature on relativistic effects in the chemistry of gold. Beside earlier more general reviews on relativistic effects [1, 2, 42], Pyykkö recently presented a comprehensive compilation on theoretical work on gold and its compounds [43–45]. Shorter reviews on the role of relativistic effects in the chemistry of gold can be found in works by Schwerdtfeger [46] and Schmidbaur et al. [47]. Gorin and Toste presented experimental and computational data on homogeneous gold catalysis including a discussion on relativistic effects [21]. Shorter, more focused reviews were also published quite recently which include computational work. Schwerdtfeger presented a highlight on new gold nanostructures and its implications to heterogeneous catalysis [48]. Crawford and Klapötke gave a review on the hydrides and iodides of gold [49]. Mohr summarized the synthesis, structures, reactivity and computational results of gold fluorides [50]. There is also an earlier review on this subject by Bartlett [51]. The gas-phase chemistry of gold has just been thoroughly reviewed by Schwarz [52]. Beside Puddephatt's early treatise of gold chemistry [9] and a number of review articles in well known book series [53–58], there is a recently edited book by Schmidbaur on progress in gold chemistry [59, 60] and a new book by Bond, Louis, and Thompson on gold catalysis [25] (see also Refs. [61–64]). It is clear that the study of relativistic effects in chemical and physical properties of atoms and molecules and the solid state has led to major advances in the understanding of the special role of gold among the other surrounding elements in the periodic table, which will be described in the following sections.

4.2
The Origin of the Relativistic Maximum at Gold Along the 6th Period of Elements in the Periodic Table

Valence electrons important for chemical reactions should move rather slowly compared with the velocity c of light, and even for a heavy element like gold we expect for the velocity of a 6s electron $v_{6s} \ll c$ and therefore negligible relativistic effects [65]. Moreover, as the inner tail of the valence orbital becomes quite small closer to the nucleus, one could assume that the relativistic perturbation operator has little effect on the innermost core region for such orbitals. In contrast, the inner 1s orbital can substantially contract relativistically as velocities of $0.5c$ or more are reached for the heavy elements [32]. It was therefore suggested that the outer ns orbitals contract because of orthogonality constraints to the inner relativistically contracting core-orbitals [2, 66], in other words, the valence orbitals are pulled in by their orthogonality tail. This very simple and intuitive picture turned out to be however completely wrong, and a more complex picture emerged [67, 68]. For example, within this simple orthogonality argument it is difficult to explain why some of the outer s-orbitals exhibit larger contractions compared with the inner ones, i.e., from the Desclaux tables [32] we obtain for the relativistic to nonrelativistic ratios of the ns orbital radii (described by $\langle r \rangle_R / \langle r \rangle_{NR}$) of neutral gold 0.880 for 1s, 0.873 for 2s, 0.900 for 3s, 0.911 for 4s, 0.907 for 5s, and 0.827 for 6s. This suggests a more subtle interplay between direct and indirect relativistic shell effects.

The most important relativistic contributions in the Breit–Pauli Hamiltonian come from the first-order relativistic mass-velocity, Darwin and spin-orbit terms. These relativistic perturbation operators act in the vicinity of the nucleus where valence s-electrons exhibit a significant part of its density. Therefore the energetic contributions to these terms all originate from the innermost K-shell region, with only small contributions coming from the L-shell and regions further out. This is in contrast to the nonrelativistic total orbital energy, where the valence shell-region contributes to about 90% [68], the region where most of the density of the orbital resides.

As the inner core shells contract due to the direct action of the relativistic perturbation operator (*direct relativistic effect* which roughly increases with the square of the nuclear charge Z), the nucleus becomes more screened. This additional screening results in a reduction of the effective nuclear charge Z_{eff} (*relativistic shielding*) and in an energetic destabilization and expansion of all orbitals (*indirect relativistic effect*) [42]. Here all radial shells contribute to the indirect relativistic orbital energy correction. This indirect effect is especially important for the outer and more diffuse d- and f-orbitals. The interplay between direct and indirect effects can however be quite subtle. Due to spin-orbit splitting the $\langle r \rangle_R / \langle r \rangle_{NR}$ is 0.995 for the lower $5d_{3/2}$ orbital in gold with a rather small energetic destabilization (orbital energies $\langle \varepsilon \rangle_R / \langle \varepsilon \rangle_{NR} = 0.947$), while the $5d_{5/2}$ orbital destabilizes but slightly expands ($\langle r \rangle_R / \langle r \rangle_{NR} = 1.049$, $\langle \varepsilon \rangle_R / \langle \varepsilon \rangle_{NR} = 0.947$). This model of course depends on how we define the orbital radius, however, if we take the outermost maximum R of the radial

4.2 The Origin of the Relativistic Maximum at Gold Along the 6th Period of Elements in the Periodic Table

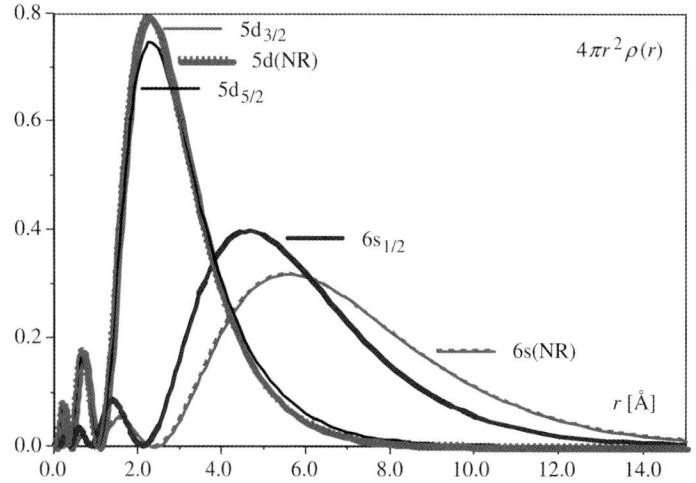

Figure 4.3 Relativistic and nonrelativistic valence 5d and 6s radial densities for neutral gold. The nonrelativistic 5d density is very close to the relativistic $5d_{3/2}$ density and shows the largest maximum.

orbital density instead we obtain $R_R/R_{NR} = 0.989$ for the $5d_{5/2}$ orbital, and 1.025 for the corresponding $5d_{5/2}$ orbital [32]. The situation is summarized in Figure 4.3.

We still have to answer why the relativistic 6s-contraction in gold is so large. First we note that orthogonality of the valence 6s orbital to the relativistically contracted core results in a very small *expansion* and not a strong contraction as often claimed [67]. Further in this picture, perturbation theory also shows that mixing in of the higher bound and continuum orbitals by the relativistic perturbation operator is responsible for the contraction. The relativistic orbital contraction is thus mainly due to the *direct action* of the relativistic perturbation operator. For s-orbitals direct effects dominate, for p-orbitals the direct and indirect effects almost cancel and for d- and f-orbitals the indirect effects dominate. Furthermore, the relativistically destabilized (expanded) d- and f-orbitals cause an *indirect stabilization* by increasing Z_{eff} (*relativistic deshielding*) for the other orbitals. Hence there is a delicate interplay between direct and indirect relativistic effects for each atomic orbital [68].

Now moving across the periodic row as shown in Figure 4.2 the direct relativistic effect smoothly increases with increasing nuclear charge Z. With filling the d-shell, the indirect stabilization increases and compensates the indirect destabilization, which has its maximum at the filled d-shell, resulting in a much larger stabilization (contraction) than the corresponding hydrogenic value, Figure 4.4. This partly explains the gold (or Group 11) maximum of relativistic effects. This is nothing other than a relativistic enhancement of the expected transition metal contraction (similar to the lanthanide contraction) [68]. Filling up the next p-shell from Group 13 to 18, the d-shell becomes more core-like and the indirect stabilization effect diminishes, Figure 4.2.

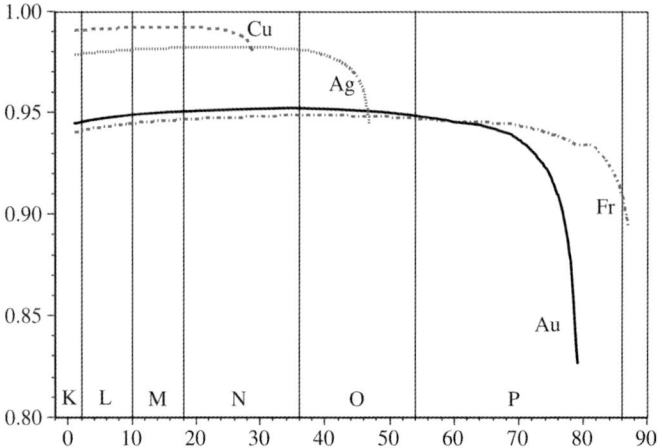

Figure 4.4 Valence relativistic s-shell contraction $\langle r \rangle_R / \langle r \rangle_{NR}$ (4s for Cu, 5s for Ag, 6s for Au and 7s for Fr). Here the ns-shell remains singly occupied and the x-axis gives the total number of electrons N with additional electrons filled in successively from the inner shells. For example $N = 3$ for Au describes the occupation $1s^2 6s^1$ and $N = 11$ the occupation $1s^2 2s^2 2p^6 6s^1$.

Figure 4.2 shows that within the 6th period of elements, Pt and Au deviate from the rather smooth trend of relativistic 6s-stabilization, in contrast to the 7th period. For these two elements we have a change in configuration from $5d^m 6s^2$ to $5d^{m+1} 6s^1$ for Pt ($m = 8$) and Au ($m = 9$), while all other transition metals in this series prefer the $5d^m 6s^2$ configuration. Moving an electron from the 6s to the 5d shell will increase Z_{eff} (additional deshielding by configurational change) and thus increases the direct relativistic effect. As a consequence, Pt and Au show a larger relativistic 6s-stabilization as expected from the smooth periodic trend [69]. Hence the maximum of relativistic effects is at gold and not at mercury. Similar arguments hold for copper and silver. Because of the extremely large and therefore dominating direct relativistic effects in the next period down (Fr to element 118), the configuration remains at $d^m s^2$ for all elements, and as a consequence the maximum of relativistic effects is shifted from Rg ($Z = 111$) to element 112 [33]. As a result of both direct and indirect effects, the $(n - 1)$d/ns gap closes and the ns/np widens thus enhancing d-participation and reducing p-participation in chemical bonding.

Finally, to end the discussion on the relativistic gold maximum, removing an electron from the d-shell, substantially reduces the relativistic s-contraction, Figure 4.4. To a much smaller extent this is also found for the Group 1 metals by removing a $(n - 1)$ p-electron as shown for Fr in Figure 4.4. However, these effects can again be quite subtle. For example for the ratio $\langle r \rangle_R / \langle r \rangle_{NR}$ we get 0.827 for Au $5d^{10} 6s^1$, 0.877 for Au$^+$ $5d^9 6s^1$, and 0.897 for Au^{2+} $5d^8 6s^1$. Even more surprising, removal of any one electron from an inner shell leads to a similar large reduction in the relativistic 6s contraction. For example, a hole in the 1s shell (Au$^+$ $1s^1 2s^2 \ldots 5d^{10} 6s^1$) yields $\langle r \rangle_R / \langle r \rangle_{NR} = 0.874$ very close to the value of the Au$^+$ $5d^9 6s^1$ configuration.

4.3 Calculations on Atomic Gold

Currently the most accurate relativistic atomic calculations at the Hartree–Fock (HF) and density functional theory (DFT) comes from numerical Dirac–Fock [70, 71] methods, and at the correlated level from numerical multi-reference Dirac–Fock [70, 72], algebraic Fock-space coupled-cluster approaches [73], and other many-body techniques [74–78]. These methods can include further contributions from the (frequency-dependent) Breit interaction [79, 80] describing the retarded relativistic interaction between two charged particles (electrons in our case) [81], and quantum electrodynamic (QED) effects such as the vacuum polarization or the electron self-energy [82–85]. Tailored basis sets for four-component calculations are basically available for the whole periodic table [86]. Dirac–Kohn–Sham calculations for atoms are also being performed [87], but here the accuracy of the different relativistic DFT methods is not comparable to wavefunction based results, and the inclusion of Breit or other QED effects within the DFT formalism is therefore not warranted [88]. Relativistic DFT becomes important in molecular calculations or in calculations for infinite systems such as surfaces and the solid state, where wavefunction based theories have serious limitations. In the last few years a number of books have been published concerning current relativistic methodology [35, 36, 39–41], which will not be discussed here. For a historical overview see Ref. [89]. In the following we discuss some of these relativistic (R) calculations and compare physical properties P to nonrelativistic (NR) results, with relativistic effects defined as $\Delta_R P = P_R - P_{NR}$, where the nonrelativistic property P_{NR} is determined by applying nonrelativistic quantum theory (the Schrödinger equation). Relativistic modifications are usually proportional to Z^2 (Z is the nuclear charge), and we may therefore use the definition for relativistic effects on properties, $\Delta_R P/P_R = (1 - P_{NR}/P_R) = \gamma_P Z^2/c^2$, with the velocity of light c set to 137.03 599 a.u. (note the different definition used in Figure 4.2). γ_P is called the *fractional relativistic correction factor* and generally varies between $+1$ and -1 [69]. For atomic orbitals, γ_P depends strongly on the two quantum numbers n and l (see discussion in the previous section and Ref. [69]). Table 4.2 contains γ_P values for the first ionization potential and electron affinity of all Group 11 elements.

Relativistic and nonrelativistic ionization potentials and electron affinities for all Group 11 elements are summarized in Figure 4.5. From Figure 4.3 it is obvious that the relativistic 6s-contraction/stabilization implies an increase in the first ionization potential ΔE_{IP} as well as the electron affinity ΔE_{EA} for all Group 11 series of elements. This is indeed the case as the results in Table 4.2 and Figure 4.5 show. The fractional relativistic correction factor γ_{IP} varies approximately between 0.5 and 0.8 for the coupled cluster values of the Group 11 elements. While Hartree–Fock (HF) underestimates relativistic effects in the ionization potential, and thus underlines the nonadditivity between electron correlation and relativistic effects, γ_{IP} is not much different to the coupled-cluster results. Note that relativistic Fock-space coupled-cluster calculations including the Breit interaction in the Coulomb gauge using a very large (21s17p11d7f) basis set already produces an ionization potential which is only 0.029 eV below the experimental value [90]. This is in the same order of magnitude

Table 4.2 Nonrelativistic (NR) and relativistic (R) ionization potentials ΔE_{IP} and electron affinities ΔE_{EA} (positive values and in eV), relativistic effects Δ_R and relativistic enhancement factors γ for the Group 11 elements of the periodic table.

Method		Cu	Ag	Au	Rg
Ionization Potential					
Exp.		7.726	7.576	9.226	—
HF	$\Delta E_{IP}(R)$	6.564	6.339	7.682	8.930
	$\Delta E_{IP}(NR)$	6.408	5.909	5.915	5.395
	$\Delta_R \Delta E_{IP}$	0.156	0.431	1.766	3.535
	γ_{IP}	0.530	0.577	0.692	0.603
CC	$\Delta E_{IP}(R)$	7.821	7.495	9.197	10.60
	$\Delta E_{IP}(NR)$	7.555	6.934	7.057	5.87
	$\Delta_R \Delta E_{IP}$	0.266	0.561	2.140	4.73
	γ_{IP}	0.530	0.763	0.702	0.680
Electron Affinity					
Exp.		1.163	1.303	2.309	—
HF	$\Delta E_{EA}(R)$	0.033	0.123	0.666	0.244
	$\Delta E_{EA}(NR)$	0.011	0.043	0.099	0.069
	$\Delta_R \Delta E_{EA}$	0.022	0.080	0.567	0.175
	γ_{EA}	14.90	5.534	2.562	1.093
CC	$\Delta E_{EA}(R)$	1.253	1.266	2.295	1.565
	$\Delta E_{EA}(NR)$	1.165	1.054	1.283	1.054
	$\Delta_R \Delta E_{EA}$	0.088	0.212	1.012	0.511
	γ_{EA}	1.568	1.424	1.327	0.490

If not otherwise stated the four-component Dirac method was used. The Hartree–Fock (HF) calculations are numerical and contain Breit and QED corrections (self-energy and vacuum polarization). For Au and Rg, the Fock-space coupled cluster (CC) results are taken from Kaldor and co-workers [4, 90], which contains the Breit term in the low-frequency limit. For Cu and Ag, Douglas–Kroll scalar relativistic CCSD(T) results are used from Sadlej and co-workers [6]. Experimental values are from Refs. [91, 92].

compared with the missing QED effects (self-energy and vacuum polarization), which lowers the ionization potential by about 0.02–0.03 eV according to Ref. [93]. We note that DFT performs well but tends to slightly overestimate the ionization potential of Au. Employing a very large contracted (8s6p6d4f3g1h)/[7s5p5d4f3g1h] valence basis set together with a scalar relativistic 19-valence electron pseudopotential for Au we obtain an ionization potential of 10.372 eV with the local spin density approximation (LSDA), 9.619 eV using the generalized gradient approximation (GGA) of Perdew and Wang (PW91) [94], and 9.441 eV using the hybrid functional B3LYP [95], which compares to the coupled clusterCCSD(T) result of 9.206 eV.

The situation for the electron affinities is more complicated. Here the HF values give rather small values at the nonrelativistic level, and most of the contribution comes from electron correlation. At the relativistic level this is still the case for Cu and Ag, but not for Au. Even though the relativistic increase in absolute values can be

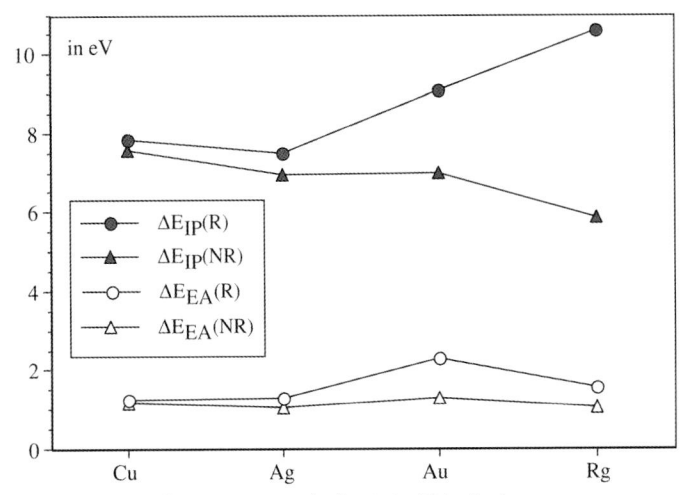

Figure 4.5 Nonrelativistic (NR) and relativistic (R) ionization potentials and electron affinities of the group 11 elements. Experimental (Cu, Ag and Au) and coupled cluster data (Rg) are from Refs. [4, 91, 92].

rather small, percentage wise it is rather large leading to an artificially large γ_{EA} value at the HF level of theory, which is substantially reduced at the correlated level. Nevertheless, relativistic effects for electron affinities can be rather large and γ_{EA} can vary substantially between the Group 11 elements. The Fock-space coupled-cluster calculations of Eliav et al. give a value only 0.014 eV below the experiment [90]. Again, the GGA functionals give quite satisfying results. Using the pseudopotential approximation for Au together with a large valence basis set as described above, we obtain an electron affinity of 3.070 eV using LSDA, 2.363 eV with the PW91 functional, and 2.232 eV using the B3LYP functional, which compares to the CCSD(T) result of 2.257 eV. Note the clearly visible anomaly in the relativistic electron affinity for Rg, which comes from a relativistic change in configuration from $6d^{10}7s^1$ to $6d^97s^2$ [4], Figure 4.5.

Eliav et al. also give nonrelativistic and relativistic excitation energies for Au [90]. Noteworthy are the spin-orbit splittings in the 5d- and 6p-shell of Au. Fock-space coupled-cluster calculations result in a $^2D_{3/2}/^2D_{5/2}$ splitting of the $5d^96s^2$ state of 1.519 eV and in a $^2P_{1/2}/^2P_{3/2}$ splitting of the $5d^{10}6p^1$ state of 0.466 eV, in excellent agreement with the experimental values of 1.522 eV and 0.473 eV respectively. The spin-orbit effects for all Group 11 elements are shown in Figure 4.6. For the $^2D_{3/2}/^2D_{5/2}$ ($5d^96s^2$) states of Au, Itano calculated the electric quadrupole moment and hyperfine constants using multi-configuration Dirac–Fock [96].

Mulliken's electronegativity is defined as the arithmetic mean of the ionization potential and the electron affinity [7], $\chi = 0.187(\Delta E_{IP} + \Delta E_{EA}) + 0.17$, where the energies are given in eV and χ is a dimensionless variable [97]. Hence for the relativistic change in χ we simply get $\Delta_R\chi = 0.187(\Delta_R\Delta E_{IP} + \Delta_R\Delta E_{EA}) = 0.07$ for Cu,

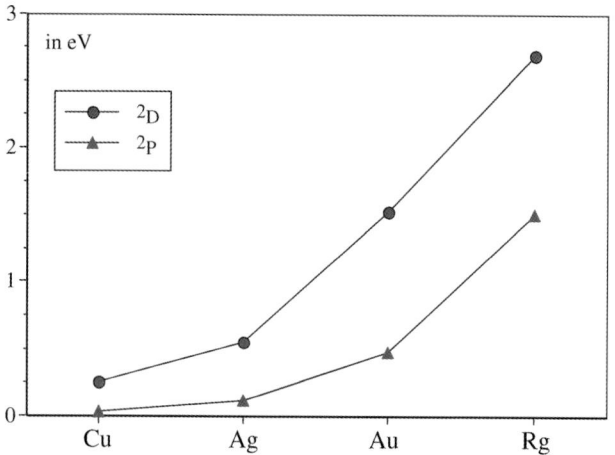

Figure 4.6 Spin-orbit splittings for the $^2D_{5/2}/^2D_{3/2}$ (d^9s^2) and $^2P_{3/2}/^2P_{1/2}$ ($d^{10}p^1$) states of the Group 11 elements. Experimental (Cu, Ag and Au) and coupled cluster data (Rg) are from Refs. [4, 91]. For the $^2P_{3/2}/^2P_{1/2}$ we used Dirac–Hartree–Fock calculations including Breit and QED corrections.

0.14 for Ag, 0.58 for Au and 1.14 for Rg using the coupled cluster data given in Table 4.2. These are huge changes for both Au and Rg leading to electronegativities of 2.37 for Au and 2.44 for Rg. This compares well with the Pauling electronegativity of gold which is 2.54. In fact, gold possesses the highest electronegativity of all metals in the periodic table (except for Rg). Thus the heaviest elements in the Group 11 series have electronegativities comparable to that of astatine (2.66) or iodine (2.2) [97], and may therefore be regarded as pseudohalides [46, 98]. In contrast, the Mulliken electronegativity is 1.83 for both Cu and Ag. It is evident that this relativistic change in χ will lead to a substantial different chemical and physical behavior of both Au and Rg as compared with its lighter congener Cu and Ag, as shown for example in Table 4.1.

Due to the relativistic 6s contraction in gold, the 6s shell becomes more compact (inert, hence the nobility of gold) and the (static dipole) polarizability α_D decreases substantially from 9.50 (NR) to 5.20 Å^3 (R) [99], Table 4.3. The relativistic enhancement factors are rather large for the polarizabilities of the neutral elements. This can easily be rationalized from the well-known dependence of α_D on the first ionization potential, which is approximately $\alpha_D \sim \Delta E_{IP}^{-2}$ [100]. Hence we expect much larger relativistic effects in dipole polarizabilities compared with ionization potentials. For the core polarizabilities the relativistic increase in the ionic radius (relativistic 5d expansion) results in a more polarizable core. Hence, we see an increase in the core polarizability from 1.54 (NR) to 1.81 Å^3 (R) for Au$^+$ (compared with Cu$^+$ with 0.9 and Ag$^+$ with 1.2 Å^3) [99]. Spin-orbit coupling is neglected in our analysis because the results shown in Table 4.2 are from scalar relativistic Douglas–Kroll calculations. Because of the additional shell expansion of the 5d$_{5/2}$ orbital due to spin-orbit coupling, we expect a further increase of the polarizability of Au$^+$. Table 4.3 also

Table 4.3 Nonrelativistic (NR) and relativistic (R) static dipole polarizabilities α_D (in Å^3), relativistic effects $\Delta_R\alpha_D$, and relativistic enhancement factors γ_α for the Group 11 elements of the periodic table.

Atom		HF	LSDA	BP86	B3LYP	CC
Positively Charged Atoms						
Cu$^+$	$\alpha_D(R)$	0.792	1.193	1.241	1.104	1.033
	$\alpha_D(NR)$	0.764	1.153	1.199	1.066	0.994
	$\Delta_R\alpha_D$	0.028	0.040	0.042	0.037	0.039
	γ_α	0.799	0.741	0.756	0.754	0.851
Ag$^+$	$\alpha_D(R)$	1.307	1.581	1.605	1.527	1.374
	$\alpha_D(NR)$	1.228	1.491	1.512	1.439	1.313
	$\Delta_R\alpha_D$	0.079	0.090	0.093	0.088	0.061
	γ_α	0.516	0.485	0.493	0.490	0.376
Au$^+$	$\alpha_D(R)$	1.823	2.110	2.151	2.063	1.809
	$\alpha_D(NR)$	1.571	1.854	1.884	1.803	1.543
	$\Delta_R\alpha_D$	0.252	0.256	0.267	0.260	0.266
	γ_α	0.416	0.365	0.374	0.379	0.444
Neutral Atoms						
Cu	$\alpha_D(R)$	10.642	5.630	5.958	6.159	6.961
	$\alpha_D(NR)$	11.579	5.980	6.340	6.585	7.509
	$\Delta_R\alpha_D$	−0.937	−0.350	−0.382	−0.427	−0.548
	γ_α	−1.966	−1.387	−1.432	−1.547	−1.758
Ag	$\alpha_D(R)$	12.387	6.938	6.488	7.198	7.74
	$\alpha_D(NR)$	15.880	7.783	8.423	8.738	9.63
	$\Delta_R\alpha_D$	−3.492	−0.845	−1.935	−1.540	−1.89
	γ_α	−2.397	−1.123	−2.536	−1.818	−2.08
Au	$\alpha_D(R)$	7.1460	4.907	5.122	5.151	5.20
	$\alpha_D(NR)$	13.243	7.311	7.831	7.929	9.50
	$\Delta_R\alpha_D$	−6.097	−2.404	−2.708	−2.779	−4.30
	γ_α	−2.567	−1.474	−1.591	−1.623	−2.49

All calculations are scalar relativistic calculations using the Douglas–Kroll Hamiltonian except for the CC calculations for the neutral atoms Ag and Au, where QCISD(T) within the pseudopotential approach was used [99]. CCSD(T) results for Ag$^+$ and Au$^+$ are from Sadlej and co-workers, and Cu and Cu$^+$ from our own work, using an uncontracted (21s19p11d6f4g) basis set for Cu [6, 102] and a full active orbital space.

shows the rather variable performance of density functionals. For example, while the relativistic polarizability at the B3LYP level is in good agreement with the more accurate coupled cluster result for neutral gold, the nonrelativistic value is underestimated by 1.57 Å^3. Hence, relativistic effects are incorrectly described at the DFT level of theory, resulting in a much smaller fractional relativistic correction factor γ_α. The polarizabilities of the negatively charged Group 11 atoms are also available from a computational study by Kellö et al. [101].

Relativistic effects in core properties like electric field gradients can be extremely large [103]. As the isotope 197-Au is widely used is Mössbauer spectroscopy [104], an accurate value for the 197-Au nuclear quadrupole moment (NQM) becomes desirable. For 197-Au the muonic measurements resulted in an NQM value of 0.547 b (1b = 10^{-28} m^2) [105]. Itano obtained 0.587 b from $^2D_{3/2}/^2D_{5/2}$ ($5d^96s^2$) excited state multi-configuration Dirac–Hartree–Fock and relativistic configuration-interaction calculations [96]. More recently, the Tel-Aviv group obtained 0.521 b from relativistic Fock-space coupled-cluster calculations with single and double excitations including the Gaunt term [106], which is perhaps the most accurate value so far.

Recently Dzuba et al. studied the binding of a positron (e^+) to silver and gold (e^+ Ag and e^+ Au) by combined relativistic configuration-interaction and many-body perturbation theory [107]. Interestingly, Ag forms a bound state with a positron with binding energy of 0.123 eV, while Au cannot bind a positron. They pointed out that in the nonrelativistic limit both systems e^+ Ag and e^+ Au are bound with binding energies of about 0.200 eV and 0.220 eV respectively. Such bound states are rather short-lived as e^+-e^- annihilation takes place. We also mention that hydrogen-like 197-Au^{78+} ($1s^1$) has been studied theoretically, as such one-electron systems provide a unique testing ground for QED of strong external fields [108]. The electronic spectrum of Au has recently been investigated for the search for the variation of the fine-structure constant in time [109].

4.4
Relativistic Methods for Molecular Calculations and Diatomic Gold Compounds

For small gold-containing molecules it is still feasible to employ the (four-component) Dirac–Coulomb Hamiltonian. Efficient algorithms to treat the Dirac equation at the correlated level are currently being developed [110], and two-component and scalar relativistic methods such as the Douglas–Kroll–Hess (DKH) [111–115] or regular approximations like ZORA [116–118] are now in use in standard program packages. Infinite order two-component (IOTC) approaches are on the rise, as they are computationally more efficient than four-component calculations, but with the same level of accuracy [119–121]. Another strategy is to treat relativistic effects perturbatively as done in direct perturbation theory (DPT) developed by Kutzelnigg and co-workers [122, 123]. Relativistic DFT at the four-component level is also widely used by various groups [124–126]. However, the main workhorse in relativistic quantum chemistry remains the relativistic pseudopotential approximation (RPPA) in its two-component (which includes spin-orbit coupling) or scalar relativistic (spin-orbit averaged) form, as the RPPA is the most economic method of sufficiently high accuracy to treat relativistic effects and correlation effects at the same time [127–131]. The importance of Breit and other QED effects for gold-containing compounds remain mostly unexplored, but they are expected to be smaller than the usual errors introduced through the many approximations applied, except perhaps for accurate spin-orbit splitting calculations in electronic excited states or in compounds containing trans-actinides.

We briefly discuss the performance of the relativisitic pseudopotential approximation with respect to all-electron methods, as this is the most widely used relativistic approximation in molecular quantum chemistry, and in solid-state calculations as well. The RPPA can be introduced in an *ad-hoc* fashion into the molecular electronic Hamiltonian (in atomic units) [132],

$$H_V = -\frac{1}{2}\sum_i^{n_V} \nabla_i^2 + \sum_{i<j}^{n_V} \frac{1}{r_{ij}} + \sum_i^{n_V}\sum_a^{N_C} \left[V_{PP}^a(r_{ai}) - \frac{Q_a}{r_{ai}} \right] + \sum_{a<b}^{N_C} \frac{Q_a Q_b}{r_{ab}} \qquad (4.1)$$

Here H_v is the valence-electron model Hamiltonian with n_V valence electrons and N_c cores (nuclei), and V_{PP} is the corresponding scalar or two-component relativistic pseudopotential for the core-valence interaction. The indices a,b run over all cores (nuclei) and i,j over all valence electrons. Q_a is the charge of core a, that is, $Q_a = Z_a - n_c^a$, n_c^a being the number of core electrons of atom a (for an all-electron treatment of a specific atom we have $Q_a = Z_a$), and the last term in Equation 4.1 describes the classical (point charge) core–core repulsion. Additional corrections to the last three sums may apply if nonfrozen core effects are taken into account. Pseudopotential calculations are of course less accurate than all-electron calculations, but they simulate the results of the latter often surprisingly well, for substantially smaller computational expenses. Relativistic effects are implicitly considered in the RPPA through the adjustment procedure to relativistic atomic all-electron calculations. This eliminates the additional treatment of more complicated relativistic perturbation operators. Breit and QED corrections can also be included in the adjustment procedure. There are a number of excellent and more specialized review articles available on pseudopotential theory and its applications [133–140]. A more concise treatment and discussion of the RPPA can be found in a recent overview by Dolg [141] or Schwerdtfeger [131]. Perhaps the most accurate and most widely used pseudopotentials are the energy consistent pseudopotentials of the Stuttgart group. It has been shown that energy-consistent pseudopotentials are size-consistent as well [142]. The smallest choice for the core definition to achieve results of reasonable accuracy seems to be the Au^{19+} core for gold, that is, leaving the 19 electrons from the (5s5p5d6s^1) space in the valence space. A nonrelativistic 19-valence electron pseudopotential for Au is also available from Ref. [143]. A newly adjusted scalar relativistic pseudopotential for gold by Figgen *et al.* is available from Ref. [144]. Accompanying Dunning-type correlation consistent basis sets for Au are available from Ref. [145] or from the Stuttgart web-site [146].

AuH and Au$_2$ serve as benchmark molecules to test the performance of various relativistic approximations. Figure 4.7 shows predictions for relativistic bond contractions of Au$_2$ from various quantum chemical calculations over more than a decade. In the early years of relativistic quantum chemistry these predictions varied significantly (between 0.2 and 0.3 Å), but as the methods and algorithms became more refined, and the computers more powerful, the relativistic bond contraction for Au$_2$ converged and is now at 0.26 Å.

Comparing the last two entries in Figure 4.7, the all-electron Douglas–Kroll coupled cluster result for $\Delta_R r_e$ is in perfect agreement with the RPPA [156, 157]. Figure 4.8 shows the relativistic effects in dissociation energies. Here, relativistic effects are very sensitive to the level of electron correlation and basis sets used. RPPA

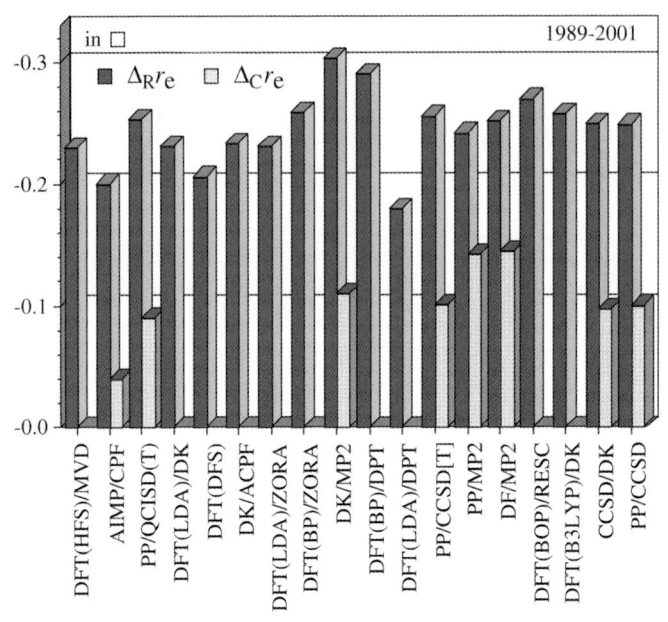

Figure 4.7 Relativistic bond contractions $\Delta_R r_e$ for Au$_2$ calculated in the years from 1989 to 2001 using different quantum chemical methods. Electron correlation effects $\Delta_C r_e = r_e(\text{corr.}) - r_e(\text{HF})$ at the relativistic level are shown on the right hand side of each bar if available. From the left to the right in chronological order: Hartree–Fock-Slater results from Ziegler et al. [147]; AIMP coupled pair functional results from Stömberg and Wahlgren [148]; EC-ARPP results from Schwerdtfeger [5]; LDA results from Häberlen and Rösch [149]; Dirac–Fock–Slater results from Bastug et al. [150]; Douglas–Kroll coupled pair functional results from Hess and Kaldor [151]; LDA and GGA ZORA results from van Lenthe et al. [152]; Douglas–Kroll MP2 results from Park and Almlöf [153]; GGA direct perturbation theory results from van Wüllen [154]; PP coupled cluster and MP2, and Dirac–Hartree–Fock results from Wesendrup et al. [155]; Relativistic elimination of the small component Becke-88+ OP DFT, Douglas–Kroll GGA and coupled cluster, and PP coupled cluster results from Hirao et al. [156, 157].

CCSD(T) calculations predict a relativistic increase in the Au$_2$ dissociation energy of 78 kJ mol^{-1} [155]. In comparison, electron correlation increases the dissociation energy by 134 kJ mol^{-1}. Spin-orbit effects increase the dissociation energy further by about 3 kJ mol^{-1} [158]. Thus we predict a $\Delta_R D_e$ value of about 80 kJ mol^{-1}. A comparison of calculated force constants reveals a similar picture, see Ref. [131].

To summarize, the RPPA is a method that can accurately describe relativistic effects, even though the relativistic perturbation operator used in the pseudopotential procedure is acting on the valence space and not the region close to the nucleus, as this is the case for the correct all-electron relativistic perturbation operator. That is, relativistic effects are completely transferred into the valence space. These effects are also completely transferable from the atomic to the molecular case as the results for Au$_2$ show. If relativistic pseudopotentials are carefully adjusted, they can produce results with errors much smaller than the errors originating from basis set incompleteness, basis set superposition or from the electron correlation procedure applied.

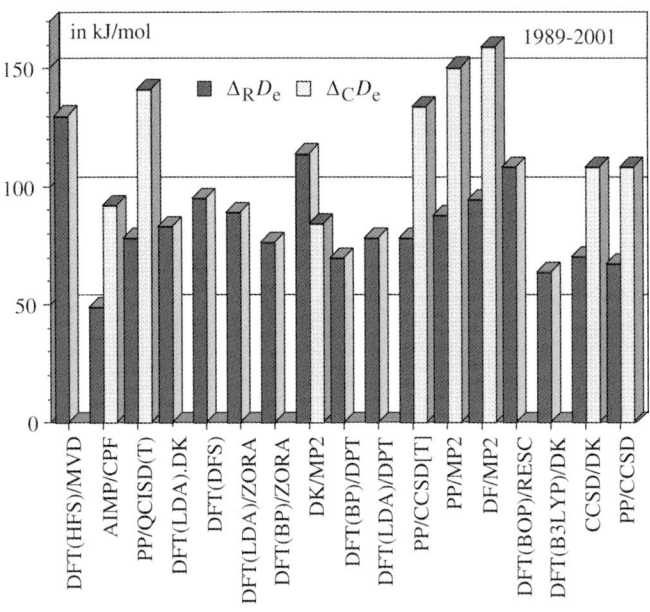

Figure 4.8 Relativistic increase $\Delta_R D_e$ in dissociation energy for Au$_2$ calculated in the years from 1989 to 2001 using a variety of different quantum chemical methods. Electron correlation effects $\Delta_C D_e = D_e(\text{corr.}) - D_e(\text{HF})$ at the relativistic level are shown on the right hand side of each bar if available. For details see Figure 4.7.

The error inherent in the RPPA is certainly much smaller than the many approximations used in DFT.

Table 4.4 shows a comparison for various properties obtained from the RPPA for Au$_2$ and its ions using a rather large contracted (8s6p6d4f3g1h)/[7s5p5d4f3g1h] valence basis set for Au and different wavefunction and density based methods. The CCSD(T) results are all in excellent agreement with experimental measurements, while the various density functionals are less accurate as one would expect [163]. A comparison between HF and CCSD(T) clearly demonstrates the importance of electron correlation effects. Also seen from Table 4.4 is that the popular MP2 approximation (second-order Møller–Plesset perturbation theory) overestimates electron correlation effects and leads to bond distances too short and dissociation energies too large for all diatomics shown in Table 4.4. We note that the RPPA results for Au$_2$ are in excellent agreement with the all-electron Douglas–Kroll CCSD(T) results (34 electrons correlated) of Hess and Kaldor who obtained $r_e = 2.488$ Å, $D_e = 2.19$ eV and $\omega_e = 187$ cm^{-1} [151], or to the recent large scale Dirac–Coulomb CCSD(T) calculations by Fleig and Visscher who obtained $r_e = 2.477$ Å, $D_e = 2.29$ eV and $\omega_e = 192$ cm^{-1} [164]. We can also compare to the Dirac (four-component level) LDA results of Wang and Liu who obtained $r_e = 2.147$ Å, $D_e = 2.791$ eV and $\omega_e = 300$ cm^{-1} [165], or to Engel and co-workers who obtained $r_e = 2.53$ Å and $\omega_e = 187$ cm^{-1} using a relativistic PW91 functional at the Dirac level [166].

Table 4.4 Spectroscopic properties for Au_2^q ($q = -1, 0, +1$) using ab-initio (Hartree Fock, HF, second-order Møller-Plesset, MP2, and coupled cluster, CCSD(T)) and DFT (local spin-density approximation, LSDA, Perdew-Wang GGA, PW91, and Becke three-parameter Lee-Yang-Parr functional, B3LYP) methods at the RPPA level of theory.

Property		HF	LSDA	PW91	B3LYP	MP2	CCSD(T)	exp.
Au_2^+	r_e	2.788	2.526	2.603	2.647	2.547	2.593	—
	D_e	1.419	3.034	2.499	2.122	2.200	2.095	2.32 ± 0.21
	ω_e	104	170	148	137	161	150	—
Au_2	r_e	2.595	2.444	2.507	2.535	2.422	2.477	2.4715
	D_e	0.813	3.023	2.361	2.007	2.680	2.272	2.29 ± 0.02
	ω_e	157	199	175	168	211	187	190.9
	ΔE_{IP}	7.101	10.361	9.481	9.326	9.903	9.384	9.16 ± 0.10
	ΔE_{EA}	0.928	2.566	1.997	1.904	1.764	1.886	2.01 ± 0.01
Au_2^-	r_e	2.764	2.541	2.626	2.670	2.513	2.591	2.582 ± 0.007
	D_e	1.107	2.519	1.996	1.679	2.168	1.902	1.92 ± 0.15
	ω_e	105	155	134	125	161	147	149 ± 10

Distances r_e are in Å, dissociation energies D_e in eV (calculated values are not corrected for the zero-point vibrational energy, harmonic frequencies ω_e in cm^{-1}, and adiabatic ionization potentials ΔE_{IP} and electron affinities ΔE_{EA} in eV. Experimental values are from Refs. [94, 159–162].

Saue and Jensen used linear response theory within the random phase approximation (RPA) at the Dirac level to obtain static and dynamic dipole polarizabilities for Cu_2, Ag_2 and Au_2 [167]. The isotropic static dipole polarizability shows a similar anomaly compared with atomic gold, that is, Saue and Jensen obtained (nonrelativistic values in parentheses) 14.2 Å^3 for Cu_2 (15.1 Å^3), 17.3 Å^3 for Ag_2 (20.5 Å^3), and 12.1 Å^3 for Au_2 (20.2 Å^3). They also pointed out that relativistic and nonrelativistic dispersion curves do not resemble one another for Au_2 [167]. We briefly mention that Au_2^{2+} is metastable at 5 eV with respect to 2 Au^+ with a barrier to dissociation of 0.3 eV [168, 169].

There are numerous calculations on diatomic gold compounds [43, 44, 170]. We only mention a few important facts on how relativity affects molecular properties in diatomic compounds. It is obvious that the relativistic 6s contraction leads to gold–ligand bond distances that are shorter, as one expects from trends down the Group 11 series of metals [171], as demonstrated above for Au_2. However, for electropositive ligands very large relativistic bond contractions are obtained yielding gold–ligand bond distances that are even smaller than for the corresponding copper–ligand bonds, Figure 4.9 [143]. It seems that this relativistic bond contraction is critically dependent on the electronegativity of the ligand, which influences the occupancy of the gold 6s orbital. Electropositive ligands (metals like Li or Na) form inter-metallic gold compounds with increased charge density on gold (M^+Au^-), as gold has an unusually high electronegativity of 2.37 due to relativistic effects. Hence relativistic effects at gold will increase even more, leading to large relativistic bond contractions. For electronegative ligands (Au^+X^-) the 6s charge density is depleted

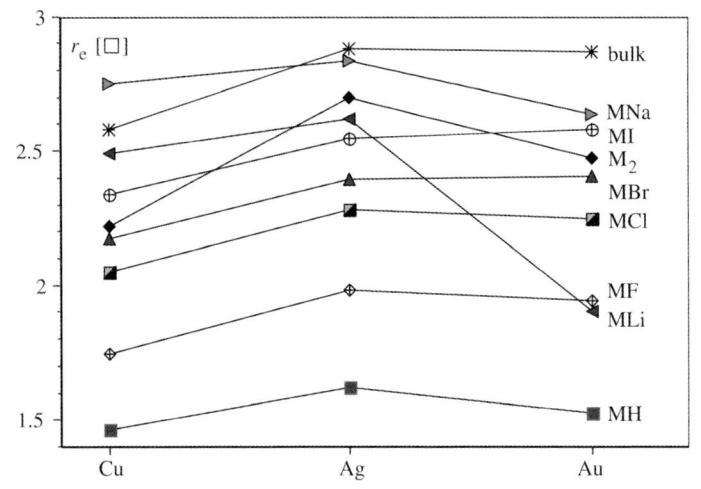

Figure 4.9 Experimental bond distances for selected diatomic Group 11 compounds (data are from Refs. [34, 159]).

and relativistic bond contractions are accordingly small. Figure 4.10 shows that for diatomic compounds the magnitude of the relativistic bond contraction correlates directly with the electronegativity of the attached ligand [143]. This leads to the empirical relationship between the relativistic bond contraction and the electronegativity of the attached ligand, $\Delta_R r_e = 0.025 \chi_L - 0.271$. Obviously, if the ligand is another heavy element with large relativistic effects, this rather crude formula is

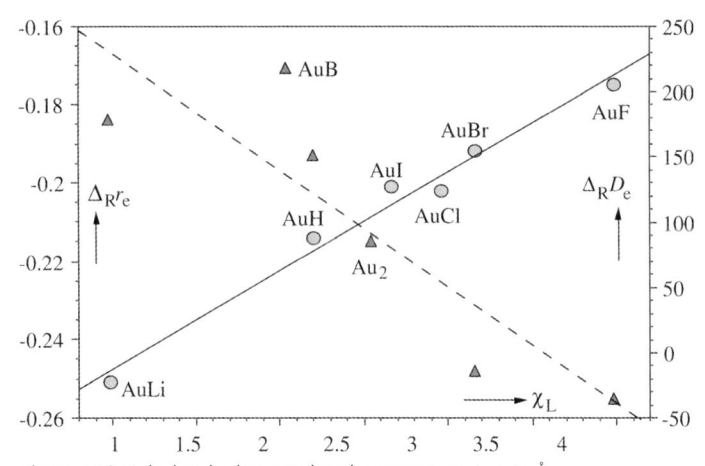

Figure 4.10 Calculated relativistic bond contractions $\Delta_R r_e$ in Å (circles and solid line, axis on the left-hand side) and relativistic change in the dissociation energy $\Delta_R D_e$ contractions (triangles and dashed line, axis on the right-hand side) for various diatomic compounds as a function of the electronegativity χ_L of the ligand.

not valid anymore. For example, for Au$_2$ this formula gives $\Delta_R r_e = -0.21$ Å compared with -0.25 Å as obtained from relativistic coupled cluster calculations.

We can use Pauling's empirical formula, which defines the Pauling electronegativity scale, for dissociation energies of a diatomic gold molecule,

$$D_e(\text{AuL}) = \frac{1}{2}\{D_e(\text{Au}_2) + D_e(L_2)\} + f\{\chi_{\text{Au}} - \chi_L\}^2 \tag{4.2}$$

Equation 4.2 consists of a covalent part (average between the dissociation energies of molecules Au$_2$ and L$_2$), and of an ionic part containing the difference in electronegativity between atom Au and ligand L. f is a simple empirical factor and is approximately 1.0 if the energies are taken in eV [143]. Due to the relativistically increased Au$_2$ dissociation energy, the covalent part of Equation 4.2 increases relativistically. The ionic part therefore determines if the Au–L bond is more or less stable due to relativistic effects. As χ_{Au} increases relativistically, the ionic contribution decreases accordingly for electronegative ligands, and increases for electropositive ones. This is indeed found for diatomic compounds as shown in Figure 4.10. In a slightly different picture, the relativistic 6s stabilization is about 2.1 eV for gold (Table 4.2). The relativistic 6s stabilization at the molecular level is critically dependent on the 6s occupation, that is, compared with the atomic value at the dissociation limit we have a larger stabilization for electropositive ligands donating electron density into the Au 6s orbital (the ionic limit is Au$^-$L$^+$), and a smaller stabilization for electronegative ligands withdrawing electron density from the Au 6s orbital (here the ionic limit is Au$^+$L$^-$). Hence, compared with the nonrelativistic case this leads to a relativistic increase in the dissociation energy for electropositive ligands and to a relativistic decrease for electronegative ligands [42]. This leads to some interesting consequences.

The Pauling formula (4.2) predicts a strong relativistic increase in the stability of intermetallic gold compounds [18], which is shown in Figure 4.11. In fact, AuLa and AuLu show a very strong bond compared with all other intermetallic dimers. Only few other transition metal dimers with multiple bonding character achieve higher dissociation energies, for example Mo$_2$ with 4.20 ± 0.2 eV [172], Nb$_2$ with 5.2 ± 0.1 eV [173], or IrLa with 5.9 ± 0.1 eV [173]. We mention that transition metal gold dimeric compounds were recently investigated by Wu using DFT [174]. If one goes along the isoelectronic series AuPt$^-$, Au$_2$ and AuHg$^+$, one obtains very similar bond distances and dissociation energies, and similar relativistic effects [155]. Here the relativistic increase in the dissociation energy is however largest for AuPt$^-$ as the negative charge increases substantially the 6s population at gold. Hirao and co-workers investigated AuPt by four-component MP2 and multi-reference CI and compared with CuPt and AgPt [175].

An interesting intermetallic gold compound is CsAu, where Au is bound to the most electropositive metal Cs ($\chi_{\text{Cs}} = 1.2$). As a result we obtain an ionic bonding situation for the semiconductor Cs$^+$Au$^-$ and not a metallic bond as one expects (two metals do not necessarily form a metallic bond!). Figure 4.12 nicely shows that the HOMO consists mainly of Au(6s) with much smaller Cs(6s) and Au(d$_\sigma$) participation, while the LUMO is mainly Cs(6s). Saue and co-workers calculated relativistic

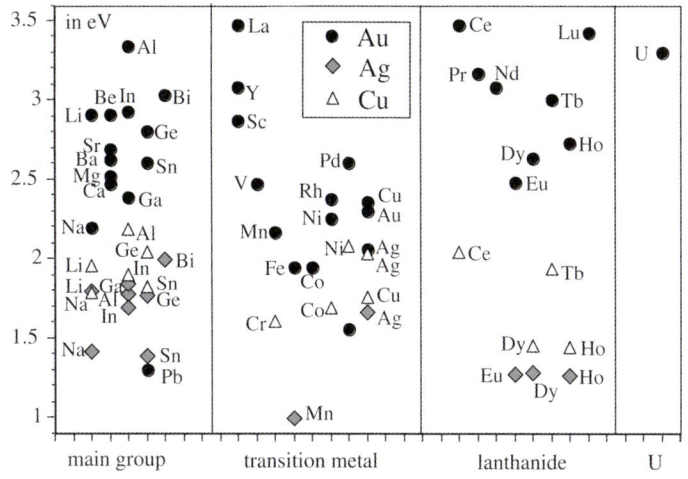

Figure 4.11 Comparison of experimental dissociation energies of gold intermetallic compounds with those of copper and silver from mass spectroscopic measurements by Gingerich and co-workers [18, 159, 173, 176] arranged in group order according to the periodic table.

effects at the Dirac CCSD(T) level of theory of $\Delta_R r_e = -0.308$ Å, $\Delta_R D_e = 1.18$ eV (an almost 100% increase due to relativistic effects) [177]. Moreover, the (Mulliken) charge on Cs increases significantly from 0.66 at the nonrelativistic level to 0.96 at the relativistic level of theory [177], thus increasing the ionicity in the Au–Cs bond. Belpassi *et al.* studied the whole alkali auride series from LiAu to CsAu with similar results for the charge transfer to gold [178]. As a result, Cs^+Au^- can be dissolved in liquid ammonia, like several other ionic alkali or alkaline earth halides [49], and the coordination compound $AuCs \cdot NH_3$ has also been isolated recently by Jansen and co-workers [179]. As a further curiosity, $AuBa^-$ consists of two interacting closed $6s^2$ shell atoms with a rather large dissociation energy of 1.48 eV [180]. It is therefore the strongest closed-shell interaction predicted so far and a result of an increased charge induced dipole interaction at a relativistically decreased bond distance [180]. For comparison, the s^2-p^6 interaction between Au^- and Xe is much weaker (0.05 eV) [181].

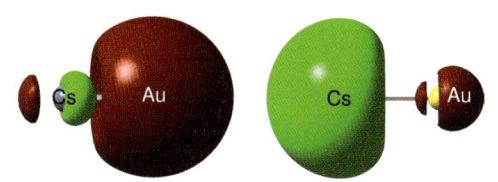

Figure 4.12 Isosurfaces of the highest occupied molecular orbital (HOMO) on the left and lowest unoccupied molecular orbital (LUMO) on the right for Cs^+Au^-.

For the most electronegative ligand, fluorine, we expect a relativistic destabilization in the Au−F bond, which was indeed determined to be −0.36 eV at the coupled cluster level [182, 183]. Nevertheless, AuF has a sufficiently high dissociation energy of about 3.17 eV and has been identified recently in the gas phase [184]. In solution or in the solid state it would disproportionate to metallic Au and compounds of Au^{3+} (AuF_3 for the solid). However, a carbene-stabilized Au(I) fluoride was synthesized only very recently (see discussion in the next section) [185].

Anomalies in force constants in diatomic gold compounds are well documented and have been identified as due to relativistic effects [42, 46, 159], that is diatomic gold compounds have relativistically increased vibrational frequencies. In fact, relativistic effects are so strong for gold-ligand stretching force constants that relativistic effects often compensate the increase in the reduced mass going down the Group 11 series of compounds, thus leading to anomalies in the vibrational frequencies as well. For example, the harmonic frequency ω_e (anharmonic correction $\omega_e x_e$) is 199.6 cm^{-1} (0.96 cm^{-1}) for CuBi, 152.1 cm^{-1} (0.41 cm^{-1}) for AgBi, and 157.7 cm^{-1} (0.25 cm^{-1}) for AuBi. For comparison, if a lighter element is attached to gold we obtain 1941.3 cm^{-1} (37.5 cm^{-1}) for CuH, 1759.9 cm^{-1} (34.1 cm^{-1}) for AgH, and 2305.0 cm^{-1} (43.1 cm^{-1}) for AuH. And for a more electronegative ligand like chlorine we get 415.3 cm^{-1} (1.6 cm^{-1}) for CuCl, 343.5 cm^{-1} (1.2 cm^{-1}) for AgCl, and 382.8 cm^{-1} (1.3 cm^{-1}) for AuCl [159]. Hence, anomalies are observed for both electronegative and electropositive ligands on gold. For a detailed discussion see Ref. [143].

The increased electronegativity for gold will also have significant influence on electric properties such as dipole or higher moments. For dipole moments of diatomic gold compounds relativistic effects have been discussed in detail in Ref. [143]. Guichemerre et al. investigated the Group 11 fluorides, chlorides and bromides using multi-reference configuration interaction (MRCI) including spin-orbit coupling at the relativistic pseudopotential level and obtained accurate spectroscopic properties including dipole moments [186]. Sadlej performed accurate calculations for the dipole moment of AuH [187] and Salek et al. investigated the frequency-dependent dipole polarizability of AuH [188]. Relativistic effects decrease the dipole moment in AuH from 3.1 to 1.6 Debye. Interestingly, the parallel dipole polarizability remains virtually unchanged because of relativistic effects, that is, nonrelativistically one obtains 45.8 a.u. and relativistically 44.1 a.u. [187] (all calculations have been carried out at the experimental AuH equilibrium distance [159]). Sadlej and co-workers also investigated relativistic effects in the electric quadrupole moment in AuH [189]. The nonrelativistic quadrupole moment at the CCSD + T level of theory is −0.18 a.u., while mass-velocity plus Darwin calculations gave a value of +2.00 a.u. [189].

The inclusion of (nonrelativistic) property operators, in combination with relativistic approximation schemes, bears some complications known as the picture-change error (PCE) [67, 190, 191] as it completely neglects the unitary transformation of that property operator from the original Dirac to the Schrödinger picture. Such PCEs are especially large for properties where the inner (core) part of the valence orbital is probed, for example, nuclear electric field gradients (EFG), which are an important ingredient in the nuclear quadrupole coupling constant NQCC [103, 192, 193]. For AuCl it has been demonstrated that the PCE leads to totally erroneous field gradients [192], and the PCE can even be larger than the relativistic effect itself for EFGs. An alternative approach has therefore been devised based on modeling the nuclear quadrupole moment (NQM) perturbation via point charges (the PCNQM model) [194, 195]. Using the PCNQM model, Schwerdtfeger *et al.* [196] recently attempted to re-determine the 197-Au NQM via the electronic route, by EFG calculations for the diatomic gold halides using all-electron Douglas–Kroll coupled cluster calculations. However, reliable values for the EFG turned out to be surprisingly difficult to calculate. Even more disturbing, density functionals lead to unreliable EFGs, and a large range of molecules is needed to obtain a reasonable NQM [197, 198]. Very recently, Belpassi *et al.* [199] carried out Dirac–Coulomb-Gaunt Hartree–Fock calculations including electron correlation at the coupled cluster level (CCSD-T) for AuF and AuH, as well as XeAuF, KrAuF, ArAuF, and OCAuF, and obtained 0.510 ± 0.015 b in good agreement with the atomic value of Kaldor and co-workers [106] as discussed above. Hence the muonic value of 0.547 b [105] remains currently unchallenged. The importance of relativistic effects in calculations of EFGs can be nicely demonstrated for AuF where at the nonrelativistic HF level an EFG at Au of -5.64 a.u. is obtained compared with -4.99 a.u. at the Dirac–Coulomb HF level [196]. Electron correlation then brings this value further up by 4.77 a.u. to a value close to zero [196], which makes the accurate determination of the 197-Au NQM from known NQCC data almost impossible. Lantto and Varra also calculated large relativistic effects for the 131-Xe NQCC in XeAuF by [200, 201].

In a recent paper David and Restrepo showed that scalar relativistic and spin-orbit effects lead to very large changes in the nuclear magnetic shielding constants $\sigma(Au)$ in AuF [202]. At the nonrelativistic B3LYP level they obtained $\sigma(Au) = 8627$ ppm, while scalar relativistic effects gave $\sigma(Au) = 12\,325$ ppm and the inclusion of spin-orbit effects at the Dirac level gave a final value of $\sigma(Au) = 31\,611$ ppm [202]. That spin-orbit effects are important in nuclear magnetic shielding constant calculations was already demonstrated by Kaupp and co-workers [203]. Finally we mention relativistic calculations of AgH and AuH in a cylindrical harmonic confinement potential by Lo and Klobukowski [204].

4.5
Calculations on Inorganic and Organometallic Gold Compounds

For larger gold compounds the main theoretical method applied is the relativistic pseudopotential approximation together with DFT [131]. This method is available in almost all molecular program packages. Another widespread procedure used for larger gold compounds is the zero-order regular approximation (ZORA) for scalar relativistic effects within the general gradient approximation in DFT [117, 118]. The relativistic stabilization of the 6s shell together with the relativistic destabilization of the 5d shell leads to a substantially decreased (increased) 5d/6s (6s/6p) gap as demonstrated in Figure 4.13. As a result, 5d-participation in gold–ligand bonding

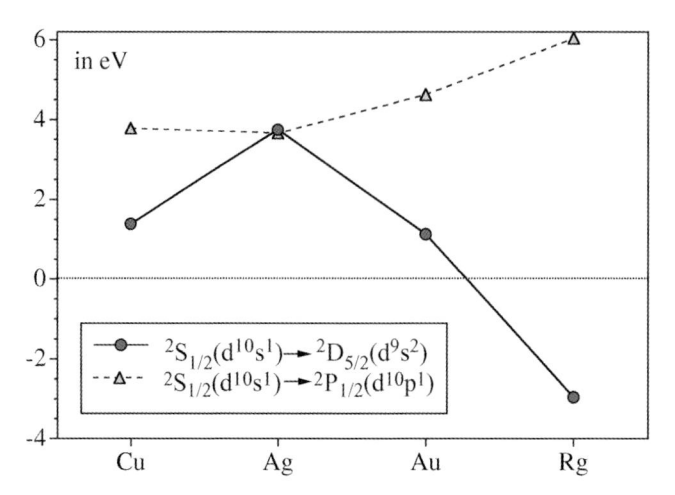

Figure 4.13 Excitation energies for the s-d and s-p gaps of the Group 11 elements. Experimental (Cu, Ag and Au) and coupled cluster data (Rg) are from Refs. [4, 91]. For the s-p gap of Rg we used Dirac–Hartree–Fock calculations including Breit and QED corrections.

becomes more pronounced. As mentioned before, for Rg the $^2D_{5/2}(6d^97s^2)$ state is below the $^2S_{1/2}(6d^{10}7s^1)$ state. The availability of the d-electrons in chemical bonding leads to a stabilization of the higher oxidation states $+$ III and $+$ V in gold, and even more so for Rg.

Relativistic effects in gold halide complexes in the oxidation state $+1$ for gold have been studied intensively in the past [205]. Recently, Mishra studied the photodetachment spectra of mixed halide complexes, XAuY$^-$ (X,Y = Cl, Br and I) [206]. Figure 4.14 shows the stability of all Group 11 fluorine complexes in the higher oxidation states $+$ III and $+$ V towards decomposition into the lower oxidation states [207]. At the nonrelativistic level the stability of the oxidation state $+$ III for the fluorides decreases with increasing nuclear charge of the central atom with a slight anomaly for gold, which is most likely due to the lanthanide contraction effect. This trend is reversed for both gold and eka-gold due to relativistic effects. The stability for the highest oxidation state $+5$ increases at both levels of theory. However, CuF$_6^-$ and AgF$_6^-$ are thermodynamically unstable, and for both gold and Rg we again see a substantial relativistic stabilization. Only gold is so far known to form a compound in the oxidation state $+$ V, AuF$_6^-$. Entropy effects will shift the equilibrium further towards decomposition. The calculations also reveal a substantially increased 5d-participation in the MF bond when going from Cu to Rg because of relativistic effects [207]. Note that the highest oxidation state of gold stands at $+$ V. The gold compound in the oxidation state $+$ VII, AuF$_7$, which has been claimed to exist by Timakov et al. [208], has recently been shown by Riedel and co-workers to be a simple complex between F$_2$ and AuF$_5$ [209, 210].

For gold compounds in the high oxidation state (e.g., AuCl$_4^-$, AuF$_6^-$, AuF$_3$ etc.,) we find rather small bond contractions as the electronegative ligands reduce the 6s

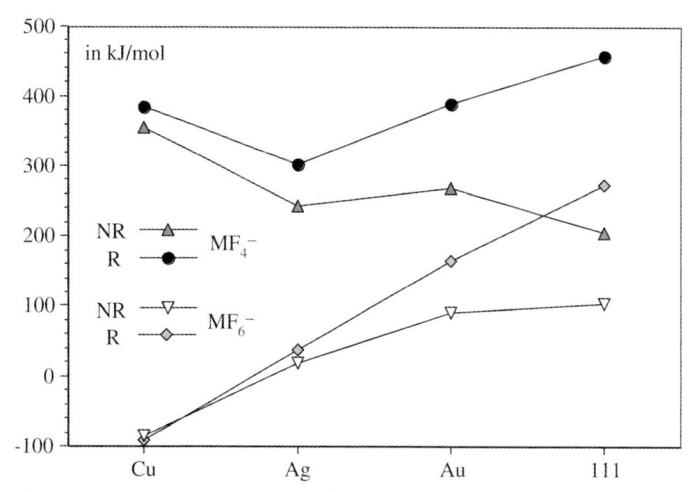

Figure 4.14 Decomposition energies for the Group 11 fluorides in the oxidation state $+5$ ($MF_6^- \rightarrow MF_4^- + F_2$) and $+3$ ($MF_4^- \rightarrow MF_2^- + F_2$). All data are from Ref. [207].

density at Au and 5d participation becomes more important [11]. As a consequence the crystal structure of $Cs_2[AuCl_2][AuCl_4]$ shows one short Au–Cl distance in the $[AuCl_2]^-$ unit and one longer distance in the $[AuCl_4]^-$ unit [211], which reverses at the nonrelativistic level of theory [11]. AuF_3 has been predicted to undergo a Jahn–Teller distortion from the planar trigonal (D_{3h}) structure to the T-shaped (C_{2v}) structure [11], which was confirmed by gas-phase electron diffraction measurements by Hargittai and co-workers [212]. For AuH_3 the situation is slightly different. The most stable form of AuH_3 is the Y-shaped structure and not the T-shaped due to the rather strong bond between the two hydrogen atoms, best described as the complex $(H_2)AuH$ [213–215]. Andrews and co-workers identified this complex recently, as well as $(H_2)AuH_3$ [216]. Relativistic effects for gold-ligand stretching force constants are very large for all oxidation states similar to the diatomic gold compounds [207].

The strong tendency to undergo gold–gold interaction is called aurophilicity [59, 217–219], and is related to relativistic effects in gold [12, 46, 220]. As described above, gold undergoes especially strong interactions with electropositive ligands (metals) [18] donating electron density to the relativistically contracted and stabilized gold 6s-orbital, thus enhancing metallophilic interactions. Table 4.3 shows a large relativistic increase in the Au^+ dipole polarizability, which can increase the dispersive type of interaction between two Au^+ (d^{10}) cores substantially. Such aurophilic interactions can be as large as 30 kJ mol^{-1}. In contrast, cuprophilic interactions are much weaker (roughly 1/3 of the aurophilic interaction) [221]. Pyykkö and co-workers demonstrated that aurophilic interactions in the model dimer $[PH_3AuCl]_2$ are of dispersive nature enhanced substantially by relativity [12, 222–228]. Hartree–Fock gave a repulsive interaction and electron correlation is solely responsible for the interaction between the two gold atoms. The aurophilic interaction is proportional to $\sim r^{-6}\alpha^2_{Au^+}$ at long-range (for α_{Au^+} see Table 4.3), and thus follows the London type dispersion force (r being the distance between the two gold atoms) [227]. Furthermore, if the occupied Au 5d orbitals are omitted in the active orbital space the entire attraction disappears [222]. Werner and co-workers provided a detailed analysis [228, 229]. The aurophilic interaction correlates with the measured bond lengths, and three groups proposed empirical formulae [12, 19, 230] [for example, in Ref. [19] the formula $D_e = 1.27 \times 10^6 \exp(-0.035 r_e)$ for the aurophilic binding energy is given with D_e in kJ mol^{-1} and the Au–Au distance r_e in pm]. Pyykkö and Zlaeski-Ejgierd recently performed basis set limit studies at the MP2 level for [PH$_3$AuCl]$_2$ [231]. However, the computationally efficient MP2 method substantially overestimates aurophilic interactions, and one has to utilize coupled-cluster techniques to obtain more reliable results [228, 229]. O'Grady and Kaltsoyannis came to the same conclusion, but they also pointed out that for [PH$_3$MCl]$_2$ (M = Cu, Ag, Au, Rg) the strength of the metallophilic interaction decreases (most likely monotonically) down the group 11 elements in contrast to the MP2 results, which show a monotonic increase [232]. A recent investigation on {(NHC)MCl}$_2$ (M = Ag, Au) shows stronger argentophilic than aurophilic interaction [233]. This suggests that the aurophilic interaction in terms of the dispersive bond strength is not so special after all. Density functionals like B3LYP or BP86 may underestimate metallophilic interactions but also show a weakening of the metal-metal bond in [PH$_3$MCl]$_2$ when going from M = Ag to Au [232]. The B3LYP functional was used recently by Humphrey et al. for studying aurophilic interactions in [(RNC)AuCl] and [(RNC)AuBr], R = (H$_3$C)$_3$N-BH$_2$, where the calculated Au–Au distances were too long compared with the experimental and MP2 results [234]. Runeberg et al. demonstrated that ionic contributions (from double excitations involving excitations from one gold center to the other) in the aurophilic interaction are also important [228]. This may explain why some density functionals perform reasonably well. They also showed that relativistic effects increase the aurophilic interaction by 28% in [PH$_3$AuCl]$_2$, and originate almost exclusively from the relativistic expansion of the two Au(5d) shells [228]. This all raises the question why metallophilic interactions are so common in gold complexes, but less so in corresponding copper and silver compounds. Perhaps the answer lies in the different coordination chemistry of copper, silver and gold compounds.

The tetramers of the group 11 halide phosphanes, (XMPH$_3$)$_4$ (M = Cu, Ag, or Au; X = F, Cl, Br, or I) have been investigated recently by DFT [235]. Copper and silver halide phosphane tetramers favor heterocubane (MX)$_4$ structures with PH$_3$ ligands externally attached to that cube forming X-M-PH$_3$ angles of around 120°. These structures are between 40 and 50 kJ mol^{-1} more stable than the step-cluster structures. Such structures are well known experimentally [236]. First, phosphine coordination is much stronger for gold compared with copper and silver, mostly due to relativistic effects [205]. Hence gold strongly favors linear structures. Second, there is substantially more charge transfer from the PH$_3$ lone pair to Au due to the relativistically increased electronegativity of gold. As a result, the gold atom in these phosphine compounds is negatively charged in contrast to copper and silver [235]. This makes the formation of a (AuX)$_4$ cube unfavorable. It was shown for the gold tetramers that they retain their linearity upon oligomerization thus allowing for aurophilic interactions [235].

These aurophilic interactions can also be responsible for observed chemiluminescent properties in Au(I) compounds [237, 238]. The lifetime and the luminescence quantum yield of these low-energy emissions often show strong solvent dependence. Studies on [Au_2(dpm)$_2$]$_2$, [(AuPH$_3$)$_2$(i-mnt)] and [Au$_2$(dpm)(i-mnt)] show that the aurophilic interactions are clearly correlated with electronic excited-state properties [239–241]. Lower emission energy is usually observed for shorter intermolecular Au–Au distances [242], and the excited states responsible for the emissions are attributed to ligand-to-metal–metal charge-transfer (LMMCT). In the case of gold, excitations out of the 5d shells are involved. As such systems are rather large in size, the optical properties in excited state calculations are either dealt within a simple configuration interaction including single excitations only (CIS), which basically is of Hartree–Fock quality, or within the "time-dependent" approach of density functional theory (TD-DFT). Both methods give only rough estimates of excitation energies and oscillator strengths. Furthermore, these calculations still lack the inclusion of spin-orbit effects, which will significantly change the calculated states and lifetimes. For a more complete list of excited state calculations in gold coordination compounds see Refs. [43, 44, 238].

Another interesting example of aurophilic interactions influencing electronic excited state properties is the unusual photochemical *cis* to *trans* conversion observed in dinuclear goldhalide bis(diphenylphosphino)ethylene complexes, Au$_2$X$_2$(dppee) (X = Cl, Br, I, p-SC$_6$H$_4$CH$_3$), as studied by Foley et al. [243]. We summarize their findings: The uncoordinated *cis*- and *trans*-dppee ligand shows a broad UV absorption band around 260 nm and they do not isomerize from *cis* to *trans* or vice versa under the applied conditions. In contrast to the uncoordinated species, *cis*-Au$_2$X$_2$(dppee) irreversibly isomerizes to the *trans* product, and irradiation of the *trans* product does not produce a detectable amount of *cis*-isomer. The mechanism of this irreversible photochemical reaction is related to aurophilic interactions in the excited singlet state and to relativistic effects [19]. CIS and CASPT2 calculations revealed that the first intense singlet transition consists of a charge-transfer from the halogen lone pair to the Au(5d6s6p) orbitals with strong intermixing from phosphorus (sp) orbitals. For the *cis* compound, aurophilic interactions allow for mixing with the ethylene C=C π^* orbital, which results in an extra stabilization of the first excited singlet state and a consequent red-shift in the spectrum together with an enhanced transition probability, with a shift in the transition from 180 nm to 260 nm for *cis*-Au$_2$Cl$_2$(dppee). This does not occur for the corresponding *trans* compound. Thus, irradiating at around 260 nm only accesses the excited state for the *cis* compound which can convert into the trans product as C=C π^* admixture is involved. Nonrelativistic calculations reveal only a weak bond between the dppee and the Au$_2$Cl$_2$ unit, and the compound would be unstable thermodynamically [19].

There have been a number of theoretical studies on gold coordination and organometallic compounds, which are cited in Refs. [43, 44]. We only mention a few recent activities in this area [244]. According to Parr and Pearson the chemical hardness is defined as $\eta = 0.5(\Delta E_{IP} - \Delta E_{EA})$, where the ionization potential and electron affinity (both defined as positive values) are given in eV [245]. Taking the first two ionization potentials for the Group 11 elements [4, 91] we obtain the following hardness values for the positively charged ions: 6.28 for Cu^+, 6.95 for Ag^+, and 5.64 for Au^+. For comparison $\eta = 21.1$ for Na^+, and the Group 11 metals are therefore all considered as soft Lewis acids. In fact Au^+ has the lowest hardness (therefore is the softest metal) of all three Group 11 elements. This is clearly a relativistic effect as nonrelativistic coupled cluster calculations reveal a hardness of 7.19 for Au^+ (here we took the $^1S(Au^+) \rightarrow {}^2D(Au^{2+})$ ionization potential from nonrelativistic CCSD(T) calculation which gave 21.442 eV). In a recent paper, Hancock et al. investigated the aqueous chemistry of the positively charged Group 11 metals by using DFT and showed that the softness follows the trend $Rg^+ > Au^+ > Ag^+ > Cu^+$ [246], in agreement with the values above, and both Au^+ and Rg^+ are strong Lewis acids.

As Au^+ is a soft Lewis acid, it binds preferentially to soft ligands like phosphines or sulfides. Indeed, Schröder et al. determined the binding energies of different ligands L to Au^+ as L = $PH_3 > CH_3SCH_3 > CH_3NC \approx NH_3 \approx C_2H_4 \approx CH_3CN > H_2S > CO > H_2O > C_6F_6 > Xe$ [247–249]. Relativistic calculations reveal that the most favored coordination number for gold is two with either a linear P–Au–P or P–Au–X arrangement (X is a strongly coordinating ligand like Cl or Br). Once $[Au(PR_3)_2]^+$ or $[ClAuPR_3]$ are formed, further coordination by PR_3 is relatively weak. Relativistic effects significantly enhance the binding of PR_3 to Au^+ or AuX [205], but also reduce the tendency for phosphine coordination beyond two [250]. PR_3 ligands open up the 5d shell in Au [251, 252] thus increasing their stability and enhancing aurophilic interactions. It was pointed out by Lauher and Wald that the $AuPR_3$ fragment is isolobal with the hydrogen atom [253]. While the frontier orbitals of H and $AuPR_3$ are substantially different [205], this interesting idea opens up the possibility for the syntheses of new gold species. Some spectacular examples are $[N(AuPPh_3)_4]^+$, $[P(AuPPh_3)_4]^+$ (see also Ref. [231] for recent theoretical work), $[As(AuPPh_3)_4]^+$ and $[HC(AuPPh_3)_4]^+$ [254–258]. Taking this idea to the extreme limit, one might be tempted to substitute all hydrogen atoms on a benzene ring to form $C_6(AuPR_3)_6$. In a recent paper, Räisänen et al. investigated coordination of pyridinethiols in Au(I) complexes, that is, bis(pyridine-2-thionato)gold(I) chloride and bis(pyridine-4-thionato)gold(I) chloride [259]. They found a perfectly linear S–Au–S arrangement, but with an agostic C–H $\cdots$ Au interaction leading to rather short H–Au distances between 2.8 and 2.9 Å. Very recently, Belpassi et al. studied the bonding between rare gas atoms and AuF [260]. They found that the formation of the rare gas–Au(I) bonds is accompanied by a large and very complex charge redistribution pattern affecting not only the outer valence region but also reaches deep into the core region. Ghanty studied noble-gas insertion (Kr and Xe) into the Au–X bond (X = F and OH) [261] and Zhang et al. investigated bonding in XAuOH (X = Kr, Xe) [262]. Ying and Xue studied Group 11 Xe compounds of the form MXe_n^+ (M = Cu, Ag, and Au; n = 1, 2) [263]. $AuXe^+$ and $AuXe_2^+$ show strong relativistic stabilization leading to anomalies in bond distances, vibrational frequencies and dissociation energies along the Group 11 series.

It has been speculated in the past that it might be possible to isolate the first Au(I) fluoride LAuF [182], if disproportionation into metallic gold and Au(III) can be avoided by stabilizing ligands L, such as $(PR_3)_3AuF$ [264]. This has just been achieved. Laitar et al. [185] were able to isolate a compound with an N-heterocyclic carbene ligand

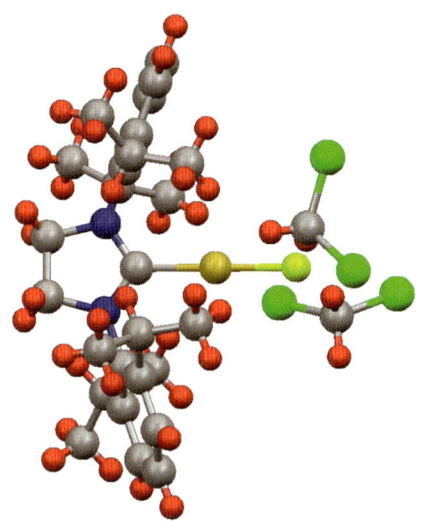

Figure 4.15 X-ray structure of 1,3-bis(2,6-diisopropylphenyl)imidazolin-2-ylidene)-gold(I) fluoride [185].

(NHC) coordinated to AuF, (NHC)AuF, see Figure 4.15. The X-ray Au–F distance is 2.028 Å and slightly larger than that of gas phase AuF which is 1.918 Å [265], and the fluorine atom is further coordinated by two CH_2Cl_2 solvent molecules. Interestingly, this compound can be seen as either a Au(I) compound with the carbene lone-pair playing exactly the same role as a phosphine ligand, or a carbon atom bond covalently to Au^+ through a double bond, >C=Au–F, which implies the oxidation state +3 for gold. A natural bond order (NBO) analysis gave a charge at Au of +0.42, which suggests Au is in the oxidation state +1. In comparison, we have a much larger charge of +0.53 at Au in AuF, which implies large charge donation from the carbene to Au. In comparison, the NBO charge of gold is +0.90 in $(CH_3)_2AuF$ and +0.63 in $H_2C=AuF$. Hence the situation here is far from clear and both descriptions may be used.

Rode and co-workers used *ab initio* QM/MM molecular dynamics simulations of solvated Au^+ in liquid ammonia [266]. They predicted a very stable linear two-coordinated $[Au(NH_3)_2]^+$. Further coordination results in a second coordination shell of about 27 NH_3 molecules which are in loose order resulting in a mean residence time of only 7.1 ps. In contrast, the mean residence time of the second hydration shell of Ag^+ (2.6 ps) is significantly shorter despite a smaller coordination number of about 17 [267, 268]. For water we find a very similar situation with two-coordinated $[Au(H_2O)_2]^+$ and fairly rigid rings in the second coordination shell, each composed of four H_2O molecules, and leading to a dumbbell like structure [269, 270]. In contrast, MP2 calculations reveal that Cu^+ has five to six water molecules in the first coordination shell [271]. Urban and Sadlej investigated the interaction of the neutral Group 11 metals with NH_3, H_2S and H_2O [272, 273], and Gourlaquen *et al.* compared H_2O binding and structure amongst the Group 11 and 12 elements with Au^+ showing large relativistic effects [274].

Table 4.5 Calculated bond distances (in Å) and dissociation energies (in eV) for linear AuCN and Au(CN)$_2^-$ (from unpublished results).

Property	AuCN			Au(CN)$_2^-$		
	HF	MP2	CCSD(T)	HF	MP2	CCSD(T)
r_e(Au–C)	2.006	1.862	1.899	2.049	1.959	2.006
r_e(C-N)	1.133	1.171	1.165	1.148	1.180	1.169
D_e(Au–CN)	3.15	4.27	3.85	3.73	4.72	4.37

For Au(CN)$_2^-$ the dissociation energy is defined as Au(CN)$_2^-$ → AuCN + CN$^-$. A scalar relativistic Stuttgart pseudopotential for Au with a uncontracted (11s/10p/7d/5f) valence basis set and a Dunning augmented correlation consistent valence triple-zeta sets (aug-cc-pVTZ) for both C and N, but with the most diffuse f function removed, was used.

There is considerable industrial interest in gold–cyanide systems, for example in gold mining, and AuCN and Au(CN)$_2^-$ have been investigated by a number of research groups in the past using both DFT and *ab initio* methods [205, 275–279]. There is no better ligand binding to gold than cyanide, and there seems to be no alternative for gold extraction than cyanide leaching. Cyanide leaching produces hazardous waste products that must be disposed of and it would be desirable to find better ligands, which are biodegradable. Table 4.5 summarizes some of our own work on gold cyanide [280], which shows that the combined binding energy of both CN groups exceeds 8 eV, which is very high. The results also show that MP2 typically overbinds. Relativistic effects are only partly responsible for these high binding energies. At the nonrelativistic CCSD(T) level we get a Au–CN dissociation energy of 3.57 eV close to the relativistic value. Moreover, it was shown that the dissociation energy for the Au(CN)$_2^-$ → AuCN + CN$^-$ reaction decreases due to relativistic effects [205]. Complexation of CN/CN$^-$ to Au$^+$ increases the C–N bond length (compare to the CCSD(T) results for the bare ligands, r_e(C–N) = 1.176 Å for CN and 1.182 Å for CN$^-$), and even more so for Au(CN)$_2^-$. We mention that Dirac–Hartree–Fock results give r_e(Au–C) = 1.997 Å and r_e(C–N) = 1.172 Å for AuCN which implies that spin-orbit coupling leads to a small decrease in the Au–C bond length. Nonrelativistic results give r_e(Au–C) = 2.167 Å and r_e(C–N) = 1.166 Å for AuCN at the CCSD(T) level, thus a relativistic bond contraction of 0.268 Å for the Au–C bond. For comparison, for CuCN and AgCN metal-CN bond distances of 1.778 Å and 2.014 Å respectively at the DFT/BP86 level of theory have been obtained by Frenking's group [276], thus revealing again a typical anomaly due to relativistic effects. For AuCN, an NBO analysis shows, besides strong ionic bonding with a bond order of about 1.2, a rather small amount of π-back-bonding from Au(5d$_\pi$) to CN(π*).

Frenking's group showed that the Group 11 isocyanides M–NC (M = Cu, Ag and Au) are less well bound compared with the corresponding cyanides M–CN [276]. They also studied CO coordination on Cu$^+$, Ag$^+$ and Au$^+$ with Au(CO)$_2^+$ being the most stable of all Group 11 dicarbonyl complexes [281]. Vaara *et al.* demonstrated the importance of relativistic effects in the 13-C NMR nuclear shielding constant in

Au(CO)$_2^+$ [282]. Very recently, Tielens et al. studied the electronic structure of charged and neutral Au–XO (X = C, N, and O) in external electric fields using DFT and coupled cluster methods [283]. Also Au(N$_3$)$_2^-$ and Au(N$_3$)$_4^-$ have been investigated theoretically [284, 285]. Interestingly, the solid state of [Me$_4$N][Au(N$_3$)$_4$] shows close alternating Au–Au contacts of 3.507 and 3.584 Å. B3LYP calculations were not able to describe these dispersive type aurophilic interactions [284]. The bonding situation in [Au(N$_3$)$_4$]$^-$ was analyzed employing the topological analysis of the electron density (AIM method) and of the electron localization function ELF [286]. Antes et al. investigated the structure and binding in ClMCO (M = Cu, Ag and Au) [287]. ClAuCO is relativistically stabilized resulting in a substantial increase of about 120 kJ mol^{-1} in the ClAu–CO dissociation energy at the CCSD(T) level of theory, which explains the unusual stability of this compound compared with its lighter congeners. Pyykkö and Runeberg recently investigated the binding of N-heterocyclic carbenes to AuCl [288]. Puzzarini and Peterson investigated multiple bonding to gold in XAuC compounds (X = F, Cl, Br and I) and found very short and strong Au–C bonds [289].

Hertwig et al. studied Au$^+$ binding to benzene [290, 291]. Au$^+$ binds preferentially in the η^1- or η^2-position, Cu$^+$ in the η^6-position, while Ag$^+$ can change its position above the whole plane of benzene virtually barrier-free. The binding energies show the typical relativistic anomaly with D_e(Cu$^+$-C$_6$H$_6$) = 2.68 eV, D_e(Ag$^+$-C$_6$H$_6$) = 1.73 eV, and D_e(Au$^+$-C$_6$H$_6$) = 2.79 eV at the CCSD(T) level of theory. From the calculated bond distances a relativistic contraction for the gold compound is also visible, that is, at the B3LYP level of theory Dargel et al. obtained for the metal-carbon distance at the η^2-position 2.123 Å for Cu$^+$, 2.415 Å for Ag$^+$, and 2.311 Å for Au$^+$ [290]. Au$^+$ adsorption on pyridine was studied by Hsu et al. [292]. Here Au$^+$ binds directly on the N-atom in the same fashion as H$^+$ does [292]. Antes and Frenking studied the structure and bonding of Group 11 methyl and phenyl compounds, MR (M = Cu, Ag or Au; R = CH$_3$ or C$_5$H$_6$) [293]. MP2 calculations predict that the M–C bond strengths of the group 11 methyl and phenyl compounds have the order Au > Cu > Ag due to relativistic effects, while the group 12 compounds M(CH$_3$)$_2$ have the order Zn > Cd > Hg [293]. Very recently, [Au(C$_2$H$_4$)$_3$]$^+$ has been synthesized, and DFT calculations reveal that the amount of Au to π*(C$_2$H$_4$) back-bonding follows the trend Au > Cu > Ag, which explains the up-field shifted 1-H and 13-C NMR data [294]. Strong back-bonding was also calculated at the DFT level for the complex [Au(bipy)(C$_2$H$_4$)](PF$_6$) [295]. Köhler et al. studied the structure and bonding in trigonal planar coordinated organogold(I) complexes stabilized by organometallic 1,4-diynes [296]. We should mention the recent progress in Au(I) and Au(III) catalyzed organic reactions [24, 297]. A number of groups of theoretical chemists are currently investigating the reaction mechanisms using quantum chemical methods [298–301]. Li and Mia published DFT calculations on Au$_5$H$_5$X hydrometal pentagons with D_{5h} planar pentacoordinate nonmetal centers (X = Si, Ge, P, S) [302]. The introduction of the nonmetal centers X introduces p aromaticity to MHX complexes.

There are only few theoretical investigations into Au(II) compounds despite the growing interest in this field [303–305]. Here we mention Barakat and Cundari's work, who studied phosphine complexation of Au(II) and their dimers [306, 307].

The most notable structure in this oxidation state is $AuXe_4^{2+}$ synthesized in Seppelt's research group [308, 309] and investigated by DFT/B3LYP by Berski et al. [310]. The relatively low stability of this oxidation state for gold (and silver) is seen as the reason for the absence of high temperature superconductors [311]. Very recently the research group headed by Craig Hill synthesized rather unusual gold-oxo complexes, $K_{15}H_2[Au(O)(OH_2)P_2W_{18}O_{68}]\cdot 25H_2O$ and $K_7H_2[Au(O)(OH_2)P_2W_{20}O_{70}(OH_2)_2]\cdot 27 H_2O$ and studied them by theoretical methods [312]. They were not able to assign a definite oxidation state for Au, which needs further careful investigation. Fackler and co-workers recently studied dinuclear and tetranuclear Au(II)-Nitrogen complexes of the form $[Au_2(hpp)_2Cl_2]$ [313].

We finally mention a recent proposal for using a gold compound in the search of parity violation (PV) effects in molecules, an electroweak symmetry breaking effect causing a very small energy difference (in the mHz range) between enantiomers in chiral compounds [314]. Such PV effects should be measurable by any absorption or emission spectroscopy provided high resolution is reached, and there are several attempts to find this tiny energy difference for the first time. As PV effects scale like Z^n ($n = 5$ in most cases), the search for suitable molecules concentrates around heavy element as chiral centers or attached on a chiral center. Compounds showing reasonably large PV effects in the C-F stretching mode accessible to the CO_2 laser frequency range are $(PR_3)Au(CHFClBr)$ or $(PR_3)Au(CHFClI)$ [315].

4.6
Calculations on Gold Clusters

Gold forms a wide variety of cluster compounds quite distinct from the other two Group 11 members or from the Group 1 metals of the periodic table [48, 316, 317]. The color of these compounds ranges from yellow to orange, red and even green (a mixture of such cluster compounds is available under the trade name Nanogold) [30]. Gold clusters are stabilized by phosphine ligands [318], and the Au–Au bonding in such clusters is further stabilized by relativistic effects as outlined above. Properties of bare gold clusters have been studied intensively in the past both by experimental and theoretical methods [43, 44, 48, 319, 320], mostly by using DFT, and we can only describe a few important recent results. We leave out cluster simulations using many-body potentials, like the Gupta-potential, the glue potential or other schemes [321–327], which are used for example for melting of large gold clusters or in nucleation studies [328–332], as these potentials do not correctly describe the many-body effects in gold clusters [333].

Au_3 in D_{3h} symmetry is $^2E'$ state and undergoes a Jahn–Teller distortion into the $^2B_2$ and $^2A_1$ states with C_{2v} symmetry with a typical Mexican hat topology [334]. However, spin-orbit coupling completely quenches this symmetry breaking effect leading to a high symmetry D_{3h} structure, and splitting the electronic states into $E_{3/2}$ and $E_{5/2}$ states in double-group symmetry [335, 336]. The relativistic bond contraction is 0.22 Å for Au_3 and already smaller compared with Au_2 [334]. Au_3^+ adopts a trigonal planar structure (D_{3h}, $^1A_1$), while Au_3^- adopts a linear structure ($^1\Sigma_g^+$) [334].

It came as a surprise, however, that Au_6 prefers a trigonal planar D_{3h} structure in the gas phase [16] and is not octahedral as one might assume, suggesting that gold clusters do not follow the usual pattern of typical Lennard-Jones, Morse or Gupta systems, which all favor a maximum number of close atom–atom contacts. The preferred planarity of small gold cluster compounds is due to relativistic effects [17]. The search for global minimum structures of neutral or charged gold clusters $Au_n^{0\pm 1}$ is still an active area of research. Wang et al. find planar structures (2D) up to $n=6$ from DFT/PW91 calculations, more compact spherical structures (3D) starting $n=16$, and flat cage-like structures in-between [337]. 2D triangulated motifs up to Au_{20} were studied by Zhao et al., and a fitting formula is provided for static polarizabilities and dipole moments for an arbitrary size [338]. However, Fernández found 2D structures up to Au_{11} using DFT/GGA (the structure of Au_{10} is shown in Figure 4.16), with the transition from 2D to 3D happening at Au_{12} [339]. The lowest energy structure for Au_8 was also found to be planar [340]. Au_8 was also investigated by Diefenbach and Kim using coupled cluster techniques, and the basis set incompleteness was also investigated [341]. In contrast, the 2D/3D structural transition for Cu and Ag occurs at $n=6$. Similar trends are found for the cationic and anionic species [16, 339, 342, 343]. They also studied trends for the cohesive energy, ionization potentials and electron affinities. Zhao et al. studied the electronic spectrum mixed Au_mAg_n clusters with $m+n=8$ [344]. Charged gold clusters up to Au_{12}^- and Au_{13}^+ have been studied in Ahlrich's group [345, 346], and for up Au_{14}^- to by Häkkinen et al. [347]. The simulated thermally weighted photoabsorption spectra showed that Au_7^- is dominated by planar structures whilst Cu_7^- and Ag_7^- give predominantly three-dimensional arrangements. Häkkinen et al. studied these negatively charged clusters [17] by relativistic DFT and showed that nonrelativistic Au_7^- behaves similarly to the analogs copper and silver [17]. Koskinen et al. studied the dynamics of Au_n^- ($n=11-14$) and showed the co-existence of 2D and 3D structures at finite temperatures [348]. Wang et al. found by photoelectron spectroscopy and DFT calculations that Au_{20} adopts an unusual tetrahedral structure with a rather large HOMO-LUMO gap (Figure 4.16) [349, 350], which has been recently confirmed by Aprà et al. [351]. In contrast, Cu_{20} and Ag_{20} show amorphous-like 3D

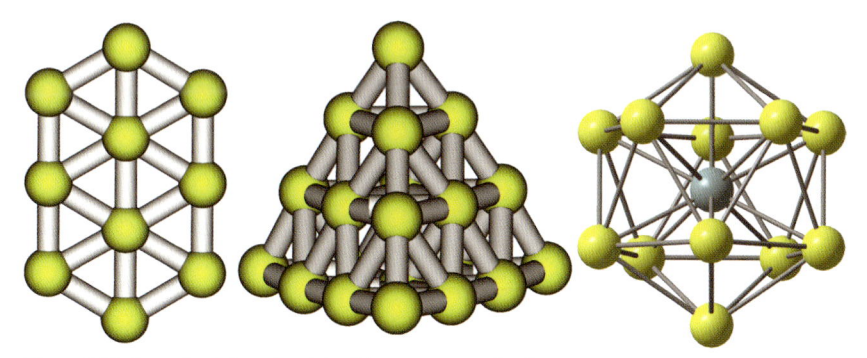

Figure 4.16 From the left to the right: Planar structure for Au_{10}, tetrahedral structure for Au_{20} and icosahedral structure for $W@Au_{12}$.

structures [339]. These differences are all attributed to relativistic effects, which substantially enhances s-d hybridization in gold [339]. Fa *et al.* studied gold clusters up to Au_{26}, arguing that the 2D/3D transition occurs between $n = 13$ and 15 [352]. Some studies of larger clusters are also available [353–357]. Idropo *et al.* studied the optical absorption spectra and dipole polarizabilities of gold clusters up to Au_{20} [358–360]. Pyykkö's group showed recently by DFT and MP2 calculations that Au_{72} could exist as a hollow nanosphere and can be characterized as an *I*-symmetric "golden fullerene" [361]. Doubly negative charged species, Au_n^{2-}, have been studied for example by Yannouleas and Landman [362] and Walter and Häkkinen [363].

Besides these many cluster studies, it is currently not known at what approximate cluster size the metallic state is reached, or when the transition occurs to solid-state-like properties. As an example, Figure 4.17 shows the dependence of the ionization potential and electron affinity on the cluster size for the Group 11 metals. We see a typical odd-even oscillation for the open/closed shell cases. Note that the work-function for Au is still 2 eV below the ionization potential of Au_{24}. Another interesting fact is that the Au_n ionization potentials are about 2 eV higher than the corresponding Cu_n and Ag_n values up to the bulk, which has been shown to be a relativistic effect [334]. A similar situation is found for the Group 11 cluster electron affinities [334].

While most studies so far concentrate on pure gold nanoclusters of different structure and size, the discovery of icosahedral mixed gold clusters $M@Au_{12}$, like $W@Au_{12}$ or $Mo@Au_{12}$ as shown in Figure 4.16, by Wang and co-workers [364], as predicted earlier by Pyykkö and Runeberg [365], opens up the field to new nanosized gold materials for applications in many areas like catalysis, with the central atom "impurity" being used to fine-tune electronic properties. These cluster types are

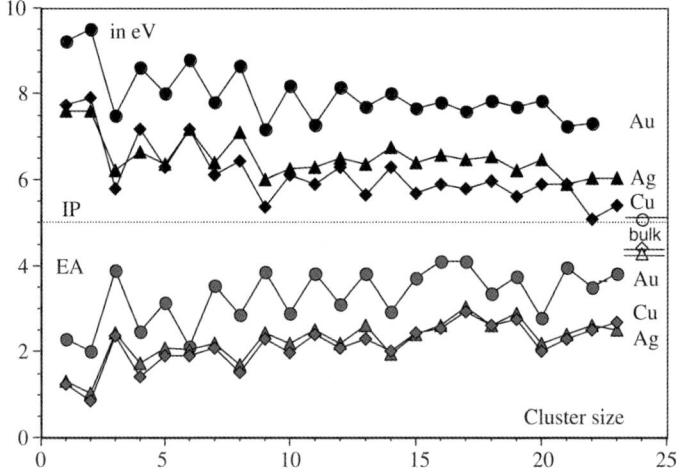

Figure 4.17 Ionization potentials (IP) and electron affinities (EA) of Group 11 clusters M_n up to $n = 23$ (in eV). The bulk metal work-functions for the (100) plane are also shown on the left hand side in open symbols. Experimental values from Refs. [370–374].

stable as they fulfill the 18-electron rule, and the peripheral gold atoms are close enough (about 2.85 Å) to undergo stabilizing metallic interactions. Furthermore, the covalent M-Au interactions are strengthened by relativistic effects [18], as explained above. For the HOMO-LUMO gap one obtains 3.0 eV at the B3LYP level of theory for W@Au$_{12}$ underpinning the unusual stability of this cluster [365, 366]. The finite-temperature dynamics of W@Au$_{12}$ using DFT-based Born–Oppenheimer molecular dynamics suggest surface-melting between 366 and 512 K [367]. Si@Au$_{16}$ was studied by Walter and Häkkinen [363], and M@Au$_{12}^-$ (M = V, Nb and Ta) by Zhai and Wang [368].Very recently Wang et al. reported a joint experimental/theoretical study of group IV atom-doped gold anion MAu$_{16}^-$ (M = Si, Ge, Sn) clusters [369].

Coating of magnetic clusters by gold atoms for passivation and stabilization of small metal particles is currently investigated in Rösch's research group [375]. They studied magnetic clusters of Ni$_6$Au$_n$ (n = 0, 8, 32) and Ni$_{13}$Au$_n$ (n = 0, 6, 8, 14, 24, 30, and 42) at the all-electron scalar-relativistic DFT level. MAu$_6$ and MAu$_6^-$ clusters (M = Ti, V, and Cr) were studied by Li et al. [376]. Pd- and Pt-doped small gold clusters were investigated using DFT by Sahu et al. [377] and Ge et al. [378]. Density functional and photo- fragmentation studies on Au$_n$M$_k$ (M = Sc, Ti, Cr, Fe) were performed by Janssens et al., and dopant-dependent differences in atomization energy, atomic-orbital occupancies and local magnetic moments were addressed [379]. Liu et al. studied AuTl and Au$_2$Tl$_2$ using a ARPP/CCSD(T) approach [380]. Bonačić-Koutecký et al. used DFT to study the structural and electronic properties of bimetallic silver–gold clusters [381]. Zhai et al. studied Au$_2$H$^-$ as a possible impurity in gold clusters [382]. Häkkinen et al. carried out DFT calculations on phosphine- and thiolate-protected gold nanoclusters, Au$_{39}$(PH$_3$)$_{14}$Cl$_6$ and Au$_{38}$(SCH$_3$)$_{24}$ [383]. For Au$_{38}$(SCH$_3$)$_{24}$ ring-like (AuSCH$_3$)$_4$ units protecting a central Au$_{14}$ core was predicted. A number of smaller thiolated gold clusters have been studied as well by this group [384]. These beautiful structures are illustrated in Figure 4.18. Au$_{13}$(SR)$_n$ (R = H, CH$_3$; n = 4, 6, 8) has been investigated by Rösch's group [385]. Very recently, Sharma et al. studied binding of Au$_6$ to size-expanded DNA bases using B3LYP [386].

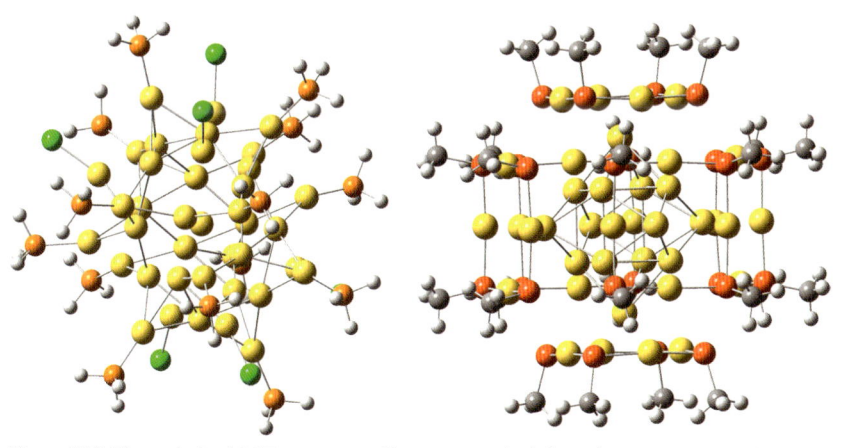

Figure 4.18 The optimized DFT structures of Au$_{39}$(PH$_3$)$_{14}$Cl$_6$ (left) and Au$_{38}$(SCH$_3$)$_{24}$ (right) [383].

4.7
Calculations on Infinite Systems: from Surfaces to the Solid State of Gold

When moving from an atom or molecule to infinite systems like the solid state, the quantum chemical description changes radically. In almost all cases one relies on the adiabatic approximation, that is one works within the Born–Oppenheimer approximation (as for most of the molecular calculations). This is a very good approximation indeed for heavy atom systems with high atomic mass. Here relativistic corrections are more important, which are treated in a similar way to molecular program codes. However, most codes work with plane-wave functions (with the notable exception of CRYSTAL [387]), and here the use of ultra-soft relativistic pseudopotentials [388] is the most popular choice in surface and solid-state calculations implemented in programs like VASP [389]. Not too long ago there was a strong belief in the physics community that spin-orbit effects were the most important relativistic effects in heavy-element-containing solids, as they directly influence the number and position of electronic bands in band structure calculations [390]. Such effects are more important in metals or covalently bonded systems because spin-orbit effects are quenched in strong ionic fields. However, Christensen and Seraphin showed early on that even scalar relativistic effects for bulk gold lead to a completely different band structure as compared with the nonrelativistic case [391]. In fact, Loucks pointed out as early as in 1966 that APW calculations for tungsten show that for most of the experimental results the relativistic values are in better agreement than the nonrelativistic ones [392]. Moreover, one can argue that the wavefunction is more compressed in the solid state compared with free atoms or molecules, thus increasing the density near the nucleus where relativistic operators act. It is interesting that even this situation seems to be modeled extremely well by the relativistic pseudopotential approach for solid-state or surface chemistry and physics applications.

In solid-state calculations one faces a number of serious drawbacks (or challenges), not usually encountered in the calculation of atoms or small molecular systems: (i) Different structures are often separated by very small energies, for example in polymorphs of molecular crystals (multiple minima problem). It is often difficult to know if the method applied predicts the correct energetic ordering of the different solid-state structures [393]. (ii) Electron correlation in the solid state is difficult to treat, especially for metallic systems which are highly multi-configurational (free electron gas) [394]. One speaks of strongly correlated systems [395]. For metallic systems, LDA often gives reasonable results for solid-state properties. (iii) Density functionals are problematic in correctly describing transition metal compounds (e.g., Mott insulators) and a parametrized Hubbard Hamiltonian is often added (the so-called DFT + U approach) [396]. (iv) Density functionals are problematic for weak interactions, that is, for Van der Waals/dispersive type of interactions [397]. (v) The vibrational problem becomes more difficult to treat (phonon branches, specific heat calculations) [398]. (vi) Electronic excitations are difficult to treat theoretically and one requires a many-body treatment, for example, a RPA + GW + Bethe-Salpeter approach [399]. (vii) Spin-polarized states are often not well described by DFT, for example, in broken symmetry states, or for magnetic properties [400]. At the moment, all so-called *first-principles methods* including DFT within the relativistic pseudopotential approach (DFT + PP) face these difficulties and one should be aware of this. Nevertheless, there have been many applications of the DFT + PP method in surface or solid-state calculations involving gold, and we will discuss a few of these applications.

Christensen and Seraphin pointed out that the shifts and splittings obtained from augmented plane wave (APW) calculations for *fcc* Au due to relativistic effects are of the same order of magnitudes as the gaps [391]. The calculated interband edge in the optical spectrum is calculated as 2.38 eV in agreement with experimental values. This compares to 3.7 eV for silver. Romaniello and de Boeij calculate the onset of interband transitions in gold at 3.5 eV using scalar relativistic TD-DFT, compared with 1.9 eV at the nonrelativistic level [401]. The low absorption edge for gold is a relativistic effect and seen as the primary reason for the yellow color of gold. Furthermore, for the bulk metals the nearest neighbor distances are r(Au–Au) ≈ (Ag–Ag), Figure 4.9. Takeuchi *et al.* calculated nonrelativistic and relativistic properties for bulk gold [402], which are summarized in Table 4.6. They also pointed out that the energy differences of the *fcc* to the *hcp* and *bcc* structures are very small. Philipsen and Baerends obtained similar results using DFT/GGA [403]. They also give spin-orbit effects for *fcc* Au which increases the cohesive energy by 5% and decreases the lattice constant by 0.4%, while the bulk modulus changes only little. Theileis and Bross showed that four-component LDA calculations lead to lattice constants for *fcc* Au which are too small compared with the experiment [404]. Souvatzis *et al.* determined the equation of state, thermal expansion coefficient, and Hugoniot for *fcc* Au using the LDA within the full-potential linear muffin-tin orbital (FP-LMTO) method including spin-orbit coupling [398]. A comparison between experimental and DFT/GGA results for *fcc* Au at high pressures shows rather large deviations [405]. Söderlind predicts a phase change to *hcp* at about 1.5 Mbar (70% compression of gold) and a phase change to *bcc* at pressures higher than 4 Mbar [406, 407]. Kirchhoff *et al.* studied the dynamical properties of gold using the tight-binding method [408]. They derived phonon-dispersion and density-of-states curves at finite temperature, and studied the temperature dependence of the lattice constant.

A list of recent solid-state calculations is given in Refs. [43–45]. We mention only a few of the most recent results discussing relativistic effects. Christensen and Kolar revealed very large relativistic effects in electronic band structure calculations for CsAu

Table 4.6 Lattice constants a, volume V, cohesive energy E_{coh} and bulk modulus B for *fcc* gold from nonrelativistic and relativistic pseudopotential DFT calculations (from Ref. [402]).

Au (*fcc*)	a [Å]	V [Å^3]	E_{coh} [eV]	B [Mbar]
nonrel.	4.314	20.072	2.27	0.996
rel.	4.104	17.282	3.46	1.790
exp.	4.079	16.967	3.81	1.732

Experimental values from Refs. [34, 413].

(which adopts the ionic CsCl structure), with nonrelativistic CsAu being a metal, while relativistic CsAu is a semiconductor with a gap of 2.6 eV [409]. The charge transfer of about 0.6e from Cs to Au [409, 410] is supported by XPS studies [411]. Wang and Zunger showed that relativistic effects change the enthalpy of formation such that CuAu becomes a stable metallic solid in contrast to CuAg [412].

Turning to ionic gold compounds, the unusual chain-like tetragonal structures of AuCl, AuBr and AuI with linear AuX_2 units and rather short Au–Au distances are due to relativistic effects [414]. Nonrelativistically, the gold halides would adopt cubic structures (rocksalt, zinc-blende or wurtzite strucures) similar to AgCl [415]. Doll et al. found that at the Hartree–Fock (HF) level the cubic and chain-like structures for AuCl are almost degenerate in energy, and electron correlation stabilizes the chain-like AuCl which offers an alternative explanation [416]. However, for AuBr and AuI even at the HF level of theory the chain-like structure is energetically preferred [415]. The predicted thermodynamic instability of solid AuF is also due to relativistic effects [182]. However, Kurzydłowski and Grochala just showed that solid AuF is stable with respect to decomposition to AuF_3 and metallic gold at pressures higher than 22.6 GPa [417]. They suggest that AuF could survive a careful low-temperature decompression to ambient pressures. The most stable solid-state structure predicted is orthorhombic with infinite one-dimensional AuF chains and an Au–F–Au angle of 143°. The smallest Au–Au distance is 2.85 Å between the chains, suggesting strong aurophilic interactions [417]. The aforementioned compound $Cs_2[AuCl_2][AuCl_4]$ undergoes a phase transition above 5 GPa where the Au–Cl distances all become equal, which is most likely due to a change in oxidation state to Au (II) in $CsAuCl_3$ [418]. This has been investigated by DFT, but no definite conclusion was reached about different phases and structures [419]. Stampfl's group recently predicted from DFT/GGA calculations that unknown Au_2O adopts a metallic phase, whereas known Au_2O_3 is a semiconductor [420]. Very recently, Pyykkö and co-workers predicted new sheet-like structures for Group 11 cyanides, M-CN, containing six-membered C_6N_6 rings connected by Group 11 metals [421, 422]. These structures are comparable in energy to the experimentally known hexagonal packed infinite chain (−M−CN−M−CN−) structures.

There has been increasing interest in the study of infinite one- and two-dimensional systems of gold, which range from gold chains to gold (or gold doped) surfaces [423], the latter mainly to study the catalytic activity of gold and related nanosystems. Gold is one of the very few metallic elements that can be used at the nanoscale due to its resistance to oxidation. The list of recent theoretical work is therefore too large to be reviewed in detail here (see Refs. [43, 44, 61, 62, 64, 424]). For example, CO or O_2 adsorption on neutral or charged gold nanoclusters as model systems for gold-doped surfaces have been studied intensively in the past [25, 60, 425–427]. The exact mechanism of catalytic processes on gold-doped metal-oxide surfaces is still not accurately known, as the (electronic) structure of such surfaces including the location of the catalytic center is still a matter of intense debate [428–430]. Moreover, the application of various DFT approximations to weak adsorption, like CO on neutral or charged gold clusters or surfaces, remains questionable as a recent study shows [431]. For the following discussion, we include adsorption studies on finite sized gold clusters as well.

Corma et al. pointed out that H_2 dissociation on a gold surface requires low-coordinated gold found in nanoparticles or in line defects [432]. Wang and Gong studied the interaction of H_2, CO and O_2 with Au_{32} using GGA/DFT [433]. Gordon's group published a series of papers on O_2 adsorption on neutral and charged gold clusters [434–436]. They found that the bond between O_2 and gold is strong if the cluster has an odd number of electrons, and is weaker if the number of electrons in the cluster is even. This group also studied H_2 and propene adsorption on gold [437, 438]. The strength of adsorption decreases with increasing number of electrons in the gold cluster, that is, $Au_n^+ > Au_n > Au_n^-$, which is related to charge transfer from the propene molecule to the gold surface [438]. Further, negatively charged gold clusters with an odd number of electrons bind O_2 more strongly than isoelectronic neutral clusters with an odd number of electrons. All clusters can bind two O_2 molecules [434]. Kryachko and Remacle studied NH_3 adsorption on charged and neutral gold clusters [439]. They showed that the cage-like Au_{32} of I_h symmetry shows a higher chemical inertness than the amorphous Au_{32} of C_1 symmetry (which is more stable at higher temperatures) with respect to the interaction with H_2, CO and O_2. Cummings and co-workers studied the self-assembly of 1,4-benzenedithiolate/ tetrahydrofuran on Au(1 1 1) surface using two-body interaction potentials in Monte Carlo simulations [440, 441]. Ford et al. studied the adsorption energetics of methanethiolate and benzenethiolate monomers and dimers on Au(1 1 1) using periodic DFT [442]. The dissociation barrier of dimethyl disulphide was estimated to be between 0.3 and 0.35 eV. Single transition metal atoms including gold on oxygen defects at the MgO(0 0 1) surface were studied theoretically by Neyman et al. [443]. Au deposited on graphite (0 0 0 1) was studied by Akola and Häkkinen [444]. Yang et al. carried out DFT calculations on supported Au on TiO_3/Mo(1 1 2) [445]. They calculated a red shift of 321 cm^{-1} in the CO vibrational frequencies when chemisorbed on the gold sites. Oxidation of CO on Au_8 clusters deposited on a MgO surface was studied by Yoon et al. using DFT [446]. Bonačić-Koutecký et al. studied the CO adsorption on $Au_mO_n^-$ ($m = 2, 3; n = 1-5$) [447]. Ge et al. [378] studied CO adsorption on Pt-doped gold clusters. Shi and Stampfl performed density-functional theory calculations to investigate the adsorption of O_2 at the Au(1 1 1) surface [448]. They found that atomic oxygen adsorbs weakly on the surface and is barely stable with respect to molecular oxygen. We note that the anomaly in the O_2 chemisorption energy in the Group 11 series of metals as shown in Table 4.1, which is most likely a relativistic effect (note the destabilizing effects of electronegative ligands to gold as discussed above). Okumura et al. studied H_2O and O_2 adsorption on gold clusters [449]. Adsorption of methanol on charged and neutral gold clusters was studied by Marx and co-workers using Car–Parrinello ab initio molecular dynamics simulations [450, 451]. They found that for Au_n^+ the C–O stretching mode of the adsorbed methanol molecule increases discontinuously as the underlying cluster structure changes from 2D to 3D. In the neutral species the CO molecule is weakly bound and much less sensitive to changes in coordination number and cluster structure. Li and Hamilton studied the binding of H_2S to Au_4 [452], and Krüger et al. investigated the binding of CH_3SH to neutral and charged gold clusters [453]. They also investigated the detaching mechanism of thiolates from copper and gold clusters [454]. They found a pronounced and systematic weakening of the S—C bond by over 1 eV upon increasing the cluster size of Cu_n, which is not observed for the corresponding gold clusters, which is a relativistic effect. CO oxidation on anionic and cationic gold oxide clusters have been studied very recently by Castleman, Bonačić-Koutecký and co-workers [455]. Kaxiras and co-workers studied chlorine bonding on Au(111) and found that the bonding of Cl to gold is primarily of covalent character [456]. This is due to the relativistic increase of the gold electronegativity [5].

The chemistry of superheavy elements has made some considerable progress in the last decade [457]. As the recently synthesized elements with nuclear charge 112 (eka-Hg), 114 (eka-Pb) and 118 (eka-Rn) are predicted to be chemically quite inert [458], experiments on these elements focus on adsorption studies on metal surfaces like gold [459]. DFT calculations predict that the equilibrium adsorption temperature for element 112 is predicted 100 °C below that of Hg, and the reactivity of element 112 is expected to be somewhere between those of Hg and Rn [460, 461]. This is somewhat in contradiction to recent experiments [459], and DFT may not be able to simulate accurately the physisorption of element 112 on gold. More accurate wavefunction based methods are needed to clarify this situation. Similar experiments are planned for element 114.

One-dimensional gold nanowires are also the subject of recent intense theoretical activity, ever since Kondo and Takayanagi discovered unexpected helical structures [462, 463]. Calculations range from single gold-chains (linear, zig-zag or Peierls distorted) [464, 465], to gold nanotubes [466, 467], gold nanowires [468–475] and nanorings [476]. Nanostrips and nanorings of gold-glued polyauronaphthyridines (or related compounds) were predicted by Pyykkö and Zaleski-Ejgierd using DFT/GGA [231, 477]. Sun *et al.* found that under appropriate experimental conditions it may still be possible to synthesize a metastable form of a gold nanoring [476]. Gold nanowires with light-element impurities have also been investigated using DFT by Skorodumova *et al.* [478]. They find that light impurities cause a significant charge redistribution in monatomic gold wires, thus altering bonding and affecting the wire structural properties. Geng and Kim investigated AuZn and AuMg nanowires [479]. Ning *et al.* investigated the conductance of a Au/1,4-diaminobenzene/Au molecular junction using a nonequilibrium Green's function formalism together with DFT [480]. The transmission coefficient at the Fermi level was calculated to be 0.028, in reasonable agreement with recent break junction experiments (0.0064). And finally, Zhang *et al.* studied surface-enhanced Raman scattering (SERS) of nicotinic acid on gold nanowires using both experimental and DFT studies, and found a significantly enhanced Raman spectrum of nicotinic acid upon interaction with gold [481].

4.8
Summary

We were only able to skim through the many applications of the computational chemistry of gold in inorganic, organometallic, and physical chemistry, and in surface science as well as solid-state physics. The chemistry and physics of gold is currently undergoing enormous progress and rapid change not seen a decade or two ago, as gold was previously regarded as an inferior catalyst compared with other catalytically active metals. The nanoscience of gold has grown into one of the most active research areas in chemistry and physics. One may even talk about a new renaissance in gold chemistry. We have seen that the physics and chemistry of gold is clearly dominated by relativistic effects as pointed out early on by Pekka Pyykkö and co-workers [1, 2], thus giving gold and its compounds some unique chemical and physical properties. On the theoretical side, the enormous progress in computer technology together with the development of computationally efficient relativistic methods such as the relativistic pseudopotential method within DFT now makes it possible to tackle very large systems including nanosized clusters, surfaces and the solid state with reasonable accuracy. A future challenge for computational chemists will be to treat such systems dynamically, modeling catalytic reactions in real time, or accurately simulating melting of clusters and the solid state from first-principles quantum theoretical methods. In summary, gold is alive and well. *Aurum scientiaque potestas sunt.*

Acknowledgment

This work was supported by the Marsden Fund managed by the Royal Society of New Zealand. Our thanks go to Detlev Figgen, Behnam Assadollahzadeh, Reuben Brown, Jon K. Laerdahl and Pekka Pyykkö for helpful discussions and suggestions, and to Wojciech Grochala for letting us know in advance of the solid state results for gold fluoride. We do not claim that our review is comprehensive, and we apologize from possible omissions of any related and important work. A more complete list of references can be found in Pyykkö's work [43–45].

References

1 Pyykkö, P. and Desclaux, J.P. (1979) Relativity and the periodic system of elements. *Accounts of Chemical Research*, **12**, 276–281.

2 Pyykkö, P. (1978) Relativistic quantum chemistry. *Advances in Quantum Chemistry*, **11**, 353–409.

3 Fricke, B. (1975) Superheavy elements a prediction of their chemical and physical properties. *Structure & Bonding*, **21**, 89–144.

4 Eliav, E., Kaldor, U., Schwerdtfeger, P., Hess, B. and Ishikawa, Y. (1994) The Ground State Electron Configuration of Element 111. *Physical Review Letters*, **73**, 3203–3206.

5 Schwerdtfeger, P. (1991) Relativistic and Electron Correlation Contributions in Atomic and Molecular Properties. Benchmark Calculations on Au and Au_2. *Chemical Physics Letters*, **183**, 457–463.

6 Neogrády, P., Kellö, V., Urban, M. and Sadlej, A.J. (1997) Ionization Potentials and Electron Affinities of Cu, Ag, and Au: Electron Correlation and Relativistic Effects. *International Journal of Quantum Chemistry*, **63**, 557–565.

7 Mulliken, R.S. (1934) A New Electroaffinity Scale; Together with Data on Valence States and on Valence Ionization Potentials and Electron Affinities. *Journal of Chemical Physics*, **2**, 782–793.

8 Mulliken, R.S. (1935) Electronic Structures of Molecules XI. Electroaffinity, Molecular Orbitals and Dipole Moments. *Journal of Chemical Physics*, **3**, 573–585.

9 Puddephat, R.J. (1978) *The Chemistry of Gold*, Elsevier, Amsterdam.

10 Schwerdtfeger, P. (1989) Relativistic effects in gold chemistry. II. The stability of complex halides of Au(III). *Journal of the American Chemical Society*, **111**, 7261–7262.

11 Schwerdtfeger, P., Boyd, P.D.W., Brienne, S. and Burrell, A.K. (1992) Relativistic effects in gold chemistry. IV. Au(III) and Au(V) compounds. *Inorganic Chemistry*, **31**, 3411–3422.

12 Pyykkö, P. (1997) Strong closed-shell interactions in inorganic chemistry. *Chemical Reviews*, **97**, 597–636.

13 Schmidbaur, H. (1990) The Fascinating Implications of New Results in Gold Chemistry. *Gold Bulletin*, **23**, 11–21.

14 Schmidbaur, H. (2000) The Aurophilicity Phenomenon: A Decade of Experimental Findings, Theoretical Concepts and Emerging Applications. *Gold Bulletin*, **33**, 3–10.

15 Bravo-Pérez, G., Garzón, I.L. and Novaro, O. (1999) Ab initio study of small gold clusters. *Journal of Molecular Structure: THEOCHEM*, **493**, 225–231.

16 Häkkinen, H. and Landman, U. (2000) Gold clusters (Au_N, $2 \leq N \leq 10$) and their anions. *Physical Review B - Condensed Matter*, **62**, R2287–R2290.

17 Häkkinen, H., Moseler, M. and Landman, U. (2002) Bonding in Cu, Ag, and Au Clusters: Relativistic Effects, Trends, and Surprises. *Physical Review Letters*, **89**, 033401-1–033401-4.

18 Schwerdtfeger, P. and Dolg, M. (1991) Anomalous high gold-metal bond stabilities: Relativistic Configuration Interaction Calculations for AuLa and AuLu. *Physical Review A*, **43**, R1644–R1646.

19 Schwerdtfeger, P., Bruce, A.E. and Bruce, M.R.M. (1998) Theoretical Studies on the Photochemistry of the cis to trans Conversion in Dinuclear Goldhalide Bis (diphenylphosphino)ethylene Complexes. *Journal of the American Chemical Society*, **120**, 6587–6597.

20 Wing-Wah Yam, V. and Chung-Chin Cheng, E. (2007) Photochemistry and Photophysics of Coordination Compounds: Gold. *Topics in Current Chemistry*, **281**, 269–309.

21 Gorin, D.J. and Toste, F.D. (2007) Relativistic effects in homogeneous gold catalysis. *Nature*, **446**, 395–403.

22 Hashmi, A.S.K. (2004) Homogeneous Catalysis by Gold. *Gold Bulletin*, **37**, 51–65.

23 Hashmi, A.S.K. (2003) Homogeneous Gold-Catalysts and Alkynes: A Successful Liaison. *Gold Bulletin*, **36**, 3–9.

24 Hashmi, A.S.K., Ata, F., Bats, J.W., Blanco, M.C., Frey, W., Hamzic, M., Rudolph, M., Salathé, R., Schäfer, S. and Wölfle, M. (2007) Homogeneous Gold-Catalysts and Alkynes: A Successful Liaison. *Gold Bulletin*, **40**, 31–35 and references therein.

25 Bond, G.C., Louis, C. and Thompson, D.T. (2006) *Catalysis by Gold*, Catalytic Science Series, vol. 6, Imperial College Press, London.

26 Haruta, M., Kobayashi, T., Sano, H. and Yamada, N. (1987) Novel Gold Catalysts for the Oxidation of Carbon Monoxide at a Temperature far Below 0°. *Chemistry Letters*, **16**, 405–408.

27 Haruta, M. (2003) When Gold Is Not Noble: Catalysis by Nanoparticles. *Chemical Record*, **3**, 75–87.

28 Ishida, T. and Haruta, M. (2007) Gold Catalysts: Towards Sustainable Chemistry. *Angewandte Chemie International Edition*, **46**, 7154–7156.

29 Haruta, M. (2007) New Generation of Gold Catalysts: Nanoporous Foams and

Tubes—Is Unsupported Gold Catalytically Active? *ChemPhysChem*, **8**, 1911–1913.
30 Cortie, M.B. (2004) The Weird World of of Nanoscale Gold. *Gold Bulletin*, **37**, 12–19.
31 Burks, R. (2007) The Riches Of Gold Catalysis. *Chemical & Engineering News*, Sept. 24, 87–91.
32 Desclaux, J.P. (1973) Relativistic Dirac-Fock expectation values for atoms with $Z=1$ to, $Z=120$. *Atomic Data and Nuclear Data Tables*, **12**, 311–406.
33 Schwerdtfeger, S. and Seth, M. (1998) Relativistic Effects of the Superheavy Elements, in *Encyclopedia of Computational Chemistry*, Vol. 4 (eds P.R. von Schleyer, P.R. Schreiner, N.L. Allinger, T. Clark, J. Gasteiger, P.A. Kollman and H.F. Schaefer III), John Wiley & Sons, Inc., New York, pp. 2480–2499.
34 *CRC handbook of chemistry and physics*, (1977) CRC Press, Cleveland, Ohio.
35 Schwerdtfeger, P. (ed.) (2002) *Relativistic Electronic Structure Theory. Part 1: Fundamentals*, Elsevier, Amsterdam.
36 Schwerdtfeger, P. (ed.) (2004) *Relativistic Electronic Structure Theory. Part 2: Applications*, Elsevier, Amsterdam.
37 Hess, B.A. (ed.) (2003) *Relativistic Effects in Heavy-Element Chemistry and Physics*, John Wiley & Sons, Ltd, Chichester.
38 Hirao, K. and Ishikawa, Y. (eds) (2004) Recent Advances in Relativistic Molecular Theory, in *Recent Advances in Computational Chemistry*, Vol. 5, World Scientific, London.
39 Dyall, K.G. and Fœgri, K. Jr (2007) *Introduction to Relativistic Quantum Chemistry*, Oxford University Press, Oxford.
40 Grant, I.P. (2007) *Relativistic Quantum Theory of Atoms and Molecules. Theory and Computation*, Springer, New York.
41 Reiher, M. and Wolf, A. (2008) *Relativistic Quantum Chemistry*, Wiley-VCH Verlag GmbH, Weinheim.
42 Pyykkö, P. (1988) Relativistic Effects in Structural Chemistry. *Chemical Reviews*, **88**, 563–594.
43 Pyykkö, P. (2004) Theoretical Chemistry of Gold. *Angewandte Chemie International Edition*, **43**, 4412–4456.
44 Pyykkö, P. (2005) Theoretical Chemistry of Gold. II. *Inorganica Chimica Acta*, **358**, 4113–4130.
45 Pyykkö, P. (2008) Theoretical Chemistry of Gold. III. *Chemical Society Reviews*, **37**, 1967–1997.
46 Schwerdtfeger, P. (2002) Relativistic Effects in Properties of Gold. *Heteroatom Chemistry*, **13**, 578–584.
47 Schmidbaur, H., Cronje, S., Djordjevic, B. and Schuster, O. (2005) Understanding gold chemistry through relativity. *Chemical Physics*, **311**, 151–161.
48 Schwerdtfeger, P. (2003) Gold Goes Nano – From Small Clusters to Low-Dimensional Assemblies. *Angewandte Chemie International Edition*, **42**, 1892–1895.
49 Crawford, M.-J. and Klapötke, T. (2002) Hydrides and Iodides of Gold. *Angewandte Chemie International Edition*, **41**, 2269–2271.
50 Mohr, F. (2004) The Chemistry of Gold-fluoro compounds: A continuing challenge for Gold. *Gold Bulletin*, **37**, 164–169.
51 Bartlett, N. (1998) Relativistic Effects and the Chemistry of Gold. *Gold Bulletin*, **31**, 22–25.
52 Schwarz, H. (2003) Relativistic Effects in Gas-Phase Ion Chemistry: An Experimentalist's View. *Angewandte Chemie International Edition*, **42**, 4442–4454.
53 Puddephatt, R.J. (1982) *Comprehensive Organometallic Chemistry*, Vol. 2 (ed. Y. Wilkinson), Pergamon, Oxford, pp. 765–821.
54 Puddephatt, R.J. (1987) *Comprehensive Coordination Chemistry*, Vol. 2 (ed. G. Wilkinson), Pergamon, Oxford, pp. 861–923.

55 Puddephatt, R.J. and Vittal, J.J. (1994) *Encyclopedia of Inorganic Chemistry*, vol. 3 (ed. R.B. King), John Wiley & Sons, Ltd, Chichester, pp. 1320–1331.

56 Schmidbaur, H. (1980) *Gmelin Handbuch der Anorganischen Chemie*, 8th edn (ed. A. Slawisch), Springer-Verlag, Berlin.

57 Schmidbaur, H. (1994) *Encyclopedia of Inorganic Chemistry*, Vol. 3 (ed. R.B. King), John Wiley & Sons, Ltd, Chichester, pp. 1332–1340.

58 Grohmann, A. and Schmidbaur, H. (1995) Gold, in *Comprehensive Organometallic Chemistry II*, vol. 3, Pergamon, Oxford, pp. 1–56.

59 Schmidbaur, H. (1995) Ludwig Mond Lecture. High-Carat Gold Compounds. *Chemical Society Reviews*, **24**, 391–400.

60 Schmidbaur, H. (ed.) (1999) *Gold. Progress in Chemistry, Biochemistry and Technology*, John Wiley & Sons, Ltd, Chichester.

61 Thompson, D. (1999) New Advances in Gold Catalysis Part II. *Gold Bulletin*, **32**, 12–19.

62 Bond, G.C. and Thompson, D.T. (1999) Catalysis by Gold. *Catalysis Reviews-Science and Engineering*, **41**, 319–388.

63 Thompson, D. (1998) New Advances in Gold Catalysis. *Gold Bulletin*, **31**, 111–118.

64 Bond, G.C. (2002) Gold: a relatively new catalyst. *Catalysis Today*, **72**, 5–9.

65 See discussion of Dirac's famous statement that relativistic effects are not important for chemical relevant systems: Kutzelnigg, W. (2000) *Theoretical Chemistry Accounts*, **103**, 182–186.

66 Balasubramanian, K. and Pitzer, K.S. (1987) Relativistic Quantum Chemistry. *Advances in Chemical Physics*, **67**, 287–319.

67 Baerends, E.J., Schwarz, W.H.E., Schwerdtfeger, P. and Snijders, J.G. (1990) Relativistic atomic orbital contractions and expansions: magnitudes and explanations. *Journal of Physics B- Atomic Molecular and Optical Physics*, **23**, 3225–3240.

68 Schwarz, W.H.E., van Wezenbeek, E., Baerends, E.J. and Snijders, J.G. (1989) The origin of relativistic effects of atomic orbitals. *Journal of Physics B: Atomic Molecular and Optical Physics*, **22**, 1515–1530.

69 Siekierski, S., Autschbach, J., Seth, M., Schwerdtfeger, P. and Schwarz, W.H.E. (2002) The dependence of relativistic effects on the electronic configurations in the atoms of the d- and f-block elements. *Journal of Computational Chemistry*, **23**, 804–813.

70 Dyall, K.G., Grant, I.P., Johnson, C.T., Parpia, F.A. and Plummer, E.P. (1989) GRASP: A general-purpose relativistic atomic structure program. *Computer Physics Communications*, **55**, 425–456.

71 Parpia, F.A., Froese-Fischer, C. and Grant, I.P. (1996) GRASP92: A package for large-scale relativistic atomic structure calculations. *Computer Physics Communications*, **94**, 249–271.

72 Mayo, R., Ortiz, M., Parente, F. and Santos, J.P. (2007) Experimental and theoretical transition probabilities for lines arising from the 6p configurations of Au II. *Journal of Physics B: Atomic Molecular and Optical Physics*, **40**, 4651–4660.

73 Landau, A., Kaldor, U. and Eliav, E. (2001) Intermediate Hamiltonian Fock-space coupled-cluster method. *Advances in Quantum Chemistry*, **39**, 171–188.

74 Sapirstein, J.P. (1998) Theoretical methods for the relativistic atomic many-body problem. *Reviews of Modern Physics*, **70**, 55–76.

75 Lindgren, I. and Morrison, J. (1986) *Atomic Many-Body Theory*, Springer-Verlag, Berlin.

76 Vilkas, M.J., Ishikawa, Y. and Koc, K. (1998) Quadratically convergent multiconfiguration Dirac-Fock and multireference relativistic configuration-interaction calculations for many-electron systems. *Physical Review E*, **58**, 5096–5110.

77 Ishikawa, Y. and Koc, K. (1997) Relativistic many-body perturbation calculations for open-shell systems. *Physical Review A*, **56**, 1295–1304.

78 Rodrigues, G.C., Ourdane, M.A., Bieron, J., Indelicato, P. and Lindroth, E. (2001) Relativistic and many-body effects on total binding energies of cesium ions. *Physical Review A*, **63**, 012510-1–012510-10.

79 Gorceix, O. and Indelicato, P. (1988) Effect of the complete Breit interaction on two-electron ion energy levels. *Physical Review A*, **37**, 1087–1094.

80 Lindroth, E. and Mårtensson-Pendrill, A.-M. (1989) Further analysis of the complete Breit interaction. *Physical Review A*, **39**, 3794–3802.

81 Jackson, J.D. (1975) *Classical Electrodynamics*, John Wiley & Sons, Inc., New York.

82 Indelicato, P. and Lindroth, E. (1992) Relativistic effects, correlation, and QED corrections on Kα transitions in medium to very heavy atoms. *Physical Review A*, **46**, 2426–2436.

83 Mohr, P.J. (1992) Self-energy correction to one-electron energy levels in a strong Coulomb field. *Physical Review A*, **46**, 4421–4424.

84 Mohr, P.J., Plunien, G. and Soff, G. (1998) QED corrections in heavy atoms. *Physics Reports-Review Section of Physics Letters*, **293**, 227–369.

85 Flambaum, V.V. and Ginges, J.S.M. (2005) Radiative potential and calculations of QED radiative corrections to energy levels and electromagnetic amplitudes in many-electron atoms. *Physical Review A*, **72**, 052115-1–052115-13.

86 Fægri, K. (2004) Even tempered basis sets for four-component relativistic quantum chemistry. *Chemical Physics*, **311**, 25–34.

87 Anton, J., Fricke, B. and Schwerdtfeger, P. (2005) Non-collinear and collinear four-component relativistic molecular density functional calculations. *Chemical Physics*, **311**, 97–103.

88 Perdew, J.P. and Cole, L.A. (1982) On the local density approximation for Breit interaction. *Journal of Physics C*, **15**, L905–L908.

89 Schwerdtfeger, P. and Thierfelder, C. (2006) Relativistic Quantum Chemistry – A Historical Overview, in *Trends and Perspectives in Modern Computational Science*, Vol. 6 Lecture Series on Computer and Computational Sciences (eds G. Maroulis and T. Simos), Brill Academic Publishers, Leiden, The Netherlands, pp. 453–460.

90 Eliav, E., Kaldor, U. and Ishikawa, Y. (1994) Open-shell relativistic coupled-cluster method with Dirac-Breit wave functions: Energies of the gold atom and its cation. *Physical Review Letters*, **49**, 1724–1729 Including newer unpublished results from this group.

91 Moore, C.E. (1958) Atomic Energy Levels, (Natl. Bur. Stand. (U. S.) Circ. No. 467, U. S. GPO Washington, D. C.

92 Hotop, H. and Lineberger, W.C. (1975) Binding energies in atomic negative ions. II. *Journal of Physical and Chemical Reference Data*, **14**, 731–750.

93 Labzowsky, L., Goidenko, I., Tokman, M. and Pyykkö, P. (1999) Calculated self-energy contributions for an ns valence electron using the multiple-commutator method. *Physical Review A*, **59**, 2707–2711.

94 Perdew, J.P. and Wang, Y. (1992) Accurate and simple analytic representation of the electron-gas correlation energy. *Physical Review B - Condensed Matter*, **45**, 13244–13249.

95 Lee, C., Yang, W. and Parr, R.G. (1988) Development of the Colle-Salvetti correlation-energy formula into a functional of the electron density. *Physical Review B - Condensed Matter*, **37**, 785–789.

96 Itano, W.M. (2006) Quadrupole moments and hyperfine constants of metastable states of Ca^+, Sr^+, Ba^+, Yb^+, Hg^+, and Au. *Physical Review A*, **73**, 022510-1–022510-11.

97 Huheey, J.E. (1978) *Inorganic Chemistry*, 2nd edn, Harper and Row, New York, pp. 167.

98 Nuss, H. and Jansen, M. (2006) [Rb([18] crown-6)$(NH_3)_3$]Au.NH_3: Gold as

Acceptor in N–H···Au⁻ Hydrogen Bonds. *Angewandte Chemie International Edition*, **45**, 4369–4371.

99 Schwerdtfeger, P. and Bowmaker, G.A. (1994) Relativistic effects in gold chemistry. V. Group 11 Dipole-Polarizabilities and Weak Bonding in Monocarbonyl Compounds. *Journal of Chemical Physics*, **100**, 4487–4497.

100 Schwerdtfeger, P. (2006) Atomic Static Dipole Polarizabilities, in *Computational Aspects of Electric Polarizability Calculations: Atoms, Molecules and Clusters* (ed. G. Maroulis), Imperial College Press, London, pp. 1–32.

101 Kellö, V., Urban, M. and Sadlej, A.J. (1996) Electric dipole polarizabilities of negative ions of the coinage metal atoms. *Chemical Physics Letters*, **253**, 383–389.

102 Kellö, V. and Sadlej, A.J. (1996) Standardized basis sets for high-level-correlated relativistic calculations of atomic and molecular electric properties in the spin-averaged Douglas-Kroll (no-pair) approximation I. Groups Ib and IIb. *Theoretica Chimica Acta*, **94**, 93–104.

103 Schwerdtfeger, P., Pernpointner, M. and Nazarewicz, W. (2004) Calculation of Nuclear Quadrupole Coupling Constants, in *Calculation of NMR and EPR Parameters: Theory and Applications*, (eds M. Kaupp M. Bühl and V.G. Malkin), Wiley-VCH Verlag GmbH, Weinheim, pp. 279–291.

104 Bowmaker, G.A. (1999) Spectroscopic methods in gold chemistry, in *Gold. Progress in Chemistry, Biochemistry and Technology*, (ed. H. Schmidbaur), John Wiley & Sons, Ltd, Chichester, pp. 841–882.

105 Powers, R.J., Martin, P., Miller, G.H., Welsh, R.E. and Jenkins, D.A. (1974) Muonic ^{197}Au: A test of the weak-coupling model. *Nuclear Physics A*, **230**, 413–444.

106 Yakobi, H., Eliav, E. and Kaldor, U. (2007) Nuclear quadrupole moment of 197Au from high-accuracy atomic calculations. *Journal of Chemical Physics*, **126**, 184305-1–184305-4.

107 Dzuba, V.A., Flambaum, V.V. and Harabati, C. (2000) Calculation of positron binding to silver and gold atoms. *Physical Review A*, **62**, 042504-1–042504-7.

108 Beier, T., Mohr, P.J., Persson, H., Plunien, G., Greiner, M. and Soff, G. (1997) Current status of Lamb shif predictions for heavy hydrogen-like ions. *Physics Letters A*, **236**, 329–338.

109 Dzuba, V.A. and Flambaum, V.V. (2008) Relativistic corrections to transition frequencies of Ag I, Dy I, Ho I, Yb II, Yb III, Au I, and Hg II and search for variation of the fine-structure constant. *Physical Review A*, **77**, 012515-1–012515-6.

110 Saue, T., Enveldson, T., Helgaker, T., Jensen, H.J.A., Laerdahl, J.K., Ruud, K., Thyssen, J. and Visscher, L. (2000) Dirac – A relativistic *ab initio* electronic structure program, Release 3.2.1.

111 Hess, B.A. (1986) Relativistic electronic-structure calculations employing a two-component no-pair formalism with external-field projection operators. *Physical Review A*, **33**, 3742–3748.

112 Wolf, A., Reiher, M. and Hess, B.A. (2002) The generalized Douglas–Kroll transformation. *Journal of Chemical Physics*, **117**, 9215–9226.

113 Reiher, M. and Wolf, A. (2004) Exact decoupling of the Dirac Hamiltonian. I. General theory. *Journal of Chemical Physics*, **121**, 2037–2047.

114 Reiher, M. and Wolf, A. (2004) Exact decoupling of the Dirac Hamiltonian. II. The generalized Douglas–Kroll–Hess transformation up to arbitrary order. *Journal of Chemical Physics*, **121**, 10945–10956.

115 Reiher, M. (2006) Douglas–Kroll–Hess Theory: a relativistic electrons-only theory for chemistry. *Theoretical Chemistry Accounts*, **116**, 241–252 and references therein.

116 Chang, C., Pélissier, M. and Durand, P. (1986) Regular Two-Component Pauli-Like Effective Hamiltonians in Dirac Theory. *Physica Scripta*, **34**, 394–404.

117 van Lenthe, E., Baerends, E.J. and Snijders, J.G. (1993) Relativistic regular two-component Hamiltonians. *Journal of Chemical Physics*, **99**, 4597–4610.

118 van Lenthe, E., Snijders, J.G. and Baerends, E.J. (1996) The zero-order regular approximation for relativistic effects: The effect of spin–orbit coupling in closed shell molecules. *Journal of Chemical Physics*, **105**, 6505–6516 and references therein.

119 Barysz, M. and Sadlej, A.J. (2002) Infinite-order two-component theory for relativistic quantum chemistry. *Journal of Chemical Physics*, **116**, 2696–2704.

120 Kedziera, D. (2005) Convergence of approximate two-component Hamiltonians: How far is the Dirac limit. *Journal of Chemical Physics*, **123**, 074109-1–074109-5.

121 Ilias, M. and Saue, T. (2007) An infinite-order two-component relativistic Hamiltonian by a simple one-step transformation. *Journal of Chemical Physics*, **126**, 064102-1–064102-9.

122 Kutzelnigg, W. (1989) Perturbation theory of relativistic corrections 1. The non-relativistic limit of the Dirac equation and a direct perturbation expansion. *Zeitschrift für Physik D*, **11**, 15–28.

123 Kutzelnigg, W. (1990) Perturbation theory of relativistic corrections 2. Analysis and classification of known and other possible methods. *Zeitschrift für Physik D*, **15**, 27–50.

124 Engel, E., Dreizler, R.M., Varga, S., Fricke, B. and Hess, B.A. (2001) *Relativistic Effects in Heavy-Element Chemistry and Physics*, John Wiley & Sons, Inc., New York.

125 Anton, J., Fricke, B. and Engel, E. (2004) Noncollinear and collinear relativistic density-functional program for electric and magnetic properties of molecules. *Physical Review A*, **69**, 012505-1–012505-10.

126 Anton, J., Jacob, T., Fricke, B. and Engel, E. (2002) Relativistic Density Functional Calculations for Pt_2. *Physical Review Letters*, **89**, 213001-1–213001-4.

127 Hafner, P. and Schwarz, W.H.E. (1978) Pseudo-potential approach including relativistic effects. *Journal of Physics B*, **11**, 217–233.

128 Kleinman, L. (1980) Relativistic norm-conserving pseudopotential. *Physical Review B - Condensed Matter*, **21**, 2630–2631.

129 Bachelet, G.B. and Schlüter, M. (1982) Relativistic norm-conserving pseudopotentials. *Physical Review B - Condensed Matter*, **25**, 2103–2108.

130 Pitzer, R.M. and Winter, N.W. (1991) Electronic-Structure Methods for Heavy-Atom Molecules. *International Journal of Quantum Chemistry*, **40**, 773–780.

131 Schwerdtfeger, P. (2003) Relativistic Pseudopotentials, in *Theoretical Chemistry and Physics of Heavy and Superheavy Elements*, Vol. 11 (eds U. Kaldor and S. Wilson), Progress in Theoretical Chemistry and Physics, Kluwer, Dordrecht, pp. 399–438. in references therein.

132 Szasz, L. (1985) *Pseudopotential Theory of Atoms and Molecules*, John Wiley & Sons, Inc., New York.

133 Barthelat, J.C. and Durand, Ph. (1978) Recent Progress of Pseudo-Potential Methods in Quantum Chemistry. *Gazzetta Chimica Italiana*, **108**, 225–236.

134 Krauss, M. and Stevens, W.J. (1984) Effective potentials in molecular quantum chemistry. *Annual Review of Physical Chemistry*, **35**, 357–385.

135 Christiansen, P.A., Ermler, W.C. and Pitzer, K.S. (1985) Relativistic effects in chemical systems. *Annual Review of Physical Chemistry*, **36**, 407–432.

136 Ermler, W.C., Ross, R.B. and Christiansen, P.A. (1988) Spin-Orbit Coupling and Other Relativistic Effects in Atoms and Molecules. *Advances in Quantum Chemistry*, **19**, 139–182.

137 Pickett, W.E. (1989) Pseudopotential Methods in Condensed Matter Applications. *Physics Reports-Review Section of Physics Letters*, **9**, 115–198.

138 Huzinaga, S. (1995) 1994 Polanyi Award lecture: Concept of active electrons in chemistry. *Canadian Journal of Chemistry*, **73**, 619–628.

139 Frenking, G., Antes, I., Böhme, M., Dapprich, S., Ehlers, A.W., Jonas, V., Neuhaus, A., Otto, M., Stegmann, R., Veldkamp, A. and Vyboishikov, S.F. (1996) Pseudopotential Calculations of Transition Metal Compounds. Scope and Limitations, in *Reviews in Computational Chemistry*, Vol. 8 (eds K.B. Lipkowitz and D.B. Boyd), VCH, New York, pp. 63–144.

140 Pyykkö, P. and Stoll, H. (2000) Relativistic pseudopotential calculations, 1993-June 1999. in *R.S.C. Specialist Periodical Reports, Chemical Modelling, Applications and Theory*, Vol. 1, pp. 239–305.

141 Dolg, M. (2002) *Relativistic Effective Core Potentials* (ed. P. Schwerdtfeger), Elsevier, Amsterdam.

142 Schwerdtfeger, P., Fischer, T., Dolg, M., Igel-Mann, G., Nicklass, A., Stoll, H. and Haaland, A. (1995) The Accuracy of the Pseudopotential Approximation. I. An Analysis of the Spectroscopic Constants for the Electronic Ground States of InCl and InCl$_3$. *Journal of Chemical Physics*, **102**, 2050–2062.

143 Schwerdtfeger, P., Dolg, M., Schwarz, W.H.E., Bowmaker, G.A. and Boyd, P.D.W. (1989) Relativistic effects in gold chemistry. I. Diatomic gold compounds. *Journal of Chemical Physics*, **91**, 1762–1774.

144 Figgen, D., Rauhut, G., Dolg, M. and Stoll, H. (2005) Energy-consistent pseudopotentials for group 11 and 12 atoms: adjustment to multi-configuration Dirac–Hartree–Fock data. *Chemical Physics*, **311**, 227–244.

145 Peterson, K.A. and Puzzarini, C. (2005) Systematically convergent basis sets for transition metals. II. Pseudopotential-based correlation consistent basis sets for the group 11 (Cu, Ag, Au) and 12 (Zn, Cd, Hg) elements. *Theoretical Chemistry Accounts*, **114**, 283–296.

146 http://www.theochem.uni-stuttgart.de/pseudopotentials/clickpse.en.html.

147 Ziegler, T., Baerends, E.J., Snijders, J.G., Ravenek, W. and Tschinke, V. (1989) Calculation of bond energies in compounds of heavy elements by quasi-relativistic approach. *The Journal of Physical Chemistry*, **93**, 3050–3056.

148 Strömberg, D. and Wahlgren, U. (1990) First-order relativistic calculations on Au$_2$ and Hg$_2^{2+}$. *Chemical Physics Letters*, **169**, 109–115.

149 Häberlen, O.D. and Rösch, N. (1992) A scalar-relativistic extension of the linear combination of Gaussian-type orbitals local density functional method: application to AuH, AuCl and Au$_2$. *Chemical Physics Letters*, **199**, 491–496.

150 Bastug, T., Heinemann, D., Sepp, W.-D., Kolb, D. and Fricke, B. (1993) All-electron Dirac–Fock–Slater SCF calculations of the Au$_2$ molecule. *Chemical Physics Letters*, **211**, 119–124.

151 Hess, B.A. and Kaldor, U. (2000) Relativistic all-electron coupled-cluster calculations on Au$_2$ in the framework of the Douglas–Kroll transformation. *Journal of Chemical Physics*, **112**, 1809–1813.

152 van Lenthe, E., Baerends, E.J. and Snijders, J.G. (1994) Relativistic total energy using regular approximations. *Journal of Chemical Physics*, **101**, 9783–9792.

153 Park, C. and Almlöf, J.E. (1994) Two-electron relativistic effects in molecules. *Chemical Physics Letters*, **231**, 269–276.

154 van Wüllen, C. (1995) A relativistic Kohn–Sham density functional procedure by means of direct perturbation theory. *Journal of Chemical Physics*, **103**, 3589–3599.

155 Wesendrup, R., Laerdahl, J.K. and Schwerdtfeger, P. (1999) Relativistic Effects in Gold Chemistry. VI. Coupled Cluster Calculations for the Isoelectronic Series AuPt$^-$, Au$_2$ and AuHg$^+$. *Journal of Chemical Physics*, **110**, 9457–9462.

156 Suzumura, T., Nakajima, T. and Hirao, K. (1999) Ground-state properties of MH, MCl, and M_2 (M=Cu, Ag, and Au) calculated by a scalar relativistic density functional theory. *International Journal of Quantum Chemistry*, **75**, 757–766.

157 Han, Y.-K. and Hirao, K. (2000) On the transferability of relativistic pseudopotentials in density-functional calculations: AuH, AuCl, and Au_2. *Chemical Physics Letters*, **324**, 453–458.

158 Lee, H.-S., Han, Y.-K., Kim, M.C., Bae, C. and Lee, Y.S. (1998) Spin-orbit effects calculated by two-component coupled-cluster methods: test calculations on AuH, Au_2, TlH and Tl_2. *Chemical Physics Letters*, **293**, 97–102.

159 Huber, K.P. and Herzberg, G. (1979) *Molecular Spectra and Molecular Structure Constants of Diatomic Molecules*, Van Nostrand, New York.

160 Bishea, G.A. and Morse, M.D. (1991) Spectroscopic studies of jet-cooled AgAu and Au_2. *Journal of Chemical Physics*, **95**, 5646–5659.

161 Cheeseman, M.A. and Eyler, J.R. (1992) Ionization potentials and reactivity of coinage metal clusters. *The Journal of Physical Chemistry*, **96**, 1082–1087.

162 Ho, J., Ervin, K.M. and Lineberger, W.C. (1990) Photoelectron spectroscopy of metal cluster anions: Cu_n^-, Ag_n^-, and Au_n^-. *Journal of Chemical Physics*, **93**, 6987–7002.

163 Naveh, D., Kronik, L., Tiago, M.L. and Chelikowsky, J.R. (2007) Real-space pseudopotential method for spin-orbit coupling within density functional theory. *Physical Review B - Condensed Matter*, **76**, 153407-1–153407-4.

164 Fleig, T. and Visscher, L. (2005) Large-scale electron correlation calculations in the framework of the spin-free dirac formalism: the Au_2 molecule revisited. *Chemical Physics*, **311**, 63.

165 Wang, F. and Liu, W. (2005) Benchmark four-component relativistic density functional calculations on Cu_2, Ag_2, and Au_2. *Chemical Physics*, **311**, 63–69.

166 Varga, S., Engel, E., Sepp, W.-D. and Fricke, B. (1999) Systematic study of the Ib diatomic molecules Cu_2, Ag_2, and Au_2 using advanced relativistic density functionals. *Physical Review A*, **59**, 4288–4294.

167 Saue, T. and Jensen, H.J.Aa. (2003) Linear response at the 4-component relativistic level: Application to the frequency-dependent dipole polarizabilities of the coinage metal dimers. *Journal of Chemical Physics*, **118**, 522–536.

168 Mukherjee, S., Pastor, G.M. and Bennemann, K.H. (1990) Theoretical study of the metastability of Au_2^{2+} clusters. *Physical Review B - Condensed Matter*, **42**, 5327–5330.

169 Ortiz, G. and Ballone, P. (1991) Metastability of doubly charged transition-metal dimers in density-functional theory. *Physical Review B - Condensed Matter*, **44**, 5881–5884.

170 Wang, F. and Li, L. (2002) A singularity excluded approximate expansion scheme in relativistic density functional theory. *Theoretical Chemistry Accounts*, **108**, 53–60.

171 Snijders, J.G. and Pyykkö, P. (1980) Is the relativistic contraction of bond lengths an orbital contraction effect? *Chemical Physics Letters*, **75**, 5–8.

172 Gupta, S.K., Atkins, R.M. and Gingerich, K.A. (1978) Mass spectrometric observation and bond dissociation energy of dimolybdenum, $Mo_2(g)$. *Inorganic Chemistry*, **17**, 3211–3213.

173 Gingerich, K.A. (1980) Experimental and Predicted Stability of Diatomic Metals and Metallic Clusters. *Faraday Symposia of the Chemical Society*, **14**, 109–125.

174 Wu, Z.J. (2005) Theoretical study of transition metal dimer AuM (M = 3d, 4d, 5d element). *Chemical Physics Letters*, **406**, 24–28.

175 Abe, M., Mori, S., Nakajima, T. and Hirao, K. (2005) Electronic structures of PtCu, PtAg, and PtAu molecules: a Dirac four-component relativistic study. *Chemical Physics*, **311**, 129–137.

176 Gingerich, K.A. (1980) *Current Topics in Materials Science*, Vol. 6 (ed. E. Kaldis), North-Holland, New York, pp. 345.

177 Fossgaard, O., Gropen, O., Eliav, E. and Saue, T. (2003) Bonding in the homologous series CsAu, CsAg, and CsCu studied at the 4-component density functional theory and coupled cluster levels. *Journal of Chemical Physics*, **119**, 9355–9363.

178 Belpassi, L., Tarantelli, F., Sgamellotti, A. and Quiney, H.M. (2006) The Electronic Structure of Alkali Aurides. A Four-Component Dirac-Kohn-Sham Study. *The Journal of Physical Chemistry A*, **110**, 4543–4554.

179 Mudring, A.-V., Jansen, M., Daniels, J., Kramer, S., Mehring, M., Prates Ramalho, J.P., Romero, A.H. and Parinello, M. (2002) Cesiumauride Ammonia (1/1), $CsAu.NH_3$: A Crystalline Analogue to Alkali Metals Dissolved in Ammonia? *Angewandte Chemie International Edition*, **41**, 120–124.

180 Wesendrup, R. and Schwerdtfeger, P. (2000) Extremely strong s^2-s^2 closed shell interactions. *Angewandte Chemie International Edition*, **39**, 907–910.

181 Pyykkö, P. (1995) Predicted chemical bonds between rare gases and Au^+. *Journal of the American Chemical Society*, **117**, 2067–2070.

182 Schwerdtfeger, P., McFeaters, J.S., Stephens, R.L., Liddell, M.J., Dolg, M. and Hess, B.A. (1994) Can AuF be synthesized? A theoretical study using relativistic configuration interaction and plasma modelling techniques. *Chemical Physics Letters*, **218**, 362–366.

183 Laerdahl, J.K., Saue, T. and Faegri, K. Jr (1997) Direct relativistic MP2: properties of ground state CuF, AgF and AuF. *Theoretica Chimica Acta*, **97**, 177–184.

184 Schröder, D., Hrušak, J., Tornieporth-Oetting, I.C., Klapötke, T.M. and Schwarz, H. (1994) Neutral Gold(I) Fluoride Does Indeed Exist. *Angewandte Chemie International Edition*, **33**, 212–214.

185 Laitar, D.S., Müller, P., Gray, T.G. and Sadighi, J.P. (2005) A Carbene-Stabilized Gold(I) Fluoride: Synthesis and Theory. *Organometallics*, **24**, 4503–4505.

186 Guichemerre, M., Chambaud, G. and Stoll, H. (2002) Electronic structure and spectroscopy of monohalides of metals of group I-B. *Chemical Physics*, **280**, 71–102.

187 Sadlej, A.J. (1991) The dipole moment of AuH. *Journal of Chemical Physics*, **95**, 2614–9622.

188 Salek, P., Helgaker, T. and Saue, T. (2005) Linear response at the 4-component relativistic density-functional level: application to the frequency-dependent dipole polarizability of Hg, AuH and PtH_2. *Chemical Physics*, **311**, 187–201.

189 Kellö, V. and Sadlej, A.J. (1991) Quadrupole moments of CuH, Ag, and AuH. A study of the electron correlation and relativistic effects. *Journal of Chemical Physics*, **95**, 8248–8253.

190 Kellö, V. and Sadlej, A.J. (1998) Picture change and calculations of expectation values in approximate relativistic theories. *International Journal of Quantum Chemistry*, **68**, 159–174.

191 Barysz, M. and Sadlej, A.J. (1997) Expectation values of operators in approximate two-component relativistic theories. *Theoretica Chimica Acta*, **97**, 260–270.

192 Pernpointner, M., Schwerdtfeger, P. and Hess, B.A. (2000) Accurate electric field gradients for the coinage metal chlorides using the PCNQM method. *International Journal of Quantum Chemistry*, **76**, 371–384.

193 Malkin, I., Malkina, O.L. and Malkin, V.G. (2002) Relativistic calculations of electric field gradients using the Douglas–Kroll method. *Chemical Physics Letters*, **361**, 231–236.

194 Pernpointner, M., Seth, M. and Schwerdtfeger, P. (1998) A Point-Charge Model for the Nuclear Quadrupole Moment. Accurate Coupled-Cluster, Dirac-Fock, Douglas-Kroll and Nonrelativistic Hartree-Fock Calculations

for the Cu and F Electric Field Gradients in CuF. *Journal of Chemical Physics*, **108**, 6722–6738.

195 Kellö, V. and Sadlej, A.J. (2000) The point charge model of nuclear quadrupoles: How and why does it work. *Journal of Chemical Physics*, **112**, 522–526.

196 Schwerdtfeger, P., Bast, R., Gerry, M.C.L., Jacob, C.R., Jansen, M., Kellö, V., Mudring, A.V., Sadlej, A.J., Saue, T., Söhnel, T. and Wagner, F.E. (2005) The quadrupole moment of the $3/2^+$ nuclear ground state of ^{197}Au from electric field gradient relativistic coupled cluster and density functional theory of small molecules and the solid state. *Journal of Chemical Physics*, **122**, 124317-1–124317-9.

197 Thierfelder, C., Schwerdtfeger, P. and Saue, T. (2007) ^{63}Cu and ^{197}Au Nuclear Quadrupole Moments from Four-Component Relativistic Density Functional Calculations using Exact Long-Range Exchange. *Physical Review A*, **76**, 034502-1–034502-4.

198 Belpassi, L., Tarantelli, F., Sgamellotti, A., Götz, A.W. and Visscher, L. (2007) An indirect approach to the determination of the nuclear quadrupole moment by four-component relativistic DFT in molecular calculations. *Chemical Physics Letters*, **442**, 233–237.

199 Belpassi, L., Tarantelli, F., Sgamellotti, A., Quiney, H.M., van Stralen, J.N.P. and Visscher, L. (2007) Nuclear electric quadrupole moment of gold. *Journal of Chemical Physics*, **126**, 064314-1–064314-7.

200 Lantto, P. and Vaara, J. (2006) Calculations of nuclear quadrupole coupling in noble gas–noble metal fluorides: Interplay of relativistic and electron correlation effects. *Journal of Chemical Physics*, **125**, 174315-1–174315-7.

201 Vaara, J. (2007) Theory and computation of nuclear magnetic resonance parameters. *Physical Chemistry Chemical Physics*, **9**, 5399–5418.

202 David, J. and Restrepo, A. (2007) Relativistic effects on the nuclear magnetic shielding in the MF (M=Cu, Ag, Au) series. *Physical Review A*, **76**, 052511-1–052511-5.

203 Kaupp, M., Malkina, O.L., Malkin, V.G. and Pyykkö, P. (1998) How do spin-orbit induced heavy-atom effects on NMR chemical shifts function? Validation of a simple analogy to spin-spin coupling by DFT calculations on some iodo compounds. *Chemistry - A European Journal*, **4**, 118–126.

204 Lo, J.M.H. and Klobukowski, M. (2007) Relativistic calculations on the ground and excited states of AgH and AuH in cylindrical harmonic confinement. *Theoretical Chemistry Accounts*, **118**, 607–622.

205 Schwerdtfeger, P., Boyd, P.D.W., Burrell, A.K., Robinson, W.T. and Taylor, M.J. (1990) Relativistic effects in gold chemistry. III. Gold(I) Complexes. *Inorganic Chemistry*, **29**, 3593–3607.

206 Mishra, S. (2007) Theoretical Calculation of the Photodetachment Spectra of XAuY$^-$ (X, Y=Cl, Br, and I). *The Journal of Physical Chemistry A*, **111**, 9164–9168.

207 Seth, M., Cooke, F., Pelissier, M., Heully, J.-L. and Schwerdtfeger, P. (1998) The Chemistry of the Superheavy Elements II. The Stability of High Oxidation States in Group 11 Elements. Relativistic Coupled Cluster Calculations for the Fluorides of Cu, Ag, Au and Element 111. *Journal of Chemical Physics*, **109**, 3935–3943.

208 Timakov, A.A., Prusakov, V.N. and Drobyshevskii, Y.V. (1986) *Doklady Akademii Nauk SSSR*, **291**, 125–128.

209 Himmel, D. and Riedel, S. (2007) After 20 Years, Theoretical Evidence That "AuF$_7$" Is Actually AuF$_5\cdot$F$_2$. *Inorganic Chemistry*, **46**, 5338–5342.

210 Riedel, S. and Kaupp, M. (2006) Has AuF$_7$ Been Made? *Inorganic Chemistry*, **45**, 1228–1234.

211 Eijnhoven, J.C.M.T. and Verschoor, G.C. (1987) Redetermination of the crystal structure of Cs$_2$AuAuCl$_6$. *Materials Research Bulletin*, **9**, 1667–1670.

212 Réffy, B., Kolonits, M., Schulz, A., Klapötke, T.M. and Hargittai, M. (2000) Intriguing Gold Trifluorides – Molecular Structure of Monomers and Dimers: An Electron Diffraction and Quantum Chemical Study. *Journal of the American Chemical Society*, **122**, 3127–3134.

213 Bayse, C.A. and Hall, M.B. (1999) Prediction of the Geometries of Simple Transition Metal Polyhydride Complexes by Symmetry Analysis. *Journal of the American Chemical Society*, **121**, 1348–1358.

214 Bayse, C.A. (2001) Interaction of Dihydrogen with Gold (I) Hydride: Prospects for Matrix-Isolation Studies. *The Journal of Physical Chemistry A*, **105**, 5902–5905.

215 Balabanov, N.B. and Boggs, J.E. (2001) Lowest Singlet and Triplet States of Copper, Silver, and Gold Trihydrides: an ab Initio Study. *The Journal of Physical Chemistry A*, **105**, 5906–5910.

216 Wang, X. and Andrews, L. (2002) Infrared Spectra and DFT Calculations for the Gold Hydrides AuH, (H_2)AuH, and the AuH_3 Transition State Stabilized in (H_2) AuH_3. *The Journal of Physical Chemistry A*, **106**, 3744–3748.

217 Scherbaum, F., Grohmann, A., Huber, B., Krüger, C. and Schmidbaur, H. (1988) "Aurophilicity" as a Consequence of Relativistic Effects: The Hexakis (triphenylphosphaneaurio)methane Dication $[(Ph_3PAu)_6C]_2^+$. *Angewandte Chemie International Edition*, **27**, 1544–1546.

218 Schmidbaur, H., Graf, W. and Müller, G. (1988) Weak Intramolecular Bonding Relationships: The Conformation-Determining Attractive Interaction between Gold(I) Centers. *Angewandte Chemie International Edition*, **27**, 417–419.

219 Jansen, M. (1987) Homoatomic d10-d10 Interactions: Their Effects on Structure and Chemical and Physical Properties. *Angewandte Chemie International Edition*, **26**, 1098–1110.

220 Pyykkö, P. (2002) Relativity, gold, closed-shell interactions and $CsAu.NH_3$. *Angewandte Chemie International Edition*, **41**, 3573–3578.

221 Hermann, H.L., Boche, G. and Schwerdtfeger, P. (2001) Metallophilic Interactions between Closed-Shell Copper(I) Molecules – A Theoretical Study. *Chemistry - A European Journal*, **7**, 5333–5342.

222 Pyykkö, P., Li, J. and Runeberg, N. (1994) Predicted ligand dependence of the Au (I)..Au(I) attraction in $(XAuPH_3)_2$. *Chemical Physics Letters*, **218**, 133–138.

223 Pyykkö, P. and Zhao, Y.-F. (1991) Ab initio calculations on the $(ClAuPH_3)_2$ dimer with relativistic pseudopotential: Is the "aurophilic attraction" a correlation effect? *Angewandte Chemie International Edition*, **30**, 604–605.

224 Li, J. and Pyykkö, P. (1992) Relativistic pseudo-potential analysis of the weak Au (I)..Au(I) attraction. *Chemical Physics Letters*, **197**, 586–590.

225 Pyykkö, P., Schneider, W., Bauer, A., Bayler, A. and Schmidbaur, H. (1997) An ab initio study of the aggregation of LAuX molecules and $[LAuL]^+[XAuX]^-$ ions. *Chemical Communications*, (12), 1111–1112.

226 Pyykkö, P., Runeberg, N. and Mendizabal, F. (1997) Theory of the d^{10}-d^{10} closed-shell attraction. I. Dimers near equilibrium. *Chemistry - A European Journal*, **3**, 1451–1457.

227 Pyykkö, P. and Mendizabal, F. (1997) Theory of the d^{10}-d^{10} closed-shell attraction. II. Long-distance behaviour and non-additive effects in dimers and trimers of type $(X-Au-L)_n$ (X=Cl, I, H; L=-PH_3, PMe_3, -NCH). *Chemistry - A European Journal*, **3**, 1458–1465.

228 Runeberg, N., Schütz, M. and Werner, H.-J. (1999) The aurophilic attraction as interpreted by local correlation methods. *Journal of Chemical Physics*, **110**, 7210–7215.

229 Magnko, L., Schweizer, M., Rauhut, G., Schütz, M., Stoll, H. and Werner, H.-J.

(2002) A comparison of metallophilic attraction in (X–M–PH$_3$)$_2$ (M=Cu, Ag, Au; X=H, Cl). *Physical Chemistry Chemical Physics*, **4**, 1006–1013.
230 Perreault, D., Drouin, M., Michel, A., Miskowski, V.M., Schaefer, W.P. and Harvey, P.D. (1992) Silver and Gold Dimers. Crystal and Molecular Structures of Ag$_2$(dmpm)$_2$Br$_2$ and [Au$_2$(dmpm)$_2$](PF$_6$)$_2$ and Relation between Metal-Metal Force Constants and Metal-Metal Separations. *Inorganic Chemistry*, **31**, 695–702.
231 Pyykkö, P. and Zaleski-Ejgierd, P. (2008) From nanostrips to nanorings: the elastic properties of gold-glued polyauronaphthyridines and polyacenes. *Physical Chemistry Chemical Physics*, **10**, 114–120.
232 O'Grady, E. and Kaltsoyannis, N. (2004) Does metallophilicity increase or decrease down group 11? Computational investigations of [Cl–M–PH$_3$]$_2$ (M=Cu, Ag, Au, [111]). *Physical Chemistry Chemical Physics*, **6**, 680–687.
233 Ray, L., Shaikh, M.M. and Ghosh, P. (2008) Shorter Argentophilic Interaction than Aurophilic Interaction in a Pair of Dimeric {(NHC)MCl}$_2$ (M=Ag, Au) Complexes Supported over a N/O-Functionalized N-Heterocyclic Carbene (NHC) Ligand. *Inorganic Chemistry*, **47**, 230–240.
234 Humphrey, S.M., Mack, H.-G., Redshaw, C., Elsegood, M.R.J., Young, K.J.H., Mayerand, H.A. and Kaskad, W.C. (2005) Variable solid state aggregations in a series of (isocyanide)gold(I) halides with the novel trimethylamine-isocyanoborane adduct. *Dalton Transactions*, (3), 439–446.
235 Schwerdtfeger, P., Krawczyk, R.P., Hammerl, A. and Brown, R. (2004) A Comparison of Structure and Stability between the Group 11 Halide Tetramers M$_4$X$_4$ (Cu, Ag, Au; X=F, Cl, Br and I) and the Group 11 Chloride and Bromide Phosphanes (XMPH$_3$)$_4$. *Inorganic Chemistry*, **43**, 6707–6716.

236 Melnik, M. and Parish, R.V. (1986) Classification and analysis of gold compounds on the basis of their x-ray structural and mössbauer spectroscopic data. *Coordination Chemistry Reviews*, **70**, 157–257.
237 Gade, L.H. (1997) Hyt was of Gold, and Shon so Bryghte...: Luminescent Gold(I) Compounds. *Angewandte Chemie International Edition*, **36**, 1171–1173.
238 Yam, V.W.-W. and Lo, K.K.-W. (1999) Luminescent polynuclear d^{10} metal complexes. *Chemical Society Reviews*, **28**, 323–334.
239 Zhang, H.-X. and Che, C.-M. (2001) Aurophilic Attraction and Luminescence of Binuclear Gold(I) Complexes with Bridging Phosphine Ligands: ab initio Study. *Chemistry - A European Journal*, **7**, 4887–4893.
240 Pan, Q.-J. and Zhang, H.-X. (2003) Aurophilic attraction and excited-state properties of binuclear Au(I) complexes with bridging phosphine and/or thiolate ligands: An ab initio study. *Journal of Chemical Physics*, **119**, 4346–4352.
241 Pan, Q.-J. and Zhang, H.-X. (2003) Ab initio Study on Luminescence and Aurophilicity of a Dinuclear [(AuPH$_3$)$_2$(i-mnt)] Complex (i-mnt isomer-Malononitriledithiolate). *European Journal of Inorganic Chemistry*, **23**, 4202–4210.
242 Forward, J.M., Bohmann, D., Fackler, J.P. Jr and Staples, R.J. (1995) Luminescence Studies of Gold(I) Thiolate Complexes. *Inorganic Chemistry*, **34**, 6330–6336.
243 Foley, J.B., Bruce, A.E. and Bruce, M.R.M. (1995) An Unprecedented Photochemical Cis to Trans Isomerization of Dinuclear Gold(I) Bis(diphenylphosphino)ethylene Complexes. *Journal of the American Chemical Society*, **117**, 9596–9597.
244 Gimeno, M.C. and Laguna, A. (2004) *Comprehensive Coordination Chemistry II*, Vol. 6 (ed. C. McCleverty), *et al.* Elsevier, Amsterdam, pp. 911–1145.
245 Parr, R.G. and Pearson, R.G. (1983) Absolute Hardness: Companion

Parameter to Absolute Electronegativity. *Journal of the American Chemical Society*, **105**, 7512–7516.

246 Hancock, R.D., Bartolotti, L.J. and Kaltsoyannis, N. (2006) Density Functional Theory-Based Prediction of Some Aqueous-Phase Chemistry of Superheavy Element 111. Roentgenium (I) Is the "Softest" Metal Ion. *Inorganic Chemistry*, **45**, 10780–10785.

247 Schröder, D., Schwarz, H., Hrušak, J. and Pyykkö, P. (1998) Cationic gold(I) complexes of xenon and of ligands containing the donor atoms oxygen, nitrogen, phosphorus, and sulfur. *Inorganic Chemistry*, **37**, 624–632.

248 Schröder, D., Hrušak, J., Hertwig, R.H., Koch, W., Schwerdtfeger, P. and Schwarz, H. (1995) Experimental and Theoretical Studies in Gold(I) Complexes Au-(L)$^+$ (L=H$_2$O, CO, NH$_3$, C$_2$H$_4$, C$_3$H$_6$, C$_4$H$_6$, C$_6$H$_6$, C$_6$F$_6$). *Organometallics*, **14**, 312–316.

249 Hrušak, J., Hertwig, R.H., Schröder, D.H., Schwerdtfeger, P., Koch, W. and Schwarz, H. (1995) Relativistic Effects in Cationic Gold(I)-Ligand Complexes: A Comparative Study of Ab Initio Pseudopotential and Density Functional Methods. *Organometallics*, **14**, 1284–1291.

250 Schwerdtfeger, P., Hermann, H.L. and Schmidbaur, H. (2003) The Stability of the Gold – Phosphine Bond. A Comparison with other Group 11 Elements. *Inorganic Chemistry*, **42**, 1334–1342.

251 Rösch, N., Görling, A., Ellis, D.E. and Schmidbaur, H. (1989) Aurophilicity as Concerted Effect: Relativistic MO Calculations on Carbon-Centered Gold Clusters. *Angewandte Chemie International Edition*, **28**, 1357–1359.

252 Görling, A., Rösch, N., Ellis, D.E. and Schmidbaur, H. (1991) Electronic structure of main-group-element-centered octahedral gold clusters. *Inorganic Chemistry*, **30**, 3986–3994.

253 Lauher, J.W. and Wald, K. (1981) Synthesis and structure of triphenylphosphinegold-dodecacarbonyltricobaltiron ([FeCo$_3$(CO)$_{12}$AuPPh$_3$]): a trimetallic trigonal-bipyramidal cluster. Gold derivatives as structural analogs of hydrides. *Journal of the American Chemical Society*, **103**, 7648–7650.

254 Slovokhotov, Yu.L. and Struchkov, Yu.T. (1984) X-ray crystal structure of a distorted tetrahedral cluster in the salt [(Ph$_3$P)$_4$Au$_4$N]$^+$BF$_4^-$. Geometrical indication of stable electronic configurations in post-transition metal complexes and the magic number 18-e in centred gold clusters. *Journal of Organometallic Chemistry*, **277**, 143–146.

255 Zeller, E., Beruda, H., Kolb, A., Bissinger, P., Riede, J. and Schmidbaur, H. (1991) Change of coordination from tetrahedral gold-ammonium to square-pyramidal gold-arsonium cations. *Nature*, **352**, 141–143.

256 Bachman, R.E. and Schmidbaur, H. (1996) Isolation and Structural Characterization of [P(AuPPh$_3$)$_5$][BF$_4$]$_2$ via Cleavage of a P-P Bond by Cationic Gold Fragments: Direct Evidence of the Structure of the Elusive Tetrakis [phosphineaurio(I)] phosphonium(+) Cation. *Inorganic Chemistry*, **35**, 1399–1401.

257 Li, J. and Pyykkö, P. (1993) Structure of E(AuPH$_3$)$_4^+$, E=N, P, As: T$_d$ or C$_{4v}$? *Inorganic Chemistry*, **32**, 2630–2634.

258 Schmidbaur, H., Gabba, F.P., Schier, A. and Riede, J. (1995) Hypercoordinate Carbon in Protonated Tetraauriomethane Molecules. *Organometallics*, **14**, 4969–4971.

259 Räisänen, M.T., Runeberg, N., Klinga, M., Nieger, M., Bolte, M., Pyykkö, P., Leskelä, M. and Repo, T. (2007) Coordination of Pyridinethiols in Gold(I) Complexes. *Inorganic Chemistry*, **46**, 9954–9960.

260 Belpassi, L., Infante, I., Tarantelli, F. and Visscher, L. (2008) The Chemical Bond between Au(I) and the Noble Gases. Comparative Study of NgAuF and NgAu$^+$ (Ng: Ar, Kr, Xe) by Density Functional and Coupled Cluster Methods. *Journal of the*

American Chemical Society, **130**, 1048–1060.
261 Ghanty, T.K. (2005) Insertion of noble-gas atom (Kr and Xe) into noble-metal molecules (AuF and AuOH): Are they stable? *Journal of Chemical Physics*, **123**, 074323-1–074323-7.
262 Zhang, P.X., Zhao, Y.F., Hao, F.g.Y. and Li, X.Y. (2008) Bonding analysis for NgAuOH (Ng=Kr, Xe). *International Journal of Quantum Chemistry*, **108**, 937–944.
263 Xin-Ying, L. and Xue, C. (2008) Ab initio study of MXe_n^+ (M = Cu, Ag, and Au; n = 1, 2). *Physical Review A*, **77**, 022508-1–022508-5.
264 Mathieson, T., Schier, A. and Schmidbaur, H. (2000) Tris[(triphenylphosphine)gold(I)] oxonium Dihydrogentrifluoride as the Product of an Attempted Preparation of [(Triphenylphosphine)gold(I)] Fluoride. *Zeitschrift für Naturforschung. Teil B: Chemie, Biochemie, Biophysik, Biologie*, **55**, 1000–1004.
265 Evans, C.J. and Gerry, M.C.L. (2000) Confirmation of the Existence of Gold(I) Fluoride, AuF: Microwave Spectrum and Structure. *Journal of the American Chemical Society*, **122**, 1560–1561.
266 Armunanto, R., Schwenk, C.F. and Rode, B.M. (2004) Gold(I) in Liquid Ammonia: Ab Initio QM/MM Molecular Dynamics Simulation. *Journal of the American Chemical Society*, **126**, 993–99354.
267 Hofer, T.S., Tran, H.T., Schwenk, C.S. and Rode, B.M. (2004) Characterization of dynamics and reactivities of solvated ions by ab initio simulations. *Journal of Computational Chemistry*, **25**, 211–217.
268 Armunanto, R., Schwenk, C.F. and Rode, B.M. (2003) Structure and Dynamics of Hydrated Ag (I): Ab Initio Quantum Mechanical-Molecular Mechanical Molecular Dynamics Simulation. *The Journal of Physical Chemistry A*, **107**, 3132–3138.
269 Armunanto, R., Schwenk, C.F., Tran, T. and Rode, B.M. (2004) Structure and Dynamics of Au^+ Ion in Aqueous Solution: Ab Initio QM/MM MD Simulations. *Journal of the American Chemical Society*, **126**, 2582–2587.
270 Reveles, J.U., Calaminici, P., Beltrán, M.R., Köster, A.M. and Khanna, S.N. (2007) H_2O Nucleation around Au^+. *Journal of the American Chemical Society*, **129**, 15565–15571.
271 Sukrat, K. and Parasuk, V. (2007) Importance of hydrogen bonds to stabilities of copper–water complexes. *Chemical Physics Letters*, **447**, 58–64.
272 Urban, M. and Sadlej, A.J. (2000) Core correlation effects in weak interactions involving transition metal atoms. *Journal of Chemical Physics*, **112**, 5–8.
273 Granatier, J., Urban, M. and Sadlej, A.J. (2007) Van der Waals Complexes of Cu, Ag, and Au with Hydrogen Sulfide. The Bonding Character. *The Journal of Physical Chemistry A*, **111**, 13238–13234.
274 Gourlaquen, C., Piquemal, J.-P., Saue, T. and Parisel, O. (2005) *Journal of Computational Chemistry*, **27**, 142–156.
275 Veldkamp, A. and Frenking, G. (1993) Theoretical studies of organometallic compounds. 6. Structures and bond energies of $M(CO)^{n+}$, MCN, and $M(CN)^{2-}$ (M = silver, gold; n = 1–3). *Organometallics*, **12**, 4613–4622.
276 Dietz, O., Rayón, V.M. and Frenking, G. (2003) Molecular Structures, Bond Energies, and Bonding Analysis of Group 11 Cyanides TM(CN) and Isocyanides TM(NC) (TM=Cu, Ag, Au). *Inorganic Chemistry*, **42** 4977–4984.
277 Tadjeddine, M. and Flament, J.P. (1999) Analysis of a nonlinear optical response of CN- ions adsorbed on metal electrode: tentative interpretation by means of ab initio molecular calculations. *Chemical Physics*, **240**, 39–50.
278 Liap, M.-S., Lu, X. and Zhang, Q.-E. (1998) Cyanide adsorbed on coinage metal electrodes: A relativistic density functional investigation. *International Journal of Quantum Chemistry*, **67**, 175–185.

279 Lee, D.-k., Lim, I.S., Lee, Y.S., Hagebaum-Reignier, D. and Jeung, G.-H. (2007) Molecular properties and potential energy surfaces of the cyanides of the groups 1 and 11 metal atoms. *Journal of Chemical Physics*, **126**, 244313-1–244313-9.

280 Kendall, R.A., Dunning, T.H. Jr and Harrison, R.J. (1992) Electron affinities of the first-row atoms revisited. Systematic basis sets and wave functions. *Journal of Chemical Physics*, **96**, 6796–6806.

281 Lupinetti, A.J., Jonas, V., Thiel, W., Strauss, S.H. and Frenking, G. (1999) Trends in Molecular Geometries and Bond Strengths of the Homoleptic d^{10} Metal Carbonyl Cations $[M(CO)_n]^{x+}$ ($M^{x+} = Cu^+, Ag^+, Au^+, Zn^{2+}, Cd^{2+}, Hg^{2+}$; $n = 1$–6): A Theoretical Study. *Chemistry - A European Journal*, **5**, 2573–2583.

282 Vaara, J., Malkina, O.L., Stoll, H., Malkin, V.G. and Kaupp, M. (2001) Study of relativistic effects on nuclear shieldings using density-functional theory and spin–orbit pseudopotentials. *Journal of Chemical Physics*, **114**, 61–71.

283 Tielens, F., Gracia, L., Polo, V. and Andrés, J. (2007) A Theoretical Study on the Electronic Structure of Au-XO$^{(0,-1,+1)}$ (X=C, N, and O) Complexes: Effect of an External Electric Field. *The Journal of Physical Chemistry A*, **111**, 13255–13263.

284 Klapötke, T.M., Krumm, B., Galvez-Ruiz, J.-C. and Nöth, H. (2005) Highly Sensitive Ammonium Tetraazidoaurates(III). *Inorganic Chemistry*, **44**, 9625–9627.

285 Beck, W., Klapötke, T.M., Klüfers, P., Kramer, G. and Rienäcker, C.M. (2001) X-Ray Crystal Structures and Quantum Chemical Calculations of Tetraphenylarsonium Tetraazidoaurate(III) and Azido (triphenylphosphine)Gold(I). *Zeitschrift für Anorganische und Allgemeine Chemie*, **627**, 1669–1674.

286 Afyon, S., Höhna, P., Armbrüster, M., Baranov, A., Wagner, F.R., Somer, M. and Kniep, R. (2006) Azidoaurates of the Alkali Metals. *Zeitschrift für Anorganische und Allgemeine Chemie*, **632**, 1671–1680.

287 Antes, I., Dapprich, S., Frenking, G. and Schwerdtfeger, P. (1996) The Stability of Group 11 ClMCO complexes (M=Cu, Ag, Au). *Inorganic Chemistry*, **35**, 2089–2096.

288 Pyykkö, P. and Runeberg, N. (2006) Comparative Theoretical Study of N-Heterocyclic Carbenes and Other Ligands Bound to Au^I. *Chemistry - An Asian Journal*, **1**, 623–628.

289 Puzzarini, C. and Peterson, K.A. (2005) Multiple bonds to gold: a theoretical investigation of XAuC (X=F, Cl, Br, I) molecules. *Chemical Physics*, **311**, 177–186.

290 Dargel, T.K., Hertwig, R.H. and Koch, W. (1999) How do coinage metal ions bind to benzene? *Molecular Physics*, **96**, 583–591.

291 Hertwig, R.H., Hrušák, J., Schröder, D., Koch, W. and Schwarz, H. (1995) The metal-ligand bond strengths in cationic gold(I) complexes. Application of approximate density functional theory. *Chemical Physics Letters*, **236**, 194–200.

292 Hsu, H.-C., Lin, F.-W., Lai, C.-C., Su, P.-H. and Yeh, C.-S. (2002) Photodissociation and theoretical studies of the Au$^+$–(C$_5$H$_5$N) complex. *New Journal of Chemistry*, **26**, 481–484.

293 Antes, I. and Frenking, G. (1995) Structure and Bonding of the Transition Metal Methyl and Phenyl Compounds MCH$_3$ and MC$_6$H$_6$ (M=Cu, Ag, Au) and M(CH$_3$)$_2$ and M(C$_6$H$_6$)$_2$ (M=Zn, Cd, Hg). *Organometallics*, **14**, 4263–4268.

294 Dias, H.V.R., Fianchini, M., Cundari, T.R. and Campana, C.F. (2008) Synthesis and Characterization of the Gold(I) Tris (ethylene) Complex [Au(C$_2$H$_4$)$_3$][SbF$_6$]. *Angewandte Chemie International Edition*, **47**, 556–559.

295 Cinellu, M.A., Minghetti, G., Cocco, F., Stoccoro, S., Zucca, A., Manassero, M. and Arca, M. (2006) Synthesis and properties of gold alkene complexes. Crystal structure of [Au(bipy^{oXyl})(η^2-CH$_2$=CHPh)](PF$_6$) and DFT calculations on the model cation [Au(bipy)(η^2-CH$_2$=CH$_2$)]$^+$. *Dalton Transactions*, (48), 5703–5716.

296 Köhler, K., Silverio, S.J., Hyla-Kryspin, I., Gleiter, R., Zsolnai, L., Driess, A., Huttner, G. and Lang, H. (1997) Trigonal-Planar-Coordinated Organogold(I) Complexes Stabilized by Organometallic 1,4-Diynes: Reaction Behavior, Structure, and Bonding. *Organometallics*, **16**, 4970–4979.

297 Bongers, N. and Krause, N. (2008) Golden Opportunities in Stereoselective Catalysis. *Angewandte Chemie International Edition*, **47**, 2178–2181.

298 Nieto-Oberhuber, C., Pérez-Galán, P., Herrero-Gómez, E., Lauterbach, T., Rodríguez, C., López, S., Bour, C., Rosellón, A., Cárdenas, D.J. and Echavarren, A.M. (2008) Gold(I)-Catalyzed Intramolecular [4+2] Cycloadditions of Arylalkynes or 1,3-Enynes with Alkenes: Scope and Mechanism. *Journal of the American Chemical Society*, **130**, 269–279.

299 Shi, F.-Q., Li, X., Xia, Y., Zhang, L. and Yu, Z.-X. (2007) DFT Study of the Mechanisms of In Water Au(I)-Catalyzed Tandem [3,3]-Rearrangement/Nazarov Reaction/[1,2]-Hydrogen Shift of Enynyl Acetates: A Proton-Transport Catalysis Strategy in the Water-Catalyzed [1,2]-Hydrogen Shift. *Journal of the American Chemical Society*, **129**, 15503–15512.

300 Kovács, G., Ujaque, G. and Lledós, A. (2008) The Reaction Mechanism of the Hydroamination of Alkenes Catalyzed by Gold(I)-Phosphine: The Role of the Counterion and the N-Nucleophile Substituents in the Proton-Transfer Step. *Journal of the American Chemical Society*, **130**, 853–864.

301 Correa, A., Marion, N., Fensterbank, L., Malacria, M., Nolan, S.P. and Cavallo, L. (2008) Golden Carousel in Catalysis: The Cationic Gold/Propargylic Ester Cycle. *Angewandte Chemie International Edition*, **47**, 718–721.

302 Li, S.-D. and Miao, C.-Q. (2005) M_5H_5X (M=Ag, Au, Pd, Pt; X=Si, Ge, P, S): Hydrometal Pentagons with D_{5h} Planar Pentacoordinate Nonmetal Centers. *The Journal of Physical Chemistry A*, **109**, 7594–7597.

303 Laguna, A. and Laguna, M. (1999) Coordination chemistry of gold(II) complexes. *Coordination Chemistry Reviews*, **193/195**, 837–856.

304 Walker, N., Wright, R., Barren, P., Murrell, J. and Stace, A. (1997) Synthesis of Au(II) Fluoro Complexes and Their Structural and Magnetic Properties. *Journal of the American Chemical Society*, **119**, 1020–1026.

305 Elder, S., Lucier, G.M., Hollander, F.J. and Bartlett, N. (2001) Synthesis of Au(II) Fluoro Complexes and Their Structural and Magnetic Properties. *Journal of the American Chemical Society*, **123**, 4223.

306 Barakat, K., Cundari, T. and Omary, M. (2003) Jahn-Teller Distortion in the Phosphorescent Excited State of Three-Coordinate Au(I) Phosphine Complexes. *Journal of the American Chemical Society*, **125**, 14228–14229.

307 Barakat, K. and Cundari, T. (2005) Chemical and photophysical properties of Au^I, Au^{II}, Au^{III}, and Au^I-dimer complexes. *Chemical Physics*, **311**, 3–11.

308 Seidel, S. and Seppelt, K. (2000) Xenon as a Complex Ligand: The Tetra Xenono Gold(II) Cation in $AuXe_4^{2+}(Sb_2F_{11}^-)_2$. *Science*, **290**, 117–118.

309 Pyykkö, P. (2000) Noblesse Oblige. *Science*, **290** 64–65.

310 Berski, S., Latajka, Z. and Andrés, J. (2002) The nature of the Au–Rg bond in the $[AuRg_4]^{2+}$ (Rg=Ar, Kr and Xe) molecules. *Chemical Physics Letters*, **356**, 483–489.

311 Larsson, S. (2004) Superconductivity in Copper, Silver, and Gold Compounds. *Chemistry - A European Journal*, **10**, 5276–5283.

312 Cao, R., Anderson, T.M., Piccoli, P.M.B., Schultz, A.J., Koetzle, T.F., Geletii, Y.V., Slonkina, E., Hedman, B., Hodgson, K.O., Hardcastle, K.I., Fang, X., Kirk, M.L., Knottenbelt, S., Kögerler, P., Musaev,

D.G., Morokuma, K., Takahashi, M. and Hill, C.L. (2007) Terminal Gold-Oxo Complexes. *Journal of the American Chemical Society*, **129**, 11118–11133.

313 Mohamed, A.A., Mayer, A.P., Abdou, H.E., Irwin, M.D., Pérez, L.M. and Fackler, J.P. Jr (2007) Dinuclear and Tetranuclear Gold–Nitrogen Complexes. Solvent Influences on Oxidation and Nuclearity of Gold Guanidinate Derivatives. *Inorganic Chemistry*, **46**, 11165–11172.

314 Crassous, J., Chardonnet, C., Saue, T. and Schwerdtfeger, P. (2005) Recent experimental and theoretical developments towards the observation of parity violation (PV) effects in molecules by spectroscopy. *Organic and Biomolecular Chemistry*, **3**, 2218–2224.

315 Bast, R. and Schwerdtfeger, P. (2003) Parity Violation Effects in the C-F Stretching Mode of Heavy Atom Containing Methyl Fluorides. *Physical Review Letters*, **91**, 23001-1–23001-3.

316 Jadzinsky, P.D., Calero, G., Ackerson, C.J., Bushnell, D.A. and Kornberg, R.D. (2007) Structure of a Thiol Monolayer-Protected Gold Nanoparticle at 1.1 Å Resolution. *Science*, **318**, 430–433.

317 Cleveland, C.L., Landman, U., Schaaff, T.G., Shafigullin, M.N., Stephens, P.W. and Whetten, R.L. (1997) Structural Evolution of Smaller Gold Nanocrystals: The Truncated Decahedral Motif. *Physical Review Letters*, **79**, 1873–1876.

318 Schwerdtfeger, P. and Boyd, P.D.W. (1992) The Role of Phosphine Ligands in Gold Cluster Chemistry. Relativistic SCF calculations on Au_2 and $Au_2(PH_3)_2$. *Inorganic Chemistry*, **31**, 327–329.

319 Bishop, K.J.M. and Grzybowski, B.A. (2007) Nanoions: Fundamental Properties and Analytical Applications of Charged Nanoparticles. *ChemPhysChem*, **8**, 2171–2176.

320 de Heer, W.A. (1993) The physics of simple metal clusters: experimental aspects and simple models. *Reviews of Modern Physics*, **65**, 611–676.

321 Michaelian, K., Rendón, N. and Garzón, I.L. (1999) Structure and energetics of Ni, Ag, and Au nanoclusters. *Physical Review B - Condensed Matter*, **60**, 2000–2010.

322 Garzón, I.L., Michaelian, K., Beltrán, M.R., Posada-Amarillas, A., Ordejón, P., Artacho, E., Sánchez-Portal, D. and Soler, J.M. (1998) Lowest Energy Structures of Gold Nanoclusters. *Physical Review Letters*, **81**, 1600–1603.

323 Soler, J.M., Beltrán, M.R., Michaelian, K., Garzón, I.L., Ordejón, P., Sánchez-Portal, D. and Artacho, E. (2000) Metallic bonding and cluster structure. *Physical Review B - Condensed Matter*, **61**, 5771–5780.

324 Chui, Y.H., Grochola, G., Snook, I.K. and Russo, S.P. (2007) Molecular dynamics investigation of the structural and thermodynamic properties of gold nanoclusters of different morphologies. *Physical Review B - Condensed Matter*, **75**, 033404-1–033404-4.

325 Järvi, T.T., Kuronen, A., Meinander, K. and Nordlund, K. (2007) Contact epitaxy by deposition of Cu, Ag, Au, Pt, and Ni nanoclusters on (100) surfaces: Size limits and mechanisms. *Physical Review B - Condensed Matter*, **75**, 115422-1–115422-9.

326 Li, T.X., Ji, Y.L., Yu, S.W. and Wang, G.H. (2000) Melting properties of noble metal clusters. *Solid State Communications*, **116**, 547–550.

327 Negreiros, F.R., Soares, E.A., de Carvalho, V.E. and Bozzolo, G. (2007) Atomistic modeling of Au-Ag nanoparticle formation. *Physical Review B - Condensed Matter*, **76**, 245432-1–245432-5.

328 Cleveland, C.L. and Luedtke, W.D. (1999) and Uzi Landman, Melting of gold clusters. *Physical Review B - Condensed Matter*, **60**, 5065–5077.

329 Grochola, G., Snook, I., Chui, D. and Russo, S.P. (2006) Exploring the effects of different immersion environments on the growth of gold nanostructures. *Molecular Simulation*, **32**, 1255–1260.

330 Grochola, G., Russo, S.P. and Snook, I. (2007) On morphologies of gold

nanoparticles grown from molecular dynamics simulation. *Journal of Chemical Physics*, **126**, 164707-1–164707-8.

331 Mendez-Villuendas, E. and Bowles, R.K. (2007) Surface Nucleation in the Freezing of Gold Nanoparticles. *Physical Review Letters*, **98**, 185503-1–185503-4.

332 Chui, Y.H., Snook, I.K. and Russo, S.P. (2007) Visualization and analysis of structural ordering during crystallization of a gold nanoparticle. *Physical Review B - Condensed Matter*, **76**, 195427-1–195427-6.

333 Hermann, A., Krawczyk, R.P., Lein, M., Schwerdtfeger, P., Hamilton, I.P. and Stewart, J.J.P. (2007) Convergence of the many-body expansion of interaction potentials: From van der Waals to covalent and metallic systems. *Physical Review A*, **76**, 013202-1–013202-10.

334 Wesendrup, R., Hunt, T. and Schwerdtfeger, P. (2000) Relativistic Coupled Cluster Calculations for Neutral and Singly Charged Au_3-Clusters. *Journal of Chemical Physics*, **112**, 9356–9362.

335 Rusakov, A.A., Rykova, E., Scuseria, G.E. and Zaitsevskii, A. (2007) Importance of spin-orbit effects on the isomerism profile of Au_3: An ab initio study. *Journal of Chemical Physics*, **127**, 164322-1–164322-5.

336 Guo, R., Balasubramanian, K., Wang, X. and Andrews, L. (2002) Infrared vibronic absorption spectrum and spin – orbit calculations of the upper spin–orbit component of the Au_3 ground state. *Journal of Chemical Physics*, **117**, 1614–1620.

337 Wang, J., Wang, G. and Zhao, J. (2002) Density-functional study of Au_n (2-20) clusters: Lowest-energy structures and electronic properties. *Physical Review B - Condensed Matter*, **66**, 035418-1–035418-6.

338 Zhao, J., Yang, J. and Hou, J.G. (2003) Theoretical study of small two-dimensional gold clusters. *Physical Review B - Condensed Matter*, **67**, 085404-1–085404-6.

339 Fernández, E.M., Soler, J.M., Garzón, I.L. and Balbás, L.C. (2004) Trends in the structure and bonding of noble metal clusters. *Physical Review B - Condensed Matter*, **70**, 165403-1–165403-14.

340 Grönbeck, H. and Broqvist, P. (2005) Comparison of the bonding in Au8 and Cu8: A density functional theory study. *Physical Review B - Condensed Matter*, **71**, 073408-1–073408-4.

341 Diefenbach, M. and Kim, K.S. (2006) Spatial Structure of Au_8: Importance of Basis Set Completeness and Geometry Relaxation. *The Journal of Physical Chemistry. B*, **110**, 21639–21642.

342 Xing, X., Yoon, B., Landman, U. and Parks, J.H. (2006) Structural evolution of Au nanoclusters: From planar to cage to tubular motifs. *Physical Review B - Condensed Matter*, **74**, 165423-1–165423-6.

343 Koskinen, P., Häkkinen, H., Seifert, G., Sanna, S., Frauenheim, Th. and Moseler, M. (2006) Density-functional based tight-binding study of small gold clusters. *New Journal of Physics*, **8**, 9–11.

344 Zhao, G., Sun, J. and Zeng, Z. (2007) Absorption spectra and electronic structures of Au_mAg_n ($m+n=8$) clusters. *Chemical Physics*, **342**, 267–274.

345 Gilb, S., Weis, P., Furche, F., Ahlrichs, R. and Kappes, M.M. (2002) Structures of small gold cluster cations ($Au_n^+ n<14$): Ion mobility measurements versus density functional calculations. *Journal of Chemical Physics*, **116**, 4094–4101.

346 Furche, F., Ahlrichs, R., Weis, P., Jacob, C., Gilb, S., Bierweiler, T. and Kappes, M.M. (2002) The structures of small gold cluster anions as determined by a combination of ion mobility measurements and density functional calculations. *Journal of Chemical Physics*, **117**, 6982–6990.

347 Häkkinen, H., Yoon, B., Landman, U., Li, X., Zhai, H.-J. and Wang, L.-S. (2003) On the Electronic and Atomic Structures of Small Au_N^- ($N=4$–14) Clusters:, A Photoelectron Spectroscopy and Density-Functional Study. *The Journal of Physical Chemistry A*, **107**, 6168–6175.

348 Koskinen, P., Häkkinen, H., Huber, B., von Issendorff, B. and Moseler, M. (2007) Liquid-Liquid Phase Coexistence in Gold Clusters: 2D or Not 2D? *Physical Review Letters*, **98**, 015701-1–015701-4.

349 Li, J., Li, X., Zhai, H.-J. and Wang, L.-S. (2003) Au20: A Tetrahedral Cluster. *Science*, **299**, 864–867.

350 Kryachko, E.S. and Remacle, F. (2007) The magic gold cluster Au_{20}. *International Journal of Quantum Chemistry*, **107**, 2922–2934.

351 Aprà, E., Ferrando, R. and Fortunelli, A. (2006) Density-functional global optimization of gold nanoclusters. *Physical Review B - Condensed Matter*, **73**, 205414-1–205414-5.

352 Fa, W., Luo, C. and Dong, J. (2005) Bulk fragment and tubelike structures of Au_N (N = 2–26). *Physical Review B - Condensed Matter*, **72**, 205428-1–205428-4.

353 Fernández, E.M., Soler, J.M. and Balbás, L.C. (2006) Planar and cagelike structures of gold clusters: Density-functional pseudopotential calculations. *Physical Review B - Condensed Matter*, **73**, 235433-1–235433-8.

354 Magyar, R.J., Mujica, V., Marquez, M. and Gonzalez, C. (2007) Density-functional study of magnetism in bare Au nanoclusters: Evidence of permanent size-dependent spin polarization without geometry relaxation. *Physical Review B - Condensed Matter*, **75**, 144421-1–144421-7.

355 Häkkinen, H., Moseler, M., Kostko, O., Morgner, N., Hoffmann, M.A. and Issendorff, B. v. (2004) Symmetry and Electronic Structure of Noble-Metal Nanoparticles and the Role of Relativity. *Physical Review Letters*, **93**, 093401-1–093401-4.

356 Häkkinen, H. and Moseler, M. (2006) 55-Atom clusters of silver and gold: Symmetry breaking by relativistic effects. *Computational Material Science*, **35**, 332–336.

357 Jalbout, A.F., Contreras-Torres, F.F., Pérez, L.A. and Garzón, I.L. (2008) Low-Symmetry Structures of Au_{32}^Z (Z = +1, 0, −1) Clusters. *The Journal of Physical Chemistry A*, **112**, 353–357.

358 Wang, J.L., Yang, M.L., Jellinek, J. and Wang, G.H. (2006) Dipole polarizabilities of medium-sized gold clusters. *Physical Review A*, **74**, 023202-1–023202-5.

359 Idrobo, J.C., Walkosz, W., Yip, S.F., Öğüt, S., Wang, J. and Jellinek, J. (2007) Static polarizabilities and optical absorption spectra of gold clusters (Au_n, n = 2–14 and 20) from first principles. *Physical Review B - Condensed Matter*, **76**, 205422-1–205422-12.

360 Wang, J.L., Jellinek, J., Zhao, J., Chen, Z.F., King, R.B. and Schleyer, P.V. (2005) Hollow Cages versus Space-Filling Structures for Medium-Sized Gold Clusters: The Spherical Aromaticity of the Au_{50} Cage. *The Journal of Physical Chemistry A*, **109**, 9265–9269.

361 Karttunen, A.J., Linnolahti, M., Pakkanen, T.A. and Pyykkö, P. (2008) Icosahedral Au_{72}: a predicted chiral and spherically aromatic golden fullerene. *Chemical Communications*, (4), 465–467.

362 Yannouleas, C. and Landman, U. (2000) Decay channels and appearance sizes of doubly anionic gold and silver clusters. *Physical Review B - Condensed Matter*, **61**, R10587–R10589.

363 Walter, M. and Häkkinen, H. (2006) A hollow tetrahedral cage of hexadecagold dianion provides a robust backbone for a tuneable sub-nanometer oxidation and reduction agent via endohedral doping. *Physical Chemistry Chemical Physics*, **8**, 5407–5411.

364 Li, X., Boggavarapu, K., Li, J., Zhai, H.-J. and Wang, L.-S. (2002) Experimental Observation and Confirmation of Icosahedral $W@Au_{12}$ and $Mo@Au_{12}$ Molecules. *Angewandte Chemie International Edition*, **41**, 4786–4789.

365 Pyykkö, P. and Runeberg, N. (2002) Icosahedral WAu_{12}: A Predicted Closed-Shell Species, Stabilized by Aurophilic Attraction and Relativity and in Accord with the 18-Electron Rule. *Angewandte*

Chemie International Edition, **41**, 2174–2176.

366 Xie, J.R.H., Cheung, C.F. and Zhao, J.J. (2006) Tuning Optical Absorption and Emission of Sub-Nanometer Gold-Caged Metal Systems M@Au_{14} by Substitutional Doping. *Journal of Computational and Theoretical Nanoscience*, **3**, 312–314.

367 Manninen, K., Pyykkö, P. and Häkkinen, H. (2005) A small spherical liquid: A DFT molecular dynamics study of WAu_{12}. *Physical Chemistry Chemical Physics*, **7**, 2208–2211.

368 Zhai, H.J., Li, J. and Wang, L.S. (2004) Icosahedral gold cage clusters: M@Au_{12}^- (M=V, Nb, and Ta). *Journal of Chemical Physics*, **121**, 8369–8374.

369 Wang, L.-M., Bulusu, S., Huang, W., Pal, R., Wang, L.-S. and Zeng, X.C. (2007) Doping the Golden Cage Au_{16}^- with Si, Ge, and Sn. *Journal of the American Chemical Society*, **129**, 15136–15137.

370 Morse, M.D., Hopkins, J.B., Langridge-Smith, P.R.R. and Smalley, R.E. (1983) Spectroscopic studies of the jet-cooled copper trimer. *Journal of Chemical Physics*, **79**, 5316–5328.

371 Cheshnovsky, O., Pettiette, C.L. and Smalley, R.E. (1988) *Ion and Cluster Spectroscopy* (ed. J.P. Maier), Elsevier, Amsterdam.

372 Jackschath, C., Rabin, I. and Schulze, W. (1992) Electron Ionisation potentials of the gas phase silver clusters, Ag_n, $n<36$. *Zeitschrift für Physik D*, **22**, 517–520.

373 Rabin, I., Schulze, W. and Jackschath, C. (1992) Electron impact ionisation potentials of gold and silver clusters, Me_n $n<22$. *Berichte Bunsengesellschaft für Physikalische Chemie*, **96**, 1200–1204.

374 Knickelbein, M.B. (1992) Electronic shell structure in the ionization potentials of copper clusters. *Chemical Physics Letters*, **192**, 129–134.

375 Krüger, S., Stener, M. and Rösch, N. (2001) Relativistic density functional study of gold coated magnetic nickel clusters. *Journal of Chemical Physics*, **114**, 5207–5215.

376 Li, X., Kiran, B., Cui, L.F. and Wang, L.S. (2005) Magnetic Properties in Transition-Metal-Doped Gold Clusters: M@Au6 (M=Ti,V,Cr). *Physical Review Letters*, **95**, 253401-1–253401-4.

377 Sahu, B.R., Maofa, G. and Kleinman, L. (2003) Density-functional study of palladium-doped small gold clusters. *Physical Review B - Condensed Matter*, **67**, 115420-1–115420-4.

378 Ge, Q., Song, C. and Wang, L. (2006) A density functional theory study of CO adsorption on Pt–Au nanoparticles. *Computational Material Science*, **35**, 247–253.

379 Janssens, E., Tanaka, H., Neukermans, S., Silverans, R.E. and Lievens, P. (2004) Electron delocalization in $Au_N X_M$, (X=Sc, Ti, Cr, Fe) clusters: A density functional theory and photofragmentation study. *Physical Review B - Condensed Matter*, **69**, 085402-1–085402-9.

380 Liu, F., Zhao, Y., Li, X. and Hao, F. (2007) Ab initio study of the structure and stability of $M_n Tl_n$ (M=Cu, Ag, Au; $n=1$, 2) clusters. *Journal of Molecular Structure (THEOCHEM)*, **809**, 189–194.

381 Bonačić-Koutecký, V., Burda, J., Mitrić, R., Ge, M., Zampella, G. and Fantucci, P. (2002) Density functional study of structural and electronic properties of bimetallic silver – gold clusters: Comparison with pure gold and silver clusters. *Journal of Chemical Physics*, **117**, 3120–3131.

382 Zhai, H.-Ji., Kiran, B. and Wang, L.-S. (2004) Observation of Au_2H^- impurity in pure gold clusters and implications for the anomalous Au-Au distances in gold nanowires. *Journal of Chemical Physics*, **121**, 8231–8236.

383 Häkkinen, H., Walter, M. and Grönbeck, H. (2006) Divide and Protect: Capping Gold Nanoclusters with Molecular Gold-Thiolate Rings. *The Journal of Physical Chemistry B*, **110**, 9927–9931.

384 Grönbeck, H., Walter, M. and Häkkinen, H. (2006) Theoretical Characterization of Cyclic Thiolated Gold Clusters. *Journal of*

the American Chemical Society, **128**, 10269–10275.

385 Genest, A., Krüger, S., Gordienko, A.B. and Rösch, N. (2004) Gold-Thiolate Clusters: A Relativistic Density Functional Study of the Model Species $Au_{13}(SR)_n$, R=H, CH_3 $n = 4, 6, 8$. *Zeitschrift für Naturforschung*, **59b**, 1585–1599.

386 Sharma, P., Singh, H., Sharma, S. and Singh, H. (2007) Binding of Gold Nanoclusters with Size-Expanded DNA Bases: A Computational Study of Structural and Electronic Properties. *Journal of Chemical Theory and Computation*, **3**, 2301–2311.

387 Dovesi, R., Saunders, V.R., Roetti, C., Orlando, R., Zicovich-Wilson, C.M., Pascale, F., Civalleri, B., Doll, K., Harrison, N.M., Bush, I.J., D'Arco, Ph. and Llunell, M. (2006) CRYSTAL06, University of Torino, Torino.

388 Singh, D.J. and Nordstrom, L. (eds) (2006) *Planewaves, Pseudopotentials, and the LAPW Method*, Springer, New York.

389 Kresse, G. and Joubert, D. (1999) From ultrasoft pseudopotentials to the projector augmented-wave method. *Physical Review B - Condensed Matter*, **59**, 1758–1775.

390 Czycholl, G. (2004) *Solid State Physics*, Springer-Verlag, Berlin.

391 Christensen, N.E. and Seraphin, B.O. (1971) Relativistic Band Calculation and the Optical Properties of Gold. *Physical Review B - Condensed Matter*, **4**, 3321–3344.

392 Loucks, T.L. (1966) Relativistic Electronic Structure in Crystals. II. Fermi Surface of Tungsten. *Physical Review*, **143**, 506–512.

393 Dronskowski, R. (2005) *Computational Chemistry of Solid State Materials*, Wiley-VCH Verlag GmbH, Weinheim.

394 Vetere, V., Monari, A., Bendazzoli, G.L., Evangelisti, S. and Paulus, B. (2008) *Journal of Chemical Physics*, **128**, 024701-1–024701-8.

395 Fulde, P. (1995) *Electron Correlations in Molecules and Solids*, Springer-Verlag, Berlin.

396 Anisimov, V.I., Zaanen, J. and Andersen, O.K. (1991) Band theory and Mott insulators: Hubbard U instead of Stoner I. *Physical Review B - Condensed Matter*, **44**, 943–954.

397 Dion, M., Rydberg, H., Schröder, E., Langreth, D.C. and Lundqvist, B.I. (2004) Van der Waals Density Functional for General Geometries. *Physical Review Letters*, **92**, 246401-1–246401-4.

398 Souvatzis, P., Delin, A. and Eriksson, O. (2006) Calculation of the equation of state of fcc Au from first principles. *Physical Review B - Condensed Matter*, **73**, 054111-1–054111-6.

399 Mahan, G.D. (2000) Many Particle Physics, in *Series: Physics of Solids and Liquids*, 3rd edn. Springer-Verlag, Berlin.

400 Hermann, A., Vest, B. and Schwerdtfeger, P. (2006) Density functional study of α-$CrCl_2$: structure, electronic and magnetic properties. *Physical Review B - Condensed Matter*, **74**, 224402-1–224402-7.

401 Romaniello, P. and de Boeij, P.L. (2005) The role of relativity in the optical response of gold within the time-dependent current-density-functional theory. *Journal of Chemical Physics*, **122**, 164303-1–164303-6.

402 Takeuchi, N., Chan, C.T. and Ho, K.M. (1989) First-principles calculations of equilibrium ground-state properties of Au and Ag. *Physical Review B - Condensed Matter*, **40**, 1565–1570.

403 Philipsen, P.H.T. and Baerends, E.J. (2000) Relativistic calculations to assess the ability of the generalized gradient approximation to reproduce trends in cohesive properties of solids. *Physical Review B - Condensed Matter*, **61**, 1773–1778.

404 Theileis, V. and Bross, H. (2000) Relativistic modified augmented plane wave method and its application to the electronic structure of gold and platinum. *Physical Review B - Condensed Matter*, **62**, 13338–13346.

405 Dewaele, A., Loubeyre, P. and Mezouar, M. (2004) Equations of state of six metals

above 94 GPa. *Physical Review B - Condensed Matter*, **70**, 094112-1–094112-8.

406 Söderlind, P. (2002) Comment on "Theoretical prediction of phase transition in gold". *Physical Review B - Condensed Matter*, **66**, 176201-1–176201-2.

407 Ahuja, R., Rekhi, S. and Johansson, B. (2001) Theoretical prediction of a phase transition in gold. *Physical Review B - Condensed Matter*, **63**, 212101-1–212101-3.

408 Kirchhoff, F., Mehl, M.J., Papanicolaou, N.I., Papaconstantopoulos, D.A. and Khan, F.S. (2001) Dynamical properties of Au from tight-binding molecular-dynamics simulations. *Physical Review B - Condensed Matter*, **63**, 195101-1–195101-7.

409 Christensen, N.E. and Kollar, J. (1983) Electronic Structure of CsAu. *Solid State Communications*, **46**, 727–730.

410 Watson, R.E. and Weinert, M. (2003) Charge transfer in gold-alkali-metal systems. *Physical Review B - Condensed Matter*, **49**, 7148–7154.

411 Wertheim, G.K., Bates, C.W. Jr and Buchanan, D.N.E. (1979) Electronic Structure of CsAu. *Solid State Communications*, **30**, 473–475.

412 Wang, L.G. and Zunger, A. (2003) Why are the $3d$-$5d$ compounds CuAu and NiPt stable, whereas the 3d-4d compounds CuAg and NiPd are not. *Physical Review B - Condensed Matter*, **67**, 092103-1–092103-4.

413 Moruzzi, V.L., Janak, J.F. and Williams, A.R. (1978) *Calculated Electronic Properties of Metals*, Pergamon, New York.

414 Söhnel, T., Hermann, H.L. and Schwerdtfeger, P. (2001) Towards the Understanding of Solid State Structures: From Cubic to Chain-Like Arrangements in Group 11 Halides. *Angewandte Chemie International Edition*, **40**, 4381–4385.

415 Söhnel, T., Hermann, H.L. and Schwerdtfeger, P. (2005) Solid State Density Functional Calculations for the Group 11 Monohalides. *The Journal of Physical Chemistry. B*, **109**, 526–531.

416 Doll, K., Pyykkö, P. and Stoll, H. (1998) Closed-shell interaction in silver and gold chlorides. *Journal of Chemical Physics*, **109**, 2339–2345.

417 Kurzydłowski, D. and Grochala, W. (2000) Elusive AuF in the Solid State As Accessed via High Pressure Comproportionation. *Chemical Communications*, (9), 1073–1075.

418 Denner, W., Schulz, H. and D'Amour, H. (1979) *Acta Crystallographica. Section A, Crystal Physics, Diffraction, Theoretical and General Crystallography*, **35**, 360.

419 Winkler, B., Pickard, C.J., Segall, M.D. and Milman, V. (2001) Density-functional study of charge disordering in $Cs_2Au(I)Au(III)Cl_6$ under pressure. *Physical Review B - Condensed Matter*, **63**, 214103-1–214103-4.

420 Shi, H., Asahi, R. and Stampfl, C. (2007) Properties of the gold oxides Au_2O_3 and Au_2O: First-principles investigation. *Physical Review B - Condensed Matter*, **75**, 205125-1–205125-8.

421 Zaleski-Ejgierd, P., Hakala, M. and Pyykkö, P. (2007) Comparison of chain versus sheet crystal structures for the cyanides MCN (M=Cu–Au) and dicarbides MC_2 (M=Be–Ba, Zn–Hg). *Physical Review B - Condensed Matter*, **76**, 094104-1–094104-9.

422 Hakala, M. and Pyykkö, P. (2006) Gold as intermolecular glue: a predicted triaurotriazine, $C_3Au_3N_3$, isomer of gold cyanide. *Chemical Communications*, (27), 2890–2892.

423 Boscoboinik, J.A., Plaisance, C., Neurock, M. and Tysoe, W.T. (2008) Monte Carlo and density functional theory analysis of the distribution of gold and palladium atoms on Au/P(111) alloys. *Physical Review B - Condensed Matter*, **77**, 045422-1–045422-6.

424 Meyer, R., Lemire, C., Shaikhutdinov, Sh.K. and Freund, H.-J. (2004) Surface Chemistry of Catalysis by Gold. *Gold Bulletin*, **37**, 72–124.

425 Chau, T.-D., de Bocarme, T.V., Kruse, N., Wang, R.L.C. and Kreuzer, H.J. (2003) Formation of neutral and charged gold carbonyls on highly facetted gold

nanostructures. *Journal of Chemical Physics*, **119**, 12605–12610.
426 Ding, X., Li, Z., Yang, J., Hou, J.G. and Zhu, Q. (2004) Adsorption energies of molecular oxygen on Au clusters. *Journal of Chemical Physics*, **120**, 9594–9600.
427 Molina, L.M. and Hammer, B. (2005) Oxygen adsorption at anionic free and supported Au clusters. *Journal of Chemical Physics*, **123**, 161104-1–161104-5.
428 Frondelius, P., Honkala, K. and Häkkinen, H. (2007) Adsorption of gold clusters on metal-supported MgO: Correlation to electron affinity of gold. *Physical Review B - Condensed Matter*, **76**, 073406-1–073406-4.
429 Walter, M., Frondelius, P., Honkala, K. and Häkkinen, H. (2007) Electronic Structure of MgO-Supported Au Clusters: Quantum Dots Probed by Scanning Tunneling Microscopy. *Physical Review Letters*, **99**, 096102-1–096102-4.
430 Häkkinen, H., Abbet, S., Sanchez, A., Heiz, U. and Landman, U. (2003) Structural, Electronic, and Impurity-DopingEffects in Nanoscale Chemistry: Supported Gold Nanoclusters. *Angewandte Chemie International Edition*, **42**, 1297–1300.
431 Schwerdtfeger, P., Lein, M., Krawczyk, R.P. and Jacob, Ch. (2008) The adsorption of CO on charged and neutral Au and Au_2. A comparison between wave-function based and density functional theory. *Journal of Chemical Physics*, **128**, 124302-1–124302-10.
432 Corma, A., Boronat, M., González, S. and Illas, F. (2007) On the activation of molecular hydrogen by gold: a theoretical approximation to the nature of potential active sites. *Chemical Communications*, (23), 3371–3373.
433 Wang, Y. and Gong, X.G. (2006) First-principles study of interaction of cluster Au_{32} with CO, H_2, and O_2. *Journal of Chemical Physics*, **125**, 124703-1–124703-12.
434 Mills, G., Gordon, M.S. and Metiu, H. (2002) The adsorption of molecular oxygen on neutral and negative Au_n clusters ($n = 2$–5). *Chemical Physics Letters*, **359**, 493–499.
435 Varganov, S.A., Olson, R.M., Gordon, M.S. and Metiu, H. (2003) The interaction of oxygen with small gold clusters. *Journal of Chemical Physics*, **119**, 2431–2537.
436 Mills, G., Gordon, M.S. and Metiu, H. (2003) A study of the reactions of molecular hydrogen with small gold clusters. *Journal of Chemical Physics*, **118**, 4198–4205.
437 Varganov, S.A., Olson, R.M., Gordon, M.S., Mills, G. and Metiu, H. (2004) A study of the reactions of molecular hydrogen with small gold clusters. *Journal of Chemical Physics*, **120**, 5169–5175.
438 Chrétien, S., Gordon, M.S. and Metiu, H. (2004) Binding of propene on small gold clusters and on Au(111): Simple rules for binding sites and relative binding energies. *Journal of Chemical Physics*, **121**, 3756–3766.
439 Kryachkoa, E.S. and Remaclea, F. (2007) The gold-ammonia bonding patter ns of neutral and charged complexes $Au_m^{0,\pm1}$-$(NH_3)_n$. I. Bonding and charge alternation. *Journal of Chemical Physics*, **127**, 194305-1–194305-11.
440 Zhao, X., Leng, Y. and Cummings, P.T. (2006) Self-Assembly of 1,4-Benzenedithiolate/Tetrahydrofuran on a Gold Surface: A Monte Carlo Simulation Study. *Langmuir*, **22**, 4116–4124.
441 Leng, Y.S., Dyer, P.J., Krstic, P.S., Harrison, R.J. and Cummings, P.T. (2007) Calibration of chemical bonding between benzenedithiolate and gold: the effects of geometry and size of gold clusters. *Molecular Physics*, **105**, 293–300.
442 Forda, M.J., Hoft, R.C. and Gale, J.D. (2006) Adsorption and dimerisation of thiol molecules on Au(111) using a Z-matrix approach in density functional theory. *Molecular Simulation*, **32**, 1219–1225.
443 Neyman, K.M., Inntam, C., Matveev, A.V., Nasluzov, V.A. and Rösch, N. (2005) Single d-Metal Atoms on F_s and F_s^+

Defects of MgO(001): A Theoretical Study across the Periodic Table. *Journal of the American Chemical Society*, **127**, 11652–11660.

444 Akola, J. and Häkkinen, H. (2006) Density functional study of gold atoms and clusters on a graphite 0001 surface with defects. *Physical Review B - Condensed Matter*, **74**, 165404-1–165404-9.

445 Yang, S., Phillips, J.M. and Ouyang, L. (2006) Density functional calculation of the geometric and electronic structure of a (1×1) and a (1×3) supported gold system: $Au/TiO_3/Mo(112)$. *Physical Review B - Condensed Matter*, **74**, 245424-1–245424-11.

446 Yoon, B., Häkkinen, H., Landman, U., Wörz, A., Antonietti, J.-M., Abbet, S., Judai, K. and Heiz, U. (2005) Charging Effects on Bonding and Catalyzed Oxidation of CO on Au_8 Clusters on MgO. *Science*, **307**, 403–407.

447 Bonačić-Koutecký, V., Mitrić, R., Bürgel, C. and Schäfer-Bung, B. (2006) Cluster properties in the regime in which each atom counts. *Computational Material Science*, **35**, 151–157.

448 Shi, H. and Stampfl, C. (2007) First-principles investigations of the structure and stability of oxygen adsorption and surface oxide formation at Au(111). *Physical Review B - Condensed Matter*, **76**, 075327-1–075327-14.

449 Okumura, M., Haruta, M., Kitagawa, Y. and Yamaguchi, K. (2007) Theoretical Study of H_2O and O_2 Adsorption on Au Small Clusters. *Gold Bulletin*, **40**, 40–44.

450 Rousseau, R. and Marx, D. (2000) The interaction of gold clusters with methanol molecules: Ab initio molecular dynamics of $Au_n^+CH_3OH$ and Au_nCH_3OH. *Journal of Chemical Physics*, **112**, 761–769.

451 Rousseau, R., Dietrich, G., Krückeberg, S., Lutzenkirchen, K., Marx, D., Schweikhard, L. and Walther, C. (1998) Probing cluster structures with sensor molecules: methanol adsorbed onto gold clusters. *Chemical Physics Letters*, **295**, 41–46.

452 Li, G.P. and Hamilton, I.P. (2006) Complexes of small neutral gold clusters and hydrogen sulphide: A theoretical study. *Chemical Physics Letters*, **420**, 474–479.

453 Krüger, D., Fuchs, H., Rousseau, R., Marx, D. and Parinello, M. (2001) Interaction of short-chain alkane thiols and thiolates with small gold clusters: Adsorption structures and energetics. *Journal of Chemical Physics*, **115**, 4776–4786.

454 Konôpka, M., Rousseau, R., Stich, I. and Marx, D. (2004) Detaching Thiolates from Copper and Gold Clusters: Which Bonds to Break? *Journal of the American Chemical Society*, **126**, 12103–12111.

455 Bürgel, C., Reilly, N.M., Johnson, G.E., Mitrić, R., Kimble, M.L., Castleman, A.W. and Bonačić-Koutecký, V. (2008) Influence of Charge State on the Mechanism of CO Oxidation on Gold Clusters. *Journal of the American Chemical Society*, **130**, 1694–1698.

456 Baker, T.A., Friend, C.M., and Kaxiras E., (2008) Nature of Cl Bonding on the Au (111) Surface: Evidence of a Mainly Covalent Interaction. *Journal of the American Chemical Society*, **130**, 3720–3721.

457 Schädel, M. (2003) *The Chemistry of Superheavy Elements*, Kluwer Academic, Dordrecht.

458 Schädel, M. (2006) Chemistry of Superheavy Elements. *Angewandte Chemie International Edition*, **45**, 368–401.

459 Eichler, R., Aksenov, N.V., Belozerov, A.V., Bozhikov, G.A., Chepigin, V.I., Dressler, R., Dmitriev, S.N., Gäggeler, H.W., Gorshkov, V.A., Haensssler, F., Itkis, M.G., Lebedev, V.Y., Laube, A., Malyshev, O.N., Oganessian, Yu.Ts., Petruschkin, O.V., Piguet, D., Rasmussen, P., Shishkin, S.V., Shutov, A.V., Svirikhin, A.I., Tereshatov, E.E., Vostokin, G.K., Wegrzecki, M. and Yeremin, A.V. (2007) Chemical characterization of element 112. *Nature*, **447**, 72–75.

460 Pershina, V., Bastug, T., Sarpe-Tudoran, C., Anton, J. and Fricke, B. (2004) Predictions of adsorption behaviour of the superheavy element 112. *Nuclear Physics A*, **734**, 200–203.

461 Sarpe-Tudoran, C., Pershina, V., Fricke, B., Anton, J., Sepp, W.-D. and Jacob, T. (2003) Adsorption of super-heavy elements on metal surfaces. *European Physical Journal D*, **24**, 65–67.

462 Kondo, Y. and Takayanagi, K. (2000) Synthesis and Characterization of Helical Multi-Shell Gold Nanowires. *Science*, **289**, 606–608.

463 Kondo, Y., Ru, Q. and Takayanagi, K. (1999) Thickness Induced Structural Phase Transition of Gold Nanofilm. *Physical Review Letters*, **82**, 751–754.

464 Skorodumova, N.V. and Simak, S.I. (2000) Spatial configurations of monoatomic gold chains. *Computational Material Science*, **17**, 178–181.

465 Häkkinen, H., Barnett, R.N., Scherbakov, A.G. and Landman, U. (2000) Nanowire Gold Chains: Formation Mechanisms and Conductance. *The Journal of Physical Chemistry B*, **104**, 9063–9066.

466 Zhou, J. and Dong, J. (2007) Vibrational properties of single-walled gold nanotubes from first principles. *Physical Review B - Condensed Matter*, **75**, 155423-1–155423-7.

467 Yang, X. and Dong, J. (2005) Geometrical and electronic structures of the (5, 3) single-walled gold nanotube from first-principles calculations. *Physical Review B - Condensed Matter*, **71**, 233403-1–233403-4.

468 da Silva, E.Z., Novaes, F.D., da Silva, A.J.R. and Fazzio, A. (2004) Theoretical study of the formation, evolution, and breaking of gold nanowires. *Physical Review B - Condensed Matter*, **69**, 115411-1–115411-11.

469 Skorodumova, N.V. and Simak, S.I. (2003) Stability of gold nanowires at large Au-Au separations. *Physical Review B - Condensed Matter*, **67**, R121404-1–R121404-4.

470 Krüger, D., Fuchs, H., Rousseau, R., Marx, D. and Parinello, M. (2002) Pulling Monatomic Gold Wires with Single Molecules: An Ab Initio Simulation. *Physical Review Letters*, **89**, 186402-1–186402-4.

471 Krüger, D., Rousseau, R., Fuchs, H. and Marx, D. (2003) Towards "Mechanochemistry": Mechanically Induced Isomerizations of Thiolate–Gold Clusters. *Angewandte Chemie International Edition*, **42**, 2251–2253.

472 Landman, U., Luedtke, W.D., Salisbury, B.E. and Whetten, R.L. (1996) Reversible Manipulations of Room Temperature Mechanical and Quantum Transport Properties in Nanowire Junctions. *Physical Review Letters*, **77**, 1362–1365.

473 Barnett, R.N., Häkkinen, H., Scherbakov, A.G. and Landman, U. (2004) Hydrogen Welding and Hydrogen Switches in a Monatomic Gold Nanowire. *Nano Letters*, **4**, 1845–1852.

474 Pu, Q., Leng, Y., Tsetseris, L., Park, H.S., Pantelides, S.T. and Cummings, P.T. (2007) Molecular dynamics simulations of stretched gold nanowires: The relative utility of different semiempirical potentials. *Journal of Chemical Physics*, **126**, 144707-1–144707-6.

475 Li, C., Pobelov, I., Wandlowski, T., Bagrets, A., Arnold, A. and Evers, F. (2008) Charge Transport in Single Au | Alkanedithiol | Au Junctions: Coordination Geometries and Conformational Degrees of Freedom. *Journal of the American Chemical Society*, **130**, 318–326.

476 Sun, Q., Wang, Q., Jena, P., Note, R., Yu, J.-Z. and Kawazoe, Y. (2004) Metastability of a gold nanoring: Density-functional calculations. *Physical Review B - Condensed Matter*, **70**, 245411-1–245411-5.

477 Pyykkö, P., Hakala, M.O. and Zaleski-Ejgierd, P. (2007) Gold as intermolecular glue: a theoretical study of nanostrips based on quinoline-type monomers. *Physical Chemistry Chemical Physics*, **9**, 3025–3030.

478 Skorodumova, N.V., Simak, S.I., Kochetov, A.E. and Johansson, B. (2007)

Ab initio study of electronic and structural properties of gold nanowires with light-element impurities. *Physical Review B - Condensed Matter*, **75**, 235440-1–235440-4.

479 Geng, W.T. and Kim, K.S. (2003) Linear monatomic wires stabilized by alloying: *Ab initio* density functional calculations. *Physical Review B - Condensed Matter*, **67**, 233403-1–233403-4.

480 Ning, J., Li, R., Shen, X., Qian, Z., Hou, S., Rocha, A.R. and Sanvito, S. (2007) First-principles calculation on the zero-bias conductance of a gold/1,4-diamino-benzene/gold molecular junction. *Nanotechnology*, **18**, 345203-1–345203-6.

481 Zhang, L., Fang, Y. and Zhang, P. (2008) Experimental and DFT theoretical studies of SERS effect on gold nanowires array. *Chemical Physics Letters*, **451**, 102–105.

5
Luminescence and Photophysics of Gold Complexes
Chi-Ming Che and Siu-Wai Lai

5.1
Introduction

The photophysical properties of closed-shell d^{10} gold(I) complexes have been widely investigated over several decades, particularly regarding correlation of the intriguing emission properties of polynuclear gold(I) complexes with aurophilic attraction [1]. The first report on the photoluminescence of $[Au_2(\mu\text{-dppm})_2]^{2+}$ (dppm = bis(diphenylphosphino)methane) was independently documented by Fackler [2] and Che [3], and the strong and long-lived phosphorescence at $\lambda_{max} = 593$ nm ($\tau = 21$ μs, $\phi = 0.31$) in solution [2] was attributed to the Au–Au bonded excited state. Subsequently, polynuclear gold(I) phosphine complexes involving Au(I)···Au(I) interactions have been extensively studied and were found to display long-lived emissions in the visible region that are usually assigned to Au–Au bonded excited states [4, 5].

In general, Au(I)–Au(I) interactions would modify the 5d → 6s/6p transition to a lower-energy 5dσ* → 6pσ transition, where 5dσ* and 6pσ refer to the antibonding combination of $5d_{z^2}$ and the bonding combination of 6s/6p$_z$ orbitals, respectively, and the Au–Au axis is defined as the z axis [6]. Assignments for the 5dσ* → 6pσ transitions of the $[Au_2(\text{diphosphine})_2]^{2+}$ and $[Au_2(\text{diphosphine})_3]^{2+}$ systems have received support from resonance Raman experiments [7]. The ^{3}[6pσ, 5dσ*] excited states of $[Au_2(\text{diphosphine})_2]^{2+}$ system were found to be powerful reductants [8], and can mediate light-induced C–X bond cleavage from benzyl halides [9].

The discriminatory emission properties between two-coordinate d^{10} gold(I) complexes and their respective three-coordinate counterparts have been demonstrated in the literature [6, 10–13]. As discussed in the later sections, Che and coworkers have rationalized that the extraordinarily large Stokes shift of the visible emission of $[Au_2(\text{diphosphine})_2]^{2+}$ from the [5dσ* → 6pσ] transition is due to the exciplex formation of the excited state with solvent or counterions [6]. Fackler [14–16] reported the photophysical properties of monomeric $[AuL_3]^+$ complexes, which show visible luminescence with large Stokes shifts (typically 10 000 cm^{-1}), suggesting significant excited-state distortion. Gray *et al.* [10] examined the spectroscopic properties of

Gold Chemistry: Applications and Future Directions in the Life Sciences. Edited by Fabian Mohr
Copyright © 2009 WILEY-VCH Verlag GmbH & Co. KGaA, Weinheim
ISBN: 978-3-527-32086-8

the three-coordinate [Au$_2$(dcpe)$_3$](PF$_6$)$_2$ (dcpe = bis(dicyclohexylphosphino)ethane) complex without any Au(I)–Au(I) interaction, and concluded that upon electronic excitation, shortening of the Au–P bonds results in stabilization of the $^3[(5d_{x^2-y^2}, 5d_{xy})(6p_z)]$ excited state, which consequently leads to an intense emission at λ_{max} 508 nm ($\tau = 21.1$ μs, $\phi = 0.80$) with large Stokes shift from the respective absorption maximum. Omary and coworkers [17] concluded through computational calculations that the excited-state distortion of luminescent, three-coordinate mononuclear [Au(PR$_3$)$_3$]$^+$ complexes is due to Jahn–Teller symmetry change from the trigonal planar ground-state geometry to the excited-state T-shaped structure. Eisenberg and coworkers explored the reversible change in photoluminescence characteristics between two- and three-coordinate gold(I) complexes as a signaling mechanism for the detection of volatile organic compounds [18]. Dramatic color change and enhanced luminescence were induced upon exposure of linear chain gold(I) dimer compounds to solvent vapor. *It is apparent that perturbation of the coordination environment at two-coordinate Au(I) site(s) is the key factor which determines the diverse and complex luminescent properties of gold(I) complexes.*

5.2
Spectroscopic Properties of Gold(I) Complexes

5.2.1
Mononuclear Gold(I) Complexes

The photoluminescence properties of discrete, mononuclear two-coordinate gold(I) complexes mainly originate from electronic transitions of the following: (i) intraligand, (ii) metal-to-ligand charge transfer, (iii) ligand-to-metal charge transfer, (iv) ligand-to-ligand charge transfer, all of which are affected by ligand coordination at Au(I) [19–21]. Their emission properties are sensitively perturbed by aurophilic interaction and/or additional ligand coordination to afford three-coordinate Au(I) species. Fackler and Schmidbaur reported a change in intermolecular Au···Au separation upon protonation of [(TPA)AuCl] (TPA = 1,3,5-triaza-7-phosphaadamantane), which in turn alters the emission properties [22]. In the solid form, the Au···Au distance in [(TPA)AuCl] is 3.092(1) Å (Figure 5.1), which lengthens to 3.322 (1) Å in the protonated species, [(TPAH)AuCl]Cl. Both unprotonated and protonated species are non-emissive in solution at room temperature. At 77 K, solid [(TPAH)AuCl]Cl luminesces yellow ($\lambda_{max} = 596$ nm), while the unprotonated species having a shorter Au···Au separation intensely luminesces red with $\lambda_{max} = 674$ nm (see Figure 5.2 for temperature-dependent emission of [(TPA)AuCl]). The red shift of the emission band of [(TPA)AuCl] apparently coincides with increased aurophilic interaction between molecules, which destabilizes the HOMO and decrease the HOMO–LUMO gap [23].

Besides intermolecular Au···Au interactions, visible metal-centered emission of mononuclear Au(I) complexes is usually associated with trigonal, non-centrosymmetric coordination at Au(I). For example, mononuclear two-coordinate bis

5.2 Spectroscopic Properties of Gold(I) Complexes

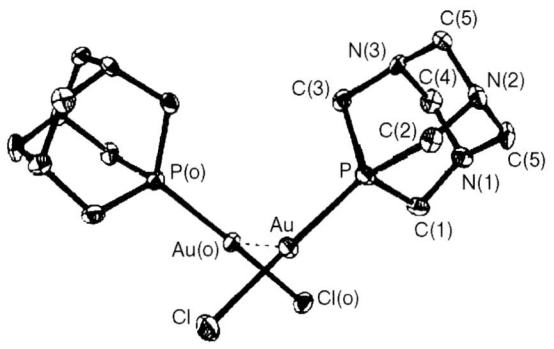

Figure 5.1 Thermal ellipsoid drawing of [(TPA)AuCl]·CH$_3$CN in 50% probability. Reproduced with permission from [22]. Copyright (1995) American Chemical Society.

(phosphine) gold(I) complexes, [Au(PR$_3$)$_2$]$^+$ (R = Ph, ^{n}Bu) were reported to be non- or weakly emissive; no emission was observed from [Au(P(^{n}Bu)$_3$)$_2$]BPh$_4$, and the weak emission from single crystals of [Au(PPh$_3$)$_2$]PF$_6$ is attributed to intraligand $^1(\pi\pi^*)$ fluorescence of the aryl groups. Addition of excess PPh$_3$ to [Au(PPh$_3$)$_2$]PF$_6$ in acetonitrile resulted in intense yellow luminescence at 512 nm with lifetime of 10 µs (Figure 5.3) [14]. The absence of any intermolecular Au···Au interaction, as deduced from the X-ray crystal structure of [Au(PPh$_3$)$_3$]BPh$_4$, precludes assignment of the emission to metal–metal bonded excited state. Three-coordinate Au(I) with loss of centrosymmetric structure, as compared to linear two-coordinate Au(I), has increased absorptivity for the (5d,6s) → 6p electronic transition. This favors

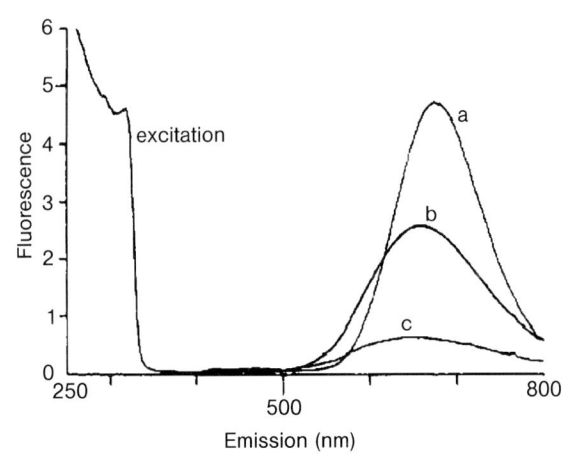

Figure 5.2 Temperature-dependent emission spectra of [(TPA)AuCl]: (a) 78 K; (b) 200 K; (c) 250 K. Emission is quenched at room temperature. The excitation spectrum was recorded at 78 K. Reproduced with permission from [22]. Copyright (1995) American Chemical Society.

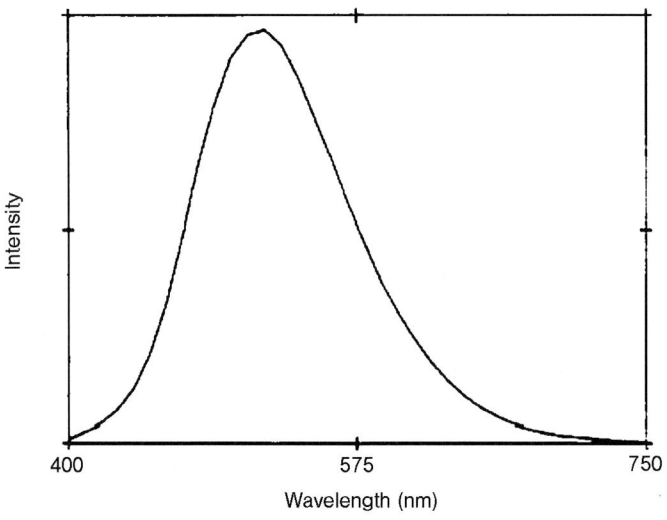

Figure 5.3 Room-temperature emission spectrum of [Au(PPh$_3$)$_3$]$^+$ in acetonitrile. Reproduced with permission from [14]. Copyright (1992) American Chemical Society.

the formation of metal-centered $^3[(5d_{x^2-y^2}, 5d_{xy})(6p_z)]$ excited state, which is responsible for the intense visible luminescence. The three-coordinate water-soluble [Au(TPA)$_3$]Cl complex was synthesized and is luminescent both in aqueous solution (547 nm) and in the solid state (533 nm) [15]. The long-lived and large Stokes shift (~2000–3000 cm^{-1}) for [Au(TPA)$_3$]Cl was attributed to a triplet metal-centered [6p$_z$ → (5d$_{x^2-y^2}$, 5d$_{xy}$)] transition.

Laguna [24] reported that molecules of [Au(PPh$_2$C≡CH)I] aggregate into dimers with gold–gold contacts of 3.0625(9) Å. A solid sample of [Au(PPh$_2$C≡CH)I] shows an emission maximum at 530 nm with excitation maximum at ~300 nm, whereas an emission at λ_{max} ~ 530 nm with a shoulder at 483 nm is observed when the excitation maximum is changed to 335 nm. The higher energy emission at 483 nm was attributed to the phosphine intraligand transition(s), and the lower-energy emission to electronic excited states with metal–metal interactions. Dependence of the emission properties of the [Au(TPA)Br] solid on excitation wavelength has been observed; the low-energy emission at 647 nm dominates at 78 K when the excitation wavelength is 320 nm, whereas a high-energy, structured emission centered at 450 nm is detected upon excitation at 340 nm [23a]. The former was assigned to a metal-centered triplet excited state, and the latter to a singlet halide-to-metal charge transfer excited state.

5.2.2
Di- and Polynuclear Gold(I) Complexes with Metal–Metal Interactions

Since the first report on the photoluminescence of [Au$_2$(μ-dppm)$_2$]$^{2+}$ [2, 3], di- and polynuclear gold(I) phosphine complexes have received considerable attention, particularly those displaying close Au(I)···Au(I) contacts and long-lived emissions in the visible region. Che and coworkers showed that the short gold(I)···gold(I) distances (2.9389(9)–3.0756(6) Å) of $[Au_2(\mu\text{-dcpm})_2]^{2+}$ (dcpm = bis(dicyclohexylphosphino)methane) salts may not necessarily play a decisive role in determining the emission energy, as the auxiliary counter anion or solvent can dramatically affect the photophysical properties [6a,b]. Similar observations for doubly-bridged dinuclear gold(I) derivatives containing an unsymmetric diphosphine ligand, $[Au_2X_2(\mu\text{-P}^iPr_2CH_2PPh_2)_2]$ (X = Cl$^-$, Br$^-$, I$^-$) and $[Au_2(\mu\text{-P}^iPr_2CH_2PPh_2)_2]A_2$ (A = CF$_3$SO$_3^-$, ClO$_4^-$) were reported by Laguna [25].

The crystal structures of $[Au_2(\mu\text{-dppm})_2]^{2+}$ and $[Au_2(\mu\text{-dcpm})_2]^{2+}$ complexes reveal close Au(I)···Au(I) separations of 2.8 – 3.1 Å [2, 3, 6a, b, 26], which signify weak metal–metal interactions. These $[Au_2(P\cap P)_2]^{2+}$ complexes (P$\cap$P = dppm, dcpm) show an intense absorption band at 270–310 nm, which are assigned to 5dσ^* → 6pσ transitions. The 5dσ^* → 6pσ assignment of the 277-nm absorption band of $[Au_2(\mu\text{-dcpm})_2]^{2+}$ was supported by resonance Raman experiments, which revealed a fundamental Au–Au stretch vibrational frequency of 88 cm^{-1} with overtones using 282.4 nm excitation (Figure 5.4) [7a]. The complex $[Au_2(dmpm)_3]^{2+}$ (dmpm = bis(dimethylphosphino)methane) also exhibits an intense absorption band at 256 nm, attributable to 5dσ^* → 6pσ transition. This assignment has also been supported by resonance Raman experiments [7b], revealing the intense fundamental (and overtones) of a 79 cm^{-1} vibrational mode assigned to Au–Au stretching frequencies using 252.7 nm excitation. For $[Au_2(\mu\text{-dmpp})_2]^{2+}$ with the

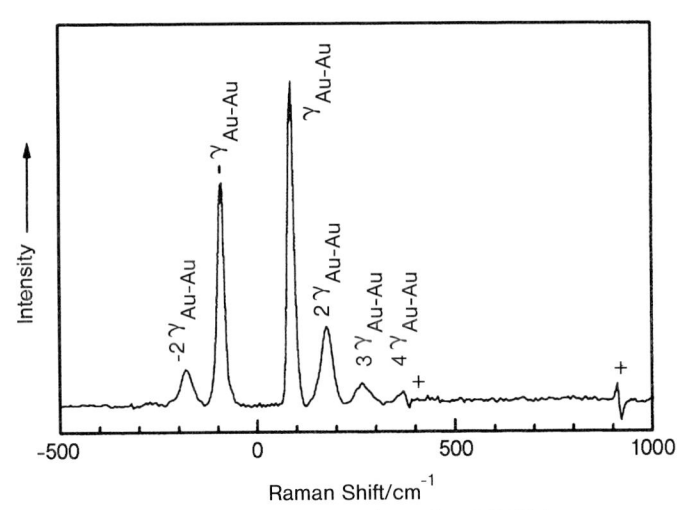

Figure 5.4 Resonance Raman spectrum of $[Au_2(dcpm)_2](ClO_4)_2$ in acetonitrile solution at room temperature taken with 282.4 nm excitation, after intensity corrections and subtractions of the Rayleigh line, glass bands, and solvent bands. Reproduced with permission from [7a]. Copyright (1999) American Chemical Society.

dimethylphosphinopyridine (dmpp) ligand, the Au···Au distance is 2.776 Å [27]. The complex [Au$_2$(μ-dppm)$_2$](ClO$_4$)$_2$ displays luminescence with λ_{max} at 575 nm in acetonitrile [2, 3, 9, 28]. Importantly, this emission is substantially red-shifted relative to the spin- and dipole-allowed ^{1}(5dσ* → 6pσ) electronic transition at 290 nm, with an apparent Stokes shift of 17 100 cm^{-1} (2.1 eV). This is an extraordinarily large Stokes shift, particularly when compared to those (6000–8000 cm^{-1}) reported for the phosphorescence of binuclear d^8–d^8 compounds relative to their ^{1}(ndσ* → (n+1)pσ) absorptions [29]. The 575-nm emission of [Au$_2$(μ-dppm)$_2$]$^{2+}$ was assigned to originate from the ^{3}[5dσ*, 6pσ] excited state.

Che reported the synthesis, crystal structures and photoluminescence of [Au$_2$(μ-dcpm)$_2$]X$_2$ (X = ClO$_4^-$, PF$_6^-$, CF$_3$SO$_3^-$, Au(CN)$_2^-$, Cl$^-$, SCN$^-$, I$^-$) [6a, b]. The bridging dcpm ligand was chosen since its intraligand transition(s) occurs at a much higher energy than the 5dσ* → 6pσ transition, and there would be no intraligand ππ* emission in the 400–700 nm spectral region. As determined by X-ray crystallography, the Au–Au distances in these Au(I) complexes (X = ClO$_4^-$: 2.9389(9), Au(CN)$_2^-$: 2.9876(5), Cl$^-$: 2.9925(2), SCN$^-$: 2.9821(7), I$^-$: 3.0756(6) Å) are within the range expected for weak Au–Au interactions [30]. For [Au$_2$(μ-dcpm)$_2$]Cl$_2$ (Figure 5.5), one of the two gold atoms engages in a close Au–Cl contact (2.7755(9) Å), which is shorter than other Au–anion contacts in the series (Au···I = 2.9960(7); Au···SCN = 3.011(3);

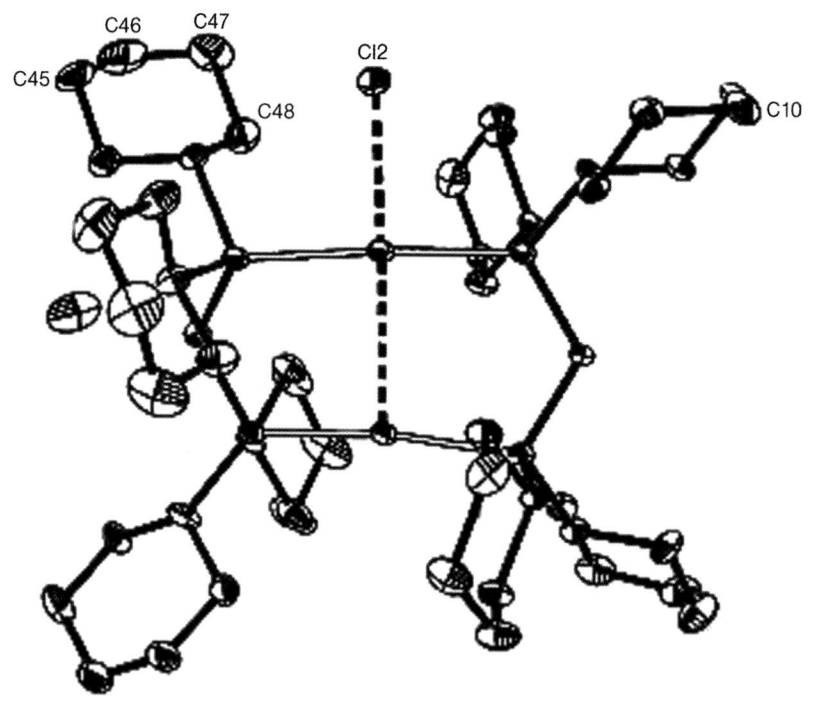

Figure 5.5 Perspective view of cation in [Au$_2$(μ-dcpm)$_2$]Cl$_2$. Reproduced with permission from [6b]. Copyright (2001) Wiley-VCH.

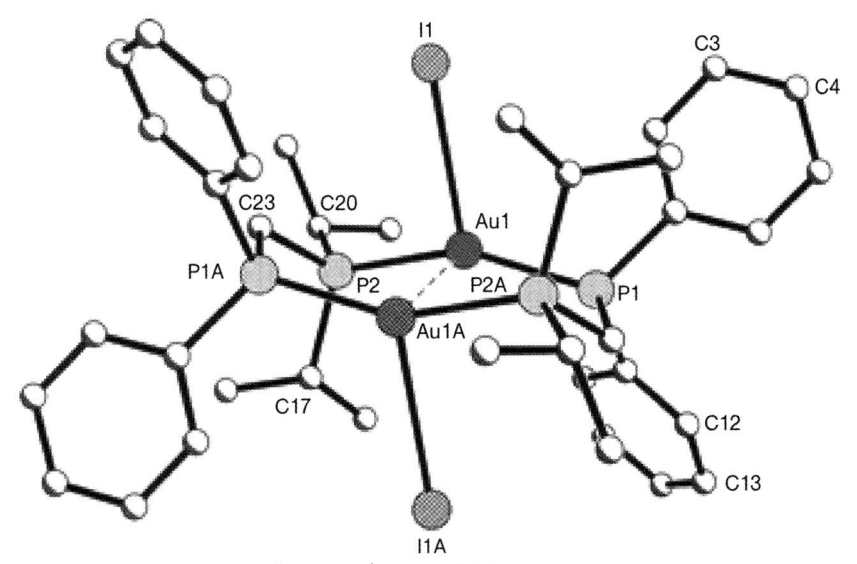

Figure 5.6 Perspective view of [Au$_2$I$_2$(μ-P^{i}Pr$_2$CH$_2$PPh$_2$)$_2$]. Reproduced with permission from [25]. Copyright (2003) Royal Society of Chemistry.

Au···Au(CN)$_2$ = 3.33(1); Au···OClO$_3$ = 3.36(2) Å. As discussed in Section 5.2.3, variations in gold–anion contacts have a significant impact upon the solid-state emission of [Au$_2$(μ-dcpm)$_2$]X$_2$ complexes.

The crystal structure of the dinuclear gold(I) iodide complex with an unsymmetric diphosphine ligand, [Au$_2$I$_2$(μ-P^{i}Pr$_2$CH$_2$PPh$_2$)$_2$] displays three-coordinate gold(I) ions in a T-shaped geometry, with a Au(I)–Au(I) distance of 2.9931(6) Å and Au–I distance of 3.0999(6) Å (Figure 5.6). The triflate derivative, [Au$_2$(μ-P^{i}Pr$_2$CH$_2$PPh$_2$)$_2$](CF$_3$SO$_3$)$_2$ contains two-coordinate gold centers and a gold–gold distance of 2.9838(5) Å [25]. They are intensely photoluminescent in the visible region (λ_{max} = 459–513 nm) at both room and low temperature in the solid state, but there is apparently no correlation between emission energies and gold–gold distances. The emission maximum of the [Au$_2$(μ-P^{i}Pr$_2$CH$_2$PPh$_2$)$_2$](CF$_3$SO$_3$)$_2$ solid featuring non-coordinating CF$_3$SO$_3$$^-$ anions occurs at 482 and 459 nm at 298 and 77 K respectively. These spectroscopic data are similar to those of the perchlorate analog. However, the emission λ_{max} of [Au$_2$I$_2$(μ-P^{i}Pr$_2$CH$_2$PPh$_2$)$_2$] with coordinated iodide ligands is red-shifted to 513 (298 K) and 499 nm (77 K) respectively. Indeed, the room temperature 513-nm emission of the [Au$_2$I$_2$(μ-P^{i}Pr$_2$CH$_2$PPh$_2$)$_2$] is similar to the 558-nm emission of [Au$_2$(dmpm)$_3$]$^{2+}$ salt having a comparable Au–Au distance of 3.050(1) Å, the emissive excited state of which was assigned to ^{3}[5dσ*, 6pσ] [7b].

A series of dinuclear gold(I)-carbene complexes of imidazolium-linked cyclophanes and related acyclic bis(imidazolium) salts have been synthesized and their spectroscopic properties were examined by Baker and coworkers [31]. X-ray structural analysis of the cation in **1** and **2** (Scheme 5.1) revealed intramolecular Au···Au contacts of 3.5425(6) and 3.0485(3) Å respectively. The electronic absorption

Scheme 5.1

spectrum of **2** in aqueous solution reveals an intense band near 200 nm with a prominent shoulder near 220 nm, which comes from intraligand transitions of the arene groups. Two additional prominent absorption bands were observed for **2** at 259 and 308 nm (Figure 5.7). Both **1** and **2** are luminous in the solid state, but only **2** is emissive in solution at room temperature. The emission spectrum of **2** (λ_{ex} = 260 nm) shows a sharp, high-energy emission band at 400 nm, and a less intense and broad low-energy emission band at 780 nm (Figure 5.7). The excitation spectra of both high- and low-energy emission bands are almost identical and resemble the absorption spectrum. Photoluminescence of **2** in both the solid state

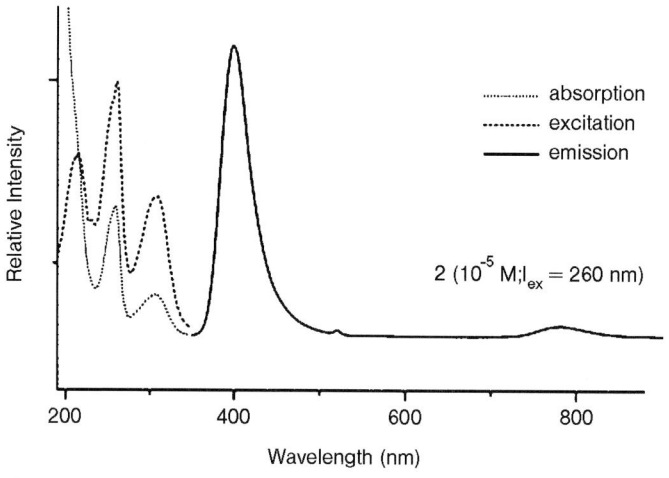

Figure 5.7 Room-temperature electronic absorption, excitation and emission spectra for **2** in aqueous solution. The excitation spectrum of **2** was recorded by monitoring emission at 400 nm. Reproduced with permission from [31]. Copyright (2004) Royal Society of Chemistry.

and solution are strongly dependent on the complex anion. Complex **2** with the shortest Au···Au contact in the series was found to emit intensely at 400 nm, whereas the other complexes with longer Au···Au distances are non-emissive in solutions at room temperature. The 400-nm emission of **2** could be assigned to $^3[5d\sigma^*, 6p\sigma]$ excited state, although perturbation of this excited state due to exciplex formation cannot be excluded.

Intermolecular aurophilic interactions usually lead to aggregation of gold(I) complexes with close gold(I)···gold(I) contacts, to afford polynuclear and supramolecular assemblies with novel structural features and intriguing luminescent properties [32]. A two-dimensional array of molecules of luminescent polynuclear gold(I) complexes with Au(I)···Au(I) distances of 2.96–3.13 Å has been reported [33]. A one-dimensional gold(I) polymer, $[\{Au_2(2,6\text{-bis(diphenylphosphino)pyridine})(C\equiv CPh)_2\}_\infty]$ with intermolecular Au(I)···Au(I) separations of 3.252(1) Å was reported to be strongly emissive with λ_{max} at 500 nm in the solid state at room temperature [34].

The luminescent trinuclear gold(I) complex $[(8\text{-QNS})_2Au(AuPPh_3)_2]BF_4$ (8-QNS = quinoline-8-thiolate) (Figure 5.8a), with intramolecular gold(I)···gold(I) distances of 3.0952(4) and 3.0526(3) Å, was reported to form the novel hexanuclear supermolecule $\{[(8\text{-QNS})_2Au(AuPPh_3)_2]\}_2(BF_4)_2$, featuring close intermolecular gold(I)···gold(I) distances of 3.1135(3) Å (Figure 5.8b) [35]. Solvent dependence of the emission for the dinuclear complex $[(AuPPh_3)_2(8\text{-QNS})]BF_4$ was previously communicated [36], and $[(8\text{-QNS})_2Au(AuPPh_3)_2]BF_4$ emits with peak maximum at about 440 and 636 nm in CH_2Cl_2, but only the high-energy emission at about 450 nm is observed in CH_3CN. The long-lived emission at 636 nm (16.2 µs) in CH_2Cl_2 solution is quenched by CH_3CN and CH_3OH (Figure 5.9) with quenching rate constants of 1.00×10^5 and $3.03 \times 10^4 \, s^{-1} \, M^{-1}$, respectively. Disruption of gold(I)···gold(I) interactions due to scrambling of $[AuPPh_3]^+$ units was suggested to account for these observations.

Catalano reported the luminescent two-coordinate Au(I) dimer, $[Au_2(dpim)_2]^{2+}$ (dpim = 2-(diphenylphosphino)-1-methylimidazole) [37]. Upon initial formation of $[Au_2(dpim)_2]^{2+}$, the presence of a small amount of Cl^- impurity was found to lead to an orange emission ($\lambda_{max} = 548$ nm, $\lambda_{ex} = 336$ nm) (solid lines in Figure 5.10). However, upon crushing, heating, or recrystallization, the complex luminesces blue ($\lambda_{max} = 483$ nm, $\lambda_{ex} = 368$ nm) (dashed lines in Figure 5.10). The crystal structure of $[Au_2(dpim)_2](BF_4)_2 \cdot 2CH_3CN$ reveals intramolecular aurophilic interaction, with a Au(I)–Au(I) distance of 2.8174(10) Å. The dependence of the emission properties of $[Au_2(dpim)_2]^{2+}$ upon the counter anion resembles the findings for $[Au_2(dcpm)_2]^{2+}$ complexes reported by Che and coworkers [6a]. These results are consistent with the hypothesis proposed by Che, that the luminescent behavior of Au(I)–Au(I) dimers can be significantly affected by exciplex formation with anions or solvent in the local environment. Thus, the unusual orange/blue emission by $[Au_2(dpim)_2]^{2+}$ is likely a result of the inclusion/exclusion of chloride ions in the crystal structure.

The Au–Au distances of $[Au_3(\mu\text{-dpmp})_2Cl_2]Cl$ (dpmp = bis-(diphenylphosphinomethyl)phenylphosphine) are 2.946(3) and 2.963(3) Å, which are shorter than those (3.0137(8) and 3.0049(8) Å) in $[Au_3(\mu\text{-dpmp})_2]^{3+}$ [38]. The intramolecular Au–Au

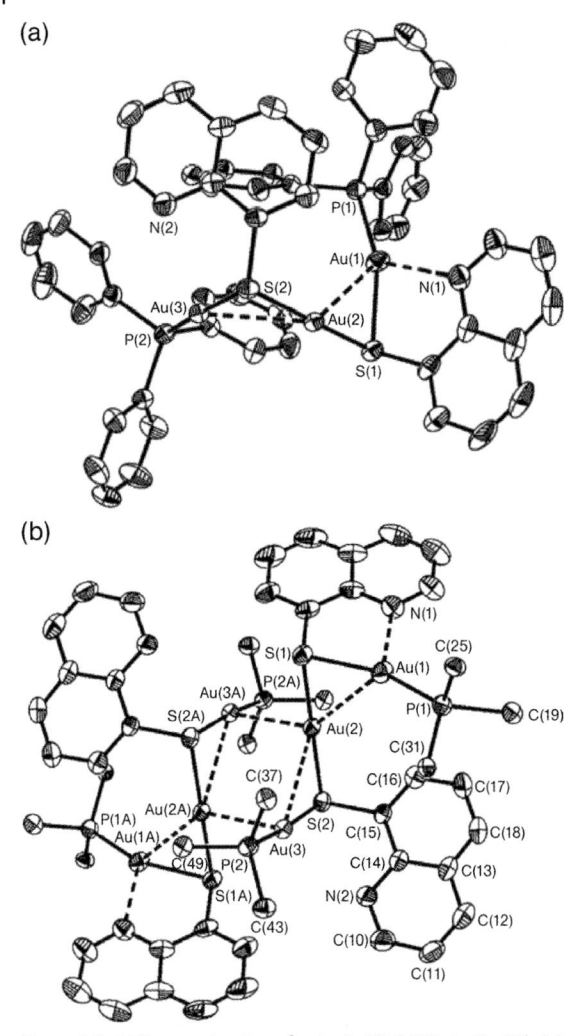

Figure 5.8 (a) Perspective view of cation in [(8-QNS)$_2$Au(AuPPh$_3$)$_2$]BF$_4$ (Au(1)···Au(2) 3.0952(4) Å, Au(2)···Au(3) 3.0526(3) Å). (b) Dimeric aggregate leading to the formation of hexanuclear gold(I) supermolecule {[(8-QNS)$_2$Au(AuPPh$_3$)$_2$]}$_2$(BF$_4$)$_2$ with intermolecular Au(2)···Au(3A) contact of 3.1135(3) Å. Reproduced with permission from [35]. Copyright (2003) American Chemical Society.

distances in [Au$_3$(μ-dpmp)$_2$I$_2$]I are 2.952(1) and 3.020(1) Å [4b]. The chloro derivative displays an intense absorption band at 356 nm with a shoulder at 326 nm, while the iodo analog shows absorption maximum at 320 nm with shoulders at 375 and 420 nm. These spectral features are attributed to gold(I)–gold(I) interactions, which cause a red-shift of the metal-centered 5d → 6s/6p transition(s). The room-temperature emission spectrum of [Au$_3$(μ-dpmp)$_2$I$_2$]I solid displays a

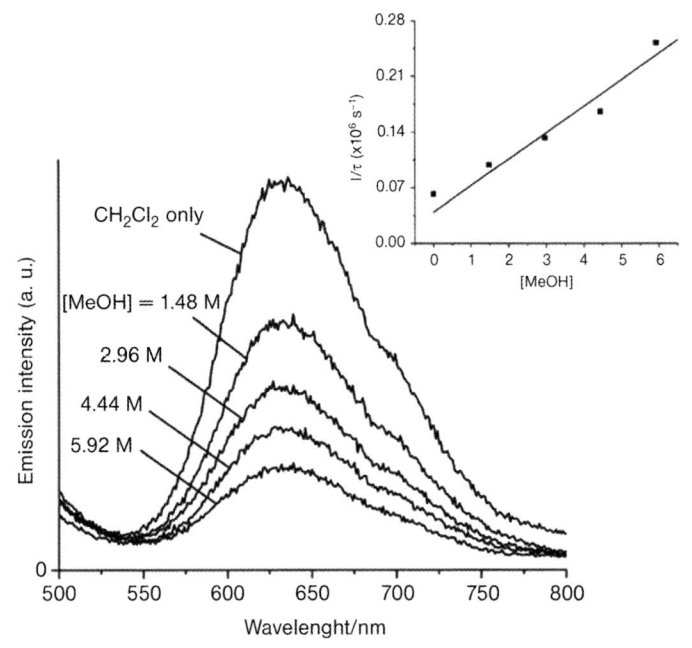

Figure 5.9 Spectral changes upon addition of CH_3OH to a CH_2Cl_2 solution of $[(8\text{-QNS})_2Au(AuPPh_3)_2]BF_4$ (7.72×10^{-5} M) with excitation at 320 nm. Inset: Changes in lifetime according to Stern-Volmer equation. Reproduced with permission from [35]. Copyright (2003) American Chemical Society.

low-energy emission at 575 nm, while $[Au_3(\mu\text{-dpmp})_2Cl_2]Cl$ shows a relatively weak emission in the visible region. The low-energy emission of $[Au_3(\mu\text{-dpmp})_2I_2]I$ was previously assigned to originate from excited states with $I^- \rightarrow$ phosphine (π^*) charge-transfer character, whereas that of $[Au_3(\mu\text{-dpmp})_2Cl_2]Cl$ was attributed to Au(d) $\rightarrow$ phosphine (π^*) transition that is perturbed by metal–metal interaction(s). These original spectroscopic assignments may require re-consideration and we propose that structural distortion of the 5d $\rightarrow$ (6s, 6p) excited state modified by Au(I)–Au(I) and/or Au(I)-halide interactions can account for these emissions.

The trinuclear $[Au_3(\mu\text{-dpmp})_2](SCN)_3$ complex with Au–Au distances of 3.0137(8) and 3.0049(8) Å and tetranuclear $[Au_4(\mu\text{-dpmp})_2(SCN)_2][SCN]Cl$ complex with Au–Au distances of 3.057(1)–3.150(1) Å were prepared and their photophysical properties examined [38]. The UV/Vis absorption spectrum of $[Au_3(\mu\text{-dpmp})_2]^{3+}$ shows a broad absorption with λ_{max} at 326 nm ($\varepsilon = 3.16 \times 10^4$ M^{-1} cm^{-1}) and a weaker shoulder at ~350 nm. The 326-nm absorption band and 350-nm shoulder were attributed to spin-allowed 5d$\sigma^* \rightarrow$ 6pσ and 5d$\delta^* \rightarrow$ 6pσ transitions respectively. Complex $[Au_4(\mu\text{-dpmp})_2(SCN)_2]^{2+}$ displays a 5d$\sigma^* \rightarrow$ 6pσ transition that is red-shifted to 350 nm. In acetonitrile, complex $[Au_3(\mu\text{-dpmp})_2]^{3+}$ displays at room temperature an intense photoluminescence at 600 nm with an emission lifetime of 3.7 µs (Figure 5.11). In view of the relatively small difference in emission energy from

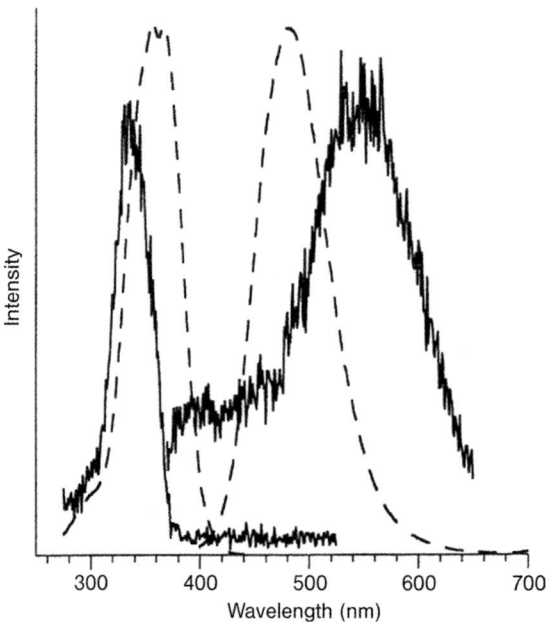

Figure 5.10 Normalized solid-state excitation (left) and emission (right) spectra of orange [Au$_2$(dpim)$_2$]$^{2+}$ (solid line) and blue [Au$_2$(dpim)$_2$]$^{2+}$ (dashed line), at room temperature. Reproduced with permission from [37]. Copyright (2003) American Chemical Society.

that of [Au$_2$(μ-dppm)$_2$]$^{2+}$ (λ_{em} 570 nm), and the large Stokes shift of the emission energy from the 5dσ* → 6pσ transition energy, the emitting electronic state for [Au$_3$(μ-dpmp)$_2$]$^{3+}$ is unlikely to be $^3[(5d\sigma^*)^1(6p\sigma)^1]$. This emission was attributed to originate from low-lying $^3[(5d\delta^*)^1(6p\sigma)^1]$ state.

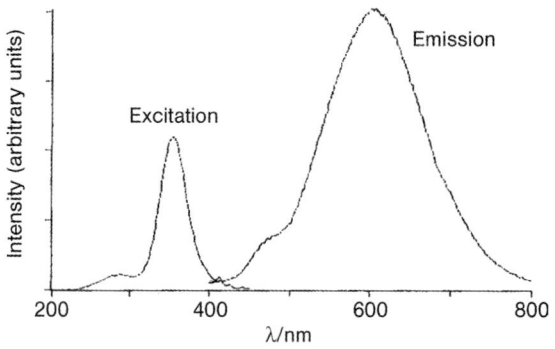

Figure 5.11 Emission and excitation spectra of [Au$_3$(μ-dpmp)$_2$](SCN)$_3$ in degassed acetonitrile at room temperature. Reproduced with permission from [38]. Copyright (1993) Royal Society of Chemistry.

Balch reported the polynuclear ((diphenylphosphino)methyl)phenylarsine (dpma) bridged gold(I) complexes, [Au$_4$(μ-dpma)$_2$Cl$_2$]X$_2$ (X = PF$_6^-$, NO$_3^-$) and [(μ-dpma)(AuCl)$_3$] having intramolecular Au–Au contacts of 2.965(1)–3.110(2) Å, and 3.131(1) and 3.138(1) Å respectively [39]. Laguna reported a series of trinuclear and tetranuclear gold(I) complexes [(μ-dpmp)(AuX)$_3$] (X = Cl$^-$, C$_6$F$_5^-$), [Au$_3$(μ-dpmp)$_2$](CF$_3$SO$_3$)$_3$, and [Au$_4$(μ-dpmp)$_2$Cl$_2$](CF$_3$SO$_3$)$_2$. The complex [Au$_4$(μ-dpmp)$_2$Cl$_2$](CF$_3$SO$_3$)$_2$ displays a rhomboidal geometry for the gold atoms with gold–gold distances of 3.1025(11) and 3.1059(14) Å [40]. The trinuclear gold(I) complexes [(μ-dpmp)(AuC$_6$F$_5$)$_3$] and [Au$_3$(μ-dpmp)$_2$](CF$_3$SO$_3$)$_3$ show absorption λ_{max} at ~310 and 350 nm respectively. The absorption λ_{max} of tetranuclear [Au$_4$(μ-dpmp)$_2$Cl$_2$](CF$_3$SO$_3$)$_2$ is at ~300 nm. The 350-nm band of [Au$_3$(μ-dpmp)$_2$](CF$_3$SO$_3$)$_3$ is characteristic of the 5dσ* → 6pσ transition of a linear Au(I)–Au(I)–Au(I) array. The luminescence of these complexes is related to gold(I)–gold(I) interactions, and is most intense for trinuclear derivatives (Figure 5.12). The emission of [Au$_3$(μ-dpmp)$_2$](CF$_3$SO$_3$)$_3$ at about 600 nm is reassigned here to originate from a three-coordinate Au(I) site in the excited state, as found in the analogous derivative, [Au$_3$(μ-dpmp)$_2$(SCN)$_3$] [38].

The trinuclear gold(I) complexes [(AuX)$_3$(μ-triphos)] (triphos = bis(2-diphenylphosphinoethyl)phenylphosphine; X = Cl$^-$, Br$^-$, I$^-$, C$_6$F$_5^-$), where X = Br$^-$ and I$^-$, were reported to emit at room temperature with emission energies significantly red-shifted from that of the free phosphine ligand (2477 cm^{-1} for Br$^-$ with λ_{ex} 350 nm; 4482 cm^{-1} for I$^-$ with λ_{ex} 385 nm) [41]. The chloro, pentafluorophenyl, and 1,3,5-tris(trifluoromethyl)phenyl derivatives do not emit, even at λ_{ex} < 300 nm. The emission properties

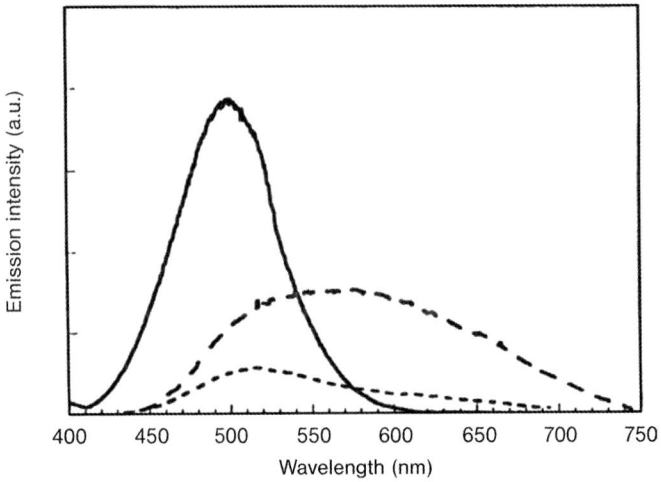

Figure 5.12 Emission spectra of [(μ-dpmp)(AuC$_6$F$_5$)$_3$] (——, λ_{ex} = 370 nm), [Au$_3$(μ-dpmp)$_2$](CF$_3$SO$_3$)$_3$ (– – – –, λ_{ex} = 420 nm), and [Au$_4$(μ-dpmp)$_2$Cl$_2$](CF$_3$SO$_3$)$_2$ (······, λ_{ex} = 370 nm) in the solid state at 300 K. Reproduced with permission from [40]. Copyright (1998) American Chemical Society.

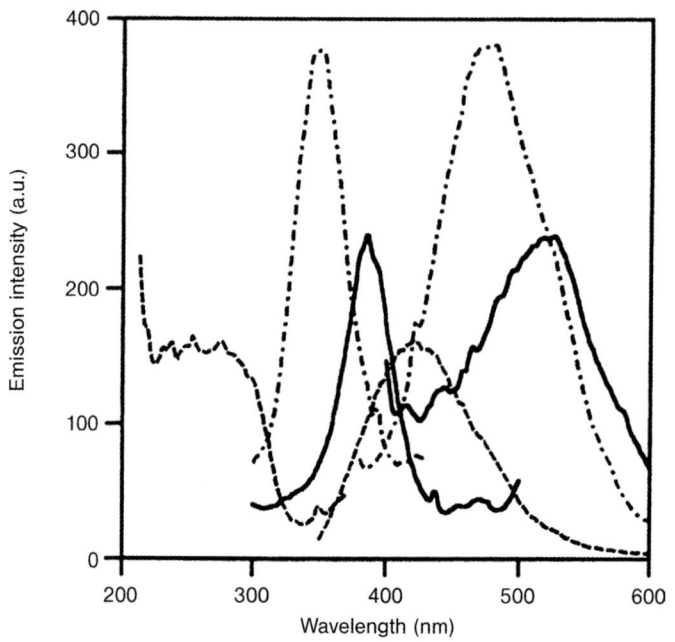

Figure 5.13 Solid-state excitation and emission spectra of triphos (------), [(AuBr)$_3$(μ-triphos)] (•—•—•—), and [(AuI)$_3$(μ-triphos)] (———) at room temperature. Reproduced with permission from [41]. Copyright (2001) American Chemical Society.

of [(AuX)$_3$(μ-triphos)] are dependent upon the halide anions (Figure 5.13), presumably via stabilization of a three-coordinate Au(I) ion in the excited state, as described for other trinuclear gold(I) derivatives [4b].

Self-assembly of gold(I) complexes by weak aurophilic interactions have afforded structurally intriguing supramolecular aggregates, including dimers, trimers, higher oligomers, and one-, two- and three-dimensional polymers [42]. In many instances, these polynuclear d^{10} complexes display interesting photoluminescence properties affected by the aurophilic interactions [43], where a metal–metal bonded excited state is typically invoked [42]. Upon introduction of appropriate functional units, these aggregates can display potential applications as luminescent sensors or catalysts. Examples of chemosensory applications for gold(I) complexes will be discussed in Section 2.4.

5.2.3
High-Energy ^{3}[5dσ*, 6pσ] Excited State Versus Visible Metal-Anion/Solvent Exciplex Emission

For di- and polynuclear gold(I) complexes with two Au(I) centers held in close proximity, a lower energy 5dσ* → 6pσ transition that is red-shifted in energy from its mononuclear counterpart is a spectroscopic signature [2, 7, 44]. Excitation to the 5dσ* → 6pσ transition gives rise to a ^{3}[5dσ*, 6pσ] excited state having a formal metal–metal single bond, [6a, 7a]. Recent studies by Che and coworkers argued that the visible emissions of [Au$_2$(dcpm)$_2$]$^{2+}$ salts, with intramolecular Au(I)···Au(I) separations of 2.92–3.02 Å, arise from exciplex formation as a result of complexation between solvent/counterion and the ^{3}[5σ*, 6σ] excited-state [6a]. Che also demonstrated that the solid-state emissions of [Au$_2$(dcpm)$_2$]$^{2+}$ salts are affected by the counterion, with the emission maximum at 368 nm for the ClO$_4^-$ salt and 530 nm for the I$^-$ complex [6a, b].

[Au$_2$(dcpm)$_2$]X$_2$ complexes with weakly interacting counterions (X = ClO$_4^-$, PF$_6^-$, CF$_3$SO$_3^-$, Au(CN)$_2^-$) exhibit similar UV/Vis absorption spectra in acetonitrile. They show an intense absorption band at 277 nm ($\varepsilon = 2.6$–2.9×10^4 M^{-1} cm^{-1}) and a weak shoulder at ~315 nm ($\varepsilon \approx 400$ M^{-1} cm^{-1}), which are assigned to ^{1}(5dσ* → 6pσ) and ^{3}(5dσ* → 6pσ) transitions respectively. The UV/Vis absorption spectrum of the iodide salt (X = I$^-$) differs significantly. Namely, [Au$_2$(dcpm)$_2$I$_2$] displays an intense absorption band at λ = 275 nm, which has a similar band shape and energy to the 277 nm absorption band of the [Au$_2$(dcpm)$_2$]$^{2+}$ ion, and can be assigned to the ^{1}(dσ* → pσ) transition. At a concentration of 6.0×10^{-5} M in CH$_3$CN, [Au$_2$(dcpm)$_2$I$_2$] shows additional intense absorptions with peak maxima at 323 ($\varepsilon = 6136$ M^{-1} cm^{-1}) and 365 nm ($\varepsilon = 1344$ M^{-1} cm^{-1}) (Figure 5.14). These two absorption bands are attributed to the coordination of I$^-$ to [Au$_2$(dcpm)$_2$]$^{2+}$ in the ground state.

The [Au$_2$(dcpm)$_2$]X$_2$ complexes (X = ClO$_4^-$, PF$_6^-$, CF$_3$SO$_3^-$, Au(CN)$_2^-$) display intense photoluminescence with λ$_{max}$ at 360–368 nm in the solid state at room temperature and in glassy solutions at 77 K. Solid samples of [Au$_2$(dcpm)$_2$](ClO$_4$)$_2$ and [Au$_2$(dcpm)$_2$](PF$_6$)$_2$ individually display a weak visible emission at λ$_{max}$ = 564 and 505 nm, respectively, but a similar emission is not detected for [Au$_2$(dcpm)$_2$]

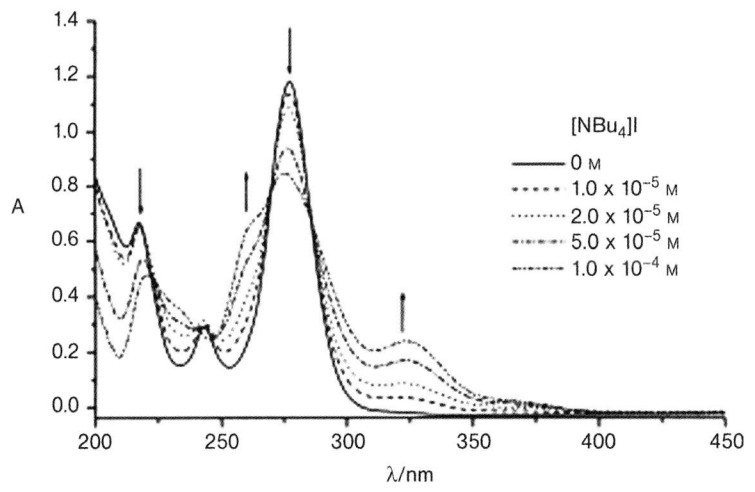

Figure 5.14 UV/Vis spectral changes of [Au$_2$(dcpm)$_2$](ClO$_4$)$_2$ in degassed acetonitrile at room temperature as a function of [NBu$_4$]I concentration. Reproduced with permission from [6b]. Copyright (2001) Wiley-VCH.

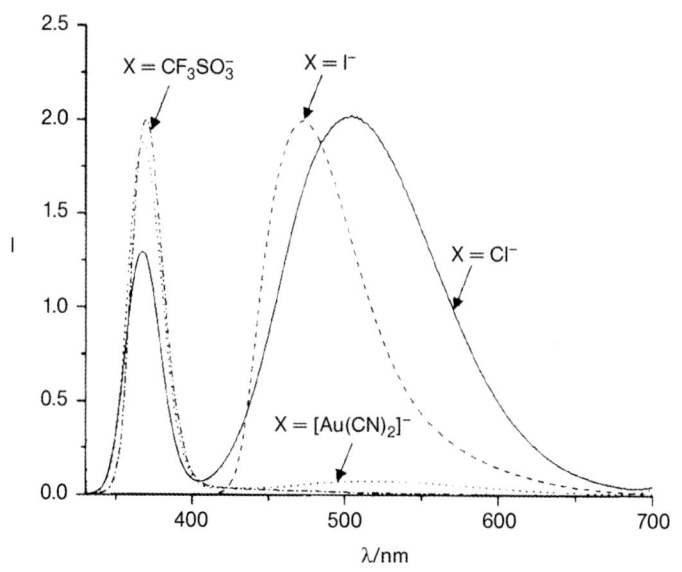

Figure 5.15 Room-temperature solid-state emission spectra of [$Au_2(dcpm)_2$]X_2 (X = $CF_3SO_3^-$, $Au(CN)_2^-$, Cl^-, and I^-) with excitation at λ = 280 nm. I = intensity Reproduced with permission from [6b]. Copyright (2001) Wiley-VCH.

(CF_3SO_3)$_2$ (Figure 5.15). Both the high- (360–370 nm) and low- (500–570 nm) energy emissions have lifetimes in the microsecond time regime, thus implying triplet-parentage excited states. The high-energy emission was ascribed to intrinsic phosphorescence from the ^{3}[5dσ*, 6pσ] excited state. The Stokes shift between the ^{1}[5dσ*, 6pσ] absorption at 278 nm and ^{3}[5dσ*, 6pσ] emission at 368 nm for [$Au_2(dcpm)_2$](ClO_4)$_2$ is 8930 cm^{-1}, which compares well with that (7793 cm^{-1}) observed for the well-established [$Pt_2(P_2O_5H_2)_4$]$^{4-}$ system [29].

A crystalline sample of [$Au_2(dcpm)_2$](SCN)$_2$ shows only one intense emission with λ$_{max}$ at 465 nm at room temperature. Crystalline samples of [$Au_2(dcpm)_2$]I_2 and [$Au_2(dcpm)_2$]I_2·0.5CH_2Cl_2 emit at 473 and 486 nm, respectively at room-temperature (Figure 5.15). Since the two complexes have virtually the same bonding parameters, the inclusion of CH_2Cl_2 molecules into the crystal structure of [$Au_2(dcpm)_2$]I_2 account for the slight red shift in the emission energy. The low-energy solid-state emission of [$Au_2(dcpm)_2$]I_2 can be attributed to the excited state associated with the three-coordinate AuP_2I moiety, which is consistent with complex formation ([$Au_2(dcpm)_2$]$^{2+}$ + 2I$^-$) as determined by UV/Vis spectrophotometry. For [$Au_2(dcpm)_2$]X_2 with X = ClO_4^-, PF_6^-, $CF_3SO_3^-$, the Au···anion distances are significantly longer than that observed for X = I$^-$, and consequently the effect of the counteranion on the ^{3}[5dσ*, 6pσ] excited state of [$Au_2(dcpm)_2$]$^{2+}$ is diminished.

The Au···anion distance of [$Au_2(dcpm)_2$][$Au(CN)_2$]$_2$ lies between those with non-coordinating (ClO_4^- or PF_6^-) anions and that of the I$^-$ salt. The visible emission of the [$Au_2(dcpm)_2$][$Au(CN)_2$]$_2$ solid at λ = 515 nm is more intense than the visible emission of solid samples of ClO_4^- and PF_6^- salts of [$Au_2(dcpm)_2$]$^{2+}$ at 564 and 505 nm respectively, but is not as intense as the 473 nm emission of the [Au$_2$(dcpm)$_2$]I$_2$ solid (Figure 5.15). A crystalline sample of [Au$_2$(dcpm)$_2$]Cl$_2$ displays both high-energy UV (λ_{max} = 366 nm) and low-energy visible emissions (λ_{max} = 505 nm) at room temperature (Figure 5.15). The X-ray crystal structure of [Au$_2$(dcpm)$_2$]Cl$_2$ shows a close Au$\cdots$Cl contact for one of the two gold atoms. The ratios of the intensities of the high-energy UV to low-energy visible emissions were found to correspond to differences in the Au$\cdots$anion contact distances, that is, Au$\cdots$Cl 2.7755(9); Au$\cdots$I = 2.9960(7); Au$\cdots$SCN = 3.011(3); Au$\cdots$Au(CN)$_2$ = 3.33(1); Au$\cdots$OClO$_3$ = 3.36(2) Å.

[Au$_2$(dcpm)$_2$]X$_2$ complexes in degassed acetonitrile solutions exhibit intense phosphorescence with λ_{max} ranging from 490 to 530 nm (Figure 5.16), which can be attributed to adduct formation between the triplet excited state and solvent/counterion. The triplet-state difference absorption spectra of [Au$_2$(dcpm)$_2$]X$_2$ (X = ClO$_4^-$, PF$_6^-$, CF$_3$SO$_3^-$, Au(CN)$_2^-$) in acetonitrile show an absorption peak maximum at 350 nm plus a shoulder/absorption maximum at 395–420 nm, the relative intensities of which are dependent upon the halide ion present in solution. Binding of the ^{3}[5dσ^*, 6pσ] excited state with halide species (Y$^-$), to give [Au$_2$(dcpm)$_2$Y]$^{+*}$ would account for the lower-energy absorption maximum in the triplet-state difference absorption spectra. For the iodide anion, [Au$_2$(dcpm)$_2$]$^{2+}$ undergoes photoinduced electron-transfer to give the radical anion I$_2^-$.

Che and coworkers have studied the effect of halide ions on the emission. Initial addition of a small amount of [NBu$_4$]Y to [Au$_2$(dcpm)$_2$](CF$_3$SO$_3$)$_2$ in acetonitrile led to enhancement of the emission intensity and red-shifted emission λ_{max}, from 508 to 510 (Cl$^-$), 514 (Br$^-$) and 530 (I$^-$) nm. Further addition of [NBu$_4$]Y resulted in a decrease in emission intensity. It is reasonable to assign the absorption band

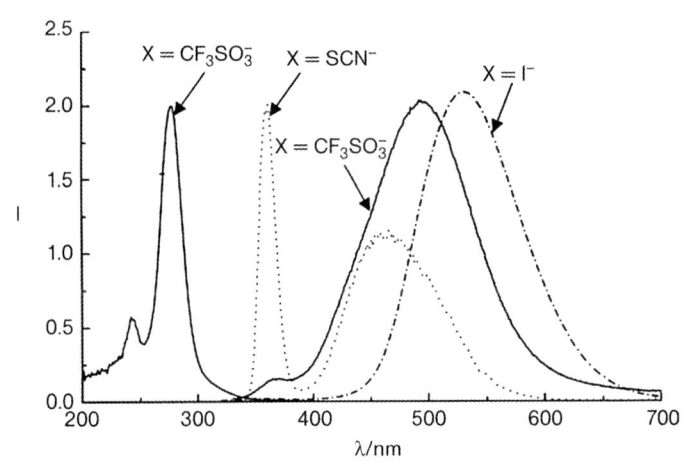

Figure 5.16 Excitation (left, emission monitored at 500 nm) and emission (right) spectra of [Au$_2$(dcpm)$_2$]X$_2$ (X = CF$_3$SO$_3^-$ and I$^-$), with λ_{ex} at 280 nm, in degassed acetonitrile at room temperature, and emission spectrum of [Au$_2$(dcpm)$_2$](SCN)$_2$, with λ_{ex} at 280 nm, in EtOH/MeOH (1:4 v/v) at 77 K. Reproduced with permission from [6b]. Copyright (2001) Wiley-VCH.

observed for [Au$_2$(dcpm)$_2$]I$_2$ at 365 nm and the emission in solution at 530 nm to the same electronic excited state with different spin multiplicities [singlet versus triplet]. This spectral assignment results in a reasonable Stokes shift of 8500 cm^{-1}. Degassed acetonitrile solutions of [Au$_2$(dcpm)$_2$](ClO$_4$)$_2$, [Au$_2$(dcpm)$_2$](PF$_6$)$_2$, and [Au$_2$(dcpm)$_2$](CF$_3$SO$_3$)$_2$ give an emission at ~500 nm. We attribute these emissions to adduct(s) formed between the ^{3}[5dσ*, 6pσ] excited state and the solvent, that is exciplex formation, as there is no evidence for complexation between the non-coordinating anions and [Au$_2$(dcpm)$_2$]$^{2+}$ in the ground state. Computational calculations revealed that the excited states of the three-coordinate [AuL$_3$]$^+$ complexes is T-shaped as a result of excited-state Jahn–Teller distortion, which is in contrast to their ground-state trigonal planar geometry [17]. The singlet-triplet energy gap decreases and leads to longer emission wavelengths, thus accounting for the large experimentally observed Stokes shift. This proposed Jahn–Teller excited-state distortion was also supported by the computational results for the three-coordinate dinuclear Au(I) complex [Au$_2${(Ph$_2$Sb)$_2$O}$_3$]$^{2+}$ [45].

Gray and coworkers [10] reported the emission properties of the dinuclear three-coordinate Au(I) complex [Au$_2$(dcpe)$_3$]$^{2+}$. Each Au atom is bound to one chelating dcpe and one bridging dcpe, and each AuP$_3$ unit has a distorted trigonal planar geometry with no intra- or intermolecular Au(I)–Au(I) interaction. The electronic absorption spectrum of [Au$_2$(dcpe)$_3$](PF$_6$)$_2$ in acetonitrile exhibits a weak absorption at 370 nm ($\varepsilon = 300$ M^{-1} cm^{-1} per AuP$_3$ unit), which is attributed to a [(5d$_{xy}$, 5d$_{x^2-y^2}$) → 6p$_z$] transition. [Au$_2$(dcpe)$_3$](PF$_6$)$_2$ emits at $\lambda_{max} = 501$ nm in solid state and at $\lambda_{max} = 508$ nm in acetonitrile solution at room temperature ($\tau = 21.1$ (5) μs, $\Phi = 0.80(5)$) (Figure 5.17). The excitation spectrum of [Au$_2$(dcpe)$_3$](PF$_6$)$_2$ is the same as its absorption spectrum. The large Stokes shift between the 370-nm absorption and 508-nm emission indicates a severe distortion of the AuP$_3$ structure in the excited state.

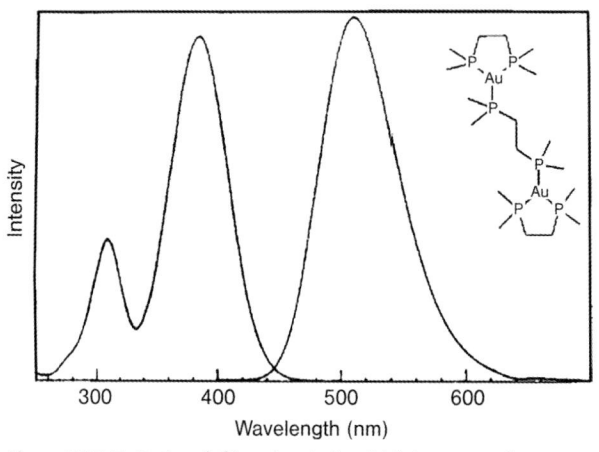

Figure 5.17 Excitation (left) and emission (right) spectra of [Au$_2$(dcpe)$_3$](PF$_6$)$_2$ in acetonitrile at room temperature. Reproduced with permission from [10]. Copyright (1992) American Chemical Society.

Our previous *ab initio* calculation on $[Au_2(H_2PCH_2PH_2)_2]^{2+}$ predicted that its $^3[5d\sigma^*, 6p\sigma]$ excited state emits at 331 nm in the absence of interaction between Au(I) and solvent molecule or neighboring anion in the excited state. This calculated emission energy is much higher than the experimental 575-nm emission of $[Au_2(dppm)_2](ClO_4)_2$ in acetonitrile. Thus, we contend that the $^3[5d\sigma^*, 6p\sigma]$ excited state readily forms exciplexes with acetonitrile molecule(s). The $^3A_u(s_\sigma) \rightarrow {}^1A_g(d_{\sigma^*})$ transition (phosphorescence) of the $[Au_2(H_2PCH_2PH_2)_2\cdot(CH_3CN)_2]^{2+*}$ exciplex was calculated at 557 nm, which matches well with the 575-nm emission of $[Au_2(dppm)_2](ClO_4)_2$ recorded in acetonitrile [46]. Based on the computational study, we propose that coordination of acetonitrile to Au(I) in the $^3[5d\sigma^*, 6p\sigma]$ excited state is primarily responsible for the visible emission of acetonitrile solutions of $[Au_2(diphosphinc)_2]^{2+}$.

5.2.4
Chemosensory Applications

The photoluminescent properties of polynuclear d^{10} metal complexes are affected by subtle structural and/or environmental changes, and this phenomenon has been used to develop chemosensory applications. Several literature examples in this area are discussed in this section. It should be noted that the emission properties of polynuclear gold(I) complexes can be very complex and some of the spectral assignments given in the literature may be open to re-interpretation. Eisenberg reported an unusual luminescence behavior attributed to the structural change of a dimeric gold(I) dithiocarbamate complex $[Au(S_2CN(C_5H_{11})_2)]_2$, induced by exposure to volatile organic compounds (VOCs) in the solid state [18]. The crystal structure of the orange, luminescent DMSO solvate of $[Au(S_2CN(C_5H_{11})_2)]_2$ shows stacks of discrete dimers forming infinite chains of Au atoms with short intermolecular Au $\cdots$ Au contacts (2.9617(7) Å) along the c axis (Figure 5.18a). The orange CH_2Cl_2 cast film of $[Au(S_2CN(C_5H_{11})_2)]_2$ exhibits intense emission at 630 nm upon excitation at 460 nm (Figure 5.18b). When the film was heated at 50 °C for 10 s, the orange color faded and the resultant pale yellow film became non-emissive. The structure of the colorless, non-emissive form of $[Au(S_2CN(C_5H_{11})_2)]_2$ shows discrete, non-interacting dimers with the shortest intermolecular Au $\cdots$ Au distance being 8.135 Å. If this solid was exposed to vapors of polar aprotic solvents, for example, acetone, CH_3CN, CH_2Cl_2, and $CHCl_3$, it became orange again and regained its emissive nature. The inset of Figure 5.18b depicts the high-energy absorption band (294 nm) of the dried film of $[Au(S_2CN(C_5H_{11})_2)]_2$ isosbestically splitting into two peak maxima at 284 and 331 nm (sh) with concomitant emergence of a peak at 465 nm upon exposure to acetone vapor at 15- or 30-s intervals. Balch and coworkers observed that when the colorless trimeric $[Au_3(CH_3N=COCH_3)_3]$ complex was irradiated with long-wavelength UV light (366 nm), followed by subsequent contact with chloroform, an intense yellow luminescence at 552 nm developed [47]. The crystal structure of $[Au_3(CH_3N=COCH_3)_3]$ reveales stacks of molecules with an intermolecular Au(I) $\cdots$ Au(I) distance of 3.346(1) Å, which is within the range for aurophilic bonding interaction. The emission was assigned to an excited state arising from intermolecular metal–metal interactions.

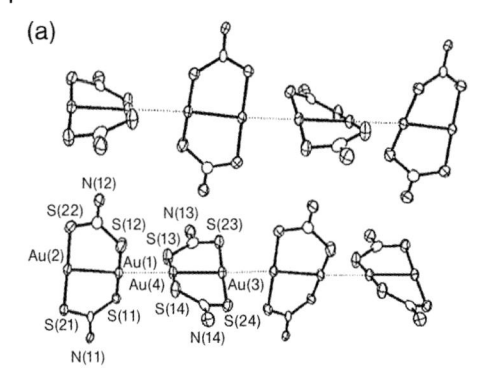

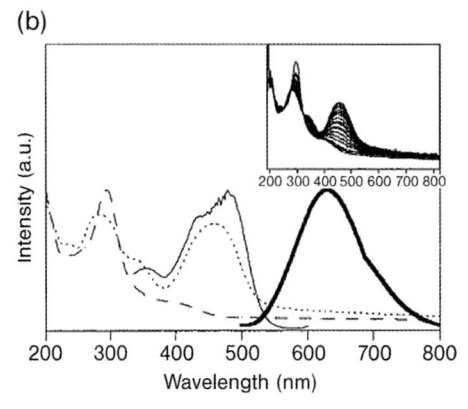

Figure 5.18 (a) Drawing of the [Au(S$_2$CN (C$_5$H$_{11}$)$_2$)]$_2$ molecules in DMSO with two repeating units of chains in the structure viewed perpendicular to the stacking axis. Thermal ellipsoids are drawn at the 50% probability level. The pentyl moieties have been omitted for clarity. (b) UV/Vis absorption (······), emission (—) (λ_{ex} = 460 nm), and excitation (——) (λ_{em} = 630 nm) spectra at room temperature of an orange CH$_2$Cl$_2$ cast film of [Au(S$_2$CN(C$_5$H$_{11}$)$_2$)]$_2$ on a quartz disk; absorption spectrum (– – – –) of heated/dried film of [Au(S$_2$CN(C$_5$H$_{11}$)$_2$)]$_2$. Inset: Absorption spectra of dried film of [Au(S$_2$CN(C$_5$H$_{11}$)$_2$)]$_2$ upon direct exposure to acetone vapor for 15- or 30-s intervals. Reproduced with permission from [18]. Copyright (1998) American Chemical Society.

Yam reported using gold–gold interactions in dinuclear gold(I) complexes to induce changes in spectroscopic properties as a luminescent probe for potassium ions [48]. Complexes [Au$_2$(P∩P)(S-benzo[15]crown-5)$_2$] (P∩P = dppm, dcpm) were synthesized, and upon addition of K$^+$ ions, the emission spectra of the former showed a decrease in emission intensity at ∼502 nm (τ < 0.1 µs) with concomitant formation of a long-lived and low-energy emission band at 720 nm (τ = 0.2 µs). The ionic radius of K$^+$ is too large to fit into the cavity of benzo[15]crown-5, so the K$^+$ ion tends to sandwich between two crown ethers and bring two Au(I) centers into close proximity. A red shift in emission energy was assigned to thiolate-to-gold charge transfer transition modified by Au(I)···Au(I) interaction. Similar findings were also observed for the dcpm derivative.

Scheme 5.2 Equilibrium between [Au(P∩P)]$^+$ and PPh$_3$ in solution.

Perturbation of metal–metal bonded excited states, through interaction with neighboring solvent molecules or anions which leads to exciplex formation, gives rise to emission in the visible region. The sensitivity of the photoluminescence of two-coordinate d^{10} metal complexes to metal-ligand coordination provides an entry to new classes of luminescent sensory materials for substrate-binding reactions. The formation of strongly emission three-coordinate Au(I) as a consequence of metal-substrate coordination in the course of chemosensory processes was illustrated by Che and coworkers [12b]. The two-coordinate [Au(P∩P)]$^+$ (P∩P = 1,8-bis(diphenylphosphino)-3,6-dioxaoctane) complex is non-emissive in solution. However, upon addition of one equivalent of PPh$_3$, an intense emission at λ_{max} = 500 nm is "switched on" (Scheme 5.2), and ground-state complexation of [Au(P∩P)]$^+$ by PPh$_3$ to give the emissive [Au(P∩P)PPh$_3$]$^+$ species was proposed and demonstrated (Figure 5.19). Yam and coworkers [49] reported that the reaction of the gold(I) alkynyl complexes

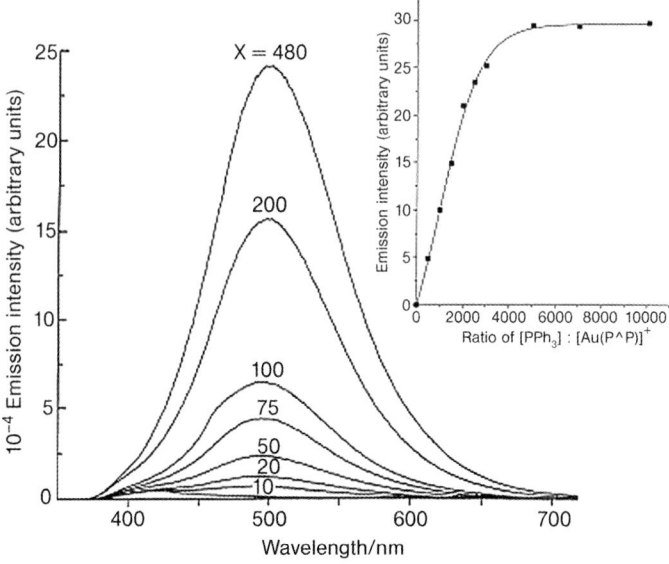

Figure 5.19 Room temperature emission spectra of [Au(P∩P)](ClO$_4$) in the presence of PPh$_3$ in degassed acetonitrile solution. Molar ratio of PPh$_3$:[Au(P∩P)]$^+$ = X:1. Inset: a plot of the emission intensity of [Au(P∩P)(PPh$_3$)]$^+$ vs. X:1 (concentration of [Au(P∩P)](ClO$_4$) = 10^{-5} mol dm^{-3}). Reproduced with permission from [12b]. Copyright (1998) Royal Society of Chemistry.

[(R₃P)Au{C≡CC(=CH₂)Me}] (R = Ph, p-Tol) with copper(I) or silver(I) ions led to the formation of luminescent η^2-alkynyl mixed-metal complexes, [{η^2-(R₃P)Au{C≡CC(=CH₂)Me}}₂M]PF₆ (M = Cu⁺, Ag⁺). The non-emissive [(μ-dppf)Au₂{C≡CC(=CH₂)Me}₂] (dppf = 1,1′-bis(diphenylphosphino)ferrocene) became highly emissive at about 565–583 nm upon copper(I) or silver(I) encapsulation to give [(μ-dppf)Au₂{η^2-C≡CC(=CH₂)Me}₂M]PF₆ (M = Cu⁺, Ag⁺).

In other cases, the changes in emission properties of polynuclear d^{10} metal complexes upon exposure to volatile organic compounds were due to other factors apart from metal–metal interactions. Eisenberg [50] reported that the reversible luminescence tribochromism from dimeric gold(I) thiouracilate complexes can be related to two different structural motifs interconverted by protonation of the uracilate ligand. Fackler and coworkers [[23]b] reported a dramatic increase in emission intensity upon grinding non-emissive crystals of the linear one-dimensional chain compound [(TPA)₂Au][Au(CN)₂]. This phenomenon was attributed to the generation of lattice defects near the crystal surface. Mills and coworkers [51] reported the incorporation of luminescent three-coordinate mononuclear gold(I) complexes, namely bis{μ-(bis(diphenylphosphino)octadecylamine-P,P′)}diiododigold(I) and 1,1,1-tris(2-diphenylphosphenitomethyl)ethane gold(I) chloride, into thin plastic films as luminescent sensors for the detection of oxygen. In polystyrene film with thickness of 20 μm, these complexes exhibited response and recovery times of 23–26 and 71–100 s respectively.

5.3
Spectroscopic Properties of Gold(III) Complexes

Tremendous efforts have been devoted to photophysical studies of gold(I) systems, whereas reports on luminescent gold(III) complexes are sparse in the literature. In 1994, Che and coworkers reported the synthesis, photoluminescence and excited state properties of the cyclometallated gold(III) complexes [Au(C^N^N-dpp)Cl]X (HC^N^N-dpp = 2,9-diphenyl-1,10-phenanthroline; X = C₇H₇SO₃⁻, ClO₄⁻) [52]. The crystal structure of [Au(C^N^N-dpp)Cl]C₇H₇SO₃ consists of pairs of cations with stacked molecular planes and an intermolecular Au···Au separation of 3.6 Å. Strong absorption bands in the 222–310 nm region ($\varepsilon > 10^4$) are assigned to intraligand transitions, whereas those at $\lambda = 384$–427 nm ($\varepsilon > 10^3$) may have metal character. The [Au(C^N^N-dpp)Cl]⁺ complex is weakly emissive in solution with quantum yields of $\sim 10^{-4}$ and lifetimes of 0.4–0.7 μs. Its luminescence is insensitive to solvent polarity and the emission spectrum recorded in glassy methanol solution at 77 K displays highly structured bands with vibrational progression of ~ 1400 cm⁻¹, which is close to the skeleton vibration frequency of aryl rings. The emission maximum of [Au(C^N^N-dpp)Cl]⁺ shows a small red shift of 1400 cm⁻¹ in emission energy from that of free dpp, suggestive of a metal-perturbed intraligand $^3(\pi\pi^*)$ emissive electronically excited state (Figure 5.20).

The estimated excited-state reduction potential $E^{\circ *}(\text{Au}^{\text{III/II}})$ of 2.2 V (versus NHE) suggests that the excited state of [Au(C^N^N-dpp)Cl]⁺ is a powerful oxidant.

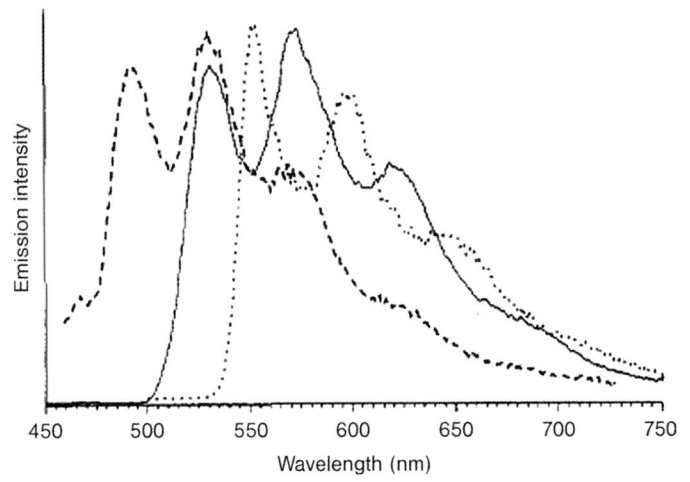

Figure 5.20 Emission spectra of dpp (– – –), [Pt(C^N^N-dpp)(NCCH$_3$)]$^+$ (······) and [Au(C^N^N-dpp)Cl]$^+$ (——) (λ_{ex} = 355 nm) in 77 K methanol glass. Reproduced with permission from [52]. Copyright (1994) American Chemical Society.

Photoinduced oxidation of 1,4-dimethoxybenzene (DMB) and tetrahydrofuran (THF) by [Au(C^N^N-dpp)Cl]$^+$ in acetonitrile upon UV/Vis irradiation have been observed. The time-resolved absorption spectrum recorded 12 μs after excitation of [Au(C^N^N-dpp)Cl]$^+$ with a laser pulse at 355 nm showed the absorption band of the DMB$^{+\bullet}$ radical cation at ~460 nm, whereas upon excitation at 406 nm in the presence of THF, a broad emission characteristic of the protonated salt of 2,9-diphenyl-1,10-phenanthroline (Hdpp$^+$) developed at 500 nm.

Further examples of emissive cyclometallated gold(III) complexes are [Au(L)Cl]$^+$ (L = tridentate carbanion of 4'-(4-methoxyphenyl)-6'-phenyl-2,2'-bipyridine) [53], as well as mono- and binuclear bis-cyclometallated gold(III) complexes, namely [Au(C^N^C)L']$^{n+}$ (C^N^C = tridentate dicarbanion of 2,6-diphenylpyridine; L' = deprotonated 2-mercaptopyridine (2-pyS$^-$), n = 0; L' = PPh$_3$ or 1-methylimidazole, n = 1) and [Au$_2$(C^N^C)$_2$(P∩P)](ClO$_4$)$_2$ (P∩P = dppm, dppe) respectively [54]. The crystal structures of the binuclear derivatives show intramolecular interplanar separations of 3.4 Å between the [Au(C^N^C)]$^+$ moieties, implying the presence of weak π-π interactions. The mononuclear complexes show absorption with vibronic structure at 380–405 nm (ε > 10^3 M^{-1} cm^{-1}), attributed to metal-perturbed intraligand transition. The absorption bands of the binuclear complexes are red-shifted compared to the mononuclear analogs, and this is attributed to intramolecular π-π interactions between C^N^C ligands in solution (Figure 5.21a).

The gold(III) complexes, [Au(C^N^C)L']$^{n+}$ and [Au$_2$(C^N^C)$_2$(P∩P)](ClO$_4$)$_2$ are emissive in acetonitrile at low temperature. The frozen-state (77 K) emission spectra of the mononuclear complexes [Au(C^N^C)L']$^{n+}$ show well-resolved vibronic structures with spacings in the 1100–1300 cm^{-1} range, which correlate with the skeletal vibrational frequency of the tridentate C^N^C ligand. By comparing the emission

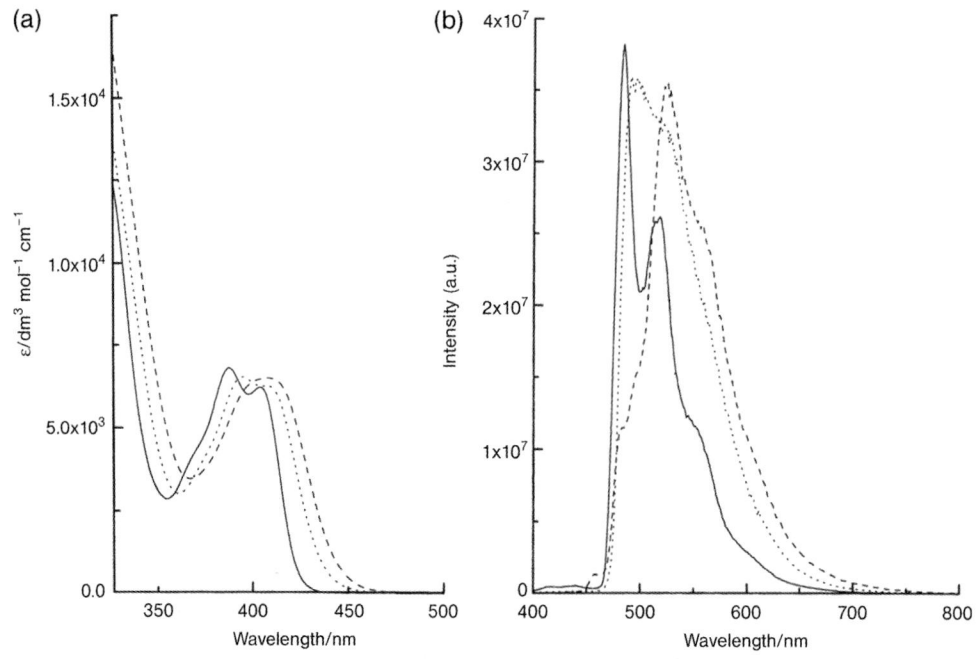

Figure 5.21 (a) 298 K UV/Vis absorption spectra, and (b) 77 K emission spectra of [Au(C^N^C)PPh$_3$]ClO$_4$ (——), [Au$_2$(C^N^C)$_2$(μ-dppm)](ClO$_4$)$_2$ (– – –) and [Au$_2$(C^N^C)$_2$(μ-dppe)](ClO$_4$)$_2$ (·····) in acetonitrile. Reproduced with permission from [54]. Copyright (1998) American Chemical Society.

spectra with those of the free and protonated forms of 2,6-diphenylpyridine, we tentatively inferred that the emission of [Au(C^N^C)L']$^{n+}$ originates from a metal-perturbed $^3\pi\pi^*$ state [52, 53]. The emission maxima of [Au$_2$(C^N^C)$_2$(dppm)](ClO$_4$)$_2$ and [Au$_2$(C^N^C)$_2$(dppe)](ClO$_4$)$_2$ are red-shifted from that of [Au(C^N^C)L']$^{n+}$, presumably due to excimeric π-π interactions (Figure 5.21b).

Yam [55] and coworkers recently reported the synthesis and electrochemical, photophysical, and computational studies of a class of luminescent cyclometallated alkynylgold(III) complexes, namely [Au(C^N^C)(C≡CR)] (HC^N^CH = 2,6-diphenylpyridine; R = C$_6$H$_5$, C$_6$H$_4$Cl-p, C$_6$H$_4$NO$_2$-p, C$_6$H$_4$OCH$_3$-p, C$_6$H$_4$NH$_2$-p, C$_6$H$_4$ (C$_6$H$_{13}$)-p, C$_6$H$_{13}$), [Au(^{t}BuC^N^C^{t}Bu)(C≡CC$_6$H$_5$)] (H^{t}BuC^N^Ct-BuH = 2,6-bis(4-tert-butylphenyl)pyridine), and [Au(C^NTol^C)(C≡CC$_6$H$_4$(C$_6$H$_{13}$)-p] (HC^NTol^CH = 2,6-diphenyl-4-p-tolylpyridine). Electrochemical studies revealed that the first oxidation wave is an alkynyl ligand-centered oxidation, while the first reduction couple was ascribed to a ligand-centered reduction of the cyclometallated ligand with the exception of [Au(C^N^C)(C≡CC$_6$H$_4$NO$_2$-p)], in which the first reduction couple was assigned to an alkynyl ligand-centered reduction. In dichloromethane solutions at room temperature, the low-energy absorption bands are assigned to π-π* intraligand (IL) transitions of the cyclometallated (C^N^C) ligand mixed with [π(C≡CR) → π*(C^N^C)] ligand-to-ligand charge transfer (LLCT) character. The

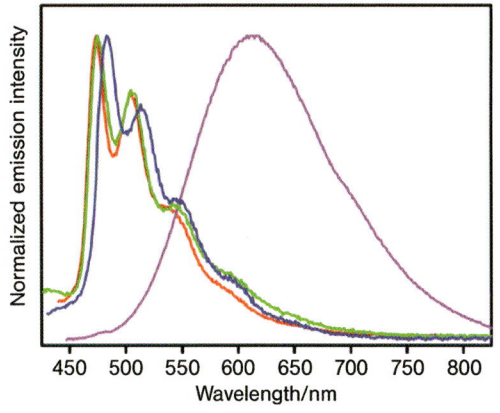

Figure 5.22 Emission spectra of [Au(C^N^C)(C≡CC$_6$H$_5$)] (red), [Au(C^N^C)(C≡CC$_6$H$_4$NH$_2$-p)] (pink), [Au(^{t}BuC^N^C^{t}Bu)(C≡CC$_6$H$_5$)] (green), and [Au(C^NTol^C)(C≡CC$_6$H$_4$(C$_6$H$_{13}$)-p] (blue) in degassed CH$_2$Cl$_2$ at room temperature. Reproduced with permission from [55b]. Copyright (2007) American Chemical Society.

low-energy emission for all complexes, with the exception of [Au(C^N^C)(C≡CC$_6$H$_4$NH$_2$-p)], were ascribed to originate from the π–π* IL transition of the cyclometallated C^N^C ligand. In the case of [Au(C^N^C)(C≡CC$_6$H$_4$NH$_2$-p)], the low-energy emission band at 610 nm was found to show a significant red shift and ascribed to a [π(C≡CC$_6$H$_4$NH$_2$) → π*(C^N^C)] LLCT excited-state origin (Figure 5.22). Applications of these luminescent cyclometallated alkynylgold(III) complexes for the fabrication of organic light-emitting diodes have been described [56].

Laguna and coworkers have reported the synthesis and spectroscopic properties of di- and trinuclear gold(III) complexes with a triphosphine ligand, [{Au(C$_6$F$_5$)$_3$}$_n$(μ-triphos)] (n = 2, 3) [41]. The absorption spectra of these gold(III) complexes in dichloromethane showed, in addition to the phenyl absorption at 232 nm, an intense absorption at about 256 nm with tailing up to 300 nm. The dinuclear gold(III) complex [{Au(C$_6$F$_5$)$_3$}$_2$(μ-triphos)] emits neither at room temperature (77K), nor low temperature, whereas the trinuclear [{Au(C$_6$F$_5$)$_3$}$_3$(μ-triphos)] complex emits at both temperatures. When compared with the spectrum of the free phosphine ligand, the excitation and emission spectra of [{Au(C$_6$F$_5$)$_3$}$_3$(μ-triphos)] are red-shifted, and the number of tris(pentafluorophenyl)gold(III) units significantly affects the luminescent properties.

5.4
Photoinduced Electron Transfer Reactions of Gold Complexes

There are few photochemical studies on dinuclear and polynuclear gold (I) and (III) complexes in the literature. The $^3[(n + 1)$pσ, ndσ*] excited states of [Au$_2$(dppm)$_2$]$^{2+}$ are powerful reductants, with an E° value of −1.6(1) V vs. SSCE (saturated sodium chloride calomel electrode) [3, 4a, 9]; however [Au$_2$(diphosphine)$_2$]$^{2+}$ systems do not

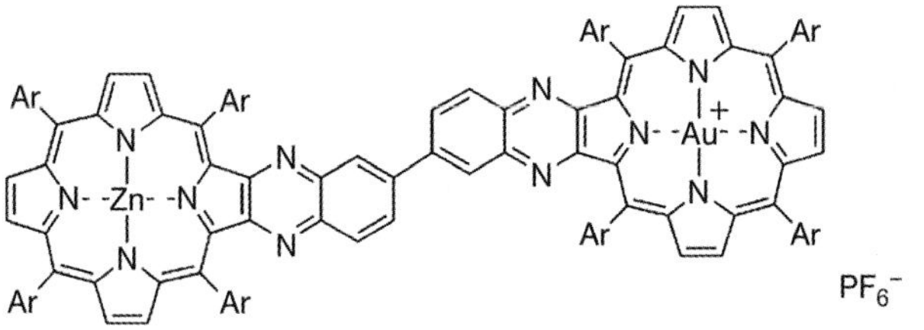

Figure 5.23 Structure of ZnPQ–AuPQ$^+$ PF$_6^-$ dyad (Ar = 3,5-$^tBu_2C_6H_3$).

react with saturated C–H bonds under photochemical conditions. [Au$_2$(μ-dppm)$_2$]$^{2+*}$ was reported to catalyze the photochemical cleavage of C–X (X = halide) bonds from benzyl halides, affording the bibenzyl C–C bond coupling product with turnover number of 172. This photochemical reaction was found to take place by an electron-transfer mechanism rather than an atom-transfer mechanism [9]. The apparent lack of reactivity of the ^{3}[5dσ*, 6pσ] excited state towards C–H activation was attributed to the fact that the form exists as a solvent/anion exciplex in solution, which renders the gold(I) less accessible towards interaction with C–H bonds.

As described in the previous section, luminescent cyclometallated gold(III) complexes are capable of oxidizing THF upon light excitation. In general, gold(III) complexes are powerful photo-oxidants. The excited states of gold(III) porphyrin complexes with lifetimes of hundreds of picoseconds or nanoseconds have proven to be strong oxidants [57]. In a recent report on the electron donor-acceptor dyad system containing linked gold(III) and zinc(II) porphyrins, ZnPQ–AuPQ$^+$ PF$_6^-$ (Figure 5.23), a long-lived charge-separated state with a lifetime of 10 μs in non-polar solvent such as cyclohexane (dielectric constant = 2.02) was observed upon light excitation [58]. The ESR spectrum of a toluene solution of ZnPQ–AuPQ$^+$ at 143 K under photo-irradiation showed a broad signal attributed to Au(II) species, revealing that the site of electron transfer is at Au(III) rather than the porphyrin ligand. The long-lived charge-separated state may result from small reorganization energy for the metal-centered electron transfer reaction of AuPQ$^+$ in non-polar solvents.

Studies on the effect of porphyrin ring substituents towards the first reduction of gold(III) porphyrins were undertaken (see Figure 5.24 for structures) [59]. Through electrochemical and spectroelectrochemical investigations in non-aqueous media, the first reduction step of (P-H)AuPF$_6$, (P-NH$_2$)AuPF$_6$ and (PQ)AuPF$_6$ were found to be metal-centered, giving a Au(II) porphyrin. However, for (P-NO$_2$)AuPF$_6$, the site of electron transfer is at the porphyrin macrocycle, giving rise to a gold(III) porphyrin π-anion radical.

5.5
Concluding Remarks

This chapter provides an overview of spectroscopic investigations into the photophysics of luminescent mono- and polynuclear gold(I) and -(III) complexes.

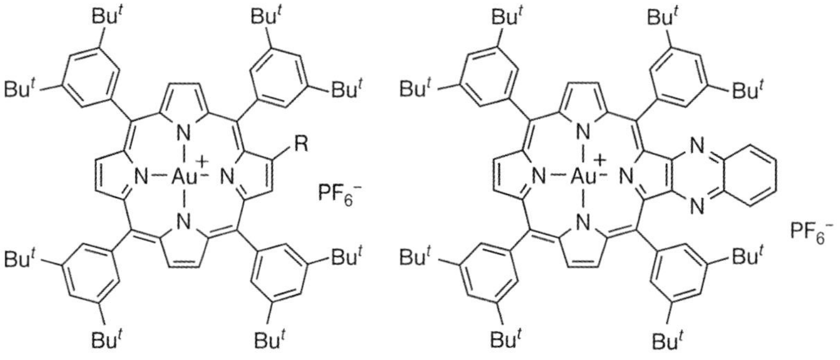

R = H,   (P-H)AuPF$_6$
R = NO$_2$,   (P-NO$_2$)AuPF$_6$
R = NH$_2$,   (P-NH$_2$)AuPF$_6$

(PQ)AuPF$_6$

Figure 5.24 Structures of hexafluorophosphate[5,10,15,20-tetrakis(3,5-di-*tert*-butylphenyl)porphyrinato]gold(III) and derivatives.

The propensity for perturbation of a two-coordinate Au(I) site either by Au(I)–Au(I) interaction or Au(I)-ligand coordination is highlighted. The existence of Au(I)–Au(I) interaction in [5dσ*, 6pσ] excited states of di- and polynuclear gold(I) complexes could be inferred by the existence of an intense red-shifted absorption attributed to the [5dσ* → 6pσ] transition, the assignment of which was substantiated by resonance Raman spectroscopy. For the di- and polynuclear Au(I) complexes, the intrinsic ^{3}[6pσ, 5dσ*] emission occurs in the UV region. Metal-ligand coordination in both ground and excited states would stabilize the ^{3}[(5d$_{x^2-y^2}$, 5d$_{xy}$), 6p$_z$] state of a three-coordinate Au(I) site, which is emissive and lower in energy than the ^{3}[6pσ, 5dσ*] excited state. Perturbation of the emission properties of gold(I) complexes due to subtle changes in the local environment, induced either by Au(I)–Au(I) interaction or Au(I)–substrate coordination, could be useful for developing chemosensory applications.

There is an impetus to develop new classes of metal photocatalysts for light-induced atom transfer and excited-state substrate-binding reactions after the seminal work on [Pt$_2$(P$_2$O$_5$H$_2$)$_4$]$^{4-}$. In this context, photoluminescent d^{10} gold(I) phosphine complexes have become a target for photochemical studies. The [Au$_2$(diphosphine)$_2$]$^{2+}$ system has vacant coordination sites at the metal atom and long-lived and emissive ^{3}[5dσ*, 6pσ] excited state in fluid solutions at room temperature. It has been established that the ^{3}[5dσ*, 6pσ] triplet excited state has a formal metal–metal single bond and is a powerful photoreductant. Photoinduced C–C bond coupling from alkyl halides by the excited state of [Au$_2$(dppm)$_2$]$^{2+}$ in the presence of sacrificial electron donors has been demonstrated. In general, gold(III) complexes are powerful photo-oxidants though intriguing photochemical reactions have yet to be unearthed.

Acknowledgements

We are grateful for financial support from the Research Grants Council of the Hong Kong SAR, China [HKU 7011/07P, HKU 7030/06P], the Chinese Academy of Sciences–Croucher Foundation Funding Scheme for Joint Laboratories, and the University Development Fund (Nanotechnology Research Institute Program).

References

1 (a) Forward, J.M., Fackler, J.P. Jr and Assefa, Z. (1999) *Optoelectronic Properties of Inorganic Compounds* (eds D.M. Roundhill and J.P. Fackler Jr), Plenum Press, New York, pp. 195–229; (b) Gade, L.H. (1997) "Hyt was of Gold, and Shon so Bryghte...": Luminescent Gold(I) Compounds. *Angewandte Chemie (International Edition in English)*, **36**, 1171–1173.

2 King, C., Wang, J.-C., Khan, M.N.I. and Fackler, J.P. Jr (1989) Luminescence and metal-metal interactions in binuclear gold(I) compounds. *Inorganic Chemistry*, **28**, 2145–2149.

3 Che, C.-M., Kwong, H.-L., Yam, V.W.-W. and Cho, K.-C. (1989) Spectroscopic properties and redox chemistry of the phosphorescent excited state of $[Au_2(dppm)_2]^{2+}$ [dppm = bis(diphenylphosphino)methane]. *Journal of the Chemical Society, Chemical Communications*, 885–886.

4 (a) Yam, V.W.-W., Lai, T.-F. and Che, C.-M. (1990) Novel luminescent polynuclear gold(I) phosphine complexes. synthesis, spectroscopy, and X-ray crystal structure of $[Au_3(dmmp)_2]^{3+}$ [dmmp = bis(dimethylphosphinomethyl)methylphosphine]. *Journal of the Chemical Society, Dalton Transactions*, 3747–3752; (b) Xiao, H., Weng, Y.-X., Wong, W.-T., Mak, T.C.W. and Che, C.-M. (1997) Structures and luminescent properties of polynuclear gold(I) halides containing bridging phosphine ligands. *Journal of the Chemical Society, Dalton Transactions*, 221–226.

5 Omary, M.A., Mohamed, A.A., Rawashdeh-Omary, M.A. and Fackler, J.P. Jr (2005) Photophysics of supramolecular binary stacks consisting of electron-rich trinuclear Au(I) complexes and organic electrophiles. *Coordination Chemistry Reviews*, **249**, 1372–1381.

6 (a) Fu, W.-F., Chan, K.-C., Miskowski, V.M. and Che, C.-M. (1999) The intrinsic $^3[d\sigma*p\sigma]$ emission of binuclear gold(I) complexes with two bridging diphosphane ligands lies in the near UV; emissions in the visible region are due to exciplexes. *Angewandte Chemie*, **111**, 2953–2955; (1999) *Angewandte Chemie (International Edition in English)*, **38**, 2783–2785; (b) Fu, W.-F., Chan, K.-C., Cheung, K.-K. and Che, C.-M. (2001) Substrate-binding reactions of the $^3[d\sigma*p\sigma]$ excited state of binuclear gold(I) complexes with bridging bis(dicyclohexylphosphino)methane ligands: emission and time-resolved absorption spectroscopic studies. *Chemistry – A European Journal*, **7**, 4656–4664; (c) Che, C.-M. and Lai, S.-W. (2005) Structural and spectroscopic evidence for weak metal–metal interactions and metal–substrate exciplex formations in d^{10} metal complexes. *Coordination Chemistry Reviews*, **249**, 1296–1309.

7 (a) Leung, K.H., Phillips, D.L., Tse, M.-C., Che, C.-M. and Miskowski, V.M. (1999) Resonance raman investigation of the Au(I)–Au(I) interaction of the $^1[d\sigma*p\sigma]$ excited state of $Au_2(dcpm)_2(ClO_4)_2$ (dcpm = bis(dicyclohexylphosphine)methane). *Journal of the American Chemical Society*,

121, 4799–4803; (b) Leung, K.H., Phillips, D.L., Mao, Z., Che, C.-M., Miskowski, V.M. and Chan, C.-K. (2002) Electronic excited states of [Au$_2$(dmpm)$_3$](ClO$_4$)$_2$ (dmpm = bis(dimethylphosphino)methane). *Inorganic Chemistry*, **41**, 2054–2059; (c) Phillips, D.L., Che, C.-M., Leung, K.H., Mao, Z. and Tse, M.-C. (2005) A comparative study on metal-metal interaction in binuclear two- and three-coordinated d^{10}-metal complexes spectroscopic investigation of M(I)-M(I) interaction in the ^{1}[dσ∗pσ] excited state of [M$_2$(dcpm)$_2$]$^{2+}$ (dcpm = bis(dicyclohexylphosphino)methane) (M = Au, Ag, Cu) and [M$_2$(dmpm)$_3$]$^{2+}$ (dmpm = bis(dimethylphosphino)methane) (M = Au, Ag, Cu) complexes. *Coordination Chemistry Reviews*, **249**, 1476–1490.

8 Che, C.-M., Kwong, H.-L., Poon, C.-K. and Yam, V.W.-W. (1990) Spectroscopy and redox properties of the luminescent excited state of [Au$_2$(dppm)$_2$]$^{2+}$ (dppm = Ph$_2$PCH$_2$PPh$_2$). *Journal of the Chemical Society, Dalton Transactions*, 3215–3219.

9 Li, D., Che, C.-M., Kwong, H.-L. and Yam, V.W.-W. (1992) Photoinduced C–C bond formation from alkyl halides catalysed by luminescent dinuclear gold(I) and copper(I) complexes. *Journal of the Chemical Society, Dalton Transactions*, 3325–3329.

10 McCleskey, T.M. and Gray, H.B. (1992) Emission spectroscopic properties of 1,2-bis(dicyclohexylphosphino)ethane complexes of gold(I). *Inorganic Chemistry*, **31**, 1733–1734.

11 Brandys, M.-C. and Puddephatt, R.J. (2001) Strongly luminescent three-coordinate gold(I) polymers: 1D chain-link fence and 2D chickenwire structures. *Journal of the American Chemical Society*, **123**, 4839–4840.

12 (a) Shieh, S.-J., Li, D., Peng, S.-M., and Che, C.-M. (1993) Synthesis, photophysical properties and crystal structure of a luminescent binuclear three-co-ordinated gold(I) complex without metal-metal interaction. *Journal of the Chemical Society, Dalton Transactions*, 195–196; (b) Chan, W.-H., Mak, T.C.W. and Che, C.-M. (1998) A two-co-ordinated gold(I) loop [Au(dpdo)]ClO$_4$ [dpdo = 1,8-bis(diphenylphosphino)-3,6-dioxaoctane] as a luminescent light switch for substrate binding reactions. *Journal of the Chemical Society, Dalton Transactions*, 2275–2276.

13 (a) Gimeno, M.C. and Laguna, A. (1997) Three- and four-coordinate gold(I) complexes. *Chemical Reviews*, **97**, 511–522; (b) Crespo, O., Gimeno, M.C., Jones, P.G., Laguna, A., López-de-Luzuriaga, J.M., Monge, M., Pérez, J.L. and Ramón, M.A. (2003) Luminescent *nido*-carborane-diphosphine anions [(PR$_2$)$_2$C$_2$B$_9$H$_{10}$]$^-$ (R = Ph, ^{i}Pr). Modification of their luminescence properties upon formation of three-coordinate gold(I) complexes. *Inorganic Chemistry*, **42**, 2061–2068.

14 King, C., Khan, M.N.I., Staples, R.J. and Fackler, J.P. Jr (1992) Luminescent mononuclear gold(I) phosphines. *Inorganic Chemistry*, **31**, 3236–3238.

15 Forward, J.M., Assefa, Z. and Fackler, J.P. Jr (1995) Photoluminescence of gold(I) phosphine complexes in aqueous solution. *Journal of the American Chemical Society*, **117**, 9103–9104.

16 Assefa, Z., Forward, J.M., Grant, T.A., Staples, R.J., Hanson, B.E., Mohamed, A.A. and Fackler, J.P. Jr (2003) Three-coordinate, luminescent, water-soluble gold(I) phosphine complexes: structural characterization and photoluminescence properties in aqueous solution. *Inorganica Chimica Acta*, **352**, 31–45.

17 Barakat, K.A., Cundari, T.R. and Omary, M.A. (2003) Jahn–Teller distortion in the phosphorescent excited state of three-coordinate Au(I) phosphine complexes. *Journal of the American Chemical Society*, **125**, 14228–14229.

18 Mansour, M.A., Connick, W.B., Lachicotte, R.J., Gysling, H.J. and Eisenberg, R. (1998) Linear chain Au(I) dimer compounds as environmental sensors: A luminescent switch for the detection of volatile organic

compounds. *Journal of the American Chemical Society*, **120**, 1329–1330.

19 (a) Kutal, C. (1990) Spectroscopic and photochemical properties of d^{10} metal complexes. *Coordination Chemistry Reviews*, **99**, 213–252; (b) Chao, H.-Y., Lu, W., Li, Y., Chan, M.C.W., Che, C.-M., Cheung, K.-K. and Zhu, N. (2002) Organic triplet emissions of arylacetylide moieties harnessed through coordination to $[Au(PCy_3)]^+$. Effect of molecular structure upon photoluminescent properties. *Journal of the American Chemical Society*, **124**, 14696–14706; (c) Lu, W., Zhu, N. and Che, C.M. (2003) Polymorphic forms of a gold(I) arylacetylide complex with contrasting phosphorescent characteristics. *Journal of the American Chemical Society*, **125**, 16081–16088; (d) Bayón, R., Coco, S. and Espinet, P. (2005) Gold liquid crystals displaying luminescence in the mesophase and short F · · · F interactions in the solid state. *Chemistry – A European Journal*, **11**, 1079–1085.

20 (a) Yam, V.W.-W., Lo, K.K.-W. and Wong, K.M.-C. (1999) Luminescent polynuclear metal acetylides. *Journal of Organometallic Chemistry*, **578**, 3–30; (b) Yam, V.W.-W. and Lo, K.K.-W. (1999) Luminescent polynuclear d^{10} metal complexes. *Chemical Society Reviews*, **28**, 323–334.

21 (a) Uang, R.-H., Chan, C.-K., Peng, S.-M. and Che, C.-M. (1994) Luminescent metallomacrocycles from gold(I) and silver(I) complexes of 2,7-bis (diphenylphosphino)-1,8-naphthyridine (L) and Crystal Structure of $[Au_2K(L)_3]$ $[ClO_4]_3 \cdot CH_2Cl_2 \cdot 2MeOH \cdot 0.5\ H_2O$. *Journal of the Chemical Society, Chemical Communications*, 2561–2562; (b) Hong, X., Cheung, K.-K., Guo, C.-X. and Che, C.-M. (1994) Luminescent organometallic gold(I) complexes. Structure and photophysical properties of alkyl-, aryl- and μ-ethynylene gold(I) complexes. *Journal of the Chemical Society, Dalton Transactions*, 1867–1871; (c) Li, D., Hong, X., Che, C.-M., Lo, W.-C. and Peng, S.-M. (1993) Luminescent gold(I) acetylide complexes. Photophysical and photoredox properties and crystal structure of $[\{Au(C\equiv CPh)\}_2$ $(\mu\text{-}Ph_2PCH_2CH_2PPh_2)]$. *Journal of the Chemical Society, Dalton Transactions*, 2929–2932; (d) Che, C.-M., Wong, W.-T., Lai, T.-F. and Kwong, H.-L. (1989) Novel luminescent binuclear gold(I) isocyanide complexes. Synthesis, spectroscopy, and X-ray crystal structure of $Au_2(dmb)(CN)_2$ (dmb = 1,8-di-isocyano-*p*-menthane). *Journal of the Chemical Society, Chemical Communications*, 243–244.

22 Assefa, Z., McBurnett, B.G., Staples, R.J., Fackler, J.P. Jr, Assmann, B., Angermaier, K. and Schmidbaur, H. (1995) Syntheses, structures, and spectroscopic properties of gold(I) complexes of 1,3,5-Triaza-7-phosphaadamantane (TPA). Correlation of the supramolecular Au · · · Au interaction and photoluminescence for the species (TPA)AuCl and [(TPA-HCl)AuCl]. *Inorganic Chemistry*, **34**, 75–83.

23 (a) Assefa, Z., McBurnett, B.G., Staples, R.J. and Fackler, J.P. Jr (1995) Structures and spectroscopic properties of gold(I) complexes of 1,3,5-Triaza-7-phosphaadamantane (Tpa). 2. Multiple-state emission from (TPA)AuX (X = Cl, Br, I) complexes. *Inorganic Chemistry*, **34**, 4965–4972; (b) Assefa, Z., Omary, M.A., McBurnett, B.G., Mohamed, A.A., Patterson, H.H., Staples, R.J. and Fackler, J.P. Jr (2002) Syntheses, structure, and photoluminescence properties of the 1-Dimensional chain compounds $[(TPA)_2Au][Au(CN)_2]$ and $(TPA)AuCl$ (TPA = 1,3,5-Triaza-7-phosphaadamantane) *Inorganic Chemistry*, **41**, 6274–6280.

24 Bardají, M., Jones, P.G. and Laguna, A. (2002) Acetylenephosphino gold(I) derivatives: Structure, reactivity and luminescence properties. *Journal of the Chemical Society, Dalton Transactions*, 3624–3629.

25 Bardaji, M., Jones, P.G., Laguna, A., Villacampa, M.D. and Villaverde, N. (2003) Synthesis and structural characterization of luminescent gold(I) derivatives with an

unsymmetric diphosphine. *Dalton Transactions*, 4529–4536.
26. (a) Schmidbaur, H., Wohlleben, A., Schubert, U., Frank, A. and Huttner, G. (1977) Gold complexes of diphosphinomethanes. 2. Synthesis and crystal-structure of 8-membered heterocycles of gold(I) with Au-Au interaction. *Chemische Berichte* **110**, 2751–2757; (b) Wang, J.-C., Khan, M.N.I. and Fackler, J.P. Jr (1989) Structure of Bis (tetraphenyldiphosphinomethane) digold(I) Dinitrate. *Acta Crystallographica.* **C45**, 1482–1485; (c) Porter, L.C., Khan, M.N.I., King, C. and Fackler, J.P. Jr (1989) Structure of the Bis[bis (diphenylphosphino)methane]digold(I) Cation in [Au$_2$(dppm)$_2$](BF$_4$)$_2$. *Acta Crystallographica.* **C45**, 947–949.
27. Inoguchi, Y., Milewski-Mahrla, B. and Schmidbaur, H. (1982) Liganden für extrem kurze metall-metall-kontakte in goldkomplexen. *Chemische Berichte*, **115**, 3085–3095.
28. Khan, M.N.I., King, C., Heinrich, D.D., Fackler, J.P. Jr and Porter, L.C. (1989) Syntheses and Crystal Structures (No Au-H Interactions) of Luminescent [Au$_2$(dppm)$_2$][BH$_3$CN]$_2$ and of [Au$_2$(dppm)$_2$(I)][Au(CN)$_2$] and [Au$_2$(dppm)$_2$(S$_2$CNEt$_2$)][BH$_3$CN]. *Inorganic Chemistry*, **28**, 2150–2154.
29. Roundhill, D.M., Gray, H.B. and Che, C.-M. (1989) Pyrophosphito-bridged diplatinum chemistry. *Accounts of Chemical Research*, **22**, 55–61.
30. Schmidbaur, H. (1995) Ludwig Mond lecture. High-carat gold compounds. *Chemical Society Reviews*, **24**, 391–400.
31. Barnard, P.J., Baker, M.V., Berners-Price, S.J., Skelton, B.W. and White, A.H. (2004) Dinuclear gold(I) complexes of bridging bidentate carbene ligands: Synthesis, structure and spectroscopic characterisation. *Dalton Transactions*, 1038–1047.
32. Tzeng, B.-C., Li, D., Peng, S.-M. and Che, C.-M. (1993) Photoluminescent properties and molecular structure of [{Au (PPh$_3$)}$_2$(μ-bbzim)] and [{Au(PPh$_3$)}$_4$ (μ-bbzim)][ClO$_4$]$_2$ (bbzim = 2, 2′-Bibenzimidazolate). *Journal of the Chemical Society, Dalton Transactions*, 2365–2371.
33. (a) Tzeng, B.-C., Cheung, K.-K., and Che, C.-M. (1996) Photoluminescent two-dimensional gold(I) polymers bearing a macrocyclic tetraaza cavity. *Chemical Communications*, 1681–1682; (b) Tzeng, B.-C., Che, C.-M. and Peng, S.-M. (1997) Luminescent gold(I) supermolecules with trithiocyanuric acid. Crystal structure, spectroscopic and photophysical properties. *Chemical Communications*, 1771–1772.
34. Shieh, S.-J., Hong, X., Peng, S.-M. and Che, C.-M. (1994) Synthesis and crystal structure of a luminescent one-dimensional phenylacetylide-gold(I) polymer with 2,6-bis(diphenylphosphino) pyridine as ligand. *Journal of the Chemical Society, Dalton Transactions*, 3067–3068.
35. Tzeng, B.-C., Yeh, H.-T., Huang, Y.-C., Chao, H.-Y., Lee, G.-H. and Peng, S.-M. (2003) A Luminescent Supermolecule with Gold(I) Quinoline-8-thiolate: Crystal Structure, Spectroscopic and Photophysical Properties. *Inorganic Chemistry*, **42**, 6008–6014.
36. Tzeng, B.-C., Chan, C.-K., Cheung, K.-K., Che, C.-M. and Peng, S.-M. (1997) Dramatic solvent effect on the luminescence of a dinuclear gold(I) complex of quinoline-8-thiolate. *Chemical Communications*, 135–136.
37. Catalano, V.J. and Horner, S.J. (2003) Luminescent gold(I) and silver(I) complexes of 2-(diphenylphosphino)-1-methylimidazole (dpim): Characterization of a three-coordinate Au(I)-Ag(I) dimer with a short metal–metal separation. *Inorganic Chemistry*, **42**, 8430–8438.
38. Li, D., Che, C.-M., Peng, S.-M., Liu, S.-T., Zhou, Z.-Y. and Mak, T.C.W. (1993) Spectroscopic properties and crystal structures of luminescent linear tri- and tetra-nuclear gold(I) complexes with bis

(diphenylphosphinomethyl) phenylphosphine ligand. *Journal of the Chemical Society, Dalton Transactions*, 189–194.

39 Balch, A.L., Fung, E.Y. and Olmstead, M.M. (1990) Polynuclear ((diphenylphosphino)methyl) phenylarsine bridged complexes of gold(I). Bent chains of gold(I) and a role for Au(I)–Au(I) interactions in guiding a reaction. *Journal of the American Chemical Society*, **112**, 5181–5186.

40 Bardají, M., Laguna, A., Orera, V.M. and Villacampa, M.D. (1998) Synthesis, structural characterization, and luminescence studies of gold(I) and gold(III) complexes with a triphosphine ligand. *Inorganic Chemistry*, **37**, 5125–5130.

41 Bardají, M., Laguna, A., Vicente, J. and Jones, P.G. (2001) Synthesis of luminescent gold(I) and gold(III) complexes with a triphosphine ligand. *Inorganic Chemistry*, **40**, 2675–2681.

42 (a) Irwin, M.J., Vittal, J.J. and Puddephatt, R.J. (1997) Luminescent gold(I) acetylides: From model compounds to polymers. *Organometallics*, **16**, 3541–3547; (b) Hunks, W.J., Jennings, M.C. and Puddephatt, R.J. (2000) Self-association in gold chemistry: A tetragold(I) complex linked by both aurophilic and hydrogen bonding. *Inorganic Chemistry*, **39**, 2699–2702; (c) Brandys, M.-C., Jennings, M.C. and Puddephatt, R.J. (2000) Luminescent gold(I) macrocycles with diphosphine and 4,4'-bipyridyl ligands. *Journal of the Chemical Society, Dalton Transactions*, 4601–4606; (d) Yip, J.H.K. and Prabhavathy, J. (2001) A luminescent gold ring that flips like cyclohexane. *Angewandte Chemie*, **113**, 2217–2220; (2001) *Angewandte Chemie (International Edition in English)*, **40**, 2159–2162; (e) Lin, R., Yip, J.H.K., Zhang, K., Koh, L.L., Wong, K.-Y. and Ho, K.P. (2004) Self-assembly and molecular recognition of a luminescent gold rectangle. *Journal of the American Chemical Society*, **126**, 15852–15869.

43 Bachman, R.E., Bodolosky-Bettis, S.A., Glennon, S.C. and Sirchio, S.A. (2000) Formation of a novel luminescent form of gold(I) phenylthiolate via self-assembly and decomposition of isonitrilegold(I) phenylthiolate complexes. *Journal of the American Chemical Society*, **122**, 7146–7147.

44 Rawashdeh-Omary, M.A., Omary, M.A., Patterson, H.H. and Fackler, J.P. Jr (2001) Excited-state interactions for $[Au(CN)_2^-]_n$ and $[Ag(CN)_2^-]_n$ oligomers in solution. Formation of luminescent gold–gold bonded excimers and exciplexes. *Journal of the American Chemical Society*, **123**, 11237–11247.

45 Bojan, V.R., Fernández, E.J., Laguna, A., López-de-Luzuriaga, J.M., Monge, M., Olmos, M.E. and Silvestru, C. (2005) Phosphorescent excited state of $[Au_2\{(Ph_2Sb)_2O\}_3]^{2+}$: Jahn–Teller distortion at only one gold(I) center. *Journal of the American Chemical Society*, **127**, 11564–11565.

46 Zhang, H.-X. and Che, C.-M. (2001) Aurophilic attraction and luminescence of binuclear gold(I) complexes with bridging phosphine ligands: ab initio study. *Chemistry – A European Journal*, **7**, 4887–4893.

47 Vickery, J.C., Olmstead, M.M., Fung, E.Y. and Balch, A.L. (1997) Solvent-stimulated luminescence from the supramolecular aggregation of a trinuclear gold(I) complex that displays extensive intermolecular Au···Au interactions. *Angewandte Chemie*, **109**, 1227–1229; (1997) *Angewandte Chemie (International Edition in English)*, **36**, 1179–1181.

48 Yam, V.W.-W., Li, C.-K. and Chan, C.-L. (1998) Proof of potassium ions by luminescence signaling based on weak gold–gold interactions in dinuclear gold(I) complexes. *Angewandte Chemie*, **110**, 3041–3044; (1998) *Angewandte Chemie (International Edition in English)*, **37**, 2857–2859.

49 Yam, V.W.-W., Cheung, K.-L., Cheng, E.C.-C., Zhu, N. and Cheung, K.-K. (2003)

50 Lee, Y.-A. and Eisenberg, R. (2003) Luminescence tribochromism and bright emission in gold(I) thiouracilate complexes. *Journal of the American Chemical Society*, **125**, 7778–7779.

51 (a) Mills, A., Lepre, A., Theobald, B.R.C., Slade, E., and Murrer, B.A. (1997) Use of Luminescent Gold Compounds in the Design of Thin-Film Oxygen Sensors. *Analytical Chemistry* **69**, 2842–2847; (b) Mills, A., Lepre, A., Theobald, B.R.C., Slade, E. and Murrer, B.A. (1998) Luminescent gold compounds in optical oxygen sensors. *Gold Bulletin*, **31**, 68–70.

52 Chan, C.-W., Wong, W.-T. and Che, C.-M. (1994) Gold(III) photooxidants. Photophysical, photochemical properties, and crystal structure of a luminescent cyclometallated gold(III) complex of 2,9-diphenyl-1,10-phenanthroline. *Inorganic Chemistry*, **33**, 1266–1272.

53 Liu, H.-Q., Cheung, T.-C., Peng, S.-M. and Che, C.-M. (1995) Novel luminescent cyclometallated and terpyridine gold(III) complexes and DNA binding studies. *Journal of the Chemical Society, Chemical Communications*, 1787–1788.

54 Wong, K.-H., Cheung, K.-K., Chan, M.C.-W. and Che, C.-M. (1998) Application of 2,6-diphenylpyridine as a tridentate [C^N^C] dianionic ligand in organogold(III) chemistry. Structural and spectroscopic properties of mono- and binuclear transmetalated gold(III) complexes. *Organometallics*, **17**, 3505–3511.

55 (a) Yam, V.W.-W., Wong, K.M.-C., Hung, L.-L. and Zhu, N. (2005) Luminescent

The text continues in the right column with "Syntheses and luminescence studies of mixed-metal gold(I)–copper(I) and –silver(I) alkynyl complexes. The "turning-on" of emission upon d^{10} metal ion encapsulation. *Dalton Transactions*, 1830–1835." appearing at the top of the left column before entry 50.

Syntheses and luminescence studies of mixed-metal gold(I)–copper(I) and –silver (I) alkynyl complexes. The "turning-on" of emission upon d^{10} metal ion encapsulation. *Dalton Transactions*, 1830–1835.

gold(III) alkynyl complexes: Synthesis, structural characterization, and luminescence properties. *Angewandte Chemie*, **117**, 3167–3170; (2005) *Angewandte Chemie (International Edition in English)*, **44**, 3107–3110; (b) Wong, K.M.-C., Hung, L.-L., Lam, W.H., Zhu, N. and Yam, V.W.-W. (2007) A class of luminescent cyclometallated alkynylgold(III) complexes: Synthesis, characterization, and electrochemical, photophysical, and computational studies of [Au(C^N^C)(C≡C–R)] (C^N^C = κ^3C, N,C bis-cyclometallated 2,6-diphenylpyridyl). *Journal of the American Chemical Society*, **129**, 4350–4365.

56 Wong, K.M.-C., Zhu, X., Hung, L.-L., Zhu, N., Yam, V.W.-W. and Kwok, H.-S. (2005) A novel class of phosphorescent gold(III) alkynyl-based organic light-emitting devices with tunable colour. *Chemical Communications*, 2906–2908.

57 Andréasson, J., Kodis, G., Lin, S., Moore, A.L., Moore, T.A., Gust, D., Mårtensson, J. and Albinsson, B. (2002) The gold porphyrin first excited singlet state. *Photochemistry and Photobiology*, **76**, 47–50.

58 Fukuzumi, S., Ohkubo, K., Wenbo, E., Ou, Z., Shao, J., Kadish, K.M., Hutchison, J.A., Ghiggino, K.P., Sintic, P.J. and Crossley, M.J. (2003) Metal-centered photoinduced electron transfer reduction of a gold(III) porphyrin cation linked with a zinc porphyrin to produce a long-lived charge-separated state in nonpolar solvents. *Journal of the American Chemical Society*, **125**, 14984–14985.

59 Ou, Z., Kadish, K.M., Wenbo, E., Shao, J., Sintic, P.J., Ohkubo, K., Fukuzumi, S. and Crossley, M.J. (2004) Substituent effects on the site of electron transfer during the first reduction for gold(III) porphyrins. *Inorganic Chemistry*, **43**, 2078–2086.

6
Gold Compounds and Their Applications in Medicine
Elizabeth A. Pacheco, Edward R.T. Tiekink, and Michael W. Whitehouse

6.1
Introduction

Metallic gold has been admired for its inert and aesthetic qualities throughout human history, but only in the last few decades has there been much interest in gold biochemistry, arising in the most part from the use of gold-based pharmaceuticals. The modern medicinal use of gold finds its beginnings in the pioneering work of German physician and microbiologist Robert Koch (Nobel laureate in Medicine, 1905). He reported that gold cyanide exhibited antibacterial effects against tubercle bacilli [1]. This discovery led to subsequent experimentation with various gold derivatives for the treatment of tuberculosis. In 1929, a French physician, Jacques Forestier, reported the anti-arthritic activity of gold complexes [2, 3]. After World War II, the Empire Rheumatism Council in the UK initiated double-blind multicenter trials that showed that gold drugs might provide an effective treatment for many (not necessarily all) patients suffering from rheumatoid arthritis. In spite of these favorable findings, research into the biochemistry of gold was not pursued extensively until the 1970s, allied to a push to study the emerging field of bioinorganic chemistry. Since that time, research has continued to include not only the prolonged use of gold complexes to treat rheumatoid arthritis, but has also expanded to investigate its possible utility as antitumor, anti-HIV agents, and so on. This chapter focuses on the chemistry and biological disposition of the gold complexes currently used medicinally. The clinical properties, that is, efficacy and side effects, of these drugs are fully discussed elsewhere [1].

6.2
The Aqueous Chemistry of Gold Compounds

Gold has oxidation states ranging from $-I$ to $+V$. Gold(0) is metallic gold, the form most commonly thought of as "gold." Gold(I) and gold(III) are the forms used in most biological gold research. They are stable in aqueous solution, rendering the gold(I) and gold(III) states suitable for medicinal purposes. Other known oxidation states,

−I, +II, +IV, and +V are of interest for chemical reasons but not important for medical purposes, as they are probably not involved in biological processes.

6.2.1
Structures of Gold(I) and Gold(III) Complexes

Gold(I) has a $5d^{10}$ outer shell electron configuration. Its complexes are similar to those of silver(I), copper(I), and mercury(II) with two-, three-, or four-coordination exhibited when coordinated with ligands, the latter two coordination motifs occurring less frequently than for copper(I) or silver(I). The gold to ligand bond lengths vary from donor atom to donor atom but for a given ligand, the bond lengths of gold(I) will generally increase as the coordination number increases.

Two-coordinate gold(I) complexes are the most common forms of gold(I) encountered in biological systems. An example of the tendency of gold(I) to favor two coordination is $[Au(CN)_2]^-$. Only di-cyano gold has been successfully isolated while di-, tri-, and tetra-cyano complexes have been isolated for both copper(I) and silver(I). Two-coordinate gold(I) complexes are made with ligands that will coordinate to a soft metal, that is, containing soft donor atoms. The complexes can be neutral, positively- or negatively-charged, depending on the ligands and have bond angles that are usually virtually linear. Some ligands of pharmacological interest include thiolates, thioethers, phosphines, cyanide, and alkyl groups as well nitrogen donor ligands such as amines and heterocycles.

Three-coordinate gold(I) and four-coordinate gold(I) complexes are much rarer. Three-coordinate gold complexes generally contain at least one neutral ligand such as phosphine and have trigonal planar geometry. Four-coordinate gold(I) complexes exhibit tetrahedral coordination often with some distortion due to steric hindrance.

Gold(III) has a $5d^8$ outer shell electron configuration. It is isoelectronic with platinum(II) and rhodium(I) ions, and forms almost entirely four-coordinate, square-planar complexes [4–7]. The gold(III) to ligand bond lengths are generally shorter than for comparable gold(I) to ligand bond distances. A wide array of ligands forms stable complexes with the gold in the +III oxidation state. Because of this variety of ligands, gold(III) complexes are able to exhibit a wide range of physical and chemical properties. The overall charge of gold(III) complexes can vary considerably. For example, with four neutral ligands, such as NH_3, the resulting complex has a +3 charge while with four negative ligands, such as CN^-, the charge is −1. Variance of charge by ligand exchange reactions occurring *in vivo* can greatly influence the hydrophilicity and the lipophilicity of gold(III) complexes and hence their stability and biodistribution.

6.2.2
Oxidation–Reduction Reactions

Both gold(I) and gold(III) are easily reduced to elemental gold.

$$Au^{1+} + e^- = Au(0) \qquad E_0 = +1.68 \text{ volts} \tag{6.1}$$

$$Au^{3+} + 3e^- = Au(0) \quad E_0 = +1.42 \text{ volts} \tag{6.2}$$

Due to the large, positive values of E_0, both gold(I) and gold(III) are capable of being reduced by mild reducing agents. Gold(III) is reduced under biological conditions to gold(I) or gold(0) by many natural occurring reductants, of which thiols, thioethers, and disulfides are examples.

The reduction potentials are dependent upon ligation of the gold species. Stabilization of either gold oxidation state can occur with suitable ligand choice [8, 9] (CyS$^-$ is cysteinato):

$$\begin{aligned} AuCl_2^- + 1e^- &= Au(0) + 2Cl^- & E_0 &= +1.15 \text{ volts} \\ AuBr_2^- + 1e^- &= Au(0) + 2Br^- & E_0 &= +0.959 \text{ volt} \\ Au(SCy)_2^- + 1e^- &= Au(0) + 2CyS^- & E_0 &= -0.14 \text{ volt} \\ Au(CN)_2^- + 1e^- &= Au(0) + 2CN^- & E_0 &= -0.48 \text{ volt} \end{aligned} \tag{6.3}$$

$$\begin{aligned} AuCl_4^- + 3e^- &= Au(0) + 4Cl^- & E_0 &= +1.00 \text{ volt} \\ AuBr_4^- + 3e^- &= Au(0) + 4Br^- & E_0 &= +0.85 \text{ volt} \\ Au(CN)_4^- + 3e^- &= Au(0) + 4CN^- & E_0 &= -0.10 \text{ volt} \end{aligned} \tag{6.4}$$

The halide complexes of gold(I) and gold(III) have large positive reduction potentials and are still powerful oxidants. Ligation with cyanide of both gold(I) and gold(III) and ligation with cysteinato of gold(I) stabilizes their oxidation states.

6.2.3
Ligand Exchange Mechanisms

The exchange of monodentate ligands bound to gold(I) and gold(III) is generally rapid. Exchange occurs via an associative mechanism whereby gold(I) goes through a three-coordinate transition state and gold(III) goes through a five-coordinate transition state. Sometimes intermediates are formed. Ligand exchange for gold(I) bound ligands is generally faster than for gold(III) bound ligands when comparable ligands are involved. An example of this tendency is the exchange of Me$_3$P at trimethyl gold(III) and monomethylgold(I) [10]:

$$Me_3P^* + Me_3PAuMe \rightleftharpoons Me_3P^*AuMe + Me_3P \tag{6.5}$$

$$Me_3P^* + Me_3PAuMe_3 \rightleftharpoons Me_3P^*AuMe_3 + Me_3P \tag{6.6}$$

The gold(I) exchange occurs rapidly on the NMR time scale. The phosphine dependence of the signal collapse establishes an associative mechanism. By contrast, the gold(III) exchange occurs slowly on the NMR time scale.

Gold(I) ligand exchange has been determined to occur via an associative mechanism by second-order associative rate laws and negative entropies of activation for the reactions. The entropies of activation are negative for the self-exchange of thiol ligands at bis(thiolato)gold(I) in the following equilibrium [11]:

$$RS^*H + Au(SR)_2^- \rightleftharpoons RS^*AuSR^- + RSH \tag{6.7}$$

When N-acetylcysteine and thiomalate were studied in a mixed ligand system, the respective activation entropies were found to be $\Delta S^{\ddagger} = -141$ and $-151\,\mathrm{J\,mol^{-1}\,K^{-1}}$ [11]. Increasing the pH accelerated the reactions, which indicates that the thiolate RS^- is the active nucleophile.

The facile exchange of ligands in two-coordinate gold(I) species can be explained in terms of the three-coordinate transition states and intermediates. There are empty, low-lying p_x and p_y orbitals in the linear complex. The availability of these orbitals facilitates re-hybridization to a trigonal geometry necessary to accommodate the entering ligand. Consequently, many stable three-coordinate gold(I) complexes have been isolated. The stability of these complexes demonstrates that trigonal complex formation, during the interchange of ligands between two-coordinate structures, is a viable stable reaction intermediate. An example is [(Ph$_3$P)$_2$AuCl] which is the expected intermediate for the associative displacement of chloride from Ph$_3$PAuCl:

$$Ph_3PAuCl + Ph_3P \rightleftharpoons [(Ph_3P)_2Au]^+ + Cl^- \tag{6.8}$$

The intermediate [(Ph$_3$P)$_2$AuCl] has been crystallized and shows two independent molecules in the crystallographic asymmetric unit which differ slightly in bond angles and lengths. The systematic changes in bond lengths and angles along the reaction coordinate are as expected for the linear and trigonal species. As the P–Au–P bond angle increases, the Au–Cl bond length increases and concomitantly the Au–P bond lengths decrease. The Au–Cl bond length increases from 227 to 250 pm, while the Au–P bonds of the trigonal complexes are on average 10 pm longer than in the linear chloride and 5 pm than in the bis(phosphine) cation, which is consistent with their retention in the final product [12–14]. Other reaction mechanisms have been observed for organogold(I) species, but the coordination required almost certainly precludes them from functioning in physiological systems.

The exchange mechanism of square-planar gold(III) complexes with monodentate ligands have been thoroughly studied. In complexes with four monodentate ligands, the reactions are associative and often rapid. The stepwise conversion of AuCl$_4^-$ into AuBr$_4^-$ goes through a series of complexes including cis- and trans-AuBr$_2$Cl$_2^-$. All forward and reverse step of the reaction is described by second-order kinetics:

$$\text{Rate} = k_2[\text{AuL}_4][\text{X}] \tag{6.9}$$

Typically, square planar platinum(II) substitution reactions involve a solvolysis term that is not necessary for this rate law. In spite of this, the reactions of gold(III) are generally several orders of magnitude faster than their platinum(II) analogs and strongly depend on the nature of the entering nucleophile. There is a linear relationship between the free energies of activation and the free energies of the overall reaction [15]. This is consistent with the formation of five-coordinate intermediates, even though few stable penta-coordinate gold(III) species have actually been characterized crystallographically. This indicates that the reactions probably involve labile transition states instead of intermediates.

The previous reactions all involved monodentate ligands. Ligands with higher denticity often reduce the thermodynamic driving force for ligand exchange and/or increase the activation energy, causing slower reactions. This can result in greater stability of the bidentate and multidentate complexes in aqueous solution [16] compared with monodentate complexes. Kinetic studies of gold(III) reactions with ethylenediamine and related ligands show that the initial displacement of one end of the chelate is most often followed by rapid reclosure of the ring, rather than displacement of the second bond to the metal ion [15].

6.3
Medicinally Important Gold Complexes, Their Analogs and Reactions

The earliest use of gold complexes in medicine primarily involved gold thiolates, which being yellow or even gold-colored in solution, led to their designation as *chrysotherapy* (chrysos is Greek for gold).

6.3.1
Oligomeric Gold(I) Thiolates and Analogs

Gold(I) thiolates in the thiolate : gold ratio of 1 : 1 or 2 : 1 were the first chrysotherapeutic agents to be used. They are administered intramuscularly and not orally because they cannot be readily absorbed from the gut and by other tissues. Table 6.1 lists some chrysotherapy agents that will be discussed, with their formulae and Figure 6.1 shows chemical structures for some important thiolate ligands.

The ratio of available thiolate : gold will determine the structure formed. The gold (I) centers will be bridged by thiolate ions. If the gold : thiolate ratio is exactly 1 : 1 a cyclic structure will ensue. This occurs to preserve two-coordination of the gold(I). If excess thiolate is present, the structure of the formed oligomer will either be (a) alternating gold to thiolate linkages yielding an open chain with the excess thiolate capping its ends, or (b) an equilibrium between a cyclic structure and a bis-thiolato gold complex:

$$TmS\text{-}[Au(STm)]_n^- \rightleftharpoons [AuSTm]_{n-1} + Au(STm)_2^- \qquad (6.10)$$

The solubility of these oligomers is important for their practical use in medicine. In the 1 : 1 reactions, the nature of the thiolate is important since the solubility of the resulting oligomer is affected by substituents on the thiolate moiety. Polar groups

Table 6.1 Medicinally important chrysotherapy complexes. Adapted from Shaw [17].

Name	Trade Name	Formula
Gold sodium thiomalate	Myochrysine	Na_2AuSTm
Gold thioglucose	Solganol	$AuSTg$
Gold sodium thiosulfate	Sanochrysin	$Na_3Au(S_2O_3)_2$
2,3,4,6-Tetra-O-acetyl-1-thio-β-D-pyranosato-S-(triethylphosphine)gold(I)	Auranofin	$Et_3PAuSAtg$

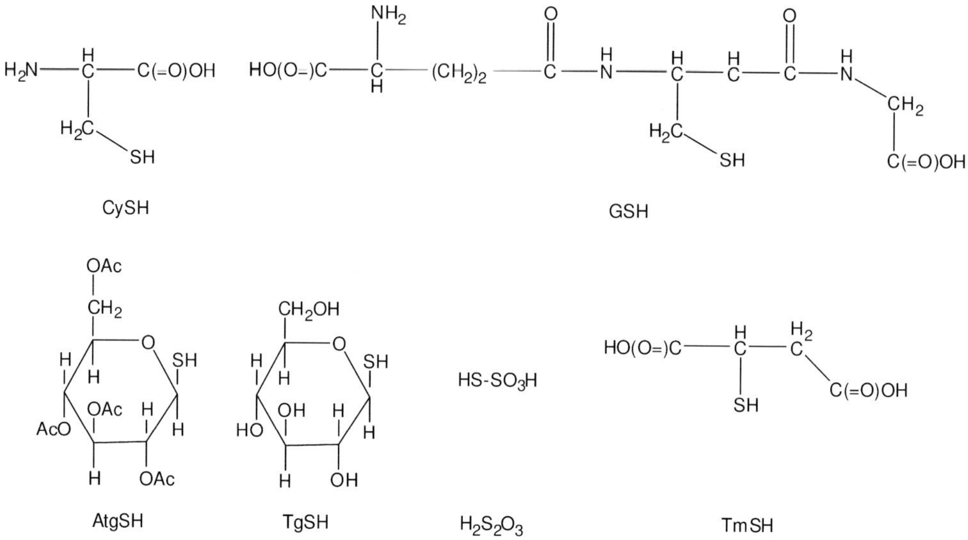

Figure 6.1 Structures of various thiolates discussed in this chapter. Abbreviations: CySH = cysteine; GSH = glutathione; AtgSH = tetra-O-2,3,4,6-acetyl-1-β-D-thioglucose; TgSH = β-1-D-thioglucose; $H_2S_2O_3$ = thiosulfuric acid; and TmSH = thiomalic acid.

such as carboxylates and hydroxyls on the thiol ligand usually increase aqueous solubility while unsubstituted aromatics and aliphatic groups will lead to relatively insoluble compounds [18]. Many complexes of gold(I) with biologically relevant ligands such as cysteine, N-acetylcysteine, glutathione, and penicillamine have been synthesized and characterized [19–23] but there are surprisingly few reports about their respective biological activities.

One of the most studied, yet least understood of the gold(I) drugs is aurothiomalate, AuSTm (Myocrysine). AuSTm is prepared with a slight excess of thiomalate [24–26] and also about a third of a mole of glycerol. The glycerol appears to be necessary in order to obtain a precipitate. Because of the excess thiomalate used, AuSTm is assumed to comprise open-chained oligomers with the excess thiomalate capping the ends. Due to its composition, AuSTm is better represented $Au_n(STm)_{n+1}^-$ or $Au_n(STm)_m$.

The first and only single crystal X-ray structure of a gold(I) thiomalate, resembling myocrysine, with the structure $Na_2CsAu_2(TmS)(TmSH)$, where TmS is the thiomalate trianion $[O_2CCH_2CH(S)CO_2]^{3-}$, has been reported [27]; see Figure 6.2. In the solid state, it is polymeric with the gold–sulfur backbone consisting of two interpenetrating spirals that have approximate fourfold helical symmetry, Figure 6.2. A racemic mixture is formed with S-thiomalate forming left-handed helices and R-thiomalate forming right-handed helices. The symmetric unit contains two different gold atoms. The Au–S distances are experimentally equivalent with Au_1–S being 228.9(8) Å and Au_2–S = 228.5(7) pm, consistent with the findings from previous EXAFS spectroscopic studies [20, 27, 28]. The bond angles of the two gold

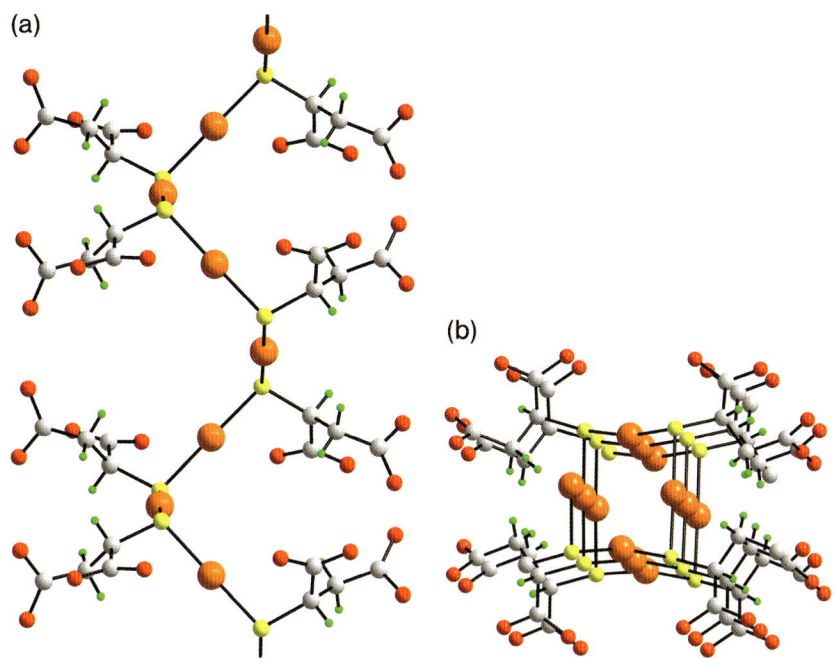

Figure 6.2 Two views of the polymeric chain in Na$_2$CsAu$_2$(TmS)(TmSH) highlighting the (a) polymeric structure and (b) the fourfold helical symmetry.

atoms differ with the first being essentially linear with a S–Au$_1$–S bond angle of 178.9(5)°. The second gold atom is distorted with a S–Au$_2$–S bond angle of 169.4(4)°. The Au–S–Au angle was found to be 99.2(3)° [27].

Soluble RSAu oligomers have been observed to form bis(thiolato) gold(I) complexes in the presence of thiols [11, 29–33]:

$$\text{RSH} + 1/n[\text{RSAu}]n \rightleftharpoons \text{Au(SR)}_2^- + \text{H}^+ \tag{6.11}$$

These reactions forming bis(thiolato) complexes provide model systems to study reactions likely to occur *in vivo* between gold drugs and proteins including various enzymes. Many thiols can cleave thiolate bridges in these oligomeric aurothiolate complexes to form monomeric bis(thiolato) complexes [11, 29–33]. The reaction is pH dependent, but is usually rapid and complete at physiological pH. The extent of the reaction depends on the affinity of the thiol for gold(I) [11]. This may not hold true though when the ability of a thiolate to bridge two gold(I) ions is similar to its affinity to coordinate terminally. An example is provided by the equilibrium constants found for Equation 6.11 when RSH = CySH [29] and TmSH [32]. For these thiols, the equilibrium constants are comparable, even though the two thiol ligands have different affinities to gold(I) [33]. A notable exception is ergothioneine which exists in a tautomeric equilibrium between a thiol (RSH) and a thione (R=S). The thione having lower affinity for gold(I) than the thiolate tautomer [29, 34].

The cysteine ligand forms an unusually insoluble gold(I) oligomer, $[(AuSCy)_n]$. The formation of this insoluble product can displace the equilibria of a number of reactions of biological significance including those with AuSTm [29, 35–38]:

$$CySH + Au(CN)_2^- + H^+ \rightarrow 1/n[CySAu]_n + 2HCN \qquad (6.12)$$

$$CySH + 1/n[AuSTm]_n \rightarrow 1/n[CySAu]_n + TmSH \qquad (6.13)$$

$$CySSCy + 1/n[AuSTm]_n \rightarrow 1/n[CySAu]_n + TmSSCy \qquad (6.14)$$

However, $[CySAu]_n$ is found to redissolve when a large excess of CySH is present. This is due to the formation of the bis(cysteinato)gold(I) complex, $[Au(SCy)_2]^-$, which at pH 7.4 is favored. This is an example of bis(thiolato)gold(I) formation.

Gold(I) thiolates also undergo reactions with disulfides. An example is 2,2-dithiobis(2-nitrobenzoic acid) (ESSE) used as a kinetic probe of sulfhydryl reactivity in metallothioneins [39–42]. ESSE is readily attacked by metal-bound thiolates. The reaction with the open chain form of AuSTm seems to occur in two stages. The first attack involves the terminal sulfides and the second attack involves the bridging thiolates [19]:

$$RS\text{-}[AuSR]_n\text{-}AuSR^- + 2ESSE \rightarrow ES^-[AuSR]_n\text{-}Au\text{-}SE^- + 2RSSE \qquad (6.15)$$

$$ES\text{-}[AuSR]_n\text{-}Au\text{-}SE^- + nESSE \rightarrow ES\text{-}[AuSE]_n\text{-}Au\text{-}SE^- + nRSSE \qquad (6.16)$$

AuSTm also reacts with cystine, the oxidized form of cysteine, forming the insoluble $(AuSCy)_n$ oligomer [36].

Although for many metal ions chelation is an important part of complex formation, tri- and tetrakis-(thiolato)gold(I) complexes and their cyanide analogs are not easily formed. Thus, chelation via traditional five- and six-membered rings is not observed with bis(thiolato)gold(I). Structural analyses of the cysteinato and penicillamine complexes of gold(I) [18, 21] showed oligomeric structures with bridging thiolates and no gold chelation via the sulfur or nitrogen atoms.

6.3.2
Bis(thiolato)Gold(I) Species

These bis(thiolato)gold(I) complexes are useful models for providing insights into the chemistry of the end products of the reactions between 1:1 gold thiolates and any excess (including other) thiol, including some reactions with proteins *in vivo*.

Aqueous solutions of the sodium salt, $Na_3[Au(S_2O_3)_2]$, are used medicinally under the generic name sanochrysin, an example of a truly "inorganic" drug, that is, containing no carbon. This bis(thiosulfato) complex of gold(I) has been crystallographically examined on multiple occasions [43, 44]. It is a linear complex where the thiolates are terminally coordinated to the gold center with a Au—S bond distance of 227 pm and a bond angle of S—S—Au of 104° [43, 44]. The Mössbauer parameters are consistent with the AuS_2 coordination environment [18, 21]. Two related complexes, $[Au(SC_6H_2iPr_3)_2]$ and $[Au(SPh)_2]^-$, with the similar $[Au(SR)_2]$ coordination, have been characterized crystallographically and found to have Au—S bond lengths of 229 and 227 pm, respectively.

6.3.3
Dithiocarbamates

Dithiocarbamate complexes of gold and their derivatives have been increasingly studied in recent years, being prepared with both gold(I) and gold(III). Complexes have been made featuring both monodentate and bidentate dithiocarbamate ligands.

Gold(I) dithiocarbamate complexes are interesting because their structures sometimes contain short intermolecular Au···Au distances [45] that can aggregate into chains also with short intramolecular Au···Au distances [46]. Recently mononuclear, dinuclear, and hexanuclear gold(I) complexes have been made with (aza-15-crown-5)dithiocarbamate. A hexanuclear complex, $[Au_6(S_2CNC_{10}H_{20}O_4)_6]$ has been characterized, that contains six gold atoms arranged in a chair-like conformation held together by a zig-zag system of bridging dithiocarbamates [46]. This is the first reported cyclohexane-like geometry in gold cluster chemistry. The Au···Au distances of 293 pm are found as well as Au···Au···Au bond angles of 117°. The molecular structure, with the hexagold layer being sandwiched between two crown-ether layers, pushes the sulfur and gold atoms apart preventing the gold atoms from aligning in a planar conformation [46].

Gold(III) dithiocarbamate complexes have also been synthesized. Stabilization of the gold(III) oxidation state by nitrogen-donating groups has focused particular interest on preparing gold(III) dithiocarbamate derivatives with similar structures to *cis*-diamminedichloroplatinum(II), cisplatin [47]. *In vitro* cytotoxicity testing have indicated that many gold(III) dithiocarbamates are sometimes up to four orders of magnitude more cytotoxic than cisplatin [47, 48]. However, gold(III) dithiocarbamates may be difficult to bring into solution which at present is a hindrance to their being used pharmacologically.

6.3.4
Auranofin and Other Phosphine(Thiolato)Gold(I) Species

The most recent gold-based anti-arthritic drug to be introduced was auranofin (Ridaura), a two coordinate gold(I) complex containing the anion derived from 2,3,4,6-tetra-O-acetyl-β-1D-thiolglucose and triethylphosphine as ligands [49]. The Au—S and Au—P bond distances are 229 and 226 pm, respectively and the S—Au—P bond angle is 174° [50]. The bond angle is slight distorted from linear due to steric effects. The bond angle Au—S—C_1 is 106° [50]. Auranofin was specifically designed to be lipophilic to ensure effective absorption from the intestine after oral administration.

As auranofin demonstrates both anti-arthritic and antitumor effects, many analogs differing in the thiolate and the phosphine residues have been characterized [18, 21, 51–53]. Like auranofin, these complexes are generally linear at the two-coordinate gold center. Di-μ-(diethylphosphinoethylthio)digold(I) is a complex that demonstrates the tendency to retain two-coordination even though ligands favorable for chelation are present [54].

Auranofin contains both a thiolate and a phosphine in the same simple monomeric unit. Because these two ligands have different reactivities, the chemical behavior of auranofin can be quite complex. When reacted with other thiols in aqueous or alcohol/aqueous solution, auranofin and its analogs undergo ligand exchanges with only the thiolate ligand. The phosphine ligand is more strongly coordinated and remains gold bound in aqueous solution:

$$Et_3PAuSAtg + RSH \rightleftharpoons Et_3PAuSR + AtgSH \qquad (6.17)$$

Like the gold(I) thiolates discussed earlier, the position of the equilibrium depends upon the affinity of the thiols for gold(I) with the established sequence being AtgSH > TgSH > GSH > TmSH. This sequence corresponds to increasing pK_{SH} [12] or by decreasing ^{31}P NMR chemical shifts for the Et_3PAuSR complex in aqueous solution [55].

In contrast to the equilibrium established with thiols, excess phosphine will displace the thiolate of auranofin or its close analogs to yield a bis(phosphine) complex.

$$Et_3PAuSAtg + Et_3P + H_2O \rightarrow (Et_3P)_2Au^+ + AtgSH + OH^- \qquad (6.18)$$

If cyanide is present, both the thiolate and the phosphine ligand can be displaced [56]. While the mixed thiol environment displacement is an equilibrium process, the right-hand reaction is much more favored in the reactions of phosphine and cyanide.

In acidic solutions auranofin undergoes deacetylation to form the thioglucose complex [57]:

$$Et_3PAuSAtg \rightarrow Et_3PAuStg + 4\,HOAc \qquad (6.19)$$

The final product after complete deacetylation contains the thioglucose ligand. This product was considered for medicinal use, but it is extremely hygroscopic giving it a short shelf-life.

There is some evidence that auranofin may also be biologically de-acetylated during its absorption from the gut [58]. It is unfortunate that so many *in vitro* studies to determine possible mechanisms for the anti-arthritic activity of auranofin have not considered (tetra) desacetyl-auranofin as the first likely "active metabolite," with its far greater hydrophilicity than the administered auranofin (which is only a pro-drug).

6.3.5
Au—S Bond Length Comparisons

Some comparisons of the Au—S bond length within some complexes are possible with those complexes having the stoichiometries $[AuSR]_6$, $[Au(SR)_2]^-$ and $[Ph_3PAuSR]$ for a single ligand, RSH = 2,4,6-tri*iso*propylphenylthiol [53, 59]. Table 6.2 compares their Au—S bond lengths and selected bond angles with those of the complexes used as anti-arthritic agents. There are no significant changes in

Table 6.2 Selected bond angles and lengths of medicinal gold(I) thiolates and 2,4,6-tri*iso*propylphenylthiolatogold(I) analogs. Adapted from Shaw [17].

Compound	D (Au—S/P)	Angle (S—Au—S/P)	Angle (Au—S—C/S)	Ref.
Oligomeric thiolates, —[—AuSR—]—:				
[Au(S$_6$H$_2$iPr$_3$)]$_6$	229	176	108	[59]
Au$_n$STm$_m$	230			[60]
Bis(thiolato)gold(I), [Au(SR)$_2$]$^-$:				
[Au(SPhiPr$_3$)$_2$]$^-$	229	176	105	[59]
[Au(S$_2$O$_3$)$_2$]	228	177	104	[59]
Phosphinegold(I)thiolate, R'$_3$PAuSR:				
Ph$_3$PAuSPhiPr$_3$	S 229, P 226	176	105	[53]
Et$_3$PAuSAtg	S 229, P 226	174	106	[50]

bond length between terminal and bridging Au—S bonds or when the ligand opposite a terminal Au—S is changed from thiolate to phosphine. More remote chemical changes in the molecular structure have apparently little effect on the thiol and sulfur cores.

6.3.6
Complexes of Diphosphine Ligands

Diphosphine ligands are a class of compounds developed from the search for new gold complexes with antitumor activity. The complexes are named for the bis (diphenylphosphino)ethane (Ph$_2$PCH$_2$CH$_2$PPh$_2$, dppe) ligand and other related "diphos" ligands they contain [61–64]. These ligands can chelate a single gold(I) ion or bridge two gold centers [65]. In serum, a rearrangement occurs to form bis (diphos)gold(I) cations from digold complexes [22]:

$$2(\mu\text{-dppe})(\text{AuX})_2 \rightarrow [\text{Au(dppe)}_2]^+ X + 3\text{AuX} \quad (6.20)$$

This demonstrates the inherent stabilization of the complex provided by chelation. The cationic complex formed is stable for 25 hours in serum and eight days in the presence of glutathione [16]. It does, however, react with copper(II) to form [Cu(dppe)$_2$]$^{2+}$. Structure–function studies have shown that diphosphines which complex copper(II) are active, and those that do not are inactive [63]. An example is [Au(dppe)$_2$]Cl, bis[1,2-bis(diphenylphosphino)ethane]gold(I) chloride, that is a cationic, four-coordinate tetrahedral gold(I) phosphine complex with chelating diphosphine ligands [66]. This complex is much more stable than the previously discussed gold(I) complexes in terms of ligand exchange. This decrease in exchange reactivity is important because it confers more stability in the presence of serum proteins, thiols, and disulfides, compared with auranofin [66]. Another potential advantage of gold(I) diphosphine complexes is the ability to use appropriate substituted diphosphine ligands to adjust the complexes' lipophilicity and hydrophilicity [66, 67]. This has led

6.3.7
Gold(I) Cyanide Complexes

Aurocyanide, also known as dicyanogold(I), [Au(CN)$_2$]$^-$, was highlighted by Robert Koch for its bacteriostatic properties [68]. [Au(CN)$_2$]$^-$ has subsequently been identified as a common metabolite of both oligomeric gold(I) thiolates and auranofin [69], present in serum and excreted in the urine. It has been proposed that [Au(CN)$_2$]$^-$ might have a role as a therapeutic agent (see below). The anion is linear and the Au–C bond length is 212 pm [70]. The cyanide ion has an extremely great affinity for gold(I) with an association constant estimated to be 10^{38} [71]. Despite this high affinity, the bound cyanides undergo rapid exchange with free cyanide ions [72], which exemplifies the difference between thermodynamic affinity and kinetic inertness.

Cyanide, *in vivo*, should react with auranofin. When the stoichiometry of cyanide is less than 1:1, the thiolate ion is displaced:

$$Et_3PAuSAtg + HCN \rightarrow Et_3PAuCN + AtgSH \tag{6.21}$$

When there is an excess of cyanide, both the thiol and the phosphine will be displaced in the next step:

$$Et_3PAuCN + HCN \rightarrow Et_3P + [Au(CN)_2]^- + H^+ \tag{6.22}$$

A study of patients being treated with aurothiomalate showed that gold was being taken up by red cells (erythrocytes) of smokers at a much greater level than by non-smokers [71]. The reason was found to be the HCN found in cigarette smoke at the level of 1700 ppm [71, 73] and readily absorbed from the lungs. This reacts with circulating serum protein-bound gold or oligomeric gold thiolates to form [Au(CN)$_2$]$^-$, which is responsible for gold uptake into red blood cells.

The equilibrium constant at pH 7.4 for the cyanolysis of AuSTm to yield [Au(CN)$_2$]$^-$ is 6.0×10^2 M^{-1}. The equilibrium constant for the competition of cyanide and cysteine for gold(I) is 1.67×10^2 M^{-1}. Although strictly these constants cannot be compared due to the different forms of oligomeric AuSTm and monomeric [Au(SCy)$_2$]$^-$, it can be shown that in the presence of equimolar quantities of cyanide and thiol, the cyano complex will be favored. Excess thiol is more likely than excess cyanide under *normal* biological conditions, so the equilibrium may be one in which cyanide complexes are decomposed by thiols.

However, the situation may be completely different in a *pathogenic* context, particularly when cyanide is locally generated within a site of inflammation, for example in arthritic joints. Here, leukocytes, recruited by inflammagenic chemotactic factors, are able to generate cyanide endogenously from plasma thiocyanate (a normal detoxication product of exogenous cyanide). Consequently, there may be a high local cyanide concentration, more than sufficient to liberate gold(I) from non cell-permeating thiolate complexes in the circulation. The [Au(CN)$_2$]$^-$ now produced locally as an "active metabolite" can then be taken up by activated leukocytes and other inflammatory cells, to down-regulate their hyperactivity by a type of Trojan-horse therapy/strategy. Within these cells, where thiols are in excess, the reverse transformation can occur that is, thiolysis. The $[Au(CN)_2]^-$ now being decomposed to liberate endocellular cyanide – a known potent intracellular toxin.

This mechanism only requires $[Au(CN)_2]^-$ to be a highly transient "metabolite", generated locally from administered gold drugs and in turn perhaps largely metabolized/decomposed locally.

$[Au(CN)_2]^-$ inhibits the oxidative burst of polymorphonuclear leukocytes and the proliferation of lymphocytes *in vitro* (both types of cells actively participate in the development and maintenance of inflammatory processes of rheumatoid disease) [71]. It is also far more toxic than gold(I) thiolates to bacteria, plants and animals.

6.3.8
Heterocyclic Carbenes

Gold(I) N-heterocyclic carbenes (NHC) have been known since 1989, but have attracted much recent interest because of some similarities between NHC and phosphine ligands [74]. Gold(I)–phosphine complexes have been studied for their antitumor properties. Gold(I)–NHC complexes are now being synthesized and studied not only as an alternative ligand to gold(I) phosphines, but also to gain more understanding of the role that phosphine ligands may play in conferring antitumor activitiy [75]. The neutral or cationic complexes are made in a variety of methods, often from easily synthesized imidazolium salts and have formulas of (NHC)AuX and $(NHC)_2AuX$ [74]. Like the diphosphine ligand class, NHC can be synthesized to have variable hydrophilic and lipophilic properties depending on ligand substitution [67, 75].

6.4
Gold–Protein Reactions and Complexes

In vivo, gold(I) binds predominately with proteins and low-molecular weight thiols. Reactions involving selenols, present for example in the enzyme glutathione peroxidase are also important. Gold(I) appears to not interact significantly with DNA or RNA. In general, gold(III) which can bind to nucleic acids (*cf.* platinum(II)) *in vitro* may not do so *in vivo* as it is probably reduced to gold(I) before it can enter cell nuclei. The next section surveys some reactions of gold compounds with selected proteins and enzymes.

6.4.1
Serum Albumin

Serum Albumin is the most abundant plasma protein in humans (and other mammals) with a concentration of about 600 µM in healthy humans. Around 80–90% of the extracellular gold in the circulation is bound to albumin and about

10–20% is bound to immunoglobulins [76–79]. Only trace amounts of free gold are present in serum. Gold is so prominently bound in this manner because albumin is the primary source of extracellular thiols in the blood, with a circulating thiolate concentration of about 400 µM. The protein is folded into three domains with each domain having two sub-domains. The tertiary structure is stabilized by 17 disulfide bonds. There are 35 cysteines overall but Cys-34 is the only cysteine present in the reduced form and rarely forms a disulfide linkage with any external ligand, for example, only perhaps after therapy with D-penicillamine. The Cys-34 site is located in a fold between helices h2 and h3 in sub-domain IA [80]. The pK_{SH} is extremely low at biological pH indicating that it is in effect deprotonated at physiological pH [81].

Cys-34 is blocked forming disulfide species, usually with glutathione or cysteine, in about 30% of circulating albumin molecules [82]. These natural disulfides formed with Cys-34 have little affinity for gold but develop avidity for gold if reductively cleaved. Gold compounds have been shown to react with Cys-34 by ligand exchange reactions [22, 23, 82–90]:

$$AlbS^- + Et_3PAuSAtg + H^+ \rightarrow AlbSAuPEt_3 + AtgSH \qquad (6.23)$$

$$AlbS^- + 1/n(AuSTm)_n \rightarrow AlbS(AuSTm)_n^- \rightarrow AlbSAuSTm^- \qquad (6.24)$$

These gold exchange reactions are prevented by prior alkylation of this protein thiol. Because of its lower affinity for gold when Cys-34 exists as a disulfide species, gold-to-sulfhydryl and gold-to-total albumin ratios are considered when analyzing data. Albumin with the Cys-34 blocked by alkylation is often used.

Using the complex AuSTm, up to six or seven equivalents of gold could be bound to the protein. Radiolabeling of the gold in AuSTm was carried out to track the fate of its associated ligand which was found to be retained upon gold binding. EXAFS studies showed that the gold retains AuS_2 coordination environment with normal Au–S bond lengths around 228–230 pm [63]. When protein is in excess, there is binding of a single AuSTm at Cys-34, but an oligomeric species binds with an excess of AuSTm present [22]. The AuS_2 coordination environment for $AlbS(AuSTm)_{6-7}$ is consistent with the formation of an open chain oligomer and not with the binding of single AuSTm groups to the histidine residues.

It was shown that ^{31}P NMR was a good tool for investigating the formation of protein adducts [84, 85]. The reaction of albumin with $Et_3PAuSAtg$ and with Et_3PAuCl was followed with ^{31}P NMR. One resonance only was observed for both complexes that corresponded to $AlbSAuPEt_3$. A further reaction of Et_3PAuCl at the histidine groups occurs, forming adducts with up to Et_3PAu^+ 17 groups bound. The reactions at the histidine groups are probably physiologically irrelevant, because they require the Cys-34 sites to be saturated first.

The albumin-binding of auranofin and a tri-*iso*-propylphosphine analog have also been analyzed. They each reacted rapidly with Cys-34 [91]. With excess auranofin, a rate-limiting rearrangement of the albumin molecule occurred which can be used to study the kinetics of the system. The rate-limiting step is the first one in a three step mechanism and involves a conformational change of the albumin where a crevice is opened surrounding Cys-34. The second step has auranofin binding to albumin with the AtgS$^-$ ligand leaving. Finally, a closure of the crevice around the gold is observed in the third step. The kinetic process is governed most likely by one of two processes or perhaps a combination of both. The first is attributed to the opening of a crevice surrounding Cys-34 [80]. The second has been described as a rearrangement of the Cys-34 residue, monitored by an allosteric effect on His-3 associated with the gold binding and disulfide formation [89]. ^{31}P NMR reveals a much faster second-order exchange of R$_3$PAu$^+$ between AtgSH and Cys-34. This is governed by the second step of the mechanism and occurs frequently with Cys-34 in the exposed position. The mechanism is completed by closure of the crevice around the bound gold.

In vivo, the albumin concentration nearly always exceeds the gold concentration by more than an order of magnitude and so, the reaction can be treated as pseudo-first order:

$$\text{rate} = k_{\text{pfo}}[\text{auranofin}] \tag{6.25}$$

The half-life is short, about 2 seconds, which is consistent with the rapid separation of AtgSH and Et$_3$PAu$^+$ in vivo [92]. This indicates that auranofin is short-lived in vivo and should preferably be studied using the albumin adducts of it (and other gold compounds) instead of the drug itself. Unfortunately, many model studies of the effects of gold thiolates described in the literature deal with their effects upon isolated cells in vitro but have neglected to include a physiological concentration of albumin. Combination with albumin effectively "buffers" the level of free (more toxic) gold ions and diminishes the likelihood of observing toxicities that are in fact artifacts, such as irreversible damage to cell membranes.

The harmless metabolite, Et$_3$PO, is formed when the phosphine of auranofin is oxidized in vivo. AlbSAuPEt$_3$ is used to model the formation of Et$_3$PO from phosphine. AlbSAuPEt$_3$ is stable for over a week in the absence of thiols [83]. However, if thiols like GSH or AtgSH are present, it is slowly oxidized over a 24 hour period [86]. The reaction is anaerobic indicating that oxygen is not a necessary oxidant [86]. However, when the reaction is carried out aerobically in ^{17}O-enriched water, Et$_3$P^{17}O is produced [87]. AlbSAuPEt$_3$ reacts rapidly with cyanide to release Et$_3$PO [93]. A reaction scheme depicting the equilibrium between the reactants Et$_3$PAu$^+$, serum albumin, and cyanide combined in varying forms has a common three-coordinate intermediate with gold coordinated to CN$^-$, PEt$_3$ and AlbS$^-$ [93]. This scheme shows that cyanide sourced from smoking or otherwise (see 6.3.7) may determine the metabolism of chrysotherapeutic agents. [Au(CN)$_2$]$^-$ anions bind to serum albumin predominantly by the formation of adducts without the displacement of cyanide [94]. The ions bind tightly to albumin independent of the oxidative state of Cys-34. The equilibrium constant values for [Au(CN)$_2$]$^-$ binding to serum albumin are similar to values for other gold complexes that bind to albumin. This indicates that albumin can act as a carrier for transporting [Au(CN)$_2$]$^-$ in the bloodstream.

6.4.2
Metallothioneins

Metallothioneins (MTs) are small proteins with an especial affinity for the binding of various heavy metals active in a wide range of reactions [95–97]. Besides their role in metabolism of the essential metals copper(I) and zinc(II), these proteins are known to protect against oxidative stress. Metallothioneins help detoxify against four general classes of cytotoxins: (i) metals accumulated from the environment such as cadmium(II) and mercury(II); (ii) reactive oxygen species (ROS); (iii) alkylating agents such as chlorambucil and streptozotazin; and (iv) therapeutic heavy metals particularly gold(I) and platinum(II) liberated from anti-arthritic and antitumor drugs [95–98]. This section considers only the reactions of MTs with gold complexes. The metallothionein proteins that are gold-bearing are sometimes called aurothioneins. The reactions of gold complexes with metallothioneins depend upon the initial ligation of the gold.

Mammalian MTs have two domains with a total of 61 amino acids, 20 being cysteine residues. Each MT can bind up to seven zinc(II) or cadmium(II) ions [98, 99]. The N-terminal β-domain of MT has an M_3SCy_9 arrangement and the C-terminal α-domain has an M_4SCy_{11} arrangement. In each of the domains is a crevice which exposes the metal-coordinated thiolates to solvent and low-molecular-weight solutes [99]. Metallothioneins react with a wide range of electrophiles including disulfides, alkylating agents and metal ions [100]. The cysteine thiolate groups exposed at the crevice may be involved in the initial attack of metallothioneins. If no metal is bound to the protein, it lacks structure and is a random coil [101].

A considerable amount of the gold that accumulates in the kidneys and liver of mammalian species is bound to MTs. This buildup of gold in the kidneys is accompanied by elevated levels of renal copper to form copper-rich, gold-bearing MTs. In cell lines that overproduce MT, there is commonly a resistance to the cytotoxic effects of gold compounds. This resistance is also seen often in parent lines that have been repeatedly exposed to gold complexes. The mechanisms of resistance include but are not limited to enhanced biosynthesis of MT [102].

There are four possible ways that gold(I) binds to MTs. Two motifs use two-coordinate binding of gold to cysteine residues. In bidentate chelation, a gold ion will be bound to two thiolates giving adducts with the general formula $(Au_{10}S_{20})$-MT. The thiolate-bridged model also involves two-coordinate gold in an AuS environment. In this model, gold ions will bind two thiolates with an alternating gold-thiolate chain forming. A general formula for this particular adduct is $(Au_{20}S_{20})$-MT. With monodentate coordination, gold is bound to one cysteine of the MT and to another exogenous ligand giving a general formula of $(LAuS)_{20}$-MT. The final proposed binding mode is termed isomorphous substitution. For isomorphous substitution to occur, gold(I) would need to displace zinc(II) or cadmium(II) and adopt a tetrathiolate coordination environment to form an adduct with the general formula (Au_7S_{20})-MT. Isomorphous substitution is not seen although the other binding modes have all been observed. If excess AuSTm is present, it will displace all M^{2+} ions to form $MT(SAuSTm)_{20}$. The thiomalate ligand is retained upon displacement. EXAFS has established an Au–S coordination number of two and bond lengths of 230 pm [20, 103]. Thus, the product entails monodentate coordination of AuSTm moieties to the cysteines [20, 103]. This is similar to the binding of AuSTm at Cys-34 of albumin.

If the thiomalate ligand is not present in excess, it is completely displaced. When gold binds, the M^{2+} ions are displaced in the gold:metal ratio of 3:2, which suggests bidentate coordination. EXAFS data shows AuS_2 coordination with Au−S bond lengths of 229 pm [20, 103]. In the presence of both cadmium and zinc, zinc is preferentially displaced which is possibly a consequence of thermodynamics as Zn_7MT and Cd_7MT react at comparable rates.

Unlike AuSTm, the phosphine-containing complex of $Et_3PAuSAtg$ does not react with M_7-MT [83, 104]. The Et_3P and AtgSH ligands have much greater affinities for gold than TmSH, which is easily displaced. When the MT protein is free of bound metal, $Et_3PAuStg$ will react [83] due to the changes in thermodynamics of the reaction. Around 80% of the Et_3P ligand and all of the AtgSH will be displaced [83]. Because some phosphine is retained, it suggests that gold is bound in the monodentate mode. Et_3PAuCl will react more readily than the thiolate, displacing zinc in preference over cadmium [105].

Metallochromic dyes have been used to study the kinetics of aurothionein due to the lack of intrinsic and distinctive chromophoric changes during reaction [105]. The reaction of the holo-protein with AuSTm occurs in two phases with each step showing first-order dependence on the MT concentration but independent of AuSTm [105]:

$$\text{rate}_{\text{holo}} = k_f[\text{MT}] + k_s[\text{MT}] \qquad (6.26)$$

The rates are essentially independent of the distribution of metal in the MT with similar rates between Zn_7−MTm $(Cd, Zn)_7$-MT, and Cd_7−MT. The values of the rate constants are $k_s = 6.9(\pm 0.9) \times 10^{-4} s^{-1}$ and $k_f = 2.7(\pm 1.2) \times 10^{-2} s^{-1}$ for the holoprotein. The slow rate constant is similar in magnitude to the first-order protein-dependent steps observed for reactions of DTNB (5,5′-dithiobis(2-nitrobenzoate)), EDTA (ethylenediamine tetraacetate), cisplatin, and other reagents, which has been attributed to a rearrangement of the protein. The fast step is more rapid by an order of magnitude, which suggests that other mechanisms are prevailing.

The α- and β-domains have been isolated in order to study their role in the two reaction phases. The slow reactions occur predominately with the β-domain while the fast reaction is associated entirely with the α-domain [106]. This pattern follows other circumstances where the α-domain is more reactive than the β-domain. However, the β-site has Cd^{2+} ions that are thermodynamically less tightly bound and more labile to inter-site exchange.

Some interesting biological consequences stem from the physiochemical properties of aurothioneins. It has been noted that typically only a third of the gold accumulated in cells and tissues is actually bound to MT, unlike cadmium which is more extensively bound. Structural changes arising from the differences in coordination environment of gold(I) ions and zinc(II) or cadmium(II) ions [103] have been invoked to explain the rapid degradation of aurothioneins (half-life of about 0.75 hour) compared with Cd_7−MT (24 hours) and Zn_7−MT (10 hours) [107, 108]. While aurothionein formation is rapid [20], the uptake of gold into the kidneys and livers of test animals is usually rather slow [106], indicating that the kinetics of gold binding to MT is not the rate-determining step for gold accumulation within organs.

6.4.3
Selenium-Dependent Glutathione Peroxidase

Glutathione-peroxidase (GSH-Pxase) is an enzyme found in erythrocytes and other tissues that has an essential selenocysteine residue involved in the catalytic decomposition of reactive oxygen species. In the erythrocyte, hydrogen peroxide is the principle reactive oxygen species available.

The effect of aurothioglucose (AuSTg) on GSH-Pxase has been studied both *in vitro* [109] and *in vivo* [110, 111]. The catalytic mechanism of the protein entails the oxidation of the active-site selenocysteine residue which is reduced in two steps by glutathione through a mixed selenium-sulfide intermediate. AuSTg competitively inhibits GSH-Pxase by forming an inactive seleno-gold complex with an inhibition constant, K_i of 2.3 µM. Thiomalate is a non-competitive inhibitor of GSH-PXase. Because of this, the ability of AuSTm to inhibit the reaction could not be determined. The glutathione complex of gold, AuSG, inhibits GSH-Pxase competitively when studied *in vitro* [109] with an inhibition constant of 2.8 ± 0.4 mM. This value is similar to the inhibition constant with AuSTg value which has a K_i of 2.4 ± 0.5 mM [110, 111]. The close similarity of these K_i values confirms a postulate of Chaudière and Tappel that the 1 mM excess of GSH used in this experimental study actually displaces thioglucose from gold, so the actual inhibitor is probably $Au(SG)_2^-$:

$$AuSTg + 2GSH \rightleftharpoons Au(SG)_2^- + TgSH + H^+ \quad (6.27)$$

Oligomeric $[AuSG]_n$ reacts with excess glutathione to form the same inhibitor $Au(SG)_2^-$. The K_i values for $Et_3PAuSAtg$ and Et_3PAuCl have also been determined being, 11.6 ± 0.8 and 10.8 ± 0.5 mM, respectively. The values are around four times larger than the values for AuSG and AuSTg, indicating that the former two (with the Et_3P ligand) are less effective inhibitors. The phosphine complexes exhibit similar K_i values although the observation that the chloride and ATgSH ligands differ greatly in affinity for gold(I) indicates both ligands are displaced in the presence of GSH, forming a phosphine-gold(I)-glutathione complex:

$$Et_3PAuX + GSH \rightleftharpoons Et_3PAuSG + X^- + H^+ \quad (6.28)$$

This type of *in vitro* study is relevant because it mimics likely physiological conditions. Red cells contain 1–2 mM GSH so the formation of glutathione complexes would be a likely consequence when auranofin metabolites enter red cells [112, 113]. Compared with typical *in vivo* gold concentrations of 10–15 µM observed in patients, the K_i values for the four gold complexes are less than anticipated. This indicates that chrysotherapy can greatly depress GSH-Pxase activity *in vivo* so the normal cellular redox balance would be displaced, favoring the accumulation of H_2O_2 and possibly GSH [3, 79, 114].

6.4.4
Gold(III) Oxidation of Insulin and Ribonuclease

Insulin and ribonuclease have been used as model systems for examining the chemical and immunological effects of their oxidation by gold(III). Being a powerful oxidant, gold(III) readily oxidizes thiols to disulfides and thioethers, for example, methionine to sulfoxides. Gold(III) can even oxidize disulfides to higher oxidation states, RSO_n^-, $n = 1, 2, 3$ [115]. Formation of gold(III) derivatives *in vivo* may be responsible for both some of the immunological-driven adverse reactions but could also help deliver some of the benefits of gold therapeutics [116–118].

Insulin, a small protein of molecular mass 6000 daltons, is composed of two chains designated A and B. There are no reduced cysteine residues in insulin, but it contains three essential disulfide bonds; two that crosslink the A and B chains, and one internal to the A chain to stabilize the overall tertiary structure. These disulfide bonds are cleaved in the presence of excess AuX_4^-, leaving A and B chains that have cysteine residues that have become oxidized to sulfonic acids [119]. With smaller amounts of AuX_4^-, a single disulfide bond will be attacked to form sulfinic acid [119]. The reaction is second order for $AuCl_4^-$ while $AuBr_4^-$ reacts too quickly for accurate monitoring.

Ribonuclease (RNase) is larger than insulin and has a number of methionine residues as well as two histidine residues at the active site [120]. RNase is shown to be inhibited when treated with $AuCl_4^-$ at pH 5.6 [121], the methionines are oxidized to sulfoxides at this pH with concomitant aggregation of the protein. The protein is partially unfolded at lower pH. At the site of unfolding, the reaction is more rapid and the loss of activity somewhat greater [121]. Gold(III) ions may also bind to the histidine catalytic center. In another study, gold(III) ions were noted to bind to both histidine and methionine residues, inhibiting RNase A activity and inducing a conformation change in the enzyme protein. However, with the addition of thiourea, which forms stable complexes with gold ions, RNase A was able to refold indicating that the gold-induced denaturation was largely reversible [120].

6.4.5
Enzyme Inhibition

Gold complexes have been extensively investigated for their ability to inhibit enzymes that is, act as cellular poisons in the course of providing therapeutic benefits [122]. Examples include effects on proteolytic (auto) digestion of joint tissue, formation of pro-inflammatory arachidonic acid metabolites (eicosanoids) and the protein kinase C signal transduction pathway. Auranofin has caused some serious gastrointestinal side effects (diarrhea, even fibrosis) attributed to an initial inhibition of gut wall ATPases. While many enzymes have been postulated as targets for gold complexes, some require mM levels of gold for IC_{50} while the maximum circulating levels of total gold *in vivo* is less than 20 µM. This circulating concentration is misleading, because most gold is "held up" in the albumin. For enzyme inhibition to be effective from gold agents, the K_i values for the enzymes must be less than 20 µM. Table 6.3 lists K_i

Table 6.3 K_i or IC_{50} values for biologically important enzymes and gold complexes. Adapted from Shaw [17].

Enzyme	Complex	K_i or IC_{50} (μM)	Ref.
Ribonuclease	$AuCl_4^-$	$IC_{50} = 40$ μM	[121, 123]
Glutathione Peroxidase	AuSTg	$K_i = 2.3$ μM	[109]
	AuSG	$K_i = 2.8$ μM	[112]
	$Et_3PAuSAtg$	$K_i = 11.6$ μM	[112]
	Et_3PAuCl	$K_i = 10.8$ μM	[112]
Myelperoxidase	AuSTm	$K_i = 180$ μM	[124]
	$Et_3PAuSAtg$	$K_i = 6.2$ μM	[124]
Cathepsin-B_1	AuSTm	$IC_{25} = 6.8$ μM	[125]
	$Et_3PAuSAtg$	$IC_{25} = 54$ μM	[125]
	$AuCl_4^-$	$IC_{25} = 20$ μM	[125]
Cathespin-D	AuSTm	$IC_{25} = 46$ μM	[125]
	$Et_3PAuSAtg$	$IC_{25} = 540$ μM	[125]
Na, K-ATPase	$Et_3PAuSAtg$	$IC_{50} = 200$ μM	[126]

or IC_{50} values for biologically important enzymes and gold complexes. Even this circulating concentration is misleading, because most serum gold is transported in the form of an albumin complex.

6.4.6
Zinc finger Proteins

The zinc finger proteins are regulatory elements for protein expression. The coordination of cysteine to zinc in these structures is analogous to that in metallothionein polypeptides. It is thought that these particular proteins might likewise react with intracellular gold. With coordination environments of $Zn(SCy)_4$ or $Zn(His)_2(Cys)_2$, they play an important role in stabilizing DNA binding elements. It has been verified that gold compounds can interact with the particular zinc finger proteins for the cJun and cFos proteins, both important regulators of immune response [127].

A study was performed using DNA binding assays to assess the interaction of gold(I) thiomalate with two different zinc finger transcription factors, TFIIIA and Sp1 [128]. Both of these factors have multiple Cys_2His_2 zinc finger domains and play a role in stimulating cell growth. Gold was found to dissociate from the thiomalate complex and the free gold(I) ions could then bind to the zinc finger structure, effectively replacing the Zn^{2+} ion. It was found that Au^{1+} ions exchanged between the zinc finger domains much more slowly than Zn^{2+} ions indicating a higher affinity for gold [129]. The zinc finger motifs are present in a wide variety of regulatory proteins and are in turn regulated by the release or binding of zinc. If gold(I) ions are able to disrupt this homeostasis due to their greater affinity, this property could also contribute to the overall therapeutic efficacy of gold(I) complexes [128].

6.4.7
Hemoglobin and Interprotein Gold Transfer ("Transauration")

Hemoglobin, the major protein of the red blood cell, has two cysteine residues near the surface (Cys-β93) and four cysteine residues that are buried (Cys-α104 and Cys-β112) in a tetrameric structure, $\alpha_2\beta_2$. When treated with Et_3PAuCl, Et_3PAu^+ was found to bind to the Cys-β93 residues of Hb as well as GSH in red cells [85]. The cysteine residues of hemoglobin are thought to have a lower affinity for gold(I) than the Cys-34 residues of albumin. This is attributed to the fact that the reactions of Et_3PAuCl are more complete than those of $Et_3PAuSAtg$.

Under physiological conditions, transfer of Et_3PAu^+ from hemoglobin to albumin has been observed. The transfer is direct and does not require a low-molecular weight thiol such as GSH as an intermediary and indicates that inter-protein transfer of gold can occur spontaneously. The rapid and efficient manner of establishing this equilibrium (for "transauration") could determine many of the effects of intracellular and extracellular chemistry of gold.

6.4.8
Mitochondrial Thioredoxin Reductase

Thioredoxin reductases, belonging to the same enzyme family as glutathione reductase, are a class of selenium-containing oxidoreductases that catalyze the NADPH-dependent reduction of thioredoxins, a family of disulfide reductases that maintain many proteins in their reduced state. Thioredoxin reductases have wide substrate specificity. This is due to a second redox-active site on a C-terminal −Cys−SeCys− (where SeCys is selenocysteine), that is not present in glutathione reductase [67]. Thioredoxins have many roles including regulation of redox signaling and acting as electron donors for some enzymes. Thioredoxin reductases may also have a pathophysiological role in some cancers and for rheumatoid arthritis making it a potential drug target [67].

Gold(I) containing drugs, particularly auranofin, are potent inhibitors of thioredoxin reductase, almost certainly reacting with the selenocysteine within the catalytic site. By contrast gold(I) is a poor inhibitor of glutathione reductase which lacks this particular amino acid [67, 129]. Inhibition of mitochondrial thioredoxin reductase may explain some of the therapeutic/toxic effects of gold(I) drugs as antitumor agents, given the emerging role of mitochondria as a key participants in the apoptotic process (inducing cell death).

Auranofin and other gold(I)-phosphines induce apoptosis via selective inhibition of mitochondrial thioredoxin reductase [67]. At submicromolar concentrations of gold(I) phosphines, in the presence of calcium ions, mitochondrial permeability transitions (MPT) are observed in rat liver cells including mitochondrial swelling and loss of membrane potential. These changes have little effect on the mitochondrial electron transport chain or glutathione reductase, but are associated with the selective inhibition of mitochondrial thioredoxin reductase [67, 129]. After treatment with Et_3PAuCl, isolated rat heptocytes showed changes in their mitochondria, decreased

ATP levels and oxygen consumption [66]. Et_3PAuCl treatment also reduced the potential difference across the inner mitochondrial membrane releasing sequestered calcium from mitochondria [66].

6.5
Physiological and Cellular Biochemistry

6.5.1
Biological Ligand Exchange

In both patients and laboratory animals, gold(I) drugs are rapidly metabolized. The carrier ligands and gold suffer different fates, not surprising in view of the lability of the gold complexes [24, 130–133]. This occurs by exchange of ligands between those which are exogenous, that is, present in the administered drugs, and the many endogenous ligands. This mechanism is indicated by the lack of free gold ions after chrysotherapy. If a dissociative mechanism was involved, then free gold ions should be present after the carrier ligands were lost and metabolized.

While the metabolism of gold(I) drugs is generally rapid, gold is generally retained *in vivo* much longer than the original dosed agent. When radiolabeled auranofin, $Et_3{}^{31}P$-^{195}Au-$^{35}SATg$, was given to dogs, each component of the drug was quickly distributed with differing fates. The half-lives of excretion for ^{31}P and ^{35}S being eight and six hours, respectively, while the half-life for excretion of gold is 20 days [132]. The same triply radiolabeled auranofin was also added to whole blood and the observation was that bond cleavage occurred much more rapidly [132]. This *in vitro* experiment demonstrated that rapid ligand exchanges occurred for each component of auranofin. Thus, within 20 minutes the gold had primarily moved into the red cells, the acetylthioglucose was mainly bound in the serum, and the phosphine distributed between the red cells, low-molecular weight species, and to a lesser extent extracellular proteins [132]. A similar study indicated the rapid displacement of radiolabeled ^{35}STm from $^{198}Au^{35}STm$ administered to rats [133] and ^{35}STg from $^{198}Au^{35}STg$ given to mice [131]. Gold evidently will remain in the organs much longer than the ^{35}S label. In patients receiving chrysotherapy, free thiomalate can be detected in the blood after the administration of gold thiomalate [24].

These experimental findings indicate that gold drugs, as administered, probably act as pro-drugs. The ligands present in the (original) chrysotherapeutic agents are quickly displaced *in vivo* by protein and low-molecular-weight thiols and are merely acting as carriers for introducing the gold into the body. The phosphine and acetylthioglucose constituents of auranofin are certainly readily displaced as these portions of the drug move freely into and out of red blood cells. Because these "carrier" ligands are so easily displaced, the actual exposure of cells and tissues to the intact original drug is very short-lived. Subsequently, protein-bound metabolites are the dominant species *in vivo*.

6.5.2
The Sulfhydryl-Shuttle Model

The hydrophilic gold(I) drugs which need to be injected such as AuSTm and AuSTg, cannot easily move into most cells. Instead, they bind to cell surfaces where they can alter membrane transport and affect the overall cell metabolism [115, 134].

The drug auranofin on the other hand, being much more lipophilic, readily enters many cells [135–137]. A mechanism named the sulfhydryl shuttle was invoked to explain the uptake and efflux of auranofin metabolites from cultured macrophages [137]. According to this shuttle hypothesis, the AtgSH ligand is displaced before entering the cell. The remaining "SAuPEt$_3$" complex is transported into the cell by a process that involves sulfhydryl-dependent membrane proteins (MSH). These MSH proteins promote the movement of Et$_3$PAu$^+$ across the cell membrane although much of the triphenylphosphine ligand may be oxidatively lost in the membrane before the gold is shuttled into the cytoplasm. Once inside the cell, Et$_3$PAu$^+$ is transferred to various proteins and low-molecular-weight sulfhydryl species from which Et$_3$P can be displaced and oxidized to Et$_3$PO by further reaction with intracellular thiols. The inward transfer (import) of gold can be blocked by alkylation of surface cysteine residues with agents such as N-ethylmaleimide (NEM) or oxidation to a disulfide by Ellman's reagent (ESSE) [135–137]. This important process is not energy-dependent and therefore not a form of active transport. Nor is it a process of simple diffusion since modifying the cell surface thiols inhibits gold uptake, implying that the process is nevertheless a mediated one [135].

Red cells provide a useful model for studying the likely movement of gold into and out of cells. It was shown that Et$_3$PAuCl moving into red cells binds to glutathione and hemoglobin [84, 85] providing ancillary evidence for the sulfhydryl shuttle. Similarly, $[Au(CN)_2]^-$ has been shown to enter red cells by the sulfhydryl shuttle and its uptake can be blocked by alkylation of membrane thiols [32].

The export of cellular gold can be done with or without Et$_3$P bound to it. If the complex still exists as Et$_3$PAu$^+$, it can move out of the cell on the same carriers that it entered. However, if Et$_3$P is oxidized, gold bound to glutathione may be removed via the same transport proteins that specifically transfer organic conjugates of GSH out of the cell. The efflux of gold from a red cell is accompanied by an efflux of glutathione [138]. AlbSAuSG, analogous to AlbSAuSTm and AlbSAuSAtg, is a ternary complex of extracellular albumin with both gold and glutathione associated. It has been suggested that AlbSAuSG could be the primary circulating metabolite generated after displacement of the carrier ligands [138].

6.5.3
Equilibration of Intra- and Extracellular Gold

Excluding gold complexes that cannot transverse the membrane such as AuSTm, intracellular gold concentrations must approach equilibrium with extracellular gold concentrations if active transport is not at work. The balance of gold between the environment of a cell and within the cell will depend on the ligands available.

In vitro Raw 264.7 macrophages and B16 melanoma showed decreased uptake of auranofin in the presence of increasing quantities of serum albumin [137, 138]. Friend leukemia cells had a reduced of uptake of $Ph_3PAu(tTP)$ (tTP = 8-thiotheophyllinate) under similar conditions [51]. Given that the cells are rather varied, the trend is general and not limited to specific cell types. The albumin in the medium also reduces the cytotoxicity of auranofin to B16 melanoma cells and of 5-thiotheophyllinato(triethylphosphine)gold(I) to Friend leukemia cells [51, 131, 139].

For gold accumulation in cells to be an equilibrium process, it should be seen that the equilibrium is reversible and that the efflux of gold should be dependent on extracellular ligands. Red cells exposed to serum albumin lose more gold than cells exposed to buffered isotonic saline [138], the enhanced efflux occurs due to the presence of extracellular binding sites with a high affinity for gold. This finding supports the concept of an equilibrium between intra- and extracellular gold.

6.5.4
Cytotoxicity and Antitumor Activity of Gold Complexes

Many gold compounds have been tested for their potential antitumor activity by examining their cytotoxicity against cell lines in culture. This avoids barriers that arise *in vivo* due to uncertain absorption, restricted distribution, and variable metabolism of drugs.

Table 6.4 lists a number of gold compounds tested against B16 melanoma *in vitro*. The results cannot be applied universally across all cell lines due to varying characteristics of individual cell types. However, in general the presence of a phosphine ligand is important for activity. Much less activity is observed for the binary thiolate complexes, or those with carbon or nitrogen donors [140, 141]. One example of this tendency was demonstrated in a study of 6-mercaptopurine (6-MP) and Ph_3PAuCl. While individually, the thiol ligand and Au-phosphine, that is, Ph_3PAuCl, were potent, the combined complex, that is, $Ph_3PAu(6-MP)$, was most

Table 6.4 Cytotoxcitiy and antitumor activity of selected gold complexes on B16 Melanocytes and P388 leukocyte. Adapted from Shaw [17].

Complex	B16 Melanoma IC_{50} (μM) *in vitro*	P388 *in vivo* ILS_{max} (%)	Ref.
$[AuSTm]_n$	60	24	[64]
$[AuSTg]_n$	166	15	[64]
$[AuSAtg]_n$	150	14	[64]
Et_3PAuCl	1	36	[64]
Et_3PAuCN	0.4	68	[64]
$Et_3PAuSAtg$	1.5	70	[64]
$Et_3PAuSTg$	2	68	[64]
$dppe(AuCl)_2$	8		[63]
$dppe(AuCl_3)_2$	15		[63]
dppe	60		[63]

potent [140]. Due to this tendency, auranofin analogs have shown much greater potential as antitumor agents than the binary gold(I) thiolates previously discussed. A difference in the two classes is that gold–phosphine complexes can enter cells while the oligomeric thiolates cannot, and therefore, do not accumulate in cells. Table 6.4 also lists some gold compounds tested against P388 leukemia cells *in vivo*. The response of the cells to treatment *in vivo* depends not only on the pharmacokinetics of the specific compound and whether it accumulates in the cell, but also whether the cells develop resistance to gold's cytotoxicity.

Currently cisplatin in the most widely used cancer drug. Because of its time-tested efficacy, it is used as a standard against which new potential cancer drugs are tested. Gold(I) and gold(III) complexes, especially those with diphosphine or dithiocarbamate ligands, have shown promise in cell line studies. Gold(I) dithiocarbamates and to a greater extent gold(III) dithiocarbamate complexes show cytotoxicity in many cisplatin-resistant cell lines. A similar trend is observed with many gold(I) and gold(III) diphosphine complexes showing activity in cisplatin-resistant cell lines. Significant *in vivo* activity was observed for $[Au(dppe)_2]^+$ in mice with P388 leukemia, M5076 reticulum cell sarcoma, and mammary adencarcinoma [141].

6.5.5
Oxidation States *in Vivo*

Gold(I) is the expected and most prevalent form of gold *in vivo*. There is yet no chrysotherapeutic agent designed to specifically deliver gold(III). Nevertheless, the ligand exchange reactions and interactions of proteins with gold complexes discussed previously all indicate that the gold in the $+1$ oxidation state is favored and under some conditions, preserved biologically. Aurosomes, that is, lysosomes that accumulate large amounts of gold, have been isolated from rats treated with gold(III) and found to contain predominately gold(I) [20, 28]. Peptide and protein methionine residues as well as disulfide bonds all reduce gold(III) to gold(I) [115, 119, 121].

Even under these conditions, the potential for gold(III) to be formed *in vivo* has been explored and predicted. After AuSTm was administered to mice for some weeks [116, 117] it was found that gold(III) *in vitro* elicited a hyper-reactive response in the popliteal lymph node assay (PLNA) although AuSTm, the gold(I) species originally given, does not. The PLNA is useful because it can distinguish between the effects of a drug or any of its metabolites which may be immunogenic. This observation leads to the conclusion that *in vivo* gold drugs can be transformed to a gold(III) metabolite that initiates some of the immunological side effects of chrysotherapy [130, 142]. Further clinical studies confirmed that T-cells from arthritis patients receiving chrysotherapy could become sensitized to gold(III) but not gold(I) [119].

Oxidation of gold(I) to gold(III) can occur during oxidative stress, particularly when HOCl is produced by myelperoxidase from activated leukocytes. HOCl can oxidize gold in AuSTm, AuSTg, auranofin, and $[Au(CN)_2]^-$ to form gold(III) complexes [71, 142, 143] by the following reactions:

$$AuSR + OCl^- \rightarrow AuCl_4^- + RSO_3^- \qquad (6.29)$$

$$\mathrm{Et_3PAuSAtg + OCl^- \rightarrow AuCl_4^- + AtgSO_3^- + Et_3PO} \qquad (6.30)$$

$[\mathrm{Au(CN)_2}]^-$ may be oxidized to $[\mathrm{Au(CN)_4}]^-$ [84] but this gold(III) complex is reduced back to $[\mathrm{Au(CN)_2}]^-$ by glutathione. However, $[\mathrm{Au(CN)_4}]^-$ is much more stable than other gold(III) complexes, and may perhaps diffuse from its site of synthesis to exert its pharmacological or toxic effects, before being reduced back to $[\mathrm{Au(CN)_2}]^-$.

There appears to be a disparity between the observation that gold is present predominantly as gold(I) *in vivo* but that an oxidative burst (associated with leukocyte activation) can generate gold(III) and that T-cells can become sensitized to gold(III) but not to the gold(I) drugs themselves. These facts can be reconciled if a redox cycle develops [94] in which the administered drugs first form a protein–gold (I)–ligand complex. Then during the oxidative burst this complex can be converted to $\mathrm{Au(III)X_4}^-$ that may be converted back to the gold(I) form by thiols, disulfides, thioethers, and so on. This cycle accounts for the observations that metals, thiols, proteins, and so on are seen in much greater concentration in tissues than is stoichiometrically expected compared with the low concentration of gold administered during chrysotherapy.

6.5.6
Immunochemical Consequences of Gold(III)

RNase has been used as a model to investigate the possible role of gold(III) on peptide presentation [111]. The immune system first subjects proteins to proteolytic digestion by antigen-presenting cells (APCs), such as macrophages in order to distinguish between "self" and "non-self." Peptide fragments generated during the digestion are then bound to the multi-histocompatibility complexes (MHC molecules) and transferred to the surface of the APC, where T-cells react to non-self-peptide sequences. In auto-immune diseases, T-cells react to self-proteins not usually presented. These proteins are called cryptic because they are not usually presented nor recognized, but are still part of the self-protein. When a mouse is challenged with normal RNase, it will develop T-cells that react against the peptide segment 74–88. However, challenging the same strain of mouse with gold(III)-treated RNase led to the establishment of several CD4$^+$ T-cell hybridoma lines responsive to gold(III)-treated RNase, but not to the native protein. When tested against different peptide segments of RNase, peptides 7–21 and 94–108 elicited responses, that is, were cryptic peptides elicited with gold(III)-treated RNase but not by native RNase. The ability of the oxidized metal species to generate cryptic peptides may both underlie some of the adverse (immunological) side effects experienced during chrysotherapy but also perhaps sustain some of the beneficial effect of gold(I) therapy.

6.5.7
Anti-HIV Activity

In the context of treating AIDS, there has been recent excitement from the research of anti-HIV properties of gold complexes. Reverse transcriptase (RTase) is an enzyme that converts viral RNA into DNA in the host cell. AuSTg, a reported RTase inhibitor [144], shows activity in cell-free extracts [141], but is unable to actually enter cells where RTase is active. $[Au(STg)_2]^-$ can be generated *in situ* from AuSTg and TgSH but appears to act in a different manner [145]. It inhibits the infection of MT-4 cells by HIV strain HL4-3 without inhibiting the RTase activity in the intact virions. Cys-532 on gp160, a glycoprotein of the viral envelope, has been identified as the critical target site [145]. $[Au(STm)_2]^-$ is also active, but oligomeric AuSTm and AuSTg, auranofin and many other compounds were found to be inactive.

$[Au(CN)_2]^-$ is taken up into H9 cells, a T-cell line susceptible to HIV infection and retards the proliferation in these cells. At low concentrations, as low as 20 ppb, it is well tolerated in arthritic patients and may serve well as an adjunct alongside existing HIV treatments [146].

6.5.8
Gold Nanoparticles

These are not strictly compounds but being charged particles they may be associated with counterions when formulated as hydrosols for medicinal use. Recently, nanoparticles have become of great interest in the scientific community for their unique properties, especially those containing gold [147]. Metallic gold (Au^0) nanoparticles are readily generated by the reduction of gold(III) complexes such as $HAuCl_4$ [139]. Many methods have been developed for attaching amino acids, peptides, proteins, and so on onto the nanoparticles in order to stabilize the particles, prevent aggregation, and even help them become more soluble.

Bulk metallic gold and its alloys appear to be relatively biologically inert, many studies showing they are not cytotoxic and do not elicit stress-induced secretion of pro-inflammatory cytokines [143]. Consequently, gold nanoparticles have been used as carriers of bioactive agents in gene delivery programs and shown to cause little damage to targeted living cells and tissues with no toxic results [143]. Gold nanoparticles (27 nm) have also proven useful as a potential anti-arthritic treatment, in one rat study showing activity 1000 times that of gold thiomalate [143]. This raises the interesting questions as to whether Au^0 is:

(i) the ultimate pharmaco-active species formed from labilized gold(I) complexes *in vivo* by dismutation, that is, $3Au(I)$ giving $2Au^0$ and $Au(III)$; or
(ii) a pro-drug, needing to be locally oxidized (again involving active leukocytes) to the actual bioregulant species (gold(I) or gold(III)) [148].

6.6
Conclusions

Gold biochemistry has seen much development in recent years. The interaction of gold complexes with proteins and mitochondria, the possible role(s) of gold(III) *in vivo*, and the potential for developing future therapies using gold nanoparticles are all exciting prospects. Various gold complexes can generate common transformation products ("metabolites") *in vivo* which still need to be studied further as they may be either the active species or be converted into it/them. How chrysotherapy "works" is still elusive after decades of study. It is a continuing challenge to see if the use of gold complexes as potential arthritic treatments can be refined, perhaps based on the concept that the present drugs are only precursors or pro-drugs that somewhat inefficiently deliver their therapeutic effects via far more potent intermediaries. Even more exciting and still needing much further study are the potential antitumor and anti-HIV activities attributed to gold complexes in the last few decades.

Yet, while we have many facts, there is still little understanding about the real significance of the various transformations that gold complexes undergo *in vivo*. It is not always helpful to consider these transformations as either useful or toxic, since some of the clinical benefits may actually derive from controlled/selective toxicities with one example being the *local* generation of $[Au(CN)_2]^-$ and potential *restricted* delivery of cyanide within an inflammatory locus. A more helpful approach may be to consider such local transformations, particularly those such as cyanolysis of the Au−S bond or oxidations of gold(I) to gold(III) that may be initiated by certain activated cell populations, for example, reactive leukocytes within an inflamed tissue (as in arthritis) or proliferating, perhaps mutated, cells in the absence of normal feedback controls (as in cancer).

References

1 Kean, W.F. and Kean, I.R.L. (2008) Clinical pharmacology of gold. *Inflammopharmacology*, **165**, 112–125.
2 Higby, G.J. (1982) Gold in medicine: a review of its use in the west before 1900. *Gold Bulletin*, **15**, 130–140.
3 Kean, W.F., Lock, C.J.L. and Howard-Lock, H. (1991) Gold complex research in medical science difficulties with experimental design. *Inflammopharmacology*, **1**, 103–114.
4 Pennerman, R.A. and Ryan, R.R. (1972) The di-aquo proton in hydrogen tetracyanoaurate(III) dehydrate. *Acta Crystallographica*, **B28**, 1629–1632.
5 Weishaupt, M. and Strähle, J. (1976) Crystal structure and vibrational spectrum of tetraamminegold(III) nitrate. *Zeitschrift fur Naturforschung*, **B31**, 554–557.
6 Jones, P.G., Guy, J.J. and Sheldrick, G.M. (1975) Tetraphenylarsonium tetrachloroaurate(III). *Acta Crystallographica*, **B31**, 2687–2688.
7 Dunand, A. and Gerdil, R. (1975) The crystal and molecular structure of ethylenediammonium bis{cis-[ethylenediaminedisulphitoaurate(III)]. *Acta Crystallographica*, **B31**, 370–374.
8 Skibsted, L.H. and Bjerrum, J. (1977) The standard electrode potentials of aqua gold ions. *Acta Chemica Scandinavica Series A: Physical and Inorganic Chemistry*, **31**, 155–156.
9 Chernyak, A. S. and Shestopalova, L.F. (1976) Study of complexes of gold(I) with cysteine in an alkaline medium. *Russian Journal of Inorganic Chemistry, English Translation*, **21**, 464–465.
10 Schmidbaur, H., Shiotani, A. and Klein, H.-F. (1971) NMR studies of ligand substitution processes at two- and four-coordinate gold atoms. *Chemische Berichte*, **104**, 2831–2837.

11 Isab, A.A. and Sadler, P.J. (1982) A carbon-13 nuclear magnetic resonance study of thiol exchange reactions of gold(I) thiomalate ("Myocrisin") including applications to cysteine derivatives. *Journal of the Chemical Society, Dalton Transactions*, (1), 135–141.

12 Baenziger, N.C., Dittemore, K.T. and Doyle, J.R. (1974) Crystal and molecular structure of chlorobis (triphenylphosphine) gold(I) hemibenzenate. *Inorganic Chemistry*, **13**, 805–810.

13 Baenziger, N.C., Bennett, W.E. and Sobroff, D.M. (1976) Chloro (triphenylphosphine)gold(I). *Acta Crystallographica*, **B32**, 962–966.

14 Khan, M., Oldham, G. and Tuck, D.G. (1981) The direct electrochemical synthesis of triphenylphosphine adducts of Group IB monohalides. *Canadian Journal of Chemistry*, **59**, 2714–2718.

15 Skibsted, L.H. (1986) Ligand substitution and redox reactions of gold(III) complexes. *Advances in Inorganic and Bioinorganic Mechanisms*, **4**, 137–183.

16 Berners-Price, S., Jarret, P.S. and Sadler, P.J. (1987) Phosphorus-31 NMR studies of $[Au_2(\mu\text{-dppe})]^{2+}$ antitumor complexes. Conversion into $[Au(dppe)_2]^+$ induced by thiols and blood plasma. *Inorganic Chemistry*, **26**, 3074–3077.

17 Shaw, C.F. III (1999) The biochemistry of gold, in *Gold Progress in Chemistry, Biochemistry, and Technology*, (ed. H. Schmidbaur), John Wiley & Sons, Ltd, Chichester, UK, pp. 260–305.

18 Al-Sa'ady, A.K.H., Moss, K., McAuliffe, C.A. and Parish, R.V. (1984) Mössbauer and nuclear magnetic resonance spectroscopic studies on "Myocrisin", "Solganol", "Auranofin", and related gold (I) thiolates. *Journal of the Chemical Society, Dalton Transactions*, (8), 1609–1616.

19 Brown, D.H., McKinley, G.C. and Smith, W.E. (1978) Gold complexes of L-cysteine and D-penicillamine. *Journal of the Chemical Society, Dalton Transactions*, (3), 199–201.

20 Elder, R.C., Tepperman, K.G., Eidsness, M.K., Heeg, M.J., Shaw, C.F. III and Schaeffer, N. A. (1983) Gold-based antiarthritic drugs and metabolites. Extended x-ray adsorption fine structure (EXAFS) spectroscopy and x-ray absorption near edge spectroscopy (XANES). *ACS Symposium Series*, **209**, 385–400.

21 Hill, D.T., Sutton, B.M., Isab, A.A., Razi, T., Sadler, P.J., Trooster, J.M. and Calis, G.H.M. (1983) Gold-197 Mossbauer Studies of Some Gold(I) Thiolates and Their Phosphine Complexes Including Certain Antiarthritic Gold Drugs. *Inorganic Chemistry*, **22**, 2936–2942.

22 Shaw, C.F. III, Schaeffer, N.A., Elder, R.C., Eidsness, M.K., Trooster, J.M. and Calis, G.H.M. (1984) Bovine serum albumin-gold thiomalate complex: gold-197 Moessbauer, EXAFS and XANES, electrophoresis, sulfur-35 radiotracer, and fluorescent probe competition studies. *Journal of the American Chemical Society*, **106**, 3511–3521.

23 Carlock, M.T., Shaw, C.F. III, Elder, R.C. and Eidsness, M.K. (1986) Reactions of Auranofin and Et_3PAuCl with Bovine Serum Albumin. *Inorganic Chemistry*, **25**, 333–339.

24 Rudge, S.R., Perrett, D., Swannell, A.J. and Drury, P.L. (1984) Circulating thiomalate after administration of disodium aurothiomalate: impurity or active metabolite? *The Journal of Rheumatology*, **11**, 150–152.

25 Isab, A.A. and Sadler, P.J. (1981) Hydrogen-1 and carbon-13 nuclear magnetic resonance studies of gold(I) thiomalate (Myocrisin) in aqueous solution. Dependence of the solution structure on pH and ionic strength. *Journal of the Chemical Society, Dalton Transactions*, (7), 1657–1663.

26 Grootveld, M.C. and Sadler, P.J. (1983) Differences between the structure of the antiarthritic gold drug 'Myocrisin' in the

solid state and in solution: a kinetic study of dissolution. *Journal of Inorganic Biochemistry*, **19**, 51–64.

27 Bau, R. (1998) Crystal structure of the antiarthritic drug gold thiomalate (Myochrysine): A double-helical geometry in the solid state. *Journal of the American Chemical Society*, **120**, 9380–9381.

28 Elder, R.C. and Eidsness, M.K. (1987) Synchrotron X-ray studies of metal-based drugs and metabolites. *Chemical Reviews*, **87**, 1027–1046.

29 Shaw, C.F. III, Schmitz, G., Thompson, H.O. and Witkiewicz, O. (1979) Bis(l-cysteinato)gold(I): chemical characterization and identification in renal cortical cytoplasm. *Journal of Inorganic Biochemistry*, **10**, 317–330.

30 Shaw, C.F. III, Eldridge, J. and Cancro, M.O. (1981) Carbon-13 NMR studies of aurothioglucose: ligand exchange and redox disproportionation reactions. *Journal of Inorganic Biochemistry*, **14**, 267–274.

31 Danpure, C.J. (1976) The interaction of aurothiomalate and cysteine. *Biochemical Pharmacology*, **25**, 2342–2346.

32 Elder, R.C., Jones, W.B., Floyd, R., Zhao, Z., Dorsey, J.G. and Tepperman, K. (1994) Myochrysine solution structure and reactivity. *Metal-Based Drugs*, **1**, 363–374.

33 Grootveld, M.C., Razi, M.T. and Sadler, P.J. (1983) Progress in the characterization of gold drugs. *Clinical Rheumatology*, **3** (Suppl. 1), 5–16.

34 Isab, A.A. (1992) The carbon-13 NMR study of the binding of gold(I) thiomalate with ergothionine in aqueous solution. *Journal of Inorganic Biochemistry*, **45**, 261–267.

35 Lewis, G. and Shaw, C.F. III (1986) Competition of thiols and cyanide for gold(I). *Inorganic Chemistry*, **25**, 58–62.

36 Danpure, C.J. and Lawson, K.J. (1977) Interaction of aurothiomalate and cystine. *Biochemical Society Transactions*, **5**, 1366–1368.

37 Brown, D.H., McKinlay, G.C. and Smith, W.E. (1977) Electronic and circular dichroism spectra of gold(I) complexes having sulfur- and phosphorus-containing ligands. *Journal of the Chemical Society, Dalton Transactions*, (19), 1874–1879.

38 Coates, G.E., Kowala, C. and Swan, J.M. (1966) Coordination compounds of Group IB metals. I. Triethylphosphine complexes of Au(I) mercaptides. *Australian Journal of Chemistry*, **19**, 539–545.

39 Reglinski, J., Hoey, S. and Smith, W.E. (1988) Exchange reactions between disulfides and myocrisin: an *in vitro* model for a mechanism in chrysotherapy. *Inorganica Chimica Acta*, **152**, 261–264.

40 Savas, M.M., Petering, D.H. and Shaw, C.F. III (1991) On the rapid, monophasic reaction of the rabbit liver metallothionein α-domain with 5,5′-dithiobis (2-nitrobenzoic acid) (DTNB). *Inorganic Chemistry*, **30**, 581–583.

41 Savas, M.M., Petering, D.H. and Shaw, C.F. III (1993) The oxidation of rabbit liver metallothionein-II by 5,5′-dithiobis (2-nitrobenzoic acid) and glutathione disulfide. *Journal of Inorganic Biochemistry*, **52**, 235–249.

42 Zhu, Z., Petering, D.H. and Shaw, C.F. III (1995) Electrostatic Influences on the Kinetics of the Reactions of Lobster Metallothioneins with the Electrophilic Disulfides 2,2′-Dithiodipyridine (PySSPy) and 5,5′-Dithiobis(2-nitrobenzoic acid) (ESSE). *Inorganic Chemistry*, **34**, 4477–4483.

43 Baggio, R.F. and Baggio, S. (1973) Crystal and molecular structure of trisodium dithiosulfateaurate(I) dehydrate, $Na_3[Au(S_2O_3)_2] \cdot 2H_2O$. *Journal of Inorganic & Nuclear Chemistry*, **35**, 3191–3200.

44 Reuben, Z., Zalken, A., Faltens, M.O. and Templeton, D.H. (1974) Crystal structure of sodium gold(I) thiosulfate dihydrate, $Na_3Au(S_2O_3)_2 \cdot 2H_2O$. *Inorganic Chemistry*, **13**, 1836–1839.

45 Fernández, E.J., López-de-Luzuriaga, J.M., Monge, M. and Olmos, E. (1998) Dithiocarbamate Ligands as Building-

Blocks in the Coordination Chemistry of Gold. *Inorganic Chemistry*, **37**, 5532–5536.

46 Arias, J., Baradaji, M. and Espinet, P. (2008) Mononuclear, dinuclear, and hexanuclear gold(I) complexes with (aza-15-crown-5)Dithiocarbamate. *Inorganic Chemistry*, **47**, 1597–1606.

47 Ronconi, L., Marzano, C., Zanello, P., Corsini, M., Miolo, G., Maccà, C., Trevisan, A. and Fregona, D. (2006) Gold(III) dithiocarbamate derivatives for the treatment of cancer: solution chemistry, DNA binding, and hemolytic properties. *Journal of Medicinal Chemistry*, **49**, 1648–1657.

48 Ronconi, L., Giovagnini, L., Marzano, C., Bettìo, F., Graziani, R., Pilloni, G. and Fregona, D. (2005) Gold Dithiocarbamate derivatives as potential antineoplastic agents: design, spectroscopic properties, and *in vitro* antitumor activity. *Inorganic Chemistry*, **44**, 1867–1881.

49 Sutton, B.M., McGusty, E., Walz, D.T. and DiMartino, M.J. (1972) Antiarthritic properties of alkylphosphinegold coordination complexes. *Journal of Medicinal Chemistry*, **15**, 1095–1098.

50 Hill, D.T. and Sutton, B.M. (1980) (2,3,4,6-Tetra-O-acetyl-1-thio-β -D-glucopyranosato-S) (triethylphosphine) gold, $C_{20}H_{34}AuO_9PS$. *Acta Crystallographica Section C: Crystal Structure Communications*, **9**, 679–686.

51 Arizti, M.P., Garcia-Orad, A., Sommer, F., Silvestro, L., Massiot, P., Chevallier, P., Gutierrez-Zorilla, J.M., Colacio, E., Martinez de Pancorbo, M. and Tapiero, H. (1991) Intracellular accumulation and cytotoxic effect of (8-thiotheophyllinate) (triphenylphosphine) gold(I) in Friend leukemia cells. *Anticancer Research*, **11**, 625–628.

52 Shaw, C.F. III, Beery, A. and Stocco, G.C. (1986) Antitumor activity of two binuclear gold(I) complexes with bridging dithiolate ligands. *Inorganica Chimica Acta*, **123**, 213–216.

53 Tong, Y.Y., Pombiero, A.J.L., Hughes, D.L. and Richards, R.L. (1995) Syntheses and properties of the gold(I) complexes with bulky thiolates $[Au(SR)]_6$ and $[Au(SR)(PPh_3)]$ ($R = C_6H_2Me_3$-2,4,6 or $C_6H_2Pr_3i$-2,4,6), and the molecular structure of $[Au(SC_6H2Pr_3i$-2,4,6)(PPh_3)]$. *Transition Metal Chemistry*, **20**, 372–375.

54 Crane, W.S. and Beall, H. (1978) The structure of a large ring gold complex. *Inorganica Chimica Acta*, **31**, L469–L470.

55 Shaw, C.F. III, Coffer, M.T., Klingbeil, J. and Mirabelli, C.K. (1988) Application of phosphorus-31 NMR chemical shift: gold affinity correlation to hemoglobin-gold binding and the first inter-protein gold transfer reaction. *Journal of the American Chemical Society*, **110**, 729–734.

56 Hormann, A.L., Coffer, M.T. and Shaw, C.F. III (1988) Reversibly and irreversibly formed products from the reactions of mercaptalbumin (AlbSH) with Et_3PAuCN and of $AlbSAuPEt_3$ with hydrocyanic acid. *Journal of the American Chemical Society*, **110**, 3278–3284.

57 Razi, M.T., Sadler, P.J., Hill, D.T. and Sutton, B.M. (1983) Proton, carbon-13, and phosphorus-31 nuclear magnetic resonance studies of (2,3,4,6-tetra-O-acetyl-1-thio-β -D-glucopyranosato-S) (triethylphosphine)gold (auranofin), a novel antiarthritic agent. *Journal of the Chemical Society, Dalton Transactions*, (7), 1331–1334.

58 Tepperman, K., Finer, R., Donovan, S., Doi, J., Ratclift, D. and Ng, K. (1984) Intestinal uptake and metabolism of auranofin, a new oral gold-based antiarthritis drug. *Science*, **225**, 430–432.

59 Schröter, I. and Strähle, J. (1991) Thiolato complexes of monovalent gold. Synthesis and structure of $[(2,4,6\text{-iso-}Pr_3C_6H_2S) Au]_6$ and $(NH_4)[(2,4,6\text{-iso-}Pr_3C_6H_2S)_2Au]$. *Chemische Berichte*, **124**, 2161–2164.

60 Elder, R.C., Ludwig, K., Cooper, J.N. and Eidsness, M.K. (1985) EXAFS and WAXS structure determination for an antiarthritic drug, sodium gold(I) thiomalate. *Journal of the American Chemical Society*, **107**, 5024–5026.

61 Berners-Price, S., Girard, G.R. and Hill, D.T. (1990) Cytotoxicity and antitumor activity of some tetrahedral bis(diphosphino)gold(I) chelates. *Journal of Medicinal Chemistry*, **33**, 1386–1392.

62 Berners-Price, S., Mirabelli, C.K. and Johnson, R.K. (1986) In vivo antitumor activity and in vitro cytotoxic properties of bis[1,2-bis(diphenylphosphino)ethane] gold(I) chloride. *Cancer Research*, **46**, 5486–5493.

63 Snyder, R.M., Mirabelli, C.K. and Johnson, R.K. (1986) Modulation of the antitumor and biochemical properties of bis(diphenylphosphine)ethane with metals. *Cancer Research*, **46**, 5054–5060.

64 Mirabelli, C.K., Johnson, R.K. and Hill, D.T. (1986) Correlation of the in vitro cytotoxic and in vivo antitumor activities of gold(I) coordination complexes. *Journal of Medicinal Chemistry*, **29**, 218–223.

65 Bates, P.A. and Waters, J.M. (1985) Structure of tetraphenylphosphonium bis(benzenethiolato)aurate(I), $[P(C_6H_5)_4]$ $[Au(C_6H_5S)_2]$. *Acta Crystallographica.*, **C41**, 862–865.

66 McKeage, M.J., Maharaj, L. and Berners-Price, S.J. (2002) Mechanisms of cytotoxicity and antitumor activity of gold(I) phosphine complexes: the possible role of mitochondria. *Coordination Chemistry Reviews*, **232**, 127–135.

67 Barnard, P.J. and Berners-Price, S.J. (2007) Targeting the mitochondrial cell death pathway with gold compounds. *Coordination Chemistry Reviews*, **251**, 1889–1902.

68 Koch, R. (1890) Ueber bacteriologische Forschung. *Deutsche Medizinische Wochenschrift (1946)*, **16**, 756–757.

69 Elder, R.C., Zhao, Z., Zhang, Y., Dorsey, J.G., Hess, E.V. and Tepperman, K. (1993) Dicyanogold (I) is a common human metabolite of different gold drugs. *The Journal of Rheumatology*, **20**, 268–272.

70 Rosenzweig, A. and Cromer, D.T. (1959) The crystal structure of $KAu(CN)_2$. *Acta Crystallographica*, **12**, 709–712.

71 Graham, G.G., Whitehouse, M.W. and Bushell, G.R. (2008) Aurocyanide, dicyano-aurate (I), a pharmacologically active metabolite of medicinal gold complexes. *Inflammopharmacology*, **16**, 126–123.

72 Pesek, J.J. and Mason, W.R. (1979) Carbon-13 magnetic resonance spectra of diamagnetic cyano complexes. *Inorganic Chemistry*, **18**, 924–928.

73 Smoking and Health Report of the Advisory Committee to the Surgeon General of the Public Health Service, PHS publication 1103, p. 53.

74 de Frémont, P., Stevens, E.D., Eelman, M.D., Fogg, D.E. and Nolan, S.P. (2006) Synthesis and characterization of gold(I) N-Heterocyclic carbene complexes bearing biologically compatible moieties. *Organometallics*, **25**, 5824–5828.

75 Baker, M.V., Barnard, P.J., Berners-Price, S.J., Brayshaw, S.K., Hickey, J.L., Skelton, B.W. and White, A.H. (2005) Synthesis and structural characterization of linear Au(I) N-heterocyclic carbene complexes: New analogues of the Au(I) phosphine drug auranofin. *Journal of Organometallic Chemistry*, **690**, 5625–5635.

76 Sadler, P.J. (1976) The biological chemistry of gold: a metallo-drug and heavy-atom label with variable valency. *Structure and Bonding*, **29**, 171–219.

77 Shaw, C.F. III (1979) The mammalian biochemistry of gold: an inorganic perspective of chrysotherapy. *Inorganic Perspective in Biology and Medicine*, **2**, 287–355.

78 Brown, D.H. and Smith, W.E. (1980) The chemistry of the gold drugs used in the treatment of rheumatoid arthritis. *Chemical Society Reviews*, **9**, 217–239.

79 Lewis, A.J. and Walz, D.T. (1982) Immunopharmacology of gold. *Progress in Medicinal Chemistry*, **19**, 1–58.

80 Carter, D. and Ho, J.X. (1994) Structure of serum albumin. *Advances in Protein Chemistry*, **45**, 153–204.

81 Lewis, S.D., Misra, D.C. and Shafer, J.A. (1980) Determination of interactive thiol ionizations in bovine serum albumin, glutathione, and other thiols by potentiometric difference titration. *Biochemistry*, **19**, 6129–6180.

82 Ni Dhubhghaill, O.M., Sadler, P.J. and Tucker, A. (1992) Drug-induced reactions of bovine serum album: proton NMR studies of gold binding and cysteine release. *Journal of the American Chemical Society*, **114**, 1117–1118.

83 Ecker, D.J., Hempel, J.C., Sutton, B.M., Kirsch, R. and Crooke, S.T. (1986) Reactions of the metallodrug auranofin [(1-thio-β-D-glucopyranose-2,3,4,6-tetraacetato-S)(triethylphosphine) gold] with biological ligands studied by radioisotope methodology. *Inorganic Chemistry*, **26**, 3139–3143.

84 Razi, M.T., Otiko, G. and Sadler, P.J. (1983) Ligand exchange reactions of gold drugs in model systems and in red cells. *ACS Symposium Series*, **209**, 371–384.

85 Malik, N.A., Otiko, G. and Sadler, P.J. (1980) Control of intra- and extra-cellular sulfhydryl-disulfide balances with gold phosphine drugs: Phosphorus-31 nuclear magnetic resonance studies of human blood. *Journal of Inorganic Biochemistry*, **12**, 317–322.

86 Coffer, M.T., Shaw, C.F., Hormann, A.L., Mirabelli, C.K. and Crooke, S.T. (1987) Thiol competition for Et_3PAuS-albumin: a nonenzymic mechanism for Et_3PO formation. *Journal of Inorganic Biochemistry*, **30**, 177–187.

87 Isab, A.A., Shaw, C.F. III and Locke, J. (1988) GC-MS and oxygen-17 NMR tracer studies of triethylphosphine oxide formation from auranofin and water-^{17}O in the presence of bovine serum albumin: an *in vitro* model for auranofin metabolism. *Inorganic Chemistry*, **27**, 3406–3409.

88 Shaw, C.F. III, Isab, A.A., Hoeschele, J.D., Starich, M., Locke, J., Schulteis, P. and Xiao, J. (1994) Oxidation of the phosphine from the auranofin analog, triisopropylphosphine(2,3,4,6-tetra-O-acetyl-1-thio-β-D-glucopyranosato-S) gold(I), via a protein-bound phosphonium intermediate. *Journal of the American Chemical Society*, **116**, 2254–2260.

89 Christodoulou, J., Sadler, P.J. and Tucker, A. (1994) A new structural transition of serum albumin dependent on the state of Cys34. Detection by 1H-NMR spectroscopy. *European Journal of Biochemistry*, **225**, 363–368.

90 Kamel, H., Brown, D.H., Ottaway, J.M. and Smith, W.E. (1977) Determination of gold in separate protein fractions of blood serum by carbon furnace atomic-absorption spectrometry. *Analyst*, **102**, 645–663.

91 Roberts, J.R., Xia, K., Schliesman, B., Parsons, D.J. and Shaw, C.F. III (1996) Kinetics and mechanism of the reaction between serum albumin and auranofin (and its isopropyl analog) *in vitro*. *Inorganic Chemistry*, **35**, 424–433.

92 Intoccia, A.P., Flanagan, T.L., Walz, D.T., Gutzait, L., Swagzdis, J.E., Flagiello, J., Hwang, B.Y.-H., Dewey, R.J. and Noguchi, H. (1983) Pharmacokinetics of auranofin in animals, in *Bioinorganic Chemistry of Gold Coordination Compounds*. (ed. B.M. Sutton and R.G. Franz), SK&F, Philadelphia, pp. 21–33.

93 Isab, A.A., Hormann, A.L., Coffer, M.T. and Shaw, C.F. III (1988) Reversibly and irreversibly formed products from the reactions of mercaptalbumin (AlbSH) with Et_3PAuCN and of $AlbSAuPEt_3$ with hydrocyanic acid. *Journal of the American Chemical Society*, **110**, 57–60.

94 Shaw, C.F. III, Schraa, S., Gleichmann, E., Grover, Y.P., Dunemann, L. and Jagarlamudi, A. (1994) Redox chemistry and $[Au(CN)_2]^-$ in the formation of gold metabolites. *Metal-Based Drugs*, **1**, 351–362.

95 Stillman, M.J. Shaw, C.F. JIII and Suzuki, K.T. (eds) (1992) *Metallothioneins: Synthesis, Structure, and Properties of Metallothioneins, Phytochelatins and Metal-*

thiolate Complexes, VCH Publishers, Weinheim, p. 442.
96 Suzuki, K.T., Imura, N. and Kimura, M. (1993) *Metallothionein III: Biological Roles and Medical Implications*, Birkhaüser Berlag, Basel, p. 470.
97 Kägi, J.H.R. and Kojima, Y. (1993) *Metallothionein II*, Birkhaüser Verlag, Basel, p. 479.
98 Robbins, A.H. and Stout, C.D. (1992) Metallothioneins: Synthesis, Structure and Properties of Metallothioneins, Phytochelatins, and Metal-Thiolate Complexes (eds M.I. Stillman, C.F. ShawIII and K.T. Suzuki), VCH Publishers, Weinheim, pp. 31–54.
99 Robbins, A.H., McRee, D.E., Williamson, M., Collett, S.A., Xuong, N.H., Furey, W.F., Wang, B.C. and Stout, C.D. (1991) Refined crystal structure of Cd, Zn metallothionein at 2.0 A resolution. *Journal of Molecular Biology*, **221**, 1269–1293.
100 Otvos, J.D., Petering, D.H. and Shaw, C.F. III (1989) Structure - Reactivity Relationships of Metallothionein, a Unique Metal-Binding Protein. *Comments on Inorganic Chemistry*, **9**, 1–35.
101 Vašák, M., Galdes, A., Hill, H.A.O., Kägi, J.H.R., Bremner, I. and Young, B.W. (1980) Investigation of the Structure of Metallothioneins by Proton Nuclear Magnetic Resonance Spectroscopy. *Biochemistry*, **19**, 416–425.
102 Shaw, C.F. and Savas, M.M. (1992) *Metallothioneins: Synthesis, Structure and Properties of Metallothioneins, Phytochelatins and Metal-Thiolate Complexes* (eds M.J. Stillman, C.F. ShawIII and K.T. Suzuki), VCH Publishers, Weinheim, pp. 145–163.
103 Laib, J.E., Shaw, C.F. III, Petering, D.H., Eiddness, M.K., Elder, R.C. and Garvey, J.S. (1985) Formation and characterization of aurothioneins: Au,Zn, Cd-thionein, Au,Cd-thionein, and (thiomalato-Au)x-thionein. *Biochemistry*, **24**, 1977–1986.

104 Shaw, C.F. III and Laib, J.E. (1986) Aurothionein formation from zinc, cadmium-thionein and chloro (triethylphosphine)gold, but not (2,3,4,6-tetra-O-acetyl-1-thio-β-D-glucopyranosato)(triethylphosphine) gold (Auranofin). *Inorganica Chimica Acta*, **123**, 197–199.
105 Shaw, C.F. III, Laib, J.E., Savas, M.M. and Petering, D.H. (1990) Biphasic kinetics of aurothionein formation from gold sodium thiomalate: a novel metallochromic technique to probe zinc $(2+)$ and cadmium$(2+)$ displacement from metallothionein. *Inorganic Chemistry*, **29**, 403–408.
106 Munoz, A., Laib, F., Petering, D.H. and Shaw, C.F. III (1999) Characterization of the cadmium complex of peptide 49-61: a putative nucleation center for cadmium-induced folding in rabbit liver metallothionein IIA. *Journal of Biological Inorganic Chemistry*, **4**, 495–507.
107 Monia, B.P., Butt, T.R., Ecker, D.J., Mirabelli, C.K. and Crooke, S.T. (1986) Metallothionein turnover in mammalian cells. Implications in metal toxicity. *The Journal of Biological Chemistry*, **261**, 10957–10959.
108 Mogilnicka, E.M. and Webb, M. (1981) Comparative studies on the distribution of gold, copper and zinc in the livers and kidneys of rats and hamsters after treatment with sodium gold-195-labeled-aurothiomalate. *Journal of Applied Toxicology*, **1**, 287–291.
109 Chaudière, J. and Tappel, A.L. (1984) Interaction of gold(I) with the active site of selenium-glutathione peroxidase. *Journal of Inorganic Biochemistry*, **20**, 313–325.
110 Hu, M.-L., Dillard, C.J. and Tappel, A.L. (1988) Aurothioglucose effect on sulfhydryls and glutathione-metabolizing enzymes: in vivo inhibition of selenium-dependent glutathione peroxidase. *Research Communications in Chemical Pathology and Pharmacology*, **59**, 147–160.

111 Baker, M.A. and Tappel, A.L. (1986) Effect of gold thioglucose on subcellular selenium distribution in rat liver and kidney. *Biological Trace Element Research*, **9**, 113–123.

112 Roberts, J. and Shaw, C.F. III (1998) Inhibition of erythrocyte selenium-glutathione peroxidase by auranofin analogs and metabolites. *Biochemical Pharmacology*, **55**, 1291–1299.

113 Roberts, J. (1993) The kinetic properties of Au(I) drug binding to serum albumin and selenium-glutathione peroxidase and their significance for rheumatoid arthritis. Ph.D. thesis, University-Milwaukee.

114 Smith, W.E. and Reglinski, J. (1991) Gold drugs used in the treatment of rheumatoid arthritis. *Perspectives on Bioinorganic Chemistry*, **1**, 183–208.

115 Shaw, C.F. III, Cancro, M.P., Witkiewicz, P.L. and Eldridge, J. (1980) Gold(III) oxidation of disulfides in aqueous solution. *Inorganic Chemistry*, **19**, 3198–3201.

116 Takahashi, K., Griem, P., Goebel, C., Gonzalez, J. and Gleichmann, E. (1994) The antirheumatic drug gold, a coin with two faces: Au(I) and Au(III), Desired and undesired effects on the immune system. *Metal-Based Drugs*, **1**, 483–496.

117 Schuhmann, D., Kubicka-Myranyi, M., Mirtcheva, J., Günther, J., Kind, P. and Gleichmann, E. (1990) Adverse immune reactions to gold. I. Chronic treatment with a gold(I) drug sensitizes mouse T cells not to gold(I), but to gold(III) and induces autoantibody formation. *Journal of Immunology*, **145**, 2132–2139.

118 Verwilghen, J., Kingsley, G.H., Gambling, L. and Panayi, G.S. (1992) Activation of gold-reactive T lymphocytes in rheumatoid arthritis patients treated with gold. *Arthritis and Rheumatism*, **35**, 1413–1418.

119 Witkiewicz, P.L. and Shaw, C.F. III (1981) Oxidative cleavage of peptide and protein disulfide bonds by gold(III): a mechanism for gold toxicity. *Journal of the Chemical Society, Chemical Communications*, (21), 1111–1114.

120 Maruyama, T., Sonokawa, S. and Matsushitaa, H. Goto, M. (2007) Inhibitory effects of gold(III) ions on ribonuclease and deoxyribonuclease. *Journal of Inorganic Biochemistry*, **101**, 180–186.

121 Isab, A.A. and Sadler, P.J. (1977) Reactions of gold(III) ions with ribonuclease A and methionine derivatives in aqueous solution. *Biochimica et Biophysica Acta*, **492**, 322–330.

122 Leibfarth, J.H. and Persellin, R.H. (1981) Mechanisms of action of gold. *Agents and Actions*, **11**, 458–472.

123 Christenson, G.M. and Olson, D.L. (1981) Effect of water pollutants and other chemicals upon ribonuclease activity in vitro. *Environmental Research*, **26**, 274–280.

124 Matheson, N.R. and Travis, J. (1983) The effect of gold compounds on human neutrophil myeloperoxidase, in *Bioinorganic Chemistry of Gold Coordination Compounds* (eds B.M. Sutton and R.G. Franz), SK&F, Philadelphia, pp. 58–66.

125 Lewis, A.J., Cottney, J., White, D.D., Fox, P.K., McNeillie, A., Dunlop, J., Smith, W.E. and Brown, D.H. (1980) Action of gold salts in some inflammatory and immunological models. *Agents and Actions*, **10**, 63–77.

126 Hardcastle, J., Hardcastle, P.T., Kelleher, D.K., Henderson, L.S. and Fondacaro, J.D. (1986) Effect of auranofin on absorptive processes in the rat small bowel. *The Journal of Rheumatology*, **13**, 541–546.

127 Handel, M.L., Watts, C.K., DeFazio, A., Day, R.O. and Sutherland, R.L. (1995) Inhibition of AP-1 binding and

transcription by gold and selenium involving conserved cysteine residues in Jun and Fos. *Proceedings of the National Academy of Sciences of the United States of America*, **92**, 4497–4501.

128 Larabee, J.L., Hocker, J.R. and Hanas, J.S. (2005) Mechanisms of aurothiomalate-Cys2His2 Zinc finger interactions. *Chemical Research in Toxicology*, **18**, 1943–1954.

129 Pia Rigobello, M., Messori, L., Marcon, G., Cinellu, M.A., Bragadin, M., Folda, A., Scutari, G. and Bindoli, A. (2004) Gold complexes inhibit mitochondrial thioredoxin reductase: consequences on mitochondrial functions. *Journal of Inorganic Biochemistry*, **98**, 1634–1641.

130 Turkall, R.M. (1979) Gold sodium thiomalate: selected aspects of metabolism and distribution. Ph. D. Thesis, University of Ohio.

131 Schwartz, H.A., Christian, J.E. and Andrews, F.N. (1960) Distribution of sulfur-35- and gold-198-labeled gold thioglucose in mice. *The American Journal of Physiology*, **199**, 67–70.

132 Intoccia, A.P., Glanagan, T.L., Walz, D.T., Gutzait, L., Swagzdis, J.E., Flagiello, J., Hwang, B.Y.-H., Dewey, R.J. and Noguchi, H. (1983) Pharmacokinetics of auranofin in animals, in *Bioinorganic Chemistry of Gold Coordination Compounds* (eds B.M. Sutton and R.G. Franz), SK&F, Philadelphia, pp. 21–33.

133 Cottrill, S.M., Sharma, H.L., Dyson, D.B., Parish, R.V. and McAuliffe, C.A. (1989) The role of the ligand in chrysotherapy: a kinetic study of gold-199- and sulfur-35-labeled myocrisin and auranofin. *Journal of the Chemical Society, Perkin Transactions II*, (1), 53–58.

134 Smith, W.E., Reglinski, J., Hoey, S., Brown, D.H. and Sturrock, R.D. (1990) Changes in glutathione in intact erythrocytes during incubation with penicillamine as detected by proton spin-echo NMR spectroscopy. *Inorganic Chemistry*, **29**, 5190–5196.

135 Snyder, R.M., Mirabelli, C.K. and Crooke, S.T. (1987) The cellular pharmacology of auranofin. *Seminars in Arthritis and Rheumatism*, **1**, 71–80.

136 Matheson, N.R. and Travis, J. (1983) The effect of gold compounds on human neutrophil myeloperoxidase, in *Bioinorganic Chemistry of Gold Coordination Compounds* (eds. B.M. Sutton and R.G. Franz), SK&F, Philadelphia, pp. 58–66.

137 Snyder, R.M., Mirabelli, C. and Crooke, S.J. (1986) Cellular association, intracellular distribution, and efflux of auranofin via sequential ligand exchange reactions. *Biochemical Pharmacology*, **35**, 923–932.

138 Shaw, C.F. III, Isab, A.A., Coffer, M.T. and Mirabelli, C.K. (1990) Gold(I) efflux from auranofin-treated red blood cells. Evidence for a glutathione-gold-albumin metabolite. *Biochemical Pharmacology*, **40**, 1227–1234.

139 Mirabelli, C.K., Johnson, R.K., Sung, C.M., Faucetter, L., Muirhead, K. and Crooke, S.T. (1985) Evaluation of the *in vivo* antitumor activity and *in vitro* cytotoxic properties of auranofin, a coordinated gold compound, in murine tumor models. *Cancer Research*, **45**, 32–39.

140 Tiekink, E.R.T. (2003) Gold Compounds in Medicine: Potential Anti-Tumour Agents. *Gold Bulletin*, **36**, 117–124.

141 Tiekink, E.R.T. (2002) Gold derivatives for the treatment of cancer. *Critical Reviews in Oncology/Hematology*, **42**, 225–248.

142 Beverly, B. and Couri, D. (1987) Role of Myeloperoxidase (MPO) in Aurothiomalate Metabolism. *Federation Proceedings*, **46**, 854.

143 Brown, C.L., Whitehouse, M.W., Tiekink, E.R.T. and Bushell, G.R. (2008) Colloidal metallic gold in not bio-inert. *Inflammopharmacology*, **16**(3), 133–137.

144 Blough, H. (1990) Organic thio compounds having a sulfur-gold linkage as specific inhibitors for retrovirus reverse transcriptase, and

their application as medicaments. Abstracts 3rd International Conf. Gold and Silver in Medicine, Manchester, UK, p.14.

145 Okada, T., Patterson, B.K., Ye, S.Q. and Gurney, M.E. (1993) Aurothiolates inhibit HIV-1 infectivity by gold(I) ligand exchange with a component of the virion surface. *Virology*, **192**, 631–642.

146 Tepperman, K., Zhang, Y., Roy, P.W., Floyd, R., Zhao, Z., Dorsey, J.G. and Elder, R.C. (1994) Transport of the dicyanogold(I) anion. *Metal-Based Drugs*, **1**, 433–444.

147 Daniel, M.-C. and Astruc, D. (2004) Gold nanoparticles: assembly, supramolecular chemistry, quantum-size-related properties, and applications toward biology, catalysis, and nanotechnology. *Chemical Reviews*, **104**, 293–346.

148 Brown, C.L., Bushell, G., Whitehouse, M.W., Agrawal, D.S., Tupe, S.G., Paknikar, K.M. and Tiekink, E.R.T. (2007) Nanogold-pharmaceutics; (i) the use of colloidal gold to treat experimentally-induced arthritis in rat models; (ii) Characterization of the gold in *Swarna bhasna* a microparticulate use in traditional Indian medicine. *Gold Bulletin*, **40**, 245–250.

7
Nanoscience of Gold and Gold Surfaces
M.B. Cortie and A. McDonagh

7.1
Introduction

This chapter focuses on the phenomena that come to the fore when working with nanoscale gold particles or structures, and their surfaces. As we will show, very small gold structures possess physical and, in some cases, chemical properties that can be quite different from those of the bulk phase. There are several existing reviews of the field [1–9] usually focused on various specialized aspects of the overall topic. The papers of Daniel and Astruc [9] and Glomm [1] are especially comprehensive. Here we provide only an overview, seeking mainly to highlight the latest information and insights on this subject. Furthermore, we will bias our review towards the chemical and physical aspects of the science of nanoscale gold and its surfaces, but will not delve into any detail into specific technological applications.

The field of nanotechnology, which links chemistry, physics, materials science, electrical engineering, and biotechnology, has catalyzed a strongly interdisciplinary theme in much of current science. Metallic gold in the form of bulk surfaces, nanoparticles, or fabricated nanostructures is widely used in this emerging technology. This is because of gold's chemical stability, useful surface chemistry, and unique optical properties, which are simultaneously associated with a very convenient range of processing technologies [6, 10, 11]. Gold is not only the most electronegative (i.e., "noble") of metals, it is also a good conductor of heat and electricity, and has a face-centered cubic (fcc) crystal structure which makes it soft and ductile. As a consequence of these properties, it is arguably easier to work with gold at the nanoscale than with any other metal. Naturally, these same attributes also make bulk gold an interesting element to exploit in niche macroscopic applications such as electronics and dental prostheses. However, additional interesting properties come to the fore in nanoscale gold structures. For example, gold's natural affinity for sulfur has stimulated much scientific interest, since it makes gold a very convenient substrate on which to assemble ordered monolayers of a wide range of organic molecules containing the –SH group [10]. These are usually termed SAMs (self-assembled monolayer)s and have been the subject of a very prolific literature since they were placed on a scientific footing by Nuzzo and Allara in 1983 [12].

Nanoscale coatings of organic molecules can also be formed on the surface of colloidal gold nanoparticles, although, in this case, the coating would not normally be called a "SAM" since it would be somewhat disordered. Such composite structures are easy to synthesize in polar and non-polar media [9] and their availability opens up a large number of interesting possibilities. The pioneering development of immunogold labeling in 1971 [13] has inspired the use of gold nanoparticles in many biological applications, including fluorescent biological labels, detection of pathogens or proteins, blood immunoassays, and DNA analyses [14–16]. Historically, there has also been considerable use of gold colloidal particles in Chinese [17] and Indian [18] traditional medicine as well as in quack Western "medicine" [9, 19]. The revolutionary work of J. Forestier set the scene for the rational use of gold thiolates for some types of rheumatoid arthritis over fifty years ago [19]. Recent developments suggest that colloidal gold particles may also have applications in modern therapeutic medicine [20]. In particular, they have been used to target and destroy cancer cells [15], macrophages [21], and pathogens [22]. In general, the idea is that a nanoparticle of gold, being immune to corrosion in physiological environments, can be safely targeted to specific cells or regions of the body, at which point some destructive action is triggered [23]. The action might involve release of a cytotoxic compound previously conjugated to the surface of the gold nanoparticle [24], or the generation of heat caused by a plasmon resonance of the nanoparticle [15].

As mentioned previously, one reason for the widespread interest in gold for nanoscale technologies is that very small, oxide-free components of it can be fabricated. If an unoxidized ("naked") metallic surface is required under ambient conditions then gold and platinum are among the few viable material options. Furthermore, since room temperature oxidation of metals normally produces an oxide layer of at least several nanometers if not microns in thickness, it follows that nanoscale electrically-conducting structures of most other metallic elements would normally be rapidly destroyed by oxidation. In this particular case, the use of gold in nanotechnology does not exploit any special nanoscale property; it is merely based on a combination of oxidation resistance and some bulk property (such electrical conductivity) scaled to the nano-domain. However, it is the appearance of new and unexpected material properties at the nanoscale that has attracted the greater interest in the field, and the situation with respect to gold is no exception. The localized plasmon resonance of gold nanoparticles is a well known example of such a unique nanoscale property, which we will discuss later in greater detail. However, it is less well known that very small clusters of gold can be semi-conducting in nature [25], photoluminescent [26] or even magnetic [27]. These attributes are completely unexpected in an element such as gold, and are the result of the onset of new and interesting physical and chemical phenomena that only appear in the smallest of structures.

7.2
Forms of Gold at the Nanoscale

7.2.1
Clusters and Nanoparticles of Less than 5 nm Diameter

We will consider the very smallest of gold nanoparticles first: clusters which consist of at most a few hundred Au atoms. An approximate relationship between number of atoms and size is plotted in Figure 7.1. The best of the currently available transmission electron microscopes can detect a particle of about 2 nm diameter, which would contain a few hundred Au atoms. Smaller clusters cannot normally be directly imaged and are usually characterized using indirect methods such as mass spectrometry [28] or Mössbauer spectroscopy [29]. Such small clusters are not stable in naked form in aqueous environments, and must, for example, be produced and collected in vacuum, or protected from aggregation by a surface layer of capping ligands. There are several examples of ligand-capped clusters, of which $Au_{55}[(C_6H_5)_3P]_{12}Cl_6$ is perhaps the best known [30]. Another cluster worth mentioning is Au_{20}, for which there is a surprisingly stable naked tetrahedral form [31]. The Au_{20} tetrahedron is estimated to be about 0.7 eV more stable than any of its isomers [32], and even the Au_{19} and Au_{21} clusters derived from it by subtracting or adding an Au atom respectively enjoy a special, although somewhat diminished, lowest energy status [33]. Such entities are probably most correctly considered to be large molecules and, as such, the question of their internal "crystalline" structure does not arise.

However, as the cluster becomes larger there comes a point at which it becomes more accurate to consider it to be a nanoparticle, possibly with an exterior coated with a layer of capping ligands. We consider a diameter of 1 nm to be a reasonable although arbitrary value for differentiating between "clusters" and "nanoparticles."

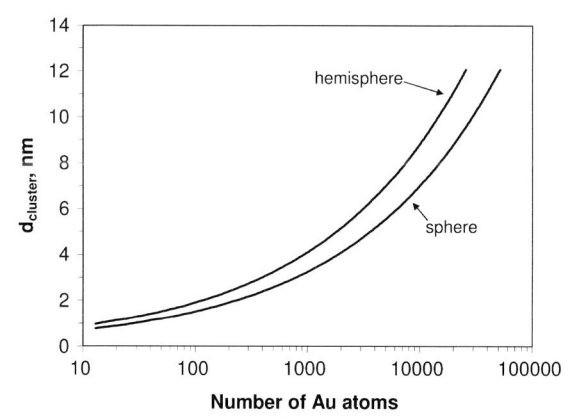

Figure 7.1 Approximate size of nominally spherical and hemispherical clusters of Au_n, Redrawn from Cortie and van der Lingen [34], Copyright (2002) Materials Australia.

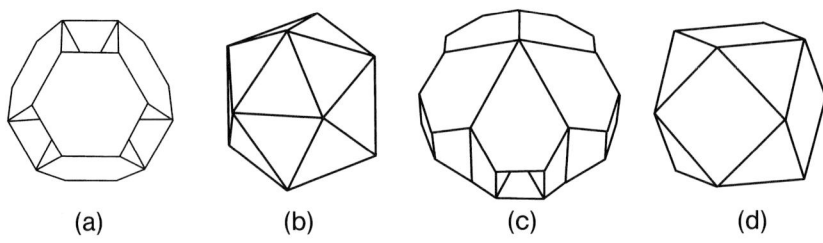

Figure 7.2 Comparison of (a) truncated octahedron, (b) icosahedron, (c) Marks decahedron and (d) cuboctahedron. Note the re-entrant facets in the truncated octahedron and the Marks decahedron. Reproduced from Cortie and van der Lingen [34], Copyright (2002) Materials Australia.

The internal atomic arrangement of the Au atoms in nanoparticles between 1 and 5 nm diameter is both complex and very interesting. Of course in the bulk, that is, macroscopic, form gold has the *fcc* structure, and macro- or even micro-crystals of gold exhibit the highly symmetrical cubic, octahedral or rhombododecahedral forms associated with this crystal structure. However, it turns out that the very smallest gold nanoparticles do not necessarily possess the *fcc* structure, a point which has attracted a fair amount of discussion in the literature [35]. The problem is amenable to various theoretical approaches and there have been several studies in which the lowest energy structures of very small nanoparticles have been predicted or determined [36–38]. To date, however, there is certainly not yet a complete agreement between calculation and experiment. Conflicting claims have been made for icosahedral or decahedral quasi-crystal structures, amorphous structures, or octahedra, cuboctahedra and truncated octahedra based on *fcc* packing (Figure 7.2). In any case, it is generally agreed that the energies of these structures are all very similar, making differentiation in numerical or experimental work quite ambiguous [36, 37]. Nevertheless, it is found that clusters or nanoparticles containing certain "magic" numbers of Au atoms are more stable than others [39]. We have already mentioned Au_{55}, while a mixture of Marks decahedra with about 70, 110 or 140 Au atoms dominate some experimental preparations [35].

The Au_{13} cluster is the smallest possible member of the sequence of icosahedral particles and, in principle, larger icosohedra containing 55, 147, 309, 561, 923 or more atoms are possible [40]. Both icosahedral and decahedral clusters have five-fold symmetry and for nano-scale particles, the icosahedron yields an efficient compromise between surface area and packing density which, however, is only about 69% [40] (compared with the 74% expected for a *fcc* structure). The Marks decahedron on the other hand has a somewhat lower strain energy [35]. In fact, cuboctahedral clusters with *fcc* packing can be assembled from the same series of magic numbers than the icosahedra [40], and, in addition, very few of these clusters (whether decahedral or octahedral) are defect-free [30, 37, 39, 41, 42], which somewhat blurs the classification of their atomic arrangement. Furthermore, structure types may change as result of a change in temperature [38, 39]. It has been shown, that icosohedral particles of up to 12 nm diameter will transform to the decahedral structure when heated [38].

7.2.2
Nanospheres

The topic of gold nanospheres attracted the interest of several famous nineteenth century scientists such as Michael Faraday, Richard Zsigmondy, and Gustov Mie [43]. Interest diminished in the mid-twentieth century although some excellent contributions were made by Turkevich [42, 44], Frens [45], and Brust [46] in that period regarding the controlled preparation of nearly monodisperse colloidal suspensions. A good review of the topic was provided by Daniel and Astruc in 2004 [9]. From a synthetic point of view the emphasis was (and still is) very much on controlling the diameter of the nanoparticles produced. Although gold nanospheres of various diameters can be separated to some extent by means such as chromatography, fractional crystallization [35], diafiltration [47], isoelectric focussing, centrifugation, or selective precipitation, it is obviously far better to directly synthesize a monodisperse population of the target size. Of course, the gold nanoparticles thus produced must be stabilized against agglomeration (reversible coalescence into loosely bound units by weak forces of attraction) or aggregation (irreversible coalescence into bound units by chemical or metallurgical bonding) [48].

From a structural point-of-view the "bulk" metallic state, that is, *fcc* lattice (with varying densities of defects such as twins and stacking faults) is generally established in gold nanoparticles of about 10 nm diameter and upwards. However, such particles still display many unusual physical properties, primarily as the result of their small size. Shrinking the size of gold particles has an important effect: it increases both the relative proportion of surface atoms and of atoms of even lower coordination number, such as edge atoms [49] and these atoms in turn are relatively mobile and reactive.

There are very many papers in the literature that address some aspect of gold nanospheres. In particular, their plasmon response (see Section 7.3.1.1) has been well studied, as has their agglomeration [50–52] and the manner in which they can be assembled into highly ordered colloidal crystals [50, 53, 54]. The latter are interesting and will be further discussed in Section 7.3.8.2. Conjugation of gold nanospheres with proteins and antibodies, for use as a stain in microscopy [55] or possibly, in medical applications [23], is another rich field.

7.2.3
Nanoshells

Hollow gold spheres or core-shell particles consisting of a gold shell on a core of some other material have recently attracted attention. This is because they have interesting and tunable optical extinction properties [56]. These can be readily calculated using Mie theory [57], and there had been some scattered early interest in these shapes as a result [58, 59], but the versatility and properties of these particles only became widely

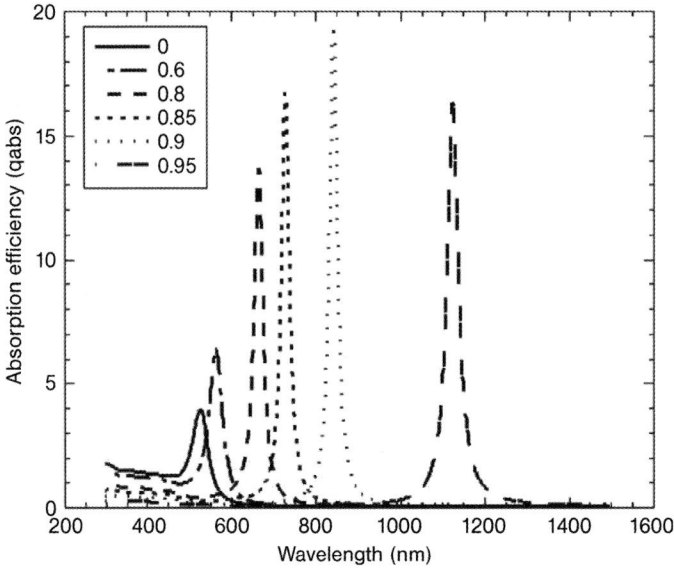

Figure 7.3 Absorption efficiencies for a 50 nm diameter nanoshell with varying aspect ratios. Reproduced from Harris [72], Copyright (2006) American Chemical Society.

appreciated after a series of landmark publications [15, 60–64] and patents [65] by the group of N. Halas in the USA. This group also coined the term "nanoshell" to describe the morphology, although some in the field prefer the term "core-shell particle" [66, 67]. The tunability of gold nanoshells is illustrated in Figure 7.3. Note that the peak optical extinction of the particles can be tuned to match the so-called tissue window [68], the part of the near IR spectrum in which our bodies are the most transparent. It is evident that, in principle, the core-shell particle morphology offers a versatile platform on which to build diverse technological applications in the medical therapeutic [15, 63] and diagnostic [62, 69] areas, as well as in spectrally selective coatings [70] or even as the basis of a "smart" particle that can regulate its own temperature [71].

There are three main wet chemical methods by which gold nanoshells have been synthesized. The first involves the formation of an Au shell on an Au_2S core [58, 60]. However, there has been controversy in the literature regarding the efficacy of this technique, and it has been claimed that, to some extent, the resulting optical properties are simply those of aggregated gold nanospheres [51, 73]. The second method involves the functionalization of some dielectric core, for example, silica [61, 67] or latex [74], so that very small gold nanospheres become attached. Thereafter the gold nanospheres are coarsened in a suitable growth solution so that they coalesce to form a continuous shell. Although this technique has been claimed in a series of patents by the Halas group [65], it is possible that they had been preceded by Mulvaney [75]. Finally, gold nanoshells or other shapes can be readily produced by galvanic reaction with cores of a more reactive metal, most commonly, silver [76, 77].

7.2.4
Nanorods

Gold nanorods represent another, highly tunable shape of gold nanoparticle that is currently receiving attention from researchers. Although they have been produced and studied since the mid 1980s [78], interest was limited until the discovery of batch methods by which they could be synthesized in high yield, and with a controlled aspect ratio in 2001 [3, 79]. More recently, continuous synthesis techniques have been reported [80], indicating that scale-up of production will not be difficult. Interest in the science and applications of these particles is now quite lively. Like nanoshells, they can be readily tuned to have peak optical extinction across a range of wavelengths (Figure 7.4) and as a result they too have been investigated as biomedical markers [81] for medical treatment against cancer [82, 83], macrophages [20, 84] or invading pathogens [22], as spectrally selective [84] or dichroic [85] coatings, or as SERS substrates [86], to name only a few of the possible applications. There is little to differentiate nanorods and nanoshells from one another as far as the efficiency of their optical response (in terms of quantity of gold used) is concerned [87], but there is now no doubt that rods can be more readily synthesized than nanoshells.

The lower symmetry of nanorods (in comparison to nanoshells) allows additional flexibility in terms of the tunability of their optical extinction properties. Not only can the properties be tuned by control of aspect ratio (Figure 7.4a) but there is also an effect of particle volume (Figure 7.4b), end cap profile (Figure 7.4c), convexity of waist (Figure 7.4d), convexity of ends (Figure 7.4e) and loss of rotational symmetry (Figure 7.4f).

7.2.5
Other Nanoparticle Shapes

While nanorods, nanoshells and nanospheres receive the bigger share of interest at present, there are several other interesting particle shapes such as triangles [89], cages or boxes [76] or semi-shells [90]. These generally have more complex optical extinction spectra due to their lower symmetry. Triangles (and spheres) have been produced by reducing Au salts with brews or mixtures of vegetable substances [91], but it is hard to see what advantage such a route offers since synthesis with pure reagents should generally give better reproducibility. Cages and cubes are made by galvanic replacement of a more active, usually silver, template [76, 92]. Finally, while semishells can be produced by wet chemistry [93], it is also convenient to prepare them by physical vapor deposition [90]. In this case both arrays [94, 95] and discrete semi-shells [90] can be produced by careful control of the deposition parameters.

7.2.6
Mesoporous Sponges

There has been some interest in mesoporous sponges of gold [96, 97]. These structures combine the electrical conductivity of gold with a very high surface area.

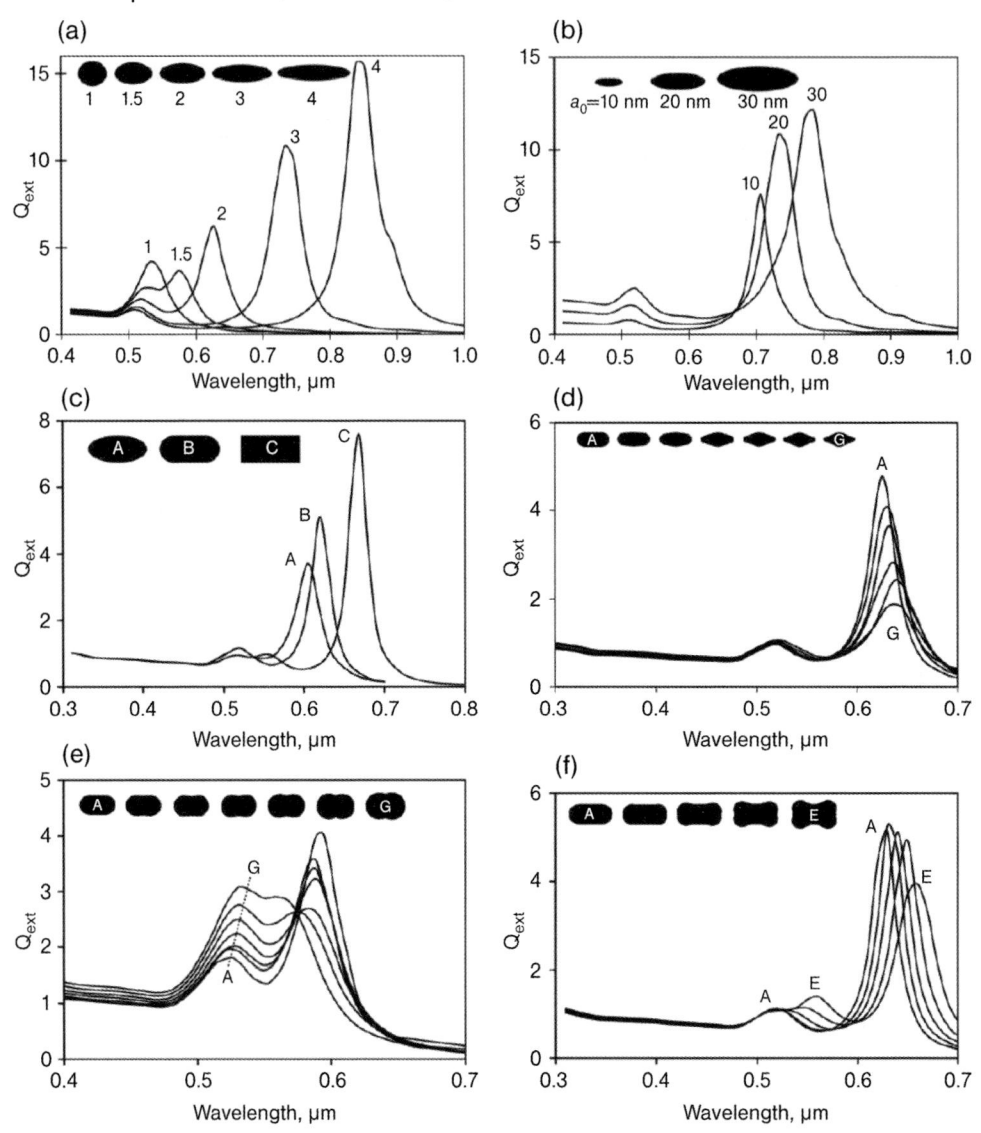

Figure 7.4 Influence of nanorod shape on its optical extinction properties, as simulated using the discrete dipole approximation. (a) different aspect ratios, fixed volume, (b) fixed aspect ratio, variable volume, (c) aspect ratio and volume fixed, variable end cap geometry, (d) convexity of waist varied, fixed volume and aspect ratio, (e) convexity of ends varied, fixed volume and aspect ratio, (f) transition from rod to "dogbone," fixed volume and aspect ratio. Reproduced with permission from Xu and Cortie [88], Copyright (2006) Wiley-VCH.

Possible applications include optical coatings [98], catalysts [99–101], substrates for Surface Enhanced Raman spectroscopy [102] or biosensor electrodes [103]. Mesoporous gold can be prepared by de-alloying a suitable precursor such as a

Figure 7.5 Two topologically distinct types of mesoporous gold sponge, each with 50 volume % gold. (a) Swiss-cheese morphology produced by de-alloying, (b) aggregated particle morphology produced by sintering of nanoparticles.

gold-silver or gold-copper alloy [98, 104] or an Au–Al intermetallic compound [99, 105], or by aggregation and partial sintering of gold nanospheres. It is important to note, however, that de-alloying produces a sponge with very different topology to that produced by the sintering of particles. This is illustrated in Figure 7.5, which shows blocks of two sponges, each with the same volume fraction of gold, but produced by the two different techniques. In general, percolation is achieved at a far lower volume fraction of gold in the sponge produced by de-alloying, and this has a dramatic influence on the optical and mechanical properties.

7.2.7
Thin Films

Films of gold with a thickness of more than about 50 nm are basically no different in their physical and chemical properties to bulk gold. One of their attributes that has proved to be very useful in a scientific sense, is that the gold crystallites tend to take up a strong (1 1 1) crystal orientation with respect to the plane of the substrate. In particular, thin films of gold on mica, prepared by physical vapor deposition followed by gentle annealing, display large, flat (1 1 1) domains that are ideal for the self-assembly of ordered monolayers of diverse organic molecules. There is an extensive literature which addresses this phenomenon, and we will provide an overview of it in Section 7.4. Films which are thinner than about 30 nm have some interesting "nanoscale" properties. One aspect that has attracted commercial attention in the past is that such films are moderately transparent to visible light, yet reasonably reflective with respect to the near infra-red. This is an ideal combination of properties for solar filters, as applied for example, to windows [106]. Simulations of the transmission of light through a series of thin films of gold applied to glass are plotted in Figure 7.6. The thinnest continuous coatings that can be produced are in the range 15 to 20 nm thick. If less than that amount of gold is applied the coating is no longer continuous and its optical properties change significantly. In particular, it develops localized plasmon resonances (see Section 7.3.1.1) with light in the 530 to 650 nm region of the spectrum, while reflectivity in the near infra-red becomes negligible [106, 107].

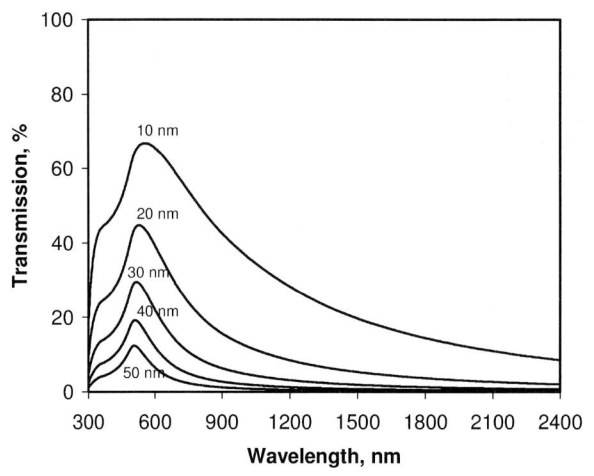

Figure 7.6 Simulations of the transmission of light through continuous thin films of gold. Data courtesy of Dr A. Maaroof, University of Technology Sydney.

7.2.8
Assemblages of Nanoparticles

7.2.8.1 Disordered Aggregates

The coalescence of colloidally suspended gold nanoparticles into agglomerates or aggregates is a nuisance for some and an opportunity for others. Generally, workers seek to avoid the collapse of their sols, a process which necessitates the use of pure starting materials and clean glassware. On the other hand, there is a color change accompanying the collapse, and the sol changes from burgundy red, through purple, blue and on to black. This has been used as the basis of a colorimetric sensor [108] for analytes as diverse as DNA [16] and lead [52]. The agglomeration process can also be detected by non-optical methods [109]. Further information may be found in some other recent publications on the subject [110].

7.2.8.2 Colloidal Crystals

If the process of agglomeration is allowed to take place slowly in a dispersion of monodisperse gold nanospheres, then highly ordered structures known as colloidal crystals, supracrystals, or artificial solids may form, Figure 7.7. These crystal-like structures may have a size of up to several microns and contain nanospheres arranged in a closest-packed configuration. In general they are believed to be *fcc* or, less often, hexagonal close packed, however in either case the "crystal" has a large number of stacking faults [50, 54, 111]. The process of colloidal assembly is driven by dispersive forces, and the Hamaker constant of Au-to-Au across the intervening medium is an important parameter. It is worth noting that the Au-water-Au, for example, has one the highest of the known Hamaker constants (40×10^{-20} J, to be compared with polystyrene-water-polystyrene of $\sim 1 \times 10^{-20}$ J [112]). It is also

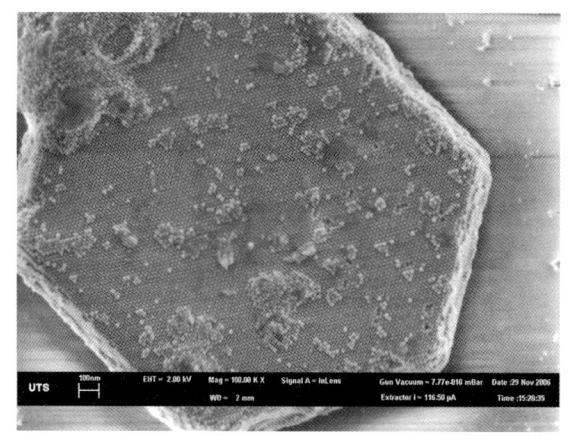

Figure 7.7 Colloidal crystal formed from oleylamine-stabilized gold nanoparticles, Reproduced with permission from Harris et al. [53b], Copyright (2007) Institute of Physics.

possible to assemble supracrystals using ligand-ligand bonding, for example, face centered, body-centered and simple cubic structures have been formed this way from gold-DNA conjugate particles [113].

7.3
Onset of New Phenomena

7.3.1
Optical

7.3.1.1 The Plasmon Resonance

The electromagnetic waves of light can only penetrate several tens of nanometers into gold before being completely attenuated. This is however sufficient to generate some very interesting phenomena. In the case of bulk gold or thin gold films, light of an appropriate frequency can be coupled onto the surface where it becomes confined as a traveling wave known as a surface plasmon polariton [114]. These can propagate some tens of microns along the surface before being dissipated [115]. While various "plasmonic" devices have been proposed to exploit this particular phenomenon, it is its application in surface plasmon resonance spectroscopy (SPR) that may be of greatest value currently, and which is the basis of many commercially available instruments. In this technique a laser beam is coupled onto the surface of a gold film, which in turn is functionalized to selectively capture an analyte molecule. The nature of the reflected beam is affected by any change in the localized refractive index in the near-field above the gold film. This technique allows the exploitation of the unique surface chemical properties of gold films to be combined with a plasmonic phenomenon to yield a sensitive technique for detecting the binding of organic molecules such as proteins [116].

A related phenomenon, localized surface plasmon resonance, occurs only in discrete nanostructures. In this case, light of an appropriate frequency causes the "free" electrons of gold (and those of a few other elements such as silver, copper, aluminum and sodium) to resonantly oscillate in the electric field of the light. As a result, much of the energy of the light at the resonant frequencies is absorbed or scattered. The resulting extinction peak in, for example, colloidal suspensions of the nanoparticles, causes them to exhibit strong and beautiful colors in transmission. Gold nanospheres of between 10 and 50 nm diameter selectively absorb the blue and green light with a peak extinction at wavelengths between 520 and 540 nm, causing suspensions or coatings of such particles to have a burgundy red color since that is, the predominant color of the light remaining. Two useful accounts of the phenomenon are available elsewhere [2, 117].

The important points to note here are that (i) spherical gold nanoparticles are the most stable form, because that shape offers the minimum surface area for a given volume, (ii) the optical extinction for nanospheres of gold peaks at around 520 nm (green light) and (iii) extinction is due to the additive effects of absorption and scattering (Figure 7.8). However, the overall optical extinction for particles of less than 100 nm diameter is dominated by absorption rather than by scattering. The absorbed light is converted to heat, which is the basis for some of the proposed technological applications of gold nanoparticles.

The principles of SPR spectroscopy described earlier can be extended to exploit the localized plasmon resonance in nanoparticles of gold [108, 119] in the technique known as localized surface plasmon resonance spectroscopy (LSPR). However, this is not yet in the commercial mainstream. Other applications of gold nanoparticles for chemical analysis depend on optical effects caused by the surface enhanced Raman effect [120], particle aggregation [16, 52], a shift in peak resonance [121], or resonant fluorescent enhancement [122]. The tunability of the optical extinction in nanorods and nanoshells has suggested their use for new medical treatments [15, 23, 82], medical imaging [64, 81], solar glazing [84, 123], solar thermal harvesting [70], spectrally and angularly selective coatings [93, 94], dichroic and polarizing filters [85], decorative applications [124], plasmonic wave-guides [125, 126], plasmonic "circuits" [126, 127], and for the basis of building metamaterials (with negative refractive index [128], "invisibility" [129], or super-lensing [130] capabilities).

7.3.1.2 Manipulation of the Plasmon Resonance

The plasmon resonance in gold nanostructures can be modulated or shifted in diverse ways, and this flexibility is part of the reason for the great interest in these systems. We have previously mentioned that the peak extinction can be controlled by the aspect ratio of gold nanoshells and nanorods. However, the absorption peak can also be red-shifted to the top of the visible by aggregating the individual spheres, with the resulting material possessing a deep blue color in transmission [51]. The wavelength at which maximum absorption occurs can be controlled by varying the interparticle spacing or, equivalently, the volume fraction of gold.

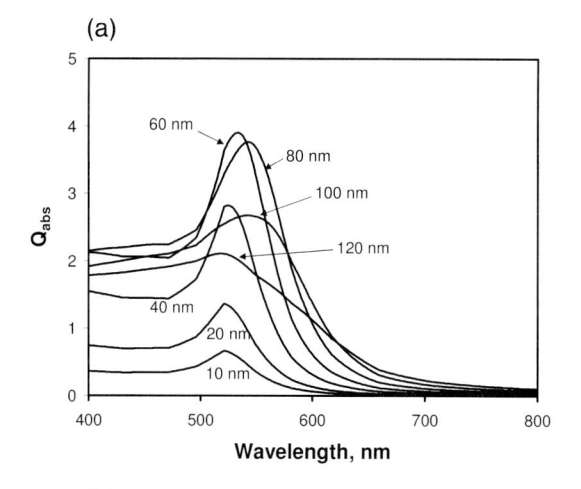

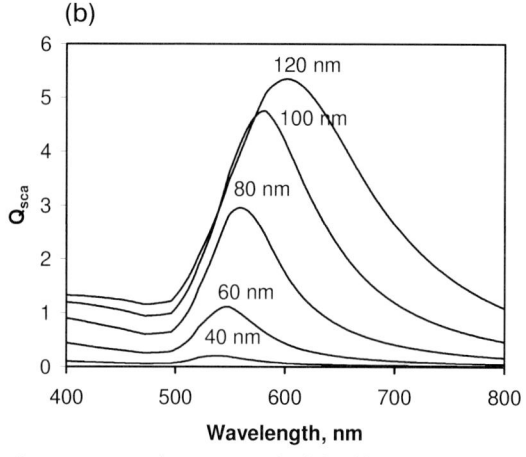

Figure 7.8 Optical properties of solid gold spheres of indicated diameters, (a) absorption, (b) scattering. Redrawn from Cortie et al. [118], Copyright (2005) Society of Photo-Optical Instrumentation Engineers.

7.3.1.3 Fluorescence and Luminescence

The light emitting properties of very small gold nanoparticles (<2 nm in diameter) has also been the subject of some interest. It has been claimed that such particles of gold behave like semi-conductor quantum dots, with an analogous fluorescence [26, 131]. On the other hand, while it is certainly true that preparations of gold-containing organic compounds fluoresce [132, 133], the origin of the light emission may lie in the molecule rather than any nanoscale gold cluster as such. Light emission from gold-containing molecules is not strictly a nanoscale phenomenon and so this will not be discussed further here. On the other hand, genuine luminescence of a different kind from naked gold surfaces and metallic nanoparticles is certainly possible, albeit with a very low efficiency [134]. The two-photon luminescence of gold nanoparticles can be quite efficient [135] and that of gold nanorods has been demonstrated as a biomedical contrast agent [136]. These are topics that would warrant further investigation.

7.3.2
Physical

7.3.2.1 Depression of the Melting Point
Very small particles of a substance have a depressed melting point relative to the bulk due to the curvature of their surfaces, which may be expressed as [137, 138]:

$$\frac{T_m(d)}{T_m(\infty)} = 1 - \frac{4}{\rho_s L}\left\{\gamma_s - \lambda_l\left(\frac{\rho_s}{\rho_l}\right)^{\frac{2}{3}}\right\}\frac{1}{d}$$

where $T_m(r)$ and $T_m(\infty)$ are respectively the melting points of a particle of diameter d and the bulk, L is the molar heat of fusion, ρ is density, γ is surface free energy, d the diameter, and the subscripts s and l refer to solid and liquid respectively. The situation for gold has been both measured and modeled, and the melting point of 1.5 nm gold spheres was found to be in the range 600 to 650 K [39, 137]. It is reasonable to extrapolate the melting point of even smaller nanospheres downwards, perhaps to as low as 470 K, or even below that [138]. Therefore, it is possible that the very small gold nanoparticles used in catalysis might, in some cases, be molten at their service temperatures- a factor that is, very seldom considered. On the other hand, at least one prediction has appeared in the literature that the melting point of a few, especially stable clusters might disobey this general trend. Au_{20}, for example, appears to be very stable, and it has been predicted that it would melt sharply at about 1200 K [139]. This implies that this cluster should be treated in a physical, thermodynamic and chemical sense as a quite distinct phase to bulk gold and in the other nanoscale structures discussed here.

7.3.3
Chemical

7.3.3.1 Heterogeneous Catalysis
It is now well known that nanoparticles and nano- or mesoporous sponges of gold have catalytic properties, and the oxidation of carbon monoxide has been particularly well studied in this regard. The reports of Hutchings [140] and Haruta [141] in the 1980s were influential in this field in so far as they described very positive results with gold, but there had even been a few older reports in the literature, see Ref. [142]. The scientific understanding of catalysis by gold is evolving rapidly and the topic has been reviewed several times in the last decade [4, 5, 8, 49, 142–144]. Here we will confine discussion to the points most salient to nanotechnology. The central question is of course the nature of the mechanism responsible for the catalytic efficacy of gold. In the case of the most studied reaction - the oxidation of CO – a widely held view is that

"activation" of the O_2 absorption and perhaps the oxidation reaction itself occurs at the perimeter of a hemispherical gold nanoparticle attached to a suitable oxide substrate, while adsorption of the CO occurs at corners, edges and other lowly coordinated sites of the nanoparticle. Both discrete gold nanoparticle and the oxide support are required to produce an enhancement of reaction kinetics in this view [142, 144]. It was claimed that the rate of CO oxidation scaled in proportion to the length of perimeter available. However, it has been shown elsewhere that improbably precise measurements of particle size and reactivity would be required to prove this point from a plot of activity versus particle size alone [34]. It was mentioned before that very small Au nanoparticles lose their metallic character, becoming semi-conductors [25]. Might this be the explanation for the catalytic efficacy of very small Au nanoparticles? In this view the gold nanoparticles become catalytically active when they acquire a particular value of band gap. The idea is intriguing but is rejected by the mainstream [144].

In any case, it is interesting to note that catalytic efficacy has been observed with nano- or mesoporous gold sponges [99–101, 145] suggesting that neither a discrete particle nor an oxide support is actually a fundamental requirement for catalysis. An alternative mechanism invokes the nanoscale structural effect noted in Section 7.2.2, and proposes that the catalytic effect of nanoscale gold structures is simply due to the presence of a large proportion of lowly-coordinated surface atoms, which would have their own, local electronic configurations suitable for the reaction to be catalyzed [34, 49, 146] A recent and readily available study by Hvolbæk et al. [4] summarizes the support for this alternate view.

7.4
Surface Chemistry of Gold

In this section, the surface chemistry of non-metals adsorbed as thin layers, films or SAMs on gold surfaces is discussed. Although attachment by a sulfur atom is by far the most predominant binding motif, many other elements may be used to bind to gold. Particular focus is given here to surface binding through atoms other than those already extensively covered in the literature.

7.4.1
Hydrogen

The hydrogen molecule does not chemisorb onto clean sintered gold surfaces at or above 78 K [147] but on unsintered films, a small amount of H_2 is chemisorbed if gold surface atoms of low coordination number are present [148]. Stobinski [149] found that H_2 can also chemisorb on thin sintered Au films if the surface is covered at low temperatures with a small amount of gold equivalent to 1–3 Au monolayers prior to H_2 exposure. This suggests a fundamental role of surface Au atoms of low coordination number in the chemisorption process. Deuterium molecules also chemisorb in a similar fashion on gold films at 78 K and isotope effects were observed during the desorption process [149]. Eberhardt et al. report molecular hydrogen adsorbed onto Au at 4 K and their observation that the molecule is very weakly bound and desorbs at about 15 K suggests a physisorbed, rather than chemisorbed, molecule [150]. Although H_2 does not generally dissociatively adsorb on gold due to the high energy barrier and relatively weak chemisorption state [151], dissociative adsorption was reported for H_2 on thin Au(1 1 1) films grown on Ir (1 1 1) [152]. This may be a function of local surface features such as defects or steps providing H_2 dissociation sites. Atomic hydrogen can be adsorbed onto gold if the molecules are pre-dissociated [153, 154], for example, by use of a hot filament at temperatures up to 150 K [155]. The adsorbed hydrogen atoms desorb to give H_2 at temperatures around 216 K.

7.4.2
Halogens

There are few reports that describe the interaction of molecular fluorine F_2 with gold surfaces. Metallic gold is quickly dissolved when exposed to F_2 at room temperature in the presence of anhydrous hydrogen fluoride containing an alkali fluoride to give the alkali salt of the AuF_4^- anion [155]. Arutyunov and Chaikin describe the interaction of fluorine atoms produced by a microwave discharge in a mixture of F_2 and He with gold to yield the species AuF_5 [156]. Exposure of a gold surface to F_2 or XeF_2 gas gives not more than a monolayer of adsorbed fluorine, which is contrast to silver surfaces where fluorine is taken up to form silver fluoride films [157]. NF_3 may be molecularly adsorbed on Au(1 1 1) surfaces but desorbs below 80 K. In monolayer regions, NF_3 is adsorbed via the N atom, directing fluorine atoms away from the surface [158].

In the early 1970s, Kishi and Ikeda reported that chlorine gas formed an adsorbed species on gold [159]. Spencer and Lambert investigated chlorine adsorption on the gold (1 1 1) surface at 298 K and suggested the formation of a surface chloride species AuCl [160]. Kastanas and Koel subsequently proposed that gaseous molecular chlorine dissociatively adsorbs to Au(1 1 1) above 120 K [161]. No molecular chlorine adsorbs at or above 120 K but rather chemisorbed chlorine adatoms are formed and not surface chloride compounds, as previously suggested. Two desorption events, at 790 and 640 K, involve desorption of $Cl_{(g)}$ and $Cl_{2(g)}$, respectively, and no desorption of AuCl species was observed. The estimated dissociation energy of the Cl–Au bond is $\sim$225 kJ mol^{-1}, which is at least 80 kJ mol^{-1} smaller than that of chlorine on Ag. Friend et al. [162] demonstrated that chlorine disperses clustered Au–O complexes on Au(1 1 1), redistributing the oxygen on the surface. The presence of chlorine alters the Au–O bonding, which may relate to the observation that chlorine has a dramatic effect on the selectivity for olefin epoxidation on Au surfaces [163]. Chlorine co-adsorbed with oxygen on Au(1 1 1) inhibits combustion, formation of organic acids, and the deposition of residual carbon. Chlorine adsorption on the reconstructed Au (0 0 1) surface changes the surface structure to a 1×1 structure at room temperature, probably as a consequence of chemical-bonding of the adsorbate [164]. Similarly, chemisorption of bromine on Au(1 0 0) surfaces reconstructs the (5×20) structure to give a (1 × 1) surface upon low bromine exposures. Thermal desorption of bromine from Au(1 0 0) occurs in atomic form with two main desorption features: a peak around 800 K and a very broad feature starting to appear at surface coverages exceeding 0.25. An adsorbate layer model with bromine atoms in the bridge positions and strongly coverage dependent lateral interactions was proposed [165].

Iodine and bromine adsorb onto Au(1 1 1) from sodium iodide or sodium bromide solutions under an applied surface potential with the surface structure formed being dependent on the applied potential [166]. The iodine adsorbate can also affect gold step edge mobility and diffusion of the Au surface. Upon deposition of a layer of disordered surface iodine atoms, the movement of gold atoms (assisted by the 2-dimensional iodine gas on the terrace) from step edges out onto terraces occurs. However, this diffusion occurs only at the step edge when an ordered adlayer is formed [167].

As well as the adsorption of halogen atoms or molecules, the adsorption of halide anions to gold surfaces has been extensively studied and a comprehensive review of the area has been published by Magnussen [168]. The degree of specific adsorption to gold surfaces increases in the order $F^- < Cl^- < Br^- < I^-$ with F^- only weakly specifically adsorbed. The presence of halide anions can also affect the electrodeposition of organic molecules such as pyridine on Au surfaces with chloride and bromide solutions suppressing the formation of ordered N-bonded pyridine layers [169].

7.4.3
Oxygen

In 1970 Macdonald and Hayes reported that oxygen adsorbs to the surface of gold powder [170]. Canning et al. [171] found that oxygen did not adsorb on clean Au surfaces for oxygen pressures up to 1.3×10^{-2} Pa and sample temperatures between 300 and 600 K. At the same time, Pireaux et al. [172] reported that chemisorption of molecular oxygen on Au(1 1 0) or Au(1 1 1) surfaces is dependent on surface impurities such as Si. Molecular or atomic oxygen adsorption does not occur on the stepped surface of a gold tip at temperatures between 300 and 350 K and the dissociative adsorption process is not activated by electric fields of the order of 6 V nm^{-1} [173]. Thus, a number of techniques to deposit O on gold have been developed, most of which are described by Meyer in a 2004 review [8]. For example, below 50 K, dioxygen physisorbs readily with a binding energy <12 kJ mol^{-1} and this physisorbed O_2 can be activated by electron or by photon impact to form chemisorbed O films [174]. Interestingly, a report by Rao et al. [175] that oxygen adsorbs molecularly on gold at low temperatures (~100 K) and also desorbs molecularly upon warming may in fact reflect the physisorption process rather than chemisorption.

Adsorption of oxygen atoms, achieved via electron-induced dissociation of nitrogen dioxide, induces restructuring of the gold surface. Figure 7.9 is illustrative. The number density of "elbows" (dislocations corresponding to a change in direction of the reconstruction) in the herringbone structure decreases with increasing atomic oxygen coverage. Small islands and serrated step edges form, presumably due to the

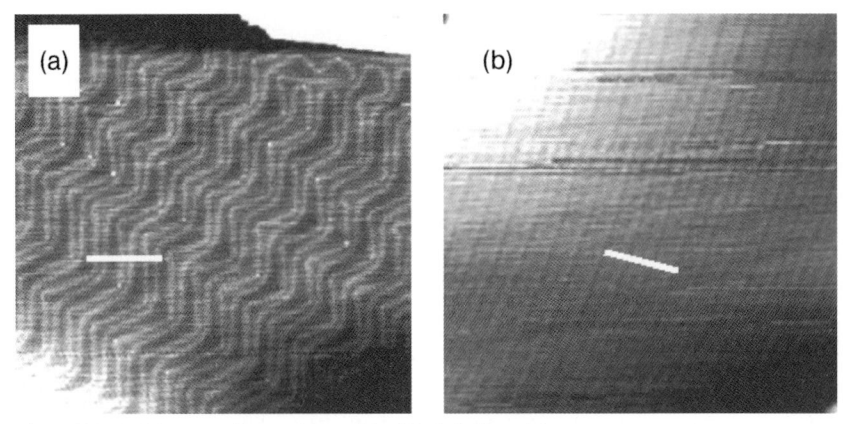

Figure 7.9 STM images of reconstructed Au(1 1 1) (left) and Au (1 1 1) covered with 0.4 ML of oxygen atoms (right). Images are 100 × 100 nm. Reproduced with permission from Min et al. [176], Copyright (2005) American Physical Society.

release of gold atoms from the "elbows" of Au(1 1 1) [176]. Clusters of gold oxides may form at high oxygen coverage by abstraction of gold atoms from terrace sites and these structural changes further affect the reactivity of the gold surfaces with respect to dissociation of [177].

In conjunction with the experimental data, DFT calculations have shown that O_2 binding to an Au(1 1 1) surface is promoted by the adsorption of Au clusters [178]. Because O_2 does not bind to the flat faces of gold surfaces, this area has quite significant implication in the area of CO catalysis. A review discussing the role of DFT in understanding CO oxidation on gold particles has been recently published [179].

7.4.4
Sulfur

The gold-sulfur interaction is the most studied. Far from being an inert surface, it has been shown that the interaction of sulfur with Au(1 1 1) surfaces has a dramatic effect on the gold structure. In a similar fashion to oxygen (see Section 7.4.3) sulfur lifts the herringbone reconstruction of Au (1 1 1) surface at relatively low coverages (<0.1 ML). At higher coverages (>0.3 ML) gold surface atoms are removed from regular terrace sites and incorporated into a growing gold sulfide phase. This process give rise to the commonly observed pit and mound surface morphology [180].

Sulfur-anchored SAMs and thin films, mostly from organosulfur precursors, have been discussed at length by a number of authors [10, 181]. SAMs of organosulfur compounds (thiols, disulfides, sulfides) form on gold substrates by spontaneous adsorption from either the liquid or the vapor phase. A number of experimental factors can affect the formation and structure of SAMs such as choice of solvent, temperature, concentration, immersion time, purity of adsorbate, oxygen concentration in solution, cleanliness, and structure of the adsorbate. Interestingly, the

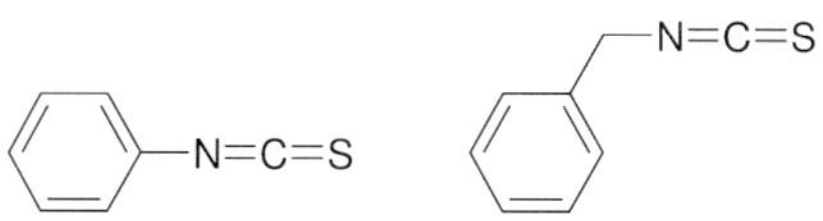

Figure 7.10 Chemical structures of phenyl isothiocyanate (left) and benzyl isothiocyanate (right).

absorption of H_2S is completely reversible. Also, unsymmetrical disulfides are useful for preparing mixed SAMs.

7.4.4.1 Isothiocyanates

SAMs can be formed from benzyl isothiocyanate or phenyl isothiocyanate molecules on gold and have upright geometries binding via the sulfur atom (see Figure 7.10). Kinetic data suggest that benzyl isothiocyanate should self-assemble faster than the phenyl analog, which has a less upright orientation relative to the surface [182]. Importantly, the single-molecule conductance of *n*-butanediisothiocyanate was found to be an order of magnitude greater than that of *n*-octanedithiol even though they both contain ten atoms (counted from sulfur to sulfur), which may have implications in molecular electronic applications. The preferential adsorption site for isothiocyanate appears to be the atop site on gold [183].

7.4.5
Selenium and Tellurium

Although the sulfur-gold bond has been most investigated, the Group 16 elements selenium and tellurium have also attracted attention and are discussed in detail here (polonium has not received attention due to its radioactivity).

Alanyalioglu *et al.* [184] showed that Se may be electrolytically deposited on Au (1 1 1) electrodes immersed in SeO_2 solutions. The selenate species is first reduced to selenite followed by a four-electron, four-proton reduction of selenite to Se. The selenium coverage corresponds to approximately 0.41 monolayers, which agrees with the scanned probe microscopy investigations on this system [185]. Ordered selenium atomic layers may be formed electrochemically on Au(1 0 0) surfaces either by direct reduction of $HSeO_3$, or by anodic or cathodic stripping of previously formed bulk Se. Coverages of 0.25, 0.33, 0,5, and 0.89 monolayers have been reported using these three methods and resulted in similar atomic layer structures on the Au(1 0 0) surface [186].

Studies of octylchalcogenate films on Au(1 1 1) by Nakamura *et al.* [186] found that the primary adsorbate species are chemisorbed chalcogenolates. At maximum coverage, octylSe and octylTe monolayers have lower coverage than the thiol analogs. The average tilt angle of the methyl group of the octylSe and octylTe monolayer is larger than that in the octylthiol molecules, possibly a consequence of the lower surface coverage of the octyl heavy chalcogenolate monolayers compared with that of the thiol analogs. In the formation of telluride monolayers, Nakamura *et al.* [187] used ditellurides (R-Te-Te-R, where R = *n*-butyl or *n*-octyl groups) to form films by autooxidation of the ditelluride. Compared with S and Se anchored films, the Te films were significantly more electronically insulating.

Self-assembled monolayers formed by immersing Au(1 1 1) surfaces into solutions of benzeneselenol and diphenyl diselenide were investigated by scanning tunneling microscopy and Auger electron spectroscopy. Chemisorption of benzeneselenol produced numerous gold islands (20–200 Å in diameter) on the surface. Upon annealing at 323 K in air, these gold islands coalesce to form larger, hexagonal-shaped islands. Chemisorption of diphenyl diselenide is qualitatively similar and upon annealing appears to produce the same surface structures observed with benzeneselenol [188]. Comparison of films on gold(1 1 1) formed from benzenethiol, benzeneselenol and diphenyl ditelluride showed that gold-sulfur and gold-selenium bonds formed in a stable manner with the gold-selenium molecular system able to carry larger current at a lower voltage than the thiol analog [189]. STM experiments on isolated bisthiol- and biselenol-terthiophene molecules inserted in SAMs of dodecanethiol on gold showed a very similar behavior for the adsorption onto the gold surfaces. The Se anchor provides a better metal–molecule electronic coupling than S [190], which was also confirmed by DFT calculations and UPS experiments [191].

Single-molecule conductance measurements of bis(acetylthio)terthiophene and bis-(acetylseleno)terthiophene molecules embedded in SAMs of octanethiol molecules were preformed with repeated analysis of a chosen target molecule. Bond fluctuations were far more prevalent in the sulfur-anchored molecules than in the selenium-anchored molecules, which may again be evidence for a more stable Se−Au bond compared with S−Au [192]. Using *ab initio* methods, Mankefors *et al.* calculated that for all physically realizable coverages, the Au–Se bond is ~0.25 eV stronger than the corresponding Au−S bond. The Se bond also displays a higher degree of metallicity [193].

Docosaneselenol, which has twenty two carbon chain, spontaneously forms an ordered SAM on gold(1 1 1) surfaces from solution [194]. Investigation into SAMs of *n*-dodecaneselenol on Au(1 1 1) using electrochemical impedance spectroscopy showed that film formation follows a two-step process. A monolayer is formed within the first minute after the electrode has been brought into contact with the deposition solution. The second step involves film reorganization and self-ordering, which can last for several hours. At small deposition times the SAM forms as many small islands of dodecaneselenol film on an Au substrate. As the time of deposition increases, the islands coalesce forming fewer, but larger, pinholes compared with short deposition times [195].

In an elegant experiment, SAMs of 4,4′-biphenylsubstituted alkaneselenolates, $CH_3(C_6H_4)_2(CH_2)nSe$ (where $n = 2$–6) were formed on gold and silver substrates (see Figure 7.11). The average tilt angle of the biphenyl moieties exhibits a systematic variation with varying number of CH_2 units in the alkyl chain. It was concluded that the substrate-Se-C angles were 104 and 180° on Au and Ag surfaces, respectively [196].

7.4.6
Nitrogen

Amines have been utilized to bind SAMs to gold surfaces and nanoparticles. Venkataraman *et al.* [197] found that in terms of molecular electronic junctions,

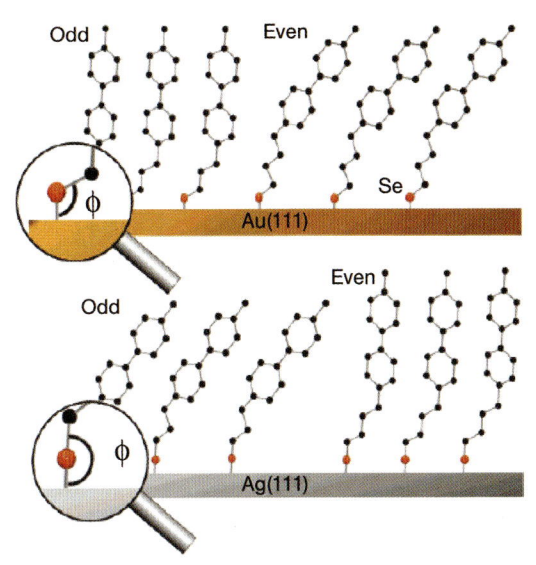

Figure 7.11 Orientation and packing for 4,4′-biphenylsubstituted alkaneselenolate molecules on Au and Ag showing the difference between an odd and even number of CH_2 groups in the alkane chain. Reproduced with permission from Shaporenko et al. [196], Copyright (2007) American Chemical Society.

amines gave more reliable conductance data than thiol analogs, which was attributed to the preferential adsorption of the nitrogen to gold adatom sites. The preference for nitrogen-containing molecules to adsorb to the atop sites of gold surfaces has also been shown using DFT calculations in the case of ammonia [198], aniline [199], and other amine compounds [200].

Chen et al. [201] investigated the single-molecule conductance of alkanes terminated with either dicarboxylic-acid, diamine, or dithiol anchoring groups. The contact resistance was again found to be dependent on the anchoring group, which varies in the order Au−S > Au−NH_2 > Au−COOH.

7.4.7
Phosphorus, Arsenic, Antimony

The triphenyl compounds of Group 15 elements (XPh_3, X = N, P, As, Sb, Bi) adsorb as monolayers on gold except in the case of bismuth where $BiPh_3$ tends to form oxygen-containing oligomeric or polymeric species on gold [202]. Uvdal et al. showed that tricyclohexylphosphine adsorbs on to gold and does so in preference to tricyclohexylphosphine oxide. Chemisorption through the phosphorus atoms is indicated such that the adsorbates donate electrons to the substrate [203]. Solution deposition of dimethylphenylphosphine (dmpp) results in a strongly bound layer on gold. The structure and binding of dmpp deposited from the gas phase in UHV are different from those obtained in toluene solution with the phosphine-metal bond stronger for solution deposited layers. Physisorbed dmpp layers desorb as intact molecules but considerable decomposition of the molecules occurs upon desorption for the UHV-deposited thin layer on gold as well as for solution-deposited films [204]. Diphenylphosphine, diphenylarsine, diphenylstilbine, as well as diphenylamine groups can be used to promote the adsorption of poly(styrene) on gold to give thin polymer layers on the surface. Upon immersion of the layers into pure solvent, the functional groups withstand desorption into pure solvent when bound to the polymer, while they are removed completely in the case of the corresponding low molecular weight compounds [205]. Adsorption of the dimethylphenyl-, methyldiphenyl- and triphenylphosphine molecules on to gold surfaces significantly changes the electron distribution in the adsorbed phosphine molecules as probed by infra-red spectroscopy. The strength of chemisorption decreases with an increasing number of phenyl groups of the phosphine. The chemisorbed phosphines appear to be prone to oxidation with IR features observed corresponding to the formation of methoxy and/or phenoxy phosphine species on the surface [206]. On gold powders, double and triple layers of dimethylphenylphosphine may be formed [207]. DFT calculations confirm that the absorption of phosphines on to gold is favorable with the Au−P surface bonds having energies of in the range 55–84 kJ mol^{-1} [208].

7.4.8
Carbon

The adsorption of CO on to gold surfaces has attracted a huge amount of interest due to the potential of catalysis and this topic has been reviewed by Meyer [8].

Kim et al. [209] have studied conjugated oligoacenes of increasing length with either thiol (-S) or isocyanide (-NC) linkers using conducting probe atomic force microscopy and UPS. In the molecules studied, the Au-CN contact was found to be more resistive than the Au-S contact. The difference is explained by noting that the highest occupied molecular orbital (HOMO) of the isocyanide series is lower in energy than the HOMO of the thiols, which leads to tunneling barrier at the point of contact between the molecule and surface in the case of the isocyanide-linked molecules.

Surface-enhanced Raman spectroscopy (SERS) showed that the alkyne group in ethynylbenzene (phenylacetylene) binds to gold surfaces [210] as well as to nanoparticles [211]. DFT calculations showed that the terminal carbon of ethynylbenzene can make a strong bond with the Au(1 1 1) surface [212], which was subsequently shown experimentally by McDonagh et al. [213] and by Gorman and co-workers [214], who attached a number of different terminal alkynes to gold surfaces as well as stabilizing gold nanoparticles using dodecyne molecules.

7.5
Conclusions

The chemical and physical properties of gold make it highly suitable for exploitation in diverse types of nanotechnologies. In particular, gold surfaces provide a convenient platform on which to assemble monolayers or other structures of diverse organic molecules. In addition, very small gold structures have a plasmon resonance with light in the visible and near infra-red parts of the electromagnetic spectrum which provides a further set of interesting capabilities. For these, and other, reasons, nanoscale gold structures lie at the heart of many new or proposed nanotechnologies. Examples include chemical analysis by Surface Plasmon Resonance, new kinds of molecular electronics devices, spectrally-selective coatings for windows, biodiagnostic stains and sensors, and new types of medical therapies.

References

1 Glomm, W.R. (2005) Functionalized gold nanoparticles for applications in bionanotechnology. *Journal of Dispersion Science and Technology*, **26**, 389–414.

2 Kelly, K.L., Coronado, E., Zhao, L.L. and Schatz, G.C. (2003) The optical properties of metal nanoparticles: the influence of size, shape, and dielectric environment. *The Journal of Physical Chemistry B*, **107**, 668–677.

3 Perez-Juste, J., Pastoriza-Santos, I., Liz-Marzan, L.M. and Mulvaney, P. (2005) Gold nanorods: Synthesis, characterization and applications. *Coordination Chemistry Reviews*, **249**, 1870–1901.

4 Hvolbæk, B., Janssens, T.V.W., Clausen, B.S., Falsig, H., Christensen, C.H. and Nørskov, J.K. (2007) Catalytic activity of Au nanoparticles. *Nano Today*, **2**, 14–18.

5 Thompson, D.T. (2007) Using gold nanoparticles for catalysis. *Nano Today*, **2**, 40–43.

6 Cortie, M.B. (2004) The weird world of nanoscale gold. *Gold Bulletin*, **37**, 12–19.

7 (a) Pyykkö, P. (2004) Theoretical chemistry of gold. *Angewandte Chemie International Edition*, **43**, 4412–4456; (b) Chen, M. and Goodman, D.W. (2006) Catalytically active gold: from nanoparticles to ultrathin films. *Accounts of Chemical Research*, **39**, 739–746.

8 Meyer, R., Lemire, C., Shaikhutdinov, S.K. and Freund, H.-J. (2004) Surface chemistry of catalysis by gold. *Gold Bulletin*, **37**, 72–124.

9 Daniel, M.-C. and Astruc, D. (2004) Gold nanoparticles: assembly, supramolecular chemistry, quantum-size-related properties, and applications toward biology, catalysis, and nanotechnology. *Chemical Reviews*, **104**, 293–346.

10 Love, J.C., Estroff, L.A., Kriebel, J.K., Nuzzo, R.G. and Whitesides, G.M. (2005) Self-assembled monolayers of thiolates on metals as a form of nanotechnology. *Chemical Reviews*, **105**, 1103–1169.

11 Stokes, N., McDonagh, A. and Cortie, M.B. (2007) Application of nanolithography to the preparation of nanoscale gold structures. *Gold Bulletin*, **40**, 310–320.

12 Nuzzo, R.G. and Allara, D.L. (1983) Adsorption of bifunctional organic disulfides on gold surfaces. *Journal of the American Chemical Society*, **105**, 4481–4483.

13 Faulk, W.P. and Taylor, G. (1971) An immunocolloid method for the electron microscope. *Immunochemistry*, **8**, 1081–1083.

14 (a) Nam, J., Thaxton, C.S. and Mirkin, C.A. (2003) Nanoparticle-based bio-bar codes for the ultrasensitive detection of proteins. *Science*, **301**, 1884–1886; (b) Zhao, X., Hilliard, L.R., Mechery, S.J., Wang, Y., Bagwe, R.P., Jin, S. and Tan, W. (2004) A rapid bioassay for single bacterial cell quantitation using bioconjugated nanoparticles. *Proceedings of the National Academy of Sciences of the United States of America*, **101**, 15027–15032.

15 Hirsch, L.R., Stafford, R.J., Bankson, J.A., Sershen, S.R., Rivera, B., Price, R.E., Hazle, J.D., Halas, N.J. and West, J.L. (2003) Nanoshell-mediated near-infrared thermal therapy of tumors under magnetic resonance guidance. *Proceedings of the National Academy of Sciences of the United States of America*, **100**, 13549–13554.

16 Elghanian, R., Storhoff, J.J., Mucic, R.C., Letsinger, R.L. and Mirkin, C.A. (1997) Selective colorimetric detection of polynucleotides based on the distance-dependent optical properties of gold nanoparticles. *Science*, **277**, 1078–1081.

17 (a) Richards, D.G., McMillin, D.L., Mein, E.A. and Nelson, C.D. (2002) Gold and its relationship to neurological/glandular conditions. *The International Journal of Neuroscience*, **112**, 31–53; (b) Huaizhi, Z. and Yuantao, N. (2001) China's ancient gold drugs. *Gold Bulletin*, **34**, 24–29.

18 Brown, C.L., Bushell, G., Whitehouse, M.W., Agrawal, D., Tupe, S., Paknikar, K. and Tiekink, E.R. (2007) Nanogoldpharmaceutics. (i) The use of colloidal gold to treat experimentally-induced arthritis in rat models; (ii) Characterization of the gold in Swarna bhasma, a microparticulate used in traditional Indian medicine. *Gold Bulletin*, **40**, 245–250.

19 Shaw, C.F. (1999) Gold-based therapeutic agents. *Chemical Reviews*, **99**, 2589–2600.

20 Pissuwan, D., Cortie, C.H., Valenzuela, S. and Cortie, M.B. (2007) Gold nanosphere-antibody conjugates for therapeutic applications. *Gold Bulletin*, **40**, 121–129.

21 Pissuwan, D., Valenzuela, S.M., Killingsworth, M.C., Xu, X.D. and Cortie, M.B. (2007) Targeted destruction of murine macrophage cells with bioconjugated gold nanorods. *Journal of Nanoparticle Research*, **9**, 1109–1124.

22 Pissuwan, D., Valenzuela, S., Miller, C.M. and Cortie, M.B. (2007) A golden bullet? Selective targeting of *Toxoplasma gondii* tachyzoites using antibody-functionalised gold nanoparticles. *Nano Letters*, **7**, 3808–3812.

23 Pissuwan, D., Valenzuela, S. and Cortie, M.B. (2006) Therapeutic possibilities of plasmonically heated gold nanoparticles. *Trends in Biotechnology*, **24**, 62–67.

24 Paciotti, G.F., Myer, L., Weinreich, D., Goia, D., Pavel, N., McLaughlin, R.E. and Tamarkin, L. (2004) Colloidal gold: a novel nanoparticle vector for tumor directed drug delivery. *Drug Delivery*, **11**, 169–183.

25 Valden, M., Lai, X. and Goodman, D.W. (1998) *Science*, **281**, 1647.

26 Zheng, J., Zhang, C.W. and Dickson, R.M. (2004) Highly fluorescent, water-soluble, size-tunable gold quantum dots. *Physical Review Letters*, **93**, 077402.

27 Yamamoto, Y., Miura, T., Suzuki, M., Kawamura, N., Miyagawa, H., Nakamura, T., Kobayashi, K., Teranishi, T. and Hori, H. (2004) Direct observation of ferromagnetic spin polarization in gold nanoparticles. *Physical Review Letters*, **93**, 116801.

28 Wallace, W.T., Wyrwas, R.B., Whetten, R.L., Mitrić, R. and Bonačić-Koutecký, V. (2003) Oxygen adsorption on hydrated gold cluster anions: experiment and theory. *Journal of the American Chemical Society*, **125**, 8408–8414.

29 Paulus, P.M., Goossens, A., Thiel, R.C., Kraan, A.M.v.d., Schmid, G. and Jongh, L.J.d. (2001) Surface and quantum-size effects in Pt and Au nanoparticles probed by ^{197}Au Mössbauer spectroscopy. *Physical Review B – Condensed Matter*, **64**, 205418.

30 Wallenberg, L.R., Bovin, J.O. and Schmid, G. (1985) On the crystal-structure of small gold crystals and large gold structures. *Surface Science*, **156**, 256–264.

31 Li, J., Li, X., Zhai, H.-J. and Wang, L.-S. (2003) Au20: A tetrahedral cluster. *Science Focus*, **299**, 864–867.

32 Wang, J., Wang, G. and Zhao, J. (2003) Structures and electronic properties of Cu_{20}, Ag_{20}, and Au_{20} clusters with density functional method. *Chemical Physics Letters*, **380**, 716–720.

33 Ford, M.J., Soulé de Bas, B. and Cortie, M.B. (2007) Stability of the tetrahedral motif for small gold clusters in the size range 16–24 atoms. *Materials Science and Engineering B*, **140**, 177–181.

34 Cortie, M.B. and van der Lingen, E. (2002) Catalytic gold nano-particles. *Materials Forum*, **26**, 1–14.

35 Cleveland, C.L., Landman, U., Schaaff, T.G. and Shafigullin, M.N. (1997) Structural evolution of smaller gold nanocrystals: the truncated decahedral motif. *Physical Review Letters*, **79**, 1873–1876.

36 Li, T.X., Yin, S.Y., Ji, Y.L., Wang, B.L., Wang, G.H. and Zhao, J.J. (2000) A genetic algorithm study on the most stable disordered and ordered configurations of Au_{38-55}. *Physics Letters A*, **267**, 403–407.

37 (a) Soler, J.M., Garzón, I.L. and Joannopoulosa, J.D. (2001) Structural patterns of unsupported gold clusters. *Solid State Communications*, **117**, 621–625; (b) Garzón, I.L., Michaelian, K., Beltrán, M.R., Posada-Amarillas, A., Ordejón, P., Artacho, E., Sánchez-Portal, D. and Soler, J.M. (1998) Lowest energy structures of gold nanoclusters. *Physical Review Letters*, **81**, 1600–1603.

38 Koga, K., Ikeshoji, T. and Sugawara, K.-i. (2004) Size- and temperature-dependent structural transitions in gold nanoparticles. *Physical Review Letters*, **92**, 115507.

39 Liu, H.B., Ascencio, J.A., Perez-Alvarez, M. and Yacaman, M.J. (2001) Melting behavior of nanometer sized gold isomers. *Surface Science*, **491**, 88–90.

40 Mackay, A.L. (1962) *Acta Crystallographica*, **15**, 916–918.

41 Cunningham, D.A.H., Vogel, W., Torres-Sanchez, R.M., Tanaka, K. and Haruta, M. (1999) *Journal of Catalysis*, **183**, 24–31.

42 Turkevich, J. (1985) Colloidal gold. Part II. Colour, coagulation, adhesion, alloying and catalytic properties. *Gold Bulletin*, **18**, 125–131.

43 (a) Sönnichsen, C. and Fritzsche, W. (2007) *100 Years of Nanoscience with the Ultramicroscope*, Shaker Verlag, Aachen; (b) Turkevich, J. (1985) Colloidal gold. Part I. Historical and preparative aspects, morphology and structure. *Gold Bulletin*, **18**, 86–91.

44 Turkevich, J., Stevenson, P.C. and Hillier, J. (1951) A study of the nucleation and growth processes in the synthesis of colloidal gold. *Discussions of the Faraday Society*, **11**, 55–75.

45 Frens, G. (1973) Controlled nucleation for the regulationof the particles size in monodisperse gold suspensions. *Nature: Physical Science*, **241**, 20–22.

46 Brust, M., Walker, M., Bethell, D., Schiffrin, D.J. and Whyman, R. (1994) *Chemical Communications*, 801–802.

47 Sweeney, S.F., Woehrle, G.H. and Hutchison, J.E. (2006) Rapid purification and size separation of gold nanoparticles via diafiltration. *Journal of the American Chemical Society*, **128**, 3190–3197.

48 (2006) E2456-06, Terminology for nanotechnology, ASTM International.

49 Lopez, N., Janssens, T.V.W., Clausen, B.S., Xu, Y., Mavrikakis, M., Bligaard, T. and Nørskov, J.K. (2004) On the origin of the catalytic activity of gold nanoparticles for low-temperature CO oxidation. *Journal of Catalysis*, **223**, 232–235.

50 Compton, O.C. and Osterloh, F.E. (2007) Evolution of size and shape in the colloidal crystallization of gold nanoparticles. *Journal of the American Chemical Society*, **129**, 7793–7798.

51 Norman, T.J., Grant, J.C.D., Magana, D., Zhang, J.Z., Liu, J., Cao, D., Bridges, F. and Duuren, A.V. (2002) Near infrared optical absorption of gold nanoparticle aggregates. *The Journal of Physical Chemistry B*, **106**, 7005–7012.

52 Lin, S.-Y., Wu, S.-H. and Chen, C.-H. (2006) A simple strategy for prompt visual sensing by gold nanoparticles: general applications of interparticle hydrogen bonds. *Angewandte Chemie International Edition*, **45**, 4948–4951.

53 (a) Sato, S., Yao, H. and Kimura, K. (2003) Equilibrium growth of three-dimensional

gold nanoparticle superlattices. *Physica E*, **17**, 521–522; (b) Harris, N., Ford, M.J., Cortie, M.B. and McDonagh, A.M. (2007) Laser-induced assembly of gold nanoparticles into colloidal crystals. *Nanotechnology*, **18**, 365301.

54 Oonishi, T., Sato, S., Yao, H. and Kimura, K. (2007) Three-dimensional gold nanoparticle superlattices: Structures and optical absorption characteristics. *Journal of Applied Physics*, **101**, 114314.

55 Hayat, M.A. (1989) *Colloidal Gold: Principles, Methods, and Applications*, Vol. 1, Academic Press, San Diego, CA.

56 Oldenburg, S.J., Averitt, R.D., Westcott, S.L. and Halas, N.J. (1998) Nanoengineering of optical resonances. *Chemical Physics Letters*, **288**, 243–247.

57 Kerker, M. and Blatchford, C.G. (1982) Elastic scattering, absorption, and surface-enhanced Raman scattering by concentric spheres comprised of a metallic and a dielectric region. *Physical Review B – Condensed Matter*, **26**, 4052–4063.

58 Zhou, H.S., Honma, I., Komiyama, H. and Haus, J.W. (1994) Controlled synthesis and quantum-size effect in gold-coated nanoparticles. *Physical Review B – Condensed Matter*, **50**, 12052–12057.

59 Neeves, A.E. and Birnboim, M.H. (1989) Composite structures for the enhancement of nonlinear-optical susceptibility. *Journal of the Optical Society of America B – Optical Physics*, **6**, 787–796.

60 Averitt, R.D., Sarkar, D. and Halas, N.J. (1997) Plasmon resonance shifts of Au-coated Au_2S nanoshells: Insight into multicomponent nanoparticle growth. *Physical Review Letters*, **78**, 4217–4220.

61 Oldenburg, S.J., Jackson, J.B., Westcott, S.L. and Halas, N.J. (1999) Infrared extinction properties of gold nanoshells. *Applied Physics Letters*, **75**, 2897–2899.

62 Hirsch, L.R., Jackson, J.B., Lee, A., Halas, N.J. and West, J.L. (2003) A whole blood immunoassay using gold nanoshells. *Analytical Chemistry*, **75**, 2377–2381.

63 O'Neal, D.P., Hirsch, L.R., Halas, N.J., Payne, J.D. and West, J.L. (2004) Photo-thermal tumor ablation in mice using near infrared-absorbing nanoparticles. *Cancer Letters*, **209**, 171–176.

64 Loo, C., Lin, A., Hirsch, L., Lee, M.H., Barton, J., Halas, N., West, J. and Drezek, R. (2004) Nanoshell-enabled photonics-based imaging and therapy of cancer. *Technology in Cancer Research & Treatment*, **3**, 33–40.

65 (a) West, J.L., Sershen, S.R., Halas, N.J., Oldenburg, S.J. and Averitt, R.D. (2002) Temperature-sensitive polymer/nanoshell composites for photothermally modulated drug delivery, US Patent 6,428,811; (b) Oldenburg, S.J., Averitt, R.D. and Halas, N.J. (2002) Metal nanoshells, US Patent 6,344,272.

66 Caruso, F. (2001) Nanoengineering of particle surfaces. *Advanced Materials*, **13**, 11–22.

67 Graf, C. and van Blaaderen, A. (2002) Metallodielectric colloidal core-shell particles for photonic applications. *Langmuir*, **18**, 524–534.

68 Weissleder, R. (2001) A clearer vision for in vivo imaging. *Nature Biotechnology*, **19**, 316–317.

69 West, J.L., Halas, N.J. and Hirsch, L.R. (2003) Optically-active nanoparticles for use in therapeutic and diagnostic methods, US Patent 6,530,944.

70 Cole, J.R. and Halas, N.J. (2006) Optimized plasmonic nanoparticle distributions for solar spectrum harvesting. *Applied Physics Letters*, **89**, 153120.

71 Cortie, M.B., Dowd, A., Harris, N. and Ford, M.J. (2007) Core-shell nanoparticles with self-regulating plasmonic functionality. *Physical Review B – Condensed Matter*, **75**, 113405.

72 Harris, N., Ford, M.J. and Cortie, M.B. (2006) Optimization of plasmonic heating by gold nanospheres and nanoshells. *The Journal of Physical Chemistry B*, **110**, 10701–10707.

73 Zhang, J.Z., Schwartzberg, A.M., Norman, T., Grant, C.D., Liu, J., Bridges,

F. and Buuren, T.v. (2005) Comment on "Gold nanoshells improve single nanoparticle molecular sensors". *Nano Letters*, **5**, 809–810.

74 (a) Zhang, J., Liu, J., Wang, S., Zhan, P., Wang, S. and Ming, N. (2004) Facile methods to coat polystyrene and silica colloids with metal. *Advanced Functional Materials*, **14**, 1089–1096; (b) Peceros, K.E., Xu, X., Bulcock, S.R. and Cortie, M.B. (2005) Dipole-dipole plasmon interactions in gold-on-polystyrene composites. *The Journal of Physical Chemistry B*, **109**, 21516–21520.

75 Mulvaney, P.C. and Liz-Marzan, L.M. (1999) Stabilized particles and methods of preparation and use thereof. US Patent 6,548,168.

76 Sun, Y. and Xia, Y. (2002) Shape-controlled synthesis of gold and silver nanoparticles. *Science*, **298**, 2176–2179.

77 Sun, Y. and Xia, Y. (2004) Mechanistic study on the replacement reaction between silver nanostructures and chloroauric acid in aqueous medium. *Journal of the American Chemical Society*, **126**, 3892–3901.

78 (a) Tierney, M.J. and Martin, C.R. (1989) *The Journal of Physical Chemistry*, **93**, 2878–2880; (b) Wiesner, J. and Wokaun, A. (1989) Anisometric gold colloids. Preparation, characterization, and optical properties. *Chemical Physics Letters*, **157**, 569–575; (c) Foss, C.A., Hornyak, G.L., Stockert, J.A. and Martin, C.R. (1992) Optical properties of composit membranes containing arrays of nanoscopic gold cylinders. *The Journal of Physical Chemistry*, **96**, 7497–7499.

79 (a) Jana, N.R., Gearheart, L.A. and Murphy, C.J. (2001) Wet chemical synthesis of high aspect ratio cylindrical gold nanorods. *The Journal of Physical Chemistry. B*, **105**, 4065–4067; (b) Nikoobakht, B. and El-Sayed, M.A. (2003) Preparation and growth mechanism of gold nanorods (NRs) using seed-mediated growth method. *Chemistry of Materials*, **15**, 1957–1962; (c) Perez-Juste, J., Liz-Marzan, L.M., Carnie, S., Chan, D.Y.C. and Mulvaney, P. (2004) Electric-field-directed growth of gold nanorods in aqueous surfactant solutions. *Advanced Functional Materials*, **14**, 571–579; (d) Jana, N.R. (2005) Gram-scale synthesis of soluble, near-monodisperse gold nanorods and other anisotropic nanoparticles. *Small*, **1**, 875–882; (e) Busbee, B.D., Obare, S.O. and Murphy, C.J. (2003) An improved synthesis of high-aspect-ratio gold nanorods. *Advanced Materials*, **15** (5), 414–416; (f) Sau, T.K. and Murphy, C.J. (2004) Seeded High Yield Synthesis of Short Au Nanorods in Aqueous Solution. *Langmuir*, **20** 6414–6420; (g) Gole, A. and Murphy, C.J. (2005) Seed-Mediated Synthesis of Gold Nanorods: Role of the Size and Nature of the Seed. *Chemistry of Materials*, **16**, 3633–3640; (h) Gou, L. and Murphy, C.J. (2005) Fine-tuning the shape of gold nanorods. *Chemistry of Materials*, **17**, 3668–3672; (j) Zweifel, D.A. and Wei, A. (2005) Sulfide-arrested growth of gold nanorods. *Chemistry of Materials*, **17**, 4256–4261; (k) Chang, J.-Y., Wu, H., Chen, H., Lingb, Y.-C. and Tan, W. (2005) Oriented assembly of Au nanorods using biorecognition system. *Chemical Communications*, 1092–1094; (l) Mieszawska, A.J. and Zamborini, F.P. (2005) Gold nanorods grown directly on surfaces from microscale patterns of gold seeds. *Chemistry of Materials*, **17**, 3415–3420.

80 Boleininger, J., Kurz, A., Reuss, V. and Sönnichsen, C. (2006) Microfluidic continuous flow synthesis of rod-shaped gold and silver nanocrystals. *Physical Chemistry Chemical Physics*, **8**, 3824–3827.

81 El-Sayed, I.H., Huang, X. and El-Sayed, M.A. (2005) Surface plasmon resonance scattering and absorption of anti-EGFR antibody conjugated gold nanoparticles in cancer diagnostics: Applications in oral cancer. *Nano Letters*, **5**, 829–834.

82 Huang, X., El-Sayed, I.H., Qian, W. and El-Sayed, M.A. (2006) Cancer cell imaging and photothermal therapy in the near-infrared region by using gold nanorods. *Journal of the American Chemical Society*, **128**, 2115–2120.

83 El-Sayed, I.H., Huang, X. and El-Sayed, M.A. (2006) Selective laser photo-thermal therapy of epithelial carcinoma using anti-EGFR antibody conjugated gold nanoparticles. *Cancer Letters*, **239**, 129–135.

84 Xu, X., Gibbons, T. and Cortie, M.B. (2006) Spectrally-selective gold nanorod coatings for window glass. *Gold Bulletin*, **39**, 156–165.

85 Cortie, M., Xu, X. and Ford, M. (2006) Effect of composition and packing configuration on the dichroic optical properties of coinage metal nanorods. *Physical Chemistry Chemical Physics*, **8**, 3520–3527.

86 Nikoobakht, B. and El-Sayed, M.A. (2003) Surface-enhanced Raman scattering studies on aggregated gold nanorods. *Journal of Physical Chemistry A*, **107**, 3372–3378.

87 (a) Jain, P.K., Lee, K.S., El-Sayed, I.H. and El-Sayed, M.A. (2006) Calculated absorption and scattering properties of gold nanoparticles of different size, shape, and composition: applications in biological imaging and biomedicine. *The Journal of Physical Chemistry B*, **110**, 7238–7248; (b) Harris, N., Cortie, M.B., Ford, M.J. and Mulvaney, P. (2008) Plasmon absorption in gold nanoparticles: shells versus rods. *Gold Bulletin*, **40**, 5–14.

88 Xu, X. and Cortie, M.B. (2006) Shape change and color gamut in gold nanorods, dumbbells and dog-bones. *Advanced Functional Materials*, **16**, 2170–2176.

89 Sajanlal, P.R. and Pradeep, T. (2008) Electric-field-assisted growth of highly uniform and oriented gold nanotriangles on conducting glass substrates. *Advanced Materials*, **20**, 980–983.

90 Liu, J., Maaroof, A.I., Wieczorek, L. and Cortie, M.B. (2005) Fabrication of hollow metal nanocaps and their red-shifted optical absorption spectra. *Advanced Materials*, **17**, 1276–1281.

91 Shankar, S.S., Rai, A., Ankamwar, B., Singh, A., Ahmad, A. and Sastry, M. (2004) Biological synthesis of triangular gold nanoprisms. *Nature Materials*, **3**, 482–488.

92 Lu, X., Au, L., McLellan, J., Li, Z.-Y., Marquez, M. and Xia, Y. (2007) Fabrication of cubic nanocages and nanoframes by dealloying Au/Ag alloy nanoboxes with an aqueous etchant based on $Fe(NO_3)_3$ or NH_4OH. *Nano Letters*, **7**, 1764–1769.

93 Charnay, C., Lee, A., Man, S., Moran, C.E., Radloff, C., Bradley, R.K. and Halas, N.J. (2003) Reduced symmetry metallodielectric nanoparticles: chemical synthesis and plasmonic properties. *The Journal of Physical Chemistry. B*, **107**, 7327–7333.

94 Liu, J., Cankurtaran, B., Wieczorek, L., Ford, M.J. and Cortie, M.B. (2006) Anisotropic optical properties of semi-transparent coatings of gold nano-caps. *Advanced Functional Materials*, **16**, 1457–1461.

95 (a) Takei, H., Himmelhaus, M. and Okamoto, T. (2002) Absorption spectrum of surface-bound cap-shaped gold particles. *Optics Letters*, **27**, 342–344; (b) Himmelhaus, M. and Takei, H. (2000) Cap-shaped gold nanoparticles for an optical biosensor. *Sensors and Actuators, B: Chemical Sensors and Materials*, **63**, 24–30; (c) Love, J.C., Gates, B.D., Wolfe, D.B., Paul, K.E. and Whitesides, G.M. (2002) Fabrication and wetting properties of metallic half-shells with submicron diameters. *Nano Letters*, **2**, 891–894.

96 (a) Cortie, M.B., Maaroof, A.I., Stokes, N. and Mortari, A. (2007) Mesoporous gold sponge. *Australian Journal of Chemistry*, **60**, 524–527; (b) Ding, Y. and Erlebacher, J. (2003) Nanoporous metals with controlled multimodal pore size

97 (a) Erlebacher, J., Aziz, M.J., Karma, A., Dimitrov, N. and Sieradzki, K. (2001) Evolution of nanoporosity in dealloying. *Nature*, **410**, 450–435; (b) Parida, S., Kramer, D., Volkert, C.A., Rösner, H., Erlebacher, J. and Weissmüller, J. (2006) Volume change during the formation of nanoporous gold by dealloying. *Physical Review Letters*, **97**, 035504.

98 Maaroof, A.I., Cortie, M.B. and Smith, G.B. (2005) Optical properties of mesoporous gold films. *Journal of Optics A: Pure and Applied Optics*, **7**, 303–309.

99 Glaner, L. and van der Lingen, E. and Cortie, M. B. (2003) Gold catalysts and methods for their preparation, Australian Patent 2003/215039.

100 Cortie, M.B., van der Lingen, E. and Pattrick, G. (2003) Catalysis and capacitance on nano-structured gold particles and sponges, in Proceedings of the Asia Pacific Nanotechnology Forum 2003, Cairns, Australia, World Scientific, Singapore, pp. 79–82.

101 Xu, C., Su, J., Xu, X., Liu, P., Zhao, H., Tian, F. and Ding, Y. (2006) Low temperature CO oxidation over unsupported nanoporous gold. *Journal of the American Chemical Society*, **129**, 42–43.

102 (a) Kucheyev, S.O., Hayes, J.R., Biener, J., Huser, T., Talley, C.E. and Hamza, A.V. (2006) Title: Surface-enhanced Raman scattering on nanoporous Au. *Applied Physics Letters*, **89**, 053102; (b) Qian, L.H., Yan, X.Q., Fujita, T., Inoue, A. and Chen, M.W. (2007) Surface enhanced Raman scattering of nanoporous gold: Smaller pore sizes stronger enhancements. *Applied Physics Letters*, **90**, 153120.

103 Mortari, A., Maaroof, A., Martin, D. and Cortie, M.B. (2007) Mesoporous gold electrodes for measurement of electrolytic double layer capacitance. *Sensors and Actuators B*, **123**, 262–268.

104 (a) Cattarin, S., Kramer, D., Lui, A. and Musiani, M.M. (2007) Preparation and characterization of gold nanostructures of controlled dimension by electrochemical techniques. *Journal of Physical Chemistry C*, **111**, 12643–12649; (b) Biener, J., Hodge, A.M., Hayes, J.R., Volkert, C.A., Zepeda-Ruiz, L.A., Hamza, A.V. and Abraham, F.F. (2006) Size effects on the mechanical behavior of nanoporous Au. *Nano Letters*, **6**, 2379–2382.

105 (a) Candy, J.P., Fouilloux, P., Keddam, M. and Takenouti, H. (1981) The characterization of porous electrodes by impedance measurements. *Electrochimica Acta*, **26**, 1029–1034; (b) Cortie, M.B., Maaroof, A., Smith, G.B. and Ngoepe, P. (2006) Nanoscale coatings of $AuAl_x$ and $PtAl_x$ and their mesoporous elemental derivatives. *Current Applied Physics*, **6**, 440–443; Cortie, M.B., Maaroof, A.I. and Smith, G.B. (2005) Electrochemical capacitance of mesoporous gold. *Gold Bulletin*, **38**, 15–23.

106 Smith, G.B., Niklasson, G.A., Svensson, J.S.E.M. and Granqvist, C.G. (1986) Noble-metal-based transparent infrared reflectors: experiments and theoretical analyses for very thin gold films. *Journal of Applied Physics*, **59**, 571–581.

107 Xu, X., Cortie, M.B. and Stevens, M. (2005) Effect of glass pre-treatment on the nucleation of semi-transparent gold coatings. *Materials Chemistry and Physics*, **94**, 266–274.

108 Hutter, E. and Fendler, J.H. (2004) Exploitation of localized surface plasmon resonance. *Advanced Materials*, **16**, 1685–1706.

109 Wang, H., Lei, C., Li, J., Wu, Z., Shen, G. and Yu, R. (2004) A piezoelectric immunoagglutination assay for *Toxoplasma gondii* antibodies using gold nanoparticles. *Biosensors & Bioelectronics*, **19**, 701–709.

110 Kim, T., Lee, C.-H., Joo, S.-W. and Lee, K. (2008) Kinetics of gold nanoparticle aggregation: Experiments and modeling. *Journal of Colloid and Interface Science*, **318**, 238–243.

111 Kimura, K., Yao, H. and Sato, S. (2006) Self-assembling of gold and silver nanoparticles at a hydrophilic/hydrophobic interface: A synthetic aspect and superstructure formation. *Synthesis and Reactivity in Inorganic, Metal-Organic, and Nano-Metal, Chemistry*, **36**, 237–264.

112 (a) Israelachvili, J. (1992) *Intermolecular & Surface Forces*, Academic Press, London; (b) Wilen, L.A., Wettlaufer, J.S., Elbaum, M. and Schick, M. (1995) Dispersion-force effects in interfacial premelting of ice. *Physical Review B-Condensed Matter*, **52**, 12426–12433.

113 (a) Crocker, J. (2008) Golden handshake. *Nature*, **451**, 528–529; (b) Aldaye, F.A. and Sleiman, H.F. (2007) Dynamic DNA templates for discrete gold nanoparticle assemblies: control of geometry, modularity, write/erase and structural switching. *Journal of the American Chemical Society*, **129** 4130–4131.

114 Barnes, W.L., Dereux, A. and Ebbesen, T.W. (2003) Surface plasmon subwavelength optics. *Nature*, **424**, 824–830.

115 Barnes, W.L. (2006) Surface plasmon–polariton length scales: a route to sub-wavelength optics. *Journal of Optics A: Pure and Applied Optics*, **8**, S87–S93.

116 Jung, L.S., Campbell, C.T., Chinowsky, T.M., Mar, M.N. and Yee, S.S. (1998) Quantitative interpretation of the response of surface plasmon resonance sensors to adsorbed films. *Langmuir*, **14**, 5636–5648.

117 (a) Bohren, C.F. and Huffman, D.R. (1998) *Absorption and Scattering of Light by Small Particles*, Wiley, New York; (b) Maier, S.A., Brongersma, M.L., Kik, P.G., Meltzer, S., Requicha, A.A.G. and Atwater, H.A. (2001) Plasmonics–A route to nanoscale optical devices. *Advanced Materials*, **13**, 1501–1505; (c) Genet, C. and Ebbesen, T.W. (2007) Light in tiny holes. *Nature*, **445**, 39–46; (d) Ozbay, E. (2006) Plasmonics: merging photonics and electronics at nanoscale dimensions. *Science*, **311**, 189–193.

118 Cortie, M., Xu, X., Zareie, H., Chowdhury, H. and Smith, G.(12th–15th Dec 2005) Plasmonic heating of gold nanoparticles and its exploitation. presented at Smart Materials, Nano-, and Micro-Smart Systems II, Sydney, Australia, SPIE, 5649, pp. 565–573.

119 (a) Barbillon, G., Bijeon, J.-L., Plain, J., Chapelle, M.L.d.l., Adam, P.-M. and Royer, P. (2007) Chemical gold nanosensors based on localized surface plasmon resonance. *Gold Bulletin*, **40**, 240–244; (b) Haynes, C.L., McFarland, A.D., Zhao, L., Duyne, R.P.V., Schatz, G.C., Gunnarsson, L., Prikulis, J., Kasemo, B. and Ka, M. (2003) Nanoparticle optics: the importance of radiative dipole coupling in two-dimensional nanoparticle arrays. *Journal of Physical Chemistry B Letters*, **107**, 7337–7342; (c) Haes, A.J. and Duyne, R.P.V. (2004) Preliminary studies and potential applications of localized surface plasmon resonance spectroscopy in medical diagnostics. *Expert Review of Molecular Diagnostics*, **4**, 527–537; (d) Yonzon, C.R., Jeoung, E., Zou, S., Schatz, G.C., Mrksich, M. and Duyne, R.P.V. (2004) A comparative analysis of localized and propagating surface plasmon resonance sensors: the binding of concanavalin A to a monosaccharide functionalized self-assembled monolayer. *Journal of the American Chemical Society*, **126**, 12669–12676.

120 (a) Yang, Y., Xiong, L., Shi, J. and Nogami, M. (2006) Aligned silver nanorod arrays for surface-enhanced Raman scattering. *Nanotechnology*, **17**, 2670–2674; (b) Lu, Y., Liu, G.L., Kim, J., Mejia, Y.X. and Lee, L.P. (2005) Nanophotonic crescent moon structures with sharp edge for ultrasensitive biomolecular detection by local electromagnetic field enhancement effect. *Nano Letters*, **5**, 119–124; (c) Orendorff, C.J., Gearheart, L.A., Jana, N.R. and Murphy, C.J. (2006) Aspect ratio

dependence on surface enhanced Raman scattering using silver and gold nanorod substrates. *Physical Chemistry Chemical Physics*, **8**, 165–170.
121 Raschke, G., Kowarik, S., Franzl, T., Sönnichsen, C., Klar, T.A. and Feldmann, J. (2003) Biomolecular recognition based on single gold nanoparticle light scattering. *Nano Letters*, **3**, 935–938.
122 Aslan, K., Leonenko, Z., Lakowicz, J.R. and Geddes, C.D. (2005) Fast and slow deposition of silver nanorods on planar surfaces: Application to metal-enhanced fluorescence. *The Journal of Physical Chemistry. B*, **109**, 3157–3162.
123 (a) Schelm S. and Smith G.B. (2003) Dilute LaB$_6$ nanoparticles in polymer as optimized clear solar control glazing. *Applied Physics Letters*, **82**, 4346–4348; (b) Xu, X., Stevens, M. and Cortie, M.B. (2004) In situ precipitation of gold nanoparticles onto glass for potential architectural applications. *Chemistry of Materials*, **16**, 2259–2266.
124 (a) Iwakoshi, A., Nanke, T. and Kobayashi, T. (2005) Coating materials containing gold nanoparticles. *Gold Bulletin*, **38**, 107–112; (b) Wagner, F.E., Haslbeck, S., Stievano, L., Calogero, S., Pankhurst, Q.A. and Martinek, P. (2000) Before striking gold in gold-ruby glass. *Nature*, **407**, 691–692.
125 (a) Maier, S.A., Kik, P.G., Atwater, H.A., Meltzer, S., Harel, E., Koel, B.E. and Requicha, A.A.G. (2003) Local detection of electromagnetic energy transport below the diffraction limit in metal nanoparticle plasmon waveguides. *Nature Materials*, **2**, 229–232; (b) Atwater, H.A., Maier, S., Polman, A., Dionne, J.A. and Sweatlock, L. (2005) The New "p-n Junction": Plasmonics enables photonic access to the nanoworld. *MRS Bulletin*, **30**, 385–389; (c) Pile, D.F.P. and Gramotnev, D.K. (2006) Adiabatic and nonadiabatic nanofocusing of plasmons by tapered gap plasmon waveguides. *Applied Physics Letters*, **89**, 041111.

126 Alù, A. and Engheta, N. (2006) Theory of linear chains of metamaterial/plasmonic particles as subdiffraction optical nanotransmission lines. *Physical Review B - Condensed Matter*, **74**, 205436.
127 Engheta, N., Salandrino, A. and Alù, A. (2005) Circuit elements at optical frequencies: nanoinductors, nanocapacitors, and nanoresistors. *Physical Review Letters*, **95**, 095504.
128 (a) Grigorenko, A.N., Geim, A.K., Gleeson, H.F., Zhang, Y., Firsov, A.A., Khrushchev, I.Y. and Petrovic, J. (2005) Nanofabricated media with negative permeability at visible frequencies. *Nature*, **438**, 335–338; (b) Zhang, S., Fan, W., Malloy, K.J., Brueck, S.R., Panoiu, N.C. and Osgood, R.M. (2005) Near-infrared double negative metamaterials. *Optics Express*, **13**, 4922–4930; (c) Smith, D.R., Pendry, J.B. and Wiltshire, M.C.K. (2004) Metamaterials and negative refractive index. *Science*, **305**, 788–792; (d) Dolling, G., Wegener, M., Soukoulis, C.M. and Linden, S. (2007) Negative-index metamaterial at 780 nm wavelength. *Optics Letters*, **32**, 53–55; (e) Shalaev, V.M. (2007) Optical negative-index metamaterials. *Nature Photonics*, **1**, 41–48.
129 (a) Schurig, D., Mock, J.J., Justice, B.J., Cummer, S.A., Pendry, J.B., Starr, A.F. and Smith, D.R. (2006) Metamaterial electromagnetic cloak at microwave frequencies. *Science*, **314**, 977–980; (b) Milton, G.W. and Nicorovici, N.A.P. (2006) On the cloaking effects associated with anomalous localized resonance. *Proceedings of the Royal Society A*, **462**, 3027–3059.
130 (a) Pendry, J.B. (2000) Negative refraction makes a perfect lens. *Physical Review Letters*, **85**, 3966–3969; (b) Blaber, M.G., Arnold, M.D., Harris, N., Ford, M.J. and Cortie, M.B. (2007) Plasmon absorption in nanospheres: a comparison of sodium, potassium, aluminium, silver and gold. *Journal of Physics B: Atomic Molecular and Optical Physics*, **394**,

184–187; (c) Melville, D. and Blaikie, R. (2005) Super-resolution imaging through a planar silver layer. *Optics Express*, **13**, 2127–2134.

131 Liu, Z., Peng, L. and Yao, K. (2006) Intense blue luminescence from self-assembled Au-thiolate clusters. *Materials Letters*, **60**, 2362–2365.

132 Fackler, J.J., Grant, T.A., Hanson, B.E. and Staples, R.J. (1999) Characterisation of luminescent, homoleptic, three coordinate, water soluble Au(I) complex of trisulfonated triphenylphosphine (TPPTS) as the caesium salt, $Cs_8[Au(TPPTS)_3]$. $5.25H_2O$. *Gold Bulletin*, **31**, 20–23.

133 (a) Bachman, R.E., Bodolosky-Bettis, S.A., Glennon, S.C. and Sirchio, S.A. (2000) Formation of a novel luminescent form of gold(I) phenylthiolate via self-assembly and decomposition of isonitrilegold(I) phenylthiolate complexes. *Journal of the American Chemical Society*, **122**, 7146–7147; (b) Arvapally, R.K., Sinha, P., Hettiarachchi, S.R., Coker, N.L., Bedel, C.E., Patterson, H.H., Elder, R.C., Wilson, A.K. and Omary, M.A. (2007) Photophysics of bis(thiocyanato)gold(I) complexes: intriguing structure-luminescence relationships. *Journal of Physical Chemistry C*, **111**, 10689–10699.

134 Link, S. and El-Sayed, M.A. (2000) Shape and size dependence of radiative, non-radiative and photothermal properties of gold nanocrystals. *International Reviews in Physical Chemistry*, **19**, 409–453.

135 Farrer, R.A., Butterfield, F.L., Chen, V.W. and Fourkas, J.T. (2005) Highly efficient multiphoton-absorption-induced luminescence from gold nanoparticles. *Nano Letters*, **5**, 1139–1142.

136 Wang, H., Huff, T.B., Zweifel, D.A., He, W., Low, P.S., Wei, A. and Cheng, J.-X. (2005) *In vitro* and *in vivo* two-photon luminescence imaging of single gold nanorods. *Proceedings of the National Academy of Sciences of the United States of America*, **102**, 15752–15756.

137 Dick, K., Dhanasekaran, T., Zhang, Z. and Meisel, D. (2002) Size-dependent melting of silica-encapsulated gold nanoparticles. *Journal of the American Chemical Society*, **124**, 2312–2317.

138 Buffat, P. and Borel, J.-P. (1976) Size effect on the melting temperature of gold particles. *Physical Review A*, **13**, 2287–2298.

139 Soulé de Bas, B., Ford, M.J. and Cortie, M.B. (2006) Melting in small gold clusters: a Density Functional molecular dynamics study. *Journal of Physics - Condensed Matter*, **18**, 55–74.

140 Hutchings, G.J. (1985) Vapor phase hydrochlorination of acetylene: Correlation of catalytic activity of supported metal chloride catalysts. *Journal of Catalysis*, **96**, 292–295.

141 (a) Haruta, M., Kobayashi, T., Sano, H. and Yamada, N. (1987) *Chemistry Letters*, **4**, 405–408; (b) Haruta, M., Sano, H. and Kobayasi, T. (1987) Method for manufacture of catalyst composite having gold or mixture of gold with catalytic metal oxide deposited on carrier, US Patent 4,698,324.

142 Bond, G.C. (2001) Gold: A Relatively New Catalyst. *Gold Bulletin*, **34**, 117–119.

143 (a) Bond, G.C. and Thompson, D.T. (2000) Gold-catalysed oxidation of carbon monoxide. *Gold Bulletin*, **33**, 41–51; (b) Hashmi, A.S.K. and Hutchings, G.J. (2006) Gold catalysis. *Angewandte Chemie International Edition*, **45**, 7896–7936.

144 Haruta, M. (2004) Gold as a novel catalyst in the 21st Century: Preparation, working mechanism and applications. *Gold Bulletin*, **37**, 27–36.

145 Jürgens, B., Kübel, C., Schulz, C., Nowitzki, T., Zielasek, V., Biener, J., Biener, M.M., Hamza, A.V. and Bäumer, M. (2007) New gold and silver-gold catalysts in the shape of sponges and sieves. *Gold Bulletin*, **40**, 142–149.

146 Grunwaldt, J.-D., Maciejewski, M., Becker, O.S., Fabrizioli, P. and Baiker, A. (1999) Comparative study of Au/TiO_2 and Au/ZrO_2 catalysts for low-temperature

CO oxidation. *Journal of Catalysis*, **186**, 458–469.

147 (a) Bond, G.C. and Thompson, D.T. (1999) Catalysis by Gold. *Catalysis Reviews: Science and Engineering*, **41**, 319–388; (b) Bond, G.C. (1972) Catalytic properties of gold. Potential applications in the chemical industry. *Gold Bulletin*, **5**, 11–13.

148 (a) Stobiński, L. and Duś, R. (1993) Molecular hydrogen chemisorption on thin unsintered gold films deposited at low temperature. *Surface Science*, **298**, 101–106; (b) Stobiński, L., Zommer, L. and Duś, R. (1999) Molecular hydrogen interactions with discontinuous and continuous thin gold films. *Applied Surface Science*, **141**, 319–325.

149 Stobiński, L. (1996) Molecular and atomic deuterium chemisorption on thin gold films at 78 K: an isotope effect. *Applied Surface Science*, **103**, 503–508.

150 Eberhardt, W., Cantor, R., Greuter, F. and Plummer, E.W. (1982) Photoemission from condensed layers of molecular hydrogen on copper and gold. *Solid State Communications*, **42**, 799–802.

151 Hammer, B. and Nørskov, J.K., (1995) Why is gold the noblest of all metals? *Nature*, **376**, (6537), 238–240.

152 Okada, M., Ogura, S., Dino, W.A., Wilde, M., Fukutani, K. and Kasai, T. (2005) Reactivity of gold thin films grown on iridium: Hydrogen dissociation. *Applied Catalysis A: General*, **291**, 55–61.

153 Pritchard, J. and Tompkins, F.C. (1960) Surface-potential measurements. Adsorption of hydrogen by Group IB metals. *Transactions of the Faraday, Society*, **56**, 540–550.

154 Sault, A.G., Madix, R.J. and Campbell, C.T. (1986) Adsorption of oxygen and hydrogen on Au(ll0)-(l × 2). *Surface Science*, **169**, 347–356.

155 Bartlett, N. (1998) Relativistic Effects and the Chemistry of Gold. *Gold Bulletin*, **31** 22–25.

156 Arutyunov, V.S. and Chaikin, A.M. (1977) Heterogeneous decay of fluorine atoms. *Reaction Kinetics and Catalysis Letters*, **6**, 169–174.

157 Loudiana, M.A., Dickinson, J.T., Schmid, A. and Ashley, E.J. (1987) Electron enhanced sorption of fluorine by silver surfaces. *Applied Surface Science*, **28**, 311–322.

158 Rzeźnicka, I.I., Lee, J. and Yates, J.T. Jr (2006) Dynamics of NF3 in a condensed film on Au(111) as studied by electron-stimulated desorption. *Surface Science*, **600**, 4492–4500.

159 Kishi, K. and Ikeda, S. (1974) X-ray photoelectron spectroscopic study of the reaction of evaporated metal films with chlorine gas. *The Journal of Physical Chemistry*, **78**, 107–112.

160 Spencer, N.D. and Lambert, R.M. (1981) Chlorine chemisorption and surface chloride formation on Au(111). *Surface Science*, **107**, 237–248.

161 Kastanas, G.N. and Koel, B.E. (1993) Interaction of chlorine with the gold (111) surface in the temperature range of 120 to 1000 K. *Applied Surface Science*, **64**, 235–249.

162 Gao, W., Zhou, L., Pinnaduwage, D.S. and Friend, C.M. (2007) Interaction of Chlorine with Au−O Surface Complexes. *Journal of Physical Chemistry C*, **111**, 9005–9007.

163 Pinnaduwage, D.S., Zhou, L., Gao, W. and Friend, C.M. (2007) Chlorine Promotion of Styrene Epoxidation on Au(111). *Journal of the American Chemical Society*, **129**, 1872–1873.

164 Iwai, H., Okadat, M., Fukutani, K. and Murata, Y. (1995) Chlorine-induced de-reconstruction on Au(001) and Cl-adsorbed layers. *Journal of Physics-Condensed Matter*, **7**, 5163–5176.

165 Bertel, E. and Netzer, F.P. (1980) Adsorption of bromine on the reconstructed Au(100) surface: LEED, thermal desorption and work function measurements. *Surface Science*, **97**, 409–424.

166 Tao, N.J. and Lindsay, S.M. (1992) *In Situ* Scanning Tunneling Microscopy Study

of Iodine and Bromine Adsorption on Au (111) under Potential Control. *The Journal of Physical Chemistry*, **96**, 5213–5217.

167 McHardy, R., Haiss, W.H. and Nichols, R.J. (2000) An STM investigation of surface diffusion on iodine modified Au (111). *Physical Chemistry Chemical Physics*, **2**, 1439–1444.

168 Magnussen, O.M. (2002) Ordered Anion Adlayers on Metal Electrode Surfaces. *Chemical Reviews*, **102**, 679–725.

169 Ikezawa, Y. and Terashima, H. (2002) Influence of halide anions on the behavior of pyridine adsorbed on Au(111) electrode. *Electrochimica Acta*, **47**, 4407–4412.

170 MacDonald, W.R. and Hayes, K.E. (1970) Comparative study of the rapid adsorption of oxygen by silver and gold. *Journal of Catalysis*, **18**, 115–118.

171 Canning, N.D.S., Outka, D. and Madix, R.J. (1984) The adsorption of oxygen on gold. *Surface Science*, **141**, 240–254.

172 Pireaux, J.J., Chtaib, M., Delrue, J.P., Thiry, P.A., Liehr, M. and Caudino, R. (1984) Electron spectroscopic characterization of oxygen adsorption on gold surfaces I. Substrate impurity effects on molecular oxygen adsorption in ultra high vacuum. *Surface Science*, **141**, 211–220.

173 Visart-de-Bocarmé, T., Chau, T.-D., Tielens, F., Andrés, J., Gaspard, P., Wang, R.L.C., Kreuzer, H.J. and Kruse, N. (2006) Oxygen adsorption on gold nanofacets and model clusters. *Journal of Chemical Physics*, **125**, 054703–054707.

174 Gottfried, J.M., Schmidt, K.J., Schroeder, S.L.M. and Christmann, K. (2002) Spontaneous and electron-induced adsorption of oxygen on Au(110)-(1 × 2). *Surface Science*, **511**, 65–82.

175 Rao, C.N.R., Kamath, P.V. and Yashonath, S. (1982) Molecularly adsorbed oxygen on metals: electron spectroscopic studies. *Chemical Physics Letters*, **88**, 13–16.

176 Min, B.K., Deng, X., Pinnaduwage, D., Schalek, R. and Friend, C.M. (2005) Oxygen-induced restructuring with release of gold atoms from Au(111). *Physical Review B – Condensed Matter*, **72**, 121410–121414.

177 Min, B.K., Alemozafar, A.R., Biener, M.M., Biener, J. and Friend, C.M. (2005) Reaction of Au(111) with sulfur and oxygen: scanning tunneling microscopic study. *Topics in Catalysis*, **36**, 77–90.

178 Mills, G., Gordon, M.S. and Metiu, H. (2003) Oxygen adsorption on Au clusters and a rough Au(111) surface: The role of surface flatness, electron confinement, excess electrons, and band gap. *Journal of Chemical Physics*, **118**, 4198–4205.

179 Chen, Y., Crawford, P. and Hu, P. (2007) Recent Advances in Understanding CO Oxidation on Gold Nanoparticles Using Density Functional Theory. *Catalysis Letters*, **119**, 21–28.

180 Biener, M.M., Biener, J. and Friend, C.M. (2007) Sulfur-induced mobilization of Au surface atoms on Au(111) studied by real-time STM. *Surface Science*, **601**, 1659–1667.

181 (a) Vericat, C., Vela, M.E., Benitez, G.A., Gago, J.A.M., Torrelles, X. and Salvarezza, R.C. (2006) Surface characterization of sulfur and alkanethiol self-assembled monolayers on Au(111). *Journal of Physics: Condensed Matter*, **18**, R867–R900; (b) Ulman, A. (1996) Formation and Structure of Self-Assembled Monolayers. *Chemical Reviews*, **96**, 1533–1554.

182 Joo, S.-W. (2006) Characterization of self-assembled phenyl and benzylisothiocyanate thin films on Au surfaces. *Surface and Interface Analysis*, **38**, 173–177.

183 Fu, M.-D., Chen, I.-W.P., Lu, H.-C., Kuo, C.-T., Tseng, W.-H. and Chen, C.-H. (2007) Conductance of Alkanediisothiocyanates: Effect of Headgroup-Electrode Contacts. *Journal of Physical Chemistry C*, **111**, 11450–11455.

184 Alanyalioglu, M., Demir, U. and Shannon, C. (2004) Electrochemical formation of Se atomic layers on Au(111) surfaces: the role of adsorbed selenate and

selenite. *Journal of Electroanalytical Chemistry*, **561**, 21–27.
185 Huang, B.M., Lister, T.E. and Stickney, J.L. (1997) Se adlattices formed on Au(100), studies by LEED, AES, STM and electrochemistry. *Surface Science*, **392**, 27–43.
186 Nakamura, T., Miyamae, T., Yoshimura, D., Kobayashi, N., Nozoye, H. and Matsumoto, M. (2005) Alkyl Chain Conformation and the Electronic Structure of Octyl Heavy Chalcogenolate Monolayers Adsorbed on Au(111). *Langmuir*, **21**, 5026–5033.
187 Nakamura, T., Yasuda, S., Miyamae, T., Nozoye, H., Kobayashi, N., Kondoh, H., Nakai, I., Ohta, T., Yoshimura, D. and Matsumoto, M. (2002) Effective Insulating Properties of Autooxidized Monolayers Using Organic Ditellurides. *Journal of the American Chemical Society*, **124**, 12642–12643.
188 Dishner, M.H., Hemminger, J.C. and Feher, F.J. (1997) Scanning Tunneling Microscopy Characterization of Organoselenium Monolayers on Au(111). *Langmuir*, **13**, 4788–4790.
189 Yokota, K., Taniguchi, M. and Kawai, T. (2007) Control of the Electrode-Molecule Interface for Molecular Devices. *Journal of the American Chemical Society*, **129**, 5818–5819.
190 Patrone, L., Palacin, S. and Bourgoin, J.P. (2003) Direct comparison of the electronic coupling efficiency of sulfur and selenium alligator clips for molecules adsorbed onto gold electrodes. *Applied Surface Science*, **212–213**, 446–451.
191 Patrone, L., Palacin, S., Charlier, J., Armand, F., Bourgoin, J.P., Tang, H. and Gauthier, S. (2003) Evidence of the Key Role of Metal-Molecule Bonding in Metal-Molecule-Metal Transport Experiments. *Physical Review Letters*, **91**, 096802.
192 Yasuda, S., Yoshida, S., Sasaki, J., Okutsu, Y., Nakamura, T., Taninaka, A., Takeuchi, O. and Shigekawa, H. (2006) Bond Fluctuation of S/Se Anchoring Observed in Single-Molecule Conductance Measurements using the Point Contact Method with Scanning Tunneling Microscopy. *Journal of the American Chemical Society*, **128**, 7746–7747.
193 Mankefors, S., Grigoriev, A. and Wendin, G. (2003) Molecular alligator clips: a theoretical study of adsorption of S, Se and S-H on Au(111). *Nanotechnology*, **14**, 849–858.
194 Samant, M.G., Brown, C.A. and Gordon, J.G. (1992) Formation of an Ordered Self-Assembled Monolayer of Docosaneselenol on Gold(111). Structure by Surface X-ray Diffraction. *Langmuir*, **8**, 1615–1618.
195 Protsailo, L.V., Fawcett, W.R., Russell, D. and Meyer, R.L. (2002) Electrochemical Characterization of the Alkaneselenol-Based SAMs on Au(111) Single Crystal Electrode. *Langmuir*, **18**, 9342–9349.
196 Shaporenko, A., Müller, J., Weidner, T., Terfort, A. and Zharnikov, M. (2007) Balance of Structure-Building Forces in Selenium-Based Self-Assembled Monolayers. *Journal of the American Chemical Society*, **129**, 2232–2233.
197 (a) Venkataraman, L., Klare, J.E., Nuckolls, C., Hybertsen, M.S. and Steigerwald, M.L. (2006) Dependence of single-molecule junction conductance on molecular conformation. *Nature*, **442**, 904–907; (b) Venkataraman, L., Klare, J.E., Tam, I.W., Nuckolls, C., Hybertsen, M.S. and Steigerwald, M.L. (2006) Single-Molecule Circuits with Well-Defined Molecular Conductance. *Nano Letters*, **6**, 458–462.
198 Bilic, A., Reimers, J.R., Hush, N.S. and Hafner, J. (2002) Adsorption of ammonia on the gold (111) surface. *Journal of Chemical Physics*, **116**, 8981–8987.
199 Li, Z. and Kosov, D.S. (2007) Nature of well-defined conductance of amine-anchored molecular junctions: Density functional calculations. *Physical Review B-Condensed Matter*, **76**, 035415.
200 Hoft, R.C., Ford, M.J., McDonagh, A.M. and Cortie, M.B. (2007) Adsorption of amine compounds on the Au(111)

201 Chen, F., Li, X., Hihath, J., Huang, Z. and Tao, N. (2006) Effect of Anchoring Groups on Single-Molecule Conductance: Comparative Study of Thiol-, Amine-, and Carboxylic-Acid-Terminated Molecules. *Journal of the American Chemical Society*, **128**, 15874–15881.

202 Steiner, U.B., Neuenschwander, P., Caseri, W.R., Suter, U.W. and Stucki, F. (1992) Adsorption of NPh_3, PPh_3, $AsPh_3$, $SbPh_3$, and $BiPh_3$ on Gold and Copper. *Langmuir*, **8**, 90–94.

203 Uvdal, K., Persson, I. and Liedberg, B. (1996) Tricyclohexylphosphine Adsorbed on Gold. *Langmuir*, **11**, 1252–1256.

204 Kariis, H., Westermark, G., Persson, I. and Liedberg, B. (1998) Infrared Spectroscopic and Temperature-Programmed Desorption Studies of Dimethylphenylphosphine Adsorbed on the Coinage Metals. *Langmuir*, **14**, 2736–2743.

205 Steiner, U.B., Caseri, W.R. and Suter, U.W. (1998) Ultrathin Layers of Substituted Poly(styrene)s on Gold and Copper. *Langmuir*, **14**, 347–351.

206 Westermark, G., Kariis, H., Persson, I. and Liedberg, B. (1999) An infrared study on the chemisorption of tertiary phosphines on coinage and platinum group metal surfaces. *Colloids and Surfaces A – Physicochemical and Engineering Aspects*, **150**, 31–43.

207 Westermark, G. and Persson, I. (1998) Chemisorption of tertiary phosphines on coinage and platinum group metal powders. An infrared reflectance absorption spectroscopic, enhanced Raman spectroscopic and surface coverage study. *Colloids and Surfaces A – Physicochemical and Engineering Aspects*, **144**, 149–166.

208 Ford, M.J., Masens, C. and Cortie, M.B. (2006) The application of gold surfaces and particles in nanotechnology. *Surface Review and Letters*, **13**, 297–307.

209 Kim, B., Beebe, J.M., Jun, Y., Zhu, X.Y. and Frisbie, C.D. (2006) Correlation between HOMO Alignment and Contact Resistance in Molecular Junctions: Aromatic Thiols versus Aromatic Isocyanides. *Journal of the American Chemical Society*, **128**, 4970–4971.

210 Feilchenfeld, H. and Weaver, M.J. (1989) Binding of alkynes to silver, gold, and underpotential deposited silver electrodes as deduced by surface-enhanced Raman spectroscopy. *The Journal of Physical Chemistry*, **93**, 4276–4282.

211 (a) Joo, S.-W. and Kim, K. (2004) Adsorption of phenylacetylene on gold nanoparticle surfaces investigated by surface-enhanced Raman scattering. *Journal of Raman Spectroscopy*, **35**, 549–554; (b) Lim, J.K., Joo, S.-W. and Shin, K.S. (2007) Concentration dependent Raman study of 1,4-diethynylbenzene on gold nanoparticle surfaces. *Vibrational Spectroscopy*, **43**, 330–334.

212 Ford, M.J., Hoft, R.C. and McDonagh, A. (2005) Theoretical study of ethynylbenzene adsorption on Au(111) and implications for a new class of self-assembled monolayer. *The Journal of Physical Chemistry. B*, **109**, 20387–20392.

213 McDonagh, A.M., Zareie, H.M., Ford, M.J. and Barton, C. (2007) Oxidation of ethynylbenzene on gold(111) surfaces. *Journal of the American Chemical Society*, **129**, 3533–3538.

214 Zhang, S., Chandra, K.L. and Gorman, C.B. (2007) Self-Assembled Monolayers of Terminal Alkynes on Gold. *Journal of the American Chemical Society*, **129**, 4876–4877.

8
Liquid Crystals Based on Gold Compounds
Silverio Coco and Pablo Espinet

8.1
Introduction

8.1.1
A Few General Concepts in Liquid Crystals

Most solid materials produce isotropic liquids directly upon melting. However, in some cases one or more intermediate phases are formed (called *mesophases*), where the material retains some ordered structure but already shows the mobility characteristic of a liquid. These materials are *liquid crystal* (LCs)(or *mesogens*) of the *thermotropic* type, and can display several transitions between phases at different temperatures: crystal–crystal transition (between solid phases), melting point (solid to first mesophase transition), mesophase–mesophase transition (when several mesophases exist), and clearing point (last mesophase to isotropic liquid transition) [1]. Often the transitions are observed both upon heating and on cooling (*enantiotropic* transitions), but sometimes they appear only upon cooling (*monotropic* transitions).

Similar behavior can occur when a crystalline network is disassembled by adding a solvent rather than by heating. These mesogens are called *lyotropic* liquid crystals and the mesophase formation shows temperature and concentration dependence. They are very important in biological systems, but have been much less studied in materials science.

Liquid crystal behavior is a genuine supramolecular phenomenon based on the existence of extended weak interactions (dipole–dipole, dispersion forces, hydrogen bonding) between molecules. For the former two to be important enough, it is usually necessary for the molecules to have anisotropic shapes, able to pack efficiently so that these weak interactions can accumulate and co-operate, so as to keep the molecules associated in a preferred orientation, but free enough to move and slide, as they are not connected by rigid bonds.

Gold Chemistry: Applications and Future Directions in the Life Sciences. Edited by Fabian Mohr
Copyright © 2009 WILEY-VCH Verlag GmbH & Co. KGaA, Weinheim
ISBN: 978-3-527-32086-8

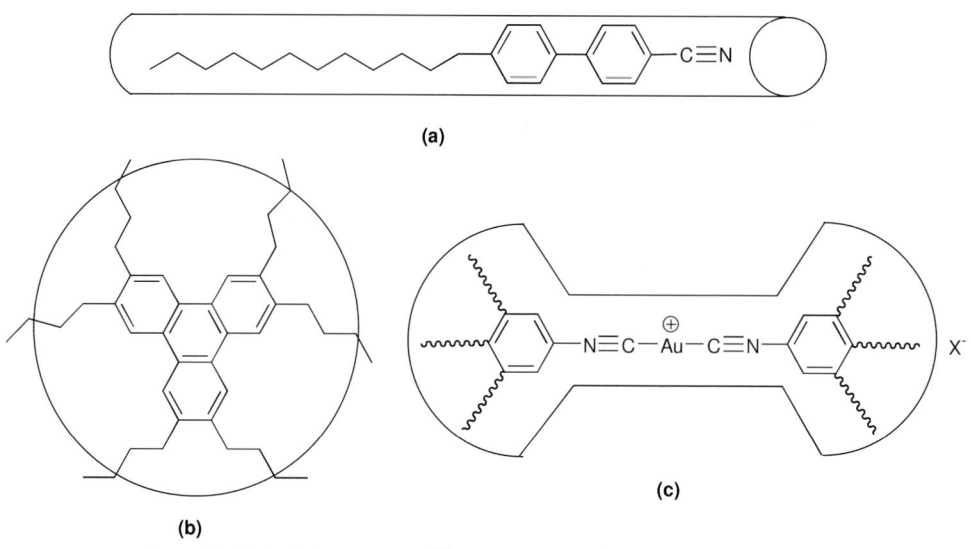

Figure 8.1 Typical shapes: (a) rod-like mesogenic molecule; (b) disk-like mesogenic molecule based on a polyaromatic core; (c) polycatenar mesogenic molecule, combining rod-like and discotic features. Note that the chains will not have the rigid representation shown in the drawing.

Considering structural factors in the molecule, thermotropic LCs can be classified into three groups: *calamitic* (formed by rod-like molecules), *discotic* (formed by disk-like molecules), and *polycatenar* mesogens (intermediate shape between rod-like and disk-like) (Figure 8.1). The last type is based on molecules with a somewhat extended calamitic central core containing several long substituents, giving rise to the name polycatenar (literally many-tailed). These types of structures determine the resulting intermolecular arrangement in the liquid crystal phase, that is, the structure of the mesophase. Polycatenar mesogens display mesomorphism characteristic of rod-like or disk-like mesogens depending on the number, length and arrangement of the chains.

The mesophases of calamitic mesogens are classified in two groups: *nematic* and *smectic*. The nematic mesophase (N) is characterized by an orientational order of the molecules that are aligned along a preferred direction (defined by a *director* n) (Figure 8.2). The molecules can slide and move in the nematic mesophase (while roughly keeping their molecular orientation) and rotate around their main axis. This is the less ordered mesophase and it is usually very fluid.

In the smectic mesophases the molecules are oriented, as in a nematic mesophase, with their principal axis roughly parallel to the director, but they are also defining layers. These layers can be perpendicular to the director, as in the *smectic A* mesophase (SmA), or tilted, as in the *smectic C* (SmC). The SmA and SmC mesophases are the less ordered and more common smectic mesophases. Other less common types of smectic mesophases are known, which differ in the degree or kind of molecular ordering within and between the layers [2].

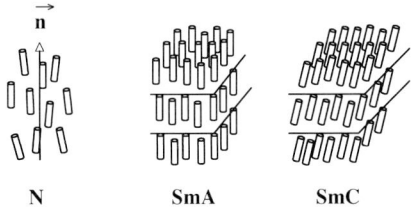

Figure 8.2 Schematic representation of the molecular arrangement in nematic (N), smectic A (SmA) and smectic C (SmC) mesophases. A cylinder represents a molecule with rod-like shape.

When the mesogenic compounds are chiral (or when chiral molecules are added as dopants) chiral mesophases can be produced, characterized by helical ordering of the constituent molecules in the mesophase. The chiral nematic phase is also called *cholesteric*, taken from its first observation in a cholesteryl derivative more than one century ago. These chiral structures have reduced symmetry, which can lead to a variety of interesting physical properties such as thermocromism, ferroelectricity, and so on.

The discotic mesophases are classified in two types: *columnar*, and *nematic discotic*. The structure of the nematic discotic mesophase (N_D, Figure 8.3, left) is similar to that of rod-like molecules, but constituted by disk-like units. In columnar mesophases, the molecules are stacked in a columnar disposition and, depending on the type of columnar arrangement, several columnar mesophases are known. The most common lattices of the columnar phases are nematic discotic (N_D), columnar nematic (N_{Col}), columnar hexagonal (Col_h), and columnar rectangular (Col_r) mesophases.

Lyotropic LCs can also be described by a simple model. Such molecules usually possess the amphiphilic nature characteristic of surfactant, consisting of a polar head and one or several aliphatic chains. A representative example is sodium stearate (soap), which forms mesophases in aqueous solutions (Figure 8.4a). In lyotropic mesophases, not only does temperature play an important role, but also the solvent, the number of components in the solution and their concentration. Depending on these factors, different types of micelles can be formed. Three representative types of micelles are presented in Figure 8.4b–d.

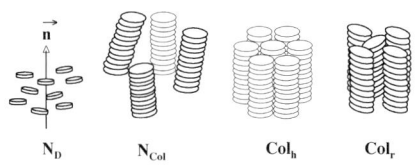

Figure 8.3 Schematic representation of the molecular arrangement in the nematic discotic (N_D), columnar nematic (N_{Col}), columnar hexagonal (Col_h), and columnar rentangular (Col_r) mesophases. A tablet represents a molecule with disk-like shape.

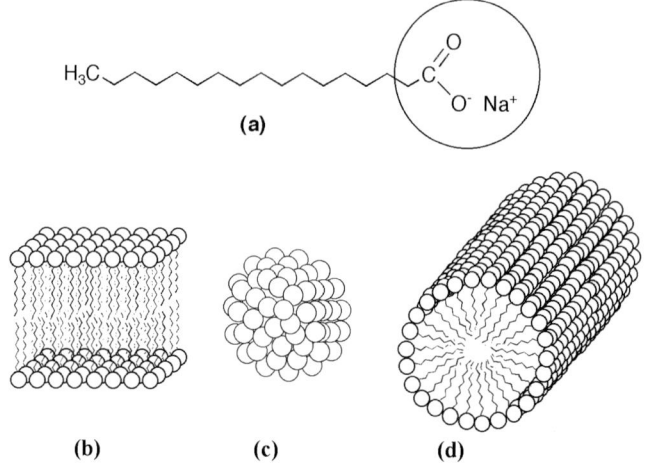

Figure 8.4 (a) A typical molecule that behaves as lyotropic liquid crystal; (b) schematic representation of a plate-shaped micelle; (c) a spherical micelle; (d) a cylindrical micelle.

Mesophase characterization of liquid crystalline materials can be achieved by a combination of different methods. The most used and classic is observation of the texture shown by the mesophases in a polarizing microscope equipped with a hot stage. The different mesophases are associated with characteristic textures, but unequivocal identification sometimes requires low angle X-ray diffraction studies on the mesophase. Differential scanning calorimetry (DSC) is also used in order to measure transition temperatures and enthalpies. Obviously large changes in ordering (e.g., melting points) are associated with large transition enthalpy values.

8.1.2
Gold, an Ideal Metal for LC Studies

Mesogenic behavior was discovered in 1888 in organic molecules [3]. Since then, the structural features favoring the appearance of organic mesogens have been reasonably well established. Usually calamitic organic mesogens consist of a rigid core (two or more conjugated rings) associated with alkyl- or alkoxychains that contribute to extend the polarizability of the molecule. The use of an extended conjugated core is a common strategy in order to increase the otherwise rather weak intermolecular interactions in organic molecules (hence their low melting points). Terminal groups that introduce dipole moments (CN, halogen) are also common, as well as systems that introduce hydrogen bonding interactions. Disk-like molecules are usually built around aromatic cores (Figure 8.1).

Metallomesogens (mesogens based on metal containing molecules) are much younger than organic LCs. Their more systematic study began in the mid 1980s [4–6]. Soon it became apparent that, when a metal is involved, the transition temperatures of the material are quite high. On the other hand, LCs can sometimes be produced with molecules having bizarre shapes, for which no mesogenic behavior would be expected in related organic molecules. Furthermore, the initial naïve ideas stressing the importance of the metal coordination geometry in determining the shape of the molecule (e.g., linear coordination for rod shaped, square-planar coordination for disk shaped), have made room for the evidence that the shape and adaptability of the bulky coordinated organic ligands very often disguise the initial geometry around the metal core, and do in fact determine the real shape of the molecule. Metal centers with coordination numbers higher than two have very complex multipolar structures and it is difficult to grasp the role of the metal in the so-called "structure-mesogenic behavior relationship." In these circumstances it was convenient to establish a simple model to understand the role of the metal atom. The linear coordination of gold(I) offers the simplest model to try to identify the effect of the metal amongst other contributions to the formation of mesophases. As a matter of fact, with very few exceptions, the papers on gold-containing mesogens deal with gold(I).

Certainly gold(I) has one disadvantage compared with some metal centers, which is the lesser thermal stability of their complexes. However, since one goal in thermotropic LCs is to achieve low temperature transitions, this should not be a main concern.

8.2
Pyridine Complexes

The first reported example of a liquid crystal based on gold complexes was the alkoxystilbazole complex **1**, which displays mesomorphic behavior between 120 and 200 °C. However, it is not thermally stable and decomposition occurs on reaching the clearing point and the exact nature of the phase was not established [7]. At this early stage, we can comment that the accumulation of electron delocalization over two rings and a double bond, which is favorable for organic mesogens, turns out to be disadvantageous when a metal is involved because it leads to high transition temperatures in conflict with the stability of the complex. However, this simple concept was not immediately understood.

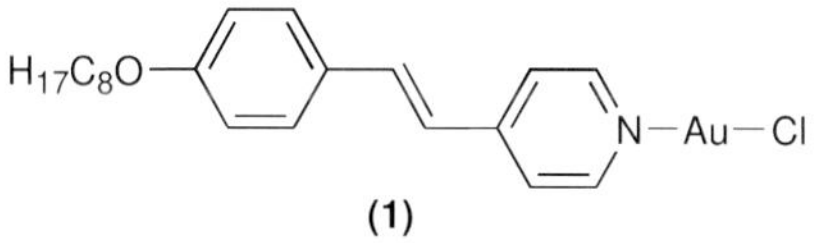

$H_{17}C_8O$—⟨⟩—=—⟨⟩—N—Au—Cl

(1)

8.3
Dithiobenzoate Complexes

Three series of mesomorphic dithiobenzoatogold(III) complexes (**2**) have been described. They show a SmA mesophase in the range 150–200 °C [8, 9]. The thermal stability of these complexes is low and the clearing transitions occur with decomposition or even at lower temperatures. The substitution of the halogens atoms (X) by methyl groups produces a decrease of both the melting temperatures and the thermal mesophase stability.

$H_{2n+1}C_nO$—⟨aryl⟩—C(=S)(S)Au(X)(X)

X = Cl, Br, CH_3 (2)

8.4
Isocyanide Complexes

Isocyanides (or isonitriles, CNR) are interesting ligands of high importance in organometallic chemistry due to their coordination ability. They form very stable complexes with many metals including gold; these display interesting and varied reactivity. The chemical behavior of the isocyanides depends on the steric and electronic features of the R group, which can be used to tune their properties and consequently those of their metal complexes. Perhaps the versatility of this ligand together with the fact that isocyanide gold(I) complexes constitute a fairly stable coordinatively simple system can explain that the majority of mesomorphic gold complexes described are based on gold(I) isocyanide derivatives. A variety of different types of mesomorphic gold isocyanide complexes are known. On the basis of the gold moiety bound to the isocyanide ligand these complexes can be classified into four types: halo-, alkynyl-, or polyfluorophenyl-gold mixed isocyanide complexes, and homoleptic bis-isocyanide cationic derivatives. Mesomorphic mixtures containing gold isocyanide complexes will be considered as an independent group.

8.4.1
Isocyanide-Halide Complexes

The first mesomorphic gold isocyanide complexes reported were the gold(I) derivatives [AuCl(CNC$_6$H$_4$COOC$_6$H$_4$OC$_n$H$_{2n+1}$)] (3a) [10]. The free isocyanides are LCs only when the alkoxy substituent is long, and the ligand with $n = 10$ displays N and SmA mesophases, between 68 and 84 °C. Coordination to gold results in an increase of the transition temperatures and gives more ordered mesophases (SmA and SmC) in the range 172–270 °C. This result shows clearly that the presence of the gold atom, which is a center containing a large number of polarizable electrons, increases molecular polarizability, resulting in stronger intermolecular interactions and higher transition temperatures. In order to hinder the molecular packing to decrease the intermolecular interactions and achieve mesomorphic behavior at lower temperatures, complexes containing a lateral chain were prepared (3b). Despite the non-mesomorphic nature of the ligands, the complexes show N and SmA phases at reduced transition temperatures (90–190 °C, depending on n and m). Longer lateral chain lengths favor the formation of nematic mesophases.

$H_{2n+1}C_nO$—⟨aryl⟩—O—C(=O)—⟨aryl, R⟩—N≡C—Au-Cl

R = H (**3a**), OC_nH_{2n+1} (**3b**)

X-ray structural analysis of the derivatives with $n=10$ and $m=1$ shows the existence of weak aurophilic gold···gold interactions. It was suggested that they could play an important role in the formation of mesophases. In this respect, studies on alkylisocyanide complexes [AuCl(CNC$_n$H$_{2n+1}$)], which show rotator phases (rotator phases are lamellar crystals which lack long-range order in the rotational degree of freedom of the molecules along their long axes), have provided evidence that aurophilic interactions can be used to induce the formation of mesophases in the absence of classical mesogenic units such aromatic rings [11].

A similar chiral chlorogold(I) compound [AuCl(CNC$_6$H$_4$COOC$_6$H$_4$OC*H-MeC$_6$H$_{13}$)] has been described, displaying ferroelectric SmC* (152 °C), and SmA (185 °C) mesophases before decomposition occurs at the clearing temperature (285 °C). However, the spontaneous polarization could not be measured with precision due to the inherent conductivity of the compound [12].

Complexes of the type [AuXL] (X = halide) where L is a promesogenic ligand (that is a ligand that is not a LC itself but has a shape or other properties that will favor the appearance of mesomorphism in the complex) can be seen as compounds where XAu is a terminal group modifying the properties of L [13]. Compared with other terminal groups, the presence of a metal produces dramatic effects. Thus the complexes [AuX(CNC$_6$H$_4$OC$_n$H$_{2n+1}$)] (**4**) give LCs (SmA mesophases for X = Cl, Br but not for X = I) between 105 and 171 °C [14]. In fact, the free isocyanides are not LCs and it was quite extraordinary at that time that mesogens were obtained from molecules containing only one aryl ring. These gold compounds showed high thermal stability and offered a very simple structure particularly suitable to study the effect of structural modifications on the liquid crystal behavior. Systematic studies have been made varying the structural and electronic properties of the isocyanide and the anionic X ligands. The influence of the metal (Cu, Au) has also been studied [15], but this will not be discussed here.

$H_{2n+1}C_nO$—⟨aryl⟩—N≡C—Au—X

n= 2, 4, 6, 8, 10; X = Cl, Br, I   (**4**)

Concerning the role of the isocyanide, incorporation of a second aryl ring in the [CNC$_6$H$_4$OC$_n$H$_{2n+1}$] system produces mesomorphic biphenylisocyanide ligands (nematic or smectic mesophases in the range of 40–85 °C), and liquid crystal complexes [AuX(CNC$_6$H$_4$C$_6$H$_4$OC$_n$H$_{2n+1}$)] (**5**) even for X = I [16].

$H_{2n+1}C_nO$—⟨aryl⟩—⟨aryl⟩—N≡C—Au—X

n= 2, 4, 6, 8, 10; X = Cl, Br, I   (**5**)

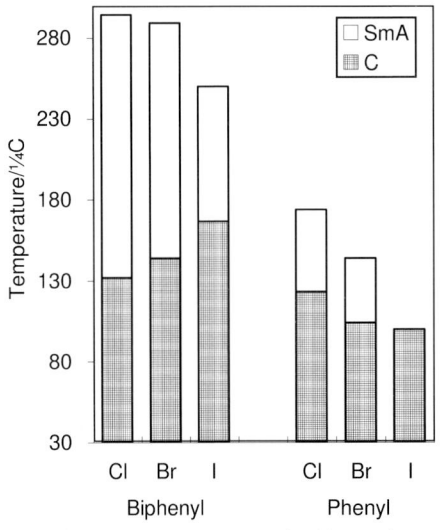

Figure 8.5 Thermal properties of gold complexes with isocyanides having a phenyl or a biphenyl core.

A closer look at the thermal behavior variation upon introduction of a second aryl ring (see Figure 8.5 for the behavior of the derivatives with a *n*-decyloxy chain) reveals very interesting features: for the phenyl isocyanide complexes the melting and clearing temperatures decrease in the order Cl > Br > I. This is also the trend of the clearing points for biphenyl isocyanide complexes, but their melting temperatures follow the opposite trend that is, I > Br > Cl.

A simple explanation of this apparently contradictory behavior can be given considering the two main contributions to the intermolecular interactions in condensed phases: (i) the dipole moment associated with the X—Au bond, and (ii) the induced dipoles associated with the polarizability of the isocyanide moiety. The variation in the X—Au dipole moment is the same for the two families (Cl > Br > I), but the contribution of the isocyanide ligand can be very different. The polarizability of the biphenyl isocyanide can be much higher than that of the phenyl isocyanide, especially when the two aryl rings of the biphenyl system are coplanar, giving extended conjugation. This difference will be attenuated if the two aryls in the biphenyl unit are perpendicular and the conjugation is broken. On this basis, the thermal properties observed can be explained as follows: the behavior observed for the melting and clearing points of the phenyl series indicates an important influence of the X—Au dipole, since both these transition temperatures increase with the value of this dipole. The clearing temperatures of the biphenyl series follow the same trend that is, they behave as the phenyl system, which suggests that in the mesophase the two aryl rings are not well conjugated. This is reasonable considering that in a fluid phase there is freedom of rotation around the aryl—aryl bond (Figure 8.6a). In contrast, the melting points of the biphenyl system follow the opposite trend to that expected from X—Au dipole control. This suggests that in the solid state the biphenyl

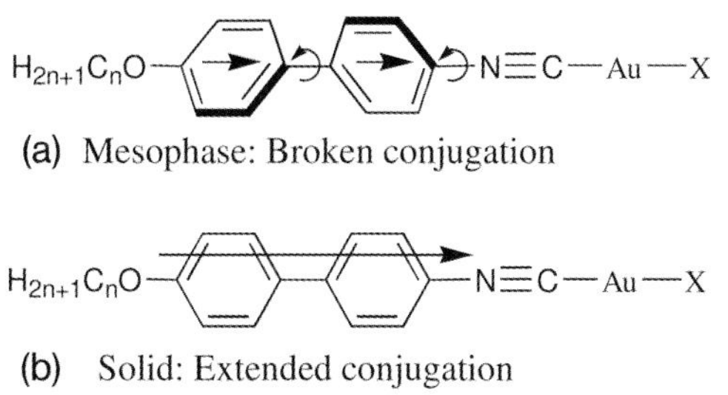

8.4 Isocyanide Complexes

$H_{2n+1}C_nO$-⟨phenyl⟩-⟨phenyl⟩-$N{\equiv}C$—Au—X

(a) Mesophase: Broken conjugation

$H_{2n+1}C_nO$-⟨phenyl⟩-⟨phenyl⟩-$N{\equiv}C$—Au—X

(b) Solid: Extended conjugation

Figure 8.6 Proposed explanation for the different influence of a biphenyl core on the melting and the clearing points.

group has become coplanar producing a higher molecular hyperpolarizability (Figure 8.6b). Moreover, the availability of electron density from the metallic center (higher for the less electronegative halogen, I > Br > Cl) controls the value of the induced dipole moments in the solid state, and seems to be the dominant factor at this point until the conjugation is broken upon melting.

This hypothesis is supported by the energy of the lowest electronic transition for the isocyanides and their complexes (Table 8.1). This energy (K band) is a practical measurement of the HOMO–LUMO separation and gives a good estimation of the relative polarizability of the molecule: the lower the energy of this band, the higher the polarizability of the molecule [17]. It can be seen that: (i) the presence of the metal produces a drastic increase in polarizability with respect to the free isocyanides; (ii) in the case of the phenyl isocyanide derivatives, the change in halogen does not produce a significant change in polarizability. Therefore, the changes in transition

Table 8.1 Relationship of the visible K band energy with the polarizability of the molecule, its variation from the free isocyanides to the different complexes, and the variation in X–Au dipolar moment.

	A	Biphenyl $E(cm^{-1})$	polarizability	$\mu(Au\text{-}X)$
Biphenyl	CN	34722		Cl
	ClAuCN	32154		Br
	BrAuCN	32051		I
	IAuCN	31047	⇓	
Phenyl	CN	40650		Cl
	XAuCN	37037	no variation for different X	Br
				I

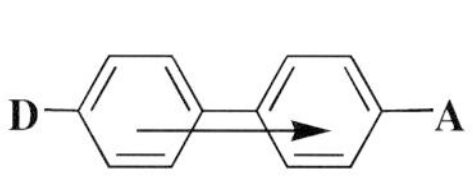

D—⟨phenyl⟩-⟨phenyl⟩—A

> The lower E, the higher the polarizability, the higher the induced dipolar moment, the higher the electrostatic interactions

A and D stand for acceptor and donor group, respectively.

temperature are not associated with the isocyanide, but rather with variations in the X–Au dipolar moment; (iii) in the biphenyl system the energy of the K band is sensitive to the halogen, hence increase in polarizability is expected upon changing the halogen in the order I > Br > Cl, as observed for the melting temperatures.

The most important conclusion of this study on gold complexes is that a strong increase in polarizability is introduced with the metal. Thus, in contrast to the usual practice for organic LCs, accumulation of conjugated aryl rings in metal complexes can lead to extremely strong intermolecular interactions, thus increasing the melting temperatures to undesirable high values.

Lateral monofluorination of the *p*-alkoxyphenyl isocyanide system [XAu(CNC$_6$H$_4$OC$_n$H$_{2n+1}$)] (X = Cl, Br, I), in ortho-(3-F) (**6a**) and meta-(2-F) (**6b**) positions relative to the alkoxy chain was also studied [18]. None of the free fluorinated ligands is a LC, but their gold complexes display mesomorphic properties. This is a typical case of a promesogenic ligand which yields a mesogen upon coordination to a metal.

$$H_{2n+1}C_nO-\underset{F}{\overset{3\quad 2}{\text{C}_6H_3}}-N\equiv C-Au-X \quad (6)$$

n = 2, 4, 6, 8, 10; X = Cl, Br, I

The complexes show SmA mesophases, except the 2-fluorohexyloxy derivative, which shows a narrow nematic range and the 2-fluoroiodo derivative, which is not a liquid crystal. The transition temperatures of the chloro complexes compared with

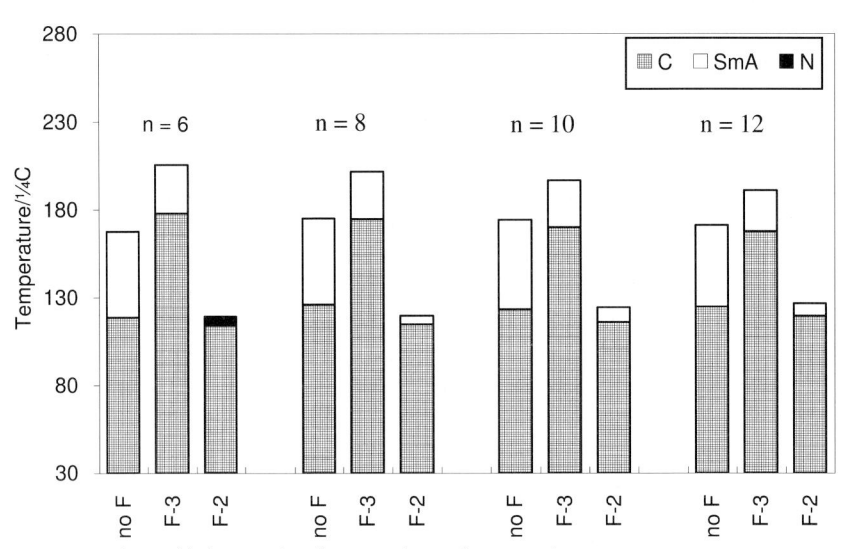

Figure 8.7 Thermal behavior of no fluorinated complexes [AuCl(CNC$_6$H$_4$OC$_n$H$_{2n+1}$)] (labeled as no F), and the corresponding fluorinated derivatives **6a** (F-3) and **6b** (F-2).

the non-fluorinated derivatives, decrease in the order 3–F > non-fluorinated > 2–F (Figure 8.7). Substitution of Cl by Br and I produces a decrease in the transition temperatures in the order Cl > Br > I, as seen above for the parent non-fluorinated complexes, according to the decrease in polarity of the Au–X bond. Rationalizing the effect of fluorination is not a simple matter, but in general, lateral fluorination causes a broadening of the molecule thus increasing the average intermolecular distances, reducing the intermolecular attractions and leading to lower transition temperatures. On the other hand, polarization effects can cause increased intermolecular interactions leading to higher transition temperatures [19]. It is clear that the position of the fluorine atom (at either the 2- or 3-position) does not significantly affect the breadth of the molecule. Thus, the 2- and 3-fluorinated derivatives will differ mainly in their polarization effects and it is not surprising that an electronegative substituent (F) *ortho* to the OR group (3-fluorination) will produce a larger effect than when it is placed in the *meta* position (2-fluorination). Consequently, the transition temperatures are expected to decrease in the order 3-F > 2-F, in agreement with the trend observed.

Very recently, mesomorphic [AuCl(CNC$_6$H$_4$O(CH$_2$)$_4$(CF$_2$)$_8$F)] (7) complexes containing an isocyanide ligand with a semifluorinated alkoxy chain have been prepared [20]. The free ligand melts at 62 °C and displays a SmA mesophase in a narrow temperature range (2 °C) as opposed to its hydrocarbon analog which is not a LC. The gold complex also displays a SmA mesophase in the range of 191–274 °C. Compared with the corresponding non-fluorinated compound, the fluorination of the terminal chain in these systems results in an increase of both melting and clearing temperatures and an enhancement of the smectic stability, as described for related organic compounds [21, 22]. Moreover, the increase of the clearing point is larger than the increase in the melting point resulting in a larger mesophase temperature range for the fluorous compound.

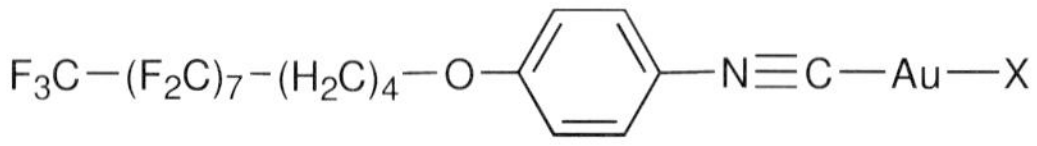

$$F_3C-(F_2C)_7-(H_2C)_4-O--N\equiv C-Au-X$$

X = Cl, Br, I   (7)

In the examples presented so far, the linear coordination of gold has given rise to rod-like molecules. The metal coordination determines the bulk shape of the molecule. However, modification of the rod-like structure of the *p*-alkoxyphenyl isocyanide system [AuX(CNC$_6$H$_4$OC$_n$H$_{2n+1}$)] is possible by addition of two alkoxy chains in the 3- and 4- positions in the phenyl ring. Complexes [AuX{CNC$_6$H$_2$-3,4,5-(OC$_n$H$_{2n+1}$)$_3$}] (8) with alkoxy chains containing six or more carbons, display columnar hexagonal mesophases at room temperature [18, 23]. This behavior is typical for discotic molecules. Although the molecules of these gold isocyanide complexes are not discotic when considered individually, they can give a disk-like rearrangement of the peripheral alkyl chains when they are considered as pairs in an antiparallel arrangement (Figure 8.8), giving rise to columnar behavior as a supramolecular property of the bulk material. Small-angle X-ray diffraction data on the mesophase support this interpretation. This does not mean that there is any *permanent* association in pairs, but that each 0.4-nm-thick disk in the column contains two molecules filling this space. It would be reasonable to assume that this arrangement might be due to aurophilic interactions, but similar behavior is found for related compounds of Ni, Cu, Pd and Mo [24, 25]. This suggests that the packing forces in a compact best space-filling arrangement are the most important factor for formation of the thermotropic mesophases in this system.

Given that the presence of three alkoxy chains in the phenyl group produces such a dramatic change in the properties of the material to the point that columnar mesophases are formed at room temperature, the structure of the aryl isocyanide ligand has been further modified to introduce more paraffinic chains, and examples of metallodendrimers containing monodendrons with an isocyanide group in the focal point, and its gold compound **9**, have been reported [26].

Although the free isocyanide dendron is not a LC, the gold dendrimer is. Compound **9** displays a cubic mesophase $Im\bar{3}m$ over a wide range of temperatures. Small-angle X-ray diffraction data of the mesophase show evidence that this dendritic molecule adopts a conical-like molecular conformation. Ten of these gold monodendrons are self-assembled into a supramolecular spherical dendrimer, and the spherical dendrimers self-organize into an $Im\bar{3}m$ cubic mesophase. A schematic model of the mesophase structure is shown in Figure 8.9.

When an organic isocyanide, CNR, has a second functional group on R, the molecule becomes a bifunctional ligand. It is obvious that the compatible groups in the molecule are limited by the reactivity of the isocyanide group. Although aqueous acids hydrolyze organic isocyanides, the synthesis of stable 4-isocyanobenzoic acid was achieved some fifty years ago [27]. In spite of the very interesting combination of these two functions, which make this molecule a promising building block in supramolecular chemistry, further studies with this isocyanide or its coordination to give metal complexes have not been reported until recently; amongst them,

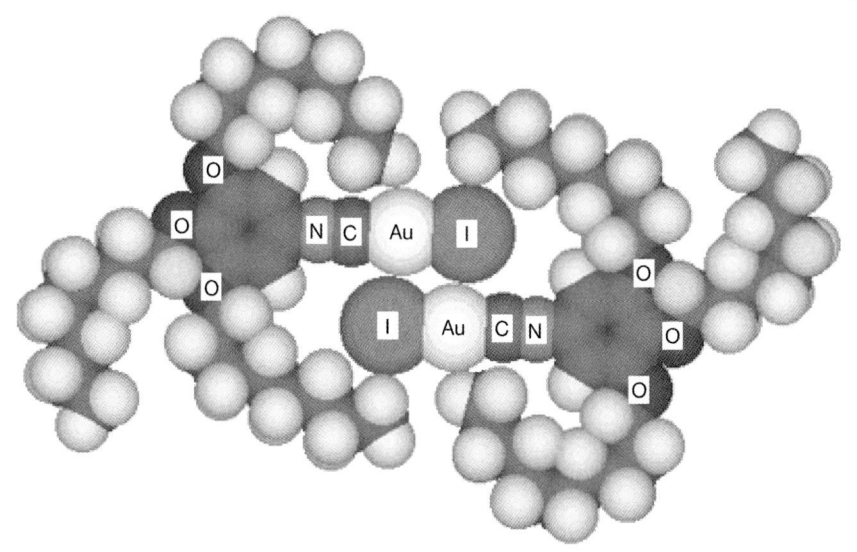

Figure 8.8 Space-filling model based on small-angle X-ray diffraction measures, showing how two hemidiscotic molecules of **8** can give rise to a disk-like shape. (Adapted from Ref. [11].)

the complexes [AuCl(CNC$_6$H$_4$COOH)] [28], and [Au(C$_6$F$_5$)(CNC$_6$H$_4$COOH)] [29]. The single crystal X-ray diffraction structure of the latter has been determined and confirms the formation of a supramolecular array in the solid state supported by hydrogen bonding.

The carboxylic acid group of the coordinated isocyanide acts as a hydrogen donor towards hydrogen acceptors. Stable hydrogen-bonded liquid crystalline metal complexes have been prepared with decyloxystilbazole (**10**). Although the gold acid derivatives used are not mesomorphic (they decompose above 200 °C without melting), and the hydrogen acceptor decyloxystilbazole only shows an ordered Smectic E phase, the [{AuCl(CNC$_6$H$_4$COOH)}·decyloxystilbazole] complex displays an enantiotropic SmA mesophase in the range of 161–187 °C. At the clearing

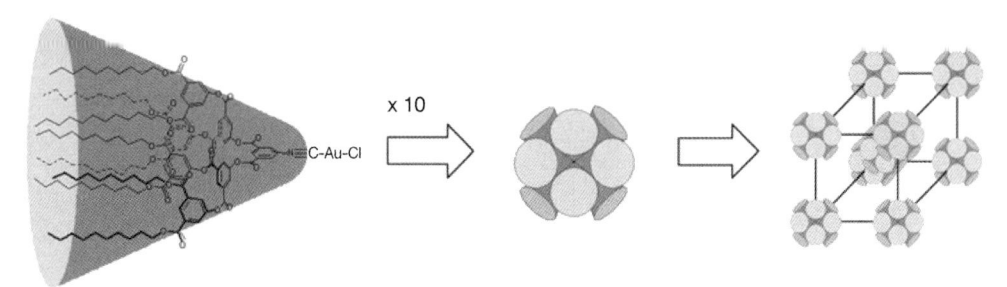

Figure 8.9 Schematic representation of a conical monodendron self-assembled into a supramolecular spherical micellar dendrimer, and then into an $Im\bar{3}m$ cubic mesophase.

transition, formation of some crystals is observed suggesting decomposition, most likely due to the thermal lability of the hydrogen bond.

$H_{21}C_{10}O$—⟨phenyl⟩—CH=CH—⟨pyridyl⟩ N---H—O—C(=O)—⟨phenyl⟩—$N\equiv C$—Au—Cl

(10)

8.4.2
Isocyanide-Alkynyl Complexes

Complexes $[Au(C\equiv C-C_6H_4-C_mH_{2m+1})(C\equiv N-C_6H_4-X)]$ (X = H, OC_nH_{2n+1}; m = 6, 8, 10, 12; n = 2, 4, 6, 8, 10) (11), prepared by reaction of $[Au(C\equiv C-C_6H_4-C_mH_{2m+1})]_n$ with the corresponding isocyanide, display SmA phases, although none of the starting materials (the free alkynes, the polymeric gold alkynyls or the free isocyanides) are LCs [30].

X—⟨phenyl⟩—$N\equiv C$—Au—$C\equiv C$—⟨phenyl⟩—C_mH_{2m+1}

(11)

X = H, OC_nH_{2n+1} (n = 2, 4, 6, 8, 10)
m = 6, 8, 10, 12

Particularly low melting points (close to 100 °C) are obtained for X = H. The melting points for the rest of the complexes, in the range of 130–170 °C, decrease regularly for equal total lengths of the molecule (same $m+n$ value) as m increases and n decreases. A sudden fall in the melting point is observed when the isocyanide lacks a *para* chain (phenyl isocyanide, n = 0) (Figure 8.10). Clearing temperatures are in the range of 150–170 °C and the transitions to the isotropic state occur with decomposition, producing gold mirrors.

Introduction of lateral chains in rod-like molecules helps, as discussed above, to lower transition temperatures and avoid the thermal decomposition of this type of

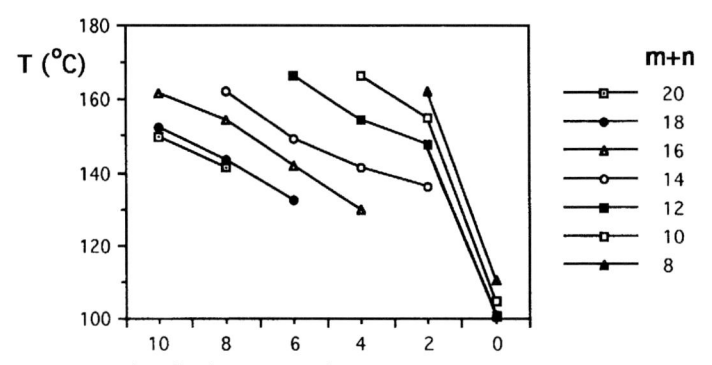

Figure 8.10 Plot of melting points of $[Au(C\equiv CC_6H_4C_mH_{2m+1})(C\equiv NC_6H_4X)]$ (X = H, OC_nH_{2n+1}) versus n. For simplicity the coordinate n = 0 is used for CNPh. (Reproduced from Ref. [28] by permission.)

complex. Complex **12**, melts at 165 °C displaying a SmA mesophase with decomposition. When a small side chain ($R = OC_2H_5$) is introduced into the isocyanide ligand (complexes **13**), the melting point decreases to 142 °C, to give the SmA mesophase, but still with some decomposition. However, for a longer side chain ($R = OC_6H_{13}$), the intermolecular interactions clearly drop and a nematic phase is found. The melting point is also lowered and no decomposition is detected at clearing [31].

$H_{2m+1}C_mO$—⟨⟩—X—⟨⟩(R)—N≡C—Au—C≡C—⟨⟩—OC_nH_{2n+1}

X = COO; R = H; m = 9; n = 5   (**12**)
X = OOC; R = OEt, OC_6H_{13}; m = 10; n = 5   (**13**)

Derivatives of aliphatic alkynes (**14** and **15**) are more thermally unstable than **12**, but they show SmA and N phases at low temperatures (below 130 °C). The type of phase and the mesophase stability depend on the length of both the terminal and the lateral chains. When both chains are elongated, the mesomorphism becomes metastable and compounds **14** display monotropic N and SmA transitions. Complexes **15**, which contains an ester group with an opposite direction to that of complexes **14**, display less stable nematic mesophases.

$H_{21}C_{10}O$—⟨⟩—X—⟨⟩(R)—N≡C—Au—C≡C—C_nH_{2n+1}

X = OOC; R = OMe, OEt, OPr; n = 4–6   (**14**)
X = COO; R = Et, OMe, OEt; n = 4   (**15**)

Liquid crystals based on aliphatic isocyanides and aromatic alkynyls (compounds **16**) show enantiotropic nematic phases between 110 and 160 °C. Important reductions in the transition temperatures, mainly in clearing points (<100 °C), are obtained when a branched octyl isocyanide is used. The nematic phase stability is also reduced and the complexes are thermally more stable than derivatives of aliphatic alkynes. Other structural variations such as the introduction of a lateral chlorine atom on one ring of the phenyl benzoate moiety or the use of a branched terminal alkyl chain produce a decrease of the transition temperatures enhancing the formation of enantiotropic nematic phases without decomposition.

$H_{17}C_8$—N≡C—Au—C≡C—⟨⟩—X—⟨⟩—OC_nH_{2n+1}

X = OOC; COO   (**16**)

Analogous complexes containing a functional alkynyl ligand [Au(C≡CR)(CNC$_6$H$_4$O(O)CC$_6$H$_4$OC$_{10}$H$_{21}$)] (R = C$_5$H$_4$N, C$_6$H$_4$NC) (**17**) have been also reported. They show SmA mesophases in the range of 144–180 °C before decomposition occurs [32].

In summary, [Au(alkynyl)(CNR)] complexes also show mesomorphic behavior but, in contrast to the halide-isocyanide derivatives, they are thermally less stable and usually decompose before or at the clearing temperatures. This thermal instability is thought to be associated with the gold-(alkynyl) bond and makes these compounds undesirable as LCs, but might make them useful as precursors for gold deposition from an oriented fluid state.

8.4.3
Isocyanide-Fluorophenyl Derivatives

When the anionic ligand is a fluorinated aryl ring (Faryl), which gives thermally stable Au-Faryl bonds, very thermally stable mesomorphic complexes are obtained. Perhalophenyl complexes [AuR(CNC$_6$H$_4$C$_6$H$_4$OC$_n$H$_{2n+1}$)] [R = C$_6$F$_5$ (**18a**), C$_6$F$_4$Br-o (**18b**), C$_6$F$_4$Br-p (**18c**); $n=4, 6, 8, 10, 12$] display a rich mesomorphism in the range of 70–220 °C. They show an N mesophase when the isocyanide has a short tail ($n=4$), N and SmA mesophases when the isocyanide has an medium tail ($n=6, 8$) and only a SmA phase for longer chains [33]. The variation in the thermal properties is quite regular with the melting and clearing transition temperatures decreasing with growing chain lengths up to $n=8$, when they become roughly constant. The decrease in transition temperatures with increasing n is small for C$_6$F$_4$Br derivatives, larger for C$_6$F$_4$Br-o and more marked for C$_6$F$_5$ complexes. As a consequence, a change is produced in the sequence of melting and clearing temperature in the order C$_6$F$_4$Br-p C$_6$F$_5$ > C$_6$F$_4$Br-o when $n=6$ and C$_6$F$_4$Br-p > C$_6$F$_4$Br-o > C$_6$F$_5$ for $n=8$ (Figure 8.11).

Comparing these complexes with the corresponding chloro complexes [AuCl(CNC$_6$H$_4$C$_6$H$_4$OC$_n$H$_{2n+1}$)], the perfluorophenyl ligand induces lower transition temperatures, shorter mesogenic ranges and an enhancement of nematic phases (Figure 8.12), according to the expected lower lateral intermolecular interactions as consequence of their larger molecular width.

Oxidative addition of I$_2$ yields the corresponding gold(III) isocyanide compounds [AuI$_2$R(CNC$_6$H$_4$C$_6$H$_4$OC$_n$H$_{2n+1}$)]. These are thermally unstable and heating leads to cleavage of the Au—R bond to give the corresponding iodogold(I) isocyanide complexes.

Similar rod-like gold(I) complexes [Au(C$_6$F$_4$OC$_m$H$_{2m+1}$)(CNC$_6$H$_4$C$_6$H$_4$-OC$_n$H$_{2n+1}$)] (**19**), with both chains in the para position of the respective aryl ring, are LCs too. They show an N phase at short chain-lengths ($n + m = 4$), N and/or SmA phases for medium tails, and both SmA and SmC phases for the longer chain-lengths ($m + n > 14$). This series of compounds constitutes one of the rare examples of metallomesogens that show photoluminescence in the mesophase as well as in the solid state and in solution [34]. The single-crystal X ray diffraction structure of [Au(C$_6$F$_4$OC$_2$H$_5$)(CNC$_6$H$_4$C$_6$H$_4$OC$_4$H$_9$)] (Figure 8.13) confirms its rod-like structure, with a linear coordination around the gold atom and reveals the absence of any Au···Au interactions. Moreover, well-defined intermolecular F$_{ortho}$···F$_{meta}$ interactions were found, which are apparently responsible for the crystal packing.

In the solid state, the complexes exhibit a yellow–green luminescence under UV irradiation (365 nm). All the emission spectra are similar and consist of three broad emissions above 370 nm (e.g., 384, 490 and 524 nm for $m=10, n=6$), while the free isocyanides (white solids) are luminescent giving one strong emission band with its maximum at about 360 nm. In dichloromethane solution both the free isocyanides and their gold complexes are luminescent too, but only one intense emission is observed for the complexes in the range of 345–387 nm (Figure 8.14). The lifetime

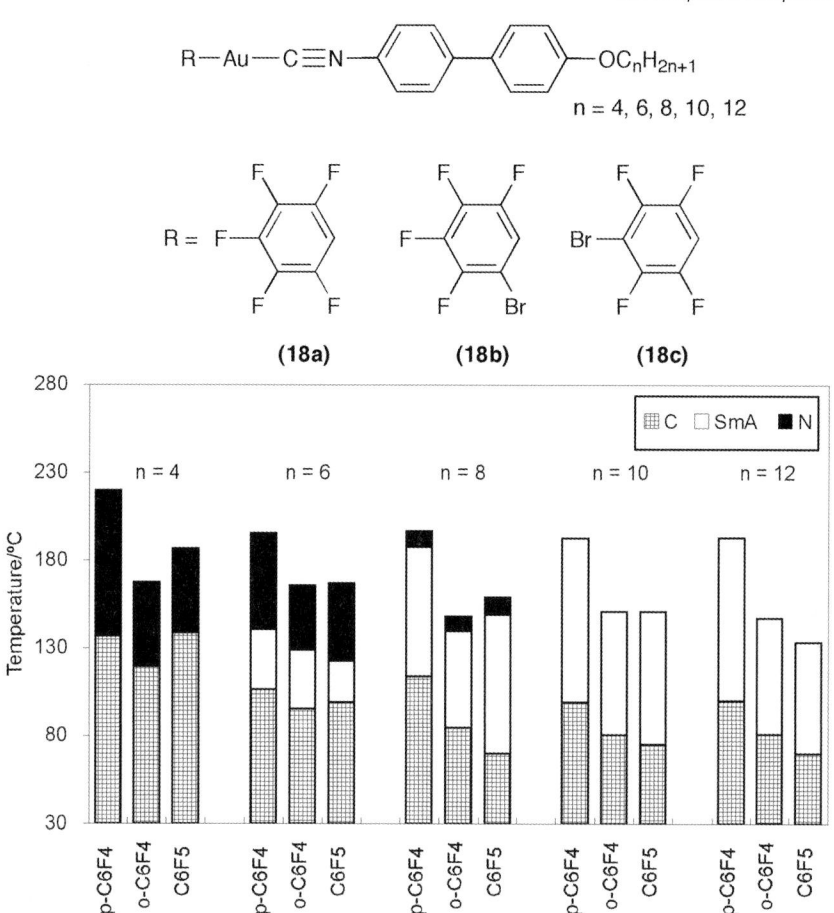

Figure 8.11 Comparison of the thermal properties of complexes **18a**, **18b** and **18c**.

values, together with the Stokes shift between the absorption and the emission bands, support that the emission at 384 nm must be a fluorescence involving intraligand localized π and π^* orbitals. In contrast, the longer lifetime of the emissions at 490 and 524 nm support a phosphorescence nature.

The luminescent behavior of [Au($C_6F_4OC_{10}H_{21}$)($CNC_6H_4C_6H_4$-OC_6H_{13})] as a function of temperature is shown in Figure 8.15. When the sample melts to the SmC mesophase (at 55.4 °C), the luminescence is not lost, but its intensity decreases noticeably and continues decreasing upon further increase of temperature. The emission practically disappears at about 130 °C. This process is reversible and the intensity of the emission is gradually recovered upon cooling. These complexes show important supercooling of the mesophase and, as a consequence, the fluid phase survives at room temperature for several hours until crystallization occurs. At the very instant that crystallization occurs, the intensity of the emission is fully recovered.

The related liquid crystalline perhalophenyl complexes [Au($C_6F_4OC_mH_{2m+1}$)($CNC_6H_4O(O)CC_6H_4OC_nH_{2n+1}$)] (**20**) have been reported. The variation in thermal

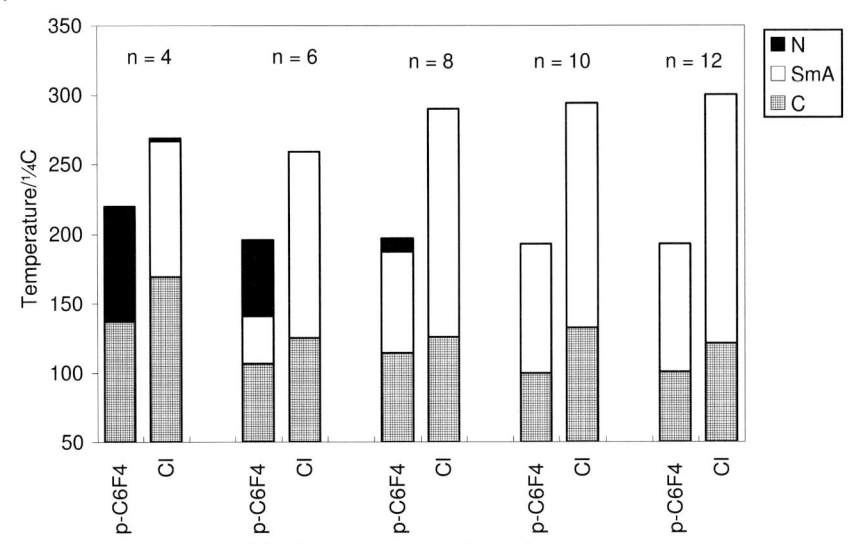

Figure 8.12 Comparison of the thermal properties of complexes [AuCl(C≡NC$_6$H$_4$C$_6$H$_4$OC$_n$H$_{2n+1}$)] (**5**) and [Au(C$_6$F$_4$Br-p)-(C≡NC$_6$H$_4$-C$_6$H$_4$OC$_n$H$_{2n+1}$-p)] (**18c**).

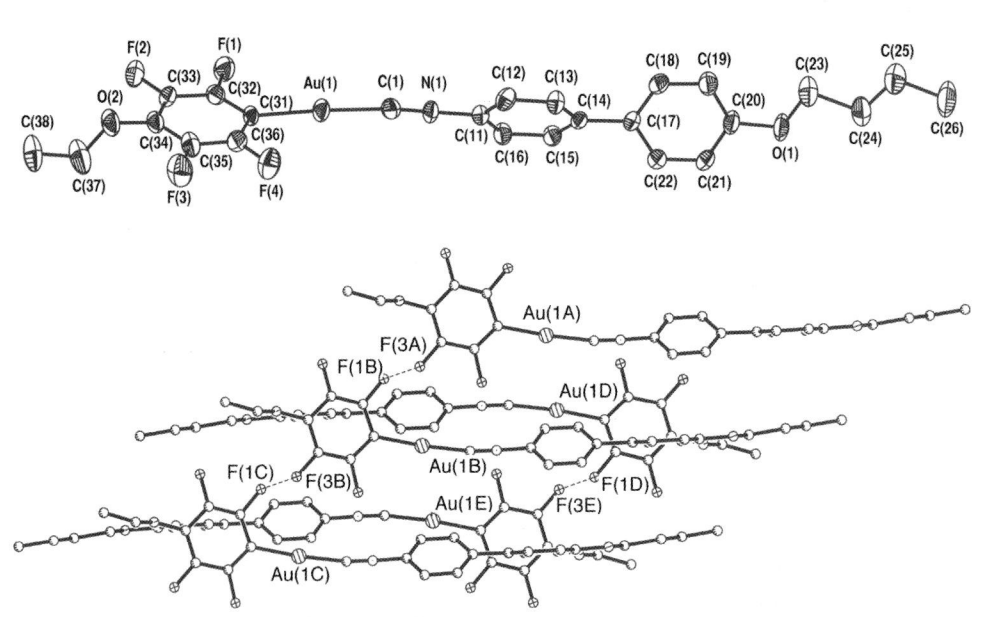

Figure 8.13 Crystal structure of [Au(C$_6$F$_4$C$_6$H$_4$OC$_2$H$_5$)(C≡NC$_6$H$_4$C$_6$H$_4$OC$_4$H$_9$)] and crystal packing representation showing some intermolecular features. (Reproduced from Ref. [32] by permission.)

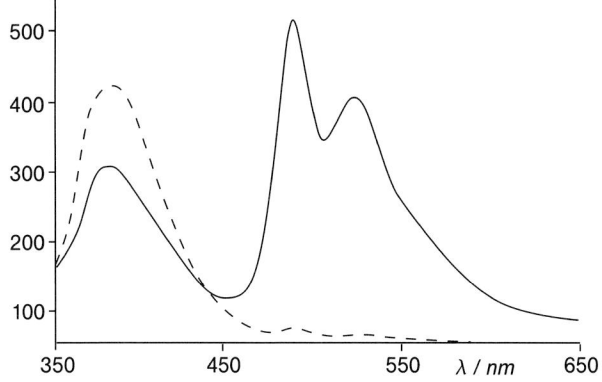

Figure 8.14 Emission spectra at 298 K of [Au(C$_6$F$_4$OC$_{10}$H$_{21}$)-(CNC$_6$H$_4$C$_6$H$_4$OC$_6$H$_{13}$)] (excitation 344 nm) in the solid state (continuous line) and in CH$_2$Cl$_2$ solution (dashed line). (Reproduced from Ref. [32] by permission.).

properties of tetrafluorophenylgold(I) complexes is very regular. With short alkoxy chains at the tetrafluorophenyl group ($m = 4$), both SmA and nematic phases are observed for shorter alkoxy substituents in the isocyanide ligand ($n = 4, 6$). For longer chains ($n = 8, 10$) only SmA phases are observed. When the alkoxy substituent at the tetrafluorophenyl group is longer ($m = 8$) both SmA and SmC phases are obtained, irrespective of the chain length of the isocyanide ligand [35].

$m = 4, 8$; $n = 4, 6, 8, 10$

(20)

The introduction of chiral centers in this type of perhalophenyl systems (**21–23**) produces chiral phases with properties absent in the corresponding achiral systems [36].

	R^1	R^2
21a	(R)-2-butyl	C$_2$H$_5$
21b	(R)-2-butyl	C$_{10}$H$_{21}$
22a	C$_2$H$_5$	(R)-2-butyl
22b	C$_{10}$H$_{21}$	(R)-2-butyl
23	(R)-2-butyl	(R)-2-butyl

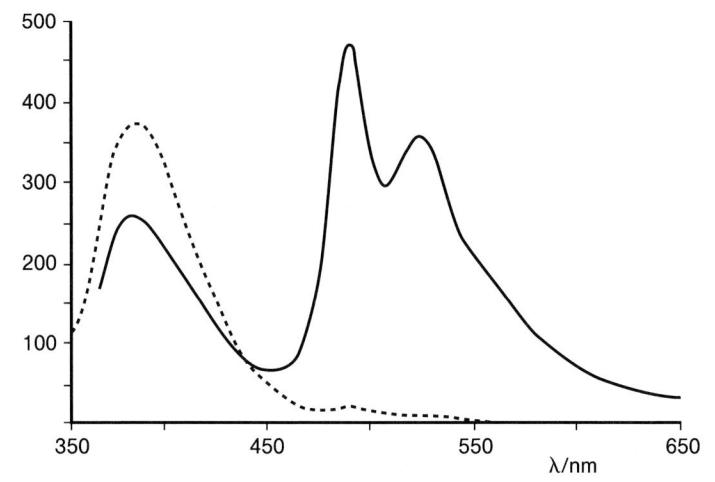

Figure 8.15 Emission spectra of [Au(C$_6$F$_4$OC$_{10}$H$_{21}$)(CNC$_6$H$_4$C$_6$H$_4$OC$_6$H$_{13}$)] (excitation 344 nm) at different temperatures (in parenthesis the corresponding phase) heating from the crystal. (Reproduced from Ref. [32] by permission.)

For the complexes containing one chiral substituent (**21** and **22**), only a SmA mesophase is observed when the length of the achiral chain is long (R^1 or R^2 = 10, **21b** and **22b** complexes), irrespective of the position of the chiral chain. For shorter alkoxy chains (R^1 or R^2 = C$_2$H$_5$, **21a** and **22a** complexes) an enantiotropic cholesteric and a monotropic SmA phase are observed (Figure 8.16). The cholesteric range is clearly higher for **21a**, where the chiral substituent is in the tetrafluorophenyl group and an unusual TGBA* phase is observed for **21a** when the SmA to

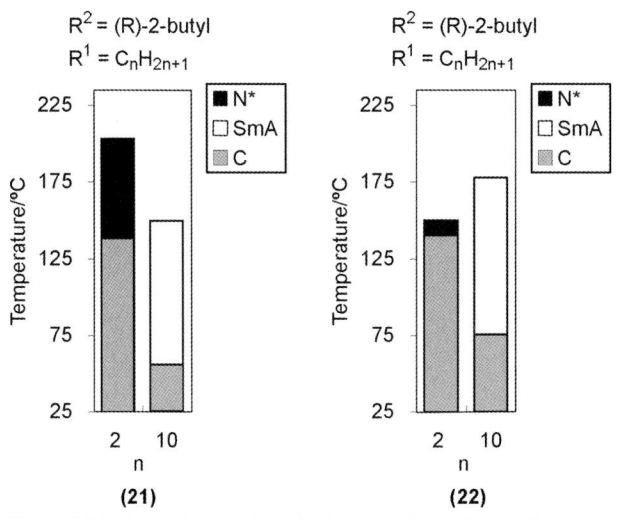

Figure 8.16 Thermal properties of chiral complexes **21** and **22**.

cholesteric transition is observed heating slowly (0.5 °C min^{-1}) from the SmA phase. Under similar conditions 22a did not show a TGBA* phase.

The introduction of a second chiral atom in the system leads to a reduction in the mesogenic properties and only a monotropic chiral nematic transition is observed for compound 23. However, when this compound is cooled down from the isotropic liquid state at a cooling rate of 0.5 °C min^{-1}, very unusual blue phases BP-III, BL-II and BP-I are observed in the range 103–88 °C. Blue phases usually require pitch values below 500 nm. Hence the pitch value of the cholesteric phase for 23 must be very short, suggesting that the packing of two chiral carbons forces a faster helical shift for successive molecules packed along the perpendicular to the director.

It is interesting that the large symmetric molecules 24 containing two gold–isocyanide fragments linked by an octafluorophenyl group give nematic phases in spite of their big molecular size. In effect, large organic molecules containing extended conjugated cores often display smectic phases because their interactions tend to associate the molecules defining aromatic and paraffinic areas, thus leading to a layer arrangement characteristic of smectic phases (Figure 8.17a). The gold derivatives, being perfectly symmetric and highly polarizable, produce large local dipoles, which lead to repulsive interactions if the molecules are packed in layers (Figure 8.17b) and favor arrangements where the aromatic and paraffinic areas are

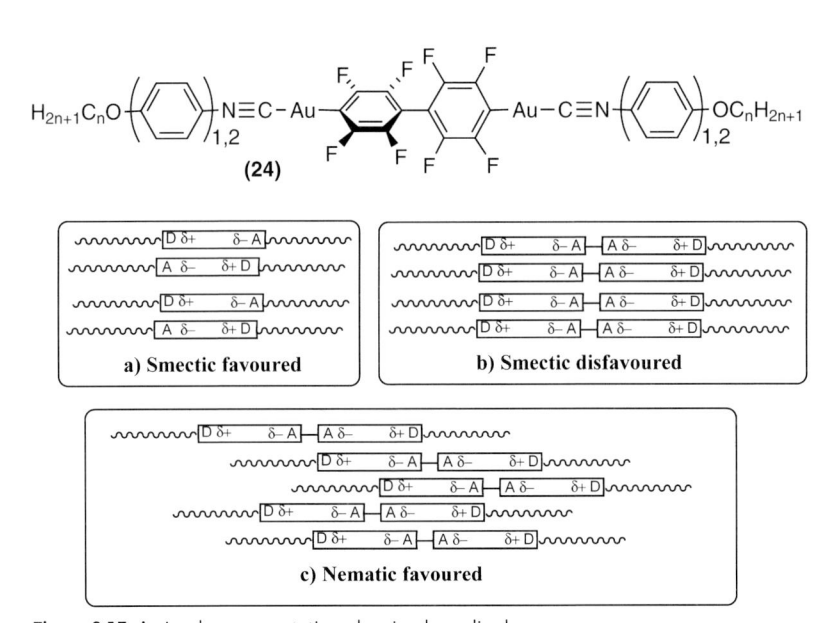

Figure 8.17 A simple representation showing how dipolar interactions can favor either discrimination of aliphatic and aromatic areas (giving rise to smectic phases), or mixing (giving rise to nematic phases). A and D stand for acceptor and donor group, respectively. (Adapted from Ref. [11].)

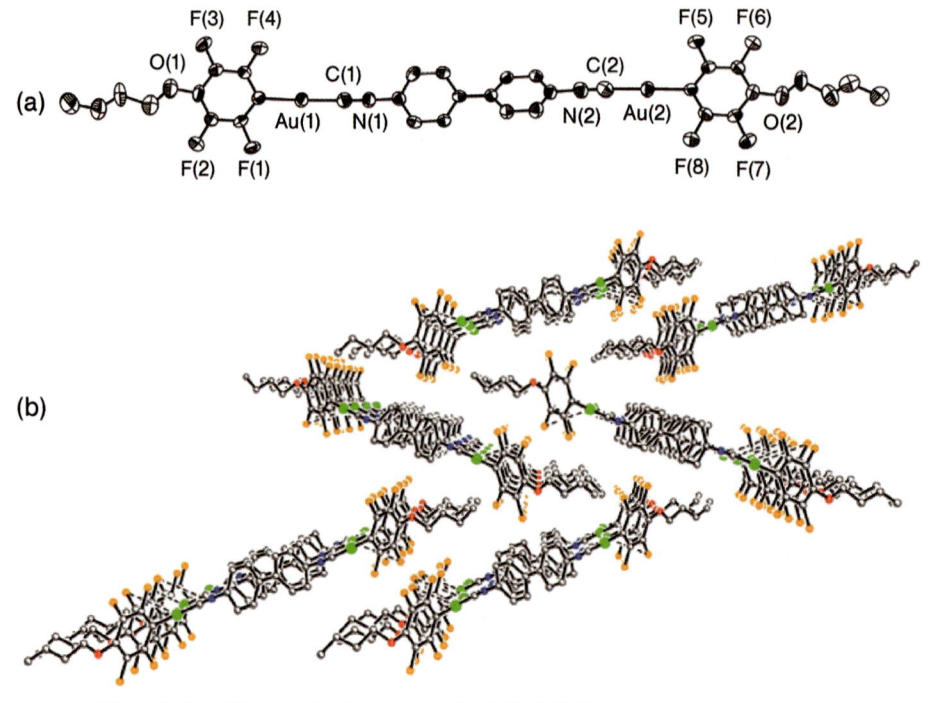

Figure 8.18 (a) X-ray molecular structure for [μ-(4,4′-CN-C$_6$H$_4$-C$_6$H$_4$-NC)-{Au(C$_6$F$_4$OC$_4$H$_9$)}$_2$] (**25**) and (b) zig-zag molecular arrangement. (Reproduced from Ref. [35] by permission.)

mixed, that is, where there is only orientational order typical of nematic mesophase (Figure 8.17c) [33].

In fact, X-ray structures of similar dinuclear compounds [μ-(4,4′-CN-R-NC)-{Au(C$_6$F$_4$OC$_n$H$_{2n+1}$)}$_2$] (R = 4-4′-biphenylene and 2,2′-dichloro-4-4′-biphenylene) containing a diisocyanide ligand bringing two gold atoms (**25**), which display only nematic mesophases do not adopt a layered packing of molecules in the solid state (Figure 8.18). This reluctance to adopt layered packing is in line with the absence of smectic phases in their mesophases [37].

As indicated above in chiral mesophases, the introduction of a functional group in mesogenic structures offers the opportunity to achieve functional LCs. With this aim, mesomorphic crown-ether–isocyanide–gold(I) complexes (**26**) have been prepared recently [38]. The derivatives with one alkoxy chain show monotropic SmC mesophases at or close to room temperature. In contrast, the complexes with three alkoxy chains behave as monotropic ($n = 4$) or enantiotropic ($n > 4$) LCs. The structure of the mesophases could not be fully elucidated because X-ray diffraction studies in the mesophase were unsuccessful and mesophase characterization was made only on the basis of polarized optical microscopy. These complexes are luminescent not only in the solid state and in solution, but also in the mesophase and in the isotropic liquid state at moderate temperatures. The emission spectra of **26a** with $n = 12$ were

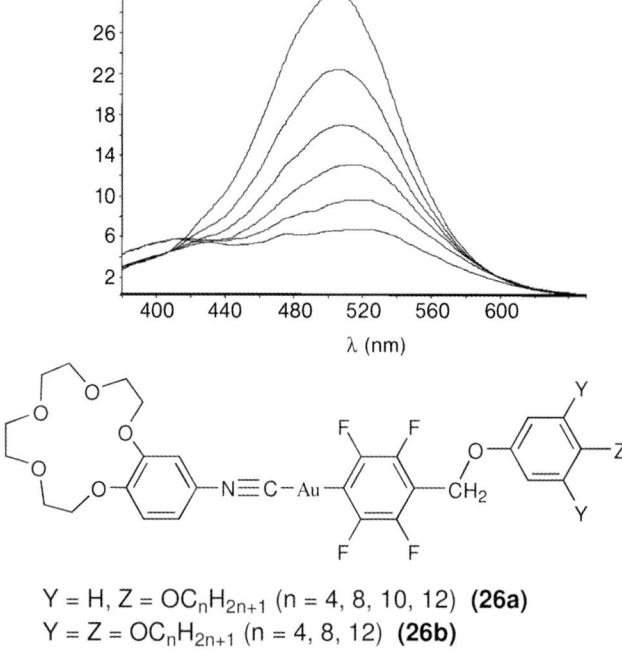

Figure 8.19 Changes in the emission spectrum of **26a** ($n = 12$) upon heating. Spectra shown at 25 (highest intensity), 35, 45, 55, 65, and 75 °C with $\lambda_{exc} = 360$ nm. Emission intensity in arbitrary units. (Reproduced from Ref. [36] by permission.)

Y = H, Z = OC$_n$H$_{2n+1}$ (n = 4, 8, 10, 12) **(26a)**
Y = Z = OC$_n$H$_{2n+1}$ (n = 4, 8, 12) **(26b)**

recorded from 25 to 75 °C (heating at 10 °C min^{-1}) (Figure 8.19). The intensity decreases when the temperature increases, but the overall profile of the spectrum is maintained even beyond the clearing transition to the isotropic liquid state (at 54.9 °C). At around 75 °C, the emission has fully disappeared. Therefore, the solid state emission persists in the liquid crystalline and isotropic liquid states, although with less intensity. Upon cooling the original emission spectrum is recovered. The 15-crown-5 group reacts with potassium perchlorate, but coordination of potassium to the crown ether destroys the mesomorphic behavior of **26**.

8.4.4
Ionic bis(isocyanide) Derivatives

Ionic LCs are interesting systems because they combine the properties of LCs with those of ionic liquids. Although alkali metal soaps were among the first thermotropic LCs to be systematically studied, ionic liquid crystalline derivatives have been reported less frequently than those based on neutral molecular and macromolecular species [39]. When the halide of [AuX(CNR)] complexes is substituted by a second isocyanide, ionic complexes [Au(CNR)$_2$][Y] [R = C$_6$H$_4$OC$_n$H$_{2n+1}$ **(27a)**,

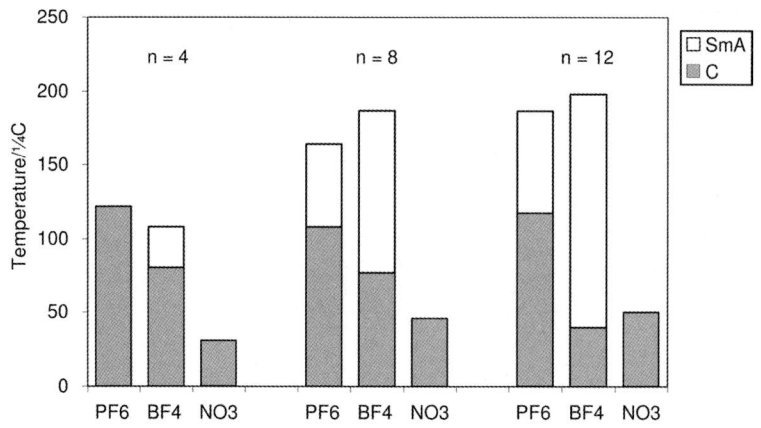

Figure 8.20 The thermal behavior of **27a**.

$C_6H_4C_6H_4OC_nH_{2n+1}$ (**27b**), $C_6H_2(3,4,5\text{-}OC_nH_{2n+1})_3$ (**26c**); $n = 4, 8, 12$; $Y = NO_3^-$, PF_6^-, BF_4^-] are obtained and different mesogenic behavior is observed depending on the isocyanide and the anion present in the molecule [40]. The thermal behavior of phenyl derivatives is summarized for $n = 12$ in Figure 8.20.

$$\left[H_{2n+1}C_nO-\!\!\left\langle\!\!\bigcirc\!\!\right\rangle\!\!-\!\!\left\langle\!\!\bigcirc\!\!\right\rangle\!\!-N\!\equiv\!C-Au-C\!\equiv\!N-\!\!\left\langle\!\!\bigcirc\!\!\right\rangle\!\!-\!\!\left\langle\!\!\bigcirc\!\!\right\rangle\!\!-OC_nH_{2n+1}\right]^+$$
$$(\mathbf{27b})$$

$$(\mathbf{27c})$$

In Figure 8.21 the thermal properties of the complexes [AuCl(CNR)] and [Au(CNC$_6$H$_4$OC$_n$H$_{2n+1}$)$_2$][PF$_6^-$] (**27a**) are compared. It can be seen that the mesogenic range is greater for neutral complexes when the chains are short (the butyloxy hexafluorophosphate derivative is not a LC). However, the increase in length of the chains affects more the ionic complexes (that have double the number of alkoxy chains) and for the derivatives with the longest chain the wider mesogenic range corresponds to the ionic complex.

The incorporation of a second phenyl ring in the system (**27b**) produces an increase of both melting and clearing temperatures, possibly arising from the larger anisotropy of polarizability of the resulting compounds (Figure 8.22) and an increase of the liquid crystal range compared with the analogous phenyl derivatives. The very high (about 220 °C) clearing temperatures of the ionic biphenylisocyanide complexes are in fact decomposition temperatures.

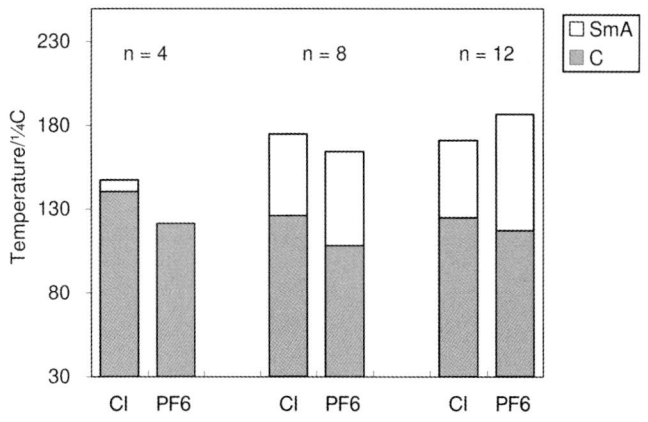

Figure 8.21 Comparison of the thermal properties of complexes [AuCl(CNR)] **(4)** and [Au(CNR)$_2$][PF$_6$] **(27a)** (R = C$_6$H$_4$OC$_n$H$_{2n+1}$).

The complexes with three alkoxy chains in the phenyl group (**27c**) can no longer be considered rod-like. Consistently, the complexes show hexagonal columnar mesophases, as confirmed by X-ray studies, some of them at room temperature (Figure 8.23). The crystallization of these gold complexes, as usually observed in complexes with a high number of alkoxy substituent and high viscosity, is not observed and a glass–liquid crystalline state is obtained on cooling the mesophase. In this series, the [Au{CNC$_6$H$_2$(3,4,5-C$_6$H$_4$OC$_n$H$_{2n+1}$)$_3$}$_2$][PF$_6$] complexes show higher clearing temperatures than the corresponding chloro complexes.

It is interesting to note the influence of the counteranions on the thermal behavior. Irrespective of the isocyanide used, all the nitrate gold derivatives show low thermal stability and undergo extensive decomposition at relatively low temperatures (only the low melting trialkoxyphenyl derivative shows liquid crystal behavior). In contrast,

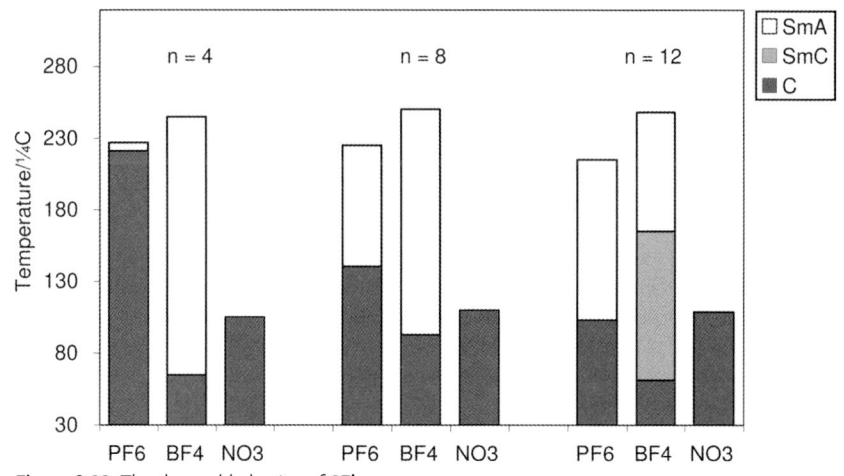

Figure 8.22 The thermal behavior of **27b**.

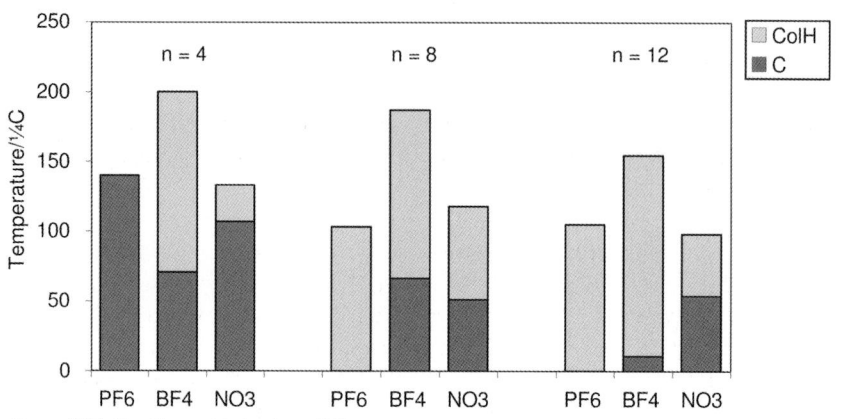

Figure 8.23 The thermal behavior of **27c**.

the corresponding silver complexes are thermally more stable and produce mesophases [40]. This thermal instability of the gold(I) compared with silver(I) nitrate derivatives is associated to the easy decomposition of the anion in the presence of many heavy metal cations [41] and can be traced back to the simple nitrates: whereas [Ag(NO$_3$)] is a commercial product, [Au(NO$_3$)] is not known [42]. It is also well established that [Au(CNR)(NO$_3$)] complexes are thermally unstable [43]. Thin-layer chromatography of heated samples of the ionic gold LCs reveals free isocyanide ligand, suggesting that the decomposition of these nitrate gold derivatives might occur on intermediates where at least one isocyanide ligand has been released. This decoordination of isocyanide should be facilitated by coordination of the nitrate anion. The probable easy formation of [Au(CNR)(NO$_3$)] neutral complexes as intermediates for the decomposition of the cationic nitrate complexes is also suggested by the observation of a covalent interaction between gold and the O atoms of the nitrate in the X-ray structure of [Au{CNC$_6$H$_3$(2,6-Me$_2$)}$_2$][NO$_3$] [43] and by the lower ν(CN) value for these [NO$_3^-$] complexes, compared with the corresponding [BF$_4^-$] or [PF$_6^-$] complexes. The lower coordinating ability of the later counteranions is supported by the absence of sub-van-der-Waals contacts between the cations and the tetrafluoroborate anions in [Au{CNC$_6$H$_3$(2,6-Me$_2$)}$_2$][BF$_4$] [44].

The influence of the "spherical" counteranions [BF$_4^-$] and [PF$_6^-$] on the thermal behavior of these gold(I) systems is quite regular. For the three kinds of isocyanide used, the general trend of melting temperatures is [PF$_6^-$] > [BF$_4^-$], but the variation of clearing temperatures is the reverse [BF$_4^-$] > [PF$_6^-$]. The two counteranions have different geometrical structures and volumes increasing in the order [BF$_4^-$] < [PF$_6^-$] [45]. This should allow for different cation–cation approximations, which determine the mesogenic behavior. Thus, the molecular arrangement in the solid phase of the [PF$_6^-$] derivatives must be more effective than that of [BF$_4^-$] compounds since the melting point observed for [PF$_6^-$] complexes is higher than for [BF$_4^-$] derivatives. The same trend has been found in hexafluorophosphate and tetrafluoroborate pyridine silver derivatives [46]. In contrast, the clearing transition is mainly due to the melting of the anion–cation arrangement in the mesophase [47]. Thus, it is not unexpected that the clearing points will be lower for bulkier the anion as experimentally found.

8.4.5
Mixtures: Liquid Crystalline "Molecular Alloys"

Applications for LCs are very demanding in temperature range, adequate response to an electric field, stability, viscosity, and so on. Generally, the desired properties do not occur in a single LC and consequently, all industrial devices use multi-component LCs. In the field of metal-containing LCs, the study of mixtures was limited to binary alkaline salt systems of carboxylic acids and related systems perhaps because a mixture of several metal-complexes is expected to bring about ligand scrambling, producing complex mixtures which are hard to study. The only exception is the study of the thermotropic behavior and phase diagrams of binary mixtures of copper and gold complexes of the type [MX(CNR)] (X = anionic ligand, R = p-alkoxyaryl group) (Scheme 8.1) [48]. The systems examined are mixtures of pairs of complexes differing in just one structural parameter, where rearrangements would not give rise to new complexes: (i) complexes [AuCl(CN$C_6H_4C_6H_4OC_nH_{2n+1}$)] changing the alkoxy length ($n=4$ and $n=12$); (ii) complexes [AuCl(CN(C_6H_4)$_m$ $OC_{12}H_{25}$)] with different numbers of aromatic rings ($m=1$ and $m=2$); and (iii) complexes differing in the metal, namely the derivative [CuCl(CN$C_6H_4OC_{12}H_{25}$)] (which is not a LC) and the mesomorphic [AuCl(CN$C_6H_4C_6H_4OC_{12}H_{25}$)]; also (iv) complexes [Au($C_6F_4OC_{10}H_{21}$)(CN$C_6H_44C_6H_4OC_6H_{13}$)] and [Au($C_6F_4OC_6H_{13}$)(CN$C_6H_4C_6H_4$

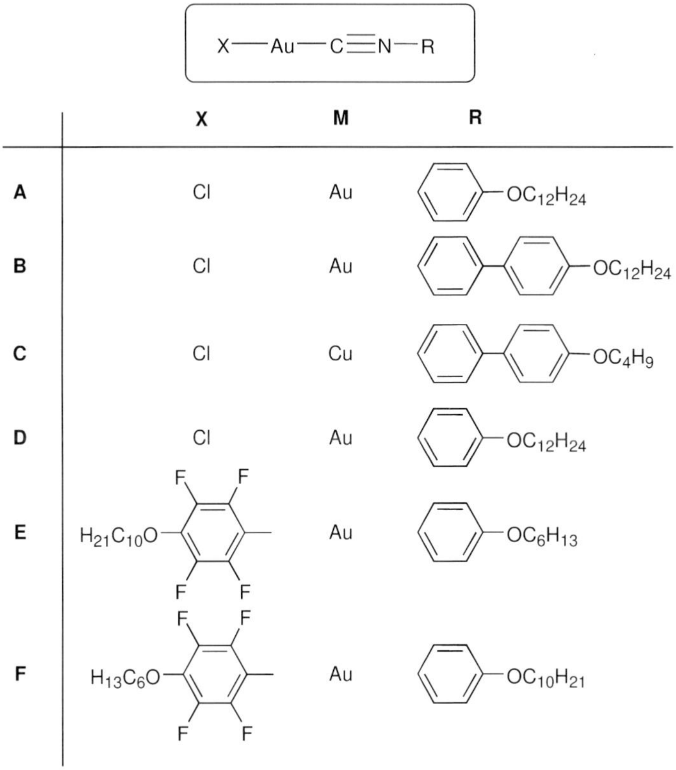

$$X-Au-C{\equiv}N-R$$

	X	M	R
A	Cl	Au	–C$_6$H$_4$–OC$_{12}$H$_{24}$
B	Cl	Au	–C$_6$H$_4$–C$_6$H$_4$–OC$_{12}$H$_{24}$
C	Cl	Cu	–C$_6$H$_4$–C$_6$H$_4$–OC$_4$H$_9$
D	Cl	Au	–C$_6$H$_4$–OC$_{12}$H$_{24}$
E	H$_{21}$C$_{10}$O–C$_6$F$_4$–	Au	–C$_6$H$_4$–OC$_6$H$_{13}$
F	H$_{13}$C$_6$O–C$_6$F$_4$–	Au	–C$_6$H$_4$–OC$_{10}$H$_{21}$

Scheme 8.1 Mixtures of gold and copper derivatives studied.

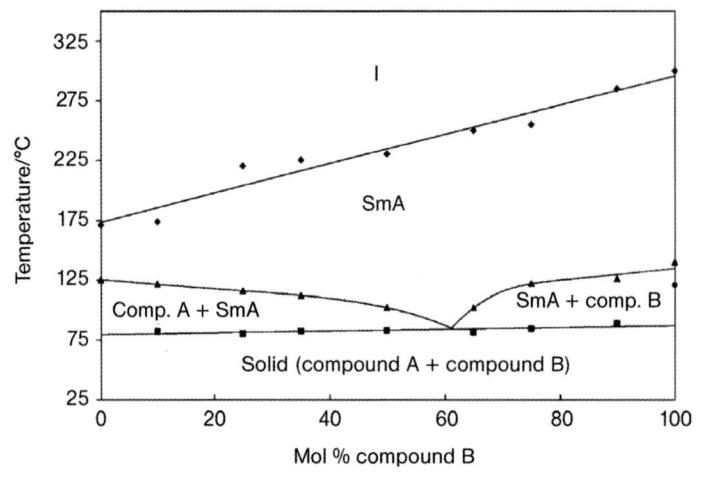

Figure 8.24 Phase diagram of the system [AuCl(CNC$_6$H$_4$OC$_{12}$H$_{25}$)] (A) + [AuCl(CNC$_6$H$_4$C$_6$H$_4$OC$_{12}$H$_{25}$)] (B). (Reproduced from Ref. [46] by permission.)

OC$_{10}$H$_{21}$)], in which rearrangement does not occur. All the mixtures studied display liquid crystal behavior with improved properties with respect to the pure components. A representative binary phase diagram and their corresponding DSC traces are presented in Figures 8.24 and 8.25 respectively, and reveal the eutectic nature of these systems.

It is interesting to note that, in contrast to other mesophases, Smectic C mesophases from room temperature to about 200 °C are obtained for mixtures (iv), which form solid solutions. The mixtures (iii) of [CuCl(CNC$_6$H$_4$OC$_{12}$H$_{25}$)] (which is not a LC) and the mesomorphic [AuCl(CNC$_6$H$_4$C$_6$H$_4$OC$_{12}$H$_{25}$)] (4), containing different metals, show mesogenic behavior for a large range of concentrations, and afford a single fluid material with an ordered homogeneous distribution of the two metal complexes. This method for producing homogeneous ordered mobile phases of different metal-containing molecules (in this case gold and copper) could be of interest for electronics and for metal deposition. These materials have been named "liquid crystalline molecular alloys" to distinguish them from heterometallic mesophases based on mesogenic heterometallic molecules, that is, where there is only one kind of molecule which contains two or more metals in its structure (the latter have been named "mixed M–M' metallomesogens").

8.5
Carbene Complexes

The reaction of isocyanide complexes with nucleophiles gives metal-carbene complexes [49], which constitute an important branch of organometallic chemistry and are effective catalyst systems for a variety of processes [50, 51].

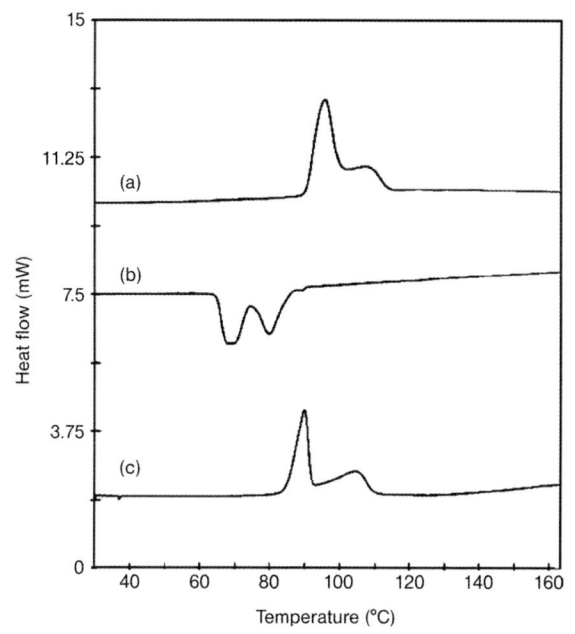

Figure 8.25 DSC scans of 75% [AuCl(CNC$_6$H$_4$OC$_{12}$H$_{25}$)] (A)-35% [AuCl(CNC$_6$H$_4$C$_6$H$_4$OC$_{12}$H$_{25}$)] (B). (a) First heating; (b) First cooling; (c) Second heating. (Reproduced from Ref. [46] by permission.)

Some liquid crystalline gold(I) carbene complexes [ClAu{C(YC$_m$H$_{2m+1}$)(NHC$_6$H$_4$CO$_2$C$_6$H$_4$OC$_n$H$_{2n+1}$)}] (Y = O, NH) (**28**), including a dinuclear gold–carbene complex where the two carbene species are linked via a hexyl diamine bridge, were prepared by reaction of gold isocyanide complexes [ClAu(CNC$_6$H$_4$CO$_2$C$_6$H$_4$OC$_n$H$_{2n+1}$)] with aliphatic alcohols and amines respectively [52, 53]. Most of them display SmA mesophases over the range of 146–160 °C and decompose in the mesophases before reaching their clearing temperatures. In solution, an isomeric mixture of two geometric isomers is observed, which, for complexes with short alkoxy substituents, could be separated by fractional crystallization. However, when the pure isomers were heated isomerization occurred to afford the isomeric mixture of E- and Z-isomers in a 3 : 1 molar ratio.

X = Cl, I; Y = OOC, COO, C≡C
Z = OC$_m$H$_{2m+1}$, NR^{1}R^2

(**28**)

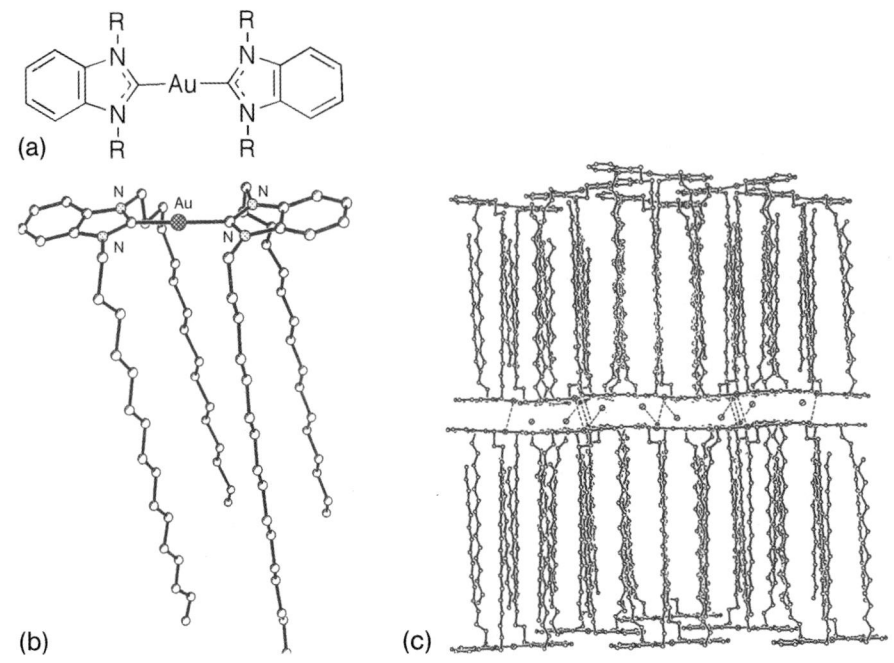

Figure 8.26 Schematic formula (a) molecular single crystal X-ray diffraction structure (b), and solid state packing (c) of the cationic gold complex [Au{bimy($C_{16}H_{33}$)}$_2$]Br·H_2O. (Reproduced from Ref. [52] by permission.)

Another interesting cationic dicarbene system is the family of dialkylbenzimidazol derivatives **29** depicted in Figure 8.26. The structure was initially reported as a mesomorphic layer structure in the range of 92–162 °C, with order in the layers (lamellar mesophases) [54]. However, examination of the XRD pattern in combination with the large clearing enthalpies suggests that these materials are highly ordered and should be better described as crystals [4]. Independent of the exact nature of these phases, this system is a good example to analyze how the ligand and the neutral or ionic nature of the complexes have a determinant influence on the molecular shape and the best space-filling arrangement beyond the initial influence of metal coordination.

From Figure 8.26a a planar molecular shape under normal conditions might be expected, but the very long chains attached to the nitrogen atoms are far from what is normal in common chemistry, and intra- and intermolecular interactions as well as the anion–cation attractions lead to the cationic structure shown in Figure 8.26b. The single crystal X-ray structure for the hexadecyl derivative shows that the cations have interdigitated their chains giving rise to paraffinic areas sandwiched between Au (bimy)$_2$ planes (bimy = 1,3-dialkylbenzimidazol-2-ylidene). These bilayers of about 22 Å thickness are stacked to give a lamellar structure with a repeating spacing of about 26 Å (Figure 8.26c). The anions (Br$^-$) and one H_2O molecule per gold atom are located between these bilayers. It is likely that this molecular arrangement is basically maintained in the phase reported as lamellar.

8.6
Complexes Containing Pyrazole-Type Ligands

8.6.1
Trinuclear Gold Pyrazolate Rings: Metallacrowns

It has been known for many years that gold pyrazolates have a trinuclear structure defining a basically hexagonal (or triangular) core (Figure 8.27) [55]. Conveniently substituted, this core can give rise to columnar organizations. Thus, using bis-alkoxy substituents in positions R^1 and R^3 or tris-alkoxy aryls (**30**), columnar hexagonal mesophases are obtained at temperatures not far from room temperature, although with a short range of liquid crystal behavior [56, 57].

(**30**)

Less substituted systems also give rise to mesomorphism. Compounds with $R^1 = R^3 =$ Me and $R^2 = C_7H_{15}, C_8H_{17}$ also display columnar hexagonal mesophases, although in this case they are monotropic. Again, the number of aliphatic chains per molecule is insufficient to fill a disk but the molecules stacked in the alternate way shown in Figure 8.27 produce disk-like dimers [58]. Similar monotropic behavior is found for an asymmetric gold trinuclear complex based on monosubsituted pyrazolate with $R^1 = R^2 =$ H and $R^3 = C_6H_4OC_{12}H_{25}$ [59].

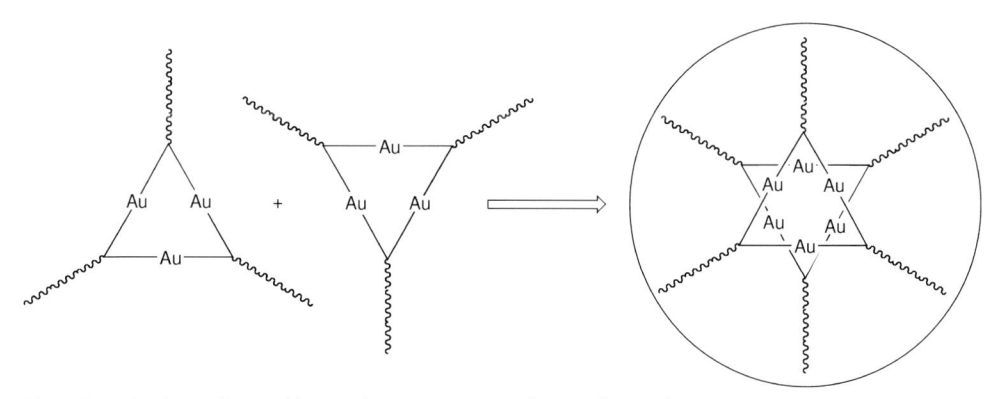

Figure 8.27 Stacking of two gold pyrazolate trimers into a discotic dimer of trimers.

8.6.2
"Mononuclear" Complexes

The simple gold 3-alkylpyrazol complex (**31**) has been reported showing a monotropic SmA mesophase in the range of 65–52 °C and luminescence in the solid state [60].

Cl–Au–N⟨pyrazole ring with NH and C$_{12}$H$_{25}$⟩ (**31**)

8.7
Ionic Imidazolium Derivatives

An example of ionic mesomorphic imidazolium cyanoaurate derivative (**32**) has been prepared displaying a SmA mesophase in the range of 66–112 °C [61].

[imidazolium–CH$_2$–C$_6$H$_4$–OC$_{12}$H$_{25}$]$^+$ (**32**)
[Au(CN)$_2$]$^-$

This species has been used as precursor of nanogold particles by electrodeposition. When the electrodeposition is induced from the isotropic state at 117 °C, the nanoparticles obtained are nanodots aggregated in a spherical-like shape. In contrast, the morphology of the nano particles prepared from the SmA mesophase at 111 °C consist of leaf-like forms interlocked in rosettes.

8.8
Liquid Crystalline Gold Nanoparticles

The preparation and study of metal nanoparticles constitutes an important area of current research. Such materials display fascinating chemical and physical properties due to their size [62, 63]. In order to prevent aggregation, metal nanoparticles are often synthesized in the presence of ligands, functionalized polymers and surfactants. In this regard, much effort has focused on the properties of nanoparticles dispersed into LCs. In contrast, the number of nanoparticles reported that display liquid crystal behavior themselves is low. Most of them are based on alkanethiolate stabilized gold nanoparticles.

Gold nanoparticles displaying anomalous melting points were first described in gold clusters of 24 Å in diameter, stabilized by chemisorbed monolayers of alkanethiol groups ($H_{2n+1}C_nS^-$, $n = 8, 12, 16$) [64]. Using DSC data and thermogravimetric analysis these transitions were assigned to partial melting of the hydrocarbon

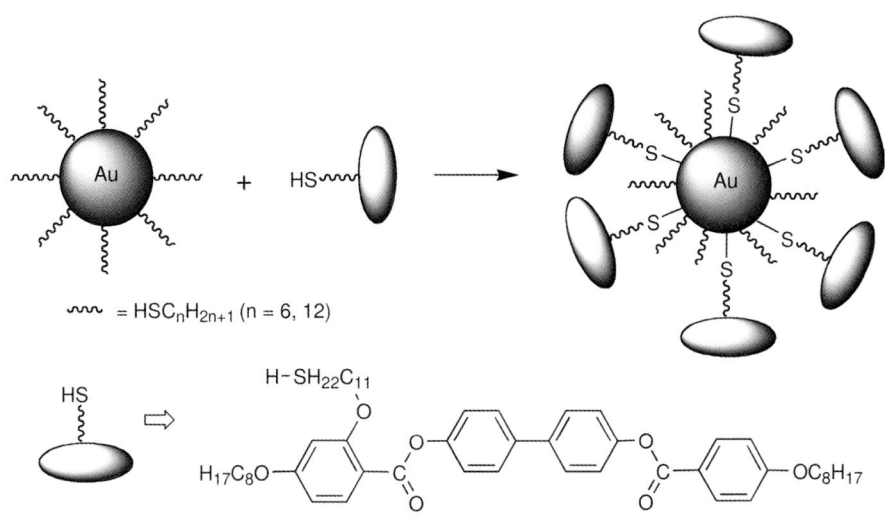

Figure 8.28 Synthesis of thermotropic nematic gold nanoparticles.

chains, but no liquid crystal behavior was described. Further studies on a similar system, formed by gold nanoparticles of 30 Å in diameter and stabilized by a monolayer of octadecanathiol groups, reported liquid crystal behavior, although the gold nanoparticles exhibit only a single phase transition assigned to an ordered-to-disordered transition of the alkanethiol monolayer at the gold nanoparticles surface [65]. The use of the liquid crystalline [10-{trans-(4-pentylcyclohexyl)phenoxy}] decane-1-thiol to stabilize gold nanoparticles gave a system that displays a mesomorphic phase in the range of 74–114 °C, but a close analysis of this behavior shows that it is due to a phase transition of the thiol moieties at the gold nanoparticle surface [66].

Liquid crystal gold nanoparticles that exhibit a thermotropic nematic phase in the bulk have been reported recently (Figure 8.28).

The nematic nanoparticles have been prepared by a two step synthetic process. First, gold nanoparticles are covered with an alkylthiol monolayer (hexyl- and dodecylthiol); in a second step, the alkylthiol-nanoparticles are reacted with the functionalized thiol mesogen in dichloromethane at room temperature to obtain the monolayer-protected liquid crystal gold nanoparticles. These materials are chemically stable and display a nematic mesophase at room temperature [67, 68]. Other examples include liquid crystal gold nanoparticles functionalized by hexaalkoxy-substituted triphenylene [69].

Several mixtures of hexanethiol capped gold nanoparticles and triphenylene based discotic LCs have been studied. These mixtures display liquid crystal behavior (columnar mesophases) and an enhancement in the DC conductivity, due to the inclusion of gold nanoparticles into the matrix of the organic LC [70]. Other studies of mixtures of gold nanoparticles with mesogens include a series of cholesteryl phenoxy alkanoates. The inclusion of the nanoparticles does not change the inherent liquid crystal properties of the cholesteryl derivative but the mesophases are thermally stabilized [71].

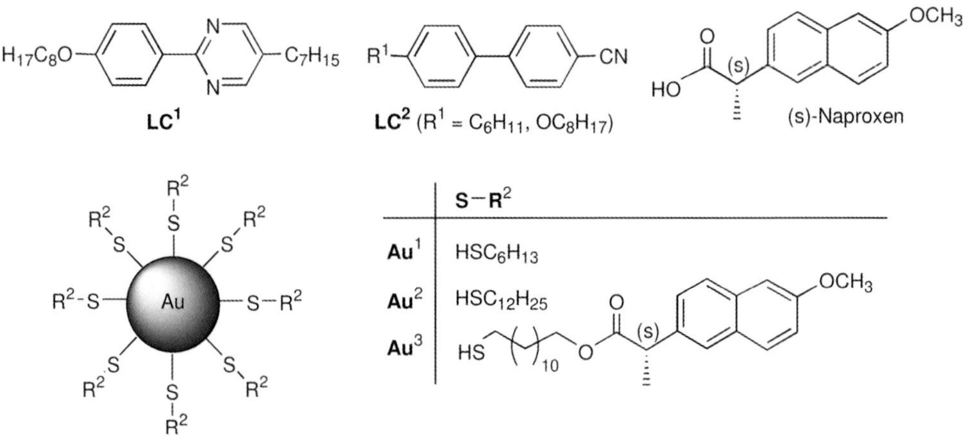

Figure 8.29 Compounds used to prepare liquid crystal mixtures containing gold nanoparticles.

Mixtures of a nematic liquid crystal (LC^1 or LC^2) with small quantities of gold nanoparticles coated with alkylthiolates (<5 wt%) including an alkylthiolate functionalized with a chiral group have been studied (Figure 8.29) [72]. All mixtures show nematic mesophases with transition temperatures and phase stability very similar to those of the liquid crystal precursors LC^1 or LC^2. The introduction of a chiral center into the mixtures (mixtures of Au^3) produce chiral nematic mesophases. A similar result is obtained in mixtures of Au^2 and LC^1 doped with the chiral dopant (s)-Naproxen.

Dendrimer-functionalized gold nanoparticles displaying thermotropic cubic mesophases (Figure 8.30) have been reported. The synthesis was carried out in two steps: the surface of the gold particles was first stabilized by dodecanethiol in order to avoid aggregation; then partial substitution of dodecanethiol chains by the dendritic thiolate group yielded the mixed thiolate gold nanoparticles. These functionalized nanoparticles show unconventional ferromagnetic behavior between 1.8 and 400 K in the bulk [73]. The exact origin of these magnetic properties is not well understood. Some theoretical results predict that gold alone could become ferromagnetic below a certain size, whilst others claim that the covalent grafting of thiols at the gold surface induces magnetism.

Lyotropic liquid crystalline nanoparticles have also been described. Concentrated solutions of gold nanorods in water in the presence of a surfactant (cetyltrimethylammonium bromide) display a nematic mesophase stable up to 200 °C [74]. The N mesophase was identified by optical microscopy by their typical "nematic droplets" texture.

8.9
Conclusions

The simple linear coordination of gold(I) offers an interesting range of opportunities in the field of metal-containing LCs. On the one hand, linear coordination is the

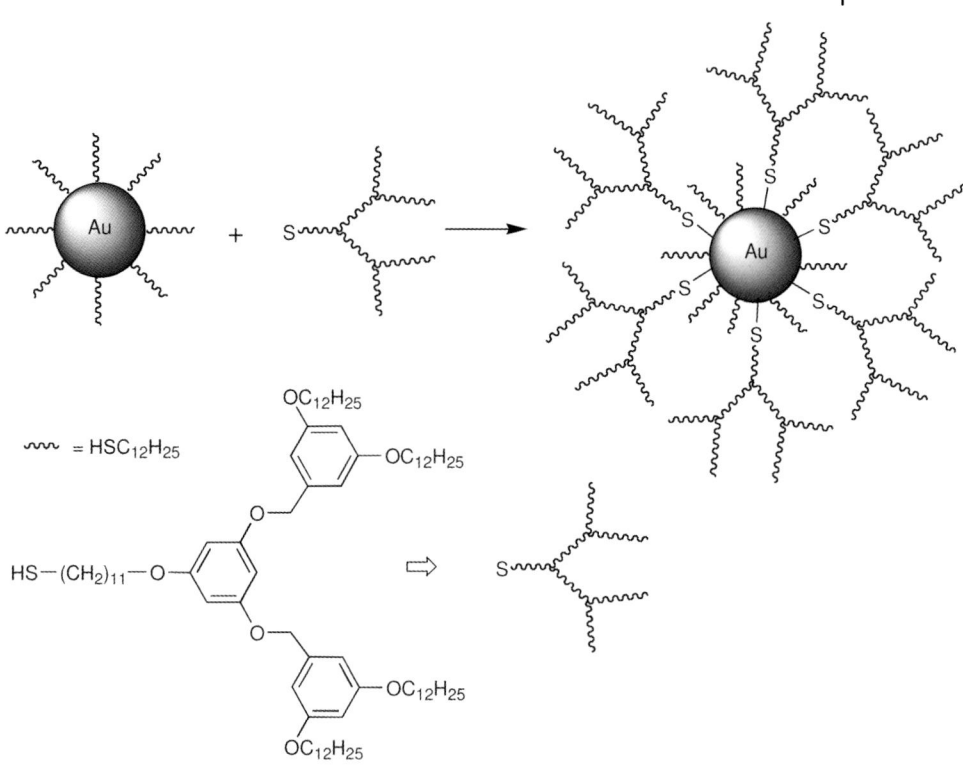

Figure 8.30 Synthesis of dendrimer-functionalized gold nanoparticles.

simplest coordination geometry to build up rod-like molecules. The gold atoms can be situated close to either an extreme or the middle of the molecule; in both cases the molecular simplicity facilitates the interpretation of the LC properties, that is the study of the structure–property relationship. On the other hand, the molecular shape is not confined to rod-like. In effect, other shapes can be built, either by playing with the ligands, or by taking advantage of the interesting and well known chemistry of polynuclear gold(I) complexes, as in the case of cyclic trimers formed with pyrazolates and related ligands.

Besides this, the remarkable properties of gold(I) compounds, which often give rise to aurophilic interactions and/or to luminescence, are of interest when these properties are transported into the liquid crystal field. Although there is much still to be studied, it is already clear that luminescence can survive in the condensed but mobile state of a mesophase, and even in the isotropic liquid state of a molten gold compound. It also seems that aurophilicity can contribute in some cases to the formation of mesophases.

Finally, gold nanoparticles functionalized with thiols have emerged as a very interesting field of research. The gold particle is used as a core to which thiols are attached and the melting of the chains can give rise to LC-like behavior on the gold surface or to real mesophases.

In summary, it appears that gold(I) will continue to be a very rich source of inspiration and experimentation, not only in liquid crystal research but in research using gold liquid crystals at the interface of gold chemistry and the material sciences.

References

1 The reader can find a comprehensive treatment on all aspects of liquid crystals in Goodby, J.W. and Gray, G.W. (eds) (1999) *Handbook of Liquid Crystals*, VCH, Weinheim.

2 More precise description of the structure of the smectic phases can be found in: (a) Leadbetter, A.J. (1987) *Thermotropic Liquid Crystals* (ed. G.W. Gray), Wiley, Chichester, Ch 1, pp. 1–27; (b) Gray, G.W. and Goodby, J.W. (1984) *Smectic Liquid Crystals; Textures and Structures*, Leonard Hill, Glasgow. (c) Dierking, I. (2003) in *Textures of Liquid Crystals*, Wiley-VCH, Weinheim.

3 Reinitzer, F. (1888) Beiträge zur Kenntniss des Cholesterins. *Monatshefte für Chemie*, 9, 421–441.

4 Espinet, P., Esteruelas, M.A., Oro, L.A., Serrano, J.L. and Sola, E. (1992) Transition Metal Liquid Crystals: Advanced materials within the reach of the coordination Chemist. *Coordination Chemistry Reviews*, 117, 215–274.

5 Serrano, J.L. (ed.) (1996) *Metallomesogens*, VCH, Weinheim.

6 Donnio, B., Guillon, D., Bruce, D.W. and Deschenaux, R. (2006) Metallomesogens in *Comprehensive Organometallic Chemistry III: From Fundamentals to Applications* (eds R.H. Crabtree and D.M.P. Mingos), Elsevier, Oxford, UK, Vol. 12 *Applications III: Functional Materials, Environmental and Biological Applications* (ed D. O'Hare), ch. 12.05, pp. 195–294.

7 Bruce, D.W., Lalinde, E., Styring, P., Dunmur, D.A. and Maitlis, P.M. (1986) Novel transition metal-containing nematic and smectic liquid crystals. *Journal of the Chemical Society, Chemical Communications*, (8), 581–582.

8 Adams, H., Bailey, N.A., Bruce, D.W., Dhillon, R., Dunmur, D.A., Hunt, S.E., Lalinde, E., Maggs, A.A., Orr, R., Styring, P., Wragg, M.S. and Maitlis, P.M. (1988) Metallo-mesogens and liquid crystals with a heart of gold. *Polyhedron*, 7, 1861–1867.

9 Adams, H., Albeniz, A.C., Bailey, N.A., Bruce, D.W., Cherodian, A.S., Dhillon, R., Dunmur, D.A., Espinet, P., Feijoo, J.L., Lalinde, E., Maitlis, P.M., Richardson, R.M. and Ungar, G. (1991) Synthesis and phase behaviour of mesomorphic transition-metal complexes of alkoxydithiobenzoates. *Journal of Materials Chemistry*, 1, 843–855.

10 Kaharu, T., Ishii, R. and Takahashi, S. (1994) Liquid-crystalline gold–isonitrile complexes. *Journal of the Chemical Society, Chemical Communications*, (11), 1349–1350.

11 Bachman, R.E., Fioritto, M.S., Fetics, S.K. and Cocker, T.M. (2001) The Structural and Functional Equivalence of Aurophilic and Hydrogen Bonding: Evidence for the First Examples of Rotator Phases Induced by Aurophilic Bonding. *Journal of the American Chemical Society*, 123, 5376–5377.

12 Omenat, A., Serrano, J.L., Sierra, T., Amabilino, D.B., Minguet, M., Ramos, E. and Veciana, J. (1999) Chiral linear isocyanide palladium(II) and gold(I) complexes as ferroelectric liquid crystals. *Journal of Materials Chemistry*, 9, 2301–2305.

13 Espinet, P. (1999) Liquid crystal made of gold. *Gold Bulletin*, 32, 127–134.

14 Coco, S., Espinet, P., Falagán, S. and Martín-Álvarez, J.M. (1995) Liquid

Crystals based on Halogold(I) Complexes with p-alkoxyphenyl isonitriles. *New Journal of Chemistry*, **19**, 959–964.

15. Benouazzane, M., Coco, S., Espinet, P. and Barbera, J. (2001) Supramolecular organization in copper(I) isocyanide complexes: Copper(I) liquid crystals from a simple molecular structure. *Journal of Materials Chemistry*, **11**, 1740–1744.

16. Benouazzane, M., Coco, S., Espinet, P. and Martín-Alvarez, J.M. (1995) Liquid Crystals based on Halogold(I) Complexes with 4-Isocyano-4′-alkoxyphenyl Derivatives. *Journal of Materials Chemistry*, **5**, 441–445.

17. Hedström, M., Benachenhou, N.S. and Calais, J.L. (1994) Nonlinear optical properties of some substituted biphenyls. *Molecular Engineering*, **3**, 329–342.

18. Coco, S., Espinet, P., Martín-Alvarez, J.M. and Levelut, A.M. (1997) Effects of Isonitrile Substituents on Mesogenic Properties of Halogoldiisonitrile Complexes: Calamitic and Discotic Liquid Crystals. *Journal of Materials Chemistry*, **7**, 19–23.

19. Bruce, D.W. and Hudson, S.A. (1994) Mesomorphic complexes of silver trifluoromethanesulfonate and silver dodecylsulfate with 2- and 3-fluoro-4-alkoxy-4-stilbazoles. *Journal of Materials Chemistry*, **4**, 479–486.

20. Dembinski, R., Espinet, P., Lentijo, S., Markowicz, M.W., Martín-Alvarez, J.M., Rheingold, A.L., Schmidt, D.J. and Sniady, A. (2008) Fluorophobic Effect in Metallomesogens – The Synthesis and Mesomorphism of Ag, Au, Cu, Fe, Pd, and Pt Fluorous Isocyanide Complexes. *European Journal of Inorganic Chemistry*, (10), 1565–1572.

21. Hird, M. and Toyne, K.J. (1998) Fluoro Substitution in Thermotropic Liquid Crystals. *Molecular Crystals and Liquid Crystals*, **323**, 1–67.

22. Tschierske, C. (1998) Non-conventional liquid crystals – the importance of micro-segregation for self-organisation. *Journal of Materials Chemistry*, **8**, 1485–1508.

23. Benouazzane, M., Coco, S., Espinet, P. and Martín-Alvarez, J.M. (1999) Liquid Crystals Based on Pseudohalogold(I) Isocyanide Complexes. *Journal of Materials Chemistry*, **9**, 2327–2332.

24. Zeng, H., Lai, C.K. and Swager, T.M. (1994) Transition Metals in Highly Correlated Discotic Phases: Designing Metallomesogens with Selected Intermolecular Organizations. *Chemistry of Materials*, **6**, 101–103.

25. Serrette, A.G. and Swager, T.M. (1994) Polar Superstructures Stabilized by Polymeric Oxometal Units: Columnar Liquid Crystals Based on Tapered Dioxomolybdenum Complexes. *Angewandte Chemie (International Edition in English)*, **33**, 2342–2345.

26. Coco, S., Cordovilla, C., Donnio, B., Espinet, P., García-Casas, M.J. and Guillon, D. (2008) Self-Organization of Dendritic Supermolecules, Based on Isocyanide–Gold(I), –Copper(I), –Palladium(II), and –Platinum(II) Complexes, into Micellar Cubic Mesophases. *Chemistry – A European Journal*, **14**, 3544–3552.

27. Samuel, D., Weinraub, B. and Ginsburg, D. (1956) Synthesis of several aromatic isocyanides. *The Journal of Organic Chemistry*, **21**, 376–377.

28. Coco, S., Espinet, E., Espinet, P. and Palape, I. (2007) Functional isocyanide metal complexes as building blocks for supramolecular materials: hydrogen-bonded liquid crystals. *Dalton Transactions*, (30), 3267–3272.

29. Coco, S., Cordobilla, C., Domínguez, C. and Espinet, P. (2008) Luminescent gold(I) metallo-acids and their hydrogen bonded supramolecular liquid crystalline derivatives with decyloxystilbazole as hydrogen acceptor. *Dalton Transactions*, **48**, 6894–6900.

30. Alejos, P., Coco, S. and Espinet, P. (1995) Liquid Crystals based on alkynylgold(I) isonitrile complexes. *New Journal of Chemistry*, **19**, 799–805.

31. Kaharu, T., Ishii, R., Adachi, T., Yoshida, T. and Takahashi, S. (1995) Liquid-crystalline

(Isonitrile)gold(I) acetylide complexes. *Journal of Materials Chemistry*, **5**, 687–692.

32 Ferrer, M., Mounir, M., Rodríguez, L., Rossell, O., Coco, S., Gómez-Sal, P. and Martín, A. (2005) Effect of the organic fragment on the mesogenic properties of a series of organogold(I) isocyanide complexes. X-ray crystal structure of [Au(CCC$_5$H$_4$N)(CNC$_6$H$_4$O(O)CC$_6$H$_4$OC$_{10}$H$_{21}$)]. *Journal of Organometallic Chemistry*, **690**, 2200–2208.

33 Bayón, R., Coco, S., Espinet, P., Fernández-Mayordomo, C. and Martín-Álvarez, J.M. (1997) Liquid Crystalline Mono and Dinuclear Perhalophenylgold(I) Isonitrile Complexes. *Inorganic Chemistry*, **36**, 2329–2334.

34 Bayón, R., Coco, S. and Espinet, P. (2005) Gold Liquid Crystals Displaying Luminescence in the Mesophase and Short F...F Interactions in the Solid State. *Chemistry – A European Journal*, **11**, 1079–1085.

35 Coco, S., Fernández-Mayordomo, C., Falagán, S. and Espinet, P. (2003) Effect of alkoxy chains on the mesomorphic properties of (p-alkoxy)tetrafluoroarylgold(I) isocyanide complexes. *Inorganica Chimica Acta*, **350**, 366–370.

36 Bayón, R., Coco, S. and Espinet, P. (2002) Twist–Grain Boundary Phase and Blue Phases in Isocyanide Gold(I) Complexes. *Chemistry of Materials*, **14**, 3515–3518.

37 Coco, S., Cordovilla, C., Espinet, P., Martín-Álvarez, J. and Muñoz, P. (2006) Dinuclear gold(I) isocyanide complexes with luminescent properties, and displaying thermotropic liquid crystalline behavior. *Inorganic Chemistry*, **45**, 10180–10187.

38 Arias, J., Bardají, M. and Espinet, P. (2008) Luminescence and Mesogenic Properties in Crown-Ether-Isocyanide or Carbene Gold(I) Complexes: Luminescence in Solution, in the Solid, in the Mesophase, and in the Isotropic Liquid State. *Inorganic Chemistry*, **47**, 3559–3567.

39 Binnemans, K. (2005) Ionic Liquid Crystals. *Chemical Reviews*, **105**, 4148–4204.

40 Benouazzane, M., Coco, S., Espinet, P., Martín-Alvarez, J. and Barberá, J. (2002) Liquid crystalline behavior in gold(I) and silver(I) ionic isocyanide complexes: smectic and columnar mesophases. *Journal of Materials Chemistry*, **12**, 691–696.

41 Jones, K. (1973) The Chemistry of Nitrogen, in *Comprehensive Inorganic Chemistry*, Vol. 2, (eds J.C. Bailar Jr, H.J. Emeléus, R. Nyholm and A.F. Trotman-Dickenson), Pergamon Press, Oxford, p. 386.

42 Johnson, B.F.G. and Davis, R. (1973) Gold, in *Comprehensive Inorganic Chemistry*, Vol. 3, (eds J.C. Bailar Jr, H.J. Emeléus, R. Nyholm and A.F. Trotman-Dickenson), Pergamon Press, Oxford, p. 153.

43 Mathieson, T.J., Langdon, A.G., Milestone, N.B. and Nicholson, B.K. (1999) Preparation and structural characterisation of isocyanide gold(I) nitrates, [Au(NO$_3$)(CNR)] (R = Et, Bu^t or C$_6$H$_3$Me$_2$-2,6); new aurophilic motifs. *Dalton Transactions*, (2), 201–208.

44 Schneider, W., Sladek, A., Bauer, A., Angermaier, K. and Schmidbaur, H. (1997) Structural Investigation of Bis(isonitrile) gold(I) Complexes. *Zeitschrift für Naturforschung B. A Journal of Chemical Sciences*, **52**, 53–56.

45 Mingos, D.M.P. and Rohl, A.L. (1991) Size and shape characteristics of inorganic molecules and ions and their relevance to molecular packing problems. *Dalton Transactions*, (12), 3419–3425.

46 Marcos, M., Ros, M.B., Serrano, J.L., Sola, M.A., Oro, L.A. and Barberá, J. (1990) Liquid-crystal behavior in ionic complexes of silver(I): molecular structure-mesogenic activity relationship. *Chemistry of Materials*, **2**, 748–758.

47 Albéniz, A.C., Barberá, J., Espinet, P., Lequerica, M.C., Levelut, A.M. López-Marcos, F.J. and Serrano, J.L. (2000) Ionic Silver Amino Complexes Displaying

Liquid Crystal Behavior close to Room Temperature. *European Journal of Inorganic Chemistry*, (2), 133–138.

48 Ballesteros, B., Coco, S. and Espinet, P. (2004) Mesomorphic Mixtures of Metal Isocyanide Complexes, Including Smectic C Mesophases at room temperature and "Liquid Crystalline Molecular Alloys". *Chemistry of Materials*, **16**, 2062–2067.

49 Bartolomé, C., Carrasco-Rando, M., Coco, S., Cordovilla, C., Espinet, P. and Martín-Álvarez, J. (2006) Structural switching in luminescent polynuclear gold imidoyl complexes by intramolecular hydrogen bonding. *Organometallics*, **25**, 2700–2703.

50 (a) McGuinness, D. and Cavell, K.J. (2000) Donor-Functionalized Heterocyclic Carbene Complexes of Palladium(II): Efficient Catalysts for C-C Coupling Reactions. *Organometallics*, **19**, 741–748; (b) Herrmann, W.A. (2002) N-Heterocyclic Carbenes: A New Concept in Organometallic Catalisis. *Angewandte Chemie (International Edition in English)*, **41**, 1290–1309.

51 Zaragoza, F. (1999) in *Metal Carbenes in Organic Synthesis*, Wiley-VCH, Weinheim.

52 Ishii, R., Kaharu, T., Pirio, N., Zhang, S.-W. and Takahashi, S. (1995) Liquid-crystalline gold(I)-carbene complexes. *Journal of the Chemical Society, Chemical Communications*, (12), 1215–1216.

53 Zhang, S.-W., Ishii, R. and Takahashi, S. (1997) Syntheses and Mesomorphic Properties of Gold(I)-Carbene Complexes. *Organometallics*, **16**, 20–26.

54 Lee, K.M., Lee, C.K. and Lin, I.J.B. (1997) A Facile Synthesis of Unusual Liquid-Crystalline Gold(I) Dicarbene Compounds. *Angewandte Chemie (International Edition in English)*, **36**, 1850–1851.

55 Minghetti, G., Banditelli, G. and Bonati, F. (1979) Metal derivatives of azoles. 3. The pyrazolato anion (and homologs) as a mono- or bidentate ligand: preparation and reactivity of tri-, bi-, and mononuclear gold(I) derivatives. *Inorganic Chemistry*, **18**, 658–663.

56 Barberá, J., Elduque, A., Giménez, R., Oro, L.A. and Serrano, J.L. (1996) Pyrazolate Golden Rings: Trinuclear Complexes That Form Columnar Mesophases at Room Temperature. *Angewandte Chemie (International Edition in English)*, **35**, 2832–2835.

57 Barberá, J., Elduque, A., Giménez, R., Lahoz, E.J., López, J.A., Oro, L.A. and Serrano, J.L. (1998) (Pyrazolato)gold Complexes Showing Room-Temperature Columnar Mesophases. Synthesis, Properties, and Structural Characterization. *Inorganic Chemistry*, **37**, 2960–2967.

58 Kim, S.J., Kang, S.H., Park, K.-M., Kim, H., Zin, W.-C., Choi, M.-G. and Kim, K. (1998) Trinuclear Gold(I) Pyrazolate Complexes Exhibiting Hexagonal Columnar Mesophases with Only Three Side Chains. *Chemistry of Materials*, **10**, 1889–1893.

59 Torralba, M.C., Ovejero, P., Mayoral, M.J., Cano, M., Campo, J.A., Heras, J.V., Pinilla, E. and Torres, M.R. (2004) Silver and Gold Trinuclear Complexes Based on 3-Substituted or 3,5-Disubstituted Pyrazolato Ligands. X-Ray Crystal Structure of cyclo-Tris{μ-[3,5-bis(4-phenoxyphenyl)-1H-pyrazolato-κN^1: κ N^2]}trigold Dichloromethane ([Au(μ-)]$_3$ CH$_2$Cl$_2$). *Helvetica Chimica Acta*, **87**, 250–263.

60 Ovejero, P., Mayoral, M.J., Cano, M. and Lagunas, M.C. (2007) Luminescence of neutral and ionic gold(I) complexes containing pyrazole or pyrazolate-type ligands. *Journal of Organometallic Chemistry*, **692**, 1690–1697.

61 Dobbs, W., Suisse, J.-M., Douce, L. and Welter, R. (2006) Electrodeposition of Silver Particles and Gold Nanoparticles from Ionic Liquid-Crystal Precursors. *Angewandte Chemie (International Edition in English)*, **45**, 4179–4182.

62 Lewis, L.N. (1993) Chemical catalysis by colloids and clusters. *Chemical Reviews*, **93**, 2693–2730.

63 Smith, G. (1992) Large clusters and colloids. Metals in the embryonic state. *Chemical Reviews*, **92**, 1709–1727.

64 Cerril, R.H., Postlethwaite, T.A., Chen, C.H., Poon, C.D., Tzerzis, A., Hutchinson, A.D., Clark, M.R., Wignall, G., Londono, J.D., Superfine, R., Calvo, M., Johson, C.S., Samulski, E.T. and Murray, R.W. (1995) Monolayers in Three Dimensions: NMR, SAXS, Thermal, and Electron Hopping Studies of Alkanethiol Stabilized Gold Clusters. *Journal of the American Chemical Society*, **117**, 12537–12548.

65 Badia, A., Demers, L., Guccia, L., Morin, F. and Lennox, R.B. (1997) Structure and Dynamics in Alkanethiolate Monolayers Self-Assembled on Gold Nanoparticles: A DSC, FT-IR, and Deuterium NMR Study. *Journal of the American Chemical Society*, **119**, 2682–2692.

66 Kanayama, N., Tsutsumi, O., Kanazawa, A. and Ikeda, T. (2001) Distinct thermodynamic behaviour of a mesomorphic gold nanoparticle covered with a liquid-crystalline compound. *Chemical Communications*, (24), 2640–2641.

67 Cseh, L. and Mehl, G.H. (2007) Structure–property relationships in nematic gold nanoparticles. *Journal of Materials Chemistry*, **17**, 311–315.

68 Cseh, L. and Mehl, G.H. (2006) The Design and Investigation of Room Temperature Thermotropic Nematic Gold Nanoparticles. *Journal of the American Chemical Society*, **128**, 13376–13377.

69 Yamada, M., Shen, Z. and Miyake, M. (2006) Self-assembly of discotic liquid crystalline molecule-modified gold nanoparticles: control of 1D and hexagonal ordering induced by solvent polarity. *Chemical Communications*, (24), 2569–2571.

70 Kumar, S., Pal, S.K. and Lakshminarayanan, V. (2005) Discotic-Decorated Gold Nanoparticles. *Molecular Crystals and Liquid Crystals*, **434**, 251–258.

71 Vemula, P.K., Mallia, V.A., Bizati, K. and John, G. (2007) Cholesterol Phenoxy Hexanoate Mesogens: Effect of meta Substituents on Their Liquid Crystalline Behavior and in Situ Metal Nanoparticle Síntesis. *Chemistry of Materials*, **19**, 5203–5206.

72 Qi, H. and Hegmann, T. (2006) Formation of periodic stripe patterns in nematic liquid crystals doped with functionalized gold nanoparticles. *Journal of Materials Chemistry*, **16**, 4197–4205.

73 Donnio, B., García-Vázquez, P., Gallani, J.-L., Guillon, D. and Terazzi, E. (2007) Dendronized Ferromagnetic Gold Nanoparticles Self-Organized in a Thermotropic Cubic Phase. *Advanced Materials*, **19**, 3534–3539.

74 Jana, N.R., Gearheart, L.A., Obare, S.O., Johson, C.J., Mann, K.J.S. and Murphy, C.J. (2002) Liquid crystalline assemblies of ordered gold nanorods. *Journal of Materials Chemistry*, **12**, 2909–2912.

Index

a
A2780 (cell lines) 55
absorption efficiency 326
acetonitrile 252–253
– degassed 260–265
acetylacetonate complexes 141
acids
– arachidonic 301
– Lewis 32, 93, 208
– small organic 30–33
acrylate, methyl 151
activation, C–H bond 48–49
addition
– alcohol 133
– cyclo- 150–151
– isocyanides 135
– oxidative, see oxidative addition
adduct formation 15
adsorption, halogens 336
aggregated particle morphology 329
aggregates
– dimeric 258
– disordered 330
AIDS 308–309
albumin, serum 295–297
alcohol addition 133
alcoholysis 74
aliphatic alkynes 371
aliphatic areas 377
alkanes, oxidation 72
alkaneselenolate molecules 341
alkoxo complexes 72–76
alkynes, aliphatic 371
alkynyl
– functional ligands 371
– isocyanide-alkynyl complexes 370–372
alkynylation, palladium-catalyzed 95
alloys, molecular 383–384

amidinate complexes 3–9, 36
amidine ligands 4
amido complexes 52
ammonium tetraazidoaurates(III) 51
anionic diphenoxide 74
anionic monodentate ligands 51–53
anionic N-donors 47–65
anionic nitrogen ligands 21
anionic pentafluorophenylgold(I)
    derivatives 103–106
anions, poly- 72
anti-arthritic activity, gold complexes 283
– gold complexes 294
anti-HIV activity 308–309
antimony 341–342
antitumor activity 293, 306–307
applications
– chemosensory 267–270
– medical 283–319
approximation
– discrete dipole 328
– relativistic pseudopotential 194–198
– ZORA 194–196, 203
aqueous chemistry, gold compounds
    283–287
arachidonic acid metabolites 301
aromatic areas 377
arsenic, surface chemistry 341–342
aryl rings 363
aryl substituent 48
arylgold chemistry 93
aryloxo complexes 73
assemblages of nanoparticles 330–331
atomic gold 189–194
[Au(C$_6$F$_5$)(Carbene)] complexes 98
[Au(C$_6$F$_5$)L] complexes 96
auranofin 286, 291–297, 300–309
aurosomes 307

Gold Chemistry: Highlights and Future Directions. Edited by Fabian Mohr
Copyright © 2009 WILEY-VCH Verlag GmbH & Co. KGaA, Weinheim
ISBN: 978-3-527-32086-8

aurothiomalate 288
Au... see also gold
azotoluene 68

b

B16 melanocytes 306
bacteriostatic properties 294
benzene 211
benzothiazole ligands 101
benzoyl peroxide 11
benzyl halides 274
benzyl imidazolates 1
benzyl isothiocyanate 339
2-benzyl pyridines 48
bidentate nitrogen ligands 21
bifunctional ligands 1, 50
– isocyanide 368
biochemistry, physiological and cellular 304–309
bioinorganic chemistry 283
biological ligand exchange 304
biologically relevant ligands 288
biphenyl core 364
bipy ligands 67–70, 75–76
bipyridyl derivatives 68
bis(amidate) ligands, chelating 62
bis(isocyanide) derivatives, ionic 379–382
bismuth complexes 117
bis(thiolato)gold(I) species 290–291
'Blue Gold' complexes 161
Bnpy ligands 52
bond angles 293
bond distances 54
– Au–S 292–293
– calculated 210
– diatomic compounds 199
– multidentate ligands 61
– O-donor ligands 67
bonds
– C–Au 94
– C–H 48–49
– metal-metal 159
– NBO analysis 209
Breit–Pauli Hamiltonian 187
bridges, halogen 131
– halogen 139
bridging hydroxides 65–66
bridging ligands 25–26
– carboxylate 77
– diphosphine 255
broken conjugation 365
Brown constant 93
Bubipy ligands 48–50
bulky groups 3

c

C–Au bonds 94
C-donor ligands 73
– pentafluorophenylgold(I) 97–100
C–H bond activation 48–49
calamitic liquid crystals 358
calculations
– atomic gold 189–194
– bond distances 210
– Dirac–Kohn–Sham 189
– gold clusters 212–215
– Hückel 160
– infinite systems 216–220
– inorganic and organometallic gold compounds 203–212
– relativistic methods 194–203
– spectroscopic properties 266
calf-thymus DNA (ct DNA) 54–56
carbamates, dithio- 291
carbenes 98, 384–387
– cyclic 135
– heterocyclic 295
carbeniates 27
– gold(I) 1
carbon
– C–Au bonds 94
– C–H bond activation 48–49
– surface chemistry 342
carboxamido substituted heterocyclic ligands 60–62
carboxylate ligands 11
– bridging 77
carcinoma cell lines, human ovarian 55
catalysis
– heterogeneous 334–335
– palladium-catalyzed alkynylation 95
catalysts, Au/TiO$_2$ 36
cation-radical salts 138
cationic pentafluorophenylgold(I) derivatives 107
cells
– membranes 297
– murine leukemia 52, 63
cellular biochemistry 304–309
C$_6$F$_5$ groups, see pentafluorophenyl...
Chagas disease 50
charge transfer 250
– electron transfer reactions 273–274
– thiolate-to-gold 268
chelate ligands 78
– bis(amidate) 62
– dcpe 266
– imidazole-containing 58
chemical nanophenomena 334–335

chemical reactions, *see* reactions
chemisorption 335
chemosensory applications 267–270
chiral centers 375
chiral chlorogold(I) compound 363
chlorine bridges 139
cholesteric range 376
chromium complexes 126–127
chrysotherapy 283–319
– complexes 286
cisplatin 53–55, 64
clusters
– calculations 212–215
– hemispherical 323
– magnetic 215
– nano-, *see* nanoclusters
– pentafluorophenyl gold 155
– spherical 323
– tetranuclear gold(I) nitrogen 15–20
CO oxidation 36
coalescence 325
coating, magnetic clusters 215
cohesive energy 217
colloidal crystals 330–331
columnar mesophases 27, 359
complexes
– acetylacetonate 141
– alkoxo 72–76
– amido 52
– antiarthritic activity 283, 294
– antitumor activity 306–307
– [Au(C6F5)(Carbene)] 98
– [Au(C6F5)L] 96
– bismuth 117
– 'Blue Gold' 161
– carbene 384–387
– chromium 126–127
– chrysotherapy 286
– copper 122
– cytotoxicity 306–307
– diamine 135
– dicationic 59
– diimine 57
– dinuclear, *see* dinuclear complexes
– diphosphine ligands 293–294
– dithiobenzoate 361–362
– gold–protein 295–304
– gold(I) amidinate 3–9, 30–33, 36
– gold(I) cyanide 294–295
– gold(I) isocyanide 97–99
– gold(III) 47–92
– hexanuclear 142
– homobridged 6
– hydroxo 65–69
– iron 123–126
– isocyanide 362
– isocyanide-alkynyl 370–372
– isocyanide-halide 362–370
– ketazine 100
– luminescence 249–281
– manganese 126
– medicinally important 287–295
– mesomorphic 367
– methoxo 75
– mixed N/O ligands 78–79
– molybdenum 126–127
– orthometallated 114
– oxaauracyclobutane 70
– oxo 69–72
– palladium 122–123
– pentafluorophenyl 93–181
– (pentafluorophenyl)gold(III) 51
– pentanuclear gold(I) 115
– perhalophenyl 375
– photoinduced electron transfer reactions 273–274
– photophysics 249–281
– polyamine gold(III) 53
– polynuclear 159–162
– pyrazole-type ligands 387–388
– pyridine 361
– pyridylamidogold(III) 59
– rhodium 123
– silver 119–122
– spectroscopic properties 250–273
– structure of gold(I)/gold(III) complexes 284
– tetranuclear 116–117
– thallium 117–119
– thermal properties 73, 364–366, 373–376, 380–382
– tin 117
– tungsten 126–127
– two-coordinate gold(I) 284
– zwitterionic 142
concomitant emergence 267
conductance measurements, single-molecule 340
conical monodendron 369
conjugation, broken/extended 365
coordination polymer 13–15
copper
– gold–copper mixtures 383
– heteronuclear complexes 122
coproportionation 158–159
core-orbitals 186
core-shell particles 325
Coulomb, Dirac–Coulomb Hamiltonian 194

counteranions 381–382
counterions 263
crown ether 378
crystallography, X-ray, see X-ray crystallography
crystals
– colloidal 330–331
– liquid, see liquid crystals
– molecular 216
ct DNA, see calf-thymus DNA
CTCs, see cyclic trinuclear complexes/compounds
cuboctahedron structure 324
cyanides
– gold(I) complexes 294–295
– iso-, see isocyanides
– mercury(II) cyanide coordination polymer 13–15
cyclic carbenes 135
cyclic ligands, polyazamacro- 63–65
cyclic methanide derivatives 138–140
cyclic trinuclear gold(I) nitrogen compounds 24–30
cycloaddition, 1,3-dipolar 150–151
cycloaurated derivatives 47
cyclometallation 49, 272
cylindrical micelle 360
cysteine 288–290, 294–305
cytokines 309
cytotoxicity, gold complexes 306–307

d

dcpe 266
deacetylation 292
decahedron structure, Marks 324
decomposition energy 205
degassed acetonitrile 260–265
dendrimers 368–369, 390–391
density functional theory (DFT) modeling 15, 18, 189–190
depression of the melting point 334
deprotonation 26
deshielding, relativistic 188
diamine complexes 135
diatomic gold compounds, molecular calculations 194–203
dicationic complexes 59
diimine complexes 57
β-diketonates 63
dimeric aggregates 258
2,6-dimethyl amidinate 8
dinuclear complexes 66–67
– gold(I) 252–262
– gold(I) amidinate 3–13
– luminescent 111–113

– oxidative-addition reactions 9–13
– pentafluorophenylgold(II) 156–159
dinuclear pentafluorophenylgold(I) derivatives 108–117
diphenoxide, anionic 74
diphenylpyrazolate 17
diphosphine ligands 293–294
– bridging 255
1,2-diphosphinobenzene 146
1,3-dipolar cycloaddition 150–151
dipolar interactions 377
dipole approximation, discrete 328
dipole polarizability, static 193
Dirac–Coulomb Hamiltonian 194
Dirac–Fock methods 189
Dirac–Kohn–Sham calculations 189–190
direct perturbation theory 194
director 358
discotic liquid crystals 358
discrete dipole approximation 328
diseases
– AIDS 308–309
– arthritis 283, 294
– Chagas 50
disk-like mesogenic molecule 358
disordered aggregates 330
disproportionation 158–159
dissociation energy 197–201
distortion, Jahn–Teller 212
dithiobenzoate complexes 361–362
dithiocarbamates 103, 291
DNA, calf-thymus 54
donor ligands
– nitrogen 47–65
– oxygen 65–79
– phosphorus 101–102
– sulfur 102–103
double alcohol addition 133
double halogen bridges 131
Douglas–Kroll Hamiltonian 193
dpp 271
droplets, nematic 390
dyes, metallochromic 299

e

electric quadrupole moment 191
electrochemistry 105
electron affinity 190–191, 214
electron transfer reactions, photoinduced 273–274
electronegative ligands 199, 202–205
electronegativity, Pauling 192
– Pauling 200
electrons, valence 186

electropositive ligands   198–200
electroweak symmetry breaking   212
elemental gold   284–285
enzymes
– inhibition   301–302
– mitochondrial thioredoxin reductase   303–304
– myelperoxidase   302
– selenium-dependent glutathione peroxidase   300
equation, Stern-Volmer   259
equilibration of intra- and extracellular gold   305–306
erythrocytes   300
ESSE   290
esters, imido   4
ether, crown   378
ethylenediamine   118
exchange mechanisms, ligands   285–287
exchange reactions   19
exciplex emission   262–267
excitation energy   204
excited-state reduction potential   270
expression, proteins   302
extended conjugation   365
extinction properties, optical   328
extracellular gold   305–306

f
facets, re-entrant   324
Faryl   372
ferrocenes   148
ferrocenyl derivatives   136
ferrocenyl methylpyrazole   125
ferro... *see also* iron
films
– octylchalcogenate   339
– thin   329–330
finger proteins, zinc   302
flexible ligands   7
fluorescence, nanoparticles   333–334
fluoroalkoxo complexes   73
fluorophenyl, isocyanide-fluorophenyl derivatives   372–379
Fock, Dirac–Fock methods   189
force constants   202
formamidinate, sodium   5
fractional oxidation states   162
functional alkynyl ligands   371

g
geometries, optimized   2
glutathione peroxidase, selenium-dependent   300

gold
– anti-arthritic activity   283, 294
– antitumor activity   306–307
– as pseudo-halide   183
– atomic   189–194
– Au–S bond distances   292–293
– Au/TiO$_2$ 36
– C–Au bonds   94
– cytotoxicity   306–307
– elemental   284–285
– gold–protein reactions and complexes   295–304
– guanidinate derivatives   21–24
– interprotein gold transfer   303
– intra- and extracellular   305–306
– luminescence   249–281
– medicinally important complexes   287–295
– metaloxide-supported   69
– mixed valence derivatives   147
– pentafluorophenyl complexes   93–181
– photophysics   249–281
– radioisotopes   57
– short contacts   120
– solid state calculations   216–220
– surface chemistry   335–342
– theoretical chemistry   183–247
– thermal properties of complexes   364–366, 373–376, 380–382
– thiolate-to-gold charge transfer   268
– 'transauration'   303
gold clusters   155
– calculations   212–215
– *see also* clusters, gold nanoscience
gold compounds
– aqueous chemistry   283–287
– chiral   363
– diatomic   194–203
– inorganic   203–212
– intermetallic   201
– ionic   218
– liquid crystals   357–396
– medical applications   283–319
– organometallic   203–212
gold-doped surfaces   218
gold nanoscience   321–356
– liquid crystals   388–390
– nanoclusters   218–219, 323–325
– nanomaterials   184–185
– nanoparticles   309, 323–325
– oleylamine-stabilized nanoparticles   331
– thermotropic nematic nanoparticles   389
gold(I)
– amidinate complexes   3–9, 30–33, 36
– benzylimidazolates   1

– carbeniates   1
– cyanide complexes   294–295
– hexanuclear supramolecule   258
– isocyanide complexes   97–99
– luminescent complexes   111–113, 118–119
– mixed-ligand nitrogen clusters   15–20
– mixed-metal trinuclear complexes   33–36
– monomethyl   285
– nitrogen chemistry   1–45
– pentafluorophenyl-, *see* pentafluorophenylgold(I)
– pentanuclear complexes   115
– pyrazolates   1
– spectroscopic properties of complexes   250–270
– structure of complexes   284
– thiolates   287–292
– trinuclear nitrogen compounds   24–28
– two-coordinate complexes   284
gold(II), theoretical chemistry   211–212
gold(III)
– complexes   47–92
– halogen derivatives   150
– immunochemical consequences   308–309
– oxidation of insulin and ribonuclease   301
– pentafluorophenyl-   127–155
– (pentafluorophenyl) complexes   51
– polyamine complexes   53
– porphyrins   64
– pyridylamido complexes   59
– spectroscopic properties of complexes   270–273
– structure of complexes   284
– trimethyl   285
gravimetric analysis, thermal   14
groups, sterically bulky   3
guanidinate   3
– gold guanidinate derivatives   21–24
Gupta potential   212

h
halide phosphanes   206
halides   274, 362–370
halogen bridges   131
halogen-gold(III) derivatives   150
halogenated solvents   10
halogens, surface chemistry   336–337
Hamiltonian
– Breit–Pauli   187
– Dirac–Coulomb   194
– Douglas–Kroll   193
Hammett constant   93
heavy elements   183
– super-   219

hemispherical clusters   323
hemoglobin   303
heterocycles   47
heterocyclic carbenes   295
heterocyclic ligands, carboxamido substituted   60–62
heterocyclic thiones   103
heterogeneous catalysis   334–335
heteronuclear pentafluorophenylgold(I) derivatives   117–127
hexanuclear complexes   142
hexanuclear gold cycle   26
hexanuclear gold(I) supramolecule   258
high-energy excited state emission   262–267
high oxidation state   204
highest occupied molecular orbital (HOMO)   15–17, 23–24
– calculated   201
– chemical nanophenomena   342
– HOMO–LUMO gap   213–215, 250
histidine containing peptides   62
hollow gold spheres   325
homobridged complexes   6
homogeneous ordered mobile phases   384
hpp ligand   21–23
Hückel calculations   160
human ovarian carcinoma cell lines   55
hydrazine   94
hydrogen
– C–H bond activation   48–49
– surface chemistry   335–336
hydroxides, bridging   65–66
hydroxo complexes   65–69
hydroxo ligands, terminal   66
hyperfine constants   191

i
icosahedron structure   324
imidazole-containing chelate ligands   58
imidazolium derivatives, ionic   388
imido ester   4
immunochemical consequences, gold (III)   308–309
in vivo oxidation states   307–308
infinite systems   216–220
inflammation   294
inhibition, enzyme   301–302
inorganic gold compounds   203–212
insulin, oxidation   301
intermetallic compounds, gold   201
interprotein gold transfer   303
intra- and extracellular gold   305–306
intracellular gold   305–306
intraligand transitions   111, 250–256, 270–273

iodine, oxidative-addition 30
ionic bis(isocyanide) derivatives 379–382
ionic gold compounds 218
ionic imidazolium derivatives 388
ionic radius 268
ionization potential 190–191, 214
iron complexes
– heteronuclear 123–126
– see also ferro...
isocyanides
– addition 135
– alkynyl complexes 370–372
– complexes 362
– fluorophenyl derivatives 372–379
– gold(I) complexes 97–99
– halide complexes 362–370
isomeric mixture 385
isothiocyanates 339

j
Jahn–Teller distortion 212
Jahn–Teller symmetry change 250
junction, molecular 219

k
ketazine complexes 100
Kohn, Dirac–Kohn–Sham calculations 189
Kroll, Douglas–Kroll Hamiltonian 193

l
laser-directed metal deposition 77
lateral monofluorination 366
lattice constants 217
leukemia cells 306
– murine 52, 63
leukocyte activation 308
Lewis acids 32
– soft 208
– strength 93
ligands, sterically bulky groups 3
ligands
– amidinate, *see* amidinate
– amidine 4
– benzothiazole 101
– bifunctional 1, 50, 368
– biological exchange 305
– biologically relevant 288
– bipy 67–70, 75–76
– Bnpy 52
– bridging 25–26, 255
– Bubipy 48–50
– C-donor, *see* C-donor ligands
– carboxylate 11, 77
– chelate, *see* chelate ligands
– cyclometallated 272
– diphosphine 293–294
– electronegative 202–205
– electronegativity 199
– electropositive 198–200
– exchange mechanisms 285–287
– flexible 7
– functional alkynyl 371
– heterocyclic 60
– hpp 21–23
– intraligand transitions 252, 256
– mixed N/O 78–79
– monodentate 47–53, 286
– multidentate 53–63
– N-donor, *see* N-donor ligands
– nitrogen 1–3
– nitrogen donor 47–65
– O-donor, *see* O-donor ligands
– oxygen donor 65–79
– P-donor, *see* P-donor ligands
– Phbipy 48–50
– phosphine 107
– polyazamacrocyclic 63–65
– polydentate 109
– polyphosphine 144
– pyrazole-type 387–388
– Rbipy 48–50
– S-donor, *see* S-donor ligands
– terminal hydroxo 66
– tetrahydrothiophene (tht) 93–161
linear response theory 198
liquid crystals 97
– calamitic 358
– discotic 358
– general concepts 357–360
– gold compound-based 357–396
– gold nanoparticles 388–390
– luminescence 373–374
– lyotropic 357, 360
– 'molecular alloys' 383–384
– perhalophenyl complexes 375
– phosphorescence 373–374
– thermotropic 357
– X-ray crystallography 377–378
lowest unoccupied molecular orbital
    (LUMO) 15–17, 23–24
– calculated 201
– HOMO–LUMO gap 213–215, 250
luminescence
– dinuclear gold(I) complexes 111–113
– gold complexes 249–281
– gold(I) complexes 118–119
– liquid crystals 373–374
– nanoparticles 333–334

– tribochromism 270
lyotropic liquid crystals 357, 360
lysosomes 307

m

macrocyclic ligands, polyaza- 63–65
macrophages 306
magnetic clusters, coating 215
mammalian MTs 298
manganese complexes, heteronuclear 126
Marks decahedron structure 324
medical applications 283–319
medicinally important gold complexes 287–295
melanocytes, B16 306
melting point depression 334
9-membered rings, trinuclear 24
membranes, cell 297
mercury(II) cyanide coordination polymer 13–15
mesogens 357–358
– metallo- 360
– see also liquid crystals
mesomorphic complexes 367
mesophases 357–359
– columnar 27
mesoporous sponges 327–329
metallochromic dyes 299
metallocycles, mixed-valence 28–29
metallodendrimers 368
metallomesogens 360
metallothioneins (MTs) 297–299
metaloxide-supported gold 69
metals
– copper, see copper
– gold, see gold
– laser-directed deposition 77
– metal-anion/solvent exciplex emission 262–267
– metal-metal bonds 159
– metal–metal interactions 252–262
– mixed-metal trinuclear complexes 33–36
– silver, see silver
methanide derivatives 138–139
methoxo complexes 75
methyl acrylate 151
methyl dithiocarbamates 103
methylammonium tetraazidoaurates(III) 51
methylimidazol units 99
mica 329
micelles 360
– micellar dendrimer 369

mitochondrial thioredoxin reductase 303–304
mitochondrial permeability transitions 303
mixed-ligand tetranuclear gold(I) nitrogen clusters 15–20
mixed-metal salts 76
mixed-metal trinuclear complexes, gold(I) and silver(I) 33–36
mixed N/O ligands 78–79
mixed-valence gold derivatives 147
mixed-valence metallocycles 28–29
mobile phases, ordered 384
model, space-filling 369
modeling, DFT 15, 18, 189
'molecular alloys' 383–384
molecular arrangement, zigzag 378
molecular calculations 194–203
molecular crystals 216
molecular junction 219
Møller–Plesset perturbation theory 197
molybdenum complexes, heteronuclear 126–127
monodendron, conical 369
monodentate ligands 286
– anionic 51–53
– neutral 47–51
monofluorination, lateral 366
monomethylgold(I) 285
mononuclear gold(I) complexes 250–252
MTs, see metallothioneins
Mulliken's recipe 183
multi-reference configuration interaction 202
multidentate ligands 53–63
– bond distances 61
multiple minima problem 216
murine leukemia cells 52, 63
myelperoxidase 302
myocrysine 288

n

N-donor ligands
– anionic 59–63
– mixed N/O ligands 78–79
– pentafluorophenylgold(I) 100–101
nanoclusters 218–219, 323–325
nanomaterials, gold 184–185
nanoparticles 309, 323–325
– assemblages 330–331
– oleylamine-stabilized 331
– thermotropic nematic 389
nanorods 327, 390
nanoscience, see gold nanoscience
nanoshells 325–326

nanospheres 325
– optical properties 333
naphthalene, octafluoro- 32–33
natural bond order (NBO) analysis 209
nematic droplets 390
nematic gold nanoparticles, thermotropic 389
nematic mesophases 358–359
neutral monodentate ligands 47–51
neutral pentafluorophenylgold(I) derivatives 95–103
nitrogen 1–45
– donor ligands 47–65
– gold(I) clusters 15–20
– ligands 1–3, 21
– surface chemistry 340–341
NMR spectra 101, 108
nuclearity 16
– gold guanidinate derivatives 21–24

o

O-donor ligands 65–79
– mixed N/O ligands 78–79
octafluoronaphthalene 32–33
octahedron structure, truncated 324
octylchalcogenate films 339
olefins 71
oleylamine-stabilized gold nanoparticles 331
oligomeric gold(I) thiolates 287–290
optical fiber sensor 121
optical nanophenomena 331–334
optical properties
– extinction 328
– nanospheres 333
optimized geometries 2
ordered mobile phases, homogeneous 384
organic acids, small 30–33
organometallic gold compounds 203–212
orthometallated complexes 114
ovarian carcinoma, human 55
oxaauracycle species 72
oxaauracyclobutane complexes 70
oxidation
– alkanes 72
– CO 36
– gold guanidinate derivatives 21–24
– insulin and ribonuclease 301
– oxidation–reduction reactions 284–285
oxidation states
– fractional 162
– high 204
– $in\ vivo$ 307–308
oxidative addition 9–13

– cyclic trinuclear gold(i)-nitrogen compounds 28–30
– iodine 30
oxo complexes 69–72
oxygen
– donor ligands 65–79
– surface chemistry 337–338

p

P-donor ligands 101–102
P388 leukemia cells 306
– murine 52, 63
palladium-catalyzed alkynylation 95
palladium complexes, heteronuclear 122–123
parity violation 212
Pauli, Breit–Pauli Hamiltonian 187
Pauling electronegativity 192, 200
pentafluorophenyl gold complexes 93–181
pentafluorophenylgold(I)
– anionic derivatives 103–106
– cationic derivatives 107
– di-/polynuclear derivatives 108–117
– heteronuclear derivatives 117–127
– neutral derivatives 95–103
pentafluorophenylgold(II) derivatives 156–162
(pentafluorophenyl)gold(III) complexes 51
pentafluorophenylgold(III) derivatives 127–155
pentanuclear gold(I) complexes 115
peptides 62, 79
perchlorate, silver 119
perhalophenyl complexes 375
periodic table 186–188
permeability transitions, mitochondrial 303
peroxidase, glutathione 300
peroxides, benzoyl 11
perturbation theory 194–197
Phbipy ligands 48–50
phenolic Schiff bases 79
phenyl core 364
phenyl isothiocyanate 339
phenyl rings 380
phosphanes, halide 206
phosphine ligands 107
phosphine(thiolato)gold(I) species 291–292
phosphorescence 249
– liquid crystals 373–374
phosphorus, donor ligands 101–102
– surface chemistry 341–342
photochemical reactions 207
photoinduced electron transfer reactions 273–274
photophysics 249–281

Ph$_3$P   286
physical nanophenomena   334
physiological biochemistry   304–309
plasmon resonance   331–333
plate-shaped micelle   360
Plesset, Møller–Plesset perturbation theory   197
polarizability   365–366
– static dipole   193
polyamine gold(III) complexes   53
polyanions   72
polyaromatic core   358
polyazamacrocyclic ligands   63–65
polycatenar mesogenic molecule   358
polydentate ligands   109
polymers   13–15, 120
– polymeric chain   289
polynuclear complexes
– gold(I)   252–262
– pentafluorophenylgold(I) derivatives   108–117
– pentafluorophenylgold(II)   159–162
polypeptide derivatives   79
polyphosphine ligands   144
porphyrins, gold(III)   64
potential
– excited-state reduction   270
– Gupta   212
– ionization   190–191, 214
– pseudo-   194–198
PPh$_3$ 251–252, 257–259, 269–272
PPN salts   106
pro-drugs   304
pro-inflammatory arachidonic acid metabolites   301
pro-inflammatory cytokines   309
proteins
– expression   302
– gold–protein reactions and complexes   295–304
– zinc finger   302
pseudo-halide   183
pseudopotential approximation, relativistic   194–198
pyrazolates, gold(I)   1
pyrazole-type ligands, complexes   387–388
pyridine complexes   361
pyridylamidogold(III) complexes   59

q

quadrupole moment, electric   191
quadrupole moment, nuclear   194, 202, 203
quantum electrodynamics (QED)   189–190
quenched emission   251, 257

r

radial density   188
radicals, cation-radical salts   138
radiolabeling   296
radiotherapeutic agents   57
Raman experiments   249
Raman spectrum   253
rate-limiting rearrangement   296
Rbipy ligands   48–50
re-entrant facets   324
reactions
– adduct formation   15
– alcohol addition   133
– alcoholysis   74
– alkynylation   95
– chemisorption   335
– co-/disproportionation   158–159
– cycloaddition   150–151
– cyclometallation   49
– deacetylation   292
– deprotonation   26
– electron transfer   273–274
– exchange   19
– gold–protein   295–304
– lateral monofluorination   366
– ligand exchange mechanisms   285–287
– medicinally important gold complexes   287–295
– oxidation, see oxidation
– photochemical   207
– rearrangement   296
– refluxing   48
– replacement   12
– substitution   134
– transmetallation   136–137
reactivity, pentafluorophenylgold(I)   105–106
rearrangement, rate-limiting   296
reductase, mitochondrial thioredoxin   303–304
reduction
– oxidation–reduction reactions   284–285
– potential   270, 285
refluxing   48
relativistic calculation methods   194–203
relativistic deshielding   188
relativistic maximum   186–188
relativistic pseudopotential approximation (RPPA)   194–198
relativistic shell contraction   183–188
replacement reactions   12
resonance
– plasmon   331–333
– Raman   249, 253
response theory, linear   198

rhodium complexes, heteronuclear 123
ribonuclease, oxidation 301
rings
– aryl 363
– phenyl 380
– trinuclear 9-membered 24
rod-like mesogenic molecule 358
RPPA, *see* relativistic pseudopotential approximation

s
S-donor ligands, pentafluorophenylgold(I) 102–103
salts
– cation-radical 138
– mixed-metal 76
– PPN 106
sandwiching 30–33
Schiff bases derivatives 78
selenium 300, 339–340
self-assembled monolayers (SAMs) 322
– sulfur-anchored 338
sensors, optical fiber 121
serum albumin 295–297
Sham, Dirac–Kohn–Sham calculations 189
shell contraction, relativistic 183–188
short gold contacts 120
shuttle model, sulfhydryl- 305
silver complexes, heteronuclear 119–122
silver perchlorate 119
silver(I), mixed-metal trinuclear complexes 33–36
simple halogen bridges 131
single alcohol addition 133
single-molecule conductance measurements 340
sintering 329
small organic acids 30–33
smectic mesophases 358–359
sodium formamidinate 5
solid state 216–220
solid state packing 386
solid state spectra 260–264
solvents
– halogenated 10
– influence on gold guanidinate derivatives 21–24
space-filling model 369
spectroscopic properties
– gold(I) complexes 250–270
– gold(III) complexes 270–273
spectroscopy, SPR 331–332
spherical clusters 323
spherical micellar dendrimer, supramolecular 369
spherical micelle 360
spin-orbit splittings 192, 196
sponges, mesoporous 327–329
static dipole polarizability 193
sterically bulky groups 3
Stern-Volmer equation 259
Stokes shift 252, 264
substitution 134
sulfhydryl-shuttle model 305
sulfur
– Au–S bond distances 292–293
– donor ligands 102–103
– surface chemistry 338–339
sulfur-anchored SAMs 338
superheavy elements 219
supramolecular entities 24, 27, 30–33
– hexanuclear gold(I) 258
– spherical micellar dendrimer 369
surface plasmon resonance (SPR) spectroscopy 331–332
surfaces
– calculations 216–220
– chemistry 335–342
– gold-doped 218
– nanoscience 321–356
Swiss-cheese morphology 329
symmetry breaking, electroweak 212

t
tautomers 289
TCNQ 31, 137
Teller
– Jahn–Teller distortion 212
– Jahn–Teller symmetry change 250
tellurium, surface chemistry 339–340
terminal hydroxo ligands 66
tetraazidoaurates(III) 51
tetrahedron structure 323
tetrahydrothiophene (tht) 93–161
tetramer 149
tetranuclear complexes 116–117
– gold(I) amidinate 3–9
tetranuclear gold(I) nitrogen clusters 15–20
thallium complexes, heteronuclear 117–119
theoretical chemistry 183–247
thermal gravimetric analysis 14
thermal properties, gold complexes 364–366, 373–376, 380–382
thermally stable complexes 73
thermotropic liquid crystals 357
thermotropic nematic gold nanoparticles 389
thin films 329–330

thiolates 268, 287–292
thiones, heterocyclic 103
thioredoxin reductase, mitochondrial 303–304
tin complexes, heteronuclear 117
'transauration' 303
transmetallation 136–137
[TR(carb)] 32–33
tribochromism, luminescence 270
trimethylgold(III) 285
trinuclear 9-membered rings 24
trinuclear complexes 3–9, 30–36
trinuclear gold(I) nitrogen compounds, cyclic 24–28
triphos 261–262, 273
trithiophosphite 102
Trojan-horse therapy/strategy 294–295
truncated octahedron structure 324
Trypanosoma cruzi 50
tungsten complexes, heteronuclear 126–127
two-coordinate gold(I) complexes 284

v
visible metal-anion/solvent exciplex emission 262–267

volatile organic compounds (VOCs) 121–122, 162–163
– chemosensors 267, 270
Volmer, Stern-Volmer equation 259

w
'whirl-wind' configuration 51

x
X-ray crystallography
– gold amidinate complexes 11–12, 22
– liquid crystals 377–378
– pentafluorophenyl gold 94–95, 102–104

y
ylides 9–11, 105

z
zero-order regular approximation (ZORA) 194–196, 203
zigzag molecular arrangement 378
zinc finger proteins 302
zwitterionic complexes 142